ONE- AND TWO-FAMILY DWELLING

Residential Electrical Systems

TWELFTH EDITION

901 Waterfall Way, Suite 602
Richardson, TX 75080-7702

 First edition published 1992
Printed in the United States of America
28 27 26 25 24 5 4 3 2 1

ISBN-13: 979-8-9921443-0-7

Photos used in this book were shot on site or at tradeshows. Use of the photos does not imply endorsement by IAEI of the manufacturers or the products. Photos without a credit line are from IAEI Archives.

NOTICE TO THE READER

This book has not been processed in accordance with NFPA Regulations Governing Committee Projects. Therefore, the text and commentary in it shall not be considered the official position of the NFPA or any of its committees and shall not be considered to be, nor relied upon as a formal interpretation of the meaning or intent of any specific provision or provisions of the 2023 edition of NFPA 70, *National Electrical Code.*©[1]

Publishers do not warrant or guarantee any of the products described herein or perform any independent analysis in connection with any of the product information contained herein. Publisher does not assume, and expressly disclaims, any obligation to obtain and include information referenced in this work.

The reader is expressly warned to consider carefully and adopt all safety precautions that might be indicated by the activities described herein and to avoid all potential hazards. By following the instructions contained herein, the reader willingly assumes all risks in connection with such instructions.

THE PUBLISHERS MAKE NO REPRESENTATIONS OR WARRANTIES OF ANY KIND, INCLUDING, BUT NOT LIMITED TO, THE IMPLIED WARRANTIES OF FITNESS FOR PARTICULAR PURPOSE, MERCHANTABILITY OR NON-INFRINGEMENT, NOR ARE ANY SUCH REPRESENTATIONS IMPLIED WITH RESPECT TO SUCH MATERIAL. THE PUBLISHERS SHALL NOT BE LIABLE FOR ANY SPECIAL, INCIDENTAL, CONSEQUENTIAL OR EXEMPLARY DAMAGES RESULTING, IN WHOLE OR IN PART, FROM THE READER'S USES OF OR RELIANCE UPON THIS MATERIAL.

[1]*National Electrical Code* and *NEC* are registered trademarks of the National Fire Protection Association, Inc., Quincy, MA 02169.

THIS BOOK CONVEYS THE INFORMATION RELATED TO EACH CHANGE AS OF JULY 1, 2022, BUT DOES NOT REFLECT ANY SUBSEQUENT APPEAL OR ACTION TAKEN BY THE NFPA STANDARDS COUNCIL.

TABLE OF CONTENTS

RESOURCES

[1]The following are registered trademarks of the National Fire Protection Association, Inc., Quincy, MA 02269.

NFPA 20, *Standard for the Installation of Stationary Fire Pumps for Fire Protection®*

NFPA 30, *Flammable and Combustible Liquids Code®*

NFPA 30A, *Code for Motor Fuel Dispensing Facilities and Repair Garages®*

NFPA 33, *Standard for Spray Application Using Flammable and Combustible Materials®*

NFPA 34, *Standard for Dipping and Coating Processes Using Flammable or Combustible Liquids®*

NFPA 54, *National Fuel Gas Code®*

NFPA 70, *National Electrical Code®* 2002, 2005, 2008, 2011, 2014, 2017, 2020, 2023.

NFPA 70-E, *Standard for Electrical Safety in the Workplace®*

NFPA 72, *National Fire Alarm Code®*

NFPA 79, *Electrical Standard for Industrial Machinery®*

NFPA 88A, *Standard for Parking Structures®*

NFPA 99, *Standard for Health Care Facilities®*

NFPA 101, *Life Safety Code®*

NFPA 220, *Standard on Types of Building Construction®*

NFPA 303, *Fire Protection Standard for Marinas and Boatyards®*

NFPA 325, *Guide to Fire Protection Hazard Properties of Flammable Liquids, Gases, and Volatile Solids contained in the NFPA "Fire Protection Guide to Hazardous Materials®–2001 edition"*

NFPA 497, *Recommended Practice for the Classification of Flammable Liquids, Gases, or Vapors and of Hazardous (Classified) Locations for Electrical Installations in Chemical Process Areas®*

NFPA 499, *Recommended Practice for the Classification of Combustible Dusts and of Hazardous (Classified) Locations for Electrical Installations in Chemical Process Areas®*

NFPA 780, *Standard for the Installation of Lightning Protection Systems®*

The following are registered trademarks of the Institute of Electrical and Electronics Engineers New York, NY 10016-5997.

ANSI C2, *National Electrical Safety Code®*

ANSI 18, *Shunt Power Capacitors®*

ANSI Z535.4, *Product Safety Signs and Labels®*

IEEE 835-1994 IEEE Standard Power Cable Ampacity Tables®

The following are registered trademarks of UL Solutions, Northbrook, IL 60062-2096.

Fire Resistance Directory®

UL Product iQ®

PREFACE

The publication "One- and Two-Family Dwelling Electrical Systems" (also known as Residential Electrical Dwellings) is designed for individuals involved in designing, installing, or inspecting electrical systems in both new and existing homes. This twelfth edition offers comprehensive information on residential electrical construction, in line with the 2023 National Electrical Code® and Part VIII of the 2024 International Residential Code®. Every effort has been made to ensure that the references and content in this book are consistent with the code editions referenced.

This publication honors the dedication of many individuals, with each edition being improved and expanded over the years. IAEI strives for this publication to remain a classic textbook on residential electrical systems, with each new edition reflecting updates and improvements in every code cycle. This edition continues those goals and will likely serve as a valuable training and instructional resource for residential electrical installations.

Who will benefit from this publication? The entire electrical and building inspection community, including electrical contractors, electricians, foremen, estimators, plans examiners, electrical utilities, and others, will find this residential guidebook very useful. It is especially valuable for combination inspectors and home inspectors who may lack the hands-on electrical experience that most electrical inspectors have.

IAEI welcomes any suggestions or comments to improve "One- and Two-Family Dwelling Electrical Systems," as your feedback is always appreciated.

— Independent Alliance of the Electrical Industry

1 General Requirements

Residential wiring is undoubtedly the most common and abundant type of electrical installation in the world. Residential occupancies are where people live, families are raised, and individuals sleep. Safe electrical installations are just as important (if not more) in residential occupancies as in any other type of occupancy.

Safe electrical installations require several key components: design, listing requirements, qualified personnel, *Code* compliance, and inspection. In fact, there are more specific requirements for one- and two-family dwellings than for any other type of occupancy referenced in NFPA 70 *National Electrical Code* (*NEC*). Anyone who designs, installs, or inspects electrical systems in one- and two-family dwellings must be thoroughly familiar with these requirements for safe installations, as found in electrical codes and product safety standards. These codes and standards must be followed carefully to provide an installation that is essentially free from electrical hazards. It is important that qualified persons perform the installation. A qualified person has the necessary skills, knowledge, and experience to ensure a safe installation. It is equally important that the installation be inspected. The inspectors are obligated to verify that applicable code rules have been followed. Knowledge of the applicable electrical code is not an option but a necessity in residential construction and wiring.

This book focuses on the requirements for electrical installations based on the 2023 *National Electrical Code* (*NEC*) and the 2024 *International Residential Code* (IRC). It includes information related to proper electrical inspections of residential wiring systems. The scope of this book covers one- and two-family dwelling electrical systems from the perspective of both the installer and the electrical inspector. A compilation of definitions of electrical terms that should be understood as they relate to the performance of grounding and bonding circuits is also included in this first chapter.

This book emphasizes the proper and consistent use of the defined terminology in both the electrical field by not using "trade slang" and the *NEC* in order to develop a common language of communication.

Structural requirements for one- and two-family dwellings are specified in the applicable building code and must be applied in conjunction with the rules outlined in the applicable electrical code. For example, NFPA 5000, *The Building Construction and Safety Code,* 2021 edition, provides structural requirements for one- and two-family dwellings. Chapter 52 specifically refers to the NEC for all electrical requirements (see 52.1).

2024 INTERNATIONAL RESIDENTIAL CODE

The 2024 International Residential Code (IRC) includes electrical provisions in Part VIII (Chapters 34 through 43), which are largely based on the 2023 edition of the *National Electrical Code* (*NEC*). The IRC adopts select *NEC* provisions that apply to typical one- and two-family dwelling construction and simplifies their application for residential use. While the IRC closely parallels the *NEC* in content and intent, it presents requirements in a streamlined and tabular format specifically tailored for residential systems.

The electrical requirements of the IRC apply only to one- and two-family dwellings and townhouses not more than three stories above grade plane in height. According to IRC Sections E3401.1 and E3401.2, the IRC governs typical wiring methods and materials encountered in residential construction, including requirements for wiring, devices, equipment, and service installations. Services covered by the IRC are limited to single-phase, 120/240-volt systems not exceeding 400 amperes. Installations exceeding these limits fall under the scope of the full *NEC*.

The IRC incorporates provisions from the NEC, granted permission by the National Fire Protection Association (NFPA), ensuring code compatibility. When specific electrical installations, materials, or methods are not addressed by Chapters 34–43, IRC Section E3401.2 directs users to the relevant NEC requirements. This guarantees comprehensive electrical safety, especially for emerging technologies or unusual systems.

Per IRC Section E3403.3, all electrical system components—including devices, materials, and equipment—must be listed and labeled by an approved agency. This requirement aligns with NEC standards, ensuring compliance with recognized testing protocols. The IRC does not set electrical rules that conflict with the *NEC*, but simplifies language for easier use in residential applications.

The IRC emphasizes accessible presentation, providing many of its requirements—such as conductor support, box fill, and wiring methods—in tabular form. This format supports easier inspection, design, and plan review. Chapter 34 sets forth general provisions for electrical work, including inspection protocols, protection of building structure, conductor identification, and the need for approved listings. These serve as foundational requirements for all subsequent chapters in Part VIII.

While many jurisdictions adopt both the IRC and the NEC, others may adopt the *NEC* in full and bypass the IRC's electrical chapters. Electrical professionals must verify local adoption and enforcement practices to ensure compliance. Despite variations in jurisdictional preferences, nearly all residential installations today are expected to comply with NEC provisions—either directly or via the IRC's adoption of NEC content.

The IRC includes a cross-reference appendix (Appendix A) linking IRC provisions to corresponding *NEC* sections, offering a valuable resource for designers, inspectors, and educators working across both codes. Where measurements are concerned, the IRC and *NEC* continue to use the International System of Units (SI) as primary, with inch-pound units provided parenthetically, following the hard conversion standards set forth in *NEC* 90.9.

THE PURPOSE OF THE CODE

Electricity is a powerful force. When under control, electricity, like fire, is a dependable servant and performs endless tasks. When uncontrolled, both electricity and fire become terrible villains that can harm people and their property.

Therefore, one of the primary purposes of the electrical code is electrical safety. An essential aspect of any safe wiring installation is conformity to the electrical code. Typically, local safety regulations, in the form of a municipal ordinance or state law, are based on the latest edition of the electrical code.

The primary purpose of the *NEC*, as outlined in Section 90.2(A), is to provide *practical or hands-on safeguards for persons and property against hazards that can arise from the use of electricity*. While installing electrical equipment in compliance with the *Code* does not guarantee that the system will be free from risk or electrical hazards, doing so reduces the likelihood of fires originating from electrical sources. Fire and electric shock hazards, often present in premises wiring systems, are minimized by proper installation and use.

Another key element in providing a safe electrical installation is to follow the manufacturer's instructions and fully comply with any limitations placed on the installation or use of the equipment. In many cases, these instructions

Photo 1.1. Typical dwelling unit at the electrical rough-in wiring stage

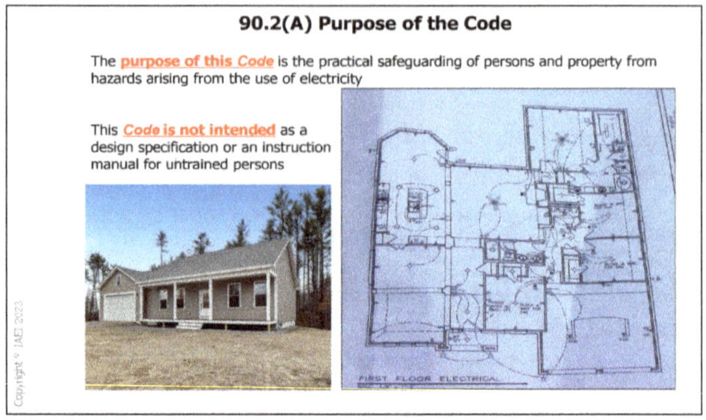

Figure 1.1. Purpose of the *Code*

accompany the product and outline requirements, such as the temperature rating of supply conductors, the minimum circuit size, the maximum rating of overcurrent protection or grounding, or the bonding needs. Compliance with these instructions is required by the Code and product standards (see *NEC* 110.3(B) or IRC E3402.3).

The *NEC* is intended to apply to premises wiring systems. Other codes govern the installation of the utility transmission, distribution, and supply systems. A premise wiring system comprises interior and exterior wiring, including outside branch circuits and feeders installed on or between buildings used for power that supplies energy to motors, lighting, control, and signal circuit wiring that are combined or used with any hardware, fittings, and wiring devices that may be either temporarily or permanently installed. The premises wiring system typically begins at the service point, located at the load end of the service drop (drip loop) or the load end of the underground service lateral, and ends at the branch circuit outlet(s).

ELECTRICAL CODES AND INSPECTION

Electrical inspection services are provided in many areas to determine conformance with the Code and other applicable safety regulations and building codes. These inspections are usually required by state, county, or municipal laws or ordinances. These jurisdictions adopt an electrical code in ordinance form, which then becomes a legally enforceable document. It is also common that these inspection agencies produce and adopt local electrical installation requirements that supplement or amend the NEC or the IRC for their jurisdiction. It is essential that the installer contact the local inspection jurisdiction before commencing an electrical installation or modification to determine which edition of the electrical code has been adopted and to obtain a copy of any local electrical amendments through adopted ordinances.

An electrical installation permit is typically required when an inspection program is in place. It is usually necessary to obtain the permit before starting the installation, and it must be posted in a visible location on the project.

INSPECTIONS REQUIRED

Several inspections are usually needed during the construction of a dwelling. Most jurisdictions require an electrical inspection before covering or "closing in" any electrical work. Some call this the "rough-in" inspection. Typically, multiple inspections are necessary during building to ensure electrical safety and code compliance. Under the 2024 IRC, Section E3403 sets the inspection and approval requirements. It states that all new electrical work, as well as any parts of existing systems affected by modifications or additions, must be inspected and approved by the building official—often a certified electrical or building inspector.

Electrical inspections to determine code compliance at dwelling units are typically performed at three critical stages of construction. These three stages of project completion are referred to as the rough-in, service, and final inspections. Checklists and cross-references provided in this textbook will provide the reader additional information relative to what is intended by the *Code* for the various aspects of one- and two-family dwelling electrical systems. A typical rough-in inspection is performed after all feeder and branch-circuit wiring, including boxes and cables, is installed and before the building finish (such as insulation and sheetrock panels) conceals any of the electrical wiring. At this stage, conductors have typically been spliced inside electrical boxes and enclosures.

The required provisions for grounding devices and equipment are in place. The wiring should not be energized at this

90.4(B) Authority Having Jurisdiction *(AHJ)*

Figure 1.2. The authority having jurisdiction is responsible for approvals of conductors, equipment, and electrical installations

point because the inspector will examine the exposed wiring inside boxes and enclosures. The installation of the electrical service, including grounding and bonding, must be inspected and approved before most electrical utilities connect the power to the electrical utility service. A final inspection is performed after all wiring is complete, including wiring devices, and after luminaires, heating and air-conditioning equipment, and appliances are installed and connected. Some inspection agencies issue a certificate of final inspection or an occupancy permit before the building or the space being added or remodeled can be occupied.

Other safeguards might be available in areas where an electrical licensing and inspection service is not needed or accessible, such as unincorporated regions outside city limits. An independent, private electrical inspector can be a practical option for property owners in these areas. For both the installer and the independent electrical inspector, the owner should request documentation confirming their qualifications for performing electrical installation and inspection. This documentation could include registration with a controlling agency, like a state agency or electric utility, passing one or more electrical certification exams, or evidence of relevant training and experience.

A comprehensive electrical safety system consists of four interdependent components:

1. Robust electrical safety codes and standards, such as the NEC and IRC.
2. Accredited electrical testing laboratories, which certify listed products.
3. Qualified, competent electrical inspectors, trained to evaluate code compliance.
4. Qualified electrical workers and contractors who perform safe, code-compliant installations.

All four elements must function together to minimize risk and ensure long-term system integrity in residential construction. The 2024 IRC, in coordination with the 2023 *NEC*, continues to serve as a foundation for safe and enforceable residential electrical practices across the United States.

APPROVAL FACTORS FOR THE AUTHORITY HAVING JURISDICTION

The authority having jurisdiction (AHJ) is responsible for enforcing the *Code* and interpreting and applying its requirements to the electrical installation under consideration. As mentioned previously, local laws or ordinances set out the requirements for electrical wiring in each jurisdiction.

An AHJ is also responsible for approving electrical equipment that will be used in the electrical installation. The Code specifies that the conductors and equipment required or permitted by the *Code* are acceptable only if approved [*NEC* 110.2 or IRC E3403.1]. Approved is defined by both the NEC and the IRC as acceptable by the authority having jurisdiction. Thus, the electrical inspection authority is responsible for determining which equipment is acceptable for the electrical project. The AHJ may be someone such as the building official, an organization, or an electrical inspector. The authority having jurisdiction (AHJ) is generally defined in the *NEC* as an organization or individual responsible for enforcing the requirements of the electrical code or adopted standard. The AHJ is also responsible for approving equipment, materials, and/or an installation or a procedure. IRC R104.1 generally designates the building official as the authority having jurisdiction.

The term "authority having jurisdiction " is broad in scope because jurisdictions and approval agencies, along with their responsibilities, vary. When public safety is the primary focus, the entity responsible may be a federal, state, local, or other regional department or individual, such as a fire chief, fire marshal, chief of a fire prevention bureau, labor department, or health department; building official; electrical inspector; or others with statutory authority. An insurance inspection department, rating bureau, or other insurance company representative may serve as the AHJ for insurance purposes. In some cases, the property owner or their designated agent assumes the role of the AHJ; at government installations, the commanding officer or departmental official may be the AHJ.

LISTED OR LABELED EQUIPMENT

Most electrical inspection authorities rely extensively on listing and labeling of electrical equipment as a core element of code compliance and enforcement. This listing is performed under the evaluation programs of qualified electrical product testing laboratories. In many jurisdictions, building codes or ordinances mandate the exclusive use of listed and labeled products in all permitted electrical work. Other jurisdictions may enforce listing requirements only when explicitly called out in the applicable code for a particular product or equipment type.

Equipment used in electrical installations should be listed or labeled by a qualified third-party electrical product testing laboratory. The Code requires electrical installations and equipment to be approved [*NEC* 110.2 or IRC E3403.1]. Inspectors are required to approve installations and generally base part of the approvals on the use of listed products and equipment.

The IRC has a mandatory requirement that all electrical materials, components, devices, luminaires, and equipment be listed [IRC E3403.3]. Equipment is also required to bear a listing label and to be installed, used, or both per the manufacturer's instructions. Product examinations should be performed by a third-party, independent, and qualified testing organization with the facilities, testing equipment, and qualified staff to perform these examinations. This organization should also perform ongoing inspections of the production of this electrical equipment [see *NEC* 110.3(C)]. The Occupational Safety and Health Administration (OSHA) recognizes qualified electrical testing laboratories that perform evaluations, testing, and certification of certain products to ensure that they meet the requirements of both the construction and general industry OSHA electrical standards. If the listing (product certification) is done under a qualified electrical testing laboratory program, this listing mark signifies that the tested and certified product complies with the requirements of one or more appropriate product safety test standards [see *NEC* 110.3(C), Informational Note].

For listed and labeled equipment, the primary role of the inspector is to ensure that the listed product is installed per the manufacturer's instructions and that the product has been tested and listed for installation and use. The *NEC* provides additional guidance in that factory-installed internal wiring or the construction of equipment listed by a qualified electrical product testing laboratory need not be inspected at the time of installation, except to detect any unnecessary alterations or damage [*NEC* 90.7].

INSTALLATION AND USE

Listed or labeled equipment is required to be installed and used in accordance with any instructions included in the listing or labeling and supplied by the manufacturer. It is important to apply these instructions properly in the field. Just because a device or a unit of electrical equipment is listed does not mean the installation of the equipment is acceptable. The product must be installed and used within the limitations of the listing. For example, if an electrical enclosure is listed for and limited to a dry location only, it cannot be installed in a damp or wet location. This would be a violation of the *Code*; in addition, it would violate a listing requirement [*NEC* 110.3(B) and IRC E3403.3].

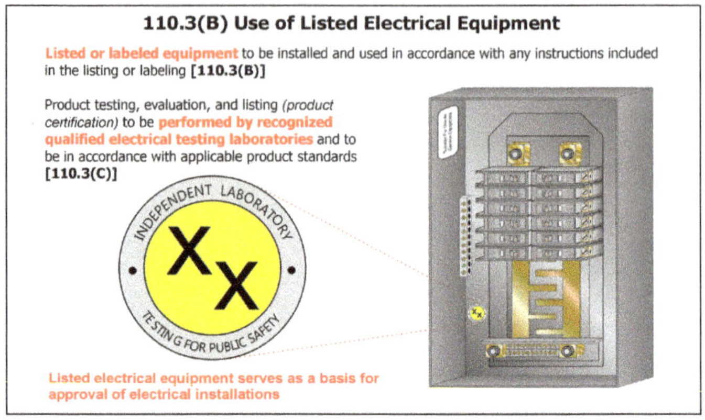

Figure 1.3. Listing marks and labels on electrical equipment

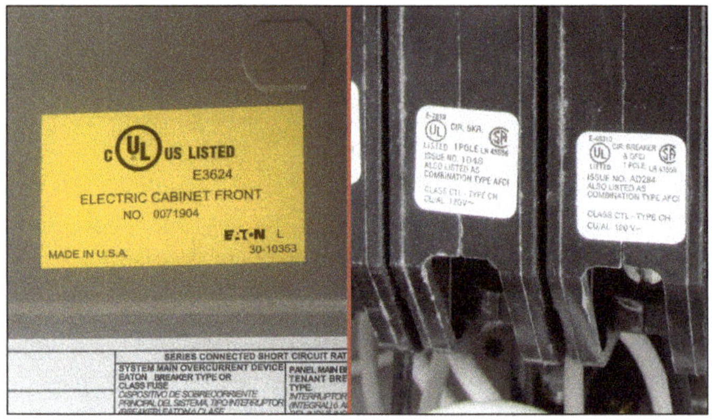

Photo 1.2. Typical listing marks and labels placed on listed electrical equipment

The manufacturer's listing requirements and the Code sections previously discussed provide essential guidelines that must be carefully followed. An additional example related to dwelling units involves recessed incandescent luminaires, which come in various styles. Some of these luminaires are suitable for direct embedding in thermal insulation (Type IC rated), while others are not (non-IC rated). Additionally, the manufacturers' instructions specify which types of trims can be installed on particular rough-in housings. Typically, an open ring-type finished recessed trim can make a recessed luminaire IC-rated, whereas a completely enclosed glass type trim generally classifies that recessed luminaire as non-IC-rated. All these markings and instructions need to be adhered to for a safe and code-compliant installation.

The evaluation, testing, and certification of electrical products must be performed by a qualified third-party testing laboratory recognized under OSHA's Nationally Recognized Testing Laboratory (NRTL) program. These laboratories verify that products meet relevant ANSI/UL, CSA, or other

Article 100: Qualified Person

Qualified Person: "One who has the skills and knowledge related to the construction and operation of the electrical equipment and installations and has received safety training on the hazards involved."

Figure 1.4. Qualified person has received safety training

nationally accepted safety standards.

Common examples of NRTLs include:

- Underwriters Laboratories (UL)
- Intertek Testing Services (ETL)
- CSA Group
- TÜV Rheinland, among others

As of 2024, OSHA recognizes 21 NRTLs, all of which maintain detailed product directories, guide information, and certification guidelines that inspectors and professionals can reference.

Many inspectors use UL Product iQ®, CSA Group listings, and other certification directories as tools to verify product suitability and correct application. These directories often include:

- Installation conditions
- Voltage and environment ratings
- Compatible components (e.g., trims, lenses, drivers)

Some authorities treat this information as equivalent to *Code* requirements, while others view it as supplemental, applicable where the code defers to listing conditions.

The terms "listed" and "labeled" are often used together, but some NRTLs may not recognize a product as fully listed unless it also bears a physical marking or label. The Authority Having Jurisdiction (AHJ) must verify the method a particular testing organization uses to indicate compliance—such as:

- Permanent listing marks
- Holographic labels
- Stamped certification numbers
- QR-code-based compliance tools

Ultimately, it is the AHJ's responsibility to ensure that the product is both listed and installed within the bounds of its listing. Failure to do so can compromise safety and invalidate inspection approvals.

QUALIFIED PERSONS

Electrical wiring and equipment should be installed by competent, qualified, and properly trained individuals. An improper installation increases the risk of fire and injury. When installed incorrectly, even the best-designed and manufactured electrical products can become hazards. Good intentions alone are not enough. Qualified and experienced individuals are trained to select equipment suited for the use and environment and install circuits capable of handling the load.

The term "qualified person" is defined identically in both the 2024 IRC (Chapter 35) and the 2023 NEC (Article 100) as:

> "One who has skills and knowledge related to the construction and operation of the electrical equipment and installations and has received safety training to recognize and avoid the hazards involved."

This definition highlights both technical competence and safety training as essential elements. It is not sufficient to be familiar with tools or equipment; a qualified person must also understand hazard recognition, code compliance, and safe work practices.

Credentialing and Licensing

In jurisdictions with formal electrical inspection and enforcement programs, individuals performing electrical work are typically required to:

- Complete an apprenticeship or vocational training program.
- Pass a competency-based written examination.
- Obtain and maintain an electrician or electrical contractor license from the state or local authority.

These licenses often have a renewal cycle of one to three years and require documented continuing education to ensure electricians stay current with code updates, technology changes, and safe work practices.

Jurisdictions without formal licensing may still require verification of competency for permitting or inspection purposes, and property owners in such areas are strongly encouraged to request proof of qualifications from anyone performing electrical work.

The Importance of Proper Training

Electrical installations become hazardous when:

- Equipment is installed contrary to the manufacturer's instructions.
- Improper materials are used for the environment or application.

- Work is performed by those lacking technical knowledge or training.

Even seemingly minor installation errors—such as improper conductor sizing, incorrect enclosure rating, or insecure terminations—can cause long-term system failure, arc flash events, or thermal hazards.

Safety Standards and Training

To meet the definition of "qualified," individuals must receive appropriate safety training, particularly for tasks involving potentially energized equipment. The recommended industry standard is NFPA 70E, *Standard for Electrical Safety* in the Workplace, which outlines:

- Electrical hazard awareness.
- Personal protective equipment (PPE) requirements.
- Safe work procedures.
- Shock and arc flash risk assessments.

While the *NEC* and IRC serve as enforceable minimum safety codes, they are not training manuals. The Code explicitly states that it is not intended to provide design specifications or serve as a substitute for formal instruction. Real-world competency is developed through:

- Structured education and apprenticeship,
- On-the-job experience under supervision,
- Ongoing professional development.

There is no substitute for the practical knowledge gained by working as an electrician, electrical designer, inspector, or engineer under qualified guidance.

MECHANICAL EXECUTION OF WORK

Despite all the rules and regulations governing the design, manufacture, testing, and installation of electrical equipment, much of the responsibility for a safe and proper installation relies on the integrity of the installer and a thorough electrical inspection. The old saying, "If it's worth doing, it's worth doing right," applies to electrical wiring and inspection. Ultimately, much of the success or failure of an installation depends on the pride and dedication of both the installer and the inspector, as well as their commitment to ensuring a safe, code-compliant setup.

ELECTRICAL EQUIPMENT INSTALLATION

The *NEC* mandates that electrical equipment be installed professionally and skillfully. Judging whether an electrical installation is done professionally, skillfully, neat, or workmanlike can be subjective. However, it is generally true that

Photo 1.3. Residential branch circuits and feeders installed at panelboards in a neat and workmanlike manner

"Good work looks good." Inspectors understand that when they see equipment like conduits, boxes, not installed plumb or cables twisted and randomly installed, other parts of the installation are likely to be flawed as well [NEC 110.12].

"Equipment" is defined by the Code as a general term that can include fittings, devices, appliances, luminaires, apparatus, machinery, and the like used as part of an electrical installation [*NEC* Article 100 or IRC Chapter 35]. Thus, the requirement that equipment be installed in a professional and skillful manner is broad and intended to be applicable to the entire electrical installation.

Here are ten basic steps that exemplify good mechanical execution of electrical work:

1. Unused openings in outlets, devices, pull and junction boxes, conduit bodies and fittings, raceways, auxiliary gutters, cabinets, equipment cases, or housings are effectively sealed with knockout seals or other materials that provide substantial protection equivalent to that of the wall of a box or piece of equipment.

2. Conductor insulation is not damaged, and wires are not nicked or damaged at terminations. Care is exercised, and proper tools are used when the insulation is stripped.

3. Cable assemblies are not kinked or excessively bent sharper than the permitted radius. Bends in cable assemblies do not have a radius that is less than five times the diameter of the cable.

4. Staples used to secure cable assemblies are not driven too tightly. In some cases, insulated types of staples may provide better protection against damage.

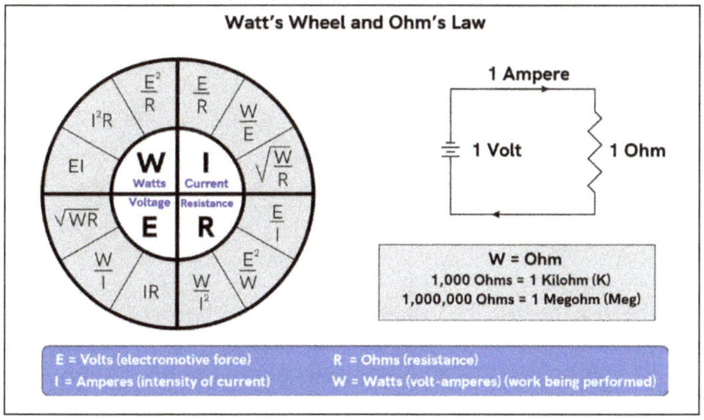

Figure 1.5. Watt's Wheel and Ohm's law

5. All terminations are made in accordance with the manufacturer's instructions provided on the equipment. Screws that engage fewer than two threads and the screws used to mount devices are not used to terminate conductors.

6. Proper tools were used to end the raceways. The scorch and burn marks on nonmetallic conduit are evidence of overheating and may indicate that permanent damage was done to the conduit.

7. Exposed wiring and equipment were protected from contamination during construction. All equipment was cleaned both inside and outside before it was energized.

8. Connections of all metal sheathing of electrical cables, raceways, and equipment are made up wrenchtight. This helps to ensure an effective fault-current path.

9. Insulation integrity for all wiring was verified. This was done to be sure that the system will not suffer from short circuits or ground faults when energized.

10. Protection against physical damage is provided for exposed electrical wiring and equipment. The potential use of the area after construction has been completed and the premises occupied was considered in making this judgment.

OHM'S LAW AND BASIC ELECTRICAL THEORY

Electrical current flowing through any electrical circuit can be compared with water under pressure flowing through a fire hose. Water flowing through a fire hose is measured in gallons per minute (GPM), and electricity flowing through a circuit is measured in amperes (A).

Water flows through a hose when pressure is exerted upon it and a valve is opened. Water pressure is measured in pounds per square inch (psi). An electrical current is moved along an electrical conductor when electrical pressure is applied to it and a path is provided for the current flow. Just as the pounds per square inch (pressure) cause gallons per minute to flow, so the volts (pressure) cause amperes (current) to flow. It takes more pressure to force the same amount of water through a small hose than through a larger hose. A small hose, with the same pressure applied to it in comparison to a larger hose, will pass much less water in a given period of time. It therefore follows that the small hose offers a greater resistance to the flow of water.

In an electric circuit, a greater electrical pressure (volts) will force a given amount of current (amperes) through a small conductor (resistance) than that required to force the same amount of current (amperes) through a larger conductor (resistance). A smaller-sized conductor will allow less current (amperes) to pass than a larger-sized conductor will if the same electrical pressure (volts) is applied to each conductor for the same period. The smaller conductor can only be assumed to offer greater resistance (ohms) than the larger conductor. Thus, we may define resistance as the "property of a body that resists or limits the flow of electricity through it." Resistance is measured in ohms—a term similar to friction in a hose or pipe.

THEORY TERMS AND DEFINITIONS

The following definitions relate to fundamental electrical theory. It is essential that installers and inspectors have a practical understanding of electrical theory. Such knowledge is often crucial in selecting the correct size of conductors for circuits with various loads.

Volt — the unit of electrical pressure — is the pressure required to force one ampere through a resistance of one ohm; abbreviated as "E," the first letter of the term *electromotive force.*

Ampere is the unit of electrical current that will flow through one ohm under a pressure of one volt in one second; abbreviated as "I," the first letter of the term *intensity of current.*

Ohm — the unit of electrical resistance — is the resistance through which one volt will force one ampere; abbreviated as "R," the first letter of the term *resistance.*

Watts is the unit of measurement of the energy flowing in an electrical circuit at any given moment. It is also the amount of work being performed in the electrical circuit. The terms *watts* or *kilowatt* have been used more commonly to express the amount of work done in the electrical circuit rather than the term *joule*. Watts is the product of multiplying volts and amperes and is sometimes referred to as *volt-amperes*. One thousand volt-amperes are referred to as one kilovolt-ampere or one kVA.

OHM'S LAW

German physicist George Simon Ohm discovered the relationship between current, voltage, and resistance in an electric circuit in 1826. He found, by experiment, that pressure equaled the product of current and resistance; this relationship is referred to as Ohm's law. This law is the practical basis on which most electrical calculations are determined. The formula may be expressed in various forms and by its use, as in the three examples shown in Figure 1.6.

If any two values are known, the third can be found by use of the formula. For example, if the resistance and the voltage are known, the current can be determined by dividing the voltage by the resistance. This can be valuable in determining the amount of current that will flow in the circuit to properly size conductors, as well as overcurrent devices.

Horsepower. Mechanical power is typically expressed in horsepower, while electrical power is expressed in watts. The term horsepower originated as the amount of work a strong London draught (draft) horse could do during a short interval. It was also used to measure the output of steam engines. One horsepower, abbreviated as "hp," equals the work required to lift 33,000 pounds one foot (33,000 foot-pounds) in one minute. This is equivalent to lifting 550 pounds one foot in one second.

Often, it is necessary to convert power from one unit to another, and the equation in Figure 1.7 is used to change horsepower to watts or watts to horsepower.

The hp formula applies to laboratory conditions as motors consume more power than they deliver. This is due to power being consumed as heat in the motor to overcome bearing friction, wind resistance, and other factors. For example, a motor of 1 hp (746 watts) may consume nearly 1,000 watts, the difference being consumed in overcoming the factors already stated. The factor of motor efficiency must be considered to determine the true power for single-phase motors.

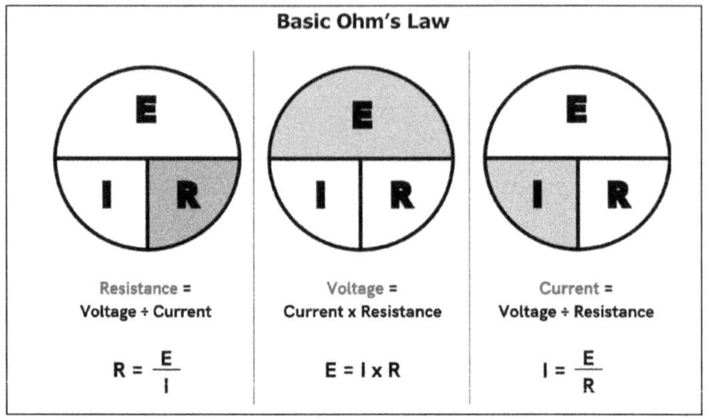

Figure 1.6. Ohm's Law basic examples and application

Power Factor and Horsepower Calculations

$$\text{Efficiency} = \frac{\text{Power Output}}{\text{Power Input}}$$

$$\text{Horsepower} = \frac{\text{Volts x Amps x Efficiency x PF}}{746}$$

$$\text{Amperes} = \frac{746 \text{ x HP}}{\text{Volts x Efficiency x PF}}$$

PF = Power Factor

746 Watts = 1 Horsepower

$$\text{Horsepower} = \frac{\text{Watts}}{746}$$

Figure 1.7. Basic power factor and hp formulas

WATT'S WHEEL

The Watt's Wheel has been included in many manuals and variations to demonstrate watts or power and its connection to the elements of Ohm's law. As shown in this text, it is accurate for DC circuits and for resistive loads in AC circuits where the power factor is close to 100 percent or unity (see Figure 1.5). Do not try to use it for motor loads, as both power factor and motor efficiency need to be considered in the formula (see Figure 1.7).

In AC circuits, we use the term impedance rather than ohms to represent the resistance of the circuit. Impedance is the total opposition to current flow in an AC circuit; it is measured in ohms. Impedance includes resistance, capacitive reactance, and inductive reactance. The last two factors are unique to AC circuits and can usually be ignored in circuits such as incandescent lighting loads and heater circuits, which consist of resistive loads. A detailed explanation of capacitive reactance and inductive reactance is beyond the scope of this text but can be found in many excellent texts on electrical theory.

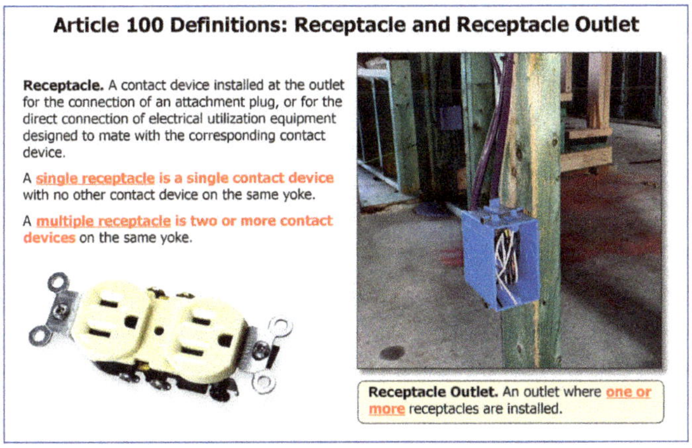

Photo 1.4. Definition of a "Receptacle" (left) and a "Receptacle Outlet" (right)

Photo 1.5. Definition: Utilization equipment (luminaire) installed at a wall-mounted lighting outlet

DEFINITIONS AND TRADE TERMINOLOGY

Both the *NEC* and IRC include some unique definitions that must be clearly understood to properly apply the rules, install, and inspect electrical systems and equipment. Each chapter in this publication will cover relevant definitions found in Article 100 of the *NEC* or Chapter 35 of the IRC. Some definitions apply to all electrical installations, while others are specific to certain parts of the electrical system. For instance, definitions related to swimming pools can be found in Article 100 of the *NEC*, with "(680)" following the IRC E4201.2 of Chapter 42 of the IRC. Let's take a closer look at some common definitions in the Code. Note that all NEC definitions were moved to Article 100 for the 2023 revision cycle.

Appliance. "Utilization equipment, generally other than industrial, that is normally built in standardized sizes or types and is installed or connected as a unit to perform one or more functions such as clothes washing, air conditioning, food mixing, deep frying, and so forth."

Branch-Circuit Overcurrent Protective Device. "A device capable of providing protection for service, feeder, and branch circuits and equipment over the full range of overcurrents between its rated current and its interrupting rating."

Device. "A unit of an electrical system, other than a conductor, that carries or controls electric energy as its principal function."

Equipment. "A general term including material, fittings, devices, appliances, luminaires, apparatus, machinery, and the like used as a part of, or in connection with, an electrical installation."

Fitting. "An accessory such as a locknut, bushing, or other part of a wiring system that is intended primarily to perform a mechanical rather than an electrical function."

Identified (as applied to equipment) "Recognizable as suitable for the specific purpose, function, use, environment, application, and so forth, where described in a particular *Code* requirement."

In Sight From (Within Sight From, Within Sight). "Equipment that is visible and not more than 15 m (50 ft) distant from other equipment is in sight from that other equipment."

Labeled. "Equipment or materials to which has been attached a label, symbol, or other identifying mark of an organization that is acceptable to the authority having jurisdiction and concerned with product evaluation, that maintains periodic inspection of production of labeled equipment or materials, and by whose labeling the manufacturer indicates compliance with appropriate standards or performance in a specified manner."

Lighting Outlet. "An outlet intended for the direct connection of a lampholder or luminaire."

Listed. "Equipment, materials, or services included in a list published by an organization that is acceptable to the authority having jurisdiction and concerned with evaluation of products or services, that maintains periodic inspection of production of listed equipment or materials or periodic evaluation of services, and whose listing states that the equipment, material, or services either meets appropriate designated standards or has been tested and found suitable for a specified purpose."

Luminaire. "A complete lighting unit consisting of a light source such as a lamp or lamps, together with the parts designed to position the light source and connect it to the power supply. It may also include parts to protect the light source or the ballast or to distribute the light. A lampholder itself is not a luminaire."

Outlet. "A point on the wiring system at which current is taken to supply utilization equipment."

Receptacle. "A contact device installed at the outlet for the connection of an attachment plug, or for the direct connection of electrical utilization equipment designed to mate with the corresponding contact device. A single receptacle is a single contact device with no other contact device on the same yoke or strap. A multiple receptacle is two or more contact devices on the same yoke or strap."

Receptacle Outlet. "An outlet where one or more receptacles are installed."

Utilization Equipment. "Equipment that utilizes electric energy for electronic, electromechanical, chemical, heating, lighting, or similar purposes."

[These definitions are from *NEC* Article 100 and/or IRC Chapter 35.]

EQUIPMENT SUITABILITY FOR THE PURPOSE AND ENVIRONMENT

One of the general requirements for electrical equipment is that it be appropriate for the operating environment. Conductors or equipment located in damp or wet areas or where exposed to other agents that could deteriorate them must be suitable for that purpose. It is the responsibility of the Authority Having Jurisdiction (AHJ) to determine the type of location involved. Ideally, the AHJ will base his or her decision on the three location definitions in NEC Article 100 or IRC Chapter 35 (see location definitions below). This decision is crucial because it will determine the types of equipment and wiring methods permitted for that location.

Equipment suitable for use in dry locations only is identified as dry locations, Type 1, or indoor use only. This equipment must always be protected against permanent damage from the weather or adverse conditions during building construction. The types of equipment for protection against environmental conditions can be found in *NEC* Table 110.28 or IRC Table E3404.4.

Photo 1.6. Is this receptacle outlet in a wet or damp location? How about the luminaire?

Photo 1.7. Receptacle outlet installed in a wet location with a in-use cover listed for a wet location

For example, nonmetallic-sheathed (Type NM) cable is suitable for locations that are normally dry. Limited exposure to moisture during construction is to be expected. However, if the ends of the installed cables become saturated with moisture, the water will "wick" up the cable. Type NM cables exposed to this type of moisture must be replaced.

Ordinary locations where electrical equipment is installed include damp, dry, and wet. It is most common to find general-use electrical equipment that is suitable for either dry or wet locations. Some luminaires are identified for damp locations. The applicable definitions from the *Code* are as follows:

Location, Damp. "Locations protected from weather and not subject to saturation with water or other liquids but subject to ***moderate degrees of moisture***. Examples of such locations include partially protected locations under canopies, marquees, roofed open porches, and like locations, and interior locations subject to moderate degrees of

Table 110.28 Types of Enclosures *(in part)*

Enclosure Type For Indoor Use

Provides a degree of protection against the following environmental conditions:	Enclosure Type Number Type 1	Type 2	Type 3
Incidental contact with the enclosed equipment	X	X	X
Falling dirt	X	X	X
Falling liquids and light splashing		X	X

Enclosure Type For Outdoor Use

Provides a degree of protection against the following environmental conditions:	Enclosure Type Number Type 3	Type 3R	Type 3S
Incidental contact with the enclosed equipment	X	X	X
Rain, snow, and sleet	X	X	X
Windblown dust	X		X

Table 1.1. Types of enclosures

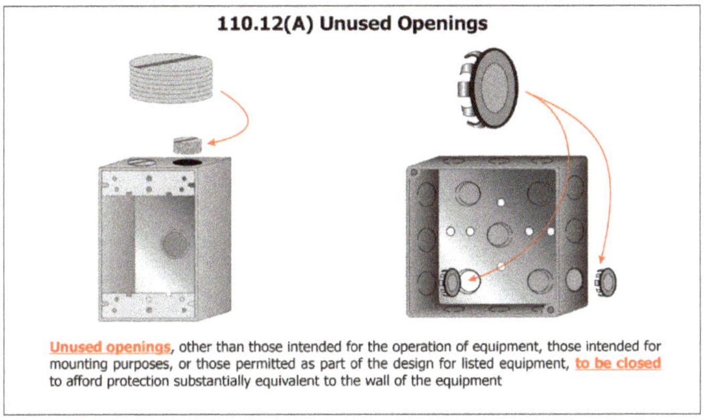

Figure 1.8. Unused openings to be sufficiently closed

moisture, such as some basements, some barns, and some cold-storage warehouses."

Location, Dry. "A location ***not normally subject to dampness or wetness.*** A location classified as dry may be temporarily subject to dampness or wetness, as in the case of a building under construction."

Location, Wet. "A location that is one or more of the following: (1) Unprotected and exposed to weather, (2) ***Subject to saturation with water or other liquids***, (3) Underground, (4) In concrete slabs or masonry in direct contact with the earth"

Determining whether a location is wet, dry, or damp requires sound, reasonable judgment from the AHJ. The AHJ must interpret the boundaries between damp and wet locations where these areas meet or overlap. For example, a dwelling with a carport roof attached to the side of the building has a damp location under the roof, as well as a part of the building wall. The section of the wall exposed to direct rain, snow, or windblown precipitation should be classified as a wet location. As mentioned earlier, the AHJ is generally responsible for determining

Photo 1.8. Unused opening in panelboard sufficiently closed with listed and identified plug

whether a location is damp, wet, or dry, based on the definitions and the product's listing and labeling by a third-party testing agency.

Suitability of equipment for a specific purpose, environment, or application is determined by a qualified testing laboratory, inspection agency, or other organization concerned with product evaluation. Such identification may include labeling or listing of a product. The package or box that contains small products, such as twist-on wire connectors or conduit fittings, may be labeled in lieu of marking the product itself.

OUTDOOR USE AND ENCLOSURE TYPES

The following helpful information is from the Underwriters Laboratories' Guide Information for Electrical Equipment Directory (The Product iQ® material) in the section on Electrical Equipment for Use in Ordinary Locations (AALZ). "In general, individual appliances and equipment have been investigated only for use indoors, in dry locations." See also the information preceding the product category or included in the individual product listing. In some cases, the title (e.g., snow movers, swimming pool luminaires) indicates the conditions for which the product has been investigated.

"Cord- and plug-connected appliances obviously intended for outdoor use, such as gardening appliances, are not intended for use in the rain, and should be stored indoors when not in use."

Electrical equipment is required to be identified for use in certain operating environments that might include damp or wet locations, locations exposed to corrosive influences, or excessive temperatures. The *Code* also gives guidance regarding protection against corrosion, where it states, in part, that corrosive conditions may be present at installations immediately adjacent to a seashore or swimming pool area. As an aid to inspection authorities, UL requires type numbers on power distribution enclosures such as cabinets and cutout boxes, enclosed panelboards or switchboards, meter sockets, enclosed circuit breakers or switches, and other equipment. Table 1.1 summarizes the intended uses of the various types of enclosures that would be applicable to one- and two-family dwelling electrical installations.

UNUSED OPENINGS

The integrity of an electrical enclosure must be preserved. Any unused openings, other than those intended for equipment operation, mounting, or permitted as part of the listed equipment design—such as circuit-breaker openings or cable and raceway knockouts—must be effectively closed to provide protection substantially equivalent to the enclosure's wall. One purpose of electrical enclosures is to contain arcs and sparks that may occur during the operation of overcurrent devices or from a short circuit or ground fault within the enclosure. This helps prevent an arc or spark from contacting a combustible surface outside the enclosure and causing a fire [NEC 110.12(A) or IRC E3404.6 and E3907.5].

Circuit breaker blank filler devices are available from circuit breaker panelboard manufacturers to cover the holes in a panelboard cover caused by removing too many circuit breaker knockouts or by using fewer or different circuit breaker locations during equipment modification. Likewise, knockout plugs are made in all common conduit sizes for filling unused knockout holes in panelboards, wireways, junction boxes, or similar enclosures. Duct tape, electrical tape, or other inferior methods are not acceptable for this purpose. Similar requirements apply for closing unused openings in junction and pull boxes and conduit bodies [NEC 408.7 or IRC E3907.5].

Photo 1.9. Unsused openings for circuit breakers in panelboard cover needs to be sufficiently closed with identified closures (filler plates)

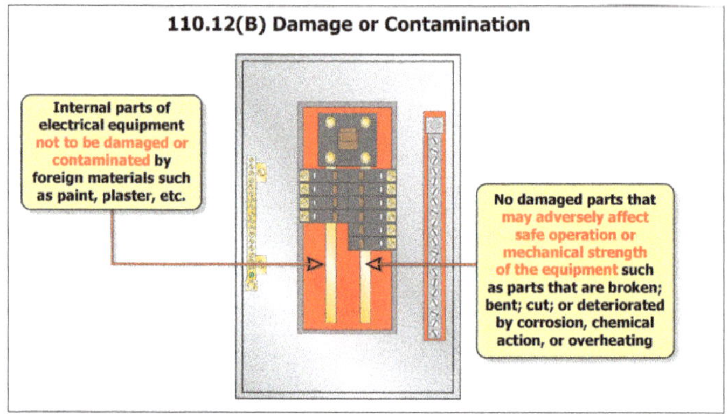

Figure 1.9. Required to maintain integrity of electrical equipment and connections

INTEGRITY OF ELECTRICAL EQUIPMENT AND CONNECTIONS

The integrity of electrical equipment must be protected from damage caused by improper actions on construction sites and the use of damaged equipment. The Code requirements regarding electrical equipment integrity state, "Internal parts of electrical equipment, including busbars, wiring terminals, insulators, and other surfaces, shall not be damaged or contaminated by foreign materials such as paint, plaster, cleaners, abrasives, or corrosive residues. There shall be no damaged parts that may adversely affect safe operation or mechanical strength of the equipment such as parts that are broken; bent; cut; or deteriorated by corrosion, chemical action, or overheating." The IRC further states that foreign debris must be removed from equipment [NEC 110.12(B) or IRC E3404.7].

CORROSION PROTECTION

The *Code* requires electrical equipment to be protected against corrosion. Metal raceways, auxiliary gutters, cable armor, boxes, cable sheathing, cabinets, enclosures (other than surrounding fences and walls), elbows, couplings, nipples, fittings, supports, and support hardware must be made of materials suitable for their installation environment. These items, along with ferrous (magnetic) raceways, auxiliary gutters, cable armor, boxes, cable sheathing, cabinets, metal elbows, couplings, fittings, supports, and support hardware, must be properly protected against corrosion both inside and outside (except at threads in joints) with an approved corrosion-resistant coating, such as zinc, cadmium, or enamel. When boxes or cabinets feature an approved system of organic coatings and are marked "Raintight," "Rainproof," or "Outdoor Type," they are allowed to be installed in outdoor locations [see NEC 300.6 or IRC Table E3801.4 for wiring methods environments].

Photo 1.10. Equipment is required to be securely mounted to the surface in a suitable manner and in a damp or wet location, provide at least 6 mm (1⁄4 in.) airspace between the enclosure and the wall

Photo 1.11. Conductor splices are to be made with connectors suitable for the purpose

Field-cut threads at joints between metal raceway sections, such as rigid or intermediate metal conduit, must be coated with an approved electrically conductive, corrosion-resistant compound. This precaution should not be taken lightly, especially when threads are cut outside the factory where the product is listed (threads cut in the field)! Experience shows that underground raceways enclosing circuit conductors, which are exposed to continuous moisture, can lead to ground-fault conditions where conductors gradually deteriorate due to insulation damage during installation. Ensuring the integrity of conduit joints is crucial for clearing ground faults within the metal raceway and when the raceway functions as the equipment grounding conductor [NEC 300.6(A)].

A qualified inspection authority will not approve conductors that show signs of damage. The proper application of the items mentioned in this chapter and the importance of correct mechanical execution of electrical work should never be overlooked.

Conductor Materials

Wire connectors or the unit containers are marked with the type of conductor material(s) as follows:

Marking ***(or equivalent)***	**For Use With**
"CU"	Copper wire only
"AL"	Aluminum wire only
"AL-CU" or "CU-AL"	Copper to copper, aluminum to aluminum, copper to aluminum but not intermixed or indirect physical contact, copper-clad aluminum to copper-clad aluminum, copper to copper-clad aluminum, aluminum to copper-clad aluminum **but not intermixed or in direct physical contact**
"AL-CU" ***(intermixed – dry locations)***	Copper to copper, aluminum to aluminum, copper to aluminum intermixed and in direct physical contact, copper-clad aluminum to copper-clad aluminum, copper to copper-clad aluminum, aluminum to copper-clad aluminum and in direct physical contact

Table 1.2. Typical markings for single-polarity wire connectors for use with all alloys of copper, aluminum, or copper-clad aluminum conductors, or all three, for the purpose of providing contact between current-carrying parts

MOUNTING OF EQUIPMENT

Electrical equipment must be securely attached to the surface where it is mounted. Using wooden plugs driven into holes in masonry, concrete, plaster, or similar materials is not allowed for this purpose, as they do not provide a durable, secure support for electrical equipment. Proper installation and support are crucial for a safe setup. Various types of anchors are available for securing equipment to masonry, concrete, plaster, and similar materials. Extra precautions are necessary when using treated lumber to support electrical equipment and enclosures. It is common to install four or more wood or sheet metal screws with these anchors or treated lumber when mounting enclosures like meter sockets and panelboards. Following standard recommendations, such as using stainless steel or "triple-dipped" galvanized fasteners, helps ensure the equipment remains securely mounted over time [NEC 110.13(A) or IRC E3404.8].

In damp or wet locations, surface-type enclosures must be mounted so as to prevent moisture or water from entering and accumulating within the cabinet or enclosure. These cabinets or enclosures must be mounted so there is at least 6 mm (1⁄4 in.) airspace between the enclosure and the wall or other supporting surface. Enclosures installed in wet locations are required to be weatherproof. For enclosures in wet locations, raceways or cables entering above the level of uninsulated live parts (such as a busbar) must use fittings listed for wet locations [*NEC* 312.2 or IRC E3907.2].

ELECTRICAL CONDUCTOR (WIRE) CONNECTIONS

An important element of any electrical installation is the electrical connections of conductors (wires). Several important requirements for a safe installation where electrical conductor(s) are terminated will be discussed in detail in the remainder of this chapter.

CONDUCTOR MATERIAL

Electrical terminations (lugs, switches, receptacles, etc.) are marked with the type of conductor material they have been investigated with and found acceptable for termination on said devices. An example of this can be found in the UL Product Spec material under the category for Wire Connectors (ZMVV). Wire conductors and connectors should be installed using the prescribed manufacturers' installation instructions. Wire connectors or the unit containers are marked with the type of conductor material(s) the terminations have been identified for use with. Except as otherwise noted on or in the shipping carton, aluminum conductors are not intended to be used in direct physical contact with copper and copper-clad aluminum conductors in the same connector. A wire connector for securing an aluminum wire in combination with a copper or copper-clad aluminum conductor, where physical contact occurs between the wires of different metals, is limited to dry locations only and is marked "AL-CU (intermixed - dry locations)."

CONDUCTOR SPLICES

The *Code* sets minimum requirements for electrical connections such as splices. Conductors must be spliced or joined using connectors approved for this purpose, or by brazing or welding. Soldered connections should be mechanically and electrically secure without solder, then soldered afterward (solder can melt during a ground fault). All splices, joints, and the free ends of conductors must be covered with insulation matching that of the conductors or with an insulating device designed for this purpose. Wire connectors or splicing devices used for direct burial applications must be listed and suitable for such use. Connections between dissimilar metals in terminal or splicing connectors are only acceptable if the connectors are rated for such intermixing and are used only in dry locations [NEC 110.14(B) or IRC E3406.11].

Additional important information on wire connectors is contained in the Product iQ® material under the category for Equipment for Use in Ordinary Locations (AALZ).

WIRE CONNECTORS

When making splices between circuit conductors, always consider using properly sized wire connectors, commonly called "wirenuts" in the field. Although some connectors display the range of wire sizes they fit, the print is usually too small to read on smaller sizes. It's important to check the smallest container for the connector to find information on how to install it and on the suitable combinations of sizes and number of conductors listed for that connector.

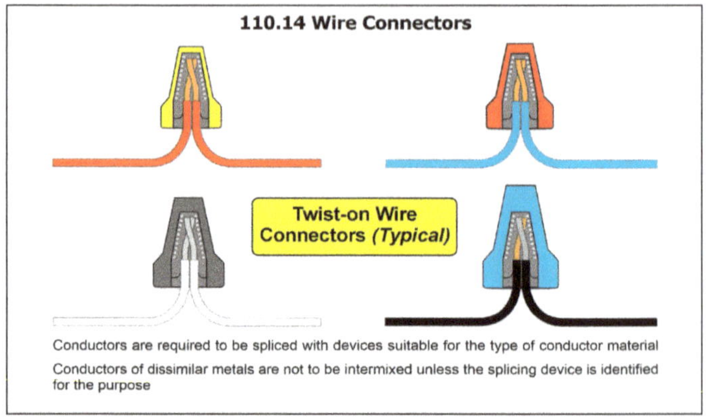

Figure 1.10. Splices with wire connectors listed for the purpose

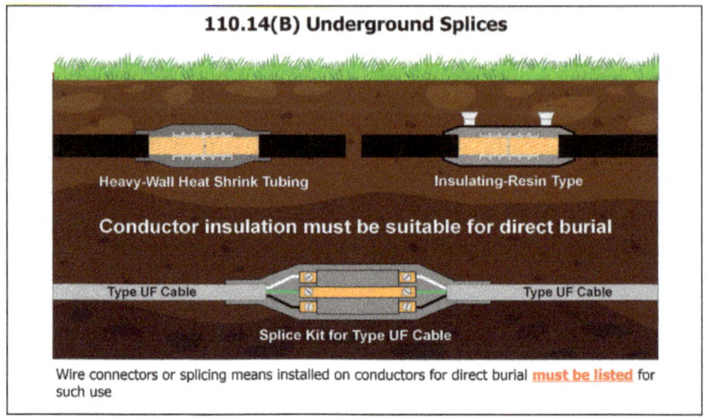

Figure 1.11. Underground splicing means must be listed for the purpose and location

Common wire connectors like wirenuts, wing nuts, and scotchlocks have insulating material inherent with the connector that has a temperature rating of 90°C or more. These connectors can be used on circuit conductors with insulation rated at 60°, 75°, and 90°C.

UNDERGROUND SPLICES

The need to make splices in underground cable runs at one- and two-family dwellings is not uncommon. These cables are often run to supply outdoor lighting, receptacles, garages, carports, storage buildings, and well or pump houses. Proper splices must be made and used where cables must be extended or repaired. Requirements for these underground splices exist in both the *Code* and in *Underwriters Laboratories' Guide Information for Electrical Equipment Directory* 2017 (The UL Product iQ® material) Product Category (ZMWQ). Also see UL Product Standard 486D.

Wire connectors or splicing means installed on conductors for direct burial must be listed for such use. In some cases, an individual wire connector is listed for the connection of

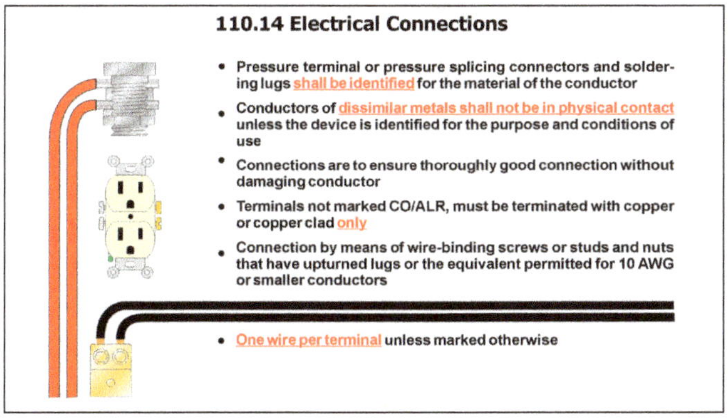

Figure 1.12. Rules for electrical connections

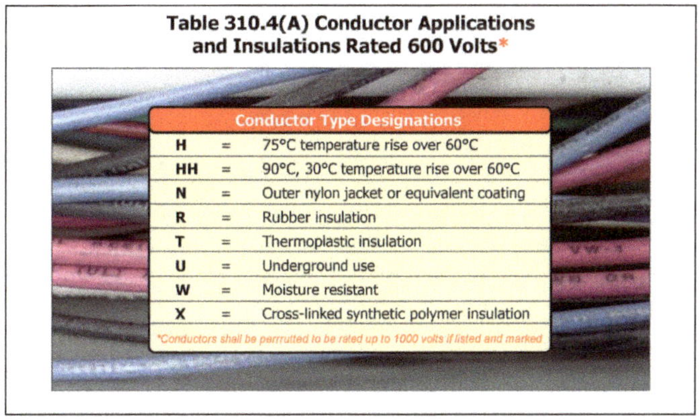

Table 310.4(A) Conductor Applications and Insulations Rated 600 Volts*

Conductor Type Designations		
H	=	75°C temperature rise over 60°C
HH	=	90°C, 30°C temperature rise over 60°C
N	=	Outer nylon jacket or equivalent coating
R	=	Rubber insulation
T	=	Thermoplastic insulation
U	=	Underground use
W	=	Moisture resistant
X	=	Cross-linked synthetic polymer insulation

*Conductors shall be permitted to be rated up to 1000 volts if listed and marked

Table 1.3. Conductor type designations

Photo 1.12. Terminals must be identified for the use and ensure a good connection of the conductor without damage to the conductor

conductors for direct burial. It is critical to note that the conductors themselves must be suitable for direct earth burial. Conductors with Type UF or Type USE insulation are typically suitable for direct burial in earth [NEC 110.14(B) or IRC E3406.11].

Type UF cable constructions encompass both single- and multiple-conductor types. Twist-on type connectors listed as suitable for connecting insulated, single-conductor Type UF cables are available. It is important to note that the outer cable sheathing insulation of Type UF cable is suitable for direct earth burial.

Twist-on type connectors cannot be used for splicing multiple-conductor Type UF cables for direct earth burial because the insulation on the individual conductors inside the cable is not suitable for direct earth burial. Obviously, the outer Type UF cable insulation must be removed to make the splice, exposing the conductor insulation. A splice kit or system must be used to splice these Type UF multiconductor cables. These kits or systems will often include both the splicing means as well as the insulating means. The two common types include heat shrink tubing or epoxy mastic and form or mold to contain the cable, splices, and mastic.

CONDUCTOR TERMINALS

Conductor connections at terminals must be made to ensure a thoroughly good connection between the terminal and the conductor without damaging the conductor. Connections at terminals can be made with pressure connectors, including the set screw types, solder lugs, or splices to flexible leads. Terminals (lug-type) and terminations for aluminum conductors must be identified for the use. Unless the terminal is identified for more than one conductor, only one conductor per terminal is allowed. Rules for panelboards clearly specify that for grounded (neutral) conductors, there is to be only one conductor under a single termination unless the terminal is listed for more than one. However, the manufacturer's markings may indicate that multiple equipment grounding conductors are permitted under the same terminal screw. Terminating the grounded conductor and the equipment grounding conductor under the same terminal screw is not permitted [*NEC* 110.14, 110.14(A), 408.41 or IRC E3406.8, E3406.10, E3706.4].

TEMPERATURE LIMITATIONS OF TERMINATIONS

Electrical conductors must be terminated in such a manner that the ampacity of the conductor is not exceeded. Temperature limitations on conductor terminations for equipment such as circuit breakers, switches, and wiring devices are a crucial requirement of the Code. The general rule is, "The temperature rating associated with the ampacity of a conductor shall be selected and coordinated so as not to exceed the lowest temperature rating of any connected termination, conductor, or device. Conductors with temperature ratings higher than specified for terminations shall be permitted to be used for ampacity adjustment, correction, or both" [NEC 110.14(C) and IRC E3705.4].

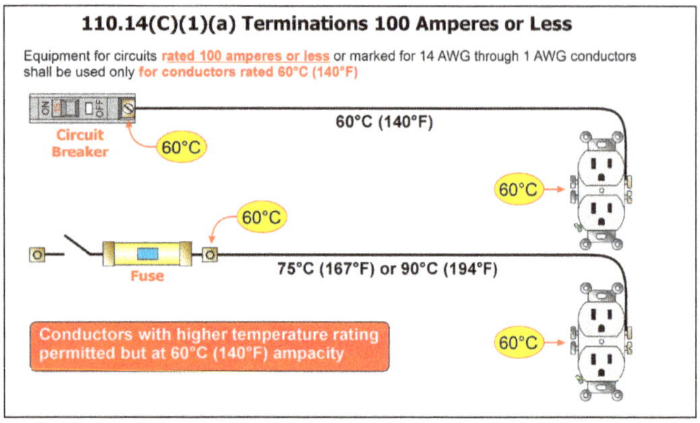

Figure 1.13. Temperature limitations for conductors in sizes 14 AWG through 1 AWG

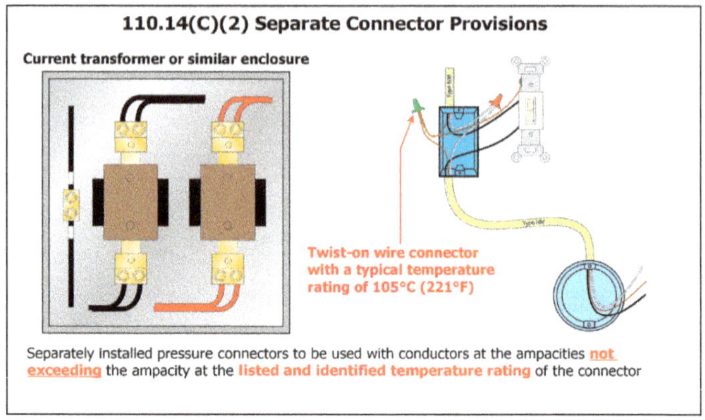

Figure 1.15. Requirements for separately installed pressure connectors

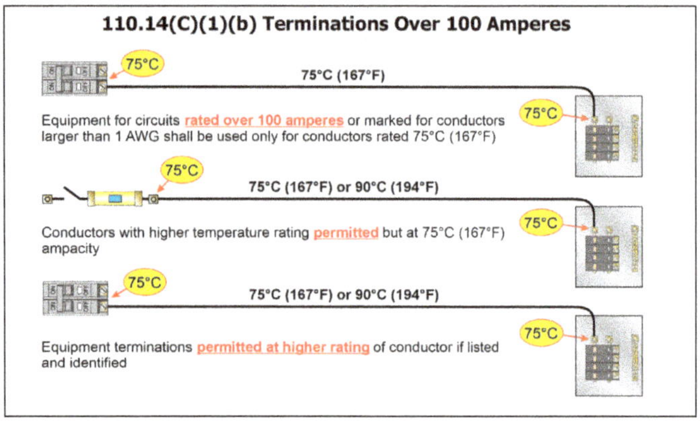

Figure 1.14. Temperature limitations for conductors larger than 1 AWG

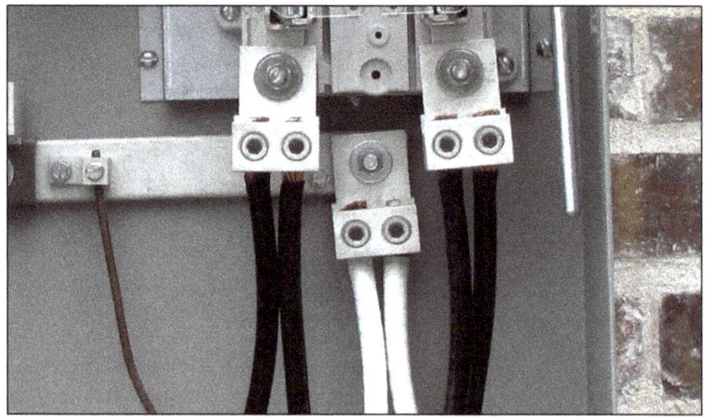

Photo 1.13. Conductor terminals for sizes 14 through 1 AWG are suitable for terminations rated at 100 amperes or less and temperature rating is limited to 60°C (140°F), unless otherwise identified. Terminals for conductors larger than 1 AWG or terminals rated over 100 amperes in equipment are rated at 75°C (167°F).

Information about electrical terminations and the temperature limitations of such terminations is covered in *NEC* 110.14, IRC 3705.4, and the UL Product iQ® material under the category (AALZ) under the subheading of Distribution and Control Equipment Terminations. All electrical terminations must follow these minimum safety requirements in all commercial, industrial, or residential applications. The following information summarizes the temperature limitation provisions.

The *Code* includes requirements for temperature limitations at conductor terminals. A temperature rating of 60°C (140°F) is the minimum starting point for temperature ratings of terminations [*NEC* 110.14(C)(1)(a)(1) or IRC E3705.4.1].

Unless the equipment is listed and identified for use with higher temperature rated conductors, electrical equipment rated 100 amperes or less, or marked for 14 AWG through 1 AWG conductors, the conductor ampacity is generally based on a 60°C (140°F) rating even though the conductors might be rated for higher temperatures, such as 75°C (167°F).

If there is one "H" in the insulation designation, the conductor has 75°C (167°F) insulation; and if there are two "HH" in the insulation designation, the conductor has a 90°C (194°F) rating.

Motors designed to meet the National Electrical Manufacturers Association (NEMA) standards are evaluated based on the 75°C (167°F) ampacity rating of conductors, even for equipment rated 100 amperes or less. However, to use this provision, all the terminations for the same conductor must be rated for 75°C (167°F), or the lower-rated 60°C (140°F) ampacity must be used.

Conductors rated lower than the equipment's designed use can cause obvious overheating problems.

An additional modification of this general requirement for 75°C (167°F) terminations is also provided. This is identical in concept to the additional provisions for 60°C (140°F) conductor terminations.

For equipment rated over 100 amperes or marked for conductors larger than 1 AWG, the conductor ampacity is based on a 75°C (167°F) rating even though conductors might

General Temperature Ratings at Terminals of Electrical Equipment

Equipment	UL Standard	Temperature Ratings
Circuit breakers	UL 489	75°C (167°F) only or 60°C (140°F) / 75°C (167°F)
Conductors *(Thermoplastic)*	UL 83	Table 310.4(A); Table 310.16
Conductors *(Thermoset)*	UL 44	Table 310.4(A); Table 310.16
Disconnect switches	UL 98	75°C (167°F) only or 60°C (140°F) / 75°C (167°F)
Receptacles/attachment plugs	UL 498	60°C (140°F) *(unless marked higher)*
Panelboards	UL 67	75°C (167°F) *(see breaker markings)**
Snap switches	UL 20	60°C (140°F) *(unless marked higher)*
Wire splicing connectors	UL 486C	105°C (221°F)

**It is important to verify the temperature limitations of the panelboard and the limitations based on the device installed in the panelboard. While the panelboard may carry the 75°C (167°F) terminal rating, the circuit breakers suitable for installation might carry only the 60°C (140°F) terminals. Many of the 12- through 50-ampere circuit breakers carry a combination of 60°C (140°F) / 75°C (167°F) temperature ratings.*

Table 1.4. General temperature ratings at terminals of electrical equipment

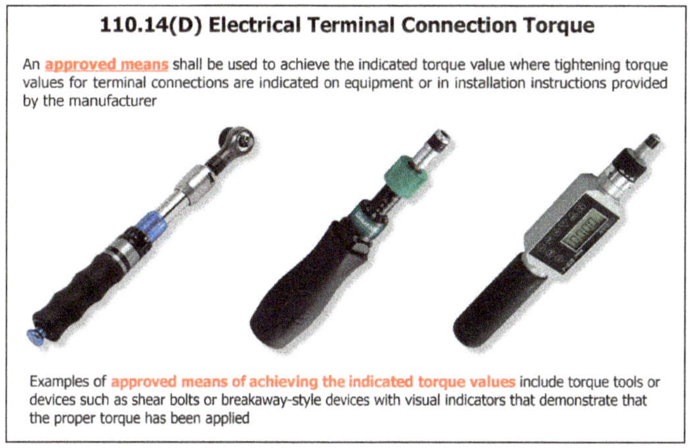

Figure 1.16. Tightening torque tools or devices required where torquing is specified on the equipment or in installation instructions

be rated for a higher temperature, such as 90°C (194°F). Conductors with insulation ratings higher than required or allowed for termination purposes are permitted for the purpose of applying correction factors, such as for conditions where more than three current-carrying conductors are installed in a raceway or cable or where the conductors are installed through ambient temperatures greater than 30°C (86°F) [*NEC* 110.14(C)(1)(b)(1) and IRC E3705.4.2].

SEPARATELY INSTALLED WIRE CONNECTORS

Separately installed pressure connectors are required to be used with conductors at ampacities not exceeding the current rating at the listed and identified temperature rating of the connector. The term separately installed is interpreted to apply to wire connections made at locations such as in outlet or junction boxes, in metering current transformer cabinets, or in wireways or auxiliary gutters. The temperature rating of the insulation and wire connector becomes the limiting factor at this point. It is quite common to find that twist-on wire connectors have a temperature rating of 105°C (221°F). Other wire connectors, commonly referred to as lugs, have a temperature rating that is identified on the connector. Remember the "weakest link" rule. It is not uncommon for separately installed pressure connectors to have a temperature rating that exceeds the temperature rating of the terminals on equipment such as circuit breakers. In this case, the terminals on the circuit breaker become the limiting factor on the entire circuit, including the separately installed connectors [*NEC* 110.14(C)(2) or IRC E3705.4.3].

TEMPERATURE RATING OF WIRE TERMINALS AND CONDUCTOR INSULATION TYPES

Manufacturers of electrical equipment may use terminals that, by themselves, in separate enclosures are suitable for higher temperatures such as 75°C (167°F) or 90°C (194°F). However, when combined with other components, such as circuit breakers or enclosed switches, these terminations become a part of the equipment in which they are installed and are subject to the overall rating imposed by the manufacturer. For additional information on general temperature ratings at electrical equipment terminals, see Table 1.4.

The NEC (Articles 310.3 and 310.4) and IRC (E3406.5) require that conductors be insulated with a material suitable for the specific application, including the environmental conditions they may be exposed to.

Acceptable insulation types for wet locations include:

- RHW, TW, THW, THHW, THWN, XHHW

Each of these types includes the letter "W", indicating wet-location approval. Other letter indicators signify specific material and thermal properties:

- "T" = Thermoplastic insulation
- "R" = Rubber insulation
- "H" = Heat-resistant (+15°C over 60°C = 75°C rating)
- "HH" = High heat-resistant (+30°C over 60°C = 90°C rating)
- "N" = Nylon jacket or equivalent protective covering

For example, THWN-2 conductors are suitable for wet and dry locations and have a 90°C rating in both environments.

A wet location is defined under the definition of location in *NEC* Article 100 or IRC Chapter 35 as follows: "Installations underground or in concrete slabs or masonry in direct contact with the earth; and locations subject to saturation with water or other liquids, such as vehicle washing areas; and in unprotected locations exposed to weather."

Where ungrounded (hot) conductors are installed for branch circuits or feeders at one- and two-family dwelling units, they are required to be insulated with material of composition and thickness that is recognized by the *Code* as electrical insulation (see the *NEC* or IRC definition of conductor). Conductors must not only be suitable for use in wet locations but, in addition, where exposed to the direct rays of the sun, they are required to be listed for sunlight resistance or listed and marked "sunlight resistant." Most manufacturers will include sunlight-resistant markings where the product is suitable for this purpose.

CONDUCTOR TERMINATIONS AND TORQUE

The connection of circuit conductors at the pressure terminal or pressure splicing devices must be tightened according to the manufacturer's instructions to ensure a proper connection. This connection must not be made in a way that would lead to damage to the conductor, either by over-tightening or under-tightening. Many terminations and their associated equipment are marked with the tightening torque specifications. NEC Section 110.14(D) and IRC E3406.12 call for the implementation of an approved means (such as tightening torque tools) to be used to achieve the indicated torque value where tightening torque values are specified on the equipment or in installation instructions provided by the manufacturer. Examples of an "approved means" of achieving the indicated torque values include torque tools or devices, such as shear bolts or breakaway-style devices with visual indicators that demonstrate the proper torque has been achieved. Once again, this is another Code requirement where it is essential to understand who does the "approving." According to NEC Article 100 or IRC Chapter 35, "Approved" means "acceptable to the authority having jurisdiction (AHJ)."

The use of proper torque tools or devices is essential to ensure that terminations are properly made and that the equipment will function properly throughout its life cycle. Without the use of the proper torquing tools or devices, it has been demonstrated that even the most experienced electricians cannot consistently tighten the connector to the manufacturer's recommended torque value.

When listed equipment is used and conductor terminations with specified torque requirements are incorporated, the tightening torque specified becomes a requirement of the

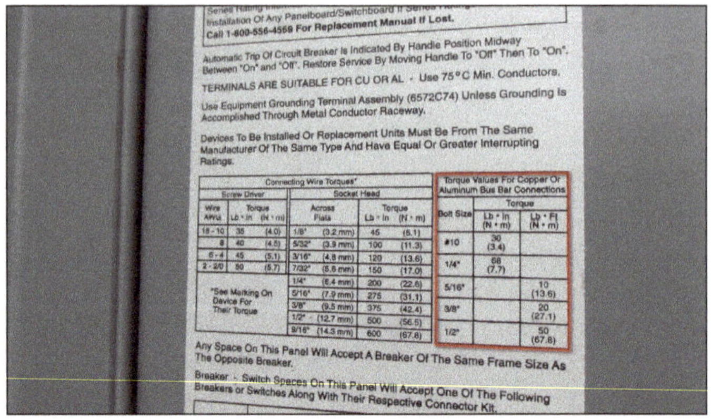

Photo 1.14. Torquing values are designated on various pieces of equipment

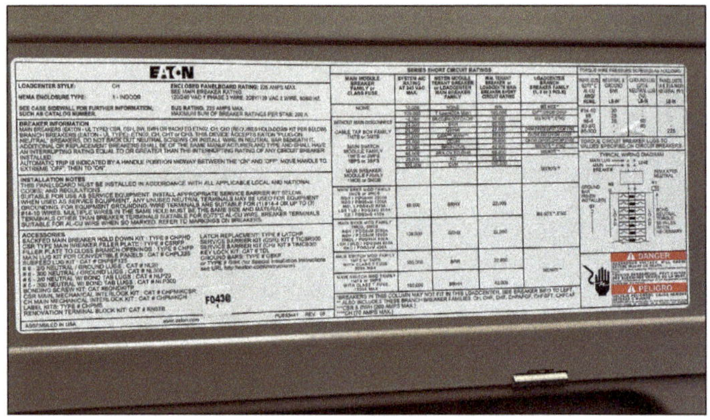

Photo 1.15. Instructions and manufacturer's specifications are found on labels inside panelboards and other type of electrical equipment

Code. The listing of the product requires the termination to comply with that torque marking. This information is often overlooked during the installation of electrical equipment.

Some examples of the different types of pressure connectors available include the following:

1. Set screw types
2. Compression terminals
3. (Crimp-type) connectors

SOLDERED CONNECTIONS

Solder connections are less common in modern electrical work. They were used in many older knob-and-tube systems. The Code still includes solder rules, but the connection must be made mechanically first. If a splice is to be soldered, it must be spliced or joined so it is both mechanically and electrically secure without solder, then soldered. For copper conductors, twisting small-sized conductors together usually counts as a mechanical splice. All splices, joints, and the free ends of conductors must be covered with insulation matching the conductor's insulation or with a device designed for that purpose. This rule applies to all circuit conductors except the grounding electrode conductor, equipment grounding

Photo 1.16. Aluminum conductors with oxide inhibiting compound. Terminations of aluminum conductors in electrical equipment must follow important guidelines to ensure suitable connections.

conductors, and bare grounded neutral service conductors in residential buildings.

EQUIPMENT MARKING

Instructions found on the inside cover of many panelboards and on other electrical equipment include information regarding proper size and type of conductor recognized for various terminals, and the proper tightening torque for the respective conductor.

Remember that listed equipment includes instructions by the manufacturer that are within the listing limitations of the product. Compliance with these markings and instructions is not an option, but a requirement [*NEC* 110.3(B) or IRC E3403.3].

ALUMINUM WIRE TERMINATIONS

To help guard against overheating and conductor terminal failures, the following procedure is recommended when connecting aluminum conductors (wire).

Step 1. Strip the wire of its insulation to the desired length without ringing or nicking the wire. Remove all coating on the wire.

Step 2. Brush the stripped portion of the conductor thoroughly with a wire brush.

Step 3. Thoroughly coat the stripped conductor with an oxide-inhibiting compound.

Step 4. Insert the conductor into the connector and tighten the connector screws securely. This operation should result in the compound oozing out from between the individual strands of the conductor. Failure of a compound to ooze out from between strands indicates insufficient coating.

Step 5. Wipe excess compound from the area adjacent to the connection because some compounds contain metallic particles and could reduce the dielectric strength of the insulating materials employed. Some compression connectors are even filled with an anti-oxidation agent to make faster and more reliable connections. Follow the manufacturer's instructions and guidelines for these terminations.

Refer to NECA/AA 104-2012, Standard for Installing Aluminum Building Wire and Cable (ANSI), for more comprehensive and detailed information on proper termination and installation methods for aluminum building wire and cables.

CONDUCTOR TERMINATION COMPOUND

Conductor termination compounds are for use on splice and termination connections of aluminum, copper-clad aluminum, and copper conductors to retard oxidation at the conductor interface. These compounds do not have a deleterious effect on the conductor metal, insulation, or equipment when used in accordance with the manufacturer's installation instructions. Refer to the product label located on the smallest unit container for specific instructions as to the proper use of the compound. If further information is required, contact the wire manufacturer.

AMPACITY OF TYPE NM CABLE

The typical and most widely used wiring method at dwelling units is by far nonmetallic-sheathed cable (Type NM cable). The insulated conductors are required to be rated for 600 volts and range from sizes 14 AWG through 2 AWG copper conductors and 12 AWG through 2 AWG aluminum conductors [*NEC* 334.104]. In addition to the insulated conductors, the cable shall have an insulated or bare conductor to be used as an equipment grounding conductor only [*NEC* 334.108].

The insulation shall be rated for 90°C (194°F) [*NEC* 334.112 or IRC E3705.4.3]. The ampacity of Type NM cable is that of 60°C (140°F) conductors in accordance with *NEC* 310.14 and *NEC* Table 310.16 or IRC Table E3705.1 [*NEC* 334.80 or IRC E3705.4.3]. The overall covering of Type NM cable shall be flame-retardant and moisture-resistant [NEC 334.116(A)].

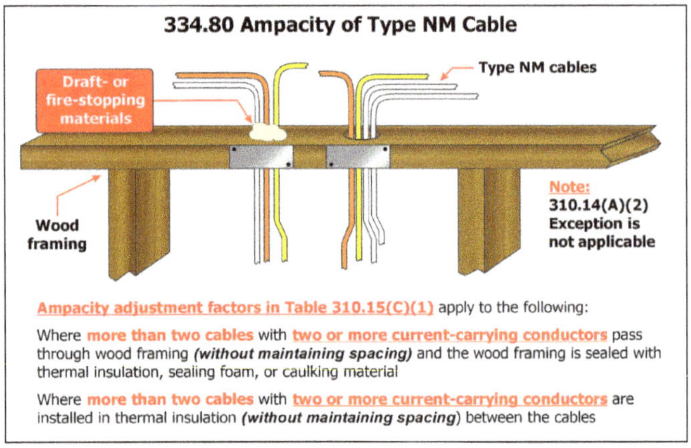

Figure 1.17. Ampacity adjustment factors for Type NM cable

Photo 1.17. Bundled cables that do not maintain spacing between cables must generally be derated

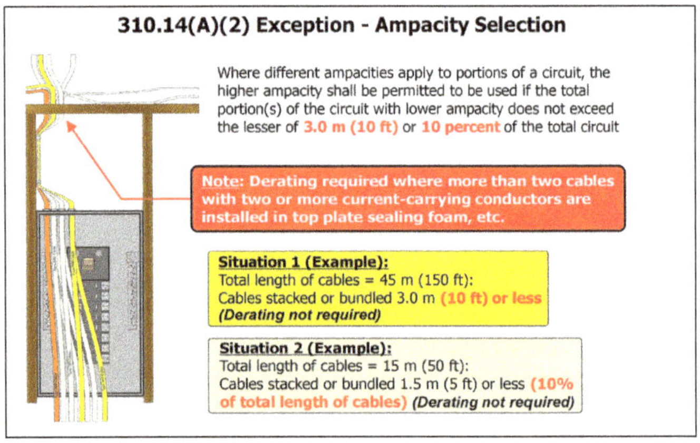

Figure 1.18. Correction factors are not required of bundled cables as shown

Table 310.16 *(in part)*

Table 310.16 Ampacities of Insulated Conductors with Not More Than Three Current-Carrying Conductors in Raceway, Cable, or Earth (Directly Buried)

Size AWG or kcmil	Temperature Rating of Conductor [See Table 310.4(A)]						Size AWG or kcmil
	60°C (140°F)	75°C (167°F)	90°C (194°F)	60°C (140°F)	75°C (167°F)	90°C (194°F)	
	Types TW, UF	Types RHW, THHW, THW, THWN,XHHW, XHWN, USE, ZW	Types TBS, SA, SIS, FEP, FEPB, MI, PFA, RHH, RHW-2, THHN, THHW, THW-2, THWN-2, USE-2, XHH, XHHW, XHHW-2, XHWN, XHWN-2, XHHN, Z, ZW-2	Types TW, UF	Types RHW, THHW, THW, THWN, XHHW, XHWN, USE	Types TBS, SA, SIS, THHN, THHW, THW-2, THWN-2, RHH, RHW-2, USE-2, XHH, XHHW, XHHW-2, XHWN, XHWN-2, XHHN	
	COPPER			ALUMINUM OR COPPER-CLAD ALUMINUM			
18*	—	—	14	—	—	—	—
16*	—	—	18	—	—	—	—
14*	15	20	25	—	—	—	—
12*	20	25	30	15	20	25	12*
10*	30	35	40	25	30	35	10*
8	40	50	55	35	40	45	8
6	55	65	75	40	50	55	6
4	70	85	95	55	65	75	4
3	85	100	115	65	75	85	3
2	95	115	130	75	90	100	2
1	110	130	145	85	100	115	1
1/0	125	150	170	100	120	135	1/0
2/0	145	175	195	115	135	150	2/0
3/0	165	200	225	130	155	175	3/0
4/0	195	230	260	150	180	205	4/0

Notes:
1. Section 310.15(B) shall be referenced for the ampacity correction factors where the ambient temperature is other than 30°C (86°F).
2. Section 310.15(C)(1) shall be referenced for more than three current-carrying conductors.
3. Section 240.4(D) shall be referenced for conductor overcurrent protection limitations, except as modified elsewhere in the *Code*.

Reproduction of NEC Table 310.16 (in part)

Table 1.5. *NEC* Table 310.16 provides ampacities of insulated conductors with not more than three current-carrying conductors in raceway, cable, or earth (directly buried)

Precautions are necessary when installing more than two Type NM cables through the same opening in wood framing members. When more than two Type NM cables containing two or more current-carrying conductors are installed through the same hole (without maintaining spacing). That hole is fire- or draft-stopped using thermal insulation, caulk, or sealing foam, the allowable ampacity of each conductor is required to be adjusted in accordance with NEC Table 310.15(C)(1), *Adjustment Factors for More Than Three Current-Carrying Conductors in a Raceway or Cable*, or IRC Table E3705.3.

For example, if six (6) 12-2 W/GR Type NM cables were installed through a single fire-caulked drilled hole in a top plate, that would equal twelve (12) current-carrying conductors. According to *NEC* Table 310.15(C)(1) or IRC Table E3705.3, those twelve (12) current-carrying conductors could only be loaded to 50 percent of their ampacity values from *NEC* Table 310.16 or IRC Table E3705.1. The provisions of *NEC* 310.14(A)(2), Exception or IRC E3701.3, Exception (two different ampacities), permits the higher ampacity to be used if higher ampacity distance is less than or equal to 3.0 m (10 ft) or 10 percent of the total circuit length. However, for adjustment factors for Type NM cable as described above, *NEC* 334.80 or IRC E3705.4.4 indicate that *NEC* 310.14(A)(2), Exception, or IRC E3701.3, Exception, does not apply in this application.

Also note that where more than two Type NM cables containing two or more current-carrying conductors are installed in contact with thermal insulation without maintaining spacing between cables, the allowable ampacity of each conductor is required to be adjusted in accordance with *NEC* Table 310.15(C)(1) as well [*NEC* 334.80 or IRC E3705.4.4].

CONDUCTOR AMPACITY AND USING THE AMPACITY TABLE(S)

Ampacity is defined as "the maximum current, in amperes, that a conductor can carry continuously under the conditions of use without exceeding its temperature rating." This definition serves as the basis for determining the amount of current that a conductor can safely carry. The ampacity tables

Table 310.15(B)(1)(1) *(in part)*
Ambient Temperature Correction Factors Based on 30°C (86°F)

For ambient temperatures other than 30°C (86°F), multiply the allowable ampacities specified in the ampacity tables by the appropriate correction factor shown below.

Ambient Temp. (°C)	Temperature Rating of Conductor 60°C	75°C	90°C	Ambient Temp. (°F)
21-25	1.08	1.05	1.04	69-77
26-30	1.00	1.00	1.00	78-86
31-35	0.91	0.94	0.96	87-95
36-40	0.82	0.88	0.91	96-104
41-45	0.71	0.82	0.87	105-113
46-50	0.58	0.75	0.82	114-122
51-55	0.41	0.67	0.76	123-131
56-60	—	0.58	0.71	132-140
61-65	—	0.47	0.65	141-149
66-70	—	0.33	0.58	150-158
71-75	—	—	0.50	159-167
76-80	—	—	0.41	168-176
81-85	—	—	0.29	177-185

Table 1.6. *NEC* Table 310.15(B)(1)(1) gives the ambient temperature correction factors where conductors are installed in other than 30°C (86 °F) temperatures

Table 310.15(C)(1) Adjustment Factors for More Than Three Current-Carrying Conductors

Number of Conductors[1]	Percent of Values in Tables 310.16 through 310.19 as Adjusted for Ambient Temperature if Necessary
4-6	80
7-9	70
10-20	50
21-30	45
31-40	40
41 and above	35

[1]Number of conductors is the total number of conductors in the raceway or cable, including spare conductors. The count to be adjusted in accordance with 310.15(E) and (F), and not to include conductors that are connected to electrical components but that cannot be simultaneously energized.

Table 1.7. *NEC* Table 310.15(C)(1) provides the adjustment factors where more than three current-carrying conductors are bundled or installed in the same raceway

of the *Code* [*NEC* Table 310.16 and IRC Table E3705.1] give the allowable ampacity a conductor is permitted to carry under the conditions of use that are identified in the table heading and the *NEC* section were the table is referenced. For electrical installations in most typical one- and two-family dwellings, the allowable ampacity tables [*NEC* 310.16 and IRC E3705.1] are used for branch circuits and feeders. Service conductors and, if present, the main power feeder are permitted to be sized in accordance with *NEC* 310.12 or IRC E3603.1.

In some cases, the allowable ampacities given in these sections must be adjusted for, or based on, the condition(s) of use. Two types of ampacity correction factors are involved here: (1) correction for ambient temperatures and (2) correction adjustment, where multiple current-carrying conductors are in the same raceway or otherwise bundled for excessive distances. These correction factors must be applied if those conditions are apparent in the installation. Conductor ampacity-adjustment rules are provided in the *Code* and must be applied [*NEC* 310.15(B)(1) and *NEC* 310.15(C)(1) or IRC E3705.2 and E3705.3].

Many different types of wiring methods are permitted for one- and two-family dwellings. Conductors are not permitted to be installed where operating temperature will be in excess of that designated for the insulation type on the conductor involved. Under no circumstances can conductors be associated together in such a way, with respect to type of circuit, the wiring method employed, or the number of conductors, that the limiting temperature of any conductor is exceeded. This means that where conductors with different temperature ratings are installed in the same raceway or as bundled cable assemblies, the temperature rating associated with the conductor having the lowest temperature rating must be used to determine the correct ampacity. The temperature rating of a conductor is the maximum temperature, at any location along its length, which the conductor can withstand over a prolonged time period without serious degradation. The correction factors at *NEC* Table 310.15(B)(1)(1) provide requirements for coordinating conductor sizes, types, allowable ampacity, ambient temperature and number of associated conductors. IRC Table E3705.2 serves the same purpose by including the ambient temperature-adjustment percentage multipliers.

The principal determinants of operating temperature that must be considered are any one or any combination of the following conditions:

1. Ambient temperature may vary along the conductor length as well as vary at different times. For example, conductors installed in areas such as an attic or on a roof are subject to high temperatures. This factor must be considered during the design stage of any electrical installation.

2. Heat generated internally in the conductor because of load current flow will have an effect on its operating temperature. This is caused from current passing through the resistance the conductor has.

3. The rate at which generated heat is dissipated into the ambient medium: Thermal insulation covering or surrounding conductors affect the rate of heat dissipation.

4. Adjacent load-carrying conductors have the dual effect of raising the ambient temperature and impeding heat dissipation. This is one reason why consideration should be given to the limitations on the number of circuits in enclosures.

AMPACITY CORRECTION AND ADJUSTMENT FACTORS

After determining the allowable ampacity for the conductors to be used, as specified in either NEC Table 310.16 or IRC Table E3705.1, one must consider two additional factors. The first factor is to compensate for ambient temperatures that exceed 86°F (30 °C). The allowable ampacities in Table 310.16 are determined based on a raceway or cable operating at an ambient temperature of 30°C (86°F). Ampacity correction factors found in NEC Table 310.15(B)(1)(1) or in IRC Table E3705.2 must be applied if the ambient temperature (where the conductors are installed) exceeds 30°C (86°F). The second factor that must be considered is where more than three current-carrying conductors are contained in the same raceway or, in the case of cables, are bundled together for a length exceeding 600 mm (24 in.). The allowable ampacities for conductors in the Table(s) are based on not more than three current-carrying conductors being in close proximity, such as in a raceway or cable.

AMPACITY CORRECTION FACTORS

The ampacity adjustment rules are provided in the *NEC* in 310.15(B)(1) and Table 310.15(B)(1)(1) and Table 310.15(C)(1). The IRC has adjustment factors included in identical code rules that are found in IRC E3705.2, E3705.3, and Tables E3705.2 and E3705.3.

Correction factors given in the allowable ampacity tables must be applied when the ambient temperature exceeds 30°C (86°F) at the location of the wiring system. Areas at a one- and two-family dwelling that may have ambient temperatures exceeding 30°C (86°F) include the exterior of the dwelling, such as on the side of the structure, on the roof, or in attics. Raceways and cables installed on rooftops must also comply with NEC 310.15(B)(2). Other interior areas are not likely to have ambient temperatures exceeding these values and would not be subject to the rooftop correction factors.

It is the responsibility of the authority having jurisdiction (AHJ) enforcing the Code to examine each particular site, determine what ambient temperatures exist, and apply the values in the tables appropriately and reasonably. Temperature correction factors apply to insulated conductors, whether part of a cord or cable assembly, installed in a raceway or run as open knob-and-tube wiring.

In the following example, an adjustment in conductor ampacity is made based solely on ambient temperature. Multiply the ampacity given in the applicable allowable ampacity table by the percentage given for the ambient temperature in

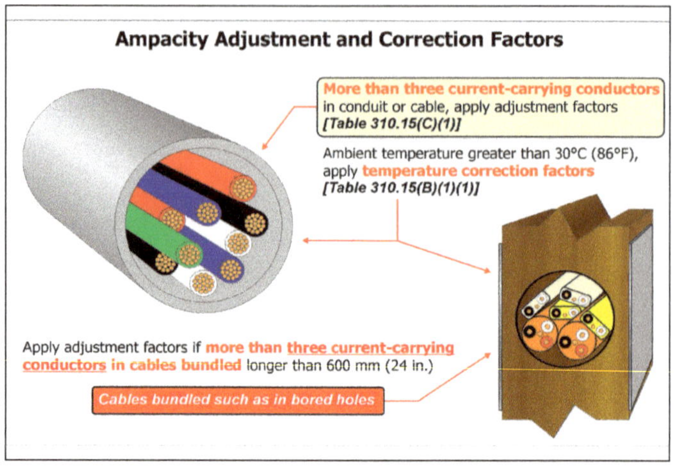

Figure 1.19. Application of correction factors for ambient temperature and for number of current-carrying conductors

which the conductor will operate. For example, a 2/0 copper conductor with 75°C insulation operates in an environment with temperatures ranging from 96°F to 104°F. NEC Table 310.16 indicates the allowable ampacity is 175 amperes. The adjustment to the allowable conductor ampacity is calculated as follows:

$$175 \times .88 = 154 \text{ amperes.}$$

The higher the ambient temperature, the less current the conductor can carry without exceeding the temperature rating of the insulation. It is not uncommon for raceways and cables to be exposed to ambient temperatures of 120°F or higher in attics, rooftops, and on the sides of buildings or other structures. Conductors installed in or on one- and two-family dwelling units are often subjected to high ambient temperatures, which may affect conductor insulation and create problems due to insulation failure.

Type NM cable is required to be manufactured with insulation having a temperature rating of 90°C. The final ampacity must be taken from the 60°C column of the allowable ampacity table. However, the derating calculations are permitted to be started from the 90°C (194°F) ampacity column of the allowable ampacity table [*NEC* Table 310.16 or IRC Table E3705.1, as allowed by *NEC* 334.80 or IRC E3705.4.4].

For example, if the ambient temperature is 108°F where Type NM cable is installed, the ampacity correction factors of *NEC* Table 310.15(B)(1)(1) or in IRC Table E3705.2 require that the ampacity shown in the 90°C column be multiplied by .87 or 87%. If the cable being installed contains 12 AWG copper conductors, the corrected ampacity is 30 x .87 = 26.1 amperes.

Now, the overcurrent protection rules in *NEC* 240.4(D)

and IRC E3705.5 must be applied. The allowable ampacities for small conductors (18 through 10 AWG) are to be per the following limitations: Overcurrent protection of 14 AWG copper not to exceed 15 amperes, 12 AWG not to exceed 20 amperes, and 10 AWG not to exceed 30 amperes, unless specifically permitted in *NEC* 240.4(E), *Tap Conductors*, or 240.4(G), *Overcurrent Protection for Specific Conductor Applications*.

In the example, even though the adjusted ampacity is 26.1 amperes, the maximum branch-circuit protective device rating is 20 amperes, unless permitted otherwise by the Code. Two common applications where the "Overcurrent Protection for Specific Conductor Applications" rule applies are for motor loads, as outlined in Article 430, and air conditioning and heat pump loads, as specified in Article 440, of the *NEC*.

AMPACITY ADJUSTMENT FACTORS

Ampacity adjustment factors are required where more than three current-carrying conductors are contained within the same raceway or cable. It also applies where cables are bundled without maintaining spacing. This adjustment factor reduces the ampacity of each conductor in a raceway or cable due to the effect of heat contributed by adjacent conductors. For example, where four through six current-carrying conductors are bundled or installed in the same raceway or cable, an 80 percent adjustment factor must be applied to the conductors. This reduces the allowable ampacity of the conductor and may require that a lower-rated overcurrent device be installed to protect the conductor at or below its reduced ampacity [*NEC* 310.15(C)(1) and IRC E3705.3].

Ampacity correction factors must be applied where single or multiconductor cables are stacked or bundled longer than 600 mm (24 in.) without maintaining spacing and are not installed in a raceway. The ampacity correction factors must be applied when installed in a raceway longer than 600 mm (24 in.). There is a provision to this rule that basically exempts this rule for Type AC and Type MC cables. Both types of cables are metal-clad. This provision does not apply to nonmetallic jacketed cables [*NEC* 310.15(C)(1)(a) and 310.15(C)(1)(d) or IRC E3705.3 and E3705.3, Exception 3].

For example, six 10-2 AWG Type NM cables are bundled longer than 600 mm (24 in.) without maintaining spacing. The ampacity of the 10 AWG copper conductor in the 90°C column of the applicable allowable ampacity table is 40 amperes. Because there are twelve current-carrying conductors in the group, their ampacity is 40 amperes × .50 or 20 amperes. These conductors would now require overcurrent protection with a rating of no more than 20 amperes. Of course, the simple solution here is to spread the cable out and not bundle it together, allowing heat to dissipate properly from these cables and conductors. Standard ampere ratings for fuses and circuit breakers are provided in the Code [NEC 240.6 and IRC E3705.6].

When more than two Type NM cables containing two or more current-carrying conductors are installed through the same hole without maintaining spacing, and that hole is fire- or draft-stopped using thermal insulation, caulk, or sealing foam, the allowable ampacity of each conductor is required to be adjusted in accordance with *NEC* Table 310.15(C)(1), Adjustment Factors for More Than Three Current-Carrying Conductors in a Raceway or Cable, or IRC E3705.3 as well [*NEC* 334.80 or IRC E3705.4.4]. This is discussed in greater detail earlier in this chapter and Chapter 11.

APPLYING BOTH CORRECTION AND ADJUSTMENT FACTORS

The *Code* makes it clear that conductors are required to be adjusted for both high ambient temperature and the number of conductors in a raceway or cable that are bundled together where these conditions exist. In this case, one determines the correction factor for ambient temperature and makes the calculation. Next, the adjustment for a number of conductors is applied. This process is often referred to in the field as a "double derate" factor.

For example, three 12-2 Type NM cables are installed in an attic with a 106°F ambient temperature. All of the phase and grounded (neutral) conductors carry current in this case. The allowable ampacity of the conductors in the ampacity table is 30 amperes. Remember that the allowable ampacity for Type NM cable is taken at 90°C for the purpose of derating. Derating the allowable ampacity for the conditions of use is performed as follows: for the high ambient temperature, a factor of .87 (87 percent) is applied (30 × .87), and a derating factor of .80 (80 percent) for the number of conductors is applied (30 × .87 × .80), which results in an adjusted allowable ampacity of 20.9 amperes.

The *Code* permits the next higher standard-rated overcurrent protective device (through 800 amperes) to be used to protect conductors from overcurrent.

Another example is where a Type NM cable having 8 AWG copper conductors is to be installed. The circuit supplies an electric range. From *NEC* Table 310.16, the conductors have an allowable ampacity of 55 amperes at 90°C. The circuit is proposed to be routed through a space where the ambient temperature is 106°F. In this case, a correction factor of .87 or 87 percent is applied (55 × .87 = 48 amperes). Next, if four two-wire cables are bundled with the 8 AWG cable for a distance longer than 600 mm (24 in.) without maintaining spacing, a 70 percent adjustment factor is applied (48 × .70 = 34 amperes). This is the new ampacity of the conductors.

The overcurrent protection rules must then be considered. The next standard overcurrent device ampere rating (35 amperes) can be used only if the circuit is other than a multioutlet branch circuit supplying receptacles for cord- and plug-connected portable loads. (In our example, the circuit supplies a range and, as such, is not a multioutlet branch circuit.) Typical circuits where the next higher ampere rating can be used include fixed appliances like water heaters, electric furnaces, dishwashers, disposals, and lighting outlet circuits.

Because a 35-ampere overcurrent device is a standard overcurrent device, it must be selected even though it may not be as common as a 40-ampere overcurrent device. Hopefully, this reduced allowable ampacity will still be adequate for the range load. Otherwise, a larger-sized cable should be selected and the ampacity recalculated. Or, once again, it may be simpler to avoid bundling the cables together!

PRODUCT SAFETY STANDARD REQUIREMENTS

It is recommended to consult the following electrical product safety guide card information and product safety standards for additional guidance on installing, operating, and using electrical equipment covered in this chapter. The four-letter code in parentheses indicates the product category in the UL Guide Information for Electrical Equipment Directory (UL Product iQ® material). A list of product standards is included in Annex A of NEC-2020 and Chapter 44 of IRC-2021. It is essential for enforcing authorities to use these references as needed to ensure compliance with the applicable requirements of the Code.

- Electrical Equipment for Use in Ordinary Locations (AALZ)
- Armored Cable (AWEZ)
- Circuit Breakers (DHJR)
- Circuit Breakers, Molded-Case, and Circuit Breaker Enclosures (DIVQ)
- Switches, Enclosed (WIAX)
- Insulating Devices and Materials (NYYV)
- Insulating Tape (OANZ)
- Panelboards (QEUY)
- Wire (ZGZX)
- Wires, Thermoset-Insulated (ZKST)
- Wires, Thermoplastic-Insulated (ZLGR)
- Wire Connectors (ZMKQ)
- Wire Connectors and Soldering Lugs (ZMVV)

REFERENCES

1. *National Electrical Code* and *NEC* are registered trademarks of the National Fire Protection Association, Inc., Quincy, MA 02169. This reprinted material is not the official position of the National Fire Protection Association, which is represented only by the standard in its entirety.

2. *International Residential Code for One- and Two-Family Dwellings* (2021). Copyright © International Code Council, Inc., June 2023.

3. UL, Northbrook, IL, *UL Guide Information for Electrical Equipment Directory* (UL Product iQ® material) unless indicated otherwise.

REVIEW QUESTIONS

1. The relationship between current, voltage, and resistance in an electric circuit was discovered by George Simon Ohm in 1726.

True _____ False _____

2. The term volt describes "the unit of electrical pressure that is required to force one ampere through a resistance of one ohm."

True _____ False _____

3. "The unit of electrical current which will flow through one ohm under a pressure of one volt in one second" is described as a watt.

True _____ False _____

4. The unit of electrical resistance described as "the resistance through which one volt will force one ampere" is called an ohm.

True _____ False _____

5. "The unit of measurement of the energy flowing in an electrical circuit at any given moment" is described as a watt.

True _____ False _____

6. Mechanical power is usually expressed in horsepower and electrical power in:

A. watts
B. ohms
C. cycles
D. farads

7. The following equation is used to change watts to horsepower:

hp = watts/746   746 watts = 1 hp

True _____ False _____

8. The purpose of the *NEC* is the practical safeguarding of persons and property from hazards arising from the use of electricity.

True _____ False _____

9. The organization responsible for recognizing qualified electrical testing laboratories that perform evaluations, testing, and certification of certain products is which of the following:

A. Authority Having Jurisdiction (AHJ)
B. Occupational Safety and Health Administration (OSHA)
C. National Electrical Manufacturers Association (NEMA)
D. National Fire Protection Association (NFPA)

10. A premises wiring system begins at the service point and extends to the outlet(s) and wiring from and including the power source to the outlets where there is no service point.

True _____ False _____

11. The authority having jurisdiction will use Section __________ in the *NEC* as a basis to approve the materials and electrical equipment that are installed in a one- and two-family dwelling electrical system.

A. 90.4
B. 110.2
C. 215.8
D. 230.42

12. Ordinary locations defined in the *NEC* and IRC include all the following EXCEPT:

A. damp location
B. dry location
C. wet location
D. moist locations

13. It is the responsibility of the ____________________ to determine the type of location involved.

A. General contractor
B. Electrical contractor
C. Authority Having Jurisdiction (AHJ)
D. National Weather Service

14. Terminals for more than one conductor and terminals used to connect ______________ shall be so identified.

A. copper
B. copper-clad aluminum
C. aluminum
D. copper or aluminum

15. Generally, the overcurrent protection shall not exceed _____ amperes for a 12 AWG copper conductor after any correction factors for ambient temperature and for number of conductors have been applied.

A. 15
B. 20
C. 25
D. 30

16. In comparing the following sizes of conductors, the ______ AWG is the smallest of the group.

A. 3
B. 4
C. 6
D. 8

17. Type THHN 8 AWG copper conductors are permitted to be installed in:

A. wet locations only.
B. dry and wet locations.
C. ambient temperatures at 90°C, 194°F.
D. PVC conduit rated at 60°C, 140°F.

18. The maximum ampacity for each of three single insulated 75°C rated 6 AWG Type THWN copper conductors in a raceway is:

A. 55 amperes.
B. 65 amperes.
C. 70 amperes.
D. 80 amperes.

19. Type XHHW-2 conductors in a dry location have a maximum insulation temperature rating of:

A. 60°C
B. 75°C
C. 90°C
D. 110°C

20. Which of the following is the maximum ampacity for three 12 AWG Type THWN aluminum conductors installed in a raceway?

A. 15 amperes
B. 20 amperes
C. 30 amperes
D. 35 amperes

21. Type THWN copper conductors are permitted to be installed in:

A. dry locations only.
B. dry and wet locations.
C. ambient temperatures at 90°C, 194°F.
D. PVC conduit rated at 60°C, 140°F.

22. Derating factors are applicable to conductors installed in raceway nipples shorter than 600 mm (24 in.).

True ______ False ______

23. 14 AWG aluminum conductors are suitable for dwelling unit branch circuits.

True ______ False ______

24. Ampacity correction factors are required to be applied where the ambient temperature exceeds:

A. 30°C (86°F)
B. 36°C (96°F)
C. 40°C (104°F)
D. 60°C (140°F)

25. The allowable ampacities of *NEC* Table 310.16 or IRC Table E3705.1 are determined at an ambient temperature of:

A. 30°C.
B. 86°F.
C. 75°C.
D. Both A and B.

26. Conductors with Type THW insulation installed in an ambient temperature of 106°F must have a correction factor of ____ percent applied.

A. 71
B. 82
C. 87
D. 88

27. Which of the following statements about the general requirements for 6 AWG conductors rated at 100 amperes or less is true?

A. They are required to be terminated at 75°C ampacity.
B. They are required to be terminated at 60°C ampacity.
C. They must have insulation rated at 60°C.
D. They must have insulation rated at 75°C.

28. Which of the following statements about the general requirements for 1/0 AWG conductors rated over 100 amperes is true?

A. They are permitted to be terminated at 75°C ampacity.
B. They are required to be terminated at 60°C ampacity.
C. They must have insulation rated at 60°C.
D. They must have insulation rated at 75°C.

29. Three 12-2 Type NM-B (with ground) cables and two 10-3 Type NM-B (with ground) cables are bundled longer than 600 mm (24 in.) without maintaining spacing. All but the equipment grounding conductors are considered to be current-carrying conductors. Which of the following adjustment factors must be applied?

A. 80 percent
B. 70 percent
C. 50 percent
D. 45 percent

30. Splicing devices for underground cables are required to be covered with insulation equivalent to that of the conductors or with an insulating device that is:

A. accessible.
B. readily accessible.
C. identified and listed for such use.
D. color-coded.

31. Unused openings, other than those intended for the operation of equipment, those intended for mounting purposes, or those permitted as part of the design for listed equipment, shall be ____________

A. welded closed.
B. sealed with silicone.
C. closed to afford protection substantially equivalent to the wall of the equipment.
D. effectively closed.

32. Which of the following statements about enclosures is true?

A. They are all suitable for outdoor use.
B. Must be identified for use in the operating environment (such as wet or damp locations).
C. Not required to be protected against permanent damage from the weather during building construction.
D. Not required to be identified in areas where exposed to gases, fumes, vapors, or liquids.

33. Which of the following statements about the integrity of electrical equipment and connections is false?

A. Cleaners or abrasives may be used to remove paint overspray from panelboard busbars.
B. Internal parts of panelboards, such as the busbars, shall be protected from contamination from foreign materials such as paint, plaster, cleaners, or abrasives.
C. Broken or damaged parts shall be repaired or replaced.
D. Parts that are damaged due to overheating shall be repaired or replaced.

34. By special permission, the authority having jurisdiction may waive specific requirements in the Code or permit alternative methods where it is ensured that equivalent objectives can be achieved by establishing and maintaining effective safety. This rule requires the AHJ to state in ___________ that the Code variation is authorized.

A. English
B. English and Spanish
C. writing
D. a court of law

35. Approved is defined by both the *NEC* and the IRC as acceptable by the______________________.

A. Consumer Product Safety Commission
B. Occupational Safety and Health Administration (OSHA)
C. Authority having jurisdiction (AHJ)
D. General Contractor

36. The determination of a damp, wet, or dry location is typically the responsibility of the ____________________ using the definitions and listing and labeling by a third-party testing agency.

A. project engineer
B. authority having jurisdiction (AHJ)
C. building contractor
D. project designer

37. Metal cabinets or enclosures installed in a damp or wet location must be installed so there is at least ________________ airspace between the enclosure and the wall or other supporting surface.

A. 3-mm (⅛-in.)
B. 10-mm (⅜-in.)
C. 1.6-mm (1/16-in.)
D. 6-mm (¼-in.)

38. A wire connector for securing an aluminum wire in combination with a copper or copper-clad aluminum conductor, where physical contact occurs between the wires of different metals, is limited to __________________ only and is marked "AL-CU."

A. dry locations
B. damp locations
C. wet locations
D. indoor locations

39. Which one of the following insulated conductors can be used in wet locations as recognized by the NEC?

A. THHN
B. XHH
C. THW
D. RHH

40. Precaution needs to be exercised when installing more than ______Type NM cables through the same opening in wood framing members.

A. 1
B. 2
C. 3
D. 4

2 Branch-Circuit, Feeder, Service, and Other Calculations

This chapter covers the methods for establishing the minimum sizes of service equipment, service entrance conductors, and feeder conductors for one- and two-family dwellings. Part A explains the proper application of the rules in the 2023 NEC. Part B provides requirements and examples for applying the provisions in the IRC 2024. There are minor differences between the two codes regarding how the sizing and calculation rules are applied; however, both aim to achieve the same overall result.

IMPORTANCE OF ELECTRICAL PLANS AND PROPER PLAN REVIEW

Many older one- and two-family homes across the country have 30- or 60-ampere services supplied by a 120-volt, two-wire, 30-ampere service drop that powers an old set of rubber-covered service entrance conductors. These conductors are often deteriorated from prolonged exposure to sunlight and the weather. The main switch enclosure may be rusted and broken, and the two-fuse, cast iron cutout box might be fitted with oversized 30-ampere Edison-base fuses. Such conditions remind us that the installer or designer did not plan for future expansion when the electrical system was first installed. Many likely believed their 30- or 60-ampere service had more capacity than they needed at the time. Over time, electric ranges, water heaters, washers, dryers, radios, televisions, and numerous other appliances have been added to these older homes. Most of these older services are severely inadequate to meet today's electrical demands.

Planning for future expansion and increases in electrical power is mentioned in the Code. However, NEC 90.8(A) (*Future Expansion and Convenience)* is advisory in nature. The Code requirements include a minimum set of requirements for electrical services, feeders, and branch circuits. Providing extra space in panelboards for future 120- and 240-volt branch circuits is always a good idea. Installing spare cable assemblies and raceways that run from the basement to the attic or other accessible areas makes it easier to add new circuits. Placing distribution centers where they can be easily serviced and maintained helps the installer meet future needs for changes and additional circuits.

A Code-compliant installation begins with careful planning. A thorough set of plans and specifications saves time and reduces the chance of a non-compliant installation. It is much easier and less expensive to modify plans than to change an installation after it is complete. While electrical plans may not always be included in the local permit process, plans and specifications are essential for ensuring the safety of occupants in the future.

Sometimes, a simple sketch on a plain piece of paper serves as a basic set of plans for an electrical installation. In other instances, the examiner may require professionally prepared plans. These electrical plans are often part of the building code permit process. They usually include a detailed diagram showing the feeder(s), the total area of the building in square feet, the total connected load, any demand factors to be applied, and the sizes and types of conductors to be installed.

The International Code Council, comprised of the former International Conference of Building Officials (ICBO), the Building Officials Congress of America (BOCA), and the Southern Building Code Congress International (SBCCI), publishes building codes that include specific requirements for electrical plan review. The International Residential Code (IRC) contains such requirements [R106, Construction Documents; R105, Permits; R109, Inspections]. Refer to IRC R106 (Construction Documents) for the specific information needed on construction documents (such as blueprints).

Ensure that you check the applicable national and local codes or ordinances for your installation. Maintaining open communication is always the best approach. Before starting a project, consult with your local authority having jurisdiction (AHJ) regarding construction permits, plans and specifications, and inspection requirements.

The methods for calculating service and feeder sizes for one- and two-family dwellings vary slightly between the NEC and the IRC. Part A of this chapter discusses load calculations based on Article 220 of the NEC. Part B includes the requirements and examples of similar calculations using the IRC 2024. Service size and voltage limitations are outlined in the IRC (E3401.2 and E3401.3).

Services for one- and two-family dwellings covered by the IRC are limited to 120/240-volt, 400-ampere, single-phase systems. It is also important to highlight that the rules in the IRC are based on the NEC. According to the language in Section E3401.2, when electrical systems, equipment, or components are not explicitly addressed in Chapters 34 through 43 of the IRC, the NEC rules apply.

CODE DEFINITION IMPORTANT TO THIS CHAPTER

The NEC and the IRC both include definitions to help apply their respective code rules. Definitions in the NEC are located in Article 100, while those related to the electrical chapters are in Chapter 35 of the IRC. Definitions in IRC Chapter 35 are directly taken from Article 100 of the NEC. The following terms are used in this chapter. If further review is needed, please refer back to Article 100 in the NEC.

- Appliance
- Bathroom
- Branch Circuit
- Branch Circuit, Appliance
- Branch Circuit, General Purpose
- Branch Circuit, Individual
- Branch Circuit, Multiwire
  Note: Electricians usually refer to this circuit as a three-wire circuit.
- Building
  Note: "Building" is not defined in Chapter 35 of the *IRC* because it exists in IRC Chapter 2 and differs slightly from the *NEC* definition.
- Clothes Closet
- Continuous Load
- Cooking Unit, Counter-Mounted
- Demand Factor
- Dwelling Unit
- Feeder
- Habitable Room
- Kitchen
- Laundry Area

PART A: CALCULATION AND SIZING REQUIREMENTS OF THE *NEC* CALCULATING SERVICE-ENTRANCE CONDUCTOR SIZES

Article 230 of the NEC outlines most of the rules for electrical service installation. For example, the minimum size of the disconnecting means for a one-family dwelling must be at least 100 amperes, 3-wire [NEC 230.79(C)]. Methods for determining the minimum size of service-entrance conductors are not included in NEC Article 230. Loads must be calculated in accordance with the rules in NEC Article 220. The Code states that the ampacity of the service-entrance conductors before any adjustment or correction factors are

Photo 2.1. Calculation(s) are required to determine the minimum size of the service equipment and service-entrance conductors

Photo 2.2. The ampacity of service conductors and feeder conductors is calculated at not less than the noncontinuous load plus 125% of the continuous load

applied must not be less than the load to be served [NEC 230.42(A)]. Typically, the ampacity of service conductors and feeder conductors should be no less than the total of non-continuous loads plus 125% of the continuous loads [NEC 230.42(A)(1)].

The *Code* also permits the service conductors to be sized based on the sum of the noncontinuous loads plus the continuous load if the service-entrance conductors terminate in an overcurrent device and assembly that are listed for operation at 100% of their rating. While not commonly found in residential electrical systems, this type of equipment is not limited to just commercial or industrial applications.

Calculating loads for service-entrance conductors must comply with the Code's requirements. The main rule is that a feeder or service load should not be less than the total of the loads on the branch circuits supplied, as determined by Part II of NEC Article 220. Demand factors allowed by Parts III or IV can be applied. Then, the minimum ampacity of the service-entrance conductors and the feeder conductors is determined from NEC 310.14 and 310.15, usually using the relevant table. NEC Table 310.16 primarily provides conductor ampacities; however, the ampacity demand factors (adjustments) in NEC 310.12 can

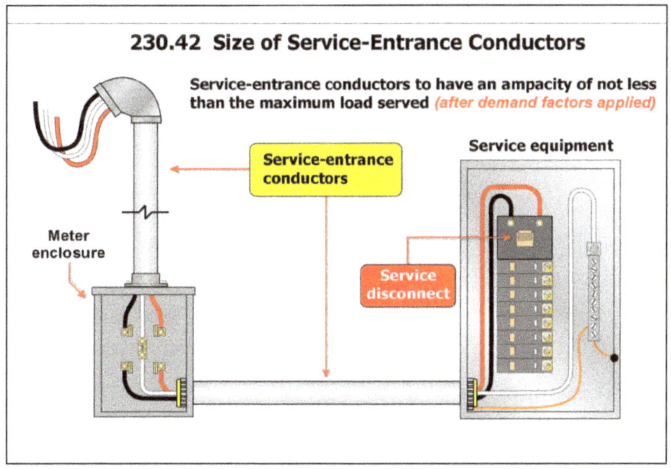

Figure 2.1. Service entrance and service-lateral conductors are to be adequate for the calculated load

be applied to services and the main power feeder (*if present*) for dwellings under certain conditions.

For a service or feeder rated between 100 and 400 amperes, the conductors supplying the entire load for a one-family dwelling, or the conductors supplying the full load for an individual unit within a two-family or multifamily dwelling, are allowed to have an ampacity of at least 83 percent of the service or feeder rating. If no adjustment or correction factors are necessary, NEC Table 310.12(A) may be used. The service or feeder overcurrent device should be calculated using the standard ampacity ratings found at NEC 240.6(A) and Table 240.6(A), along with the relevant loads. The starting point for sizing these service or main power feeder conductors is based on the ampacity ratings from NEC Table 310.16. The 83 percent adjustment for these conductors is permitted by NEC 310.12(A) (service) and (B) (feeder). It's important to note that 310.12 does not specify a different ampacity for these conductors. Instead, it provides a reduced rating based on the calculated values, which generally allows a smaller conductor to be used for dwelling units. Only the specific service and main power feeder conductors mentioned in the subsections are allowed for this application.

NEC Section 310.12 clarifies that when correction or adjustment factors are needed under 310.15(B)(1) (temperature) or 310.15(C)(1) (*number of conductors*), they can be applied to the ampacity corresponding to the conductor's temperature rating. Some Code users have questioned whether these temperature and correction factors should or even can be applied to the ampacity listed in the temperature column associated with equipment termination values or the conductor temperature rating, especially when the ampacity values allowed by 310.12 have already been used for a dwelling unit service or main power feeder.

Section 310.12 clarifies that conductors are allowed to have their ampacity adjusted based on temperature correction and adjustment factors when necessary. Service and feeder conductors sized according to 310.12 are subject to the same environmental conditions as conductors not sized under 310.12. To illustrate this, Example D7 in NEC Annex D provides two scenarios for sizing service conductors for dwelling units. One scenario is titled "With No Required Adjustment or Correction Factors," and the other is titled "With Required Temperature Correction Factor." The second example shows how to size dwelling unit service conductors using the allowed 83 percent adjustment (from 310.12(A)) combined with a temperature correction factor from Table 310.15(B)(1)(1).

FEEDER OR SERVICE CALCULATIONS

NEC Article 230 does not specify load calculation methods but instead refers to NEC Article 220 for determining the load on both feeder and service conductors. To highlight this, the title of NEC Article 220 is "Branch-Circuit, Feeder and Service Calculations." The scope of NEC Article 220 covers similar information. Part I provides general information. Part II details calculation methods used for branch-circuit load. Parts III and IV explain methods used in calculating service loads. Part V offers calculation methods for farms.

LOAD CALCULATION METHODS

The NEC permits either of the two approved methods for determining the minimum size of service conductors required for a dwelling unit. This chapter will explain and provide examples of both the Standard Calculation Method and the Optional Calculation Method. While the NEC allows both methods, the IRC offers only one method to establish these loads. These load calculations are used to determine conductor size and the minimum rating of service equipment.

We will review the general or standard load calculation in Part III of *NEC* Article 220, the optional load calculation method of Part IV, and the portion of load calculation in Part III of *NEC* Article 220 that applies to two-family dwelling services. A requirement for using the optional calculation in Part IV sets the minimum size of service-entrance conductors at 100 amperes. The *NEC* permits both methods, but the optional calculation covered in Part IV of *NEC* Article 220 usually results in a lower calculated load. We will work through both methods, as familiarity with both is important. For additional examples, refer to Informative Annex D, Examples, of the *NEC*. Information in the *NEC* annex, including these examples, is not part of the actual rules of the *Code*.

Informational Annex Example D7 Sizing of Service Conductors for Dwelling(s) [See 310.12]

With No Required Adjustment or Correction Factors

- Service conductors and the main power feeder for certain dwellings are permitted to be sized in accordance with 310.12
- If a 175-ampere service rating is selected, a service conductor is then sized as follows:
- 175 amperes × 0.83 = 145.25 amperes per 310.12
- If no other adjustments or corrections are required for the installation, then, in accordance with Table 310.16, a 1/0 AWG Cu (150 amperes) or a 3/0 AWG Al (155 amperes) meets this rating at the 75°C (167°F) column
- Note: These calculated sizes correspond to the given sizes in Table 310.12

Figure 2.2a. Reproduction of *NEC* Informative Annex D Example D7 and the requirements of 310.12 *(with no required adjustment or correction factors)*

Informational Annex Example D7 Sizing of Service Conductors for Dwelling(s) [See 310.12]

With Required Temperature Correction Factors

- Service conductors and the main power feeder for certain dwellings are permitted to be sized in accordance with 310.12
- If a 175-ampere service rating is selected, a service conductor is then sized as follows:
- 175 amperes × 0.83 = 145.25 amperes per 310.12
- Conductors to be installed in an ambient temperature of 38°C (100°F) [See Table 310.15(B)(1)]
- Using XHHW-2 conductors requires a correction factor of 0.91 to find the minimum conductor ampacity and size: 175 A x .91 = 159.25 amperes
- In accordance with Table 310.16, using the 90°C (194°F) column, a 2/0 AWG Cu or a 4/0 AWG Al would be required (with load no more than 159.25 A)

Figure 2.2b. Reproduction of *NEC* Informative Annex D Example D7 and the requirements of 310.12 *(with required temperature correction factor)*

COMMON CALCULATION REQUIREMENTS

The general requirements in Part I are shared by both calculation methods. We will review these common requirements and then discuss the requirements that apply only to the standard or only to the optional calculation method.

SELECTION OF CONDUCTORS

In the following examples, the results of the load calculations are expressed in amperes. To select conductor sizes, in most cases, we will refer to NEC Table 310.16, which provides the "Ampacities of Insulated Conductors with Not More Than Three Current-Carrying Conductors in Raceway, Cable, or Earth (Directly Buried)." Since we are determining the min-

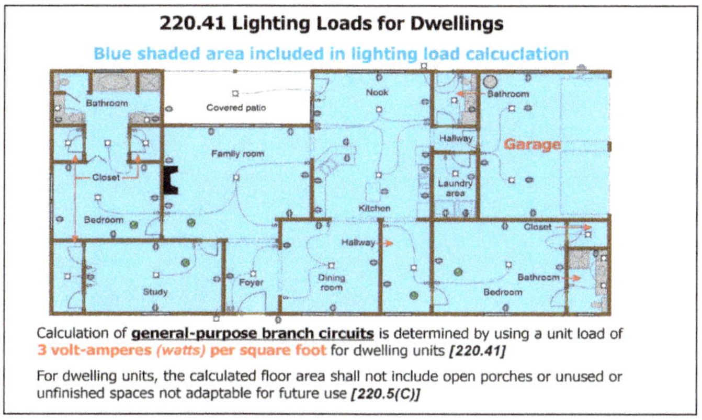

Figure 2.3. General-purpose branch circuits are determined by using a general unit load of 3 volt-amperes (watts) per square foot for one- and two-family dwelling units

imum ampacity of the service-entrance conductors, we will also refer to NEC 310.12, which provides specific sizes and types of copper and aluminum, or copper-clad aluminum, conductors permitted for dwelling-unit services and the main power feeder, if one is present.

VOLTAGES USED IN CALCULATIONS

To apply the rules in the Code uniformly, especially NEC Articles 210, 215, and 220, a nominal voltage of 120, 120/240, 240, and 208Y/120 volts is necessary for calculating the amperage load on the conductor [NEC 220.5(A) and 310.12].

FRACTIONS OF AN AMPERE

Unless a computation results in a significant fraction of an ampere (0.5 or more), such fractions will be omitted. The rounding rule for fractions is detailed in NEC 220.5(B) and is demonstrated in the examples in Informative Annex D of the NEC.

UNIT LOAD FOR BRANCH CIRCUITS AND GENERAL LIGHTING

The calculation of general-purpose branch circuits is based on using a standard unit load of 3 volt-amperes (watts) per square foot for one- and two-family dwelling units, as derived from NEC 220.41. The floor area must be measured using the outside dimensions of the dwelling unit, apartment, or any other area involved. Each floor level should be calculated separately, and then the areas should be added together to find the total floor area for the calculation.

In previous editions of the Code, the floor area included in the calculation did not require the inclusion of open porches, garages, or any unused or unfinished spaces that are not suitable for future use (see NEC 220.11 of the 2020 NEC). For the 2023 NEC, garages must be included in the calculated floor area, along with any unfinished areas that can be adapted for future use as a habitable room or occupiable space. Open porches still do not need to be included in the calculated floor area of a dwelling unit.

The decision on whether these spaces need to be included in the load calculation depends on the AHJ and how the space is shown on the plans. Building codes are used to determine if the space is suitable for human occupancy. Most unfinished areas, such as basements, are intended for storage or to house equipment like furnaces or boilers. However, the load calculation must include attics or unfinished basements that are planned for future remodeling into bedrooms, recreational rooms, or other habitable spaces [NEC 220.5(C)].

GENERAL-USE RECEPTACLE OUTLETS

General-use receptacle outlets rated at 20 amperes or less are considered outlets for general illumination in dwelling units and are allowed to be included in the general lighting load calculation. This includes receptacle outlets in bathrooms, outdoor areas, basements, and garages. No additional load is needed for these general-purpose receptacles. In commercial occupancies, general-use receptacles are calculated at 180 volt-amperes, which does not apply to dwelling units (NEC 220.41).

MINIMUM NUMBER OF GENERAL-PURPOSE LIGHTING CIRCUITS

The general branch circuits for power and lighting must be established to serve as a starting point. The other required loads will be added to this figure as the calculated load on the service grows. The minimum number of general-purpose lighting circuits required is determined by the formula provided in Figure 2.4.

Assume the dwelling has 2800 square feet, which must be included in the calculation (excluding open porches, etc.). The basic calculation for the general lighting load will be shown as indicated in Figure 2.4.

Any mix or combination of 15- or 20-ampere branch circuits can be installed as long as circuits with a capacity that meets the requirements of the general lighting load calculation are provided. In our example, four 20-ampere or five 15-ampere branch circuits, or a combination of 15- or 20-ampere branch circuits, could be used if the circuits have a total ampacity of at least 70.0 amperes. The minimum size of a branch-circuit conductor for a typical 120-volt branch circuit in a dwelling unit is 14 AWG copper or 12 AWG aluminum [NEC 310.3(A)]. This indicates that a minimum 15-ampere rated branch circuit is required for general lighting circuits. This approach determines the minimum number of general-purpose branch circuits needed for the dwelling. Of course, additional general-purpose lighting and receptacle circuits are allowed as the owner or installer wishes. When extra circuits are installed, and the load is calculated based on the volt-ampere per square foot method, no additional load is necessary in the overall service or feeder load calculation.

SMALL APPLIANCE BRANCH CIRCUITS

A minimum of two 20-ampere small appliance branch circuits must be installed to supply receptacle outlets only in areas such as the kitchen, pantry, breakfast room, dining room, or similar areas of a dwelling unit. A kitchen is defined as "an area with a sink and permanent facilities for food preparation and cooking." Although the term "permanent facilities for food preparation and cooking" is not specifically defined, there is a detailed description of which receptacle outlets must be served by these minimum two 20-ampere branch circuits. These circuits are generally required to supply small appliances in kitchens, pantries, breakfast rooms, dining rooms, or similar areas. This publication will discuss the circuit requirements in more detail later in Chapter 11,.

The code requires a minimum feeder load of 1500 volt-amperes for each of the two small appliance branch circuits needed. What happens if more than two small appliance branch circuits are installed, such as in a large dwelling? Do you need to include 1500 volt-amperes for each additional circuit? NEC 210.11(C)(1) states that "two or more 20-ampere small appliance branch circuits" must be provided for all receptacle outlets related to small appliance loads. NEC 220.52(A) mandates that the 1500-volt-ampere load be applied to those circuits that are required. Most inspection jurisdictions interpret this as requiring the load to be included only for the required two circuits, while others require a unit load of 1500 volt-amperes for each installed small appliance branch circuit. One thing is clear: the exception to NEC 220.52(A) allows the additional individual branch circuit permitted by NEC 210.52(B)(1) Exception No. 2 to be excluded from the small appliance load calculation. This exception applies to the additional small appliance branch circuit rated 15 amperes or more for appliances such as refrigeration equipment located in kitchens. Verify the requirements of your local jurisdiction before performing the load calculation.

Some single-family homes have two living areas, each with its own complete kitchen, one of which is designated as the "mother-in-law" kitchen. The Code does not allow a small appliance branch circuit to serve more than one kitchen. Therefore, each kitchen requires a minimum of two small appliance branch circuits. If there are two kitchens, at least four small appliance branch circuits must be installed. A load of 6000 volt-amperes (4 x 1500 VA) must be included in the calculation.

NEC 220.52(A) allows the 1500 volt-amperes load for each small appliance branch circuit to be combined with the general lighting load, following the demand factors in NEC 220.42 and Table 220.42(A). We will use the demand factors in the sample load calculations later in this chapter.

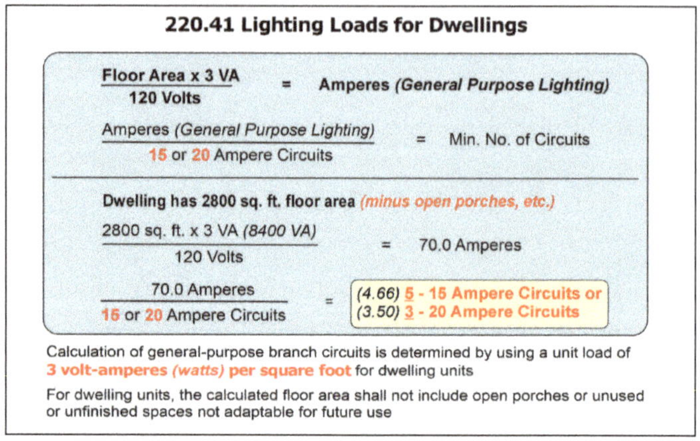

Figure 2.4. General lighting load calculation for dwelling units

Photo 2.3. General-purpose lighting circuits in a typical dwelling unit den area

LAUNDRY BRANCH CIRCUIT

A single 20-ampere laundry branch circuit is required to supply the laundry outlet(s) for a dwelling unit [NEC 210.11(C)(2)]. This circuit is in addition to those installed for general lighting and small appliances. Each laundry branch circuit must include a service or feeder load of 1500 volt-amperes [NEC 220.52(B)].

The Code does not require a laundry receptacle in dwellings other than one-family homes where they are not

Photo 2.4. Small appliance branch circuit rough wiring in kitchen half wall/island area

installed or allowed to be installed. This means that laundry receptacles might not be necessary in a two-family home that is not owned by the occupants but rented out. As a result, the 1500-volt-ampere load for each branch circuit does not need to be included in these cases. Typically, each unit of a two-family residence has a laundry equipment receptacle outlet. If so, a 1500-volt-ampere load must be allocated for each circuit.

NEC 220.52(B) allows the 1500 volt-amperes load for the laundry circuit to be added to the general lighting load, based on the demand factors in NEC Table 220.45. We use these demand factors in the load calculations shown later in this chapter.

ELECTRIC VEHICLE SUPPLY EQUIPMENT (EVSE) LOAD

If the dwelling unit has electric vehicle charging equipment installed, NEC 220.57 requires the EVSE loads to be calculated at either 7200 watts (volt-amperes) or the equipment's nameplate rating, whichever is greater.

ONE-FAMILY DWELLING CALCULATIONS

We will determine the minimum service and/or feeder ampacity for three dwellings, each with different equipment and areas. We will first perform the calculations using the standard or general method, then use the optional method.

LOAD CALCULATION EXAMPLES

We will assume the dwelling unit for which we are calculating the load has the following dimensions, appliances, or equipment (all loads are 120 volts unless indicated otherwise):

Table 220.45 General Lighting

Table 220.45 Lighting Load Demand Factor

Type of Occupancy	Portion of Lighting Load to Which Demand Factor Applies (VA)	Demand Factor (%)
Dwelling Units	First 3000 volt-amperes	100 percent
	From 3001 to 120,000 VA	35 percent
	Remainder over 120,000 VA	25 percent

The demand factors specified in Table 220.45 shall apply to that portion of the total branch-circuit load calculated for general illumination

They shall not be applied in determining the number of branch circuits for general illumination

Figure 2.5. Lighting load demand factors for dwelling units

Example No. 1

1650 — square feet of living space
1 — dishwasher, 1500VA
1 — gas or oil-fired furnace with 1/2 hp motor
1 — gas range
1 — gas clothes dryer

See Standard Calculation Example No. 1, and Optional Calculation Example No. 1 located at the end of this Chapter.

Example No. 2

2250 — square feet of living space
1 — dishwasher, 1500VA
1 — waste disposal, 1/2 hp
1 — trash compactor, 5 amperes
2 — exhaust fans, 120 VA each
1 — water heater, 4500 VA, 240-volts
1 — electric range, 11.2 kVA, 240-volts
1 — electric clothes dryer, 4.5 kVA, 240-volts
1 — electric furnace, 15 kW plus 1/2 hp fan motor, 240-volts
1 — air conditioner with 20 RLA compressor plus 1.4 A fan motor, 240-volts

See Standard Calculation Example No. 2, and Optional Calculation Example No. 2 located at the end of this Chapter.

Example No. 3

3650 — square feet of living space
1 — dishwasher, 1500VA
1 — waste disposal, 1/2 hp
1 — trash compactor, 5 amperes
4 — exhaust fans, 120 VA each
2 — water heater, 4500 VA, 240-volts
1 — electric range, 13.6 kVA, 240-volts
1 — electric clothes dryer, 5.5 kVA, 240-volts
1 — heat pump with 31.5 RLA compressor and 1.5 FLA outdoor fan motor, an indoor air handler with a total of 20 kW strip heaters plus 1/4 hp fan motor, all 240 volts
1 — 30 kW pool heater and one 3/4 hp pump, all 240-volts

See Standard Calculation Example No. 3, and Optional Calculation Example No. 3 located at the end of this Chapter.

APPLICATION OF DEMAND FACTOR

A demand factor can be applied to the general lighting load based on the calculations in (NEC 220.45). The Code also allows the load for small appliance branch circuits and the laundry circuit to be included in the demand factor calculation [NEC 220.52(A) and (B)]. The idea behind demand factors is that, typically, not all loads in the calculation will be used or turned on at the same time. As a result, the overall load on the feeder is reduced.

Refer to the Load Calculation Examples at the end of this chapter for an illustration of how to apply this general lighting load demand factor. This demand factor may only be applied when using the standard load calculation method. The load calculations determine the size of the ungrounded conductor. The minimum size of the grounded (neutral) conductor is specified in NEC 220.61. The loads used in this section can be calculated using either the standard or the optional method.

Other demand factors are provided by the NEC and apply to dwelling units (load in volt-amperes). These demand factors will be discussed in detail under the next six sections of this chapter: Other Loads, Fixed Electric Space-Heating Loads, Four or More Appliances in the Calculation, Electric Clothes Dryer Loads, Electric Range(s) Loads, and Electric Vehicle Supply Equipment (EVSE) Loads.

OTHER LOADS

The loads for outlets that supply specific appliances or other loads must use the ampere rating of the appliance or load to be served. Outlets for single motors, such as submersible pump motors or attic fan motors, must be calculated in accordance with NEC Article 430. NEC 430.22 is used for conductors supplying a single motor; conductors supplying multiple motors follow the guidelines in 430.24; and motor feeder demand factors are provided by NEC 430.26. NEC Article 440 is used to calculate the load for individual air conditioning and refrigeration units. Refer to Chapter 9 for additional information on air conditioning and heat pump equipment.

The method of calculating the load for ranges, dryers, and appliances, as well as for heating and air-conditioning equipment, differs between the standard and optional methods. We will incorporate them into each method and indicate the application of demand factors where applicable.

FIXED ELECTRIC SPACE-HEATING LOADS

The Code states that fixed electric space-heating loads must be calculated at 100% of the total connected load (NEC 220.51). It should be highlighted that NEC 424.4(B) requires the ampacity of the branch-circuit conductors to be sized at no less than 125% as a continuous load, including the heater and any motor(s).

According to NEC 220.51, the feeder or service load current rating must never be less than the rating of the largest branch circuit. This typically occurs when a significant portion of the feeder load consists of high-capacity electric space-heating equipment powered by a single branch circuit. Since the minimum branch circuit size is determined by 125% of the heater load, the feeder's load calculation must also be based on this same 125% factor. An example of this is a dwelling with two service disconnecting means, one of which supplies only electric heating loads. In such cases, a separate meter is often used to enable the customer to be billed at a lower rate for heating loads.

An exception allows the AHJ to approve a lower demand factor if the heating unit(s) operate on a duty cycle, intermittently, or if not all units run simultaneously. Contact your local electrical inspector to see if any demand factors are permitted in your area.

FOUR OR MORE APPLIANCES IN THE CALCULATION

A demand factor of 75% is allowed for the combined load of four or more appliances rated at ¼ hp or higher or 500 watts or more, fixed in place—excluding household electric cooking equipment, clothes dryers, space-heating equipment, electric vehicle charging devices, or air-conditioning units (NEC 220.53). The criterion of appliances rated ¼ hp or more, or 500 watts or above, that qualifies for inclusion in this demand factor excludes items like bathroom exhaust fans. Typically, a residential bathroom exhaust fan motor is well below these ratings. This 75% demand factor may only be applied if the same feeder or service supplies all four or more appliances. Additionally, the appliance's nameplate load must be known before applying this demand factor.

ELECTRIC CLOTHES DRYER LOADS

Electric clothes dryers installed in dwelling units must be rated at 5000 watts (volt-amperes) or the nameplate rating, whichever is greater, for each dryer served (NEC 220.54). Note that the Code does not allow a demand factor until five or more dryers are connected to the same feeder, which would not apply to a typical one- or two-family dwelling unit (NEC Table 220.54).

ELECTRIC RANGE(S) LOADS

NEC 220.55 covers electric ranges and other cooking appliances installed in dwelling units. This section includes NEC Table 220.55, which helps determine demand loads for household electric ranges, wall-mounted ovens, counter-mounted cooking units, and other household cooking appliances. Seven notes accompany the table, providing permission to use alternative methods for specific situations. Column C shows the demand load unless Note 3 states otherwise.

For an electric range with a rating of no more than 12 kW, the maximum demand may be calculated as 8 kW, as shown in Column C of NEC Table 220.55. When two ranges, each not exceeding 12 kW, are connected to the same feeder, a maximum demand of 11 kW can be used in the load calculation. If more than one cooking appliance is supplied by the same branch circuit not exceeding 50 amperes, as allowed in NEC 210.19(C), Exception No. 1, it is permitted to add the loads of up to one cooktop and two wall-mounted ovens and treat them as a single appliance for purposes of Table 220.55 (NEC Table 220.55, Note 6).

If an electric range exceeds 12 kW, the maximum demand of 8 kW in Column C must be increased by 5% for each kW or major fraction thereof that the rating exceeds 12 kW. For example, an electric range has a rating of 13.7 kW. This exceeds 12 kW by 1.7 kW, which is considered 2 kW for the purposes of the calculation. In this case, the 8 kW demand factor in Column C is increased by 5% for each kW or major fraction thereof, or 10%. The value of 8000 watts plus 10% or 8800 watts would be used in the load calculation for this 13.7 kW electric range (*NEC* Table 220.55, Note 1).

ELECTRIC VEHICLE SUPPLY EQUIPMENT (EVSE) LOAD

Provisions for electric vehicle supply equipment (EVSE) loads were added at NEC 220.57 for the 2023 NEC revision cycle. Electric vehicle supply equipment, or charging stations, has become increasingly common, even in residential units, over the past few years. EV charging infrastructure is becoming increasingly essential, and the 2026 NEC is expected to establish guidelines for load calculation requirements for EV charging. A minimum requirement of 7200 watts or volt-amps (VA) was established, based on a typical 30-amp, 240-volt, single-phase circuit. Therefore, if electric vehicle charging equipment is installed in a dwelling, this new requirement mandates that EVSE loads be calculated at either 7200 watts (volt-amps) or the equipment's nameplate rating, whichever is greater.

MINIMUM SIZE OF GROUNDED (NEUTRAL) SERVICE CONDUCTOR

The minimum size of the grounded (neutral) conductor for our service-entrance or service-lateral conductors is determined by NEC 220.61. The method of calculating the current on the grounded service-entrance conductor applies to both the standard and optional calculation methods. The purpose of this calculation is to find the maximum net unbalance of the load from the ungrounded conductors that the neutral conductor must carry.

It is common practice for electricians to connect line-to-neutral loads alternately between the two ungrounded (hot) conductors and the neutral conductor. This results in the line-to-neutral loads being fairly evenly distributed between the ungrounded (hot) conductors. If the loads could be perfectly balanced, no current would flow on the neutral conductor. If one of the ungrounded (hot) conductors fails for any reason, the neutral would then carry all the line-to-neutral loads from the other ungrounded conductor.

As shown in Figure 2.6, the neutral conductor carries only the difference in load between two ungrounded circuits connected on opposite phases of a single-phase system.

Adding all the neutral loads together at 120 volts and dividing the total by 240 volts is the most practical way to find the maximum net unbalanced current on the neutral conductor. That is the method used in this text. This approach is shown in many examples in Informative Annex D of the NEC, where the 120-volt loads, such as for general lighting, small appliances, and laundry branch circuits, are totaled and divided by 240 volts to determine the net neutral conductor load.

A panelboard schedule that displays line-to-neutral loads can be used, when available, to identify the maximum unbalanced load. This approach is illustrated in Example D1(b) in Informative Annex D and might provide slightly more precise neutral load calculations.

Our examples of load calculations using the general method demonstrate the application of the 75% demand factor for dwelling unit appliance loads with four or more appliances, as permitted in NEC 220.53, to the neutral load and the ungrounded (hot) conductors. It is logical that the same 75% demand factor should also be allowed for the corresponding neutral conductors. The demand factor assumes that not all appliances will operate simultaneously. (This 75% demand is not applied to the neutral conductors in the optional load calculation method, as the demand factor is not applied to the ungrounded [hot] conductors in that method.) Additionally, NEC Article 250 includes rules regarding the minimum size of the grounded (neutral) conductor.

NEC 250.24(D) states that the grounded service conductor (neutral conductor) must not be smaller than the size specified in NEC Table 250.102(C)(1) for grounding electrode conductors. After performing the load calculation according to NEC 220.61, compare the results with the values in NEC Table 250.102(C)(1). The larger of the two—either the conductor size determined by the calculation or the one listed in the table—must be used.

The size of the largest ungrounded service-entrance conductor is used in the application of Table 250.102(C)(1). Follow down the column with the sizes of service-entrance conductors and across the table to determine the minimum size of the grounded service conductor needed. This ensures the grounded service conductor is large enough to carry any fault current back to the source.

For a feeder or service supplying household electric ranges, wall-mounted ovens, counter-mounted cooking units, and electric dryers, the maximum unbalanced load should be considered as 70% of the load on the ungrounded conductors, as specified in NEC Table 220.55 for ranges and NEC Table 220.54 for dryers. This explains why the load calculation generally shows a 70% value based on the NEC Table 220.55 figure of 8,000 kW for ranges up to 12 kW, or 5,600 kW, rather than 70% of the nameplate rating. Calculating the grounded (neutral) conductor size based on 70% of the nameplate rating is allowed, but this results in a larger conductor.

A further demand factor of 70% is allowed for that part of the unbalanced load exceeding 200 amperes. No reduction in neutral capacity is permitted for the segment of the load that involves 2-phase wires and the neutral in a 4-wire, 3-phase, wye-connected system. This setup is sometimes used in apartment complexes supplied by two ungrounded phase conductors and a neutral from a 4-wire, 3-phase system. In such cases, the neutral will carry approximately the same current as the ungrounded phase conductors and cannot be downsized [NEC 310.15(E)(2)].

NONCOINCIDENT LOADS

Loads that do not run at the same time are considered noncoincident loads. When it is unlikely that two different loads, like heating and cooling, will operate simultaneously, it is allowed to omit the smaller load in load calculations. If a motor or air-conditioning load is part of the noncoincident load and is not the largest, 125 percent of either the motor load or the air-conditioning load (whichever is greater) must be used in the calculation (NEC 220.60).

You should consult your electrical inspector for the local interpretation of this rule as it applies to motor loads. NEC 430.24 requires using 25% of the largest motor in the load

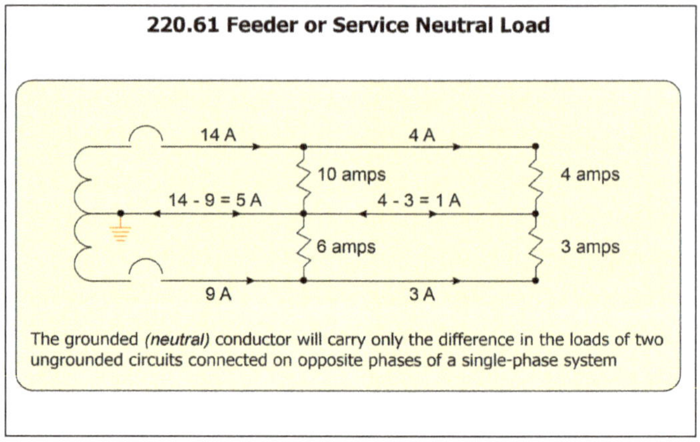

Figure 2.6. Grounded (neutral) service conductor carries unbalanced load from underground (hot) conductors

calculation. For example, a dwelling with both electric heating and air-conditioning loads, where the heating is the larger load. This means the air-conditioning load can be ignored in the load calculation. However, the air-conditioner compressor motor is still connected to the system and will almost always be the largest motor. Some inspection authorities require it to be included. Others interpret the rule that the air-conditioning motor does not need to be considered if it is smaller than the heating load.

Logic states that if the air-conditioner compressor motor load is the smaller of the two loads, it can be ignored in the load calculation. Then, the next largest motor is considered the "largest" motor for applying the 25% load rule.

OPTIONAL CALCULATION

It is allowed to calculate the feeder and service load for a dwelling unit with service-entrance or feeder conductors rated 100 amperes or more according to NEC 220.82 instead of the standard method under Part II of Article 220. As mentioned earlier, the optional method usually results in a lower load calculation than the standard method.

REQUIREMENTS SPECIFIC TO THE OPTIONAL METHOD CALCULATIONS

As noted earlier in this chapter, several elements of load calculations are the same for calculations performed using the standard and optional methods. Several conditions listed in NEC 220.82 must be met before load calculations for the optional method are allowed. These conditions include the following:

1. The dwelling unit must be served by a single 3-wire service or feeder.
2. The voltage must be 120/240 (single-phase) or 208 Y/120 (two ungrounded and one grounded conductor).

3. The service or feeder conductors must have an ampacity of 100 amperes or greater.
4. The nameplate rating must be known for all appliances that are fastened in place, permanently connected, or located on a specific circuit (ranges, wall-mounted ovens, counter-mounted cooking units, clothes dryers, water heaters, etc.).
5. The nameplate ampere or kVA rating must be known for all motors, all low-power-factor loads, and all heating and air-conditioning equipment.
6. It is permitted to use the methods in 220.61 for calculating the load on the service or feeder neutral conductor. This is the method we have followed in our calculations.

OTHER OPTIONAL CALCULATION FACTORS

Some load factors for optional calculations are the same as those in standard methods, while others are unique to the optional approach. To use the optional method, the overall calculated load should be at least 100% of the first 10 kVA plus 40% of the remaining loads.

The loads to be included as general loads are specified in NEC 220.82(B) and include the following:

1. 1500 volt-amperes for each 2-wire, 20-ampere small appliance branch circuit, and each laundry branch circuit as specified in *NEC* 210.11(C)(1) and (C)(2).

2. Three watts or volt-amperes per sq. ft.^{2} (33 sq. m^2) for general lighting and general-use receptacles. The floor area for each floor is calculated from the outside dimensions of the dwelling unit. The calculated floor area does not include open porches, garages, unused spaces, or spaces that are not adaptable for future use.

   Note to Reader: Please take note of the 2023 *NEC* change at 220.5(C). This revised section states that for dwelling units, the calculated floor area shall not include open porches or unfinished areas not adaptable for future use as a habitable room or occupiable space. Compare this revised text to the existing text at *NEC* 220.82(B)(1), which states that the calculated floor area shall not include open porches, garages, or unused or unfinished spaces not adaptable for future use. This existing language at *NEC* 220.82(B)(1) is the same language as *NEC* 220.5(C) in the previous edition of the *NEC*. These two sections should match each other, and the author of this Note to Reader suspects this is an oversight by the Code Making Panel and that this will be corrected in the next edition of the *NEC*.

3. The nameplate rating of all appliances that are fastened in place, permanently connected, or located to on a specific circuit (ranges, wall-mounted ovens, counter-mounted cooking units, clothes dryers that are not connected to the laundry branch circuit, and water heaters, etc.).

4. The nameplate ampere or kVA rating of all motors not included in 220.82(B)(2).

These loads are subject to a demand factor of 100% for the first 10 kVA (10,000 VA) and 40% for the remaining load. To this result, add the largest of the heating and air-conditioning loads from 220.82(C). Only the largest load among the following six options needs to be included.

1. 100% of the nameplate rating(s) of the air-conditioning and cooling equipment, including heat pump compressors

2. 100% of the nameplate ratings of the heat pump when the heat pump is used without any supplemental electric heating.

3. 100% of the nameplate rating(s) of the heat pump compressor and 65% of the supplemental electric heating for central space heating systems. If the heat pump compressor is prevented from operation at the same time as the supplemental heat, it does not need to be added to the supplemental heat to determine the total space heating load.

4. 65% of the nameplate rating(s) of electric space heating equipment if less than four separately controlled units.

5. 40% of the nameplate rating(s) of electric space heating equipment if four or more separately controlled units.

6. 100% of the nameplate ratings of electric thermal storage and other heating systems where the usual load is expected to be continuous at the full nameplate value. Systems qualifying under this selection

shall not be figured under any other selection in *NEC* 220.82(C).

HEAT PUMP LOADS

Example D2(c) in Annex D of the NEC offers guidance on how to apply the rules for calculating heat pump loads when supplementary strip or backup resistance heat is used. If the two loads are not operating simultaneously, the smaller load can be disregarded in load calculations. If the supplementary heat is not running at the same time as the heat pump and is the largest load, the heat pump kVA does not need to be added to the total.

Be sure to consult the control diagram or operating characteristics of the heat pump thermostat, many of which are two-stage types. The first stage activates the heat pump to heat the space to the desired temperature. When the temperature difference between the heated space and the thermostat setting exceeds one or two degrees, the thermostat engages the second stage (the resistance strip heaters). These heaters operate until the space approaches the set temperature, at which point the thermostat opens the control circuit to the heaters. The heat pump continues running until the target temperature is reached.

In colder climates, it is common for heat pump compressors to be "locked out" by an outdoor thermostat when the outdoor temperature drops to a point where the heat pump cannot produce sufficient heat from the cold air. In milder temperatures, the heat pump compressor and supplemental strip heaters often operate simultaneously.

It is also typical for the strip heaters to operate alongside the heat pump during defrost cycles and after periods of setback when the thermostat calls for heat.

As shown, it is most common for heat pumps and backup strip heaters to operate concurrently. As illustrated in Example No. D2(c) of NEC Informative Annex D, the loads of the compressor and strip heater are summed. The compressor has a 100% demand factor while a 65% demand factor is applied to the strip heaters. If the heat pump does not have strip heaters, it is assumed at 100% capacity when determining the largest of the six loads, as indicated in NEC 220.82(C).

We will assume that the heat pump and strip heaters operate at the same time and include both loads using the method specified by NEC 220.82(C)(3).

CALCULATING THE LOAD FOR A TWO-FAMILY DWELLING

Trade practice is to supply the service equipment for a two-family dwelling via a common mast or a common service lateral. A duplex meter base is usually installed when this is done. A calculation must be performed to determine the minimum size of service-entrance or service-lateral conductors needed to supply the shared loads (see Figure 2.7).

The Code allows the service or feeder load to each dwelling unit of a two-family dwelling to be calculated using more than one method. Since NEC 230.40 requires that each service drop or lateral supply only one set of service-entrance conductors, it is important that these rules be clearly understood. An exception permits one set of service-entrance conductors for each occupancy of a building with more than one occupancy, such as a two-family dwelling.

The load calculation methods permitted include:

1. A standard calculation of both feeders and service-entrance and service-lateral conductors. In this case, demand factors are applied to the individual service and service-entrance or service-lateral conductors. For example, the entire square-foot area to be served, four small appliance circuits, and two laundry circuits are included, and demand factors are applied. Also, the demand factor for two electric ranges in *NEC* Table 220.55 is applied. Standard Load Calculation (SLC) – Two-Family Dwelling, located at the end of this chapter, illustrates the standard load calculation for a new two-family dwelling unit.

2. An optional calculation of feeders to the individual dwelling units, according to *NEC* 220.80, is performed, and a standard calculation of service-entrance and service-lateral conductors, according to Part III of *NEC* Article 220, is performed.

3. An optional calculation for a new service on an existing two-family dwelling is performed under *NEC* 220.83.

We have provided an example of load calculation for the two dwelling units using the standard load calculations of Part III of NEC Article 220. We have not included an example of the optional method allowed by NEC 220.83, Existing Dwelling Unit. Refer to Chapter 16 for more information on calculating services for existing dwelling units.

The total floor area of both units is calculated similarly to that of one-family dwellings. Non-habitable parts of the building are not included. Two of every load are assumed except for small appliance circuits, where four are assumed.

The total nameplate load of appliances that are fixed in place, permanently connected, or located on a specific circuit—such as ranges, wall-mounted ovens, counter-mounted cooking units, clothes dryers, water heaters, and space heaters in both units—is included.

When standard calculations are used for both dwelling unit feeders and the service entrance or lateral conductors, up to three calculations may be needed. Two calculations are performed if the dwelling units are the same size and have identical electrical equipment. One calculation is for supplying the dwelling units. A second calculation is for the service entrance or lateral conductors that supply the total load to the building. If the dwelling units differ in size or have different electrical equipment, three calculations are required: one for each dwelling unit and a third for the service entrance or service lateral conductors.

When two dwelling units are supplied by one feeder, and the calculated load under Part III of NEC Article 220 (standard calculation) exceeds that for three identical units calculated under NEC 220.84, the lesser of the two loads may be used (NEC 220.85).

Optional calculations are not allowed for both feeders and service-entrance conductors. This would be considered applying demand factors twice to the same load. However, optional calculations can be used for the feeder conductors to individual units, while standard calculations apply to the service-entrance conductors (see optional calculations for feeders and standard calculations for service-entrance conductors).

For these calculations, we will assume the following loads for each dwelling unit.

1. an area of 1200 square feet, not counting open porches or unfinished areas not adaptable for future use as a habitable room or occupiable space,
2. laundry equipment,
3. one 10.7 kW electric range,
4. one 4.5 kW electric clothes dryer,
5. one 4.5 kW electric water heater,
6. one, 1.5 kW dishwasher,
7. one, 1/2 hp waste disposal,
8. two exhaust fans, 120 VA each, and…
9. one 9.8 kW electric heat
10. one AC unit with 20 RLA compressors plus 1.4 A fan motor, 240-volts

See the "Standard Load Calculations – Two-Family Dwelling Unit" example worksheet at the end of this chapter.

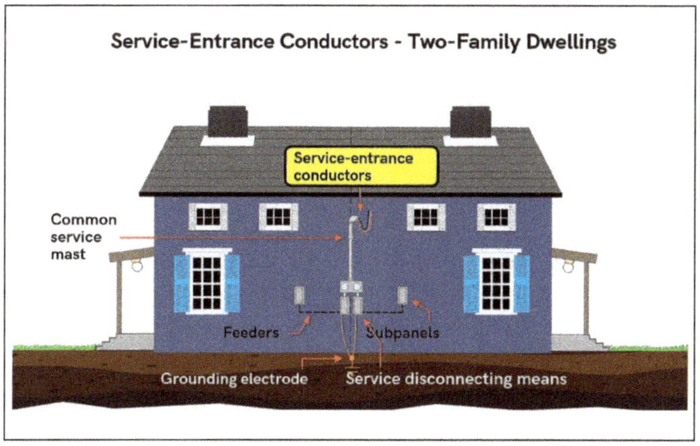

Figure 2.7. Calculating the size of service-entrance conductors supplying a two-family dwelling

PART B: CALCULATION AND SIZING REQUIREMENTS OF THE *INTERNATIONAL RESIDENTIAL CODE* GENERAL

This part of Chapter 2 discusses the requirements for sizing service equipment, service-entrance conductors, and feeder conductors according to the International Residential Code 2021 (IRC) rules. Key provisions to consider are the scope and applicability of the IRC. It is important to note that the provisions of IRC Chapters 34 through 43 are intended to apply to wiring methods and materials most commonly used in the construction of one- and two-family dwellings. Wiring methods beyond those specifically covered by the IRC must comply with the relevant requirements in NFPA 70 (the NEC) (IRC E3401.1, E3401.2).

Services within the scope of the IRC are limited to 400-ampere, single-phase, 120/240-volts. Any services exceeding this limit must comply with the applicable provisions of NFPA 70 (IRC E3401.2, E3401.3). Service equipment must be listed and bear the label of a qualified electrical testing laboratory. This equipment should be installed according to the manufacturer's instructions (IRC E3403.3). The IRC requires a dwelling unit to have only one service; no exceptions or alternatives permit additional services (IRC E3601.2).

RATINGS OF THE SERVICE EQUIPMENT AND CONDUCTORS

The minimum ampere rating of a service disconnect and the ungrounded service conductors for a single-family dwelling must be at least 100 amperes (IRC E3602.1). The minimum short-circuit current rating for the service equipment should be no less than 10,000 amperes. This is common among the service equipment available today. The equipment must be marked to show it is suitable for use as service equipment [IRC E3601.6.1].

DEFINITIONS

The definitions at the beginning of Chapter 2 also apply to this part of the chapter. Chapter 35 of the IRC contains the definitions of words and terms derived from the NEC. If terms are not defined within Chapter 35, they may be defined in IRC Section R202. An example of a difference between the IRC and NEC definitions is the word "building." In the IRC, Section R202 defines "building" as "any one- and two-family dwelling or townhouse, or portion thereof, used or intended to be used for human habitation, for living, sleeping, cooking or eating purposes, or any combination of these, or any accessory structure."

SERVICE LOAD CALCULATION

Many of the requirements in the IRC are presented in tabular form. In other words, the requirements are given in tables rather than in paragraph or section texts. The rules are generally the same as those in the NEC. The IRC requires the calculation of the load on ungrounded service conductors and devices that serve 100% of the dwelling, following the methods outlined in IRC Table E3602.2 (IRC E3602.2). The rules and methods for calculating service conductor sizes in this table are identical to those in Chapter 2 of the NEC for optional feeder and service load calculations. Feeders that serve less than 100% of the dwelling unit load must follow the calculation requirements for feeders in Chapter 37 of the IRC. Essentially, IRC Table E3602.2] includes ten steps.

The following should be summed at the beginning of the calculation.

1. The general lighting load for the dwelling unit is required to be based on 3 volt-amperes per square foot of habitable area (from *NEC* 220.41).
2. 1,500 volt-amperes are required for each small appliance and laundry circuit. (IRC E3703.2 requires a minimum of two 20-ampere branch circuits for the kitchen, and IRC E3703.3 requires a 20-ampere circuit for the laundry.)
3. The nameplate volt-ampere rating of all appliances fastened in place, permanently connected, or dedicated circuit-supplied appliances such as ranges, dryers, ovens, cooking units, and water heaters.

The first step is to add items 1 through 3 above together. Then, the applicable demand factors in the table may be

Photo 2.5. Typical two-family dwelling service lateral supplying a two-gang meter bank with dwelling service equipment and panelboards inside the building back-to-back with the meter enclosures for each dwelling

Photo 2.6. Typical service equipment at a dwelling ready for an inspection

applied. The minimum load in the calculation shall be 100% of the first 10,000 volt-amperes of the total sum, plus 40% of any portion of the remaining volt-amperes over 10,000.

After the above steps have been completed, the largest of each of the following loads is to be added to the calculation.

1. The nameplate rating of air-conditioning and cooling equipment, including heat pumps
2. The nameplate rating of electric thermal storage and other heating systems where the usual load is expected to be continuous at the full nameplate value. (Systems qualifying under this selection are not permitted to be figured into any other category in IRC Table E3602.2).
3. 65% of the nameplate rating of central electric space-heating equipment, including any supplemental integral heating used with heat pump systems.
4. 65% of the nameplate rating(s) of electrical space-heating units, if less than four separately controlled units.

TABLE E3602.2 [*NEC* 220.82(B) & (C)] MINIMUM SERVICE LOAD CALCULATION
LOADS AND PROCEDURE
3 volt-amperes per square foot of floor area for general lighting and general use receptacle outlets.
PLUS
1,500 volt-amperes multiplied by total number of 20-ampere-rated small appliance and laundry circuits.
PLUS
The nameplate volt-ampere rating of all fastened-in-place, permanently connected or dedicated circuit-supplied appliances such as ranges, ovens, cooking units, clothes dryers not connected to the laundry branch circuit and water heaters.
APPLY THE FOLLOWING DEMAND FACTORS TO THE ABOVE SUBTOTAL:
The minimum subtotal for the loads above shall be 100 percent of the first 10,000 volt-amperes of the sum of the above loads plus 40 percent of any portion of the sum that is in excess of 10,000 volt-amperes.
PLUS THE LARGEST OF THE FOLLOWING:
One hundred percent of the nameplate rating(s) of the air-conditioning and cooling equipment.
One hundred percent of the nameplate rating(s) of the heat pump where a heat pump is used without any supplemental electric heating.
One hundred percent of the nameplate rating of the electric thermal storage and other heating systems where the usual load is expected to be continuous at the full nameplate value. Systems qualifying under this selection shall not be figured under any other category in this table.
One hundred percent of nameplate rating of the heat pump compressor and 65 percent of the supplemental electric heating load for central electric space-heating systems. If the heat pump compressor is prevented from operating at the same time as the supplementary heat, the compressor load does not need to be added to the supplementary heat load for the total central electric space-heating load.
Sixty-five percent of nameplate rating(s) of electric space-heating units if less than four separately controlled units.
Forty percent of nameplate rating(s) of electric space-heating units of four or more separately controlled units.
The minimum total load in amperes shall be the volt-ampere sum calculated above divided by 240 volts.

Figure 2.8. IRC Table E3602.2 provides the minimum service load calculations

5. 40% of the nameplate rating(s) of electrical space-heating units of four or more separately controlled units.

The minimum total load in amperes must be the total of the volt-amperes calculated above divided by 240 volts. The specific sizing for the service and feeder conductors is explained in the next section.

SERVICE AND FEEDER CONDUCTOR SIZING

Once the calculation is performed using the rule found in IRC E3602.2, the minimum service and service-entrance conductor sizes can be determined using IRC E3603.1 and IRC Table E3603.1.1. The minimum grounding electrode conductor sizes are provided in IRC Table E3603.4. In summary, the minimum service size for a one-family dwelling is 100 amperes. The calculation, in accordance with IRC E3602.2 and Table E3602.2, establishes the minimum size needed beyond the 100-ampere level. Once a service size is established, IRC E3603.1 sets the minimum sizes for service-entrance conductors or the main power feeder. These requirements are the same as those in NEC 310.12. For example, if a calculated load on the service requires 251 amperes, this would call for a 300-ampere-rated service. IRC E3603.1.1 allows an 83 percent reduction ratio to be applied to the 300-ampere service rating (300 A x .83 = 249 amperes). Using IRC Table E3603.1.1, this adjusted load of 249 amperes would require a minimum copper service-entrance conductor size of 4/0 AWG insulated conductors.

According to IRC Table E3603.4, the copper grounding electrode conductor for 4/0 AWG copper service-entrance conductors must be at least 2 AWG copper. The note associated with the superscript "f" at the bottom of Table E3603.4 allows the grounding electrode conductor size to not exceed 6 AWG copper or 4 AWG aluminum for a ground rod system, or 4 AWG copper for a concrete-encased electrode, where it does not extend onto other types of electrodes that require a larger conductor size (see notes to Table E3603.4).

MINIMUM SIZE OF GROUNDED (NEUTRAL) CONDUCTOR

Similar to the NEC, the minimum size of the grounded (neutral) conductor for the service must be capable of carrying the maximum unbalanced return neutral load current, and this conductor can never be smaller than the grounding electrode conductor sizes outlined in IRC Table E3603.4. The superscript notes at the bottom of this table apply only to the grounding electrode conductor and should not be used for the minimum sizing of a grounded or neutral conductor at the service. The size of the grounded conductor cannot be smaller than what is specified by IRC E3704.3. This IRC section states that for a feeder or service supplying electric ranges, wall-mounted ovens, counter-mounted cooking units, and electric dryers, the maximum unbalanced load of the neutral conductor is to be considered as 70 percent of the load on the ungrounded conductors.

SERVICE CONDUCTORS FOR ACCESSORY BUILDINGS

Sometimes, service-entrance conductors are installed to supply additional buildings on residential property used for purposes other than human habitation, such as detached garages and workshops. The IRC specifies minimum sizing requirements for the service conductors supplying such accessory buildings or structures. The basic rule states that those service-entrance conductors must have a minimum ampacity of 60 amperes and be sized according to Chapter 37 for feeders. Two exceptions to the standard 60-ampere minimum allow the use of 8 AWG copper or 6 AWG aluminum for loads consisting of no more than two-wire circuits, or 12 AWG copper or 10 AWG aluminum for a single branch circuit (IRC E3603.2).

OVERLOAD PROTECTION

As with the requirements of the NEC, the IRC also mandates that each ungrounded service-entrance conductor must be protected against overloads (IRC E3603.3). This protection is typically provided by a main breaker or fusible switch in the service equipment, designed to limit the load on the service-entrance conductors. Overload protection usually applies to the service-entrance conductors on the line side (often the meter side) of the fuses or circuit breaker, which typically lack overcurrent protection beyond what is available on the primary side of the utility transformer. The IRC requires an overcurrent protective device in series with each ungrounded service conductor, and it must not be rated higher than the limits specified in IRC E3603.1. In other words, for a 300-ampere service, the size of the service-entrance conductors, grounding electrode conductor, and overcurrent protection all must meet the minimum sizes set forth in IRC E3603.1. Overcurrent devices are not allowed in the grounded conductor, except for circuit breaker types that disconnect all conductors, including the grounded one (IRC E3603.3.2; see Figure 2.1).

FEEDER CALCULATIONS

Feeder and branch-circuit conductor sizing is addressed in IRC Chapter 37, which discusses branch circuits and feeders. IRC Tables E3704.2(1) and E3704.2(2) are used to determine the sizes of feeders and branch circuits.

REFERENCES

1. *National Electrical Code* and *NEC* are registered trademarks of the National Fire Protection Association, Inc., Quincy, MA 02169. This reprinted material is not the official position of the National Fire Protection Association, which is represented only by the standard in its entirety.

2. *International Residential Code for One- and Two-Family Dwellings* (2021). Copyright © International Code Council, Inc., June 2023.

3. UL, Northbrook, IL, *UL Guide Information for Electrical Equipment Directory* (UL Product iQ® material) unless indicated otherwise.

TABLE E3704.2(1) [*NEC* 220.41, Table 220.42, 220.50, 220.51, 220.52, 220.53, 220.54, 220.55, and 220.60] FEEDER LOAD CALCULATION	
LOAD CALCULATION PROCEDURE	**APPLIED DEMAND FACTOR**
Lighting and receptacles: A unit load of not less than 3 VA per square foot of total floor area shall constitute the lighting and 120-volt, 15- and 20-ampere general use receptacle load. 1,500 VA shall be added for each 20-ampere branch circuit serving receptacles in the kitchen, dining room, pantry, breakfast area and laundry area.	100 percent of first 3,000 VA or less and 35 percent of that in excess of 3,000 VA.
PLUS	
Appliances and motors: The nameplate rating load of all fastened-inplace appliances 1/4 horsepower or greater, or 500 watts or greater, that are fastened in place, other than dryers, ranges, air-conditioning and space-heating equipment.	100 percent of load for three or less appliances. 75 percent of load for four or more appliances.
PLUS	
Fixed motors: Full-load current of motors plus 25 percent of the full load current of the largest motor.	
PLUS	
Electric clothes dryer: The dryer load shall be 5,000 VA for each dryer circuit or the nameplate rating load of each dryer, whichever is greater.	
PLUS	
Cooking appliances: The nameplate rating of ranges, wall-mounted ovens, counter-mounted cooking units and other cooking appliances rated in excess of 1.75 kVA shall be summed.	Demand factors shall be as allowed by Table E3704.2(2).
PLUS THE LARGEST OF EITHER THE HEATING OR COOLING LOAD	
Largest of the following two selections: 1. 100 percent of the nameplate rating(s) of the air conditioning and cooling, including heat pump compressors. 2. 100 percent of the fixed electric space heating.	

Figure 2.9. IRC Table E3704.2(1) provides feeder load calculations

TABLE E3704.2(2) (*NEC* 220.55 and Table 220.55) DEMAND LOADS FOR ELECTRIC RANGES, WALL-MOUNTED OVENS, COUNTER-MOUNTED COOKING UNITS AND OTHER COOKING APPLIANCES OVER 1¾ kVA RATING[a, b]			
NUMBER OF APPLIANCES	MAXIMUM DEMAND[b, c]	DEMAND FACTORS (percent)[d]	
	Column A maximum 12 kVA rating	Column B less than 3½ kVA rating	Column C 3½ to 8¾ kVA rating
1	8 kVA	80	80
2	11 kVA	75	65

a. Column A shall be used in all cases except as provided for in Footnote d.
b. For ranges all having the same rating and individually rated more than 12 kVA but not more than 27 kVA, the maximum demand in Column A shall be increased 5 percent for each additional kVA of rating or major fraction thereof by which the rating of individual ranges exceeds 12 kVA.
c. For ranges of unequal ratings and individually rated more than 8.75 kVA, but none exceeding 27 kVA, an average value of rating shall be computed by adding together the ratings of all ranges to obtain the total connected load (using 12 kVA for any ranges rated less than 12 kVA) and dividing by the total number of ranges; and then the maximum demand in Column A shall be increased 5 percent for each kVA or major fraction thereof by which this average value exceeds 12 kVA.
d. Over 1.75 kVA through 8.75 kVA. As an alternative to the method provided in Column A, the nameplate ratings of all ranges rated more than 1.75 kVA but not more than 8.75 kVA shall be added and the sum shall be multiplied by the demand factor specified in Column B or C for the given number of appliances.

Figure 2.10. IRC Table E3704.2(2) provides demand loads for cooking appliances

STANDARD LOAD CALCULATION - Two-Family Dwelling

Factor	Quantity	VA Ungrnd	VA Neutral
General Lighting (Sq Ft x 3 wpf)	2,400	7,200	
General Lighting Circuit Amps @ 120 V	60.0		
Install _4_ 15 Amp or _3_ 20 Amp Gen. Ltg. Circuits			
Small Appliance Circuits	4	6,000	
Laundry Circuit	2	3,000	
Total General Lighting Load		16,200	
First 3000 VA at 100 percent		3,000	
Total VA - 3000 = 13,200 VA at 35 percent		4,620	
Net General Lighting Load		**7,620**	**7,620**
Appliances			
Dishwasher	2	3,000	3,000
Disposal (1/2 hp motor)	2	2,352	2,352
Exhaust fans (120 VA each)	4	480	480
Water Heater (4,500 VA each)	2	9,000	0
Appliances rated less than ¼ hp or 500 watts		480	
4 or more Appliances Total at 75 percent		10,764	
Total Appliance Loads		**11,244**	**4,374**
Other Loads			
Range (*NEC* 220.55)(Neutral 220.61 at 70%)	(10,700)	**11,000**	**7,700**
Dryer (*NEC* 220.54)(Neutral 220.61 at 70%)	(4,500)	**10,000**	**7,000**
Electric heat (9.8 kW) (8 kW x 2)	(16,000)	**16,000**	
AC unit (21.4 A x 2)	(10,272)	0	
(*NEC* 220.60 Noncoincident Loads)			
25% of Largest Motor (*NEC* 430.24)	1,176	**294**	**294**
Total Volt-Amperes		**56,158**	**26,988**
Volt-Amperes ÷ 240 Volts = Amperes		**233.99**	**112.50**
*Minimum Conductor Size (Copper or Aluminum)		4/0 AWG (CU)	**2 AWG (CU)
Insulation Type		XHHW	XHHW

**Min. ungrounded conductor size based on standard ampacity rating of 250 A min. service rating and 83 % rating of 310.12 (207.5 A). **Grounded conductor shall not be smaller than specified in Table 250.102(C)(1) [see 250.24(C)(1)]*

STANDARD LOAD CALCULATION

Example No. 1

Factor	Quantity	VA Ungrnd	VA Neutral
General Lighting (Sq Ft x 3 wpf)	1,650	4,950	
General Lighting Circuit Amps @ 120 V	41.3		
Install 3 15 Amp or 3 20 Amp Gen. Ltg. Circuits			
Small Appliance Circuits	2	3,000	
Laundry Circuit	1	1,500	
Total General Lighting Load		9,450	
First 3000 VA at 100 percent		3,000	
Total VA - 3000 = 6,450 VA at 35 percent		2,258	
Net General Lighting Load		**5,258**	**5,258**
Appliances			
Dishwasher	1	1,500	1,500
Furnace (1/2 hp motor)	1	1,176	1,176
Appliances rated less than ¼ hp or 500 watts			
3 or less Appliances Total		**2,676**	**2,676**
4 or more Appliances Total at 75 percent		0	0
Other Loads			
Range (*NEC* 220.55)(Neutral 220.61 at 70%)		0	0
Dryer (*NEC* 220.54)(Neutral 220.61 at 70%)		0	0
25% of Largest Motor (*NEC* 430.24)	1,176	**294**	**294**
Total Volt-Amperes		**8,228**	**8,228**
Volt-Amperes ÷ 240 Volts = Amperes		**34.3**	**34.3**
*Minimum Conductor Size (Copper or Aluminum)		4 AWG (CU)	**8 AWG (CU)
Insulation Type		THW	THW

**Min. ungrounded conductor size based on 100 A min. [230.79(C)] and 83 % rating of 310.12*
***Grounded conductor shall not be smaller than specified in Table 250.102(C)(1) [see 250.24(C)(1)]*

STANDARD LOAD CALCULATION   Example No. 2

Factor	Quantity	VA Ungrnd	VA Neutral
General Lighting (Sq Ft x 3 wpf)	2,250	6,750	
General Lighting Circuit Amps @ 120 V	56.3		
Install _4_ 15 Amp or _3_ 20 Amp Gen. Ltg. Circuits			
Small Appliance Circuits	2	3,000	
Laundry Circuit	1	1,500	
Total General Lighting Load		11,250	
First 3000 VA at 100 percent		3,000	
Total VA - 3000 = _8,250_ VA at 35 percent		2,888	
Net General Lighting Load		**5,888**	**5,888**
Appliances			
Dishwasher	1	1,500	1,500
Disposal (1/2 hp motor)	1	1,176	1,176
Compactor	1	600	600
Fans (120 VA each)	2	240	240
Water Heater	1	4,500	
Appliances rated less than ¼ hp or 500 watts		240	
4 or more Appliances Total at 75 percent		5,832	
Appliance Total Loads		**6,072**	**2,637**
Other Loads			
Range (*NEC* 220.55)(Neutral 220.61 at 70%)	(11,200)	**8,000**	**5,600**
Dryer (*NEC* 220.54)(Neutral 220.61 at 70%)	(4,500)	**5,000**	**3,500**
Electric Furnace	(16,176)	**16,176**	
Air-Conditioner (21.4 A)	(5,136)	0	
(*NEC* 220.60 Noncoincident Loads)			
25% of Largest Motor (*NEC* 430.24)	1,176	**294**	**294**
Total Volt-Amperes		**41,430**	**17,919**
Volt-Amperes ÷ 240 Volts = Amperes		**172.6**	**74.7**
*Minimum Conductor Size (Copper or Aluminum)		1/0 AWG (CU)	4 AWG (CU)
Insulation Type		THWN	THWN

**Min. ungrounded conductor size based on standard ampacity rating of 175 A min. service rating and 83 % rating of 310.12 (145.3 A).   **Grounded conductor shall not be smaller than specified in Table 250.102(C)(1) [see 250.24(C)(1)]*

STANDARD LOAD CALCULATION Example No. 3

Factor	Quantity	VA Ungrnd	VA Neutral
General Lighting (Sq Ft x 3 wpf)	3,650	10,950	
General Lighting Circuit Amps @ 120 V	91.3		
Install _7_ 15 Amp or _5_ 20 Amp Gen. Ltg. Circuits			
Small Appliance Circuits	2	3,000	
Laundry Circuit	1	1,500	
Total General Lighting Load		15,450	
First 3000 VA at 100 percent		3,000	
Total VA - 3000 = 12,450 VA at 35 percent		4,358	
Net General Lighting Load		**7,358**	**7,358**
Appliances			
Dishwasher	1	1,500	1,500
Disposal (1/2 hp motor)	1	1,176	1,176
Compactor	1	600	600
Fans (120 VA each)	4	480	480
Water Heater (4,500 W each)	2	9,000	
Appliances rated less than ¼ hp or 500 watts		480	
4 or more Appliances Total at 75 percent		9,207	
Total Appliance Loads		**9,687**	**2,817**
Other Loads			
Range [NEC 220.55 (8kW x .10 = 8,800)]	(13,600)	**8,800**	**6,160**
[Neutral-NEC 220.61 (8kW x .10 x .70 = 6,160)]			
Dryer (NEC 220.54, 220.61)	(5,500)	**5,500**	**3,850**
[Neutral-NEC 220.61 (5.5 kW x .70 = 3,850)]			
Heat Pump 33 A = 20 kW b/u Heat	(28,616)	**28,616**	
30 kW Pool Heater = 3/4 hp Pump	(31,656)	**31,656**	
25% of Largest Motor (NEC 430.24)	(1,656)	**414**	**414**
Total Volt-Amperes		**92,031**	**20,479**
Volt-Amperes ÷ 240 Volts = Amperes		**383.4**	**85.3**
* **Minimum Conductor Size (Copper or Aluminum)		400 kcmil (CU)	1/0 AWG (CU)
Insulation Type		XHHW	XHHW

**Min. ungrounded conductor size based on standard ampacity rating of 400 A min. service rating and 83 % rating of 310.12 (332 A). **Grounded conductor shall not be smaller than specified in Table 250.102(C)(1) [see 250.24(C)(1)]*

Optional Load Calculation (*NEC* 220.82) Example No. 1

Factor	Quantity	VA Ungrnd	VA Neutral
General Lighting (Sq Ft x 3 wpf)	1650	4,950	4,950
General Lighting Circuit Amps @ 120 V	41.3		
Install 3 15 Amp or 3 20 Amp Gen. Lgt. Circuits			
Small Appliance Circuits	2	3,000	3,000
Laundry Circuit	1	1,500	1,500
Total General Lighting Load		9,450	9,450
First 3000 VA at 100 percent			3,000
Total VA - 3000 = 6,450 VA at 35 percent			2,258
Net General Lighting Load Neutral			5,258
Appliances (nameplate rating)			
Range (Gas)			
Dryer (Gas)			
Dishwasher	1	1,500	1,500
Furnace (1/2 hp motor)	1	1,176	1,176
Total load on ungrounded conductors		12,126	
First 10,000 VA at 100 percent		10,000	
2,126 (balance) at 40 percent		850	
Subtotal		**10,850**	
Other Loads (add largest only)	**Load**		
A/C units or heat pump compressors @ 100%		0	
Thermal storage other continuous heat @ 100%		0	
Central heat and heat strips in heat pumps @ 65%		0	
Heating units (3 or less at 65%)		0	
Heating units (4 or more at 40%)		0	
Total Volt-Amperes		**10,850**	**7,934**
Volt-Amperes ÷ 240 Volts = Amperes		**45.2**	**33.1**
*Minimum Conductor Size (Copper or Aluminum)		4 AWG CU	**8 AWG CU
Insulation Type		THW	THW

**Min. ungrounded conductor size based on standard ampacity rating of 100 A min. [230.79(C)] and 83 % rating of 310.12.* ***Grounded conductor shall not be smaller than specified in Table 250.102(C)(1) [see 250.24(C)(1)]*

Optional Load Calculation *(NEC* 220.82) Example No. 2

Factor	**Quantity**	**VA Ungrnd**	**VA Neutral**
General Lighting (Sq Ft x 3 wpf)	2,250	6,750	6,750
General Lighting Circuit Amps @ 120 V	56.3		
Install _4_ 15 Amp or _3_ 20 Amp Gen. Ltg. Circuits			
Small Appliance Circuits	2	3,000	3,000
Laundry Circuit	1	1,500	1,500
Total General Lighting Load		11,250	11,250
First 3000 VA at 100 percent			3,000
Total VA - 3000 = _8,250_ VA at 35 percent			2,888
Net General Lighting Load Neutral			5,888
Appliances (nameplate rating)			
Range (Neutral-8 kW at 70% *NEC* 220.61)	1	11,200	5,600
Dryer (Neutral-5 kW at 70% *NEC* 220.61)	1	4,500	3,500
Dishwasher	1	1,500	1,500
Waste disposal (1/2 hp)	1	1,176	1,176
Trash compactor	1	600	600
Exhaust fans (120 VA each)	2	240	240
Water heater	1	4,500	
Total load on ungrounded conductors		34,966	
First 10,000 VA at 100 percent		10,000	
_24,966_ (balance) at 40 percent		9,986	
Subtotal		**19,986**	
Other Loads (add largest only)	**Load**		
A/C units or heat pump compressors @ 100%	5,136	~~5,136~~	
Thermal storage, other continuous heat @ 100%		0	
Central heat and heat strips in heat pumps @ 65%	16,176	10,514	
Heating units (3 or less at 65%)		0	
Heating units (4 or more at 40%)		0	
Total Volt-Amperes		**30,501**	**18,504**
Volt-Amperes ÷ 240 Volts = Amperes		**127.1**	**77.1**
*Minimum Conductor Size (Copper or Aluminum)		1 AWG CU	4 AWG CU
Insulation Type		THW	THW

**Min. ungrounded conductor size based on standard ampacity rating of 150 A min. service rating and 83 % rating of 310.12 (124.5 A). **Grounded conductor shall not be smaller than specified in Table 250.102(C)(1) [see 250.24(C)(1)]*

Optional Load Calculation (*NEC* 220.82) Example No. 3

Factor	Quantity	VA Ungrnd	VA Neutral
General Lighting (Sq Ft x 3 wpf)	3,650	10,950	10,950
General Lighting Circuit Amps @ 120 V	91.3		
Install _7_ 15 Amp or _5_ 20 Amp Gen. Ltg. Circuits			
Small Appliance Circuits	2	3,000	3,000
Laundry Circuit	1	1,500	1,500
Total General Lighting Load		15,450	15,450
First 3000 VA at 100 percent			3,000
Total VA - 3000 = _12,450_ VA at 35 percent			4,358
Net General Lighting Load Neutral			**7,358**
Appliances (nameplate rating)			
Range (Neutral-*NEC* 220.61 at 70%)	1	13,600	6,160
[*NEC* 220.55 (8kW x .10 x .70 = 6,160)]			
Dryer (Neutral-*NEC* 220.61 at 70%)	1	5,500	3,850
[*NEC* 220.54 (5.5 kW x .70 = 3,850)]			
Dishwasher	1	1,500	1,500
Waste disposal (1/2 hp)	1	1,176	1,176
Trash compactor	1	600	600
Exhaust fans (120 VA each)	4	480	480
Water heater (4500 W each)	2	9,000	
30 kW pool heater + 3/4 hp pump		31,656	
Total load on ungrounded conductors		78,962	
First 10,000 VA at 100 percent		10,000	
_68,962_ (balance) at 40 percent		27,585	
Subtotal		**37,585**	
Other Loads (add largest only)	**Load**		
A/C units or heat pump compressors @ 100%		0	
Thermal storage or other continuous loads @ 100%		0	
Central heat and heat strips in heat pumps @ 65%	28,616	18,600	
Heating units (3 or less at 65%)		0	
Heating units (4 or more at 40%)		0	
		0	
Total Volt-Amperes		**56,185**	**21,124**
Volt-Amperes ÷ 240 Volts = Amperes		**234.1**	**88.0**
*Minimum Conductor Size (Copper or Aluminum)		4/0 AWG CU	**2 AWG CU
Insulation Type		XHHW	XHHW

**Min. ungrounded conductor size based on standard ampacity rating of 250 A min. service rating and 83 % rating of 310.12 (207.5 A).* ***Grounded conductor shall not be smaller than specified in Table 250.102(C)(1) [see 250.24(C)(1)]*

STANDARD LOAD CALCULATION

Factor	Quantity	VA Ungrnd	VA Neutral
General Lighting (Sq Ft x 3 wpf)			
General Lighting Circuit Amps @ 120 V			
Install _7_ 15 Amp or _5_ 20 Amp Gen. Ltg. Circuits			
Small Appliance Circuits	2	3,000	
Laundry Circuit	1	1,500	
Total General Lighting Load			
First 3000 VA at 100 percent		3,000	
Total VA - 3000 = ______ VA at 35 percent			
Net General Lighting Load			
Appliances			
Dishwasher			
Disposal			
Compactor			
Exhaust Fans			
Water Heater			
Appliances rated less than ¼ hp or 500 watts			
4 or more Appliances Total at 75 percent			
Total Appliance Loads			
Other Loads			
Range [(*NEC* 220.55)(Neutral-*NEC* 220.61)]			
Dryer (*NEC* 220.54, 220.61)			
A/C Unit			
Heat			
25% of Largest Motor (*NEC* 430.24)			
Total Volt-Amperes			
Volt-Amperes ÷ 240 Volts = Amperes			
Minimum Conductor Size (Copper or Aluminum)			
Insulation Type			

Optional Load Calculation (*NEC* 220.82)

Factor	Quantity	VA Ungrnd	VA Neutral
General Lighting (Sq Ft x 3 wpf)			
General Lighting Circuit Amps @ 120 V			
Install ___ 15 Amp or ___ 20 Amp Gen. Ltg. Circuits			
Small Appliance Circuits	2	3,000	3,000
Laundry Circuit	1	1,500	1,500
Total General Lighting Load			
First 3000 VA at 100 percent			3,000
Total VA - 3000 = _________ VA at 35 percent			
Net General Lighting Load Neutral			
Appliances (nameplate rating)			
Range *(NEC* 220.55)(Neutral 220.61 at 70%)			
Dryer *(NEC* 220.54)(Neutral 220.61 at 70%)			
Dishwasher			
Trash Compactor			
Exhaust Fans			
Water Heater			
Nameplate Amperes or kVA of Motors			
Total load on ungrounded conductors			
First 10,000 VA at 100 percent		10,000	
__________ (balance) at 40 percent			
Subtotal			
Other Loads (add largest only)	**Load**		
A/C units or heat pump compressors @ 100%			
Thermal storage or other continuous loads @ 100%			
Central heat and heat strips in heat pumps @ 65%			
Heating units (3 or less at 65%)			
Heating units (4 or more at 40%)			
Total Volt-Amperes			
Volt-Amperes ÷ 240 Volts = Amperes			
Minimum Conductor Size (Copper or Aluminum)			
Insulation Type			

REVIEW QUESTIONS

1. A room in a building for living, sleeping, eating, or cooking, but excluding bathrooms, toilet rooms, closets, hallways, storage or utility spaces, and similar areas, is defined as which of the following:

A. Living Room
B. Game Room
C. Habitable Room
D. Den

2. Computation for all required branch circuits for the general lighting load in one- and two-family dwelling units is determined by using a unit load of_____ volt-amperes per square foot.

A. 2
B. 3
C. 4
D. 5

3. General-use receptacle outlets, rated at ____ amperes or less, are considered outlets for general illumination and are permitted to be included in the general lighting load.

A. 15
B. 20
C. 25
D. 30

4. At least ____ 20-ampere small appliance branch circuits are required to be installed for the supply to receptacle outlets only in a kitchen, pantry, breakfast room, dining room, or similar area.

A. two
B. three
C. one
D. six

5. At a single-family dwelling unit, at least one ___ ampere laundry branch circuit is required to be installed for the supply of the laundry outlet(s).

A. 15
B. 20
C. 25
D. 30

6. Appliance and laundry branch circuits are required to be calculated at _____ volt-amperes each.

A. 1200
B. 1300
C. 1400
D. 1500

7. "The circuit conductors between the final overcurrent device protecting the circuit and the outlet(s)" defines a branch circuit.

True _____ False _____

8. Small-appliance branch circuits are not permitted to have any permanently connected luminaire lighting outlets that are not a part of an appliance.

True _____ False _____

9. An individual branch circuit supplies only one piece of utilization equipment.

True _____ False _____

10. "All circuit conductors between the service equipment or the source of a separately derived system and the final branch-circuit overcurrent device" defines a feeder.

True _____ False _____

11. For uniform application of *NEC* Articles 210, 215, and 220, a nominal voltage of 120, 120/240, 240, and 208Y/120 volts is required to be used for computing the ampere load on conductors at dwelling units.

True _____ False _____

12. The load for a household electric clothes dryer in a dwelling unit must be _____ watts (volt-amperes) or the nameplate rating, whichever is larger.

A. 3000
B. 4500
C. 5000
D. 8000

13. The minimum allowable number of 120-volt, 15-ampere, 2-wire general lighting branch circuits required for a residence with a calculated floor area of 21 m by 9 m (70 ft by 30 ft) is:

A. 2
B. 3
C. 4
D. 5

14. It is required that 1500 VA be added to the calculated load for a dwelling for the 20-ampere branch circuit required to serve bathroom receptacles.

True ______ False _____

15. For dwelling units, the calculated floor area shall not include any of the following EXCEPT:

A. open porches
B. garages
C. unfinished areas not adaptable for future use as a habitable room
D. unfinished areas not adaptable for future use as an occupiable space

16. Where two dwelling units are supplied by a single feeder, and the calculated load under Part III of NEC Article 220 (standard calculation) exceeds that for three identical units calculated under *NEC* 220.84, the lesser of the two loads is permitted to be used.

True ______ False _____

17. For a service or feeder rated 100 through 400 A, the service or feeder conductors supplying the entire load associated with a one-family dwelling or the service or feeder conductors supplying the entire load associated with an individual dwelling unit in a two-family or multifamily dwelling, are permitted to have an ampacity not less than __________ percent of the service or feeder rating.

A. 25
B. 43
C. 72
D. 83

18. For a service or feeder rated 100 through 400 A, the service or feeder conductors supplying the entire load associated with a one-family dwelling or the service or feeder conductors supplying the entire load associated with an individual dwelling unit in a two-family or multifamily dwelling, are permitted to have their ampacity determined by *NEC* Table 310.12(A) or IRC Table E3603.1.1 if ______________________.

A. the authority having jurisdiction (AHJ) approved the use
B. there are no objections from the authority having jurisdiction (AHJ)
C. no adjustment or correction factors are required
D. the service or feeder is rated at least 150 amperes

3 Cabinets, Panelboards, and Meter Socket Enclosures

This chapter covers installation and construction requirements for cabinets, meter socket enclosures, and panelboards as specified by the Code. It includes general guidelines for panelboards, meter sockets, and combination equipment.

A "cabinet" is an enclosure designed for either surface mounting or flush mounting. It features a frame, mat, or trim to which a swinging door or doors are attached or can be attached.

A "panelboard" refers to a single panel or a collection of panel units assembled together. It contains supply buses and automatic overcurrent protection devices. Some panelboards include switches for controlling lighting, heating, or power circuits, while others do not. They are intended to be installed in cabinets, enclosures, or cutout boxes mounted on or against walls, partitions, or other supports. Access to a panelboard is limited to the front side, as specified in NEC Article 100 or IRC Chapter 35.

The terms "meter socket" or "meter pedestal" are not explicitly defined by the Code. A meter socket or pedestal is essentially a cabinet that houses the equipment needed to mount the electric meter and connect the service-lateral and service-entrance conductors. Typically used with overhead services, a meter socket can also be suitable for underground installations. For underground service, a meter socket is designed with a wider cabinet to accommodate two conduit entries at the bottom. Meter pedestals are frequently installed for underground services as well.

A meter pedestal is constructed with an auxiliary gutter on its bottom side. The bottom of the pedestal remains open. The gutter space serves to both route and protect the service-lateral conductors as they exit the ground for connection in the pedestal. Meter sockets are typically evaluated and approved by the servicing utility. The utility may list approved metering equipment by manufacturer and model number to ensure

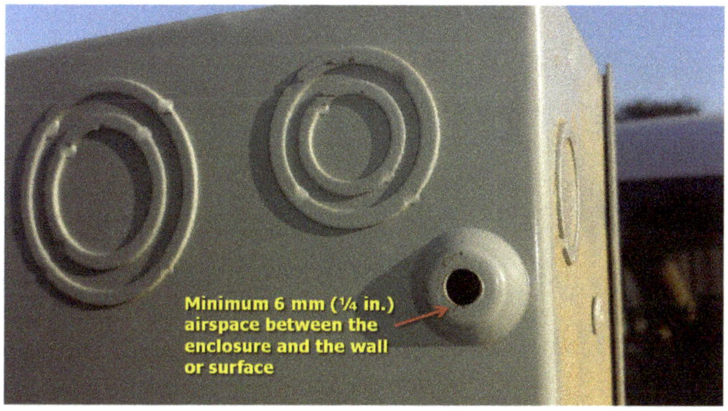

Photo 3.1. Cabinets and enclosures mounted in wet locations must be mounted with a clearance of 6 mm (¼ in.). Air space is required between the enclosure and the wall or other supporting surface.

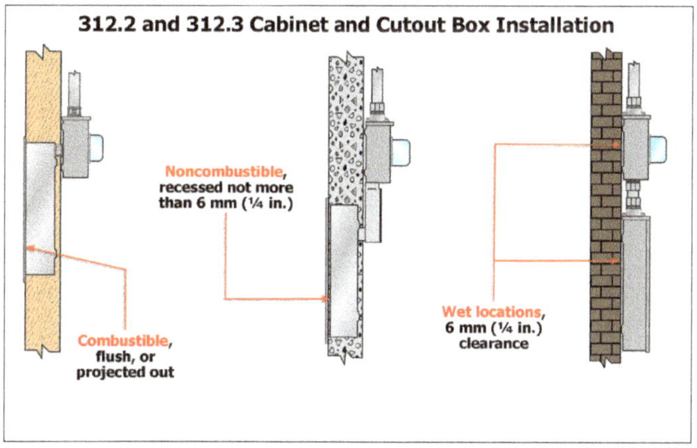

Figure 3.2. Locations of electrical enclosures relative to wet locations in relation to combustible and noncombustible surfaces

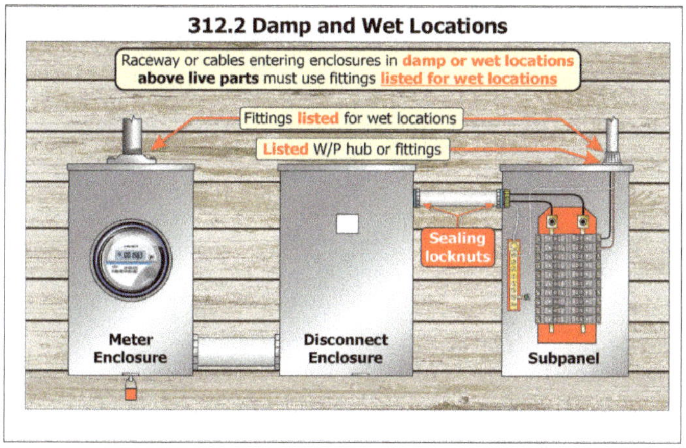

Figure 3.1. In damp or wet locations, raceways or cable enclosures entering the cabinet above live parts must use fittings listed for wet locations

Photo 3.2. Split photo of meter enclosure knockout sealing device *[outside enclosure (top) inside enclosure (bottom)]*

compatibility with the socket. Check your utility's rule book for more information on approved types of metering equipment in your area.

CABINETS INSTALLED IN DAMP AND WET LOCATIONS

In damp and wet locations, surface-mounted cabinets and meter socket enclosures must be installed to prevent moisture or water from entering and accumulating inside or behind the enclosure. A 6 mm (¼ in.) air gap is required between any metal enclosure and the wall or other support surface [NEC 312.2 or IRC E3907.2]. Enclosures designed for these areas usually have this offset feature built-in (see photo 3.1). Cabinets or meter socket enclosures installed in wet locations must be designed or protected to ensure that exposure to the weather does not affect their proper operation. Restricting water entry helps protect the internal components. Raceways and cables entering the cabinet above the level of uninsulated live parts must use special termination fittings. These fittings must be listed for wet locations and include sealing washers and bushings, hubs, and terminal adapters suitable for wet conditions [NEC 312.2 or IRC E3907.2].

Cabinets installed in concrete, tile, or other non-combustible material walls must have their front edge recessed no more than 6 mm (¼ in.) from the finished surface. When installed in wooden walls or other combustible materials, cabinets must be flush with the finished surface or extend outward from it. Surfaces like plaster, drywall, or plasterboard that are broken or incomplete need to be repaired. The maximum allowable gap around the cabinet and the finished surface is 3 mm (⅛ in.) for cabinets with a flush-type cover (NEC 312.3 and 312.4 or IRC E3907.2, E3907.3, and E3907.4).

According to NEC 110.12(A) or IRC E3404.6, unused knockout openings in cabinets or meter socket enclosures must be effectively sealed to provide protection that is roughly equal to that of the cabinet or meter socket enclosure wall. This does not include unused openings meant for equipment operation, mounting, or those allowed as part of the design for listed equipment. An unused opening above the level of live parts must be sealed with a device suitable for wet locations (see photo 3.2).

Where a metal plug or plate is used with nonmetallic cabi-

Table 3.1 Square Inches for Common Type NM Cables					
Cable (Diameter)	Area	Number of Cables	Total Area of Cables	Min. Trade Size (EMT)	Metric Designator Size (EMT)
14-2 WG (0.375)	0.110	1	0.110	½	16
		2	0.221	1	27
		3	0.331	1	27
		4	0.442	1¼	35
		5	0.552	1¼	35
		6	0.663	1½	41
14-3 WG (0.281)	0.062	1	0.062	½	16
		2	0.124	¾	21
		3	0.186	¾	21
		4	0.248	1	27
		5	0.310	1	27
		6	0.372	1¼	35
12-2 WG (0.438)	0.151	1	0.151	½	16
		2	0.301	1¼	35
		3	0.452	1¼	35
		4	0.603	1½	41
		5	0.753	1½	41
		6	0.904	2	53
12-3 WG (0.313)	0.077	1	0.077	½	16
		2	0.154	¾	21
		3	0.231	1	27
		4	0.308	1	27
		5	0.385	1¼	35
		6	0.462	1¼	35
10-2 WG (0.500)	0.196	1	0.196	¾	21
		2	0.393	1¼	35
		3	0.589	1¼	35
		4	0.785	1½	41
10-3 WG (0.375)	0.110	1	0.110	½	16
		2	0.221	1	27

Table 3.1 Square inch areas for common Type NM cables

nets or meter socket enclosures, it must be recessed at least 6 mm (¼ in.) from the outer surface.

Conductors entering cabinets or meter socket enclosures must be protected from abrasion. All openings through which conductors enter should be adequately closed. When cable is used, each cable assembly generally needs to be secured to the cabinet with fittings and connectors specifically listed for the wiring method used [NEC 312.5(C) or IRC E3907.8].

The exception to this rule applies only to cables with fully nonmetallic sheaths. Such cables may enter the top of a surface-mounted enclosure through one or more nonflexible raceways. The raceway must be at least 450 mm (18 in.) long and no longer than 3.0 m (10 ft). Additional conditions for this alternative cable installation method are as follows:

1. Each cable must be fastened within 300 mm (12 in.) of the outer end OF the raceway. Measurement is made along the cable sheath.
2. Raceway extends directly above the enclosure and does not penetrate a structural ceiling.
3. A fitting is provided on each end of the raceway to protect the cable(s) from abrasion. Listed nonmetallic fittings may be used without a bushing to meet this requirement. The fitting must remain accessible after the installation is complete.
4. Raceway is sealed or plugged at the outer end. The sealing material must be approved (acceptable to the electrical inspector). Electrical sealing putty (duct seal) or similar material can be used, provided it is installed so as to prevent access to the enclosure through the raceway. A proper seal will prevent foreign material from contaminating the live parts contained within the enclosure.
5. Cable sheath is continuous through the raceway and extends at least 6 mm (¼ in.) beyond the end of the fitting connected to the enclosure.
6. Raceway is fastened at its outer end and other points in accordance with the requirements for that type of raceway.

Photo 3.3. Type NM cable is generally required to be secured to the cabinet with fittings and connectors specifically listed for the wiring method used (as shown above)

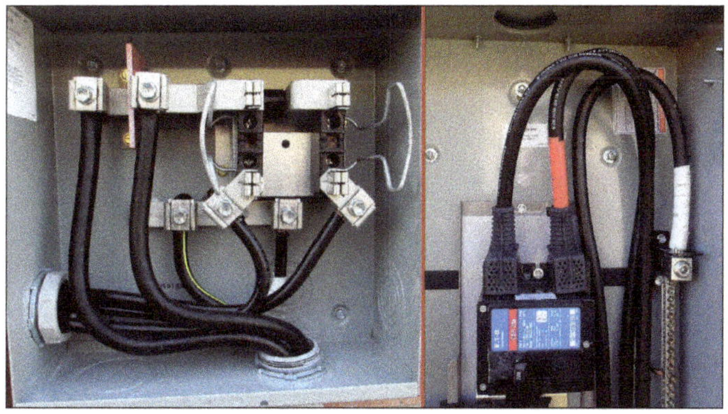

Photo 3.4. Adequate wire bending space is required in cabinets, cutout boxes, and meter socket enclosures

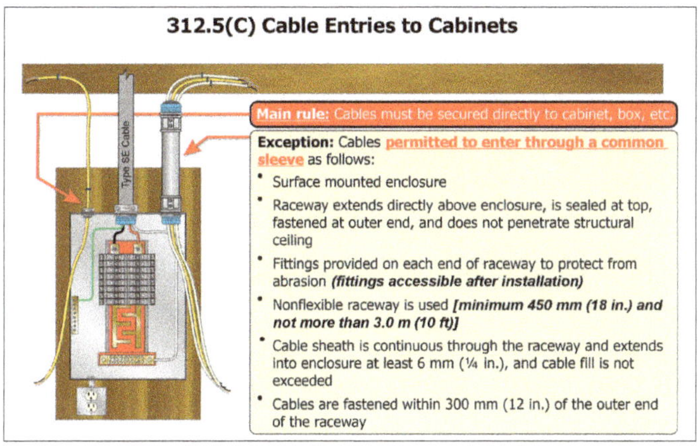

Figure 3.3 Bundled cables are permitted to enter enclosures without being individually secured to the cabinet under certain specific conditions

7. Where installed as conduit or tubing, allowable cable fill does not exceed that permitted by the *Code* for complete conduit or tubing systems.

This requirement is illustrated as follows using *NEC* Chapter 9, Tables. Three (3) 12-2 AWG and two (2) 10-2 AWG Type NM cables are to be routed through a raceway that is longer than 600 mm (24 in.). The required minimum size conduit is determined as follows:

Assume the 12-2 AWG cable has a dimension of 6 mm (¼ in.) x 13 mm (½ in.) and the 10-2 AWG cable has a dimension of 8 mm (5/16 in.) x 16 mm (5/8 in.) Note 9 to Table 1 of NEC Chapter 9 would require "For cables that have elliptical cross sections, the cross-sectional area calculation shall be based on using the major diameter of the ellipse as a circle diameter." To determine the minimum size conduit or tubing required for the number of cables to be installed, the area of the cables must be calculated and then compared to the maximum allowable fill for the conduit or tubing as specified in NEC Chapter 9, Table 4. Table 1 of NEC Chapter 9 permits one wire (cable) to fill a raceway to not more than 53% of its area, up to 31% for two wires (cables), and up to 40% for more than two wires (cables).

A metric designator 103 (4 trade size) EMT is required at minimum for this group of cables. Refer to Table 3.1 in this publication for the area in square inches for common Type NM cables. (Dimensions are based on actual measurements of cable samples. Verify the actual dimensions of your cable.)

The ampacities of multiple cable assemblies installed in a common raceway must be adjusted when there are more than three current-carrying conductors. NEC Table 310.15(C)(1) and IRC Table E3705.3 provide the adjustment factors for more than three current-carrying cables in a raceway or cable. Correction factors do not apply to conductors in short chase nipples if the length of the chase nipple does not exceed 600 mm (24 in.). The requirements for correction factors are explained in more detail in Chapter 1 of this text [NEC 310.15(C)(1) or IRC E3705.3].

WIRE BENDING SPACE AT TERMINALS

Conductors at terminals or conductors entering or leaving enclosures must be installed with sufficient wire bending space at each terminal [NEC 312.6(B) or IRC E3907.9]. This ensures connections can be made without damaging the terminal or conductor insulation. Since larger conductors are stiffer and harder to bend, more wire-bending space is needed for them.

The Code also requires that equipment and materials be installed professionally and skillfully (NEC 110.12). This includes ensuring that no damage occurs to conductors when terminating inside cabinets or other enclosures. Therefore, the Code mandates that cabinets be equipped with side and back wiring spaces, gutters, or wiring compartments [NEC 312.9, 312.101(C) and (D) or IRC E3907.9]. This requirement applies to cabinets containing devices connected to more than eight conductors within the cabinet. The conductors include those associated with branch and feeder circuits and meter loops but exclude the supply circuit conductors [NEC 312.101(C) or IRC E3907.9].

The Code requires that such cabinets include side or gutter spaces sized according to NEC 312.6. The width of the space depends on the expected routing method: a conductor will either enter the cabinet and be routed directly to the terminal or pass through the wall and then be deflected before reaching the terminal.

When the conductor does not enter or leave the enclosure through the wall opposite its terminal, the Code mandates a minimum wire deflection space, which is provided following NEC Table 312.6(A) or IRC Table E3907.9.1(2). Usually, the conductor is deflected at least 90 degrees before connecting to its terminal, requiring less wire bending space than if it entered directly across from the terminal.

A conductor that enters or leaves the enclosure through the wall opposite its terminal must include a minimum deflection space that complies with the values in *NEC* Table 312.6(B)(2) or IRC Table E3907.9.1(1). This table requires greater wire bending space than *NEC* Table 312.6(A) or IRC Table E3907.9.1(2) since it is assumed there is not enough space in the enclosure to bend the conductor before connecting it to its terminal.

A conductor is also permitted to enter or leave an enclosure through the wall opposite its terminal as long as it enters or leaves the enclosure where the gutter of the panelboard joins an adjacent auxiliary gutter that has a width that conforms to *NEC* Table 312.6(B)(2) or IRC Table E3907.9.1(1) for that conductor. This practice is typically not used in dwellings but is commonly employed in commercial construction. The IRC does not include these specific provisions for deflection space for conductors entering or leaving anything other than enclosures containing panelboards. This minimum requirement must be maintained as it can have a serious impact and result in damage to conductor insulation. Keep in mind that the IRC in Section E3401.2 indicates that the IRC does not intend to exclude any requirement for one- and two-family dwellings that are not contained within the IRC. In such cases, one must refer to the applicable provisions of the NEC, which include minimum requirements for those situations.

The enclosures for a panelboard are required to have top and bottom wire-bending space sized per Table 312.6(B)(2) *(conductors entering or leaving opposite walls)* for the largest conductor entering or leaving the enclosure. Side wire-bending space is to be in accordance with Table 312.6(A) *(conductors not entering or leaving opposite walls)* for the largest conductor to be terminated in that space. Four exceptions exist for top and bottom wire-bending spaces.

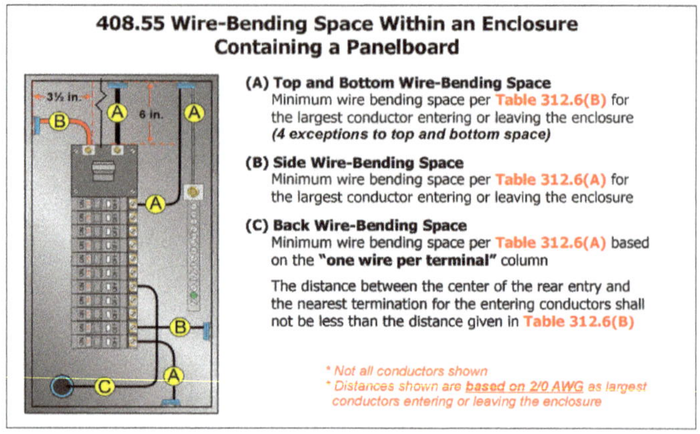

Figure 3.4. Wire bending space within cabinets containing panelboards

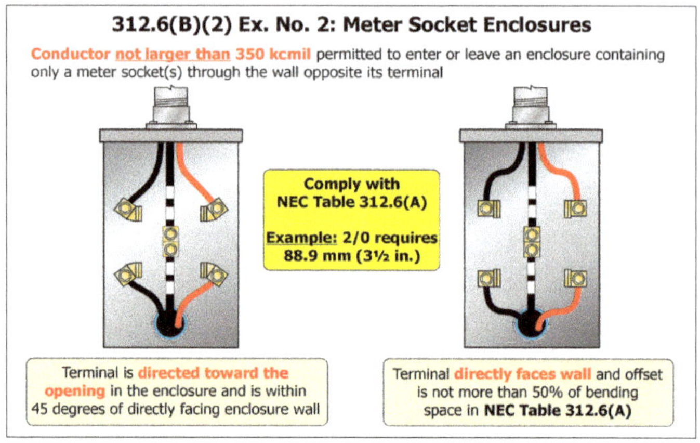

Figure 3.5. Deflection of conductors in meter socket enclosures applying *NEC* 312.6(B)(2), Ex. 2

The first two exceptions permit either the top or bottom wire-bending space to be sized in accordance with Table E3907.9.1 for a panelboard rated 225 amperes or less and designed to contain not over 42 overcurrent devices or where at least one side wire-bending space is sized in accordance with Table 312.6(B)(2) for the largest conductor to be terminated in any side wire-bending space.

The third exception permits both the top and bottom wire-bending spaces to be sized in accordance with Table 312.6(A) if the panelboard is designed and constructed for wiring using only a single 90-degree bend for each conductor and the wiring diagram shows and specifies the method of wiring that is to be used.

The fourth exception permits either the top or the bottom wire-bending space, but not both, to be sized per Table 312.6(A), where no conductors are terminated in that space. Back wire-bending space is addressed for rear entry with a removable cover on the opposite wall of the enclosure, with the distance from that wall to the cover complying with the distance required for one wire per terminal in Table 312.6(A). The distance between the center of the rear entry and the nearest

Photo 3.5. Deflection of conductors in meter socket enclosures

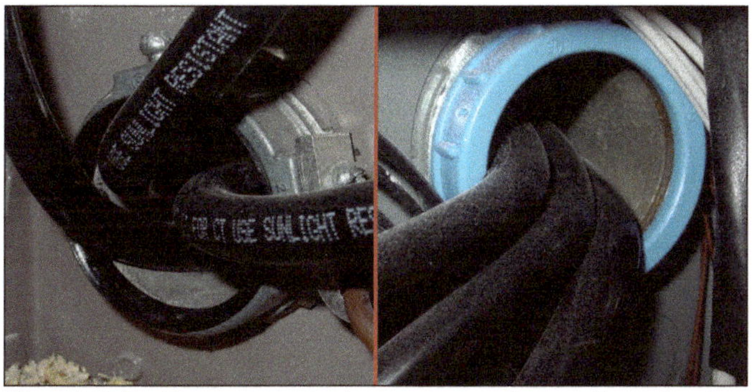

Photo 3.6. Bushings are required to protect conductors from abrasion where they enter cabinets, cutout boxes, and meter socket enclosures

termination for the entering conductors cannot be less than the distance given in Table 312.6(B)(2) [see NEC 408.55 or IRC E3907.9].

CONDUCTORS IN METER ENCLOSURES

NEC 312.6(B)(2), Exception No. 2 permits a conductor not larger than 350 kcmil to enter or leave an enclosure containing only a meter socket(s) through the wall opposite its terminal. The distance between the terminal and the wall opposite cannot be less than the distance specified in *NEC* Table 312.6(A) or IRC Table E3907.9.1(2). The terminal must be a lay-in type lug where the terminal is directly facing the enclosure wall, and the offset is not greater than 50% of the bending space specified in *NEC* Table 312.6(A) or IRC Table E3907.9.1(2). An alternative is where the terminal is directed toward the opening in the enclosure and is within a 45-degree angle or directly facing the enclosure wall (see Figure 3.5).

Per an Informational Note following *NEC* 312.6(B)(2), Exception No. 2, "Offset" is defined as the distance measured along the enclosure wall from the axis of the center line of the terminal to a line passing through the center of the opening in the enclosure.

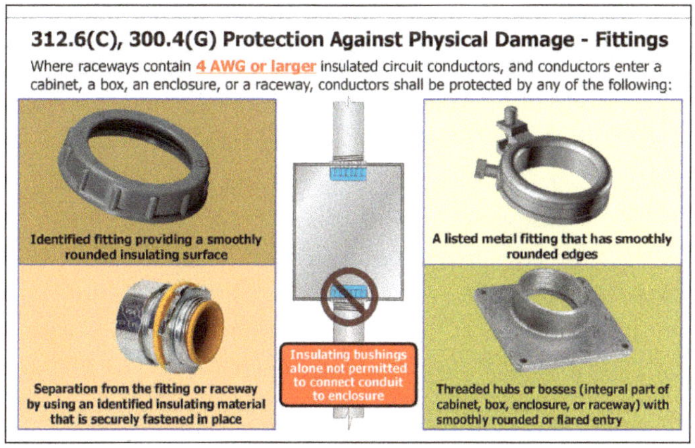

Figure 3.6. Protection of conductors from abrasion and connecting conduit to enclosures

CONDUCTOR PROTECTION FROM ABRASION

Where insulated circuit conductors of 4 AWG or larger enter a raceway in a cabinet, pull box, junction box, or auxiliary gutter, before the installation of conductors, they must be protected from abrasion in accordance with NEC 312, 6(C); 300.4(G), or IRC E3907.6 and E3906.1.1, by one of the following methods.

(1) An identified fitting providing a smoothly rounded insulating surface
(2) A listed metal fitting that has smoothly rounded edges
(3) Separation from the fitting or raceway using an identified insulating material that is securely fastened in place
(4) Threaded hubs or bosses that are an integral part of a cabinet, box, enclosure, or raceway, providing a smoothly rounded or flared entry for conductors

Insulated bushings are not required where threaded hubs or bosses that are integral to an enclosure provide a smoothly rounded or flared entry for conductors. Listed nonmetallic fittings are investigated by qualified electrical product testing laboratories to meet this requirement without the use of additional insulating bushings. Fittings are tested to ensure the edge of the fitting will not damage conductors as the conductor is pulled through the fitting.

Conduit bushings made wholly of insulating material are not permitted to be used to secure a raceway to an enclosure. A listed metal locknut would need to be used in conjunction with an insulating bushing. A metal conduit bushing with an insulated throat can be used for this purpose. An insulating fitting or insulating material is required to have a temperature rating not less than the insulation temperature rating of the installed conductors [*NEC* 300.4(G) or IRC E3906.1.1].

WIRE SPACE IN ENCLOSURES

Cabinets and meter socket enclosures must have sufficient space to accommodate all conductors installed without crowding (*NEC* 312.7).

Cabinet enclosures for switches or overcurrent devices are generally not permitted to be used as junction boxes, auxiliary gutters, or raceways for conductors that feed through or tap off to other switches or overcurrent devices. The Code permits this use where adequate space is provided within these enclosures. Conductors are not permitted to occupy more than 40% of the cross-sectional area of the wiring space at any cross-section within the enclosure.

Where conductors are spliced in these switch or overcurrent device enclosures, the conductors, splices, and taps cannot fill the wiring space at any cross-section to more than 75% of the cross-sectional area of the space. A warning label complying with *NEC* 110.21(B) must be applied to the enclosure that identifies the closest disconnecting means for any feed-through conductors (*NEC* 312.8(A) or IRC E3907.1.1).

There was a code change within *NEC* 312.8 (A) (3) to reference the bending space of conductor's size #4 AWG and larger to comply with *NEC* 314.28(A)(2) when sizing pull boxes with an angle and or U pulls, or splices.

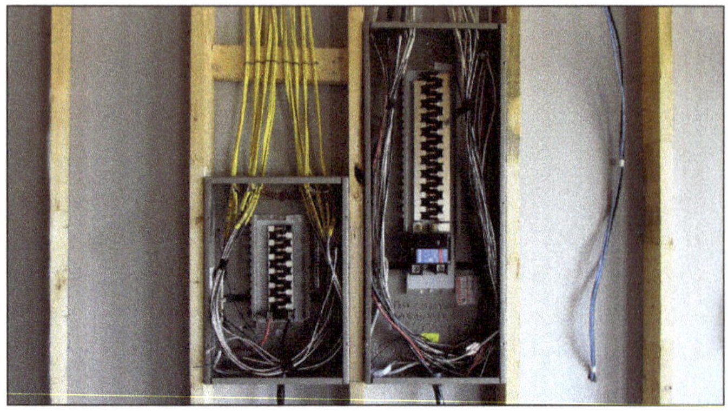

Photo 3.7. Adequate conductor (wire) space is required in enclosures

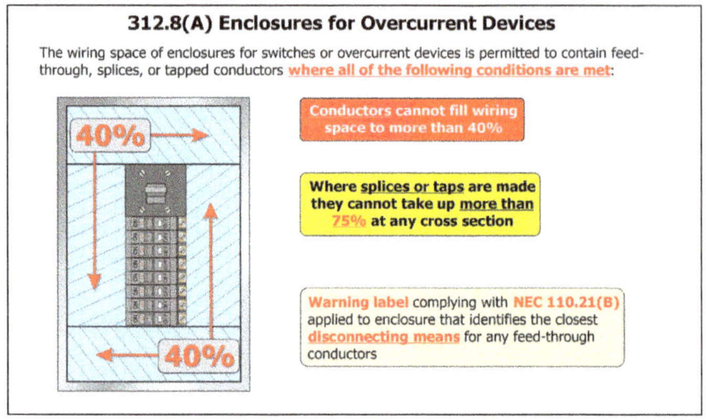

Figure 3.7. Splices and taps are generally not permitted in enclosures except where adequate space is provided

CONDUCTOR ROUTING AND INDUCTION

Where conductors carrying alternating current are installed in metal enclosures or raceways, they must be arranged to prevent heating the surrounding metal by induction. Induction is defined as "the production of an electric or magnetic state by the proximity (without contact) of an electrified or magnetized body" or "the production of an electric current in a conductor by varying the magnetic field applied to the conductor."

To accomplish this and avoid induction, all ungrounded phase conductors, along with the neutral conductor and all equipment grounding conductors when used, are typically required to be grouped by installing them in the same raceway, cable, or trench. Switch loops do not require a grounded (neutral) conductor because of the canceling effect caused by counter-electromotive force (EMF). In most cases, switches or circuit breakers cannot be used to disconnect the neutral conductor of a circuit.

MOUNTING AND LOCATION OF SWITCHES AND OVERCURRENT DEVICES

Switches and circuit breakers of the externally operable type must be installed in an enclosure approved for the intended use. Adequate wire bending space at the terminals and sufficient gutter space must be provided in switch enclosures. The minimum wire bending and gutter space requirements are listed in NEC 312.6. Refer to the earlier discussion in this chapter for additional information.

When installed in damp, wet, or outdoor locations, a switch or circuit breaker must be enclosed in a weatherproof enclosure or cabinet. Time switches, flashers, and similar devices must be either of the enclosed type or mounted in cabinet boxes or equipment enclosures. Energized parts must be protected by a barrier to prevent operator exposure when making manual adjustments or switching.

A single-throw knife switch must be wired so that the switch blades are not energized when the switch is in the open "off" position. It is standard practice to locate the line-side terminals on the top and the load side on the bottom of the switch to ensure this requirement is met. The switch must be installed in a way that gravity does not tend to close it. If a single-throw knife switch is approved for use in the inverted position, it will be equipped with a locking device to keep the switch blades in the open "off" position when set.

When installed in an enclosure, general-use switches and motor-circuit breakers must clearly show whether they are in the open "off" or closed "on" position. This position indicator must

Photo 3.8. Where operated in the vertical position, the "up" position of the handle of a circuit breaker must be in the "on" position

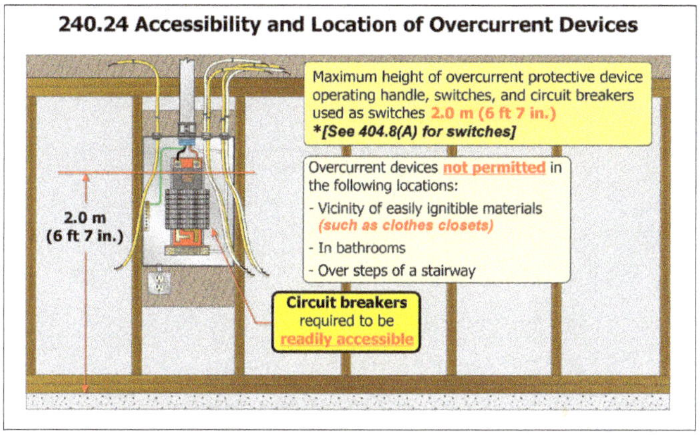

Figure 3.8. Switches and circuit breakers are not permitted to be located more than 2.0 m (6 ft 7 in.) above the floor

be placed where it is visible when accessing the external operating means. For switches or circuit breaker handles operated vertically rather than rotationally or horizontally, the "up" position of the handle must indicate "on" (NEC 404.7, 240.81, or IRC E4001.3).

All switches and circuit breakers that are used as switches must be located so they can be operated from a readily accessible location.

All switches and circuit breakers must be installed so the center of the grip of their operating handles when in their highest position, will not be more than 2.0 m (6 ft 7 in.) above the surface of the floor or other working platform [*NEC* 404.8(A), 240.24(A) or IRC E4001.6].

The Code requires overcurrent devices to be easily accessible to the occupant of a one- or two-family home. For clear reasons, overcurrent devices are prohibited from being installed near easily ignitable materials, such as in clothes closets. They are also not allowed to be located above stairway steps. The Code does not permit overcurrent devices to be installed in the bathrooms of dwelling units. There is an exception for supplementary overcurrent protective devices. These additional devices may be installed in bathrooms. Such protection is sometimes used for motors, luminaires, and water heaters (see NEC 240.24 or IRC E3705.7).

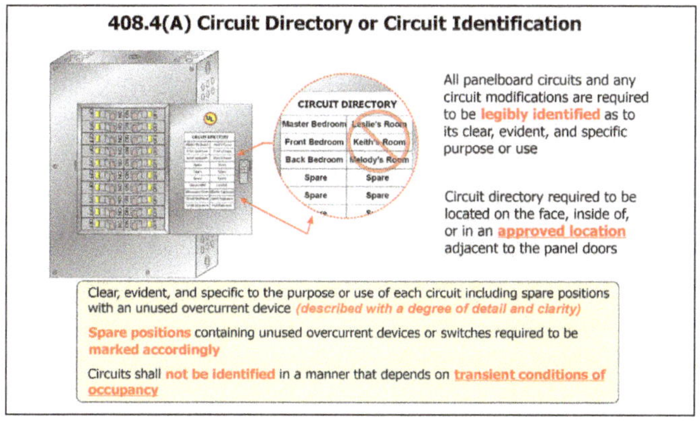

Figure 3.9. A circuit directory is required for the panelboard

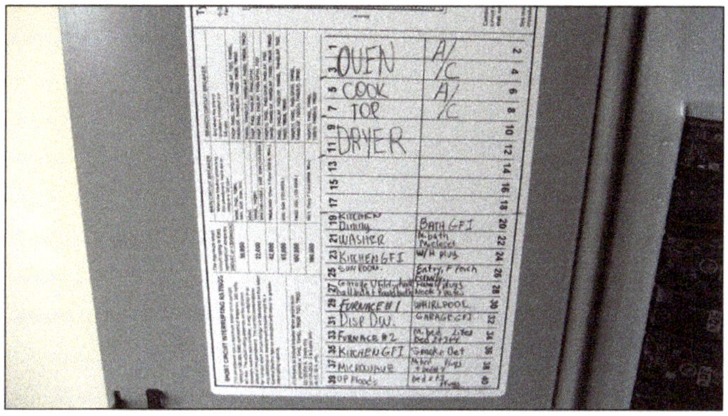

Photo 3.9. Typical circuit directory for a dwelling unit panelboard

PANELBOARDS

Requirements for panelboards and distribution boards used to control lighting and power circuits are outlined in NEC Article 408 or IRC Sections E3907 and E3706.

The rating of a panelboard cannot be less than the minimum feeder capacity required for the calculated load (*NEC* 408.30 or IRC E3706.1). A manufacturer is required to provide durable marking showing the voltage, current rating, and number of phases for which the panelboard is designed. Also required is the manufacturer's name or trademark, which must be visible after installation without disturbing the panelboard interior parts or wiring.

All panelboard circuits and any circuit modifications are required to be legibly identified as to its clear, evident, and specific purpose or use. Circuit identification requires a degree of detail, such as distinguishing each circuit from all others, resulting in clarity that is unlikely to result in confusion between circuits. Spare positions that contain unused overcurrent devices need to be indicated as well. Circuits should not be described with descriptions of transient conditions of occupancy. In other words, a circuit can't be identified as "Bill's room" or "Mary's office." This identification is required to be located on a circuit directory on the face, inside of, or in an approved location adja-

Figure 3.10. Barriers required in service panelboards

cent to the panelboard door [*NEC* 408.4(A) or IRC E3706.2].

Barriers must be placed on all service equipment, such as panelboards, to prevent inadvertent contact with uninsulated, ungrounded service busbars or service terminals by homeowners or maintenance equipment while servicing load terminations with the service disconnect in the open position.

While moving the main circuit breaker or switch to the open "off" position will de-energize the panelboard, supply terminals to this panel will remain energized and exposed. Should a metal fish tape, tool, or a loose uninsulated conductor come into contact with these exposed terminals, the result would likely be a ground fault causing an arc flash that could be lethal to personnel and expose the worker to serious burns or injuries.

One of the hurdles that had to be overcome concerning the barrier requirement for service equipment was the fact that providing such protection is more achievable for those service panelboards with a single service disconnect but less practical for service panelboards with multiple service disconnects within the same enclosure.

When this barrier requirement was initiated in the 2017 *NEC* for panelboards at *NEC* 408.3(A)(2), an exception had to accompany this rule to exempt this barrier rule for panelboards permitted by *NEC* 408.36, Ex. No. 1, 2, and 3, as it would be difficult to barrier off all "uninsulated, ungrounded service busbars or service terminals" to a panelboard with up to six means of disconnect located in a single enclosure. For the 2020 *NEC*, this exception was deleted as it was deemed unnecessary by the revisions to *NEC* 230.71 *[IRC E3601.7]*. The six disconnect rule was retained by *NEC* 230.71 *[IRC E3601.7]* but only if the "more than one disconnecting means" is located in separate enclosures *(up to six disconnecting means are no longer permitted to be in the same enclosure)*.

The revisions at NEC 230.71 [IRC E3601.7] address the challenges created by having more than one service disconnecting means in the same panelboard or other enclosure, which increases the likelihood of an incident involving energized conductors or terminal parts in the service equipment, thereby enhancing safety [NEC 230.62(C)].

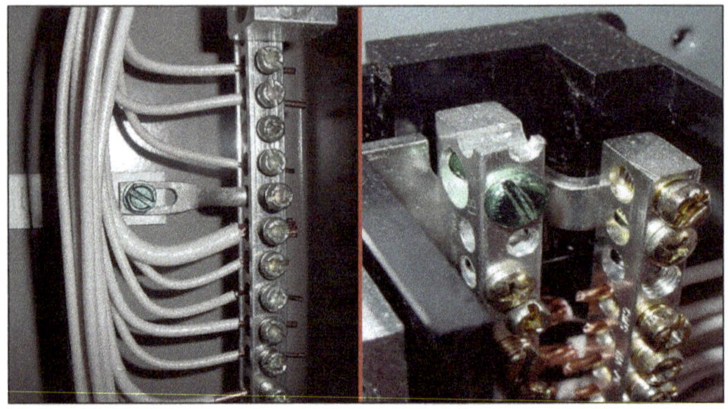

Photo 3.10. Two photos of main bonding jumpers: one in the form of a strap (left) and the other in the form of a screw (right), both connected from the grounding (neutral) conductor terminal bar to services panelboard enclosures

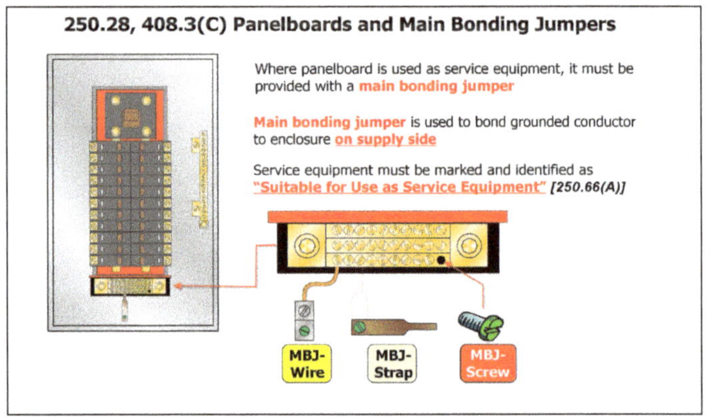

Figure 3.11. Panelboards used as service equipment must be provided with a main bonding jumper

Barriers are also required to be placed at panelboards supplied by feeder taps described at NEC 240.21(B) or transformer secondary conductors described at *NEC* 240.21(C) when the disconnecting device, to which the tap conductors are terminated, is in the open position (*NEC* 215.15). These feeder taps or transformer secondary conductors are uncommon at dwelling units but could exist. While these service equipment and feeder tap barrier requirements are not discussed in the IRC, IRC E3401.2 would require compliance with the applicable provisions of the *NEC*.

MAIN BONDING JUMPER AT PANELBOARDS

When a panelboard functions as service equipment, it must have a main bonding jumper. Usually, the main bonding jumper is a bus, strap, or screw supplied by the manufacturer, and in this case, it can be installed without field sizing requirements. It is situated inside the panelboard and connects the grounded (neutral) service conductor on its supply side to the metallic

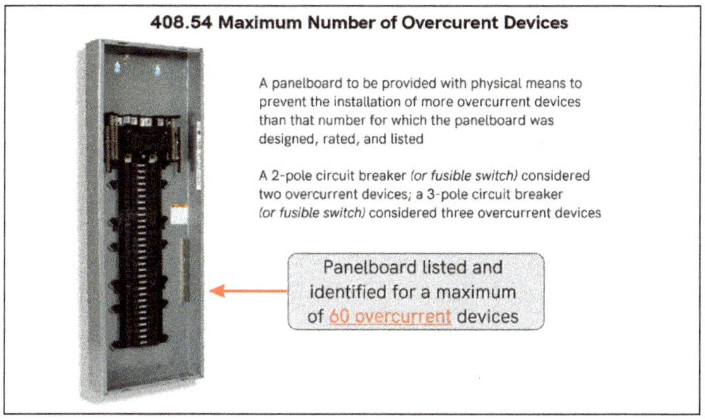

Photo 3.11. Panelboard identified for 60 overcurrent devices *(not all panelboards are limited to 42 overcurrent devices)*

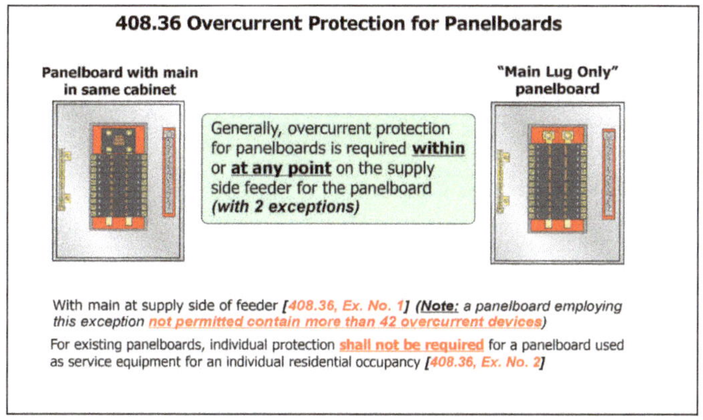

Figure 3.12. Examples of two different types of panelboard overcurrent protection

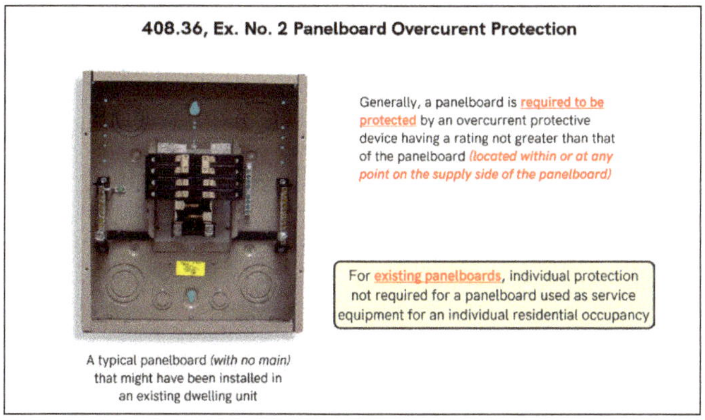

Photo 3.12. A typical panelboard could have been installed at an existing dwelling unit under *NEC* 408.36, Ex. No. 2

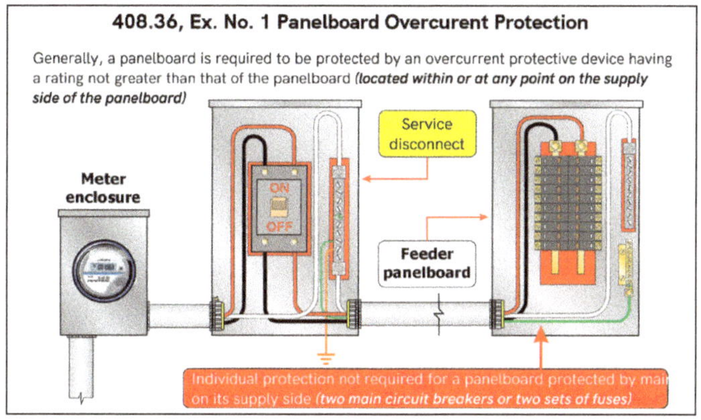

Figure 3.13. Panelboard meeting Exception No. 1 to *NEC* 408.36

frame of the panelboard [see photo 3.10 and NEC 250.28, 408.3(C) or IRC E3607.5].

INDIVIDUAL OVERCURRENT PROTECTION FOR PANELBOARDS

As previously stated, the rating of a panelboard must be equal to or greater than the minimum feeder capacity required for the calculated load. Additionally, all panelboards must be protected by an overcurrent device with a rating not exceeding that of the panelboard. This main overcurrent device must be located within the panelboard itself or at any point on the supply side of said panelboard (NEC 408.36 or IRC E3706.3). There are two exceptions to this rule, as outlined in Section 408.36 of the NEC.

The first exception excludes individual overcurrent protection for a panelboard protected on its supply side by a main overcurrent protective device, as long as that main has a combined rating not exceeding that of the panelboard. However, a panelboard built or wired under this exception cannot contain more than 42 overcurrent devices, with a 2-pole circuit breaker counting as two overcurrent devices. Before the 2008 edition of the Code, this limit of 42 overcurrent devices applied to all panelboards, not just those that comply with Exception No. 1 to NEC 408.36. Under the current edition, the number of overcurrent devices a panelboard can contain is limited only by its listing. For example, if a panelboard is listed or rated to accept 60 overcurrent devices, then it is limited to a maximum of 60 overcurrent devices when used as service equipment.

The second exception for individual protection of a panelboard applies to an existing panelboard used as service equipment at an existing residential dwelling. Before the 2020 NEC, a third exception to NEC 408.36 existed. Individual protection was not required for a panelboard used as service equipment with multiple disconnecting means in the same enclosure that was previously allowed under NEC 230.71 or IRC E3601.7 (which permits no more than six disconnects per service grouped in one location). Starting with the 2020 NEC (or the 2021 IRC), NEC 230.71(B) and IRC E3601.7 allow the service disconnecting means for each service to comprise no more than six switches or sets of circuit breakers, or a combination of both, located in separate enclosures grouped together. The provision allowing up to six disconnect means in a single enclosure was removed to align with the barrier requirement of NEC 230.62(C). With these updates to NEC

230.71(B) and IRC E3601.7, the previous third exception to NEC 408.36 is no longer necessary.

OTHER REQUIREMENTS FOR PANELBOARDS

According to most circuit breaker and panelboard manufacturers' specifications, the total load on any overcurrent device in a panelboard must not exceed 80 percent of its rating if, during normal operation, the load will continue for three hours or more unless the assembly, including the overcurrent device, is rated for continuous duty at 100 percent of its rating. This language appeared in the Code [see 384-16(d) of the 1999 NEC] but was removed during the 2002 NEC revision cycle. The 80 percent language was eliminated because feeder or branch circuit overcurrent protection is calculated at no less than the noncontinuous load plus 125 percent of the continuous load. [see NEC 210.20(A) and 215.3)]. If the feeder or branch circuit overcurrent protection is calculated correctly, there is no need to derate the overcurrent device's ampacity further. In the current Code, a panelboard must have a rating not less than the minimum feeder capacity required for the load calculated according to Part III, IV, or V of Article 220, as applicable. Loads that operate continuously for three hours or more are considered continuous loads by the Code. Continuous load is defined in NEC Article 100 and IRC Chapter 35.

Plug-in-type overcurrent protective devices or plug-in-type main lug assemblies that are back-fed must be secured in place by an additional fastener that requires other than a pull to release the device from the mounting means on the panel. The fastener used to secure the breaker to the panelboard must be identified for that specific purpose. A shock or ground-fault hazard can result from a back-fed circuit breaker becoming dislodged from its mounting on the panelboard interior busbar as the mounting bracket on the overcurrent device itself is energized due to its back-fed condition.

Without these additional fasteners holding the back-fed overcurrent device to the busbar, a back-fed circuit breaker can become dislodged by simply removing the panelboard lid due to a binding tension that can be created due to the installation of a field installed conductors to the back-fed circuit breaker. Consult the manufacturer's installation instructions or product directory for additional information [*NEC* 408.36(D) or IRC E3706.5].

Where installed in damp or wet locations, panelboards must be mounted with a 6 mm (¼ in.) airspace between the wall and support surface. Panelboards are required to be mounted in cabinets or identified enclosures. They must also be dead-front, which means having no live parts exposed to anyone on the operating side of the equipment (NEC Article 100, 408.38 (Ex) and IRC Chapter 35).

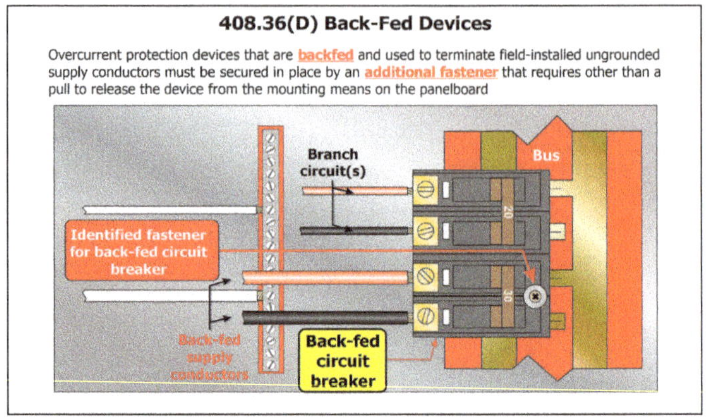

Figure 3.14. Back-fed overcurrent devices require additional fastener

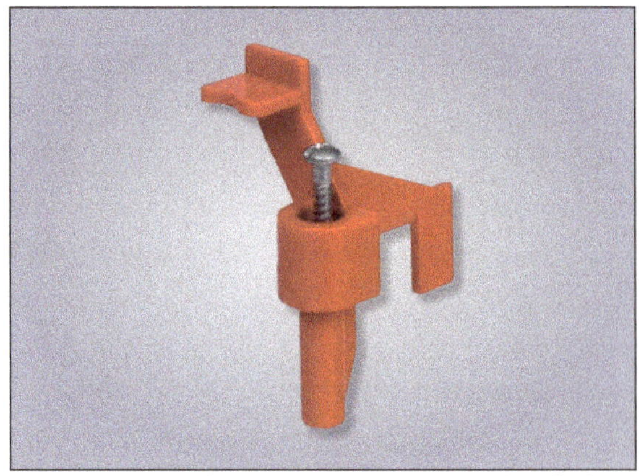

Photo 3.13. Back-fed circuit breaker with an identified fastener *(typical hold-down kit shown on left)*

Where fuses of any type are installed in panelboards, they must be installed on the load side of any switch, except where these fuses are used as service equipment [*NEC* 408.39].

Where panelboard cabinets and frames are made of metal, they must be in physical contact with each other and properly grounded by connection to an equipment grounding conductor. Where a nonmetallic raceway or cable assembly supplies the panelboard or where separate equipment grounding conductors are provided, an identified terminal bar used for all equipment grounding conductors is required to be secured to and inside the cabinet. This terminal bar is bonded to the metal cabinet and the panelboard frame. It provides a connection point for the equipment grounding conductor run with the feeder conductors supplying the panelboard (*NEC* 408.40).

Equipment grounding conductors must not be connected to a terminal bar designated for neutral conductors unless the terminal bar is specifically marked for that purpose and is located where the connection is made from the neutral conductor to a grounding electrode (NEC 408.40). The most common acceptable location is within service equipment. The connec-

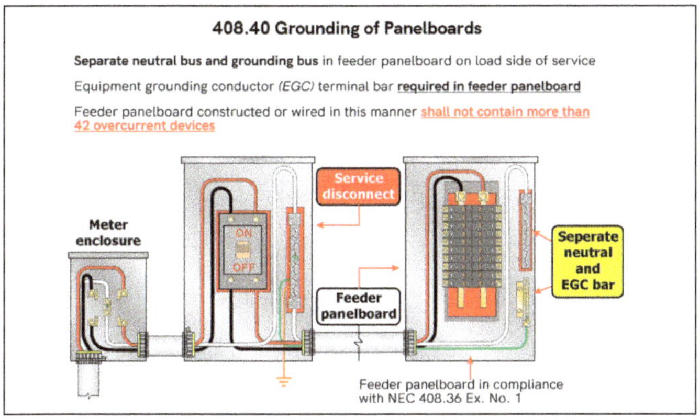

Figure 3.15. The feeder panelboard is in compliance with *NEC* 408.40 and 408.36, Ex. No. 2

Photo 3.14. Grounded (neutral) conductors terminated on neutral terminal bar in panelboard are limited to one conductor per terminal *(photo shows violation)*

tion between grounded (neutral) conductors and equipment grounding conductors is not allowed in feeder panelboards. The neutral conductor carries current, and if allowed to contact the equipment grounding conductor outside of service equipment, this circulating current could flow over grounded parts and equipment grounding conductors, potentially causing serious fire or shock hazards.

The installation rules for panelboards specify that each grounded (neutral) conductor must terminate in its own terminal and cannot be used to terminate another neutral conductor. An exception allows more than one grounded conductor to terminate under a single terminal where parallel conductors are installed, as long as they terminate in a terminal designated for multiple conductors (NEC 408.41 and 110.14 or IRC E3706.4).

As a general rule, when panelboards are installed as service equipment, the grounded service conductor (typically a neutral) is bonded to the enclosure. Where installed as a feeder panelboard (subpanel), the grounded conductor (neutral) is required to be isolated from the enclosure, and the equipment grounding conductors are connected to the enclosure [NEC 250.24(B) or IRC E3607.2]. There are instances where the Code permits a connection between the grounded (neutral) conductor and equipment grounding conductors or grounding electrode systems, but these are limited by specific Code restrictions. This connection is allowed for the existing frames of dryers and ranges (NEC 250.140). A load-side grounding connection with the grounded conductor is permitted at existing separate buildings with an existing feeder where no continuous metallic paths exist between the buildings [NEC 250.32(B)(1), Ex. No. 1 or IRC E3607.3.2]. Separate building grounding and bonding provisions are discussed in more detail in Chapter 15 of this text. The bonding and grounding of branch circuits for existing dryers and ranges are covered in Chapter 7 and Chapter 16.

PRODUCT SAFETY STANDARD REQUIREMENTS

It is recommended that the following electrical product safety guide card information and product safety standards be reviewed for additional guidance on the proper installation, operation, and use of electrical equipment covered in this chapter. The four-letter code in parentheses indicates the product category in the UL Guide Information for Electrical Equipment Directory (UL Product iQ® material).

- Equipment for Use in Ordinary Locations (AALZ)
- Cabinets and Cutout Boxes (CYIV)
- Circuit Breakers (DHJR)
- Circuit Breakers, Molded-Case, and Circuit Breaker Enclosures (DIVQ)
- Conduit Fittings (DWTT)
- Fuses (JCQR)
- Cartridge Fuses, Nonrenewable (JDDZ)
- Insulating Devices and Materials (NYYV)
- Bushings (NZMT)
- Meter Mounting Equipment (PJSR)
- Meter Fittings (PJVV)
- Meter Sockets (PJYZ)
- Meter Socket Accessories (PKAX)
- Panelboards (QEUY)
- Switches (WFXV)
- Pullout Switches – Detachable Type (WGEU)
- Enclosed Switches (WIAX)
- Knife (Switches) (WIOV)
- Switches, Molded Case (WJAZ)
- Snap Switches (WJQR)

REFERENCES

1. *National Electrical Code* and *NEC* are registered trademarks of the National Fire Protection Association, Inc., Quincy, MA 02169. This reprinted material is not the official position of the National Fire Protection Association, which is represented only by the standard in its entirety.

2. *International Residential Code for One- and Two-Family Dwellings* (2021). Copyright © International Code Council, Inc., June 2023.

3. UL, Northbrook, IL, *UL Guide Information for Electrical Equipment Directory* (UL Product iQ® material) unless indicated otherwise.

REVIEW QUESTIONS

1. In general, where ungrounded conductors of _____ AWG or larger enter a raceway, cabinet, pull box, or junction box, prior to the installation of the conductors, the conductors must be protected from abrasion by such things as an identified fitting providing a smoothly rounded insulating surface.

A. 2
B. 4
C. 6
D. 8

2. When a cabinet is to be installed in a wet location, one consideration is that the enclosure must be:

A. weatherproof
B. dust-tight
C. sealed
D. approved for a damp location

3. The minimum wire bending space and minimum width at terminals for one 1/0 AWG copper conductor (one wire per terminal) where the conductor does not leave the enclosure opposite its terminal is:

A. 114 mm (4½ in.)
B. 88.9 mm (3½ in.)
C. 63.5 mm (2½ in.)
D. 38.1 mm (1½ in.)

4. Panelboards installed in damp or wet locations must be mounted with at least 3 mm (1/8 in.) airspace between the rear of the panel and the wall or other supporting surface.

True _______ False _______

5. A cabinet or cutout box, when installed in a wall of concrete, tile, or other noncombustible material, is permitted to be recessed not more than at what distance?

A. 6 mm (¼ in.)
B. 13 mm (½ in.)
C. 16 mm (⅝ in.)
D. 19 mm (¾ in.)

6. For cabinets that have a flush-type cover, the maximum permitted gap between the finish drywall or plaster and the cabinet is:

A. No gap permitted
B. 1.6 mm (1/16 in.)
C. 3 mm (⅛ in.)
D. 6 mm (¼ in.)

7. Each grounded (neutral) conductor of a 12 AWG branch circuit shall terminate within the panelboard in an individual terminal that is not also used for another conductor.

True _______ False _________

8. Individual protection shall not be required for a panelboard protected on its supply side by two main circuit breakers having a combined rating not greater than that of the panelboard. A panelboard constructed or wired in this manner shall not contain more than _____ overcurrent devices.

A. 20
B. 36
C. 42
D. 60

9. A means for connecting equipment grounding conductors inside a metallic cabinet enclosing a panelboard must be provided where the panelboard is used with Type NM nonmetallic sheathed cable assemblies containing equipment grounding conductors.

True _______ False _______

10. Where a metal plug or plate is used with nonmetallic cabinets or meter socket enclosures, it is required to be recessed by what distance from the outer surface?

A. 0.8 mm (1/32 in.)
B. 1.6 mm (1/16 in.)
C. 3.0 mm (⅛ in.)
D. 6.0 mm (¼ in.)

11. A 2/0 kcmil conductor is permitted to enter or leave an enclosure containing only a meter socket(s) with lay-in type terminals or removable lug with integral mounting tang through the wall opposite its terminal provided a minimum of 152 mm (6 in.) wire-bending space is provided.

True _____ False _____

12. Conduit bushings made of insulating material used to protect conductors entering a raceway are permitted to be used to secure a raceway.

True _____ False _____

13. An insulated fitting or insulating material is required to have a temperature rating not less than the insulation temperature rating of the installed conductors.

True _____ False _____

14. Splices of conductors are permitted to be installed in the gutter space of a panelboard enclosure, provided all the conductors and splices do not take up more than _______ of the space.

A. 40 %
B. 50 %
C. 65 %
D. 75 %

15. Where installed in any wet location or outside a building, a surface-mounted switch or circuit breaker is required to be enclosed in a weatherproof enclosure or cabinet.

True _____ False _____

16. Where insulated, circuit conductors of 4 AWG or larger enter a raceway in a cabinet, pull box, junction box, or auxiliary gutter, the conductors are required to be protected from abrasion by which one of the following methods:

A. An identified fitting providing a smoothly rounded insulating surface.
B. A listed metal fitting that has smoothly rounded edges.
C. Separation from the fitting or raceway using an identified insulating material that is securely fastened in place.
D. Any of the above.

17. A single-throw knife switch must be placed so that gravity will not tend to close it.

True _____ False _____

18. All switches and circuit breakers must be installed with the center of the grip to their operating handles when in its highest position, will not be more than what distance above the surface of the floor or other working platform?

A. 1.8 m (6 ft 0 in.)
B. 2.0 m (6 ft 6 in.)
C. 2.0 m (6 ft 7 in.)
D. 2.3 m (7 ft 6 in.)

19. Generally, the connection between grounded (neutral) conductors and equipment grounding conductors is not permitted in feeder panelboards.

True _____ False _____

20. In a panelboard, spare positions that contain unused overcurrent devices shall be described or identified accordingly in a circuit directory that is located ________________________ .

A. on the face of the panelboard door
B. inside of the panelboard door
C. in an approved location adjacent to the panelboard door
D. any of the above

21. The rating of a panelboard cannot be less than the minimum feeder capacity required for the load computed per Parts III, IV, or V of *NEC* Article 220 as applicable.

True _____ False _____

22. All panelboard circuits and any circuit modifications are required to be legibly and permanently identified as to its clear, evident, and specific purpose or use. The circuit identification shall be described with a degree of detail and clarity such that it is ________________________.

A. unlikely to result in confusion between circuits
B. distinguished from all others
C. clearly understood
D. approved by the AHJ

23. A panelboard shall be protected by an overcurrent protective device with a rating not greater than the panelboard's. Without applying any exceptions, this overcurrent protective device shall be located at what location?

A. Anywhere the installer desires
B. Within or at any point on the supply side of the panelboard
C. Only within the panelboard
D. At least 1.5 m (5 ft) horizontally from the inside walls of the panelboard

24. Unused raceway or cable knockout openings in cabinets or meter socket enclosures must be ______________ to provide protection substantially equivalent to that of the wall of the cabinet or meter socket enclosure.

A. identified
B. covered with electrical tape
C. closed
D. welded shut

25. Plug-in-type overcurrent protective devices or plug-in-type main lug assemblies that are backfed must be secured in place by an additional fastener that requires other than a pull to release the device from the mounting means on the panelboard.

True _____ False _____

26. Where panelboard cabinets and panelboard frames are made of metal, they are required to be in physical contact with each other and are to be properly grounded by connection to an equipment grounding conductor.

True _____ False _____

27. Raceways and cables that enter the cabinet above the level of uninsulated live parts require special termination fittings. These fittings must be listed for ______________.

A. damp locations
B. wet locations
C. hazardous locations
D. all of the above

28. For cabinet or meter socket enclosures, a warning label complying with ________________ must be applied to the enclosure that identifies the closest disconnecting means for any feed-through conductors.

A. *NEC* 110.21(B)
B. *NEC* 90.3
C. *NEC* 110.24
D. *NEC* 408.3

29. Barriers are required to be placed in all ________________________ to prevent any uninsulated, ungrounded service busbar or service terminals exposed to inadvertent contact by homeowners or maintenance equipment while servicing load terminations.

A. feeder panelboards
B. service equipment
C. service panelboards
D. meter enclosures

30. The term _____________ is defined as the production of an electric or magnetic state by the proximity (without contact) of an electrified or magnetized body or the production of an electric current in a conductor by varying the magnetic field applied to the conductor.

A. induction
B. capacitance
C. reactance
D. impedance

31. Where mounted in an enclosure, General-use and motor-circuit switches and circuit breakers must clearly indicate whether they are in the open "off" or closed "on" position. This indication of the position for a switch must be in a location that is _________________________.

A. accessible to the occupant
B. close to the equipment it operates
C. visible when accessing the external operating means
D. visible only when the enclosure cover is open

32. Where switch or circuit breaker handles are operated vertically rather than rotationally or horizontally, the "up" position of the handle must be in the "off" position.

True _____ False _____

4 Installations and Inspections of Services

Requirements for electrical services are outlined in NEC Article 230 and IRC Chapter 36. It should be noted that general electrical requirements, such as electrical terminations and related matters, included in other sections of the Code, also apply to service installations. Additionally, requirements for electrical services in NEC Article 230 do not impose any limits on the size or ratings of a dwelling unit service.

The *NEC* permits services of any size or rating (including feeders of 120/208-volt "Y" systems) for dwellings. However, electrical services for dwelling units that fall under the scope of the IRC are limited in size and voltage ratings to 120/240-volt, single-phase, 400-ampere. Any electrical service beyond those ratings must follow the applicable requirements of the *NEC* [see IRC E3401.2, E3401.3(2), E3603.1, and Table E3603.1.1].

For a one-family dwelling unit, a 100-ampere, 3-wire service disconnecting means is the minimum rating required [*NEC* 230.79(C) or IRC E3602.1]. See Chapter 2 in this publication for more details on calculating the minimum rating of service conductors and equipment.

DEFINITIONS

Definitions of certain terms related to electrical services are provided in the Code. Understanding these terms is essential for correctly applying the rules for services. The following definitions are found in NEC Article 100 and IRC Chapter 35. If needed, please refer back to Article 100 in the NEC for further clarification.

- ***Premises Wiring (System)***
- ***Service***
- ***Service Cable***
- ***Service Conductors***
- ***Service Conductors, Overhead***
- ***Service Conductors, Underground***
- ***Service Drop***

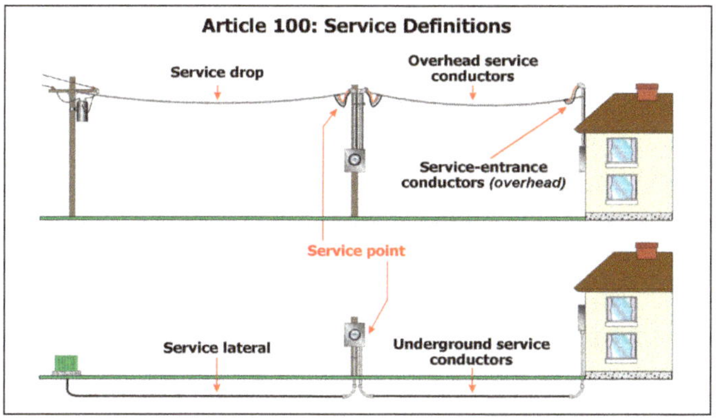

Figure 4.1. Examples of some of the service definitions

Photo 4.2. Underground service lateral supplying metering equipment

Photo 4.1. Two photos of typical underground (left) and overhead service(right) await hook-up and connection.

Photo 4.3. Service drop connects to the overhead conductors between the utility electric supply system and the service point.

- ***Service-Entrance Conductors***
- ***Service Equipment***
- ***Service Lateral***
- ***Service Point***

SERVICE EQUIPMENT, GENERAL

As defined, "service conductors" are a broad term that includes both the service drop and service-entrance conductors for overhead services, as well as service lateral and service-entrance conductors for underground systems.

The service-entrance conductors connect to the service equipment, which typically provides overload protection for them. The service equipment also offers the means to disconnect the service-entrance conductors from the building's wiring system. There may be no service entrance conductors if an underground service lateral connects directly to service equipment located outside a building. Many specific rules must be understood to install or inspect electrical services correctly.

SERVICE EQUIPMENT, GENERAL

A service drop includes the overhead conductors between the serving utility *(typically the utility electric supply system)* and the service point.

The electric utility supply company usually installs the service drop and provides the service-entrance conductors at the point where they attach to the building. This attachment point may be a porcelain insulator bolted to the structure or securely fastened to a service mast with a fitting specifically designed for this purpose (see NEC 230.27 or IRC E3604.4). The service drop is typically connected to the service-entrance conductors by the utility. The end of the service-entrance conductors is formed into a drip loop as they exit the service head (see Photo 4.3). The service head, sometimes called a weatherhead, is the fitting at the end of the service mast or cable. The drip loop relies on gravity to help prevent moisture from following the conductors and entering the raceway or cable through the weatherhead.

CLEARANCES ON BUILDINGS

The Code mandates minimum clearances from building openings and other structures such as porches, balconies, ladders, stairs, and others for service conductors installed as open conductors, like duplex or triplex service drops or multiconductor cables that lack an overall outer jacket. This

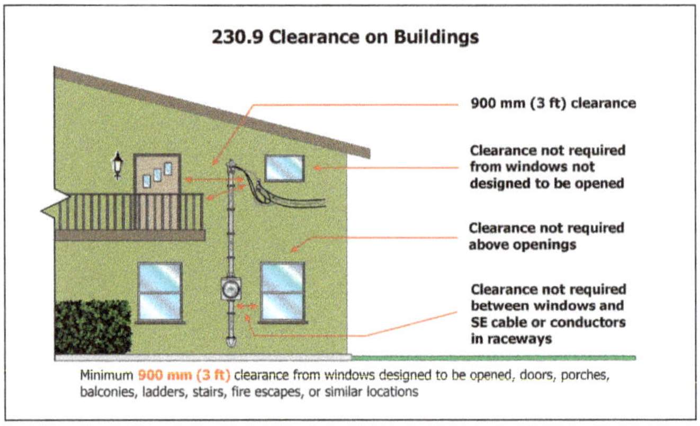

Figure 4.2. Clearances on buildings

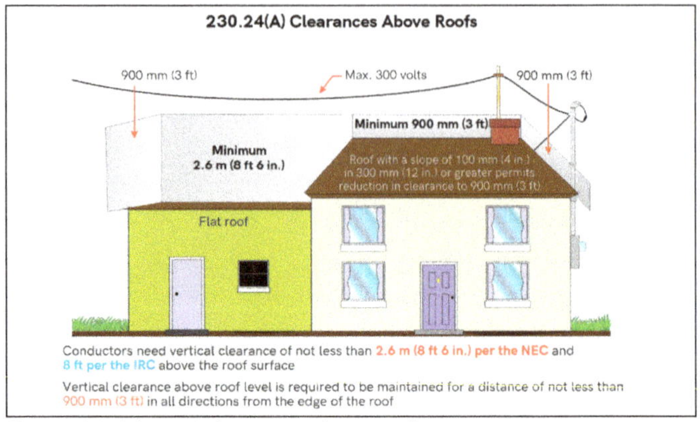

Figure 4.3. Vertical clearances above roofs

general rule does not prohibit the installation of a multiconductor cable assembly with an overall outer jacket, such as Type SE (NEC 230.9 or IRC E3604.1). Refer to Figure 4.2 for a summary of the locations on buildings where clearances apply to open service cables or conductors.

Open service conductors, including the drip loop, must be installed so they are no closer than 900 mm (3 ft) from the sides or bottom of any window intended to open, as well as from doors, porches, balconies, ladders, stairs, fire escapes, or similar locations. This clearance ensures that open service conductors stay at least "out-of-reach" from building openings or platforms. Service conductors placed above a window's top level are considered out of reach and can be closer than 900 mm (3 ft). Service conductors, such as service-entrance cable, Type SE or SER, do not need to maintain a 900 mm (3 ft) distance from building openings or platforms.

SERVICE DROP CLEARANCES ABOVE ROOFS

The vertical clearances of service drops passing over a roof and from grade are covered in NEC 230.24(A) and (B) and IRC E3604.2.1 and E3604.2.2. These rules do not apply to utility-installed service drops if they are on the line or supply side of the service point [see NEC 90.2(D)(5) or IRC E3401.3(1)]. The serving electric utility usually installs service drops. However, it is the installer's responsibility to install the electrical equipment, such as a service knob or insulator, ensuring the service drop has the proper clearance above roofs or grade. For clearance requirements of service drops over swimming pools, spas, etc., refer to NEC 230.24(D), 680.9, or IRC E4203.7 and Chapter 14.

The general rule for clearances installed above a roof is that service drops must have a minimum vertical clearance of at least 2.6 m (8 ft 6 in.) when passing over a roof. This 2.6 m (8 ft 6 in.) clearance should generally be maintained for a radius of 900 mm (3 ft) from the edge of the roof. This can be seen as a "halo" around the roof, providing an "out-of-reach" zone for someone on the roof. Five exceptions to this rule are provided (see NEC 230.24(A), Exceptions or IRC E3604.2.1 Exceptions, and figures 4.3 and 4.4).

Exception No. 1 applies to a roof surface that pedestrians or vehicles may traffic on. It could include a flat, rooftop deck used as an entertainment space.

Exception No. 2 permits a reduction of the service-drop clearance from 2.6 m (8 ft 6 in.) to 900 mm (3 ft) over the roof when the voltage between the service-drop conductors is not over 300 volts, and the roof has a slope of 100 mm (4 in.) in 300 mm (12 in.) or greater.

Exception No. 3 permits a reduction in clearance of the service-drop conductors to 450 mm (18 in.) above the roof where the:

1. voltage between conductors does not exceed 300 volts,
2. service drop passes over only the overhanging portion of the roof, and not more than 1.8 m (6 ft) of service-drop conductor passes diagonally over the roof overhang,
3. The service drop does not pass over an overhang that extends more than 1.2 m (4 ft) horizontally from the structure and
4. is secured to an electric service mast (through the roof) raceway or other approved support such as a steel or suitable wooden brace that meets the requirements of *NEC* 230.28 or IRC E3604.5. See Figure 4.4 for an illustration of service-drop clearances for a service mast that passes through the overhanging portion of the roof.

Exception No. 4 allows the final span of the service-drop

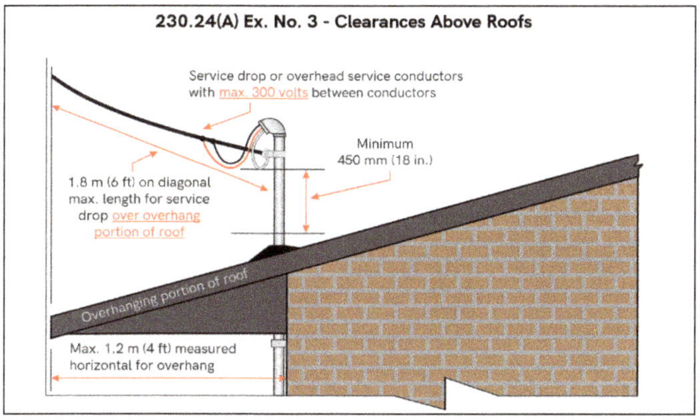

Figure 4.4. Clearance of service-drop conductors at the service mast (with exceptions)

Photo 4.4. Overhead service drop passing over a building with proper clearance

conductor to pass through the halo around the building where the service drop is attached to the side of the building. This allows the service drop to rise through the halo to end at or connect to a pole that is higher than the building.

Exception No. 5 allows reducing the clearance above the roof area to 900 mm (3 ft) when the voltage between conductors does not exceed 300 volts and the roof area is guarded or isolated. According to the definitions in NEC Article 100 or IRC Chapter 35, "guarded" means "covered, shielded, fenced, enclosed, or otherwise protected by means of suitable covers, casings, barriers, rails, screens, mats, or platforms to prevent approach or contact by persons or objects to a point of danger." The term "isolated" is defined as "not readily accessible to persons unless special means for access are used." [NEC 230.24(A) and Exceptions or IRC E3604.2.1 and Exceptions].

SUPPORTS OVER BUILDINGS

Where service conductors pass over a roof, those service conductors have to be securely supported by substantial structures. These metal structures, sometimes referred to as a "roof jack" in the field, should be adequately bonded to limit a potential shock hazard. For a grounded system, where these substantial structures are metal, they must be bonded by means of a bonding jumper and listed connector to the grounded overhead service conductor. Where practicable, such supports shall be independent of the building [see *NEC* 230.29 and IRC E3604.6 and Figure 4.5].

The bonding jumper used to accomplish this bonding is to be sized per the requirements of *NEC* 250.102 or IRC E3609.5 and *NEC* Table 250.102(C)(1) or IRC Table E3603.4 based on the size of the ungrounded service conductors. This bonding and sizing requirement is very similar to the bonding requirements for bonding a ferrous metallic raceway used to chase or enclose a grounding electrode conductor found at *NEC* 250.64(E) or IRC E3610.3.

The *Code* contains bonding requirements about "equipment for services" at *NEC* 250.92(A) [IRC E3609.2], but this provision covers "normally non-current-carrying metal parts of equipment" such as "all raceways, cable trays, cablebus framework, auxiliary gutters, or service cable armor or sheath that enclose, contain, or support service conductors." This subsection goes on to require "all enclosures containing service conductors, including meter fittings, boxes, or the like, interposed in the service raceway or armor" to be bonded together. Although this list of service-related items required to be bonded is thorough, the "substantial metal structures" described in NEC 230.29 and IRC E3604.6 do not seem to fit into any of these categories. These substantial support structures are not a raceway, as they do not act as a channel for conductors; they are auxiliary or supplemental support for overhead service conductors.

VERTICAL CLEARANCES ABOVE FINAL GRADE

Vertical clearances for service drops and overhead service conductors passing above finished grade or ground level are provided in NEC 230.24(B) or IRC E3604.2.2. The vertical clearance must be maintained at the minimum distances as follows:

> **3.0 m (10 ft)** at the point where the electric service enters the building, at the drip loop of the building's electric entrance, and above areas or sidewalks that are accessible only to pedestrians. This measurement must be taken from the final grade line or other accessible surface only for service-drop cables that are supported on and cabled together with a grounded bare messenger, which is limited to 150 volts to ground. This clearance from the ground or another accessible surface, such as a porch or sidewalk, is measured from the lowest part of the drip loop.

3.7 m (12 ft) over residential property or driveways (and those commercial areas not subject to truck traffic) where the voltage does not exceed 300 volts-to-ground.

4.5 m (15 ft) over residential property and driveways (and those commercial areas not subject to truck traffic) where the service-drop voltage to ground exceeds 300 volts. The typical voltage to ground for dwelling services is 120 volts for a 120/240-volt, 3-wire, single-phase service. The IRC does not include the 4.5 m (15 ft) clearance requirement because the voltage is limited to 240 volts for the service based on IRC E3401.2.

5.5 m (18 ft) over a public street, alley, road, or parking area that will be traveled by truck traffic, driveways on other than residential property, and other land traversed by vehicles such as cultivated, grazing, forest, and orchard areas.

It's important to consider these rules and to work cooperatively with the electric utility supply agency before installing or inspecting the location of an electric service. As previously mentioned, the service drop is usually installed by the serving utility. The installer of electrical equipment must install the equipment so the electrical utility can maintain the clearances required by the *Code* when connecting to the customer's service drop point of attachment, also known as the insulator or knob. In no case can the point of attachment be less than 3.0 m (10 ft) above grade. The utility must also follow a proscribed installation standard that includes clearances from grade and vertical and horizontal clearances from buildings and structures and on buildings. The serving utility adheres to the National Electrical Safety *Code* (NESC), which contains installation standards for the type of conductors and equipment associated with transmission, distribution, and supply systems. See Figure 4.6 for an illustration of vertical clearance from grade for various conditions.

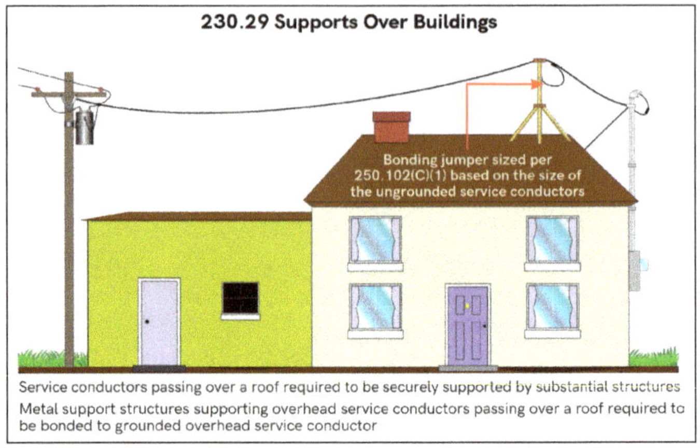

Figure 4.5. Service conductors supported by substantial structures should be adequately bonded

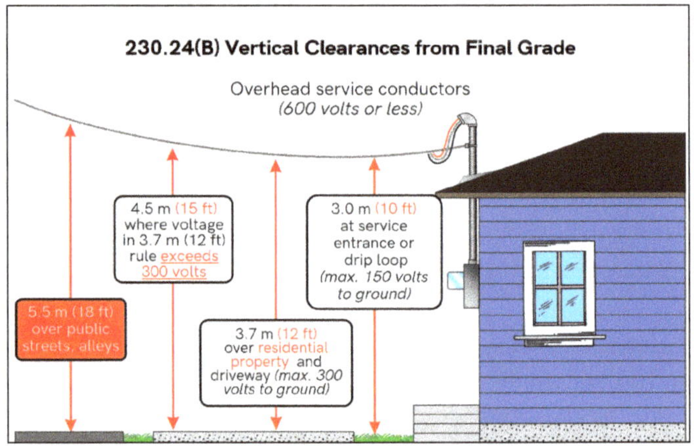

Figure 4.6. Vertical clearance from final grade for overhead service conductors

THE SERVICE MAST

Service masts used for supporting a service drop or overhead service conductors must be installed in accordance with NEC 230.28 or IRC E3604.5. This section requires that service masts used for the support of service drops must be "of adequate strength" to withstand safely the strain imposed by the service drop or be supported by braces or guys. The Code does not mandate the minimum size of conduit permitted for a service mast or the use of additional support such as guy wires. Still, it leaves the interpretation of "of adequate strength" to the authority having jurisdiction (AHJ). Many serving utility companies provide a set of regulations that detail what is required for the mast to which a service drop is to be attached. It is essential to verify with the serving utility the minimum mast size and acceptable electrical watt-hour meter and meter base locations before installing the service.

In some parts of the country where wind, rain, snow, and ice storms are prevalent, the electric utility company or local ordinances often require a metric designator 53 (2 in.) or larger steel galvanized rigid metal conduit to be used for an electric service mast, or so-called "stack" or "periscope." A roof flashing must also be installed at the roofline where the service raceway penetrates the roof to prevent damage to the roof structure by moisture and subsequent rotting of the framing materials. This flashing must be placed around openings through the roof as necessary to provide adequate drainage. This requirement is found in most major building codes for any roof penetration. See the service mast illustration in Figure 4.7.

The service mast or raceway must be equipped with a service head or weatherhead at the point of attachment to the

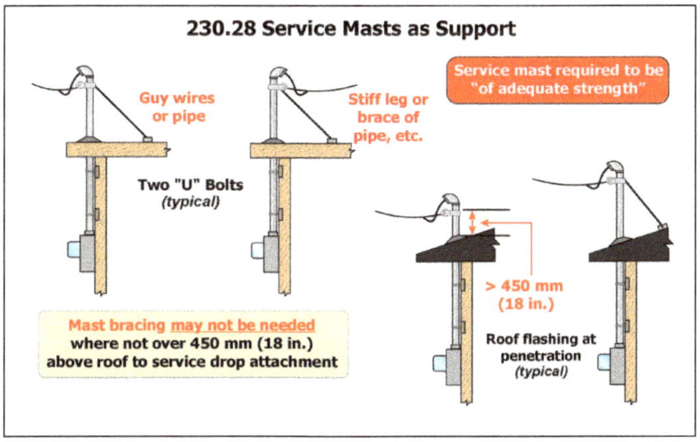

Figure 4.7. Typical bracing or guy wire supports for through-the-roof service masts. Where not "of adequate strength," a service mast must be supported by a guy wire or bracing. All raceway fittings must be identified for use with service masts when the mast is used to support the drop.

service-drop conductors. If the service head or weatherhead is installed in a wet location, it shall be listed for use in a wet location. The *Code* permits two options for Type SE cable. Either a wet location listed service head is installed on the end of the cable, or the cable is permitted to be formed into a gooseneck. The open end of the cable jacket must be sealed with a self-healing, weather-resistant, thermoplastic tape to prevent moisture from entering or accumulating within the box, conduit body, or fitting [*NEC* 230.54(A) and (B) or IRC E3605.9.1 and E3605.9.2].

Service heads and goosenecks have to be located above the point of attachment. This requirement is intended to help prevent water from dripping down and into the raceway or cable. However, this is not always practical. An exception permits the service head or gooseneck to be located not more than 600 mm (24 in.) from the point of attachment [*NEC* 230.54(C) and Exception or IRC E3605.9.3 and Exception].

The service head has multiple knock-out-type openings. The *Code* requires conductors of different potentials to be brought out through separate openings. There is an exception for jacketed multi-conductor cable, such as Type SE, as long as it has no splices.

The individual conductors that emerge from the service head must be formed into a gooseneck. The drip loop keeps gravity from drawing moisture along the conductor insulation and into the cable or raceway. The end of the drip loops connects to the service drop at a point either below the level of the service head or below the level of the end of the service-entrance cable sheath (*NEC* 230.54 or IRC E3605.9).

The Code permits only power service-drop conductors to be attached to a service mast. These rules do not permit service drops of telephone, CATV systems, or any other system to be attached to power service masts (NEC 230.28, 800.44(C), or IRC E3604.5). Outdoor antennas must never be attached to the electric service mast due to the risk of contact with energized service conductors. This restriction is found in NEC 810.12.

SERVICE-ENTRANCE CONDUCTORS SUNLIGHT RESISTANT

Service-entrance conductors exposed to direct sunlight must be sunlight resistant. This requirement applies to conductors that extend outside the service mast for connection to the service drop. To ensure sunlight resistance, the conductor insulation is made from a formula containing compounds that resist damaging sun rays. Without these compounds, some insulations can deteriorate to the point where they fall off the conductor, potentially leaving an exposed, energized (hot) conductor.

Conductors and cables exposed to direct sunlight must be of a type that is listed as "sunlight resistant" or marked as such. Individual conductors or cables may also be covered with an insulating material, such as tape or a sleeve, that is listed or marked as sunlight-resistant. Some manufacturers may abbreviate the mark as "sun res." See NEC 310.10(D) or IRC E3605.6 for the specific rules requiring this listing or marking on conductors exposed to ultraviolet rays.

Service-entrance conductors are usually required to be insulated. Here are some conditions where the *Code* will permit the grounded (neutral) service-entrance conductor to be uninsulated:

1. Bare copper installed in a raceway or part of a service cable assembly (see raceways permitted for service-entrance conductors later in this chapter).
2. Bare copper is used for direct burial installations if the copper is judged to be suitable for the conditions that will be present in the surrounding soil.
3. Bare copper used for direct burial installations without regard for soil conditions and where it is part of a cable assembly such as Type USE and specifically identified for use where buried underground.
4. Aluminum or copper-clad aluminum without individual insulation or covering where it is part of a cable assembly identified for underground use in a raceway or for direct burial.
5. Bare conductors used in an auxiliary gutter (*NEC* 230.41 Exception or IRC E3605.1).

UNDERGROUND SERVICE LATERALS

A service lateral includes the underground service conductors from the street main, including any risers on a pole or other structure or from transformers, to the service point. The service point may be the first connection to the service-entrance conductors in a terminal box or meter enclosure. This terminal box can be located either inside or outside the building wall. If there is no terminal box, meter, or other enclosure with enough space, the point of connection is considered the point where the service conductors enter the building.

PROTECTION FROM PHYSICAL DAMAGE

Service lateral and underground service conductors installed underground and emerge from grade at the point of entry into a meter base or service equipment must be protected from physical damage. This protection must be in the form of a suitable enclosure or raceway. An example of a suitable enclosure is the open-ended auxiliary gutter that makes up the bottom of a meter pedestal. The raceway must be one of those listed in NEC 230.50(B) or IRC E3605.5. Where rigid polyvinyl chloride conduit (PVC) is used for protection against physical damage, Schedule 80 PVC is required.

Schedule 40 PVC does not provide the impact and crush-resistant characteristics required for providing the protection anticipated by the *Code*. Schedule 40 PVC cannot be used in any location where protection from physical damage is required. Additional information is found in UL's General Information for Electrical Equipment under the product category Rigid Nonmetallic Schedule 40 and Schedule 80 PVC Conduit (DZYR). The guidebook indicates that Schedule 40 PVC conduit is suitable for underground use by direct burial or encasement in concrete. It is also suitable for use above ground, both indoors and outdoors, where it is exposed to sunlight and weather, and is not subject to physical damage. Schedule 80 PVC can be used in the same applications as Schedule 40 PVC. PVC conduits marked "Schedule 80" can be used where exposed to physical damage and on poles in accordance with the Code.

The raceway must extend from the minimum cover (direct burial) distance required by NEC 300.5(D) or IRC E3803.3 below grade to a point at least 2.5 m (8 ft) above the finished grade.

However, this protection does not have to extend into the ground more than 450 mm (18 in). In some cases, cables may need to be buried or covered at least 600 mm (24 in). The term "cover" is defined in Note 1 of NEC Table 300.5(A) or IRC Table E3803.1 Note e as the distance from the top of the cable or raceway to the top surface of the finished grade

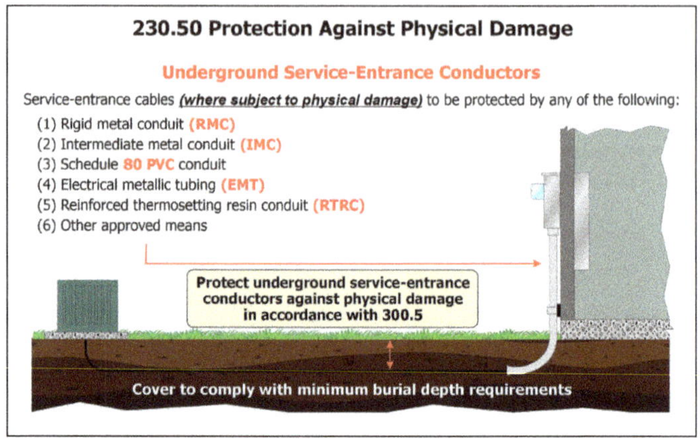

Figure 4.8. Underground service lateral conductors

(see Chapter 11 in this publication for more information on underground installations).

Burial depth tables are identical in the *NEC* and the IRC. The *NEC* minimum burial depth table is in NEC Chapter 3, Section 300.5 [Table 300.5(A)], and the IRC minimum burial depth table is in Chapter 38, Section E3803.1 (Table E3803.1).

SERVICE-ENTRANCE CABLE: TYPE USE

Cables used for underground service conductors are usually Type USE (underground service entrance) and are covered in *NEC* Article 338. These cables are usually of a single-insulated type and are often spiraled together to form a three-conductor cable, which is referred to as "URD" for "underground rural distribution." Generally, Type USE cables are limited to use underground. The primary reason for this restriction is given in the use permitted (*NEC* 330.10) of *service-entrance cable (Type USE)* and within *NEC* Article 100, where the definition indicates that Type USE cables are identified for underground use, have a moisture-resistant covering, but are not required to have a flame-retardant covering.

When using the IRC, for more specific information about service-entrance cables than what is covered in Table E3801.4 or Table E3802.1, one is required to refer to *NEC* Article 338 for more details and other installation restrictions relative to the use of these cables.

NEC 338.12(B)(2) indicates that Type USE service laterals are not permitted for aboveground installations except where they emerge outside at termination points in meter sockets or other enclosures where protected in accordance with *NEC* 300.5(D) or IRC E3803.3. This type of cable, having only a Type USE designation, is not permitted within buildings or for use as feeders, such as for supply to the interior of a mobile home. It also is not permitted within a raceway except

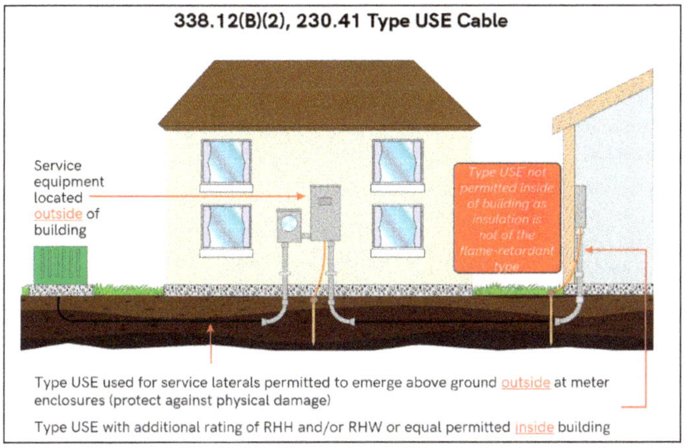

Figure 4.9. Use of Type USE cable

where the raceway is used to provide protection against physical damage or for a similar purpose. Type USE cannot be used for building wiring because the insulation is not of the flame-retardant type.

Type USE cable is also available with additional insulation designations. These designations include Type RHH and RHW (Some designations may include a suffix "-2" to indicate the cable is suitable for 90°C both wet and dry locations). These "triple-rated cables" are permitted to be used at any location the individual designation permits. For example, Type RHH is a rubber-insulated wire with insulation rated at 90°C and permitted to be used above ground in or on buildings or structures in a raceway in a dry location. Type RHW is a rubber-insulated wire that has a 75°C rating and is suitable for installation aboveground, in or on buildings or structures in a wet location in a raceway. They are permitted to be used underground due to their USE designation. As can be seen, these triple-rated cables offer a great deal of flexibility as to suitability for use in different locations. See Chapter 6 of this text for additional information on how these cables were used as feeder conductors.

The IRC does not include a separate article for each wiring method. Refer to the *NEC* for specific requirements for the various wiring methods permitted for use. The IRC includes uses permitted and uses not permitted in tabular form. IRC Chapter 38 includes three tables that provide general use and installation requirements for the type of wiring method used. Table E3801.2 provides the "Allowable Wiring Methods" with designated abbreviations; Table E3801.4 gives "Allowable Applications for Wiring Methods"; and Table E3802.1, the "General Installation and Support Requirements for Wiring Methods." This differs from the *NEC* only in that the information is provided as rules in tabular form; the rules are the same in both codes. The *NEC* installation and support requirements are found in Article 300, General Requirements, and the specifics of each wiring method are found within each wiring method article.

SERVICE-ENTRANCE CABLES: TYPES SE AND SER

A service-entrance cable, used for the supply to a 120/240-volt, three-wire, single-phase dwelling service, includes three conductors, which often consist of a bare neutral conductor and two ungrounded (hot) insulated conductors made up in the form of a cable assembly identified as Type SE. Type SE cable often has the strands of the neutral conductor wrapped spirally around the other insulated conductor to form a fairly flat cable.

A final nonmetallic jacket is then placed around the assembly. This cable is often referred to as "SEU," with the "U" referring to the type of cable that includes an uninsulated conductor along with two insulated conductors. This cable is usually flat or oval in shape. SE cable that includes an insulated grounded (neutral) conductor in addition to the uninsulated conductor used for equipment grounding purposes is often round in shape and is designated as Type SER, meaning "service-entrance round" cable [see UL Product iQ certification database (formally UL White Book) Category (TYLZ)].

Where service-entrance cables are used as service-entrance conductors, they are required to be installed in accordance with applicable requirements of *NEC* Article 230 and IRC Chapter 38 and Tables E3801.2, E3801.4, and E3802.1.

SERVICE-ENTRANCE CABLE INSTALLATION

Several fittings are made to facilitate the installation of Type SE cables for services. These include a weatherhead or service cap, watertight connectors for connection at meter bases and service equipment, and service-cable sill plates. Other fittings include one-hole and two-hole straps and one-screw and two-screw connectors. Additional information on service-entrance cable fittings can be found in the UL Product iQ certification database (formally UL White Book) in Category (TYZX).

Typically, the weatherhead is located above the point where the service drop attaches to the building. The cable is routed to the meter base location, where it is connected to a watertight connector. This connector has an oval-shaped hole and a rubber seal that compresses around the cable as the threaded nut is tightened to create a watertight connection. A cable connector is installed at the bottom of the meter base, and the cable is routed to the service equipment, where another cable connector is installed. One major advantage of

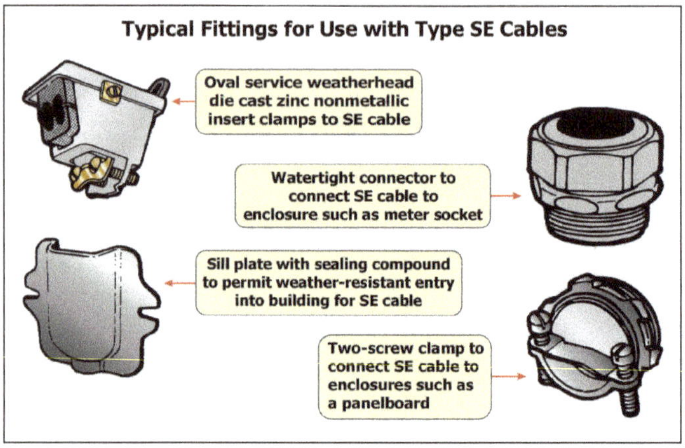

Figure 4.10. Type SE cable fittings

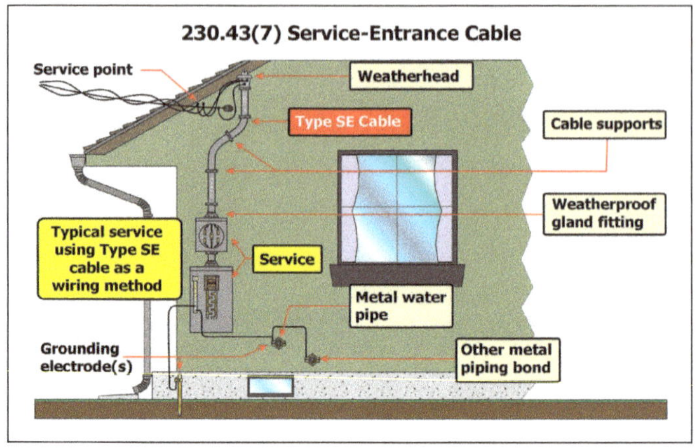

Figure 4.11. Type SE cable used as a wiring method at a dwelling unit

Type SE cable is that it can be routed around obstacles easily (see Figure 4.11 for details).

A sill plate is made to facilitate the installation of the service cable through the outside wall to the service equipment that is installed inside, sometimes in a basement. The cable enters the sill plate and makes a 90-degree bend into the wall. Care must be taken at the protective covering to ensure that the cable is not damaged where it is bent, as this would allow water to enter the cable assembly. The radius of the curve of the inner edge of any bend shall not be less than five times the diameter of the cable. The sill plate is provided with a sealing compound to seal the cable penetration into the building (see Figure 4.10).

PROTECTION OF TYPE SE CABLES FROM PHYSICAL DAMAGE

The *NEC* and the IRC permit Type SE cable assemblies to be used for service-entrance conductors. Service-entrance cable assemblies must be protected where they will be subjected to physical damage. This protection is required in exposed places such as near sidewalks, walkways, or driveways. Consideration must also be given to any object, such as a window shutter, that may cause damage to the cable assembly. This protection is accomplished by using a sleeve of rigid metal conduit, intermediate metal conduit, schedule 80 PVC conduit (sunlight-resistant if installed where exposed to the sun), electrical metallic tubing (EMT), reinforced thermosetting resin conduit (RTRC), or by any other approved means [*NEC* 230.50(B)(1) or IRC E3605.5]. Note: Type SE cable is not permitted to be run through an underground raceway (not listed for a wet location). It is suitable for use in aboveground locations only.

Where a metallic service raceway is used to enclose Type SE service-entrance conductors in this manner, it is required to be grounded. Grounding requirements for metal enclo-

Photo 4.5. Typical Type SE cable with strands of the neutral conductor wrapped spirally around two insulated conductors

Photo 4.6. Typical service disconnecting means installed and located outdoors

sures and service raceways are found in *NEC* 250.80 and IRC E3908.1. To protect the service-entrance conductors from abrasion, fittings, such as an insulated bushing, must be used on the ends of the raceway for any type that is used to enclose and protect a Type SE cable assembly. This fitting is necessary to protect the conductors from abrasion during and after their installation.

A sill plate protects the Type SE assembly at the point where the cable enters the building or basement. This protec-

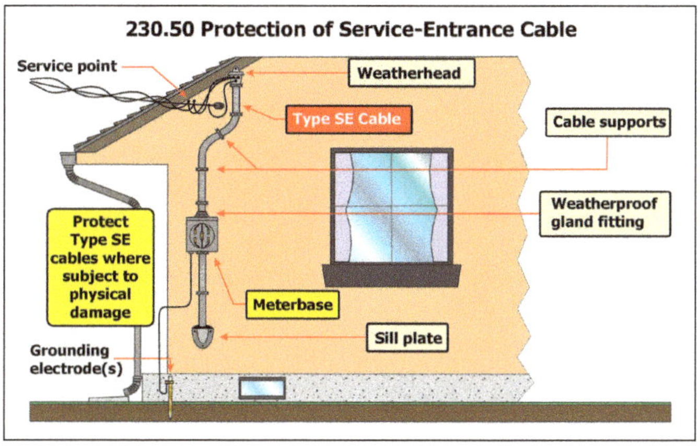

Figure 4.12. Protect Type SE cable where subject to physical damage

Photo 4.7. Service disconnecting means it is required to be readily accessible to occupants (service disconnecting means installed indoors).

tion is necessary where the service entrance cable is subject to tampering. Terrible accidents have occurred where people have damaged service entrance cables while using hacksaws, weed eaters, shovels, etc.

RACEWAYS USED FOR SERVICE-ENTRANCE CONDUCTORS

The wiring methods permitted for service-entrance conductors are found in *NEC* 230.43 or IRC E3801.4 and Table E3801.4. These sections include various wiring methods permitted to enclose service-entrance conductors. Rigid metal conduit (RMC), intermediate metal conduit (IMC), rigid polyvinyl chloride conduit (PVC), and electrical metallic tubing (EMT) are the most common types of raceways used at dwelling units. Electrical nonmetallic tubing (ENT), flexible metal conduit (FMC), liquidtight flexible metal conduit (LFMC), and liquidtight flexible nonmetallic conduit (LFNC) are also permitted to be used to enclose service-entrance conductors under certain conditions.

For flexible metal conduit and liquidtight flexible metal conduit, conditions include that the conduit is not to be longer than 1.8 m (6 ft) between the service-entrance raceways or between the service raceway and service equipment. An equipment bonding jumper or supply-side bonding jumper is required to be routed with the flexible metal and liquidtight flexible metal conduit in accordance with *NEC* 250.102(A), (B), (C), and (E) and IRC E3908.21. Where flexible metal conduit is used on the exterior of a building, the enclosed service-entrance conductors must be suitable for use in a wet location as required in *NEC* 310.10(C) or IRC Table E3801.4.

The installation must be such that water (rain or snow) will not enter into other raceways or enclosures to which the flexible metal conduit is connected. Liquidtight flexible nonmetallic conduit is also permitted to be used to enclose service-entrance conductors. Electrical nonmetallic tubing (ENT) is permitted to be used to enclose service-entrance conductors where it is concealed within walls, floors and ceilings, in dry and damp locations and in wet locations indoors. Where installed exposed, ENT is usually limited to installation inside a building. ENT is not permitted to be exposed to the direct rays of the sun unless marked "sunlight resistant."

Where a service raceway enters a building from an underground distribution system, it is required to be sealed. The seal is intended to keep moisture from entering the building through the raceway. Hazardous gases, such as radon, are also prevented from entering through the raceway if the seal is properly installed. Either or both ends of the raceway may be sealed. The sealing compound must be identified for use with the cable or conductor insulation. Some foam-based sealing compounds have been tested for use in direct contact with insulation and cables. Check the manufacturer's instructions for evidence of testing with electrical products prior to using the sealant (*NEC* 230.8 or IRC E3601.5).

Aboveground raceways are also required to be sealed where the raceway is run from outside the building to inside the building. Either or both ends of the raceway may be sealed. In this case, the intention of the seal is to prevent the circulation of warm air to a cooler area. Condensation and the resultant corrosion are inhibited by proper sealing. Some installers will use an explosionproof seal in these situations, but it is not required here. The sealant should be a material that is identified for use with cable insulation, conductor insulation, a bare conductor, a shield, or other components. Often, inert materials, such as duct seals or silicon caulk, are identified and accepted by the AHJ for this particular use [*NEC* 300.7(A) or IRC E3802.7].

SERVICE EQUIPMENT AND MAIN DISCONNECTING MEANS – LOCATION, SIZES, AND TYPES

All one- and two-family dwelling services must provide a means to disconnect all conductors installed in a building or other structure from the service-entrance conductors. Service disconnecting means must be placed in a readily accessible location. One- and two-family dwellings are required to have an emergency disconnecting means, which could include the service disconnecting means, installed in a readily accessible location on the outside of the structure. This emergency disconnect will be explained in more detail later in this chapter. The service disconnect may be located either outside or inside the building.

If located inside the building, the disconnecting means has to be located "nearest the point of entrance" from where the service conductors enter the building. The term *nearest the point of entrance* is not defined by the *Code*, and it is left to the judgment of the AHJ to determine this inside length. This section leaves room for judgment and interpretation by enforcement officials to accommodate a variety of conditions that may be encountered from one job site to another. Service-entrance conductors, which typically do not have overcurrent protection ahead of them, are allowed to enter the building or structure but are required to terminate in the service disconnect as soon as possible. This limits the length of the unprotected or unfused conductors within the building.

Service conductors are not considered to be protected from short circuits and ground faults. The only protection for these service conductors is usually located on the primary side of the serving utility transformer. Some jurisdictions place a length limitation that defines this "*nearest the point of entrance*" rule, but again, no length is specified in the *Code*. Care should be exercised here not to allow any more length than is necessary where the service-entrance conductors must enter the building [see *NEC* 230.70(A)(1) or IRC E3601.6.2].

At dwelling units, the service conductors are considered outside the building under any of the following conditions (*NEC* 230.6):

1. The conductors are installed in a suitable raceway under not less than 50 mm (2 in.) of concrete beneath a building or structure.
2. The conductors are installed in the building, but the raceway or cable is encased in not less than 50 mm (2 in.) of concrete or brick.
3. Where installed in any vault that meets the construction requirements of Article 450, Part III *(this provision would typically not apply to dwelling units)*.

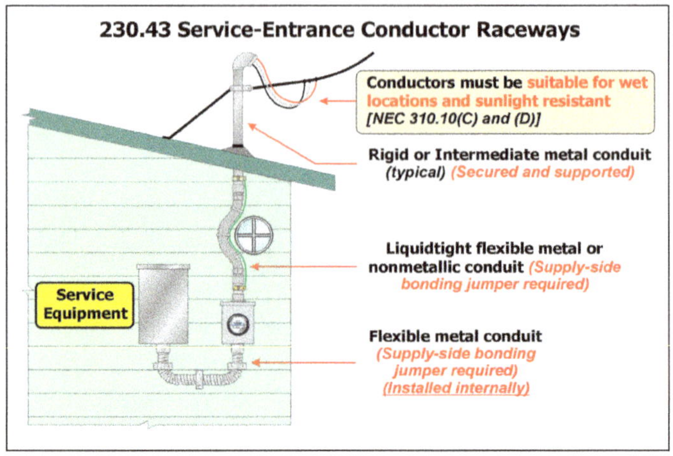

Figure 4.13. Service-entrance conductor raceways

4. The conductors are installed in conduit, and the conduit is under not less than 450 mm (18 in.) of earth beneath the building or structure.
5. Where installed within a rigid metal conduit (RMC) or intermediate metal conduit (IMC) used to accommodate the clearance requirements in *NEC* 230.24 (IRC E3604.2) and routed directly through an eave but not a wall of a building.

Note: Type SE cable is not permitted to be run through an underground raceway as it is not listed and identified as a wet location. It is suitable for use in aboveground locations only.

The disconnecting means is required to be installed where it will be readily accessible. The term *readily accessible* is defined in *NEC* Article 100 and IRC Chapter 35. It requires that the disconnecting means must be capable of being reached quickly so it may be operated, renewed, or inspected without having to use tools *(other than keys)*, climb over or under or remove obstacles, or to use portable ladders and so forth.

One- and two-family dwellings may have a service disconnecting means installed as part of the meter socket enclosure and located outside the structure. Other common locations for the service disconnecting means are in a basement or a garage on the opposite side of the wall to which the meter base is mounted. This interior location must always be nearest the point of entrance of the service-entrance conductors into the building. This phrase has been the subject of varying interpretations for many years. It should be discussed with the local electrical inspector or agency and electric supply company before the job is started.

In some cases, the local rules require the disconnecting means to be located outside the building no matter what the situation. Beginning with the 2020 *NEC* (2021 IRC), requirements were put in place demanding that an emergency disconnecting means *(which could include the service*

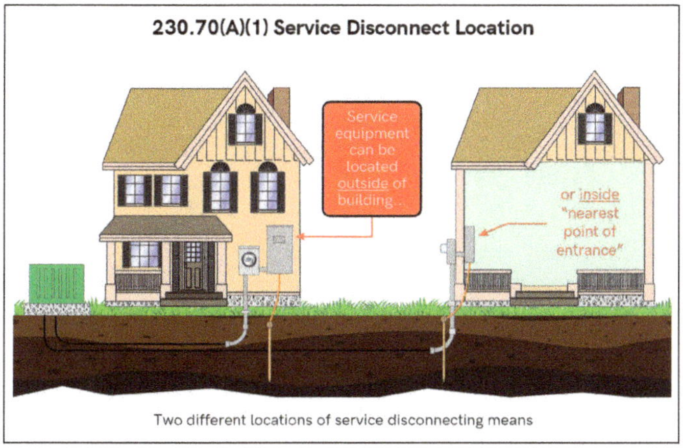

Figure 4.14. Service disconnecting means location outside the building or inside nearest the point of entry for the service conductors

Figure 4.15. Service disconnecting means or overcurrent devices are not permitted in bathrooms, including dwelling unit bathrooms

disconnecting means) for a one- or two-family dwelling be installed in a readily accessible location and located on the outside of the structure. This will be discussed in greater detail later in this chapter (see *NEC* 230.85 or IRC E3601.8).

Service disconnecting means are not permitted to be installed in bathrooms [*NEC* 230.70(A)(2) or IRC E3601.6.2]. Any type of overcurrent device *(not just service overcurrent devices)* other than supplementary overcurrent protection is not permitted to be installed in a bathroom [*NEC* 240.24(E) or IRC E3705.7(4)]. The term bathroom is defined in *NEC* Article 100 and IRC Chapter 35.

The *Code* requires overcurrent devices to be installed in readily accessible locations to the occupant of a one- or two-family dwelling. For obvious reasons, overcurrent devices are restricted from being located in the vicinity of easily ignitible material, such as in clothes closets. Overcurrent devices are also not permitted to be located over the steps of a stairway. As previously stated, the *Code* does not permit overcurrent devices to be located in bathrooms, including dwelling unit bathrooms. There is an inclusion for supplementary overcurrent protective devices.

Supplementary overcurrent devices may be installed in bathrooms. *NEC* Article 100 defines a *Supplementary Overcurrent Protective Device* as "a device intended to provide limited overcurrent protection for specific applications and utilization equipment such as luminaires and appliances. This limited protection is in addition to the protection provided in the required branch circuit by the branch-circuit overcurrent protective device." This type of supplementary protection is sometimes installed for the protection of motors, luminaires, and water heaters (see *NEC* 240.24 or IRC E3705.7).

MARKING OF SERVICE DISCONNECTING MEANS

Service equipment is required to be marked to identify it as being suitable for use as service equipment [*NEC* 230.66 or IRC E3606.4]. This marking is provided by the manufacturer of the equipment and is often a part of the information on the label affixed inside the cabinet. The marking is usually "SUITABLE FOR USE AS SERVICE EQUIPMENT," which indicates the equipment complies with the rules for disconnection of the premises wiring from the service conductors and is suitable for making grounding connections, as well as bonding the neutral conductor to the enclosure. This equipment is suitable for use as feeder panelboards or subpanels where the grounded (neutral) bar can be isolated electrically from the enclosure.

Some equipment is marked as "SUITABLE FOR USE ONLY AS SERVICE EQUIPMENT." This marking indicates that the grounded (neutral) terminal bar is bonded to the enclosure at the factory. No method exists to remove this bonding connection in the field. Therefore, this equipment cannot be used as a feeder panelboard or subpanel. In most cases, the Code does not allow the grounded (neutral) conductor to be connected to the metal parts that normally do not carry current, to equipment grounding conductors, or to be reconnected to the ground beyond the service equipment [NEC 250.24(B), 250.142(B) or IRC E3607.2].

Similar markings, such as "SUITABLE FOR USE AS SERVICE EQUIPMENT WHEN NOT MORE THAN SIX MAINS ARE INSTALLED," may also be found. This marking may appear on main lug panels without a factory-installed main breaker. A main breaker or up to six main disconnects are allowed to be installed in the field. It is very important to review the manufacturer's installation instructions before using a main lug panel as service equipment.

Often, such panels are only permitted when a main circuit breaker is installed and back-fed.

Back-fed means the supply conductors are connected to what are usually the load terminals of the breaker. Where a circuit breaker is installed in a panelboard and is back-fed, the Code requires that a hold-down mechanism be installed to secure the breaker to the enclosure. This rule also applies to field-installed main lug assemblies. The fastener must secure the breaker so that a pull on the breaker will not dislodge it. This prevents the inadvertent removal of a circuit breaker while the supply conductors to the breaker are energized. These hold-down devices are available as an accessory from the manufacturer of the circuit breaker and panelboard. [NEC 408.36(D) or IRC E3706.5]

The service disconnecting means must be permanently marked to identify it as a service disconnecting means [*NEC* 230.70(B) or IRC E3601.6.1]. Often, this marking is in the form of an adhesive label with the word "MAIN" or the words "SERVICE DISCONNECT" on it. See photo 4.8 for a typical label used to indicate the "SERVICE DISCONNECT." The service disconnecting means must be identified for the prevailing conditions. One example is a panelboard marked "Type 3R" or "Rainproof" where installed in an outdoor location [*NEC* 230.70(C)]. The rating of the panelboard is maintained by using conduit and cable fittings listed for wet locations where the raceway or cable enters the enclosure above the level of live parts (*NEC* 312.2 or IRC E3907.2). A service disconnecting means or service-rated panelboard marked "Type 1" or "Indoor Use Only" may not be used outdoors. Electrical enclosure types (such as "3R") and degrees of protection can be explained and found in *NEC* Table 110.28 or IRC Table E3404.4.

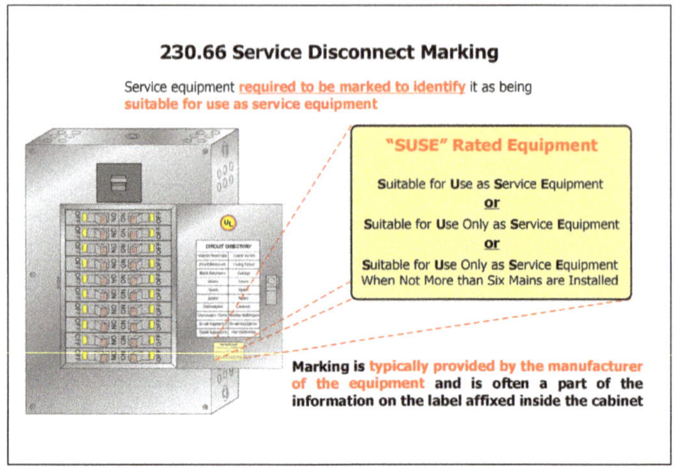

Figure 4.16. Service equipment is required to be marked as suitable for the use

Photo 4.8. Service disconnect must be clearly identified as the "service disconnect."

OUTSIDE EMERGENCY DISCONNECT(S) FOR DWELLING UNITS

Beginning with the 2023 NEC (2024 IRC), an emergency disconnecting means (which could include the service disconnecting means) for one- or two-family dwelling units was initiated. These emergency disconnects are to be installed and located on the outside of the structure in a readily accessible location on or within sight of the dwelling unit. This disconnecting means must have a short-circuit current rating equal to or greater than the available fault current. If more than one emergency disconnecting means is provided, these disconnecting means must be grouped together.

This required outdoor emergency disconnect can consist of the service disconnect(s), a properly marked meter disconnect(s), or other listed disconnect switches or circuit breakers on the supply side of each service disconnect that are suitable for use as service equipment. All of these options must be properly marked to indicate that they are service disconnects, emergency disconnects, etc., with the marking in compliance with 110.21(B). This emergency disconnecting means marking or labels must be located on the outside front of the disconnect enclosure with a red background and white text. The letters must be at least 13 mm (½ in.) high. See *NEC* 230.85(E) or IRC E3601.8 and Figure 4.17 for specific marking requirements for the emergency disconnect(s).

Where equipment for isolation of other energy source systems, such as photovoltaic (PV) power sources, emergency generators, etc., are present at the dwelling unit, and these disconnects are not located adjacent to the emergency disconnect(s), a plaque or directory identifying the location of all equipment for isolation of other energy sources is required to be located adjacent to the emergency disconnecting means

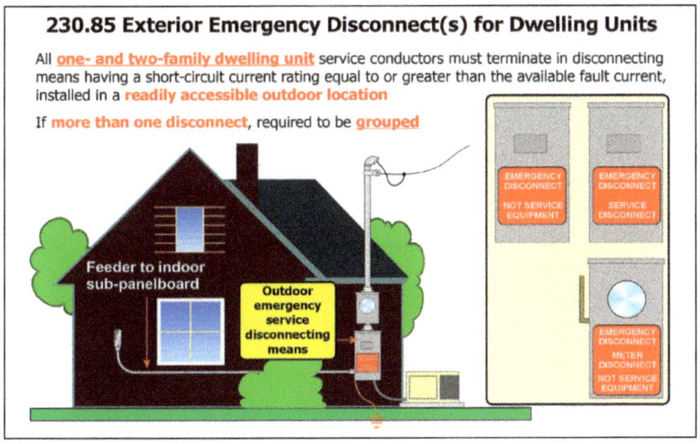

Figure 4.17. Outside emergency disconnect(s) required for one- and two-family dwellings

[*NEC* 230.85(D)].

Where service equipment is replaced at existing dwelling units, this emergency disconnecting means must also be applied. At an existing dwelling unit where only meter sockets, service entrance conductors, or related raceways and fittings are replaced, the emergency disconnect(s) requirements do not apply [*NEC* 230.85(C)].

This requirement for an outdoor emergency disconnecting means at one- and two-family dwellings primarily aims to provide first responders with an accessible outdoor emergency or service disconnect in situations such as a fire, gas leak, structural damage, or flooding. Access to the service disconnect can be very difficult for fire department personnel and first responders when it is installed indoors, such as in a basement. Furthermore, with the increased use and availability of on-site power generation, like photovoltaic (PV) systems or energy storage devices, not having easy access to safely disconnect all electrical power sources in a residence becomes even more challenging.

This new requirement will improve the safety of emergency responders. The cost of a safer environment should never be a reason to avoid implementing NEC rules and regulations. Installing an external, easily accessible disconnect allows for the safe cutoff of electrical power from outside the structure. A simple and accessible method to quickly remove electrical power is needed at dwelling units to keep firefighters and first responders safe while they work to extinguish flames and potentially save lives. They should not have to worry about shock hazards while performing their duties.

SURGE PROTECTION FOR DWELLING UNIT SERVICES

Requirements for surge protection devices (SPDs) at dwelling units were established starting with the 2023 NEC at 230.67 (IRC E3606.5). In the 2023 update, CMP-10 amended subdivision 230.67(A) by replacing the term "dwelling units" with "additional occupancies" and added a list of other locations that now need protection by a surge-protective device (SPD). These additional locations include dormitory units, guest rooms and suites in hotels and motels, as well as areas in nursing homes and limited-care facilities used exclusively for patient sleeping rooms.

A surge protection device (SPD) is a device designed to protect electrical equipment from voltage spikes. An SPD attempts to limit the voltage supplied to an electric device by either blocking or shorting to ground any unwanted voltages above a safe threshold. An SPD is defined as a protective device for limiting transient voltages by diverting or limiting surge current; it also prevents the continued flow of follow current while remaining capable of repeating these functions. It is designated in four categories.

- A Type 1 SPD is a permanently connected SPD intended for installation between the secondary of the service transformer and the line side of the service disconnect overcurrent device.
- A Type 2 SPD is a permanently connected SPD intended for installation on the load side of the service disconnect overcurrent device, including SPDs located at the branch panel.
- A Type 3 SPD is installed at the point of utilization.
- A Type 4 SPD is a component SPD, encompassing both discrete components and assemblies.

This SPD requirement mandates that all services supplying power to dwelling units include a surge-protective device (SPD). This residential SPD must be an integral part of the service equipment or installed immediately adjacent to it, unless it is provided at each subsequent level of distribution equipment downstream toward the load. The SPD must be either a Type 1 or Type 2 SPD. These SPDs are required to have a nominal discharge current rating of at least 10kA. Additionally, an SPD must be installed whenever residential service equipment is replaced.

This new SPD requirement aligns with the recognized need for surge protection to safeguard sensitive electronic systems found in most appliances and equipment used in today's modern homes. Furthermore, the increasing use of distributed energy resources (DER) within electrical systems often leads to greater opportunities or exposure for surges into the system. DER refers to smaller generation units located on the consumer's side of the service point. Examples of DERs include rooftop solar photovoltaic units, wind turbines, battery storage systems, batteries in electric vehicles used to export power back to the grid, and fuel cells.

Electronic life-saving equipment, such as fire alarm

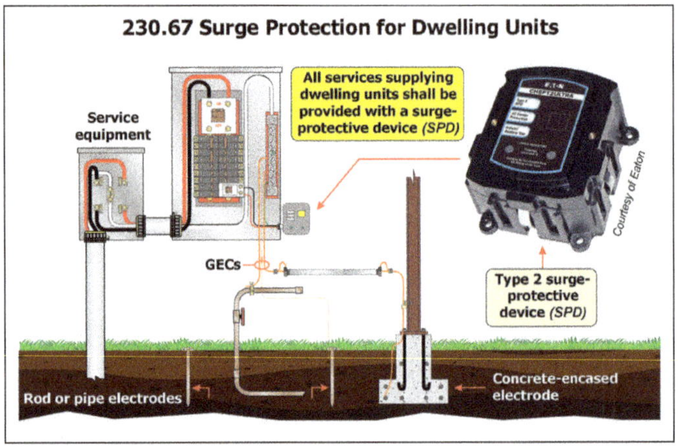

Figure 4.18. Surge protection is required on all services at dwelling units.

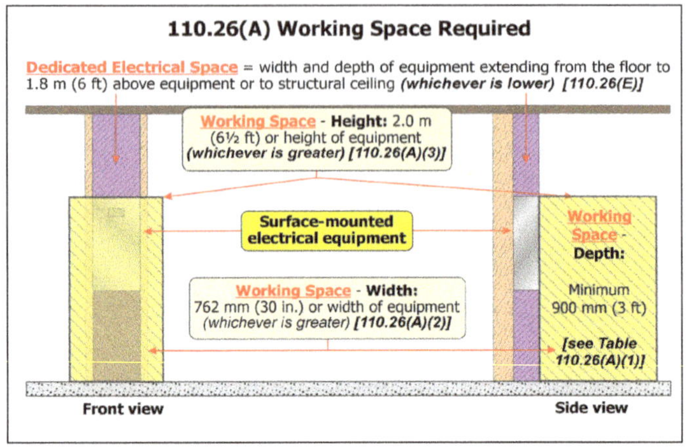

Figure 4.19. Working space and dedicated electrical space at service equipment

systems, GFCIs, AFCIs, and smoke alarms, can be damaged during a surge caused by events like lightning. In many cases, electronic devices and equipment may be harmed and become inoperable by a surge, but this damage often goes unnoticed by the homeowner. It makes sense to require SPDs at dwelling units to provide a basic level of surge protection.

WORKING SPACE AT SERVICE EQUIPMENT

Ample working space is necessary near service equipment to ensure safe performance of repairs, operation, or inspection. This chapter focuses on service equipment; however, working space is essential for any electrical equipment (not just service equipment) that is likely to require examination, adjustment, servicing, or maintenance while energized. The required indoor working space is specifically detailed in NEC 110.26 and Table 110.26(A)(1) or IRC E3405.2. A minimum width of 762 mm (30 inches), or the width of the equipment, whichever is greater, is required in front of the electrical equipment. The working space must be clear and extend from the floor or platform up to 2.0 m (6½ ft), or at least to the height of the equipment. Additionally, the workspace must allow doors on equipment to open at least 90 degrees, and open doors should not block entry or exit from the required working space.

This space is considered obstructed if the distance between the equipment's open door(s) and the opposite wall, etc., is less than 610 mm (24 in.) wide and 2.0 m (6½ ft) high. The grade, floor, or platform within the required working space must be kept clear, and the floor, grade, or platform should be as level and flat as practical throughout the entire required depth and width of the working space. This prevents a concrete mounting pad from extending out in front of the electrical equipment and then dropping off 100 mm (4 in.), 125 mm (5 in.), 150 mm (6 in.), etc., in front of the service equipment. This situation is uncommon in dwelling units.

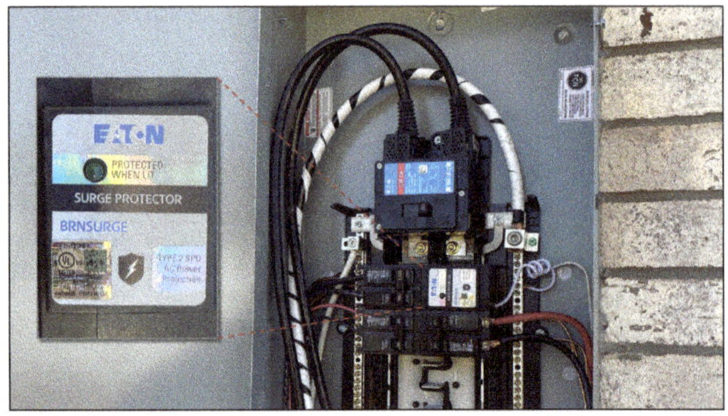

Photo 4.9. Whole-house surge protection installed at a dwelling unit

Equipment within the height limits related to the electrical installation is allowed to extend up to 150 mm (6 in.) beyond the front of the service equipment. Associated equipment may be installed either above or below the service equipment.

For example, a current transformer cabinet is 225 mm (9 in.) deep. If it is mounted below an enclosed panelboard that is 100 mm (4 in.) deep, the CT enclosure extends 125 mm (5 in.) into the working space of the panelboard, which is acceptable. A good way to describe the required working space is to imagine a large cardboard box—like a refrigerator box—with dimensions that match the working space. It should be able to sit on the floor or a working platform and slide right up to within 150 mm (6 in.) of the front face of the panelboard or service equipment. This rule applies to all electrical equipment that may need inspection, adjustment, servicing, or maintenance while energized. It is always best practice—and often required by law—to work on electrical equipment only when it is de-energized.

Table 110.26(A)(1) Working Spaces

Nominal Voltage to Ground	Minimum Clear Distance		
	Condition 1	Condition 2	Condition 3
0-150	900 mm (3 ft)	900 mm (3 ft)	900 mm (3 ft)
151-600	900 mm (3 ft)	1.0 m (3½ ft)	1.2 m (4 ft)
601-1000	900 mm (3 ft)	1.2 m (4 ft)	1.5 m (5 ft)

Condition 1 - Exposed live parts one side, no live or grounded parts, or insulated or ungrounded parts other side

Condition 2 - Exposed live parts ones side, grounded parts on the other side *(brick wall, etc.)*

Condition 3 - Exposed live parts on both sides of the work space

The requirements in Conditions 1, 2, and 3 to Table 110.26(A)(1) must be applied to establish the minimum depth of working space required in front of electrical equipment

Figure 4.20. Working space clearances as provided by *NEC* Table 110.26(A)(1)

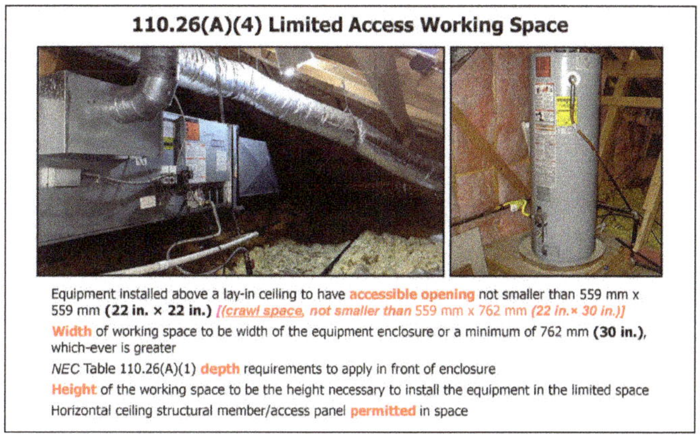

Figure 4.22. Equipment in attics and crawl spaces required to comply with "limited access" working space requirements

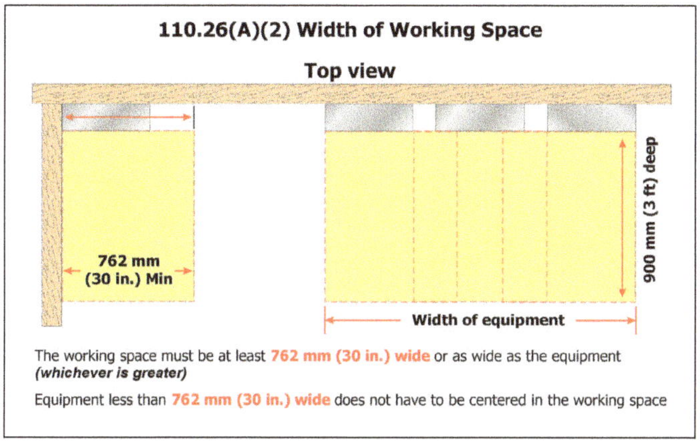

Figure 4.21. Width of working space at service equipment

Photo 4.10. Minimum working space is required for electrical equipment (including panelboard, meter base, AC units, etc.) (Note: Landscaping impeding into required working space)

At dwelling units with a typical 120/240 volt, single-phase, 3-wire service, the minimum depth of the working space required in the direction of access to the service equipment generally must be at least 900 mm (3 ft). This depth requirement is derived from NEC Table 110.26(A)(1) or IRC E3405.2 and could be more than 900 mm (3 ft) depending on the conditions and the voltages involved. This space must be clear from the floor or working platform to a height of not less than 2.0 m (6½ ft) or the height of the equipment, whichever is higher. As mentioned earlier, some have described this clear working space as an imaginary (empty) refrigerator box with the dimensions of the required working space placed on the floor correctly in front of the equipment.

HEIGHT OF WORKING SPACE

A headroom of not less than 2.0 m (6½ ft) or the height of the equipment, whichever is higher, is required [*NEC* 110.26(A)(3) or IRC E3405.2]. An exception to this rule permits headroom to be less than 2.0 m (6½ ft) for service equipment or panelboards that do not exceed 200 amperes in existing dwelling units. This exception permits the replacement of service equipment in older homes with less than the minimum headroom required by modern *Code* rules.

LIMITED ACCESS WORKING SPACE

The Code also addresses electrical equipment located in spaces with "limited access," such as above suspended ceilings, attics, or crawl spaces. Prior to the 2017 edition of the NEC, strict compliance with these working space rules in attics and crawl spaces was not feasible and, in some cases, not possible. The electrical industry has been installing and accepting equipment in such spaces for decades. Prescriptive requirements for such limited-access spaces were sorely lacking for that same time period. This limited access provision aims to provide relief for both the installer and the enforcement community. While residential construction does not typically lend itself to provisions where limited access to working spaces is the norm, such situations can occur at dwelling units.

Regarding specific requirements, the NEC does not specify a minimum size for access openings to attics. Electrical equipment installed above a lay-in ceiling must be accessible through an opening at least 559 mm x 559 mm (22 in. x 22 in.), which aligns with a standard 2 ft x 2 ft lay-in ceiling grid opening. The dimensions for a crawl space opening [559 mm x 762 mm (22 in. x 30 in.)] correspond with the relevant building codes and the spacing between standard framing members. The width of the limited-access working space must be either the width of the enclosure or a minimum of 762 mm (30 in.), whichever is greater, similar to normal working space requirements.

All doors or hinged panels must be capable of opening a minimum of 90 degrees, and the space in front of the enclosure must comply with the depth requirements of *NEC* Table 110.26(A)(1) and must be unobstructed to the floor by fixed cabinets, walls, or partitions. *NEC* 110.26(A)(1)(b) allows smaller working spaces where all exposed live parts operate at not greater than 30 volts rms, 42 volts peak, or 60 volts dc. Notice that the last list item at *NEC* 110.26(A)(4) provides relief from the general working space height requirement and permits the height of the limited access working space to be the maximum height necessary to install the equipment in the limited space.

Horizontal ceiling grid structural members or removable access panels are allowed in this space, provided that the location of any weight-bearing structural member does not result in a side reach of more than 150 mm (6 in.) while working within the enclosure. The NEC addresses limited access working space for electrical equipment at NEC 110.26(A)(4). Refer to IRC E3405.2 for the limited access working space requirements outlined by the IRC.

The IRC does address attics that contain appliances (mechanical equipment). Attics containing appliances must have an opening and a clear, unobstructed passageway that is large enough to allow the removal of the largest appliance, but not less than 762 mm (30 in.) high and 559 mm (22 in.) wide. The clear access attic opening dimensions must be at least 508 mm by 762 mm (20 in. by 30 in.) and large enough to accommodate the removal of the largest appliance (see IRC M1305.1.2).

DEDICATED ELECTRICAL SPACE —INDOORS

All indoor panelboards installed at dwelling units are required to be located in dedicated electrical spaces and protected from damage. This space is defined as the signature of the equipment, which is equal to the width and depth of the equipment and extends from the floor to a height of 1.8 m (6 ft) above the equipment or to the structural ceiling, whichever is lower. Plumbing pipes, mechanical ducts, or any other equipment foreign to the electrical system are not

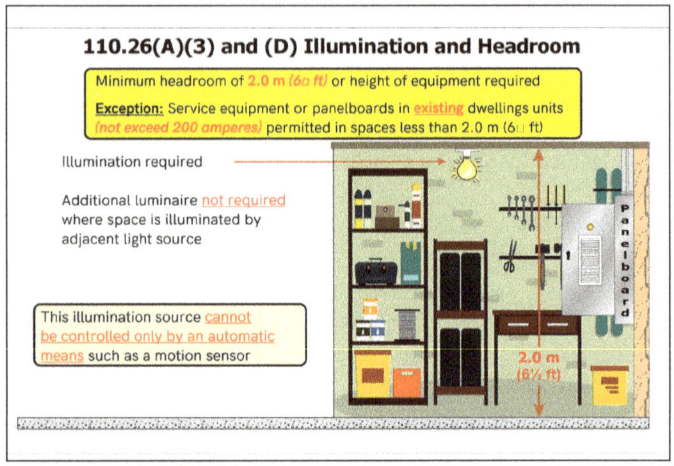

Figure 4.23. Illumination and headroom are required for service equipment.

permitted to penetrate this dedicated electrical zone or space (see purple shaded area of Figure 4.19). [*NEC* 110.26(E)(1) or IRC E3405.3].

DEDICATED ELECTRICAL SPACE—OUTDOORS

This basic dedicated equipment or electrical space is described indoors, but dedicated electrical space is also needed for outdoor installations of panelboards and electrical devices. With outdoor setups, some of the same "equipment foreign to the electrical installation" is often present, such as gas lines, water lines, irrigation equipment, phone and internet equipment, and other items not related to the electrical service. These foreign systems are typically installed outdoors above or below the electrical equipment. When placed in the outdoor dedicated electrical space, these foreign objects can block access to the electrical equipment. This issue occurs whether the electrical equipment is installed indoors or outdoors. [NEC 110.26(E)(2) or IRC E3405.4].

ILLUMINATION OF WORKING SPACE

Lighting of indoor work areas is necessary. While no minimum foot-candle level is specified, enough illumination must be provided to work safely on equipment. It is the responsibility of the Authority Having Jurisdiction (AHJ) to determine if the space is properly lit. The light source cannot be controlled solely by automatic means, such as a motion sensor, without a manual override. Additional lights dedicated to the workspace are not needed if an adjacent light source provides adequate illumination. For example, in a residential garage, if the installed lights also illuminate the service equipment workspace, this requirement is satisfied [NEC 110.26(D) or IRC E3405.7].

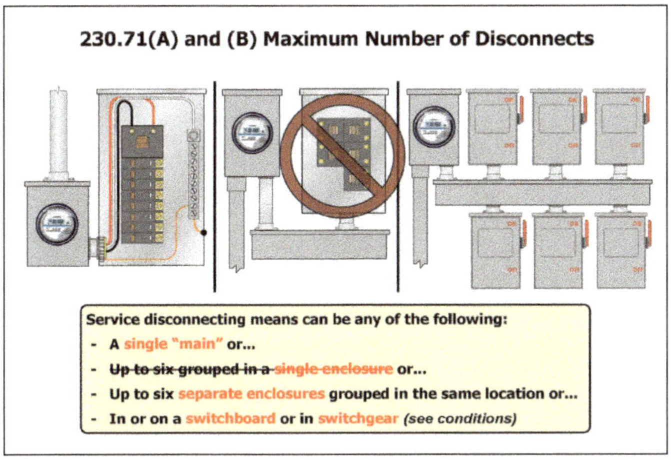

Figure 4.24. Number of service disconnecting means and grouping requirements

MAXIMUM NUMBER AND GROUPING OF DISCONNECTS

Generally, buildings or structures are permitted to have only one service. Several conditions exist at NEC 230.2 that would allow the AHJ to permit more than one service per building or structure, but typically, dwelling units are allowed only to have one service. The IRC restricts dwelling units to only one service at IRC E3601.2. A service disconnecting means is required for each service permitted by the Code. The service disconnecting means can be either a single disconnect or it is permitted to consist of not more than six switches or six circuit breakers mounted in a group of separate enclosures. No more than six service disconnects per service may be grouped in any one location. From a historical perspective, up until the 2020 NEC (2021 IRC), the Code allowed up to six means of disconnect to be located in a single enclosure or up to six separate enclosures. Permission from a Code perspective to allow these two to six means of disconnect in a single enclosure is no longer allowed [*NEC* 230.71(B) or IRC E3601.7].

Transfer switches were added at 230.71(B)(4) in the 2023 *NEC* to clarify that they must be listed for and used as service equipment. Each service disconnect is to be provided in a separate compartment. Section 230.71(B)(6) was added for motor control centers used as service equipment to limit such equipment to a maximum of two service disconnects per single motor control center with barriers required between each unit or compartment containing a service disconnect.

The probation of up to six means of disconnect to be located in a single enclosure takes into consideration the challenges created for electrical workers when encountering a panelboard with more than one service disconnecting means in the same enclosure. The six means of disconnect rule for a single enclosure made it impossible to work in service equipment when applying electrical safe work practices in accordance with NFPA 70E, *Standard for Electrical Safety in the Workplace*.

The application of an incident energy analysis for such an installation typically employs a two-second clearing time due to exposure on the line side of the multiple service disconnects, resulting in labels that prohibit justified energized work. A single service disconnect within service equipment facilitates the creation of an electrically safe work condition by opening the single service disconnect, where barriers are in place over the exposed energized conductors or terminals, to eliminate any live exposed parts.

Barriers must be provided to prevent contact with live exposed parts on switchboards, switchgear, and panelboards according to NEC 230.62(C). New wording was introduced to 230.62(C) in the 2023 NEC to clarify that barriers are required in service equipment so that no uninsulated, ungrounded busbars or terminals are exposed to accidental contact while load terminations are being serviced when the service disconnect is in the open position.

The IRC does not currently have this barrier requirement for panelboards or service equipment, so a default to the *NEC* is in order [IRC E3401.2]

A single service disconnect within service equipment facilitates the creation of an electrically safe work condition by opening the single service disconnect, where barriers are in place over the exposed energized conductors or terminals, to eliminate any live exposed parts. Barriers must be provided to eliminate live exposed parts on switchboards, switchgear, and panelboards in accordance with NEC 230.62(C). The IRC does not currently have this barrier requirement for panelboards or service equipment, so a default to the NEC is in order [IRC E3401.2]

As indicated, up to six service disconnects are permitted to be installed in up to six separate enclosures. A single main disconnect in a panelboard that is rated 100-, 150-, or 200-amperes is the most common configuration at dwelling units. The *Code* generally requires a panelboard to be protected by an overcurrent protective device with a rating not greater than that of the panelboard. This overcurrent protective device can be located within or at any point on the supply side of the panelboard. There are two exceptions to this general rule (*NEC* 408.36 or IRC E3706.3). Services between the 200- and 400-ampere ratings are commonly installed with two main breaker panelboards or two fusible switches or some combination thereof. Some electric utility or local electrical rules may require a single main disconnect. Service disconnects must be grouped at the same location. For example, in most cases, installing one service disconnect outside the dwelling and another inside is not acceptable. All have to be at the same location. This is an important safety provision for the occupant and emergency first responders at the property. Grouping the disconnecting means at one location facilitates disconnecting the building during an emergency.

Common area circuits are covered in the NEC but not mentioned in the IRC. It is uncommon for a "house" panel or common area circuits to be used in single-family homes. A house panel is more typical in a two-family dwelling where shared loads, such as a well pump or on-site waste pump, serve both occupants. Common area circuits are often used in multifamily dwellings for corridor lighting, among other things. According to IRC E3401.2, if common area circuits are present in two-family or multifamily dwelling units, the NEC rules must be followed.

NEC 230.40 Exception No. 1 permits buildings with more than one occupancy to have one set of service-entrance conductors run to each occupancy. The location of the service disconnecting means for each occupancy must be in accordance with the location provisions found in *NEC* 230.70(A) or IRC E3601.6.2. A two-family or multifamily dwelling shall be permitted to have one set of service-entrance conductors installed to supply any common area circuits covered in *NEC* 210.25(B). This allowance reflects the current practice of installing a "house panel" to supply common area branch circuits. Such circuits cannot be supplied from an individual dwelling unit panelboard.

The house panel contains the branch circuits that supply loads, such as the fire alarm system, security systems, public area lighting, shared utilization equipment (e.g., a furnace, well, or sump pump), or shared appliances (e.g., a washer or dryer). The service disconnect that supplies the house panelboard would not necessarily have to be grouped with any other service disconnect. The house panels must be readily accessible and located either outside or inside the building nearest the point of entrance of the service-entrance conductors.

SERVICE FOR OUTBUILDINGS ON THE SAME PREMISES

Several factors should be considered when deciding how to supply electrical power to an outbuilding located on the same site as the dwelling. One option is to install a single service for a dwelling and a remote building, such as a garage or workshop. In this case, two service disconnects can be installed at a common location, and a feeder can be run to each building. The service disconnects are grouped at a single common location, either inside or outside one of the buildings.

Another option is to run a set of service-entrance conductors from a common outside location to the second building. The common location might be an enclosure, such as a wireway or meter-socket enclosure. As indicated, a single service drop or lateral is allowed to supply either or both buildings. The service-entrance conductors must stay outside the buildings or structures. The service disconnecting means for each set of service-entrance conductors at any remote individual building or structure must be located either outside the building or inside, close to where the service conductors enter. The Code does not require a single disconnecting means for both buildings. However, a separate service disconnect must be installed for the conductors supplying each building (NEC 230.40 Exception No. 3).

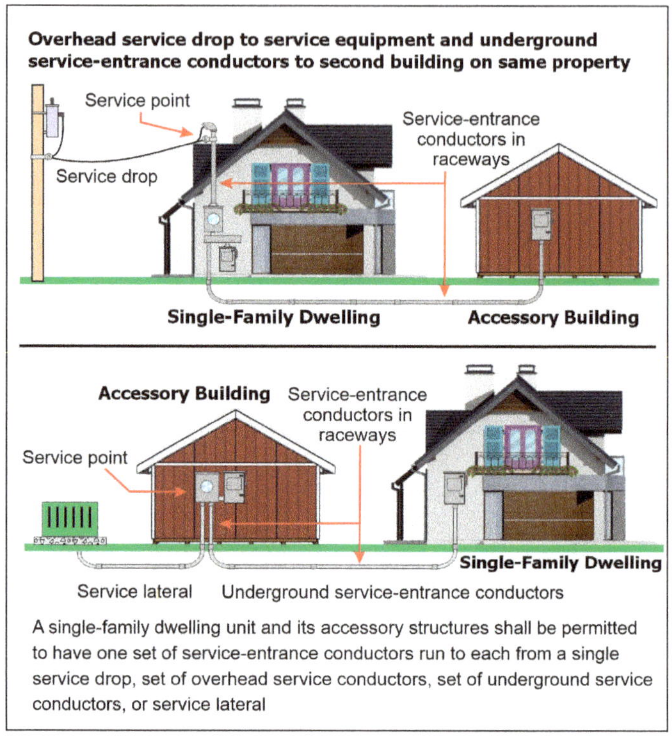

Figure 4.25. Service-entrance conductors to separate building

This is a common application in residential installations with multiple buildings on the same property. These provisions are permitted to be applied to both new and existing installations alike. In many cases, the owner does not want to request an individual service and utility meter for each building. The servicing utility normally charges a flat fee for each service and a monthly charge for each meter, even if no electricity is consumed.

Connections to the service-entrance conductors or service lateral must be made properly. Power distribution blocks, pressure connectors, and devices for splices and taps must be listed. Power distribution blocks on service conductors should be marked "suitable for use on the line side of the service equipment" or an equivalent indication. These connec-

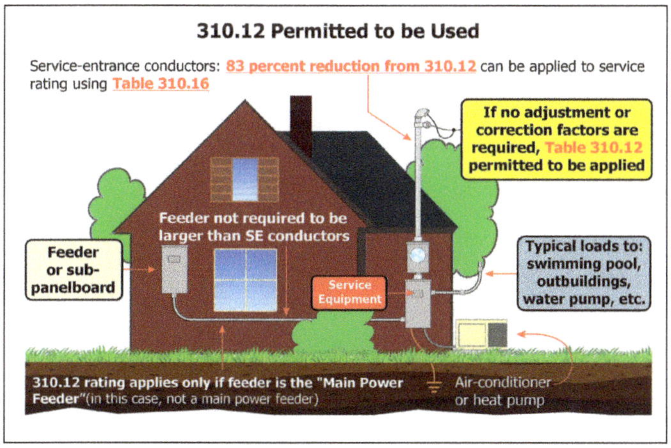

Figure 4.26. Size of service-entrance conductors and main power feeder

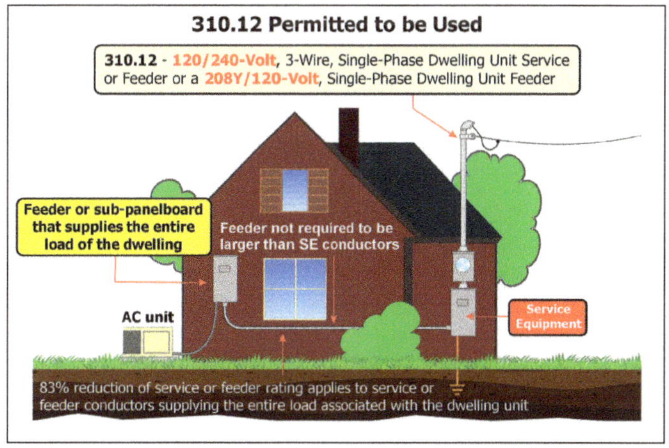

Figure 4.27. The main power feeder supplies all branch circuits, feeders, or both associated with the dwelling unit loads

tions can be made in a wireway, pull box, or inside the meter enclosure where space allows. Some meter sockets have lugs suitable for two conductors, which is ideal for making these connections (NEC 230.46 or IRC E3605.3). Chapter 15 details specific rules for required disconnects, grounding, and bonding requirements, especially when separate buildings or outbuildings are supplied by feeders or branch circuits.

SIZING OF SERVICE-ENTRANCE CONDUCTORS FOR DWELLING UNITS

A couple of methods can be used to determine the minimum rating of service-entrance conductors. The first is to perform a load calculation in accordance with *NEC* Article 220 or IRC E3602.2 and Table E3602.2. We reviewed several methods for performing load calculations in Chapter 2 of this publication. Another way of determining the minimum rating is in accordance with *NEC* 230.42(B), 230.79(C), or IRC E3602.1, where a minimum ampacity of 100 amperes is required for a one-family dwelling. This is due to the reference in IRC E3602.3 as to the rating of the service-entrance conductors not being less than the rating required for the service. The minimum rating of the service disconnecting means for a one-family dwelling is, therefore, 100 amperes according to *NEC* 230.79(C) or IRC E3602.1.

SPECIAL PROVISIONS PERMITTED FOR SIZING SERVICE-ENTRANCE CONDUCTORS FOR DWELLING UNITS

NEC 310.12 or IRC E3603.1 applies specifically to the service-entrance, service-lateral, and main power feeder conductors for dwelling units served at 120/240 volts, 3-wire, single-phase. The single-phase feeder conductors consisting of 2 ungrounded conductors and the neutral conductor from a 208Y/120-volt system. For a service or feeder rated 100 through 400 amperes, the service or feeder conductors supplying the entire load associated with a one-family dwelling, or the service or feeder conductors supplying the entire load associated with an individual dwelling, unit in a two-family or multifamily dwelling, is permitted to have an ampacity not less than 83 percent of the service or feeder rating.

The service or feeder rating would be derived using the standard ampacity ratings found at *NEC* Table 240.6(A) or IRC Table E3705.6 and the calculated loads involved. The size of these services or main power feeder conductors would then be derived from *NEC* Table 310.16 or IRC Table E3705.1. If no adjustment or correction factors are required, *NEC* Table 310.12 or IRC E3603.1.1 is permitted to be applied.

It should be noted that these sections do not assign a different ampacity for these conductors but rather provide a reduced rating when using the calculated values. Only the specific conductors addressed in the subsections are permitted to be used for this application.

These reduced calculated values are also allowed for use in the feeder conductors that serve as the main power feeder to a dwelling unit. The main power feeder is not defined in the NEC or the IRC. In the current edition of the code, it is understood as the feeder conductors supplying the entire load for a one-family dwelling or the load for an individual unit in a two-family or multifamily dwelling [see NEC 310.12 or IRC E3603.1.2].

To use this 83 percent reduction for the selection of the main power feeder, this feeder must supply *all loads* associated with the dwelling unit, which means there can only be one main feeder per dwelling unit for which this 83 percent value can be applied.

Feeder conductors supplying other buildings, such as detached garages or accessory structures, must be appropriately sized to handle the load. To calculate the minimum ampacity of these conductors, you need to know the loads [NEC 215.2(A)(2)]. Then, NEC Table 310.16 or IRC Table E3705.1 is used to determine the minimum size for these feeder conductors that are not the main power feeder (no 83 percent reduction allowed). For service conductors or a main power feeder, the sizing requirements are the same in both the NEC and the IRC, except for the service size and rating limits set by the IRC. The NEC does not specify a maximum service rating for a one- or two-family dwelling. However, NEC 310.12 is limited to 400 amperes. Any service or main power feeder exceeding these levels must meet all other applicable requirements outlined in NEC Table 310.16.

Where correction or adjustment factors are required by *NEC* 310.15(B)(1) of IRC E3705.2 (ambient temperature correction factors) or *NEC* 310.15(C)(1) or IRC E3705.3 (more than three current-carrying conductors), they are permitted to be applied to the ampacity associated with the temperature rating of these dwelling unit services or feeder conductors. See Example D7 in the Informational Annex D of the *NEC* for an example of how to apply the 83 percent reduction value to dwelling unit service and main power feeder conductors. Two examples are provided in Example D7, one example with no required adjustment or correction factor involved, and the other example shows how to apply the 83 percent reduction value along with a required temperature correction factor. Users of the *Code* will also find a table following Informational Annex D Example D7. This table is the same table that resides at *NEC* Table 310.12 or IRC Table E3603.1.1.

Where *NEC* 310.12 or E3603.1 is used to select the ungrounded (hot) service-entrance conductors, the size of the neutral service-entrance or feeder conductor is required to be no smaller than the maximum imbalance of the load. The neutral is no longer automatically permitted to be not more than two AWG (American Wire Gauge) sizes smaller than the ungrounded conductors. As can be seen in the load calculations in Chapter 2, the neutral load is usually smaller as compared to the load on the ungrounded or hot conductors. A grounded (neutral) conductor that is smaller than the ungrounded conductors is permitted where the calculation of the neutral load permits a smaller size, but the minimum size of the grounded service-entrance conductor (neutral) is not permitted to be smaller than the required grounding electrode conductor specified in either *NEC* Table 250.66 or IRC Table E3603.4.

Sizing of Service Conductors for Dwelling(s) ***[See 310.12]***

With No Required Adjustment or Correction Factors

Service conductors and the main power feeder for certain dwelling units are permitted to be sized in accordance with 310.12

If a 200-ampere service rating is selected, a service conductor is then sized as follows:

200 amperes × 0.83 = 166 amperes per 310.12

If no other adjustments or corrections are required for the installation, then, in accordance with Table 310.16, a **2/0 AWG Cu** ***(175 amperes)*** or a **4/0 AWG Al** ***(180 amperes)*** meets this rating at the 75°C (167°F) column

Note: These calculated sizes correspond to the given sizes in Table 310.12

Figure 4.28. *NEC* Section 310.12 is permitted to be used to size dwelling unit service conductors and the main power feeder

Sizing of Service Conductors for Dwelling(s) ***[See 310.12]***

With Required Temperature Correction Factors

Service conductors and the main power feeder for certain dwelling units are permitted to be sized in accordance with 310.12

If a 200-ampere service rating is selected, a service conductor is then sized as follows:

200 amperes × 0.83 = 166 amperes per 310.12

Conductors to be installed in an ambient temperature of 40°C (104°F) ***[See Table 310.15(B)(1)]***

Using XHHW-2 conductors requires a correction factor of 0.91 to find the minimum conductor ampacity and size:

200 A / .91 = 182 amperes

In accordance with Table 310.16, using the 90°C (194°F) column, a **2/0 AWG Cu** or a **4/0 AWG Al** would be required

Figure 4.29. *NEC* Section 310.12 is permitted to be used along with any required temperature correction factor to size dwelling unit service conductors and the main power feeder

OVERLOAD PROTECTION FOR SERVICE-ENTRANCE CONDUCTORS

In most cases, each ungrounded (hot) service-entrance conductor must be protected from overload by an overcurrent device installed in series with the conductor that will operate during an overload. Each ungrounded service conductor is required to have overload protection [NEC 230.90 or IRC E3603.3]. An overload occurs when equipment operates in excess of its normal, full-load rating or when a conductor exceeds its rated ampacity, which, if it persists for a sufficient length of time, would cause damage or dangerous overheating. A fault, such as a short circuit or ground fault, is not an overload.

This overcurrent protective device must generally be rated or set not higher than the conductor ampacity rating. Unlike branch circuits or feeders, service-entrance conductors are protected from overload at their load end rather than at their

Table 310.12 Single-Phase Dwelling Services and Feeders

For one-family dwellings and the individual dwelling units of two-family and multifamily dwellings, service and feeder conductors supplied by a single-phase, 120/240-volt system shall be permitted to be sized in accordance with 310.12(A) through (D). *[Single-phase feeder conductors consisting of two ungrounded conductors and the neutral conductor from a 208Y/120 volt system permitted to be sized in accordance with 310.12(A) through (C)]*

Service or Feeder Rating *(Amperes)*	Conductor *(AWG or kcmil)*	
	Copper	Aluminum or Copper-Clad Aluminum
100	4	2
110	3	1
125	2	1/0
150	1	2/0
175	1/0	3/0
200	2/0	4/0
225	3/0	250
250	4/0	300
300	250	350
350	350	500
400	400	600

Note: If no adjustment *(number of conductors)* or correction *(temperature)* factors are required, this table shall be permitted to be applied.

Figure 4.30. Reproduction of *NEC* Table 310.12 *[IRC Table E3603.1.1]*

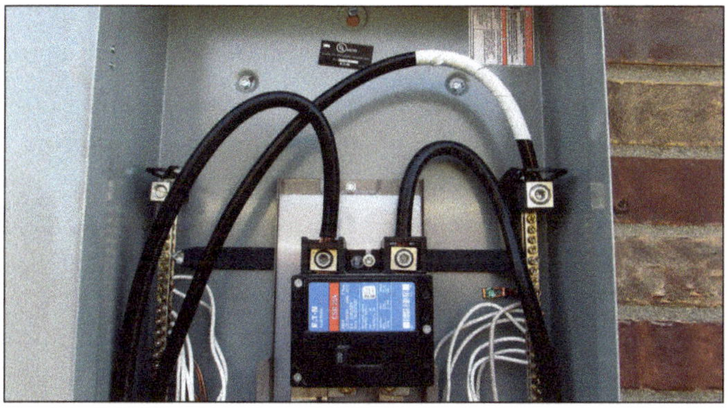

Photo 4.11 Service conductors are required to have overload protection.

230.90 Overload Protection Required

Service-entrance conductors must be provided with overload protection *(overcurrent device at load end)*

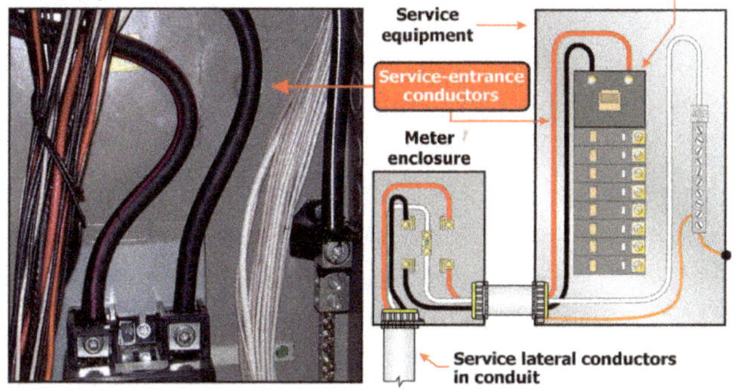

Figure 4.31. Overload protection for service-entrance conductors

supply end, as can be seen in Figure 4.31. Service-entrance conductors are not protected from overcurrents due to short circuits or ground faults. A short circuit occurs when two ungrounded conductors, or an ungrounded and a grounded conductor, come into contact with each other. A ground fault occurs when an ungrounded conductor contacts an enclosing metal raceway or other grounded part.

Short-circuit and ground-fault protection for these faults that occur on the line side of the service overcurrent protection is provided by the fuse on the line side of the electric utility transformer. This transformer line-side fuse is set to protect the transformer and provides little protection for the service-entrance conductors. This is the main reason the *Code* limits the length of service-entrance conductors inside the building.

The ungrounded service-entrance conductors must be protected at their ampacities. *NEC* 240.4(B) or IRC E3705.5.2 permits the next standard-rated overcurrent device to be used, up to 800 amperes, where the ampacity of the conductor does not match the rating of the standard circuit breaker or fuse. An example of this is a 4/0 AWG Type THW aluminum conductor, which has an allowable ampacity in *NEC* Table 310.16 or IRC Table E3705.1 of 180 amperes. It may be protected by the next standard overcurrent device rating of 200 amperes. The noncontinuous load on the conductor is limited to 180 amperes.

The IRC requires the service-entrance conductors to be protected in accordance with the specifics of IRC E3603.1. Because this section requires the minimum size conductor for the service, using the "next higher standard overcurrent protective device rating" is usually not an option, as it sometimes is in accordance with the *NEC*. It should be noted that the IRC permits this for feeders and branch circuits, as indicated in IRC E3705.5.2, but it is more restrictive for services. The exception to IRC E3603.3.1 and *NEC* 230.90(A) Exception No. 3 both permit not more than six circuit breakers or sets of fuses to serve as the overcurrent device to provide the overload protection for the service-entrance conductors. Overload protection in accordance with *NEC* 310.12 or IRC E3603.1 is permitted for single-phase dwelling unit services [*NEC* 230.90(A) Exception No. 3 or IRC E3603.3.1].

PANELBOARDS

Two types of panelboards are commonly used in residential occupancies. One type of panelboard comes with a main circuit breaker or set of fuses. The manufacturer installs the main breaker at the factory. The second type has no main breaker. This type of panelboard is commonly referred to as a "Main Lug Only" (MLO) panelboard. The latter type may be used as the service disconnecting means if marked "suitable for use as service equipment when not more than six main circuit breakers are installed" or similar marking. The construction of the panelboard is such that the number of mains is limited to the maximum number permitted by the *Code* and intended by the manufacturer.

For example, a single-phase panelboard marked "suitable for use as service equipment when not more than six main circuit breakers are installed" will have twelve poles. In other words, the construction will inherently limit the use of a maximum of six two-pole circuit breakers. Keep in mind that the Code prohibits more than one means of service disconnect from being located in a single enclosure [NEC 230.71(B) or IRC E3601.7]. A MLO panelboard being used as the service disconnecting means is only applicable for an existing dwelling unit as permitted by NEC 230.71(B), Ex. to (2), (3), (4), (5), and (6).

Where a main-lug-only (MLO) panelboard is installed, the ungrounded service or feeder conductors are connected directly to the main lugs on the load side rather than to a circuit breaker or set of fuses. The service-entrance conductors are sized for the calculated load, not necessarily for the rating of the panelboard. This (MLO) panelboard will be protected from overcurrent and overload on the supply side of the feeder with a disconnecting means consisting of two main circuit breakers or two sets of fuses having a combined rating not greater than that of the panelboard.

Also, the rating of the service-entrance conductors is not required to equal the sum of the ratings of the circuit breakers or sets of fuses. Normally, only two-pole circuit breakers are installed. The two-pole "mains" supply feeder or sub-panelboards or larger 240-volt loads such as air-conditioning equipment.

OVERCURRENT PROTECTION FOR PANELBOARDS

Panelboards must be protected by an overcurrent protective device rated no higher than the panelboard, according to NEC 408.36 or IRC E3706.3. This device can be located either within or anywhere on the supply side of the panelboard. There are two exceptions to this main rule for individual overcurrent protection.

In previous editions of the Code, individual protection was not required for a panelboard used as service equipment with multiple disconnecting means, such as a panelboard with up to six sets of circuit breakers that was previously permitted by NEC 230.71 or IRC E3601.7. As discussed earlier in this chapter, allowing up to six disconnecting means was permitted to be located in a single enclosure or in up to six separate enclosures.

Individual protection is not required at a panelboard protected on its supply side by two main circuit breakers or two sets of fuses having a combined rating not greater than that of the panelboard. An example of this would be a feeder or sub-panelboard (located in an attached garage) being supplied by a feeder from a service disconnect located on the opposite end of the dwelling. Again, caution must be exercised here because a panelboard supplied in this manner cannot contain more than 42 overcurrent devices (a 2-pole circuit breaker is considered two overcurrent devices). This scenario is the only limitation on the number of overcurrent

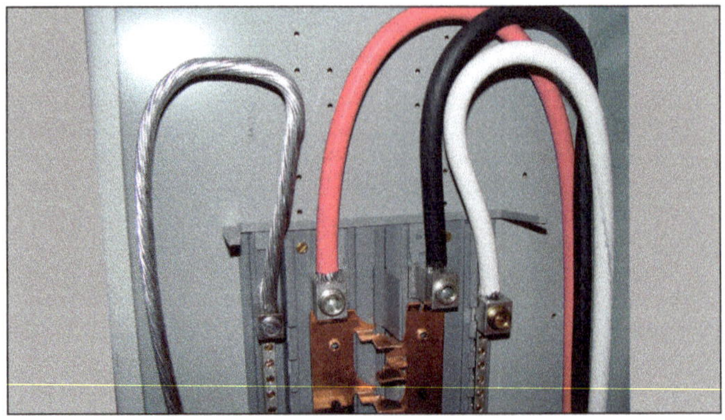

Photo 4.12 Typical main lug only (MLO) panelboard

devices that a panelboard is allowed to contain, provided the panelboard is listed to accept more. See NEC 408.36, Ex. No. 1.

The second exception to individual overcurrent protection applies to existing panelboards at individual dwellings where the panelboard functions as service equipment.

The service disconnecting means can utilize a combination of up to six main breaker panelboards, main breakers, or main fuses in separate enclosures. The two to six disconnecting means are installed adjacent to each other and are supplied from a common set of service-entrance conductors. The main service-entrance conductors are protected from overload by the calculated load on the service. The service-entrance conductors installed to the individual service disconnecting means are protected by the overcurrent device they are in series with. The service panelboard is also protected by the main breaker or fuses, which are an integral part of the panelboard serving as service equipment. A panelboard must be rated for the connected load when installed as service equipment. This panelboard must be marked "suitable for use as service equipment" when used in this manner.

EQUIPMENT CONNECTED TO THE SUPPLY SIDE OF THE SERVICE DISCONNECTING MEANS

The *Code* limits the type of equipment that is permitted on the line side of the service main. Meters, meter sockets, and meter disconnects are the most common types of equipment permitted ahead of the service disconnecting means at dwelling units. The metal meter socket or pedestal must be grounded. Refer to Chapter 5 of this book for further information on the various techniques used to ground equipment on the supply side of the main disconnect.

Interconnected alternative power sources, such as solar photovoltaic (PV) systems, fuel cell systems, and wind-powered generators, are also permitted ahead of the service main. NEC Article 690 and Chapter 18 of the IAEI's Soares

Figure 4.32. Examples of three different types of overcurrent protection for panelboards

Grounding and Bonding textbook are excellent resources for requirements and additional information on grounding solar photovoltaic (PV) systems. NEC Article 705 contains the requirements for interconnected electric power production sources. A requirement common to all such systems is the need to automatically disconnect the interconnected source from the supply conductors during a power outage. This must be done to protect utility personnel and others who may work on the supply lines.

Type 2 surge-protective devices are also permitted on the supply side of the service disconnecting means. Such devices are specified in order to protect sensitive and expensive electronic equipment from voltage surges on the supply conductors. The surge-protective devices must be installed as part of the listed equipment and provided with a disconnecting means and suitable overcurrent protection. The *Code* does not permit the surge-protective devices to be connected by placing the supply conductors to the device under the same terminal screw as the supply conductors to the main breaker or main set of fuses.

See *NEC* 230.82 for equipment that is permitted to be connected ahead of the service disconnecting means.

INSTALLING SERVICE-ENTRANCE EQUIPMENT FOR TWO-FAMILY DWELLINGS

Service equipment for two-family dwellings is usually installed using one of two methods. The first method involves grouping service disconnects at a single location, typically near the metering equipment. A feeder is run to each of the dwelling units where a subpanel or feeder panel is installed. This method is permitted by NEC 230.40 Exception No. 2.

The second method is not to group the service disconnects but to locate them in or on the individual dwelling unit. In this case, the metering equipment is at one location, which is usually required by the serving utility, and service-entrance conductors are run to each dwelling unit. *NEC* 230.40 Exception No. 1 permits this alternative method. Requirements for installing services for one-family dwellings apply for two-family dwellings as well unless specific modifications of the rules are provided.

NUMBER OF SERVICE DROPS OR LATERALS

The *Code* generally limits the number of services to a building to not more than one. Service, as defined in NEC Article 100 and IRC Chapter 35, includes the service drop and service lateral. As a result, no more than one service drop or service lateral can be run to the building unless the installation complies with one or more of the special conditions or applications at *NEC* 230.2(A) through (D). The IRC is more restrictive in this case, permitting only one service to a one- or two-family dwelling unit, with no exceptions (see IRC E3601.2).

A "building" is defined as a structure that stands alone or is separated from neighboring structures by firewalls. The NEC does not specify the types of construction that qualify as a building separation firewall, but such requirements are detailed in the IRC or other relevant building codes. When a firewall divides a structure into two separate buildings under one roof, a service—such as a service drop or service lateral—can be installed to each building.

"Building" is defined in Section R202 of the IRC as "any one- or two-family dwelling or portion thereof, including townhouses, used or intended to be used for human habitation, for living, sleeping, cooking or eating purposes, or any combination thereof, or any accessory structures." More specifics are included in the IRC definition that may assist enforcing authorities with the interpretation of some electrical code requirements.

TWO-FAMILY DWELLING SERVICES, OVERHEAD

As previously stated, the *Code* generally permits only one service drop to a building. Additional services may be installed if special conditions, special occupancies, capacity requirements, or different characteristics permit. None of these are common to one- or two-family dwelling construction. As a result, the two-family dwelling, without a firewall separating it into two separate buildings, is generally permitted to be supplied by only one service drop.

The Code generally limits a service drop or service lateral to providing no more than one set of service-entrance conductors. For buildings with multiple occupancies, such as a two-family dwelling, a separate set of service-entrance conductors is allowed for each occupancy. This rule permits the installation of two service masts with the service-entrance conductors spliced to the service-drop conductors at a common point. In this case, the service disconnecting means are not required to be grouped at one location due to the language in NEC 230.71(B) or IRC E3601.6.2. All other rules for installing service equipment—such as no more than six disconnecting means grouped at a single location—apply to the location selected for the service equipment in or on each occupancy.

A duplex meter socket may be installed at one location, and the service-entrance conductors run through each unit. This arrangement is shown in Figure 4.34. Where a duplex meter socket is provided, it must have an ampere rating for each socket that is adequate for the calculated load or service-entrance overcurrent protective device rating at the individual service disconnect. In addition, the meter socket assembly must have an overall ampere rating that is suitable for the calculated load on the service-entrance conductors supplying the service as determined by *NEC* Article 220 or IRC E3602.1 and Table E3602.2. For example, the total load for a two-family dwelling is calculated and determined to be 135 amperes. The load to each unit is calculated at 70 amperes for the first unit and 65 amperes for the second unit. The duplex meter socket should have an overall rating of at least 150 amperes, and each socket should be rated for 100 amperes. Refer to Chapter 2 of this textbook for methods and examples on calculating the load on a two-family dwelling.

Where a duplex meter socket is installed, two methods are provided for locating the service disconnecting means. The first method is to locate the service disconnecting means at the meter socket location and to run feeders to the individual dwelling units where a branch-circuit sub-panelboard is installed. The second method is to run service-entrance conductors to each unit and to locate the service disconnecting means and branch-circuit panelboard in or on the individual units. Both are common methods of serving two-family dwellings.

The latter method involves running service-entrance conductors to each unit. Each set of service-entrance conductors that arrives at its respective dwelling unit must be provided with a service disconnecting means that meets all the rules in *NEC* 230.70(A), 230.71(B), and 230.72 or IRC E3601.6.2. A single disconnect, or not more than six separate disconnects grouped at that location, would be permitted for each set of service-entrance conductors that supply each of the dwelling units of a two-family dwelling structure. An additional set is permitted to supply a "house panel." A house panel supplies loads shared by both occupants.

The former method keeps the service disconnecting means

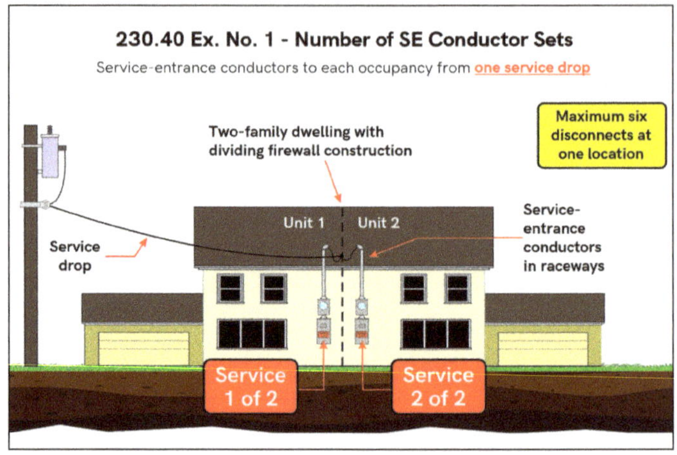

Figure 4.33. Two-family dwelling service equipment

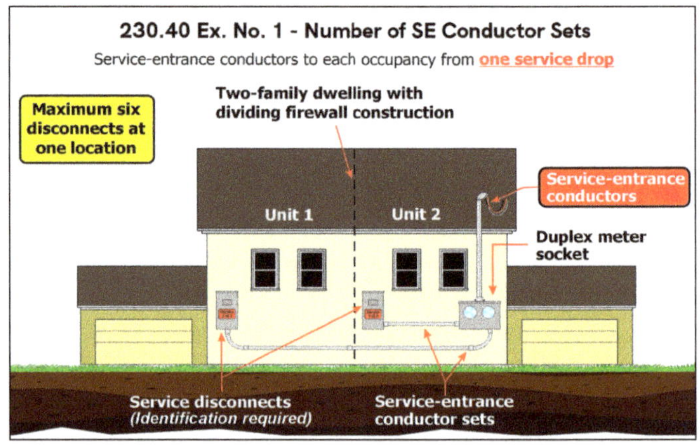

Figure 4.34. Two-family dwelling service disconnecting means located in or on the unit.

grouped at one common location. Where two (or more) sets of service-entrance conductors are derived from a single drop or lateral and the service disconnect(s) are adjacent to each other, the grouping requirements for any one location should be applied. The *Code* does not mention a distance here, and the authority having jurisdiction (AHJ) usually has the responsibility for approving the application of these locations and grouping requirements based on particular conditions.

Any service disconnect means must be located to meet the requirements found in *NEC* 230.70(A)(1) or IRC E3501.6.2. The *Code* requires the service disconnecting means to be either outside the building or inside, nearest the point of entrance of the service conductors.

TWO-FAMILY DWELLING SERVICES, UNDERGROUND

NEC 230.2 coordinates with the rule in *NEC* 230.40 that limits to one the number of service-entrance conductors that can be supplied by a service drop or service lateral. Exception No. 2 to *NEC* 230.40 permits a service lateral to supply more than one set of service-entrance conductors where two to six

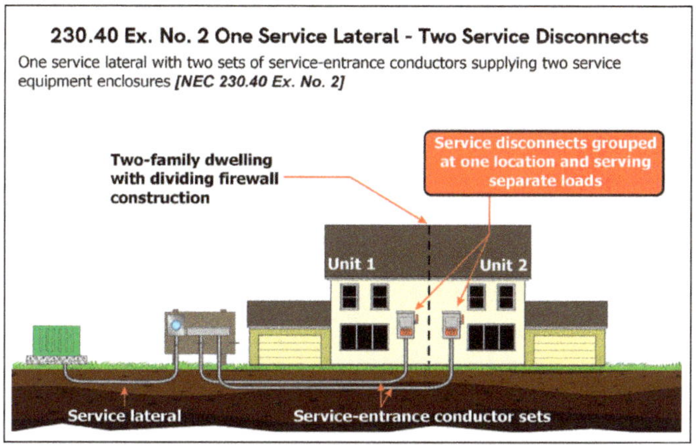

Figure 4.35. Permitted installation of service laterals

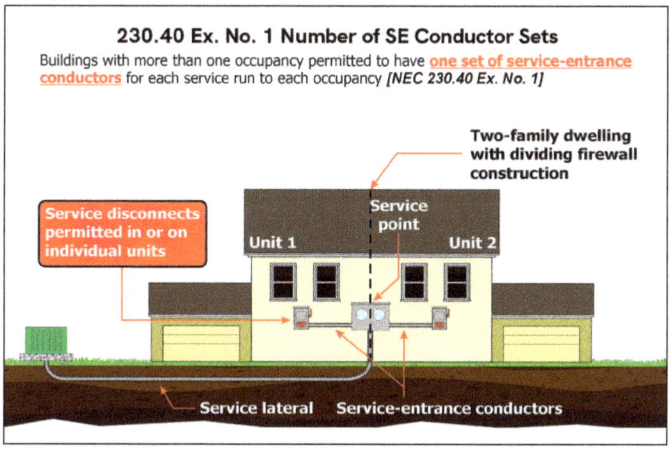

Figure 4.36. Another permitted installation of service laterals

service disconnecting means are grouped at one location and supply separate loads. The second method is not to group the service disconnects but to locate them in or on the individual dwelling unit. In this case, a splice box or metering equipment is at one location, and service-entrance conductors are run to each dwelling unit. *NEC* 230.40 Exception No. 1 permits this alternative.

Both methods are shown in Figures 4.35 and 4.36. In the first method, a separate service lateral runs to each of the building's service disconnecting means. The service disconnecting means must be grouped at one location, which is considered one service lateral by NEC 230.2.

Another option is to run a single service lateral to the building and connect it to a single enclosure. This enclosure can be a splice box, but it is usually a duplex meter pedestal. The service-lateral conductors connect to the service-entrance conductors, which then extend to each dwelling unit's service disconnecting means. The service equipment for each dwelling unit does not need to be located at a common site; it can be installed in or on the individual dwelling unit. As previously discussed for overhead services, the rules limit the maximum number of service disconnecting means to six at one location.

A two-family dwelling may be classified as two separate buildings in certain limited cases. One example is when the units share a common property line and are separated by a firewall. In this situation, the Code allows for a separate service. If two services are supplied to a single building to support two-family occupancy and they are not grouped at the same location, the Code requires an identification sign at each service point that directs people to the location of the other service within that building. This requirement may apply to installations governed by the NEC but would not apply to dwellings governed by the IRC.

PLAQUE OR DIRECTORY

Where more than one service is installed on a dwelling, the provisions in NEC 230.2, 230.2(E), and 230.40, Ex. No. 1 require a permanent plaque or directory to be placed at each service-disconnect location, denoting other service-equipment locations on the structure or building and the area served by each.

REFERENCES

1. *National Electrical Code* and *NEC* are registered trademarks of the National Fire Protection Association, Inc., Quincy, MA 02169. This reprinted material is not the official position of the National Fire Protection Association, which is represented only by the standard in its entirety.

2. *International Residential Code for One- and Two-Family Dwellings* (2024). Copyright © International Code Council, Inc., June 2023.

3. UL, Northbrook, IL, *UL Guide Information for Electrical Equipment Directory* (UL Product iQ® material) unless indicated otherwise.

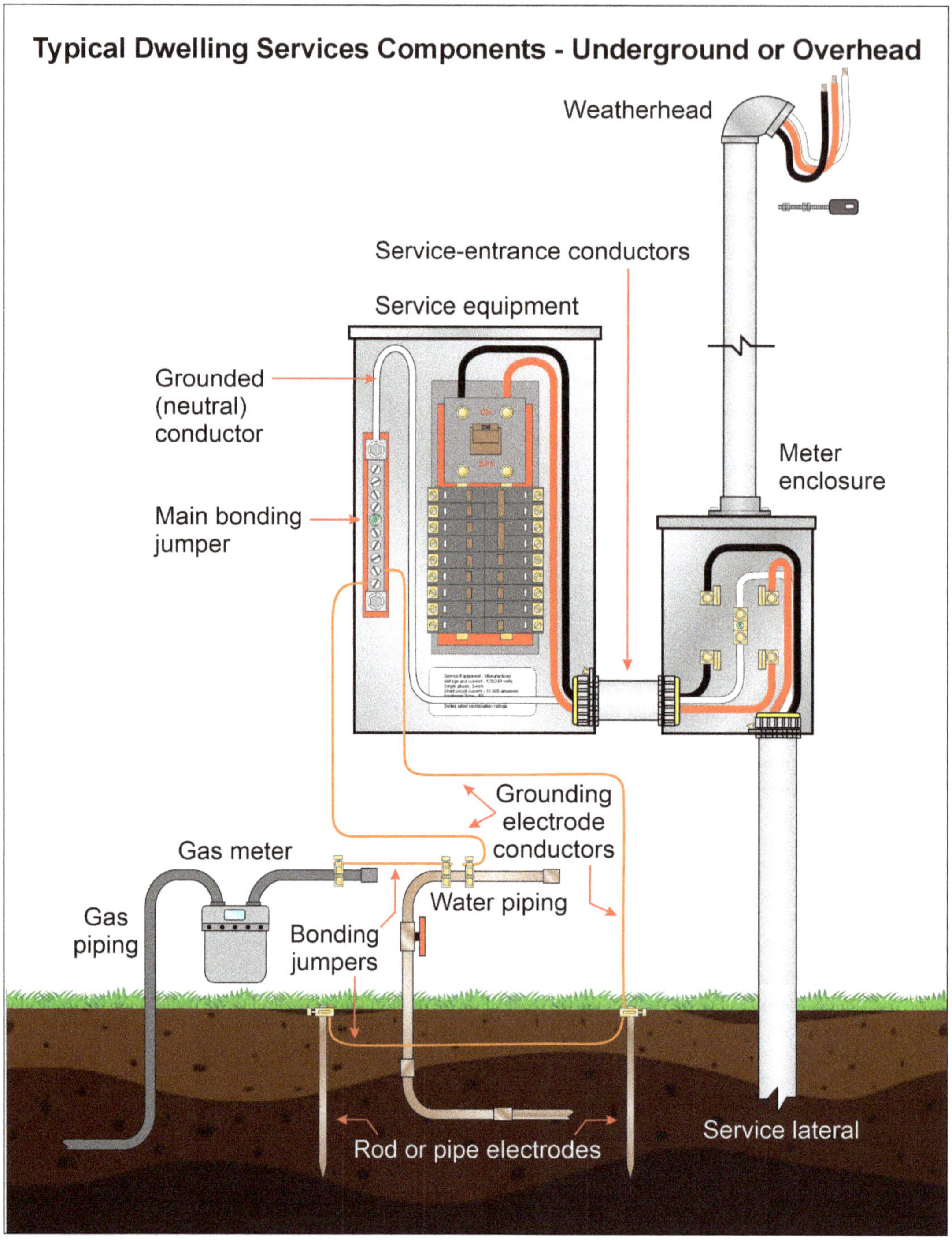

Figure 4.37. Typical dwelling unit service diagram for overhead or underground

REVIEW QUESTIONS

1. "The overhead conductors between the service point and the first point of connection to the service-entrance conductors at the building or other structure" is the definition of which of the following:

A. Service Drop
B. Service Conductors, Overhead
C. Service Conductors, Underground
D. Service Conductors, Supply-Side

2. The *NEC* requires that service masts be "of adequate strength" for any service raceway that is used for the support of service drop conductors.

True _____ False _____

3. Service conductors must have a clearance of not less than ________________ distance from the side or bottom of a window.

A. 150 mm (6 in.)
B. 475 mm (18 in.)
C. 600 mm (24 in.)
D. 900 mm (36 in.)

4. Service-entrance conductors passing over a flat roof must be at least what distance over the roof surface.

A. 900 mm (3 ft)
B. 1.8 m (6 ft)
C. 2.1 m (7 ft)
D. 2.6 m (8 ft 6 in.)

5. Service-entrance conductors are permitted to be installed within all of the following raceways except:

A. Electrical metallic tubing
B. Intermediate metal conduit
C. Surface metal raceway
D. Rigid PVC conduit

6. The minimum distance from the ground for the point of attachment for service-drop conductors is _______ above finished grade.

A. .1 m (7 ft)
B. 2.5 m (8 ft)
C. 2.7 m (9 ft)
D. 3.0 m (10 ft)

7. Where passing over a residential driveway, a 120/240 volt, 3-wire service drop must provide a minimum height above grade of:

A. 3.0 m (10 ft)
B. 3.7 m (12 ft)
C. 4.5 m (15 ft)
D. 5.5 m (18 ft)

8. Type UF cable is permitted to be used as service-entrance conductors.

True _____ False _____

9. A single-family dwelling with more than two 2-wire branch circuits must have a service disconnecting means with a rating of not less than 60 amperes.

True _____ False _____

10. Service-entrance conductors must be provided with overload protection in each:

A. Grounding electrode conductor
B. Ungrounded conductor and neutral conductor
C. Ungrounded service conductor
D. Grounding and bonding conductor

11. Where it is impracticable to locate a service head above the point of attachment of a Type SE cable, how far can it be located from the point of attachment?

A. 300 mm (12 in.)
B. 450 mm (18 in.)
C. 600 mm (24 in.)
D. 900 mm (36 in.)

12. Where service-entrance cables are installed in exposed locations (such as on a wall beside a driveway), they are required to be protected from physical damage by any of the following raceways EXCEPT:

A. Rigid metal conduit
B. Intermediate metal conduit
C. Electrical metallic conduit
D. Schedule 40 PVC conduit

13. Individual protection is not required at a panelboard protected on its supply side by two main circuit breakers or two sets of fuses other than service equipment having a combined rating not greater than that of the panelboard.

True _____ False _____

14. A reduction of clearance above the roof area for an overhead service drop to 900 mm (3 ft) is permitted where the voltage between conductors does not exceed 300 volts and the roof area is __________.

A. guarded
B. isolated
C. either A. or B.
D. flat across the entire roof surface

15. Service-entrance conductors shall be considered outside of a dwelling unit building under any of the following conditions EXCEPT:

A. Where installed under not less than 50 mm (2 in.) of concrete beneath the building.
B. Where installed inside the structure in a raceway that is encased in concrete or brick not less than 50 mm (2 in.) thick.
C. Where installed inside the structure in rigid metal conduit.
D. Where installed in conduit and under not less than 450 mm (18 in.) of earth beneath the building.

16. Two service drops are permitted to be installed to a two-family dwelling without firewalls.

True _____ False _____

17. The minimum working space at service equipment is permitted to be the width of the equipment or ________________, whichever is greater.

A. 610 mm (24 in.)
B. 762 mm (30 in.)
C. 914 mm (36 in.)
D. 900 mm (48 in.)

18. A reduction of clearance above the roof area for an overhead service drop to 900 mm (3 ft) is permitted where the voltage between conductors does not exceed 300 volts, and the roof has a slope of:

A. 100 mm in 250 mm (4 in. in 10 in.) or greater.
B. 100 mm in 300 mm (4 in. in 12 in.) or greater.
C. 100 mm in 400 mm (4 in. in 16 in.) or greater.
D. 45 degrees or greater

19. A two-family dwelling is permitted to have an additional disconnecting means to service common area loads such as lighting, joint laundry facilities, or a shared well.

True _____ False _____

20. An emergency disconnecting means for a one- or two-family dwelling is required to be installed and located on the outside of the structure. This required outdoor emergency disconnect can consist of:

A. The service disconnect(s).
B. A properly marked meter disconnect(s).
C. Other listed disconnect switches or circuit breakers on the supply side of each service disconnect that are suitable for use as service equipment.
D. Any of the above

21. A 120/240-volt, single-phase dwelling service or main power feeder shall be permitted to be sized at a minimum in accordance with:

A. *NEC* Table 310.16 (IRC Table E3705.1)
B. *NEC* 310.15(B)(7) (IRC E3603.1)
C. *NEC* Table 310.15(B)(1) (IRC Table E3705.2)
D. *NEC* 310.12 and Table 310.12 (IRC E3603.1 and Table E3603.1.1)

22. Where service conductors passing over a roof are securely supported by a substantial metal structure, the supporting structure should be bonded by means of a __________ and listed connector to the ______________________.

A. bonding jumper (grounded overhead service conductor)
B. equipment grounding conductor (metal structure)
C. bonding jumper (equipment grounding conductor)
D. main bonding jumper (grounded conductor)

23. All services delivering power to dwelling units must be provided with a surge-protective device (SPD). This residential SPD must be ___________________________ _____ unless it is supplied at each next-level distribution equipment downstream toward the load.

A. an integral part of the service equipment
B. located immediately adjacent to the service equipment
C. a Type 3 or Type 4 SPD
D. Either A. or B.

24. A device intended to provide limited overcurrent protection for specific applications and utilization equipment such as luminaires and appliances is referred to as a(n) __________.

A. supplementary overcurrent protective device
B. fuse
C. circuit breaker
D. primary overcurrent device

25. The required emergency disconnecting means at a dwelling unit marking or labels must be located on the outside front of the disconnect enclosure with ________ background and white text. The letters must be at least ______________ high.

A. blue [13 mm (½ in.)]
B. red [13 mm (½ in.)]
C. black [19 mm (¾ in.)]762 mm (30 in.)
D. red [25 mm (1 in.)]

26. Where residential service equipment is upgraded and replaced, the new service equipment is required to be provided with a surge-protective device (SPD).

True _____ False _____

27. A service disconnecting means is required for each service permitted by the *Code*. The service disconnecting means can be either a single disconnect or it is permitted to consist of not more than six switches or six circuit breakers. These two to six service disconnects are permitted to be mounted in:

A. a single enclosure
B. separate enclosures with a main service disconnecting means in each enclosure
C. separate enclosures with no main service disconnecting means in each enclosure
D. separate enclosures that are not grouped together

28. Connections to the service-entrance conductors or service lateral must be made in an acceptable manner. Power distribution blocks, pressure connectors, and devices for splices and taps are required to be listed. Power distribution blocks installed on service conductors must also be marked ______________________________ or equivalent.

A. "suitable for use on the load side of the service equipment."
B. "suitable for use with service equipment."
C. "suitable for use on the line side of the service equipment."
D. "suitable for use with feeder or service equipment."

29. A reduction of clearance above the roof area for an overhead service drop to 900 mm (3 ft) is permitted where the voltage between conductors does not exceed 300 volts, and the roof has a slope of:

A. 100 mm in 250 mm (4 in. in 10 in.) or greater.
B. 100 mm in 300 mm (4 in. in 12 in.) or greater.
C. 100 mm in 400 mm (4 in. in 16 in.) or greater.
D. 45 degrees or greater

5 Service Grounding and Bonding Requirements

GENERAL REQUIREMENTS

The requirements for grounding and bonding can be complex, but it is essential for electrical safety that minimum standards are met. It is also crucial to understand the concepts and goals of effective grounding and bonding. When you know the "how" and the "why" behind the performance requirements, applying Code rules becomes easier. Keep in mind that the same rules for grounding and bonding apply to both dwelling units and commercial settings. Although these topics can be difficult to grasp, meeting their minimum requirements is vital for safety.

Understanding the "how" and the "why" behind a performance objective makes applying Code rules easier. It's important to know that there is no separate set of grounding and bonding rules for dwelling units compared to commercial applications. When the term grounding is used in Code rules or general conversation, it should be related to the term earthed or earthing. To ground a metal object or conductor means simply to connect it to the earth. Bonding, on the other hand, involves connecting two objects together. Developing a common language for communication is crucial when applying the Code requirements to electrical systems or installations. A good understanding and proper use of the Code's definitions can help establish this shared language; the following definitions related to grounding and bonding are from NEC Article 100 and IRC Chapter 35.

DEFINITIONS APPLICABLE TO GROUNDING AND BONDING

The following definitions apply to grounding and bonding terminology at residential units. If a review is necessary, please refer to Article 100 in the NEC for additional information.

- ***Bonded (Bonding)***
- ***Bonding Conductor or Jumper***

- *Bonding Jumper, Equipment*
- *Bonding Jumper, Main*
- *Bonding Jumper, System*
- *Bonding Jumper, Supply-Side*
- *Effective Ground-Fault Current Path*
- *Energized, Likely to Become. (Likely to Become Energized)*
- *Ground*
- *Ground Fault*
- *Grounded (Grounding)*
- *Grounded Conductor*
- *Ground-Fault Current Path*
- *Grounding Conductor, Equipment (EGC)*
- *Grounding Electrode*
- *Grounding Electrode Conductor (GEC)*
- *Intersystem Bonding Termination (IBT)*
- *Neutral Conductor*
- *Neutral Point*
- *Separately Derived System*
- *Ungrounded*

PURPOSE OF GROUNDING AND BONDING — PERFORMANCE CRITERIA

The purpose of grounding and bonding electrical services and equipment is outlined in NEC 250.4. Prescriptive methods outlined in NEC Article 250 must be followed to meet the performance requirements of this NEC section. NEC 250.4 also provides descriptive and explanatory material regarding the goals of effective grounding and bonding.

NEC 250.4 consists of two sections: Part A details performance criteria for grounded systems, and Part B covers performance criteria for ungrounded systems. Since one- and two-family homes usually have grounded systems, this text will omit information related to the installation and use of ungrounded systems. These systems are used in various commercial and industrial settings and are outside the scope of this text for dwelling units. For more information on effective grounding and bonding of ungrounded electrical systems, refer to IAEI's Soares Grounding and Bonding textbook. NEC 250.4(A) highlights five main categories related to the purpose of grounding and bonding. A detailed look at each performance requirement is warranted.

GROUNDING OF ELECTRICAL SYSTEMS

Electrical systems required to be grounded are to be connected to earth in a manner that will limit the voltage imposed by lightning, line surges, or unintentional contact with higher-voltage lines and that will stabilize the voltage to earth

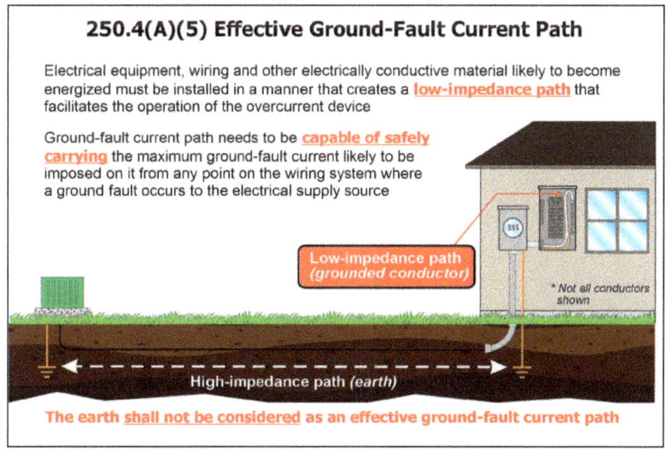

Figure 5.1. Effective low-impedance ground-fault current path must be installed

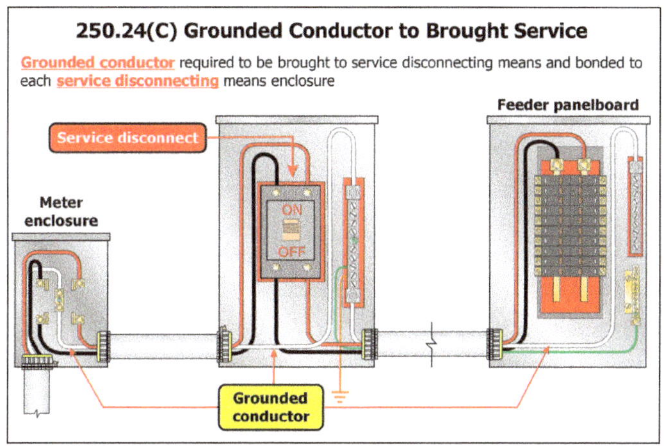

Figure 5.2. The grounded conductor must be run to the service disconnect and bonded to the service-disconnect enclosure

during normal operation [see *NEC* 250.4(A)(1)].

GROUNDING OF ELECTRICAL EQUIPMENT

Normally, non-current-carrying conductive materials enclosing electrical conductors, such as panelboards and metal raceways, and so forth, are required to be connected to the earth (grounded) to limit the voltage to ground or the difference of potential between the earth and these materials. Where the *Code* requires the electrical system to be grounded, these materials are required to be connected together and to the supply system grounded conductor. The grounded conductor for one- and two-family dwellings is usually the neutral conductor [see *NEC* 250.4(A)(2)].

BONDING OF ELECTRICAL EQUIPMENT

The *Code* requires normally non-current-carrying metallic conductive materials enclosing electrical equipment and con-

Photo 5.1. The pad-mounted transformer that supplies a single-family dwelling; the grounded conductor of this system is grounded at the transformer location and grounded at the service equipment of the dwelling

ductors to be connected together electrically in a manner that ensures effective continuity between the metallic (conductive) components and also establishes an effective ground-fault current path for any fault current likely to be imposed on them [see *NEC* 250.4(A)(3)].

BONDING OF ELECTRICALLY CONDUCTIVE MATERIALS AND OTHER EQUIPMENT

Normally, non-current-carrying electrically conductive materials, such as metal water piping, metal gas piping, and structural steel members, that are likely to become energized must be bonded to the supply system grounded conductor in a manner that establishes an effective path for ground-fault current. Bonding these metallic components plays a key role in protecting people from electrical shock hazards [see *NEC* 250.4(A)(4)].

EFFECTIVE GROUND-FAULT CURRENT PATH

Electrical equipment, wiring, and other electrically conductive materials likely to become energized must be installed in a way that creates a low-impedance circuit, allowing the overcurrent device to operate effectively. In other words, it enables the circuit breaker or fuse to quickly open the circuit during a ground fault. A ground fault is an unintentional, electrically conductive connection between an ungrounded conductor of an electrical circuit and the normally non-current-carrying conductors, metallic enclosures, metallic raceways, metallic equipment, or earth.

The ground-fault current path must be able to safely carry the maximum ground-fault current that could be imposed on it from any point on the wiring system where a ground fault might occur to the electrical supply source. The ground-fault current path is an electrically conductive route from the point of a ground fault in a wiring system through normally non-current-carrying conductors, equipment, or the earth to the electrical supply source.

Examples of common ground-fault current paths could consist of any combination of equipment grounding conductors, metallic raceways, metallic cable sheaths, electrical equipment, and any other electrically conductive material such as metal water and gas piping, steel framing members, stucco mesh, metal ducting, reinforcing steel, shields of communications cables, and the earth itself. The intentionally grounded conductor is typically the path of lowest impedance for ground fault current to return to its source during a ground fault condition.

The earth should never be used as an equipment grounding conductor or considered an effective ground-fault current path [see NEC 250.4(A)(5) or IRC E3908.4 and E3908.5].

Understanding these grounding and bonding fundamentals and knowing the difference between grounding (earthing) and bonding (connecting together) is vital for installers and inspectors to ensure safety from shock and fire hazards. This chapter will examine the specific requirements for grounding and bonding that apply to one- and two-family dwelling electrical systems.

GROUNDING SERVICE-SUPPLIED ALTERNATING-CURRENT SYSTEMS

System Grounding Connections

A premise wiring system supplied by a grounded alternating-current system is required to have (at each service) a grounding electrode system established and a connection to that system at the service with a grounding electrode conductor(s).

This grounding electrode conductor must be connected to the grounded (neutral) service conductor at any accessible point from the load end of the service drop or service lateral to and including the terminal or bus to which the grounded service conductor is connected at the service disconnecting means [*NEC* 250.24(A)(1) or IRC E3607.2] (see Figure 5.2).

Where the transformer supplying the service is located outside the building, at least one additional grounding connection must be made from the grounded (neutral) service conductor to a grounding electrode, either at the transformer or elsewhere outside the building. For one- and two-family dwellings, this transformer is usually a pole-mounted unit that supplies a service drop to the structure or is a pad-mounted unit that supplies a service lateral to the building or structure (see photo 5.1).

A grounding connection is usually not permitted to any grounded (neutral) circuit conductor on the load side of the service disconnecting means. This means that the neutral conductor may not be connected to the equipment grounding conductors or grounded enclosures except at the service equipment.

One of the few exceptions to this rule in the Code permits ranges, counter-mounted cooking units, wall-mounted ovens, and clothes dryers on existing branch circuits to be grounded by using the grounded circuit conductor. New range and dryer installations require the use of ungrounded (hot) conductors, an insulated grounded (neutral) conductor, and an equipment grounding conductor, and the use of this equipment grounding conductor for grounding purposes [NEC 250.24(B), 250.140(B) or IRC E3607.2]. Refer to Chapter 7 for further information on grounding appliances.

GROUNDED (NEUTRAL) CONDUCTOR BROUGHT TO SERVICE EQUIPMENT

Where an alternating-current system operating at less than 1000 volts is grounded at any point, the grounded (usually a neutral conductor in dwelling units) conductor is generally required to be run to and bonded to each service-disconnecting means enclosure [NEC 250.24(D)].

This conductor must be routed with the service-entrance phase conductors.

This grounded (neutral) service conductor serves a dual purpose: allowing line-to-neutral loads to be served and providing the lowest impedance path for ground-fault current during faults, until an overcurrent device clears the circuit. In this role, it offers a low-impedance connection from the service to the utility transformer so that line-to-ground faults can be cleared effectively.

If this grounded service conductor is not run to the service equipment and bonded to the enclosure, the only path for line-to-ground faults to return to the utility transformer is through the earth, which is prohibited by *NEC* 250.4 and IRC E3908.5. The earth is a high-impedance path that can leave faulted electrical equipment at the dwelling at a dangerous shock potential aboveground (see IAEI's *Soares Book on Grounding and Bonding* for additional information on this subject).

CONDUCTOR TO BE GROUNDED

For a typical dwelling unit alternating-current premise wiring system, the conductor required to be grounded is:

1. Single-phase, 2-wire: one conductor (either one).
2. Single-phase, 3-wire with a neutral conductor. The neutral conductor means that the conductor is common or neutral to both ungrounded (hot) conductors of the system or circuit.

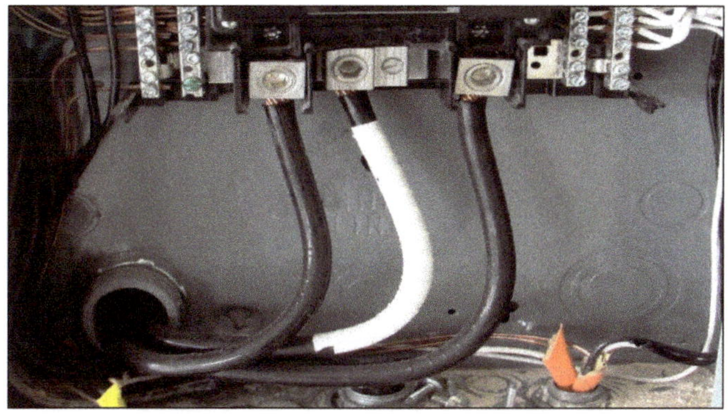

Photo 5.2. Grounded (neutral) conductor (re-identified with white tape at its termination) brought to service disconnect and bonded to disconnecting means through the main bonding jumper (green screw)

Photo 5.3. Main bonding jumper in the form of a strap from the grounded (neutral) conductor terminal bar to the enclosure

For dwelling units, the electrical utility typically performs the grounding of the electrical system either at the pole-mounted transformer or the pad-mounted transformer at grade. The installer then grounds the corresponding grounded conductor again at the building's or structure's service disconnecting means.

Grounded (neutral) conductors must be identified by the colors white or gray. An insulated grounded conductor sized at 6 AWG and smaller must be identified by a continuous color of white or gray along its entire length or by a conductor with a color other than green that is identified with three continuous white or gray stripes along its entire length. An insulated conductor sized at 4 AWG or larger is permitted to be identified the same as smaller conductors or may be identified at the time of installation by distinctive white or gray marking(s) at their terminations. White or gray phase tape or white or gray paint is often used to identify the grounded system conductor, and, where used, must encircle the insulation (*NEC* 200.6(A) and (B) or IRC E3407.1).

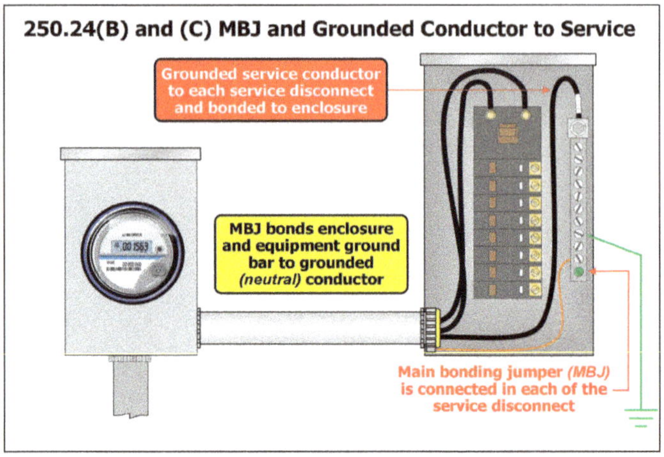

Figure 5.3. Grounded conductor required to be brought to service disconnecting means and bonded to the enclosure through the main bonding jumper

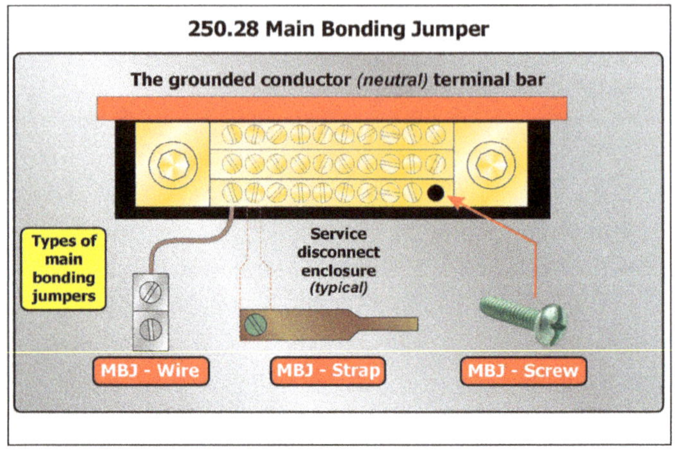

Figure 5.4. A main bonding jumper is required to be in the form of a wire, bus, screw, or similar suitable conductor

Photo 5.4. Main bonding jumper in the form of a screw provided by the manufacturer (notice the green color finish)

GROUNDING PATH TO GROUNDING ELECTRODE AT SERVICES

Connection to the earth from any electrical system or equipment is established through the grounding electrode system. The Code recognizes eight grounding electrodes, present at the building or structure, that are required to be connected together to form the grounding electrode system [see NEC 250.50 or IRC E3608.1]. The NEC includes grounded building steel in the form of metal in-ground support structures as one of these electrodes, while the IRC does not [NEC 250.52(A) or IRC E3608]. The main reason for this exclusion in the IRC is that an in-ground support structure is usually not included in the construction of most one- and two-family dwellings. However, when it is part of the construction, it must be included in the grounding electrode system if the structural metal framing members qualify as a grounding electrode per NEC 250.52(A)(2).

A grounding electrode conductor must be installed and used to connect the equipment grounding conductors, the service equipment enclosures, and the grounded (neutral) service conductor to the grounding electrode [NEC 250.24(E) or IRC E3607.4]. Requirements for grounding electrodes and conductors are explained in more detail later in this chapter.

MAIN BONDING JUMPER

The main bonding jumpers, supply-side bonding jumpers, and equipment bonding jumpers must be made of copper, aluminum, copper-clad aluminum, or other corrosion-resistant materials. A main bonding jumper must be in the form of a wire, bus, screw, or a similar suitable conductor.

A properly sized main bonding jumper is usually included as part of a listed panelboard. Some manufacturers provide a strap, while others supply a screw. When the main bonding jumper is a screw, it must have a visible green finish with the screw installed [NEC 250.28(B)], to clearly identify it as the main bonding jumper. Other manufacturers provide bonding straps of various designs. Machine screw-type fasteners or thread-forming screws that engage at least two threads or are secured with a nut are allowed to be used as the main bonding jumper or for attaching a bonding strap to the enclosure. Only the bonding screw supplied by the manufacturer with the service equipment should be used (NEC 250.8 and IRC E3406.14).

In the IRC at E3607.5, there is no color identification requirement for the screw used as the main bonding jumper in the service equipment. However, it should be emphasized that the IRC requires all equipment to be listed (IRC E3403.3). Product Standard UL 67 requires panelboards used and equipped and identified as "suitable for use as service equipment" to include a green screw for use as the main bonding jumper with the equipment. This provision is applicable where the main bonding jumper is a screw.

An unspliced main bonding jumper is required to be

installed to connect the equipment grounding conductor and the service-equipment enclosure to the system grounded (neutral) conductor of the electric utility system inside the service-equipment or service-conductor enclosure [*NEC* 250.24(C) or IRC E3607.5]. Sometimes, it is necessary to install a main bonding jumper in the field as a wire; this is common where a gutter or wireway-style service is installed. The sizing of a wire-type main bonding jumper would be based on *NEC* Table 250.102(C)(1) or IRC Table E3603.4. The majority of electrical service equipment used for one- and two-family dwellings is listed as either "Suitable for Use as Service Equipment" or "Suitable for Use Only as Service Equipment." These types of listed equipment include grounding and bonding provisions that allow conformance to the bonding requirements of the *Code* at the service.

THE GROUNDING TARGET AT THE SERVICE

For grounded electrical systems, four conductors or components must come together in the service equipment for effective grounding and bonding. These four conductors connected within the service equipment are sometimes referred to as the "grounding target." The four conductors are the grounded service conductor (neutral conductor), main bonding jumper, grounding electrode conductor(s), and equipment grounding conductor(s). In most cases, if these four components are present at the service disconnect and connected properly, the electrical service is effectively grounded and bonded (see Figures 5.5 and 5.6). Think of the service equipment as a big bull's eye, and you have four arrows to shoot at that (service) target!

Figure 5.5. The "grounding target" at service equipment

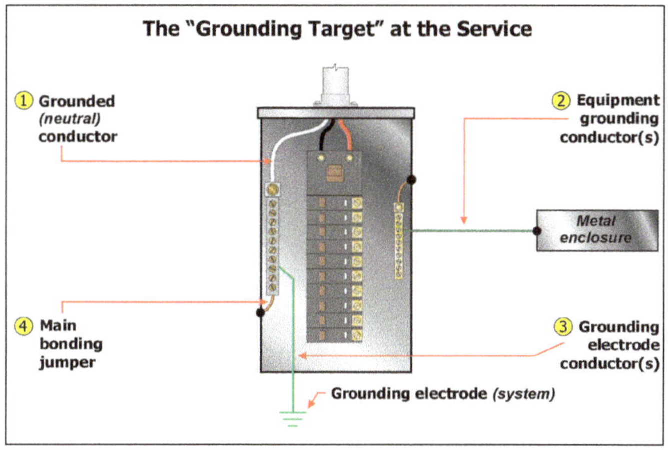

Figure 5.6. The grounding target is the service disconnect

USE OF GROUNDED (NEUTRAL) CIRCUIT CONDUCTOR FOR GROUNDING EQUIPMENT

A grounded (neutral) circuit conductor is permitted to be used to ground non-current-carrying metal parts of equipment, raceways, and other enclosures on the supply (line) side of the service disconnecting means [*NEC* 250.142(A)]. This concept is illustrated in Figure 5.7, where the neutral conductor bonds the meter base and service equipment together. In most cases, the neutral conductor terminal bar in a listed meter socket is bonded to the enclosure at the factory (see Photo 5.6).

The neutral conductor then serves as the bonding conductor between the meter socket and the service enclosure. Generally, it is not permitted to use the system-grounded conductor (often a neutral conductor) for bonding or grounding equipment downstream from the service equipment [*NEC* 250.24(B), 250.142(B), or IRC E3908.7].

Installers and inspectors should take care to verify that neutral-to-ground connections are not made on the load side of the service disconnect to avoid any parallel path for neutral (return) current back to the service and source. Grounding connections of the neutral conductor have become more restricted and limited in recent editions of the *Code*. The primary concern is to keep normal neutral return current flowing on the path it was intended for (neutral conductor) and not on other metallic paths (equipment grounding conductors, metal cabinets and enclosures, etc.), which could create a shock hazard from circulating currents on metal cabinets, etc. Only those neutral-to-ground connections specifically permitted for *existing* installations of ranges and dryers are allowed to be connected on the load side of a service disconnecting means. See *NEC* 250.140 for allowance of the use of the grounded conductor for grounding of existing ranges and dryers.

BONDING REQUIREMENTS AT SERVICE EQUIPMENT

All non-current-carrying metal parts of the service raceway, the service-equipment enclosures containing service conductors (including meter fittings, boxes, or the like) interposed

Photo 5.5. Can you identify the four "arrows" of the grounding target in the photo on the left?

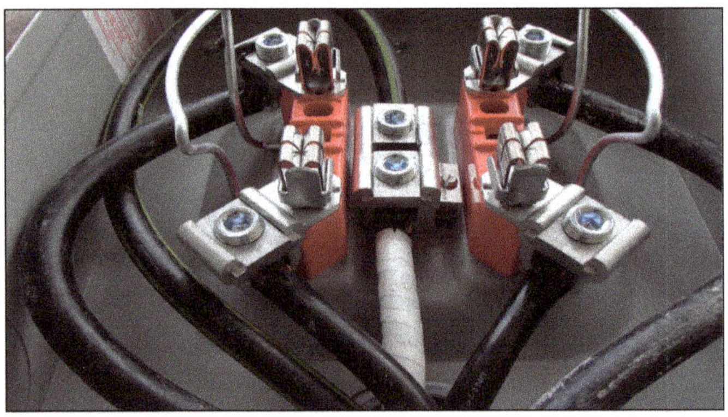

Photo 5.6. Grounded (neutral) conductor bonded to meter socket enclosure on the supply side of the service disconnecting means.

in the service raceway or armor [*NEC* 250.92(A) and IRC E3609.2], and any metallic raceway or armor enclosing a grounding electrode conductor as permitted or required in *NEC* 250.64 or IRC E3610.2 or E3610.3 must be effectively bonded together. Acceptable methods of bonding will be discussed in the following paragraphs.

The reason bonding these service enclosures is so important is that there is no overcurrent protection on the line (supply) side of them. Service conductors are protected from overloads only (not short circuits) on their load end. If an enclosure or raceway containing service conductors becomes isolated and a ground fault occurs, it would stay at a voltage level above ground and could pose a shock hazard. Effective bonding provides a low-impedance path for ground-fault current back to the source, which is usually the power company transformer. The goal is to produce enough current to cause the primary fuses of the utility transformer to open.

METHODS OF BONDING SERVICE EQUIPMENT

Bonding *(connected together)* of metallic enclosures and raceways at the service must be done in an effective and intentionally constructed manner. Service-entrance conductors are often installed in metallic raceways and enclosures up to the service disconnecting means (main). These service-entrance conductors are not protected by a fuse or circuit breaker other than any utility primary device ahead of the transformer. This bonding must be done to ensure a stouter, more rigid, and effective manner for the duration of any faulted condition. The primary purpose of bonding is to connect metal objects together both electrically and mechanically. Bonding requirements on the load side *(past the first means of disconnect)* of the service disconnecting means are less restrictive than on the supply side because there is an overcurrent device ahead of the load side that serves to open the faulted condition on the system. There are sizing differences between the supply-side bonding jumpers on the line side as compared to equipment bonding jumpers on the load side. Bonding on the line side of the service disconnect uses the size of the service-entrance conductors to establish the minimum size of the supply-side bonding jumper [*NEC* 250.102(C), Table 250.102(C)(1)]. Bonding on the load side is based on the overcurrent protection supplying the circuit [*NEC* 250.102(D), Table 250.122 or IRC E3908.20, Table E3908.13]. This section focuses on the bonding requirements at the service equipment (line side).

Electrical continuity at service equipment on the line side is to be ensured by any of the methods specified below.

1. Bonding of the service equipment to the grounded (neutral) service conductor with a listed pressure connector, listed clamp, other listed means, or by an applicable method described at *NEC* 250.8(A). Connection devices or fittings that depend solely upon solder are not permitted for this type of connection.
2. Threaded couplings, threaded entries, or listed threaded hubs on enclosures with joints that are made up wrenchtight where rigid metal conduits and intermediate metal conduits are installed.
3. Threadless couplings and connectors made up tight for metal raceways and metal-clad cables.
4. Other listed devices, such as bonding-type locknuts, bushings, or bushings with bonding jumpers.

These bonding methods for service equipment can be found at *NEC* 250.92(B) or IRC E3609.4.

Bonding jumpers are required to be installed around concentric or eccentric knockouts in enclosures that are punched or otherwise formed so as to impair the electrical connection from the installed raceway or cable to the enclosure.

It should be understood that ordinary locknuts or bushings are not recognized for use where bonding of service equipment is involved; bonding-type locknuts and bonding-type bushings must be used. Obviously, locknuts are used on the opposite side of bushings or grounding locknuts to make a solid mechanical connection. Where a concentric or eccentric knockout is encountered, a bonding bushing and jumper must be used at the service equipment to ensure adequate bonding.

These methods ensure the proper bonding of equipment to the grounded (neutral) service conductor and comply with NEC 250.8 or IRC E3406.14.1. Where threaded couplings or threaded bosses are encountered, their joints must be wrenchtight where rigid or intermediate metal conduit is used. Threadless couplings and connectors used for raceways, such as rigid metal conduit, intermediate metal conduit, and electrical metallic tubing, enclosing service-entrance conductors must be made up tight as well.

SIZING SUPPLY-SIDE BONDING JUMPERS ON SUPPLY SIDE OF SERVICE EQUIPMENT

Where a bonding bushing is used to bond raceways or other equipment within the enclosure for the service, disconnecting means and a supply-side bonding jumper must be installed on the line side of the service. The supply-side bonding jumper cannot be smaller than the sizes given in either *NEC* Table 250.102(C)(1) or IRC E3609.5 and Table E3603.4 [*NEC* 250.102(C)(1) or IRC E3609.5]. These tables are based on the size of the largest ungrounded service-entrance conductor and not on the ampere rating of the service disconnecting means.

For example, where a 200-ampere service has 4/0 AWG aluminum service-entrance conductors installed, the minimum size of the supply-side bonding jumper is 4 AWG copper or 2 AWG aluminum. *NEC* Table 250.66 is used for sizing grounding electrode conductors. *NEC* Table 250.102(C)(1) is used to size supply-side bonding jumpers, as well as main bonding jumpers. IRC Table E3603.4 is used for sizing grounding electrode conductors, supply-side bonding jumpers, and main bonding jumpers.

INTERSYSTEM BONDING AND GROUNDING REQUIREMENTS FOR COMMUNICATION SYSTEMS

An accessible means external to the service equipment enclosure for connecting intersystem bonding and grounding conductors is required by the *Code*. This intersystem bonding is required for the bonding of "limited energy" systems such as communications, optical fiber cables, radio and TV, and CATV circuits together with the main electrical service grounding and bonding system. This intersystem bonding

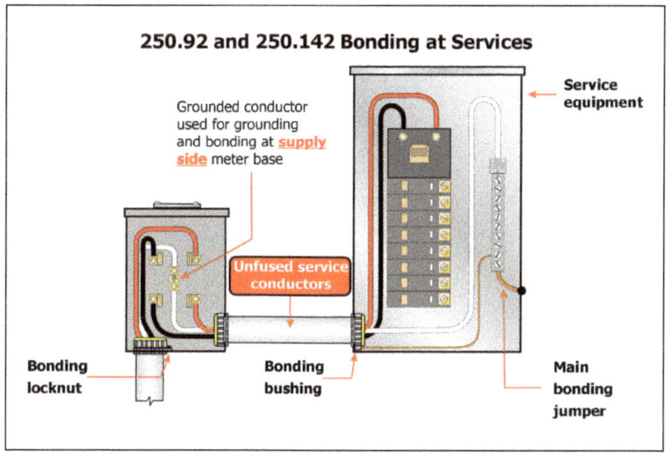

Figure 5.7. Using the grounded conductor for grounding and bonding equipment (supply side of the service disconnect) and a bonding bushing in the service disconnect for bonding at the service equipment

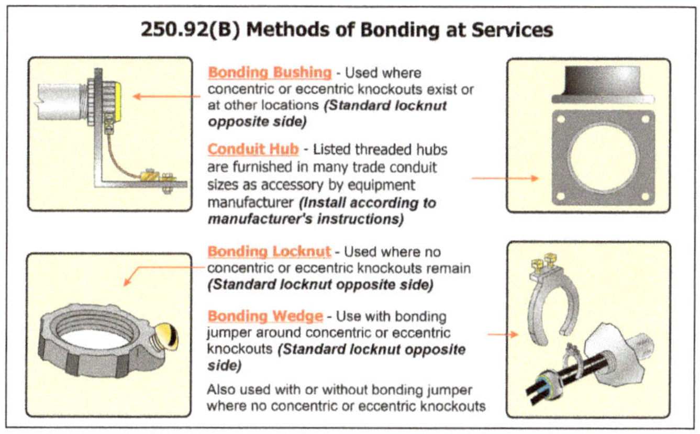

Figure 5.8. Typical fittings acceptable for bonding on the supply side of the service disconnect

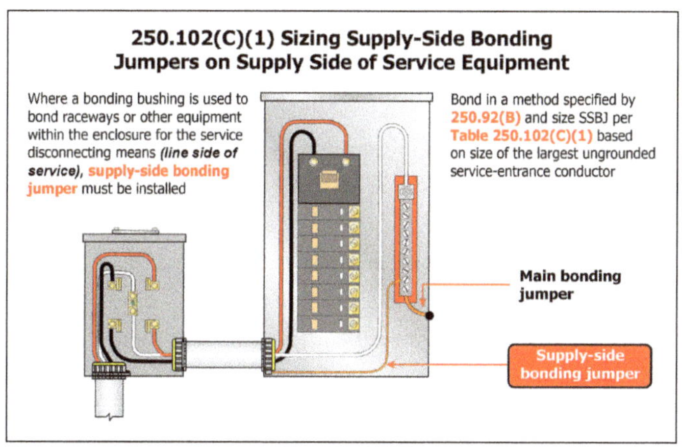

Figure 5.9. Sizing of supply-side bonding jumpers

Photo 5.7. Bonding a conduit nipple that contains service-entrance conductors on the supply side of the service disconnect between the meter base and the service disconnect enclosure

Photo 5.8. Supply-side bonding jumper installed on the line side of service equipment

termination is required at the disconnecting means for any additional buildings or structures as well as at the main electrical service disconnecting means. In a similar fashion, the disconnecting means permitted by *NEC* 550.32(A) for mobile homes is also permitted as the point of intersystem bonding. The *Code* gives two methods to achieve intersystem bonding of communication systems. The first is described at *NEC* 250.94(A) or IRC E3609.3.1. This first method pertains to a listed intersystem bonding termination (IBT) device. The second method involves connections to an aluminum or copper busbar [*NEC* 250.94(B) or IRC E3609.3.2].

Suppose a listed intersystem bonding termination (IBT) device is used. In that case, the IBT is required to be accessible for connection and inspection and must consist of at least three intersystem bonding termination points solely for communication bonding conductors. This device shall not interfere with the opening of any service or metering equipment enclosures. These IBT devices should be listed as grounding and bonding equipment.

The listed IBT should be securely mounted and electrically connected to an enclosure for the service equipment, to the meter enclosure, to the building disconnecting means, or an exposed nonflexible metallic service raceway, or be mounted at one of these enclosures and be connected to the enclosure or the grounding electrode conductor with a minimum 6 AWG copper conductor. When located at the disconnecting means for a separate building or structure, the listed IBT must be securely mounted and electrically connected to the metallic enclosure for the separate building or structure disconnecting means. The listed IBT can also be mounted at the disconnecting means for the separate building or structure and connected to the metallic enclosure or the grounding electrode conductor with a 6 AWG or larger copper conductor.

In existing buildings or structures, an accessible means external to enclosures for connecting intersystem bonding and grounding electrode conductors is permitted at the service equipment, and the disconnecting means for any additional buildings or structures [*NEC* 250.94(A) Exception]. This alternative method to the intersystem bonding of communication circuits at existing buildings or structures is not addressed in the IRC. In existing buildings or structures where any intersystem bonding and grounding electrode conductors for communication systems exist, installation of the listed IBT is not required. At existing buildings or structures where any intersystem bonding and grounding electrode conductors for communication systems exist, any of the following accessible means are permitted to be employed:

1. Exposed nonflexible metal service raceways
2. Exposed grounding electrode conductor
3. Approved means for external connection, such as a 6 AWG copper bonding jumper or grounding electrode conductor not shorter than 150 mm (6 in.) and bonded to the metallic service raceway or service equipment enclosure with a listed and identified fitting.

The second method of achieving intersystem bonding described at *NEC* 250.94(B) or IRC E3609.3.2 involves a connection to an aluminum or copper busbar not less than 6 mm thick × 50 mm wide (¼ in. thick × 2 in. wide) of sufficient length to accommodate at least three terminations for communication systems in addition to other connections. Unlike with the listed IBT, other bonding jumpers (such as a bonding jumper for bonding of a gas piping system) can be terminated at this “other means” of IBT bonding. The IBT busbar must be securely fastened and installed in an accessible location. Connections to this IBT busbar must be made by the use of a listed connector. If aluminum busbars are used, the installation must also comply with *NEC* 250.64(A) or IRC E3610.2 for such things as terminating bare aluminum or copper-clad aluminum grounding electrode conductors within 450 mm (18 in.) of the earth. This busbar must be connected to the grounding electrode system by a conductor that is as large as any conductor connected or terminated on the busbar itself.

For specific *Code* references related to bonding for communication systems and intersystem bonding terminations, see figures 5.10 and 5.11 and *NEC* 250.94 or IRC E3609.3. Intersystem bonding of communication systems is discussed further in Chapter 17 of this publication.

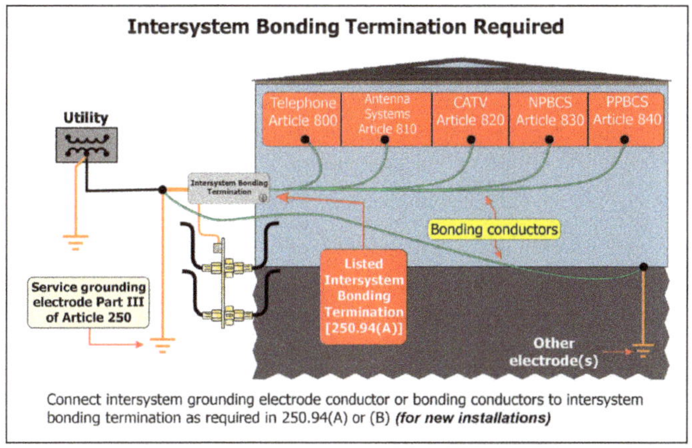

Figure 5.10. Intersystem grounding and bonding termination requirement

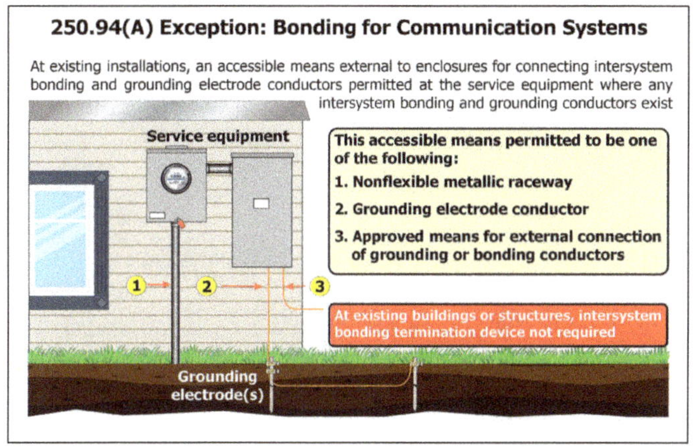

Figure 5.11. Intersystem grounding and bonding at existing buildings

Photo 5.9. Required intersystem bonding termination (with and without cover installed)

CONNECTION TO EARTH — THE GROUNDING ELECTRODE SYSTEM

Grounding electrodes provide the essential function of connecting the electrical system and electrical equipment to the earth. The earth is typically considered to be at zero voltage potential. In other words, the earth's ability to carry or conduct electricity. The earth is a conductor of electricity, but it is a poor conductor of electricity because of its dissipation factor. Typically, when electrical current is introduced into the earth, that current simply dissipates (scatters) out over the earth rather than following a nice, straight, clean path like current will do if it were introduced onto a conductor. In some cases, the grounding electrode(s) serves to ground the electrical system to earth. In other instances, the electrode(s) is used to connect non-current-carrying metallic portions of electrical equipment to the earth. In both situations, the primary purpose of the grounding electrode(s) is to maintain the electrical equipment at the *same* voltage potential as the earth voltage potential present where the grounding electrode(s) is located.

As stated, another essential function of the grounding electrode(s) is to dissipate overvoltages into the earth (see Figure 5.13). These overvoltages can be caused by high-voltage conductors being accidentally connected to the lower-voltage system, such as by a failure in a utility transformer or by an overhead conductor dropping onto the lower-voltage conductor. Overvoltages can also be caused by lightning. With more equipment (such as home appliances) containing microprocessors, this becomes increasingly important. *NEC* 250.24(E) or IRC E3607.4 requires "… the equipment grounding conductors, the service-equipment enclosures, and, if the system is grounded, the grounded service conductor to be connected to the grounding electrode(s)…" The conductor used to make this connection is the grounding electrode conductor.

A "grounding electrode" is defined as a conducting object that creates a direct connection to the earth (see NEC Article 100 or IRC Chapter 35). To establish a true understanding of what constitutes a grounding electrode, the definition must be used in conjunction with the list of electrodes identified in NEC 250.52(A) or IRC E3608.1. It can clearly be seen in this list of grounding electrodes that a grounding electrode can be a device or other conducting object, such as a building footing or metal well casing, that establishes and maintains a direct connection to the earth. The effectiveness of the connection is relative and is a variable item.

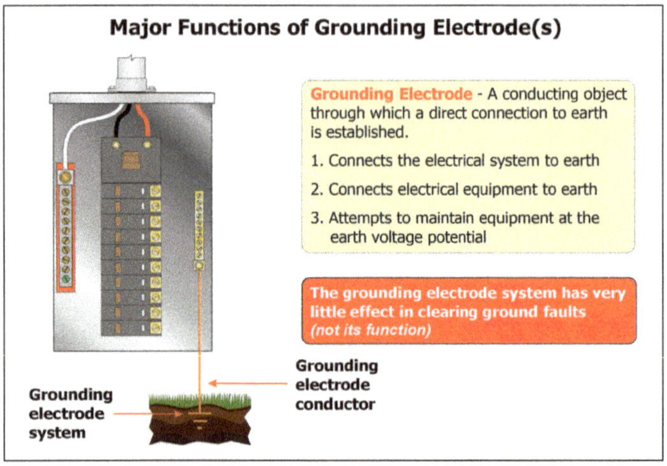

Figure 5.12. Major functions of the grounding electrode system

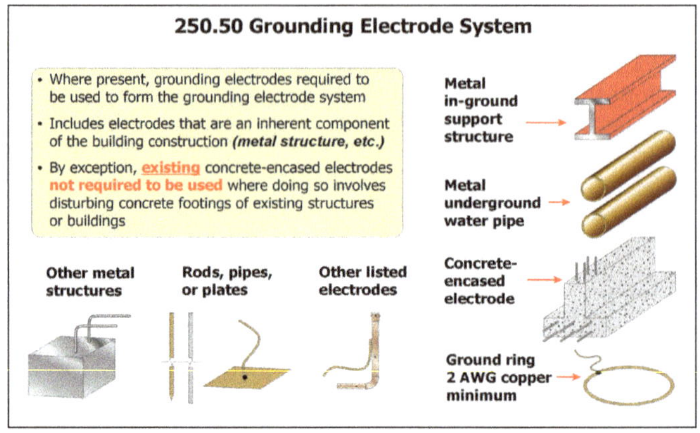

Figure 5.15. Grounding electrodes that are present must be used in the grounding electrode system

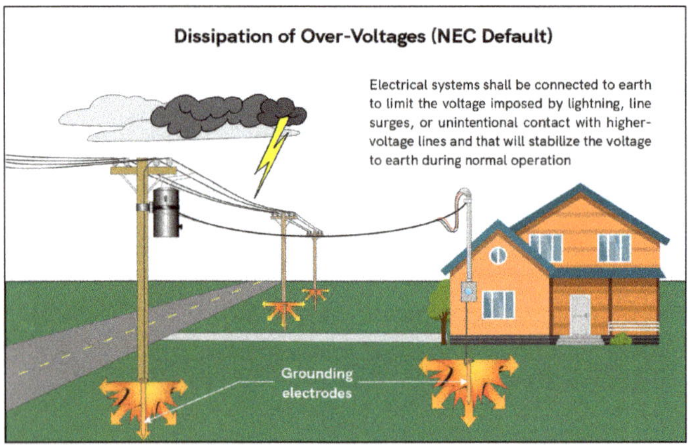

Figure 5.13. Grounding electrodes serve to dissipate overvoltages

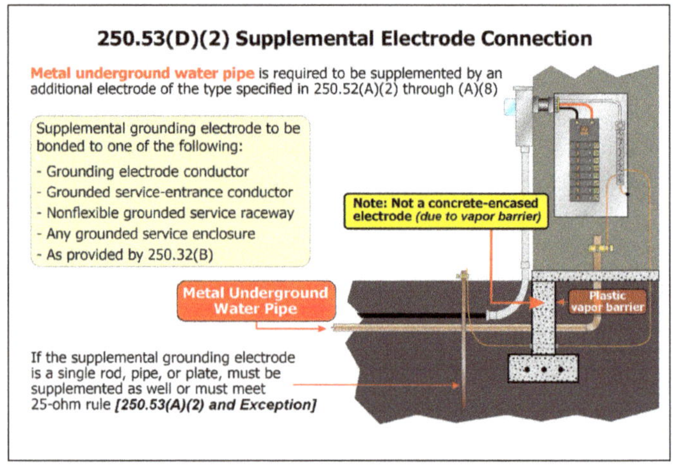

Figure 5.16. A supplemental grounding electrode is required for an underground metal water piping electrode.

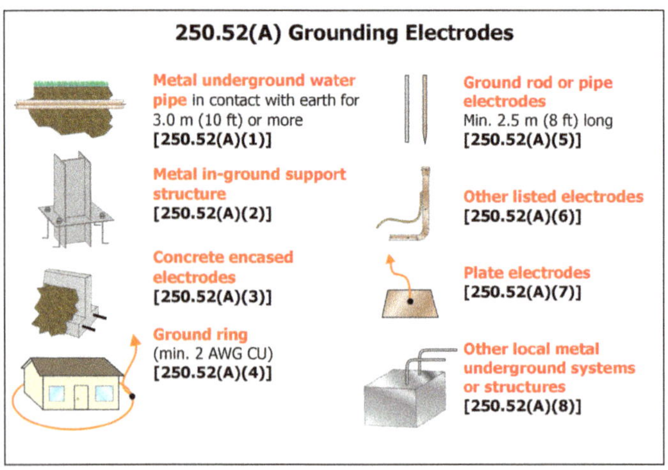

Figure 5.14. Conducting objects comprising the grounding electrode system

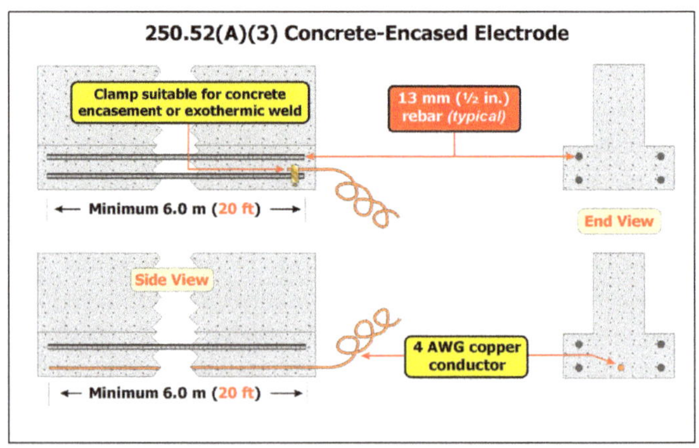

Figure 5.17. Concrete-encased electrodes must be encased in at least 50 mm (2 in.) of concrete, located horizontally or vertically within foundations or footings that are in direct contact with the earth. The electrode has to consist of at least 6.0 m (20 ft) or more of 13 mm (1/2 in.) rebar or a minimum 4 AWG bare copper conductor.

The resistance in the connection between an electrode and the earth will vary based on such things as soil conditions, electrode depth, type of electrode, and seasonal conditions or geographical location(s).

Building construction usually determines what the grounding electrode system will be. Some electrodes are inherent to the construction of one- and two-family dwellings. The *Code* requires that all grounding electrodes that are present at the job site be bonded together to form the grounding electrode system (*NEC* 250.50 or IRC E3608.1). This is required at each building or structure served on the premises. Grounding electrodes that must be bonded together *(if present at the job site)* and used as the grounding electrode system include the following:

Notice that all of these potential grounding electrodes have conditions that they must meet in order to qualify as a grounding electrode.

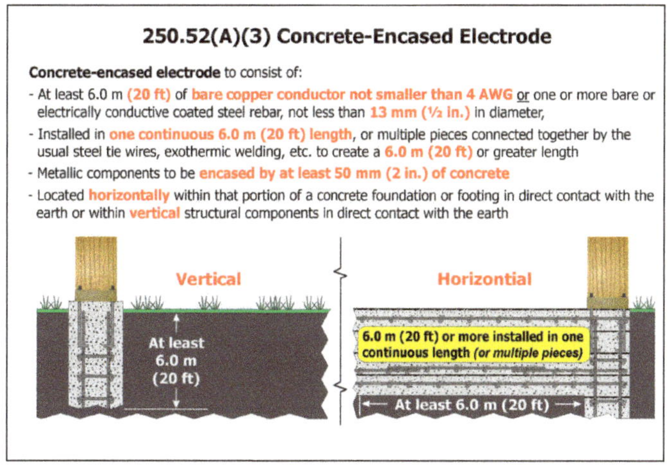

Figure 5.18. Qualifications for a concrete-encased electrode

Underground metal water pipes in contact with the earth for 3.0 m (10 ft) or more. This requirement includes metal piping for both hot and cold water. In many parts of the country, it is common to build houses with a concrete slab on grade with metal cold and hot water piping systems just below the slab in contact with the earth. If the metal pipe is brought into the building, it must be used as a grounding electrode if it is in contact with the earth for 3.0 m (10 ft) or more. Underground metal waterlines that are outside but do not enter the building may, but are not required to, be used as a grounding electrode. Water pipe electrodes must be supplemented by an additional electrode, such as a ground rod or any of the other permitted grounding electrodes [*NEC* 250.53(D)(2) and IRC E3608.1.1.2]. A grounding rod is probably the most commonly used electrode for augmenting the water pipe electrode, but is not the only electrode permitted for this purpose [*NEC* 250.52(A)(1) or IRC E3608.1.1].

The supplemental electrode is permitted to be bonded to the grounding electrode conductor, the grounded service-entrance conductor, any nonflexible grounded service raceway, or any grounded service enclosure. The most common location for the connection is the neutral conductor terminal bar at the service [*NEC* 250.53(D)(2) or IRC E3608.1.1.2].

The connection of the grounding electrode conductor to a metal water pipe electrode must be made within the first 1.52 meters (5 feet) of where the water pipe enters the building. This measurement is taken along the water piping. The same rule applies to the supplemental electrode connection to the water pipe. Typically, electricians do not have full control of the water piping system, and many unqualified persons may modify the water system or perform repairs with nonmetallic materials, which can isolate the required grounding electrode conductor or supplemental grounding electrode connection [NEC 250.68(C)(1) or IRC E3608.1.1.1].

A metal in-ground support structure that is effectively grounded must also be included in the grounding electrode system. This metal in-ground support structure of a building is an effective electrode only where 3.0 m (10 ft) or more of a single structural metal member is in direct contact with the earth (with or without concrete encasement). Grounded metal in-ground support structures are not typically found in residential one- and two-family dwellings; however, it is not impossible for them to be part of a residential building. Isolated metal structural elements, such as I-beams or support columns, are not considered a metal in-ground support structure as they are not typically in contact with the earth for 3.0 m (10 ft) or more. The NEC requires metal in-ground support structures to be used regardless of the type of occupancy [NEC 250.52(A)(2)]. The IRC does not include metal frames of buildings or metal in-ground support structures of a building as one of the grounding electrodes to be used in the grounding electrode systems, so, if present, defaulting to the NEC would be necessary in this situation.

Concrete-encased electrodes are being employed with increasing frequency as a grounding electrode due to their effectiveness as a grounding electrode and due to the increasing reduction of metal water piping systems in favor of plastic water lines, particularly in residential construction. When employing the structural metallic components (such as rebar) in the concrete-encased electrode process, they must consist of at least 13 mm (½ in.) or larger diameter rods that are 6.0 m (20 ft) or longer encased by at least 50 mm (2 in.) of concrete and located horizontally or vertically within that portion of a concrete foundation, footing, or pier that is in direct contact with the earth. This electrode is perhaps the electrode with the most conditions tied to it in order to qualify as a grounding electrode. All of the conditions described above must be met in order for the concrete-encased electrode to be considered as an effective electrode in the grounding electrode system.

Concrete-encased electrodes and their reinforcing bars or

Photo 5.10. Potential concrete-encased electrode must meet all conditions of *NEC* 250.52(A)(3)

Photo 5.11. Connection from the grounding electrode conductor to an extension from a concrete-encased electrode must be accessible

rods are typical elements of building construction inherent to the majority of one- and two-family dwellings. Where concrete-encased electrodes are present and used, an additional grounding electrode is *not required* to supplement this grounding electrode. Just like the rest of the grounding electrodes described by *NEC* 250.52(A) or IRC E3608, the *Code* does not require a concrete-encased electrode to be installed, but if it is present during construction, it must be used in the grounding electrode system.

To effectively utilize this grounding electrode, enforcing jurisdictions must take a proactive approach and install the concrete-encased electrode before the slab or foundation is poured during the construction process.

To ensure proper compliance with this important Code requirement, cooperation is essential between the electrical contractor, the builder, and the authority having jurisdiction (AHJ). Instead of reinforcement rods, 6.0 m (20 ft) or more of 4 AWG copper wire can be installed in the horizontal or vertical structural element prior to pouring to create a concrete-encased electrode if desired.

This 4 AWG or larger copper conductor must also meet the other conditions for a concrete-encased electrode of encasement by at least 50 mm (2 in.) of concrete and located horizontally or vertically within that portion of a concrete foundation, footing, or pier that is in direct contact with the earth. This wire-type electrode is sometimes installed where rebar is not used in the foundation or footing or in addition to the reinforcing steel [*NEC* 250.52(A)(3) or IRC E3608.1.2].

It has become a commonly accepted practice to extend a rebar-type extension from a concrete-encased electrode out of the footing or foundation before the slab or foundation is poured. This is accomplished by using another piece of rebar connected to a qualifying concrete-encased electrode and "stubbed-up" out of the poured concrete to provide an accessible connection point above the slab.

At a later date, the electrician can make the grounding electrode conductor connection after the foundation has been poured and cured. *NEC* 250.68(C)(3) or IRC E3611.5 gives this permission for the reinforcing rod or bar to be extended from the qualifying concrete-encased electrode location within the concrete to an accessible location above the concrete for connection of a grounding electrode conductor.

It is also common here to install a cover plate or door at the wall covering to make the connection of the ground clamp to the extension accessible (see photo 5.11). It is important to note here that the extension or "stub-up" is not part of the concrete-encased electrode. A structural component meeting all of the conditions of *NEC* 250.52(A)(3) or IRC E3608.1.2 must be present for the extension to be connected to the same. The "stubbed-up" rebar adds to or takes away nothing from the structural component that qualifies as a concrete-encased electrode (see Figure 5.19).

This "sub-up" extension must not come out of the concrete where it would be subject to corrosion, such as stubbing out sideways at the footer into the surrounding soil. The rebar extension must come out of the concrete above the slab or foundation wall or out of the side of the foundation wall in a location where moisture or other corrosive elements would not be present to cause deterioration of the rebar extension. The additional rebar "stub-up" section must be continuous with the grounding electrode rebar or be connected to the grounding electrode rebar and connected by the usual steel tie wires, exothermic welding, welding, or other effective means. Also, this rebar extension cannot be used as a conductor to interconnect other electrodes of grounding electrode systems. The ideal installation is to stub up vertically in the wall space or cavity where the service panelboard will be installed. This allows access and also protects the ground clamp or fitting required by *NEC* 250.10 or IRC E3611.6. A ground clamp used with any type of structural rebar must be listed for application with rebar (see photo 5.11).

Although the Code does not require a concrete-encased electrode for the grounding system (unless it qualifies and is present), some inspection jurisdictions mandate the installation and use of one by local ordinance because it is usually built into the construction and has proven to be cost-effective and highly efficient. In certain areas, a concrete-encased electrode has also been identified as the most effective type, capable of maintaining its resistance to earth in specific soil conditions.

MADE GROUNDING ELECTRODES

Where none of the electrodes that are inherent to the construction of the building - such as a metal underground water pipe in direct contact with the earth for 3.0 m (10 ft) or more, grounded metal in-ground support structure of the building, or concrete-encased electrode - are present, one or more electrodes must be installed and used. Where practicable, electrodes are to be embedded below the permanent moisture level. Grounding electrodes must have nonconductive coatings, such as paint or enamel, removed [*NEC* 250.53(A)(1) or IRC E3608.3]. Where more than one electrode is used, each electrode of one grounding system (including those used for air terminals for lightning protection systems) is not to be less than 1.8 m (6 ft) from any other electrode of another grounding system. Where two or more electrodes are effectively bonded together, they are treated as a single electrode [*NEC* 250.53(B) or IRC E3608.3].

The Code includes a list of grounding electrodes that must be used for grounding if present on the job site. This list describes each electrode with details and descriptions. Other local metal underground systems or structures, such as piping systems, well casings, and underground tanks, may also be used as grounding electrodes [NEC 250.52(A)(5), (7), (8) or IRC E3608.1.4 and E3608.1.5].

Some electrodes are not allowed for grounding, such as aluminum electrodes, metal underground gas piping, and the structural steel of a permanently installed in-ground swimming pool [NEC 250.52(B) or IRC E3608.5, E3608.6, and E3608.7].

MADE GROUNDING ELECTRODES PERMITTED FOR GROUNDING

Rod and pipe electrodes must be at least 2.5 m (8 ft) in length. Electrodes for pipes or conduits cannot be smaller than metric designator 21 (¾ in.) trade size. Pipe or conduit electrodes of iron or steel must have their outer surface galvanized or be otherwise metal-coated for corrosion protection.

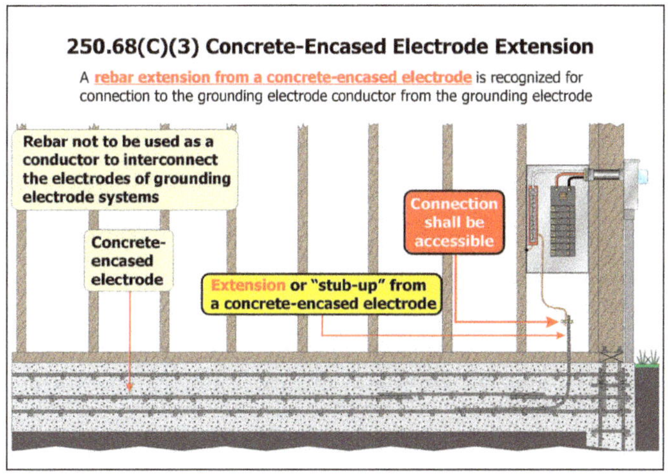

Figure 5.19. A rebar extension from a concrete-encased electrode is permitted

Iron or steel rod electrodes must be at least 15.87 mm (⅝ in.) in diameter. Nonferrous or stainless steel and copper or zinc-coated steel rods, or their equivalent, cannot be less than 15.87 mm (⅝ in.) in diameter unless listed for small diameter.

As previously mentioned, all grounding electrodes must meet specific installation requirements or conditions to qualify as a grounding electrode. This is usually not an issue for electrical installers when using metal underground water pipes or concrete-encased electrodes, as these are typically installed as part of the building's construction, but they still must satisfy certain requirements or conditions. When installing a "made" electrode, such as a rod or pipe electrode, it must also comply with the minimum requirements for grounding electrode installation. A single rod, pipe, or plate electrode must be supplemented by an additional electrode, like another rod, pipe, or plate. Additionally, existing electrodes—such as a concrete-encased electrode, metal in-ground support structures of a building, or a qualifying underground metal water pipe—can also serve to supplement these rod, pipe, or plate electrodes [NEC 250.53(A)(2) or IRC E3608.4].

The rod or pipe electrode must be installed so that at least 2.44 m (8 ft) of its length is in contact with the soil unless rock conditions are encountered. If a rocky condition prevents driving the electrode to a depth of 2.5 m (8 ft), the rod or pipe electrode may be driven at an oblique angle not exceeding 45 degrees from the vertical. If it is still not possible to obtain the minimum driven contact length of 2.44 m (8 ft), it is then permitted to be buried in a trench that is at least 750 mm (30 in.) deep. With the driven ground rod, the upper end of the electrode is required to be flush with or below ground level unless the aboveground end and the grounding electrode conductor attachment are protected against physical damage while still maintaining the required 2.44 m (8 ft)

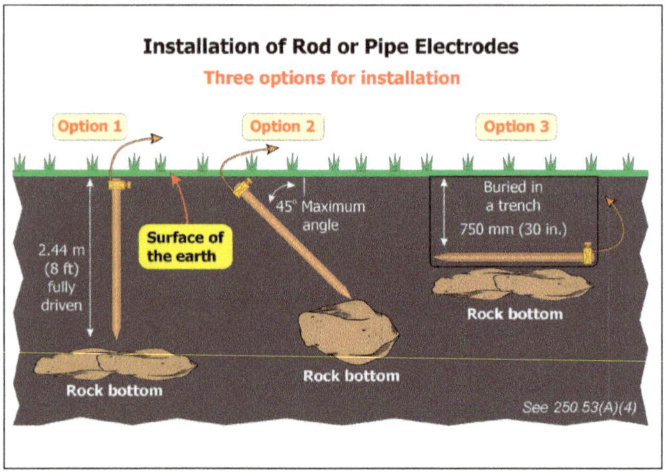

Figure 5.20. Installation of rod or pipe grounding electrodes

length in contact with the earth [NEC 250.53(A)(4) or IRC E3608.1.4.1]. See NEC 250.52(A)(5), 250.53(A)(4), or IRC E3608.1.4 for details concerning qualifying conditions and installation of rod or pipe grounding electrodes.

A plate electrode must expose not less than 0.186 m^2 (2 ft^2) of surface to exterior soil. Plate electrodes are to be installed not less than 750 mm (30 in.) below the surface of the earth. Iron or steel plate electrodes must be at least 6.4 mm (¼ in.) in thickness. Electrodes of nonferrous metal must be at least 1.5 mm (0.06 in.) in thickness. A plate electrode might be chosen because of the earth or soil conditions (such as locations where ledge or solid rock is encountered). The rules in *NEC* 250.52(A)(7) and 250.53(A)(5) or IRC E3608.1.5 would apply to plate electrodes.

If a **ground ring electrode** is chosen for installation, it must encircle the building or structure completely and be a minimum of 6.0 m (20 ft) in length. Ground rings are not commonly used in dwelling units, as they are not inherent to the construction of residential buildings and are labor-intensive. They are installed to establish a grounding electrode connection if needed. The ground ring, installed and used, must be of copper wire not smaller than 2 AWG and buried at least 750 mm (30 in.) deep [NEC 250.52(A)(4) or IRC E3608.1.3]. In addition to these grounding electrodes, all grounding electrodes, such as ground rods or ground plates, must be bonded to the grounding electrode system and thus become part of a single system.

A metal well casing is an example of an **"other local metal underground system or structure"** as it is an underground metal structure and is permitted to be used as a grounding electrode by the *NEC*. The water piping from the well to the building is often nonmetallic. The well casings are usually made from steel pipe and may be driven several hundred feet into the ground. The well casing is not an underground metal water pipe and, as such, does not require a supplemental electrode [*NEC* 250.52(A)(8)]. Other listed grounding electrodes, such as chemically treated ground rods, are also permitted to be used to establish a grounding electrode system. Still, again, these are not common in dwelling unit construction [*NEC* 250.52(A)(6) or IRC E3608.1.6].

ROD, PIPE, OR PLATE ELECTRODES REQUIRED TO BE SUPPLEMENTED

Rod, pipe, and plate electrode installation requirements are provided in *NEC* 250.53(A) or IRC E3608.3. Individually installed grounding electrodes of the rod, pipe, or plate types are required to be supplemented by an additional electrode, which could be another rod, pipe, or plate electrode [*NEC* 250.53(A)(2) or IRC E3608.4]. By exception, a single rod, pipe, or plate electrode can be installed without augmentation if the single electrode has a resistance to earth of 25 ohms or less. When supplemented by an additional electrode or a single electrode having a resistance of 25 ohms or less, no additional grounding electrodes are required to be installed regardless of the resulting resistance *[see NEC 250.53(A)(2), Exception or IRC E3608.4, Exception]*.

Where multiple rod, pipe, or plate electrodes are installed to meet these supplemental requirements, they are not permitted to be less than 1.8 m (6 ft) apart [*NEC* 250.53(A)(3)]. The paralleling efficiency of rods longer than 2.5 m (8 ft) is improved by spacing them apart greater than 1.8 m (6 ft). Trade practice is to space the rods as far apart as the rod is long. For example, 2.5 m (8 ft) long rods might be installed 2.5 m (8 ft) apart.

Many inspection authorities require that a single rod, pipe, or plate electrode be tested to verify that the resistance is 25 ohms or less. Experience has shown that a single ground rod in many areas cannot achieve the maximum 25-ohm resistance due to the surrounding soil conditions. Please note that when electrodes are supplemented as required by the Code, no resistance testing is necessary.

The principles involved in earth-resistance testing are complex, and detailed information can be found in Chapter 7 of the IAEI's *Soares Book on Grounding and Bonding* textbook. Another resource is the manufacturer of the rod electrode itself. The manufacturer often provides a training manual and step-by-step instructions for testing the resistance of the electrodes.

A COMMON GROUNDING ELECTRODE SYSTEM

A grounding electrode system can include one or multiple electrodes. When there are several electrodes, they must be bonded together to act as a single grounding system. If an AC system is connected to a grounding electrode at or in a building, the Code requires that the same "common grounding electrode" be used to ground conductors, enclosures, and equipment within or on that building. This means it's not acceptable to install one grounding electrode for the electrical service or system and another for grounding equipment like a hydromassage tub motor, or to have different grounding electrodes connected to service equipment and a feeder panelboard. When two or more services supply a single building or structure and are connected to a grounding electrode, the same grounding electrode system must be used for all of them. Two or more electrodes that are effectively bonded are considered a single or common grounding electrode in this context (NEC 250.58 or IRC E3607.6).

SIZE OF GROUNDING ELECTRODE SYSTEM BONDING JUMPERS

All grounding electrodes that are present are to be bonded together to form the grounding electrode system. It should be noted that the conductor used to bond the grounding electrodes together is a bonding jumper and not a grounding electrode conductor. As a result, there is no requirement for the bonding jumper to be installed at "one continuous length without a splice" as the grounding electrode conductor itself is required. It is permitted to connect the bonding jumper to the various grounding electrodes individually. Where the ground clamp is suitable, the bonding jumper is permitted to pass through the clamp unbroken and connect to other grounding electrodes in a "daisy chain" fashion.

The bonding jumper installed to connect multiple grounding electrodes together for the system must be sized in accordance with NEC Table 250.66 or IRC Table E3603.4 (with applicable notes) [NEC 250.53(C) or IRC E3608.2]. Table 5.1 in this chapter lists the minimum size of grounding electrode conductors or bonding jumpers required to bond the electrodes together and form the grounding electrode system.

In some cases, the bonding conductor is permitted to be smaller than the size shown in Table 5.1. For example, a size 8 AWG copper conductor is permitted to be installed for a 100-ampere service where it is protected from physical damage.

The grounding electrode conductor or bonding jumper to the concrete-encased electrode, ground ring, and made electrodes of rods, pipes, or plates is permitted to be the size as shown at *NEC* 250.66(A), (B), and (C) or the notes to IRC Table E3603.4 only where the grounding electrode conductor or bonding jumper does not extend on to other types of electrodes that require a larger size of conductor. For example, if the bonding jumper starts at the concrete-encased electrode and is routed to the water pipe or metal in-ground support structure, it must be sized for the grounding electrode conductor required for the water pipe or building steel. This "daisy-chain extension" aspect will be discussed in greater detail later in this chapter.

Where the interior metal water pipe is used to bond other electrodes, such as concrete-encased electrodes, building in-ground support structures, or a ground ring together, bonding connections must be made within the first 1.5 m (5 ft) as measured along the water piping, from where the water pipe enters the building. At a dwelling setting, the metal water pipe is not permitted to be used for bonding the grounding electrodes together beyond this first 1.5 m (5 ft) [*NEC*

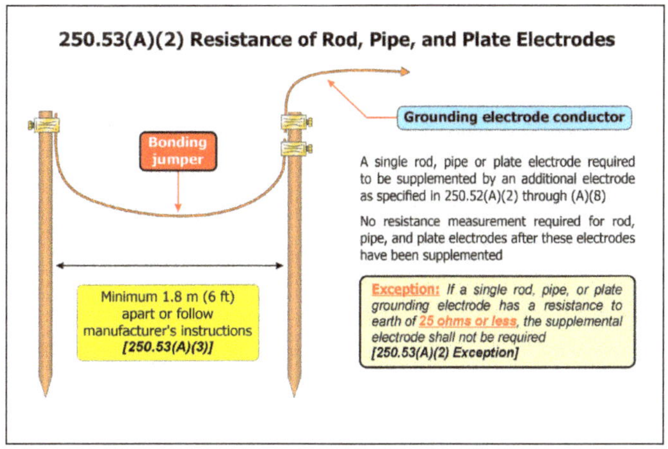

Figure 5.21. Rod, pipe, or plate electrodes are required to be supplemented

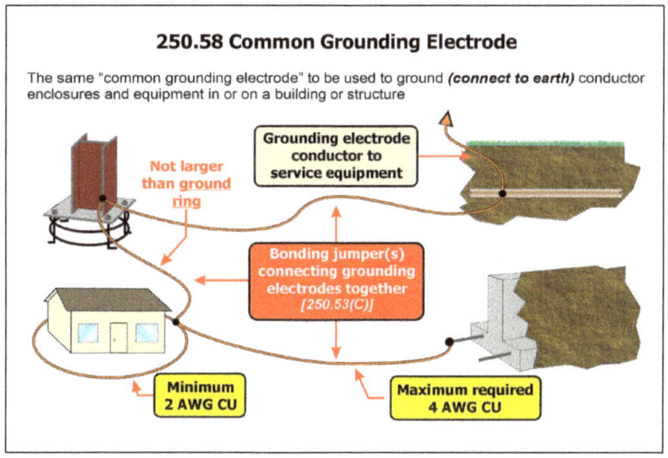

Figure 5.22. Grounding electrode conductor and bonding jumpers to multiple grounding electrodes

Sizing Grounding Electrode Conductors

Electrode	Sizing Requirement
Water Pipe Electrode	*NEC* Table 250.66 or IRC Table E3603.4
Metal In-Ground Support Structure	*NEC* Table 250.66 (Not Covered in IRC)
Concrete-Encased Electrode	*NEC* Table 250.66 or (4 AWG copper) if in compliance with *NEC* 250.66(B) or IRC E3607.4*
Ground Ring	*NEC* Table 250.66 or (2 AWG copper) if in compliance with *NEC* 250.66(C) or IRC E3607.4*
Rod, Pipe, or Plate Electrodes	*NEC* Table 250.66 or (6 AWG copper) if in compliance with *NEC* 250.66(A) or IRC E3607.4*
Other Listed Electrodes	*NEC* Table 250.66 or IRC Table E3603.4

*__IRC E3608.1__ covers the rules for establishing the grounding electrode system. Sizing requirements for grounding electrode conductors are covered at **IRC E3607.4**, which refers to **IRC Table E3603.4**. Taking a closer look at the notes at the bottom of **IRC Table 3603.4**, notice **note "f"** at the bottom of the table, which indicates that grounding electrode conductors that do not extend on to other types of electrodes that require a larger size of conductor, sizing is permitted to be a maximum 6 AWG copper or 4 AWG aluminum for rod or pipe electrodes and a maximum 4 AWG copper for concrete-encased electrodes.

Table 5.1 Sizing Grounding Electrode Conductors/Bonding Jumpers *[See NEC 250.53(C) or IRC E3608.2]*

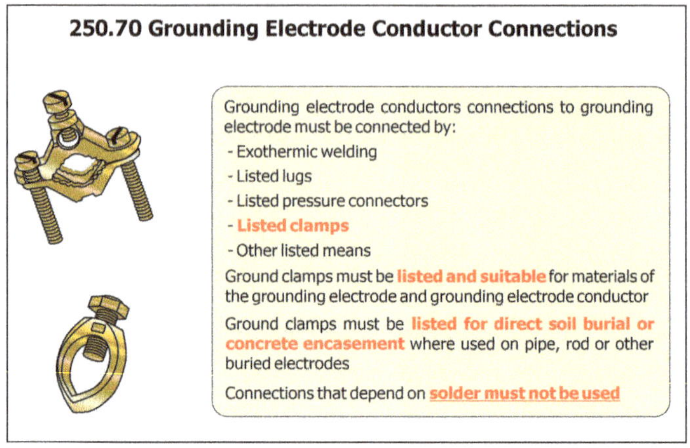

Figure 5.24. Various types of ground clamps

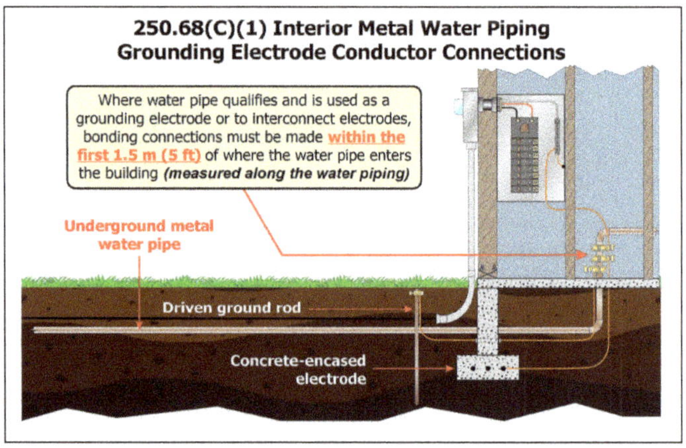

Figure 5.23. Use of interior water pipe electrode for bonding other grounding electrodes together

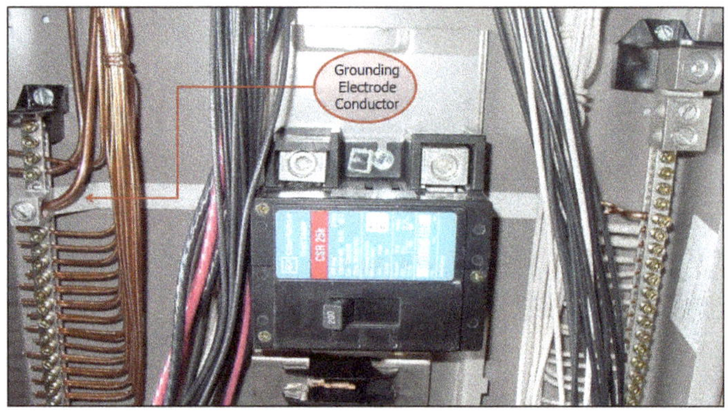

Photo 5.12. The grounding electrode conductor connected to the grounding terminal bar and the grounded (neutral) conductor terminal bar at the service equipment

250.68(C)(1) or IRC E3608.1.1.1]. A grounding electrode conductor and/or bonding jumper(s) should be run from one or more grounding electrodes directly to the service neutral point. Rebar is not permitted to be used as a conductor to interconnect the electrodes of grounding electrode systems.

GROUNDING ELECTRODE CONDUCTOR TO GROUNDING ELECTRODES

A grounding electrode conductor must be used to connect the grounding electrode system to the grounded (neutral) terminal bar in the service equipment (see photo 5.12).

The grounding electrode conductor(s) are permitted to be run to any of the grounding electrodes that make up the grounding electrode system if it is sized properly. For example, a 200-ampere service uses a metal water pipe as an electrode, and it is supplemented with two driven ground rods. It would not be acceptable to run a 6 AWG copper grounding electrode conductor from the service to the ground rod(s) and from the ground rod(s) to the metal water pipe electrode, since the grounding electrode conductor to the metal water pipe electrode is required to be a 4 AWG copper conductor. It would be acceptable to run a 4 AWG copper conductor from the service to the metal water pipe electrode and then a 6 AWG or larger copper conductor from the metal water pipe electrode to the ground rods since neither of these grounding electrode conductors or bonding jumpers does not extend on to other types of electrodes that require a larger size of conductor [*NEC* Table 250.66 and 250.66(A) or IRC Table E3603.4 and footnotes]. The language at *NEC* 250.66(A) makes it clear that the connection to single or multiple rods, pipes, or plates is considered the same as one grounding electrode system in this application.

If it is more convenient, grounding electrode conductors can be run to the grounded (neutral) bar in the service enclosure individually from each grounding electrode rather than bonding or "daisy chaining" them together with a single bonding jumper. Electrically, the same goal and objective is accomplished by using either method, and both meet the minimum provisions required by the *Code* [*NEC* 250.64(F)].

GROUNDING ELECTRODE CONDUCTOR CONNECTIONS

The grounding electrode conductor(s) are required to be connected to the grounding electrode by exothermic welding, listed lugs, listed pressure connectors, listed clamps, or other listed means. Connections that depend on solder must never be used. The ground clamp or connection process must be listed for the material of the grounding electrode and the grounding electrode conductor to prevent corrosion. Where used on a pipe, ground rod, or other buried electrode, the ground clamp must be listed for direct burial or concrete encasement. More than one conductor is not permitted to be connected to the grounding electrode by a single clamp or fitting unless the clamp or fitting is specifically listed for the connection of more than one conductor (*NEC* 250.70 or IRC E3611.1).

When connecting to a grounding electrode, a bolted clamp made of cast bronze or brass is usually used. Remember that there are no pipe fittings, pipe plugs, or other screw-in devices suitable or approved for attaching a grounding electrode conductor. For indoor communication purposes only, a listed sheet-metal-strap-type ground clamp with a rigid metal base that sits on the electrode should be used, with a strap made of appropriate material and dimensions so it will not stretch during or after installation. The UL Product Spec Category (KDER) states: "Strap-type ground clamps are not suitable for attachment of the grounding electrode conductor of an interior wiring system to a grounding electrode." These clamps should be used solely for connecting communication circuits to electrodes.

Photo 5.13. Listed grounding clamp for attachment of grounding electrode conductor to ground rod (suitable for direct burial)

Photo 5.14. Grounding clamp listed for connection of grounding electrode conductors to pipes

GROUNDING ELECTRODE CONDUCTORS

Grounding electrode conductor material must be made of copper, aluminum, or copper-clad aluminum. The material selected must be resistant to any corrosive condition existing at the installation or must otherwise be suitably protected against corrosion [*NEC* 250.62]. Bare aluminum or copper-clad aluminum grounding electrode conductors are not permitted to be used where in direct contact with the earth, masonry, or where subject to corrosive conditions. Where used outside, aluminum or copper-clad aluminum grounding conductors are not permitted to be terminated within 450 mm (18 in.) of the earth [*NEC* 250.64(A) or IRC E3610.2].

GROUNDING ELECTRODE CONDUCTOR SPLICES

The grounding electrode conductor can be solid or stranded, insulated, covered, or bare, and generally must be installed in one continuous length without a splice or joint. The *Code* permits grounding electrode conductors to be spliced only by means of an "irreversible compression-type connector that is listed as grounding and bonding equipment" or by the exothermic welding process. Sections of busbars are also permitted to be connected together to form a grounding electrode conductor.

Irreversible compression-type connectors are commonly called crimp sleeves and are installed using specially designed compression tools. Some of these tools are hydraulic. It is crucial to use the same manufacturer's compression tool and sleeves to ensure a proper connection.

When using the exothermic welding process, it is important to select the correct form designed for the conductor size and to follow all of the manufacturer's instructions [NEC 250.64(C) or IRC E3610.1] (see Figure 5.25).

GROUNDING ELECTRODE CONDUCTOR TAPS

A grounding electrode conductor (GEC) can be installed by several methods. The most common method is to install a single GEC (sized based on the largest service-entrance conductor) from the grounding electrode to the service dis-

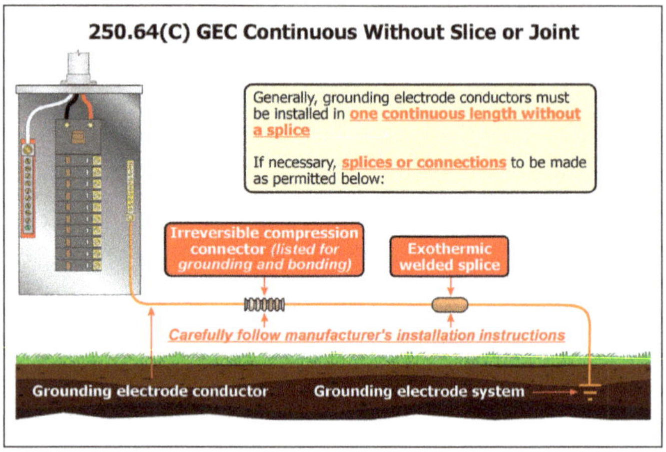

Figure 5.25. Grounding electrode conductors are generally required to be installed in one continuous length without a splice, but splicing is permitted under specific conditions

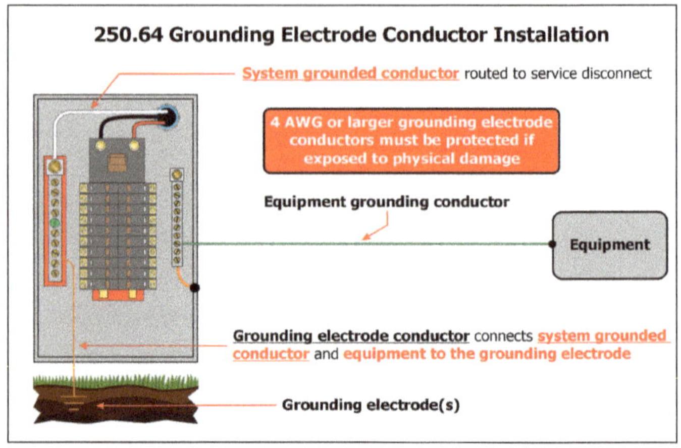

Figure 5.27. Installation of the grounding electrode conductor(s)

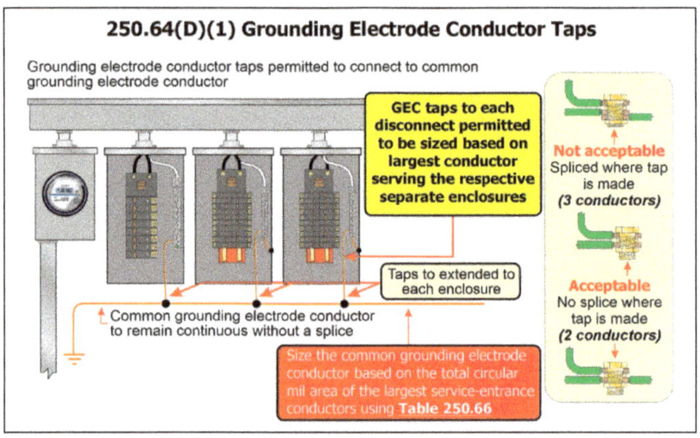

Figure 5.26. Taps from common grounding electrode conductor are permitted to be extended to each service disconnect individually.

Table 250.66 Grounding Electrode Conductor for Alternating-Current Systems

Size of Largest Ungrounded Service-Entrance Conductor or Equivalent Area for Parallel Conductors *(AWG/kcmil)*		Size of Grounding Electrode Conductor *(AWG/kcmil)*	
Copper	**Aluminum or Copper-Clad Aluminum**	**Copper**	**Aluminum or Copper-Clad Aluminum**
2 or smaller	1/0 or smaller	8	6
1 or 1/0	2/0 or 3/0	6	4
2/0 or 3/0	4/0 or 250	4	2
Over 3/0 through 350	Over 250 through 500	2	1/0
Over 350 through 600	Over 500 through 900	1/0	3/0
Over 600 through 1100	Over 900 through 1750	2/0	4/0
Over 1100	Over 1750	3/0	250

Table 5.2. Reproduction of *NEC* Table 250.66. Grounding Electrode Conductor for Alternating-Current Systems

connect. The *Code* also permits the connection of taps to the grounding electrode conductor. This method of installing the GEC is common where two or more service disconnecting means enclosures are grouped at the same location.

Typically, a common grounding electrode conductor is run from the grounding electrode to the vicinity of the multiple service disconnecting means. Each such tap conductor is required to extend from the common grounding electrode conductor to inside each enclosure. The size of the common grounding electrode conductor must be based on the total circular mil area of the largest service-entrance conductors using *NEC* Table 250.66 or IRC Table E3603.4. However, GEC taps to each disconnect are permitted to be sized based on the largest conductor serving the respective separate enclosures (see Figure 5.26). It is not common to see this method of grounding electrode conductor taps at dwelling units, but it is not impossible either. One might see this method at larger dwelling units with more than one service disconnecting means.

The sizing of the common grounding electrode conductor, shown in Figure 5.26, is based on the size of the service-entrance conductors. For example, let's say the service-entrance conductors are 2/0 AWG copper. By referring to *NEC* Table 250.66 or IRC Table E3603.4 for sizing grounding electrode conductors, one finds the minimum size would be 4 AWG copper or 2 AWG aluminum or copper-clad aluminum.

Each of the taps must be sized in accordance with the applicable table as well. Still, their size is based individually upon the largest service-entrance conductor serving the individual enclosure.

For example, if each enclosure is supplied with 2 AWG copper service-entrance conductors, the size of each GEC tap would be not less than 8 AWG copper or 6 AWG aluminum or copper-clad aluminum.

The GEC tap conductors must be connected to the common grounding electrode conductor in such a manner that the common grounding electrode conductor remains without a splice or joint. This does not require that the common grounding electrode conductor extend unbroken into one of the service enclosures. This means that the common grounding electrode conductor cannot be spliced at the point where the tap is being made [*NEC* 250.64(D)(1)]. This common GEC and taps method is not addressed in the IRC and, therefore, would be permitted by default to the *NEC* (see IRC E3401.2).

GROUNDING ELECTRODE CONDUCTOR INSTALLATION

The grounding electrode conductor (GEC) is one of the four conductors needed to reach the "grounding target" at the service disconnecting means.

The Code specifies requirements for installing grounding electrode conductors. It is crucial that the GEC is protected from physical damage, so it remains intact and continuous when it needs to perform its intended function. It must be securely fastened to the surface it is run on and protected from physical damage accordingly.

A 4 AWG or larger copper, copper-clad aluminum, or aluminum grounding electrode conductor must be protected if it is exposed to physical damage. A 6 AWG grounding electrode conductor that is free from exposure to physical damage is permitted to be run on the surface of the building construction without protection.

If exposed to physical damage, it must be protected by installation in rigid metal conduit (RMC), intermediate metal conduit (IMC), Schedule 80 rigid nonmetallic conduit (PVC), reinforced thermosetting resin nonmetallic conduit (RTRC-XW), electrical metallic tubing (EMT), or cable armor. Grounding electrode conductors smaller than 6 AWG must be protected from physical damage. Grounding electrode conductors are permitted to be installed on or through framing members [*NEC* 250.64(B) or IRC E3610.2].

Aluminum or copper-clad aluminum grounding electrode conductors are not permitted to be used when they are in direct contact with masonry or the earth or where they are subject to corrosive conditions. Where used outside, aluminum or copper-clad aluminum grounding electrode conductors must not be terminated within 450 mm (18 in.) of the earth. Aluminum or copper-clad aluminum terminations made within outdoor enclosures that are listed and identified for the environment are permitted to be terminated within 450 mm (18 in.) of the bottom of those listed enclosures [*NEC* 250.64(A) or IRC E3610.2].

SIZING OF GROUNDING ELECTRODE CONDUCTORS

The sizing of the grounding electrode conductor(s) (GEC) for an alternating-current system must be at least that speci-

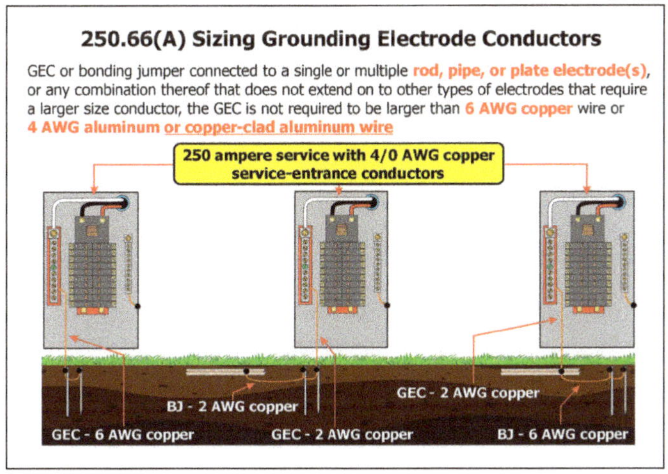

Figure 5.28. Maximum size 6 AWG copper or 4 AWG aluminum GEC to rod, pipe, or plate electrode if GEC does not extend on to other types of electrodes that require a larger size of conductor

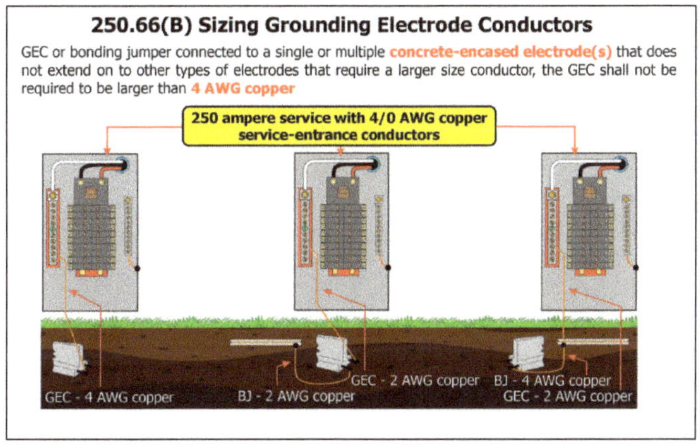

Figure 5.29. Maximum size 4 AWG copper GEC to concrete-encased electrode if GEC does not extend onto other types of electrodes that require a larger size of conductor

fied in NEC Table 250.66 or IRC Table E3603.4. The size of the GEC is based on the size of the largest service-entrance conductor, not on the rating of the circuit breaker or fuse in the service equipment, such as for equipment grounding conductors. To use the table, determine the size of the service-entrance conductor and identify whether it is made of copper, copper-clad aluminum, or aluminum. Follow the column for the size and type of conductor down until the size of the largest service-entrance conductor is found. Move to the right to find the copper, copper-clad aluminum, or aluminum grounding electrode conductor required. Note that neither the ampere rating of the service equipment nor the ampere rating of a main breaker or fuse is used to determine the size of the GEC. The ampere rating of the service equipment, the main circuit breaker, or the fuse could be used to determine the size of the largest service-entrance conductors.

For example, if the size of the service-entrance conductors is 400 kcmil copper, the grounding electrode conductor

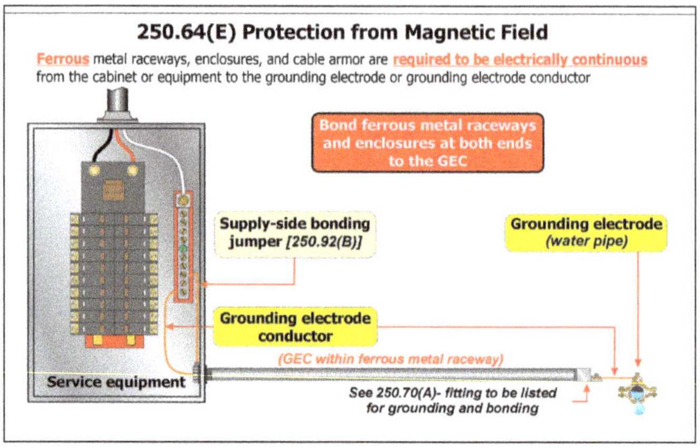

Figure 5.30. Protection from magnetic fields where grounding electrode conductor is installed in ferrous metallic raceways

must be either 1/0 AWG copper or 3/0 AWG aluminum or copper-clad aluminum.

Specific sizes of GEC for connection to rod, pipe, or plate electrodes; concrete-encased electrodes; or a grounding electrode ring that do not extend onto other types of electrodes that require a larger size conductor are provided at *NEC* 250.66(A), (B), and (C) or IRC Table E3603.4 and table notes as follows:

- If the grounding electrode conductor or bonding jumper connected to a single or multiple rod, pipe, or plate electrode(s), or any combination thereof, does not extend onto other types of electrodes that require a larger size conductor, the grounding electrode conductor is not required to be larger than 6 AWG copper or 4 AWG aluminum or copper-clad aluminum wire.
- If the grounding electrode conductor or bonding jumper connected to a single or multiple concrete-encased electrode(s) does not extend onto other types of electrodes that require a larger size of the conductor, the grounding electrode conductor is not required to be larger than a 4 AWG copper wire.
- If the grounding electrode conductor or bonding jumper connected to a ground ring does not extend onto other types of electrodes that require a larger size of the conductor, the grounding electrode conductor is not required to be larger than the conductor used for the ground ring.

BONDING METAL ENCLOSURES FOR GROUNDING ELECTRODE CONDUCTORS

It has become somewhat of a common practice to enclose a grounding electrode conductor (GEC) in a raceway or enclosure to protect it from physical damage. If the raceway or enclosure is a ferrous metal, the *Code* requires the metallic enclosure to be protected from magnetic fields. This is accomplished by bonding ferrous metallic raceways and enclosures for grounding electrode conductors to the GEC itself. Bonding of such ferrous metal raceways or enclosures between the service equipment and the grounding electrode is done in a manner that puts the ferrous metallic enclosure and the grounding electrode conductor in parallel with each other.

During events such as ground faults and lightning strikes, the amount of current flowing from the grounding electrode conductor to earth can fluctuate significantly, with relatively high amounts of current. The ferrous metallic enclosure around the GEC increases the impedance and will restrict the current from flowing to earth. This is often referred to as the choke effect in the electrical field because, without this bonding, more current will flow on the raceway than on the grounding electrode conductor itself. Bonding metal raceways and enclosures to the GEC at both ends minimizes the impedance and allows the conduit to provide an additional path over which current can flow. Electrical theory dictates that the majority of the electrical current will flow through the conduit, and a minimum amount will flow through the contained grounding electrode conductor [NEC 250.64(E) or IRC E3610.3].

Ferrous metal enclosures for GEC are required to be electrically continuous from the point of attachment to cabinets or equipment to the grounding electrode. The ferrous metal conduit must also be securely fastened to the ground clamp or fitting. Where the ferrous metal raceway is not physically continuous, it must be made electrically continuous by bonding. The bonding jumper must be at least the same size as the GEC itself. Keep in mind that protection from physical damage can be provided by Schedule 80 PVC conduit, and since the material is nonconductive, bonding is obviously not required. Bonding is also not required for non-ferrous metal raceways, such as those made of brass or aluminum. This is a common trade practice when the GEC requires protection from physical damage or a raceway is specified for other reasons, such as aesthetics.

For more detailed information concerning metal raceways carrying current and compliance with *NEC* 250.64(E) or IRC E3610.3, look for the Grounding and ElectroMagnetic Interference (GEMI) Research referred to as the "GEMI" test and software published by the Steel Tube Institute *[www.steeltubeinstitute.org]*.

BONDING OF METAL WATER PIPING SYSTEMS

Metal water piping systems that do not qualify as a grounding electrode itself, either installed in or attached to a building or structure, are required to be bonded to the service equipment enclosure, the grounded (neutral) conductor at the service, the grounding electrode conductor where of sufficient size, or to the one or more grounding electrodes used, if the grounding electrode conductor or bonding jumper to the grounding electrode is of sufficient size. If this metal water piping system qualifies as a grounding electrode per the requirements of *NEC* 250.52(A)(1) or IRC E3608.1.1, it will be bonded to the service equipment through the grounding electrode system.

The metal water piping system bonding jumper must be sized in accordance with the size of the service-entrance conductor, by reference to NEC Table 250.102(C)(1) or IRC Table E3603.4, just as though this bonding conductor were being connected to the metal water pipe as a grounding electrode conductor. This bonding jumper shall not be required to be larger than 3/0 copper, 250 kcmil aluminum, or copper-clad aluminum. The bonding of metal water piping has become even more important as more nonmetallic water piping is being used underground to supply dwellings and then transitioning to metal or copper once inside the building. Obviously, metal water piping is conductive. Electrocutions and many incidents of electric shock occurred when a metal water pipe became accidentally energized and was contacted by a person who completed the ground-fault circuit back to the source or to the earth. This bonding applies to both supply and metal drain lines. The installation of the bonding jumper to metal water pipes must be in accordance with NEC 250.64(A), (B), and (E), or IRC E3610.2 and E3610.3 (see Figure 5.31).

Points of attachment of the bonding jumper to the metal piping must be accessible. In some cases, an access panel is installed to keep these connection points accessible [NEC 250.104(A) or IRC E3609.6 and IRC E3609.7]. The clamp shown in photo 5.15 needs to be accessible.

The requirement calling for the points of attachment of the bonding jumper to be accessible originated when it was found that the common practice in some parts of the country of using a welder to thaw out frozen water lines resulted in property damage. It was not uncommon for a fire to start when the metallic water pipe or grounding electrode conductors got red-hot while they carried welder current to adjacent properties through metallic water lines. This is especially a problem where more than one dwelling is supplied from a common power transformer, and the dwellings are also supplied by a common metallic water pipe. The welder circuit can then be completed in both directions.

An alternative method for bonding metallic water piping systems in buildings incorporating multiple occupancies is

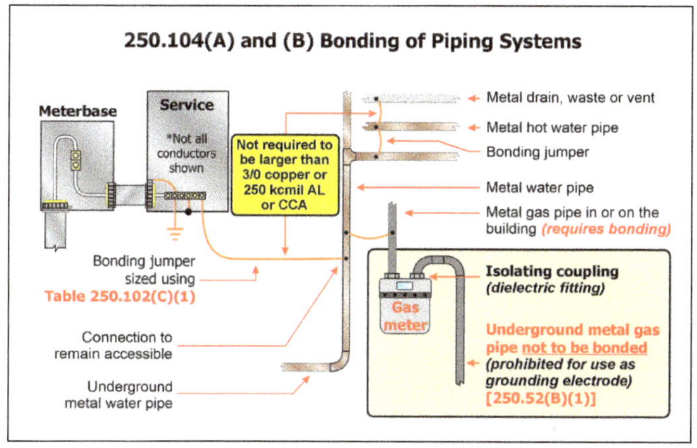

Figure 5.31. Water piping system bonding and other piping system bonding

Photo 5.15. A water pipe bonding clamp is required to be accessible and located not more than 1.52 m (5 ft) from where the water enters the building.

permitted under certain restrictive provisions. This method for bonding the interior metallic water piping system in each unit is permitted where the underground portion of the system cannot be used as a grounding electrode. An example application is where the common water service to a two-family dwelling is a nonmetallic water pipe. The interior water piping is made up of metallic pipes or tubing, such as copper or galvanized steel. The *Code* permits the interior metal water piping system for each occupancy to be bonded to the panelboard enclosure (other than at service equipment) supplying that occupancy. In this case, the size of this bonding jumper must be in accordance with *NEC* 250.102(D), which refers to *NEC* Table 250.122 or IRC Table E3908.13, based on the rating of the overcurrent protection that supplies the panelboard for that individual occupancy. These coordinates and are consistent with the minimum size of equipment grounding conductor required to be installed with the feeder from the service equipment to a feeder panelboard.

Bonding Metal Gas Piping or Other Piping Systems

Metal gas piping systems and other metallic piping systems that are likely to become energized must be bonded

Photo 5.16. The gas piping system is bonded on the customer side of the gas meter; underground gas piping is prohibited from being used as a grounding electrode

Table 250.122 *(in part)*

Minimum Size Equipment Grounding Conductors for Grounding Raceway and Equipment

Rating or Setting of Automatic Overcurrent Device in Circuit Ahead of Equipment, Conduit, etc., Not Exceeding *(Amperes)*	Size *(AWG or kcmil)*	
	Copper	Aluminum or Copper-Clad Aluminum
15	14	12
20	12	10
60	10	8
100	8	6
200	6	4
300	4	2
400	3	1
500	2	1/0
600	1	2/0
800	1/0	3/0
1000	2/0	4/0
1200	3/0	250
1600	4/0	350

Note: Where necessary to comply with 250.4(A)(5) or 250.4(B)(4), the equipment grounding conductor shall be sized larger than given in this table.

**See installation restrictions in 250.120*

Table 5.3. Size equipment grounding conductors per *NEC* Table 250.122

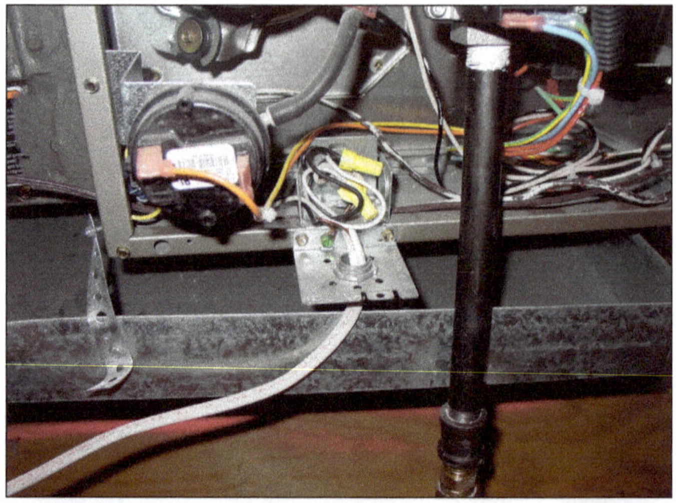

Photos 5.17 and 5.18. The equipment grounding conductor of the branch circuit supplying the gas appliance can serve as the bonding means for the gas piping and meets the minimum requirements of the *Code*.

(see photo 5.16). The Code requires this bonding jumper to connect to the service equipment enclosure, the grounded conductor at the service, the grounding electrode conductor when it is of sufficient size, or to one or more of the grounding electrodes used (if the grounding electrode conductor or bonding jumper to the grounding electrode is adequately sized) for the grounding electrode system [NEC 250.104(B) or IRC E3609.7].

The equipment grounding conductor for the equipment supply circuit that is likely to energize the piping system is permitted to serve as the bonding means to meet the minimum requirements of the Code. NEC Table 250.122 or IRC Table E3908.13 must be used for sizing this bonding jumper or equipment grounding conductor. The authority having jurisdiction (AHJ) is responsible for determining which circuit would most likely energize the piping system. It is reasonable to assume that the branch circuit supplying the equipment (such as a gas furnace) would be the circuit most likely to energize the piping system if it were to become energized.

The Code requires an equipment grounding conductor (EGC) to provide a fault return path and to ground the appliance. The EGC is sized based on the rating of the branch-circuit overcurrent protective device, as specified in NEC Table 250.122 or IRC Table E3908.13. The EGC would logically be of adequate size to bond the gas pipe connected to the appliance, such as a gas furnace.

The objective of this requirement is to maintain the gas piping system at the same voltage potential as any other metal piping system, should it become energized (see photos 5.17 and 5.18). The most likely source of this potential problem is a gas appliance such as a range, oven, furnace, boiler, or other heating appliance that also uses electrical energy.

Some electricians route the bonding conductor to the metal water pipe using a clamp on the gas pipe. Some manufacturers of flexible gas tubing recommend in their instructions that a 6 AWG bonding jumper be routed back to the service to properly bond their product. This is not a requirement of the NEC. Compliance with NEC 250.104(B) or IRC E3609.7, as mentioned above, is applicable in this case. The Code clearly states that the underground part of a metal gas pipe should not be used as a grounding electrode, but the aboveground part (inside or on the building) must be bonded. Understanding the difference between grounding and bonding is crucial for proper compliance with this Code requirement.

Nonmetallic pipe is commonly used as the underground gas supply to dwellings. An isolating bushing or coupling at the gas meter accomplishes isolation between the underground and aboveground portions of the metallic gas pipe. Because of this, the aboveground portion of the gas piping can be bonded without the underground piping being a part of the grounding electrode system.

Other metal piping systems, such as oil piping and refrigerant piping for air-conditioning systems, also fall under the requirements for "other metal piping" bonding. However, as indicated above, the minimum requirements of the Code are usually met by establishing a supply-circuit equipment grounding conductor connection to the equipment. This is a minimum requirement, and many jurisdictions may modify or amend it through local amendments or regulations.

Bonding all piping and metal air ducts within the premises will provide an additional layer of safety. This will maintain the same electrical potential for all metal parts across the various systems, thereby reducing shock hazards. For the air ducts, etc., this is a recommendation and not a Code requirement [NEC 250.104(B) Informational Note No. 1].

GROUNDING UNDERGROUND METAL ELBOWS

The *Code* requires that all metal enclosures for service enclosures and equipment be connected to the grounded system conductor. However, the *Code* exempts the bonding of a metal elbow and metal components that are installed in an underground installation of nonmetallic conduit, such as rigid polyvinyl chloride conduit (PVC), where it is isolated from possible contact by a minimum earth cover of 450 mm (18 in.) to any part of the elbow or metal component.

Installing metal elbows for pulling in an underground service raceway of rigid polyvinyl chloride conduit (PVC) is a common practice to prevent damage to the PVC raceways during cable-pulling operations. This approach is also allowed where other raceways containing conductors other than service conductors use this installation method [NEC 250.80, Exception, 250.86, Ex. No. 3(1) or IRC E3908.1 Ex. No. 2].

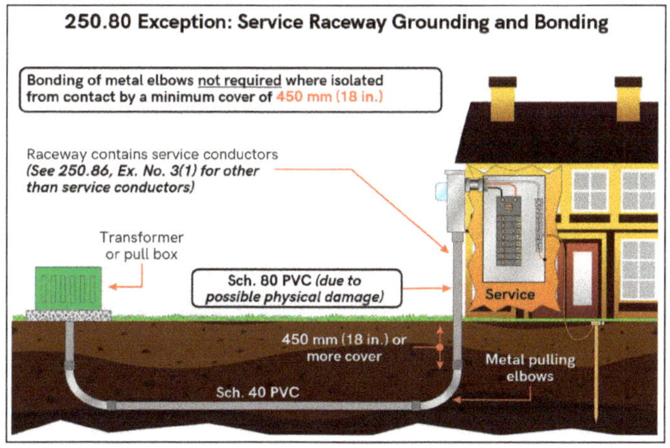

Figure 5.32. Bonding of a metal elbow and metal components that are installed in an underground installation of nonmetallic conduit is required unless isolated from possible contact by a minimum earth cover of 450 mm (18 in.)

EQUIPMENT REQUIRING GROUNDING

The Code requires that the exposed non-current-carrying part of many types of equipment be grounded by connection to the equipment grounding conductor. Any equipment installed within 2.5 m (8 ft) vertically or 1.5 m (5 ft) horizontally of ground or grounded metal objects that are likely to become energized must be grounded. Similarly, equipment located in wet or damp locations and not isolated or located in electrical contact with metal must be grounded. Electrical equipment with exposed metal parts must be grounded where the equipment is supplied by a metal-clad, metal-sheathed, or metal raceway or any other wiring method that provides an equipment grounding conductor (NEC 250.110 or IRC E3908.2). There is an exception for short sections of metal raceways used to support cables or provide protection from physical damage (NEC 250.86, Ex. No. 2 or IRC E3908.1, Ex. No. 2).

Motors, well pumps operated by motors, and metal well casings must be grounded by connecting them to the equipment grounding conductor, regardless of voltage. Many types of cord- and plug-connected equipment commonly used in homes are also required to be grounded (EGC connection). The extensive list of residential equipment listed in NEC 250.114(3) includes larger appliances such as refrigerators, freezers, air conditioners, clothes washing and drying machines, dishwashers, waste disposers, computers, sump pumps, and aquarium equipment. It also covers hand-held, stationary, fixed motor-operated tools, and outdoor motor-operated appliances like hedge clippers, lawnmowers, snow blowers, and wet scrubbers.

The Code permits listed tools, appliances, and equipment

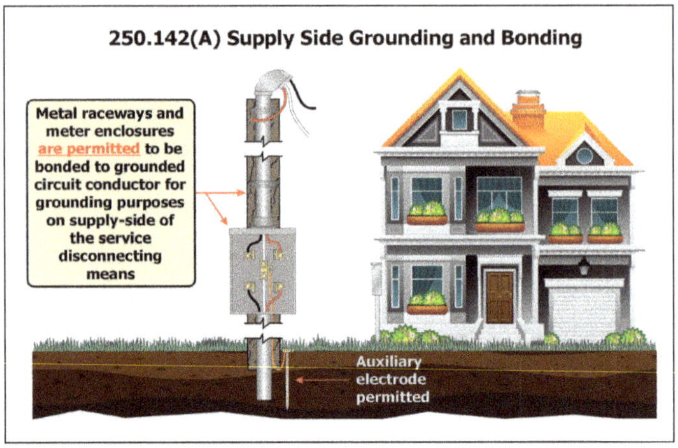

Figure 5.33. Bonding and grounding of remote metering equipment

not to be connected to the equipment grounding conductor where a system of double insulation protects the equipment. Double-insulated equipment must be distinctly marked (NEC 250.114, Exception).

Requirements for equipment grounding conductors are covered in detail Chapter 6 (for feeders) and Chapter 11 (for branch circuits). Information relative to the acceptable types of equipment grounding conductors, identification means, and sizing requirements are provided in those chapters.

BONDING AND GROUNDING REMOTE METERING EQUIPMENT

The *Code* requires all equipment containing service conductors to be grounded and bonded (*NEC* 250.80; 250.92 and IRC E3609.2; E3908.1). This includes meter enclosures, metal raceways, and metal weatherheads that are installed remotely from service equipment, typically on a yard pole. It is quite common to find this arrangement on rural premises where the term "*maypole*" has been used to describe this typical installation. This is due to the case where the electric utility serves the top of the pole with a service drop. Either the service utility or the owner then installs from one to several service drops to buildings in a circular fashion.

It is important that the grounded service (neutral) conductor be connected to the meter enclosure. This will provide a low-impedance path from any line-to-ground fault that may occur in the remote metering equipment back to the source, which is typically the utility transformer. The use of the grounded service conductor for bonding equipment on the line side of the service disconnecting means is permitted by the *Code* for this purpose [*NEC* 250.142(A) or IRC E3908.6].

It is a serious mistake to connect only a grounding electrode conductor from the meter housing to a grounding electrode, such as a ground rod installed near the meter equipment. It is very unlikely that enough current will flow through the earth (high-impedance path) to operate the transformer line-side fuse. Besides, earth return paths for fault current are prohibited by the *Code* [*NEC* 250.4(A)(5) or IRC E3908.5]. Where a ground rod or other electrode, such as a plate or pipe, is installed at the remote meter equipment, it is considered an *auxiliary grounding electrode*. Auxiliary grounding electrodes are permitted to be installed in addition to the required safety bonding and grounding connections.

There is a Residential Electrical Inspection Checklist in Appendix B, which includes a grounding and bonding checklist. This checklist helps the reader correctly apply the Code requirements for grounding and bonding in single- and two-family dwelling electrical systems.

GROUNDING REQUIREMENTS AT SEPARATE BUILDINGS OR STRUCTURES (OUTBUILDINGS)

For new installations of a grounded system, feeders and branch circuits to separate outbuildings must employ an equipment grounding conductor system. This means that the feeder or branch circuit to the outbuilding must provide a means to ground equipment in the outbuilding by providing an equipment grounding conductor that is run with the supply conductors to the outbuilding. Any installed grounded conductor cannot be connected to the equipment grounding conductors or the established grounding electrode system at the second building. A grounding electrode system must be established at the outbuilding if there is no existing grounding electrode system present at the outbuilding [*NEC* 250.32(A) and (B) or IRC E3607.3].

There is an exception to *NEC* 250.32(B), which applies to existing installations only, that permits the use of a grounded (neutral) circuit conductor for grounding purposes at the outbuilding. This rule is found in the IRC at E3607.3.2. This exception permits the grounded (neutral) conductor to be grounded again at the disconnecting means for the additional building or structure *only* if all the following conditions are met:

1. An equipment grounding conductor is not run with the supply to the second building or structure.
2. There are no continuous metallic paths bonded to the grounding system in each building or structure involved (metallic wiring method, metallic water or gas lines, phone lines, etc.).
3. Ground-fault protection of equipment has not been installed on the common AC service, which typically would not be applicable at dwelling units. See

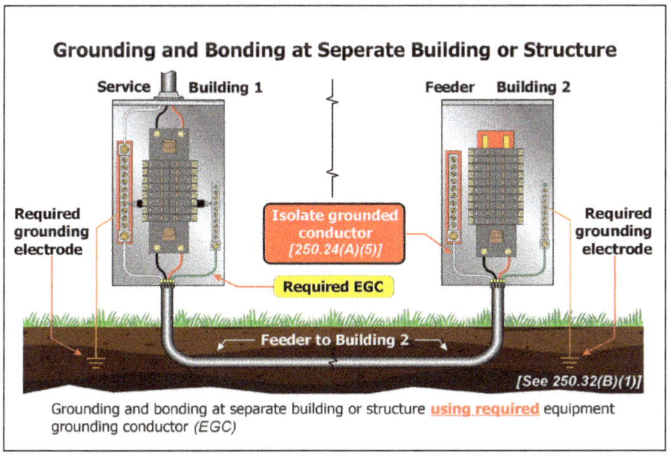

Figure 5.34. Grounding and bonding at an outbuilding supplied by a feeder

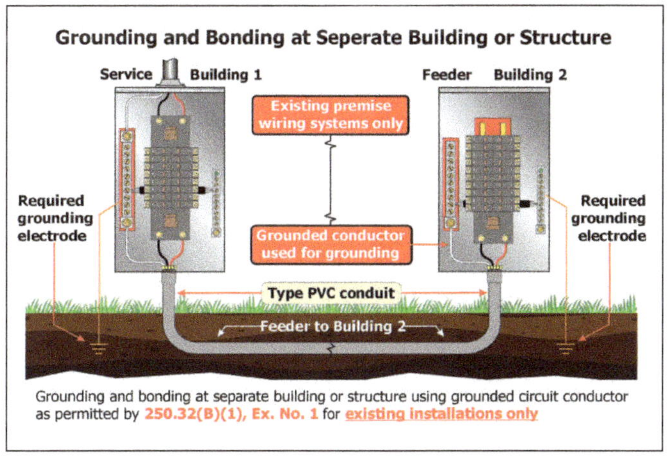

Figure 5.35. Grounding and bonding at an existing outbuilding supplied by an existing feeder

Chapter 15 of this publication for more information pertaining to wiring methods and grounding and bonding at accessory buildings or structures.

PRODUCT SAFETY STANDARD REQUIREMENTS

It is recommended to consult the following electrical product safety guide card information and product safety standards for additional guidance on the proper installation, operation, and use of electrical equipment discussed in this chapter. The four-letter code in parentheses indicates the product category in the UL Guide Information for Electrical Equipment Directory (UL Product Spec material).

- Equipment for Use in Ordinary Locations (AALZ)
- Grounding and Bonding Equipment (KDER)
- Grounding and Bonding Equipment
- Communication (KDSH)

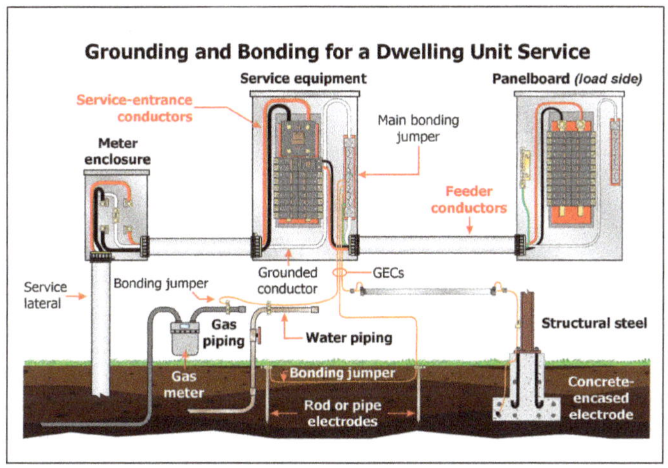

Figure 5.36. Service Grounding and Bonding Diagram

REFERENCES

1. *National Electrical Code* and *NEC* are registered trademarks of the National Fire Protection Association, Inc., Quincy, MA 02169. This reprinted material is not the official position of the National Fire Protection Association, which is represented only by the standard in its entirety.

2. *International Residential Code for One- and Two-Family Dwellings* (2021). Copyright © International Code Council, Inc., June 2023.

3. UL, Northbrook, IL, *UL Guide Information for Electrical Equipment Directory* (UL Product iQ® material) unless indicated otherwise.

REVIEW QUESTIONS

1. A conductor installed on the supply side of a service or within a service equipment enclosure(s), or for a separately derived system, that ensures the required electrical conductivity between metal parts required to be electrically connected is the definition of which of the following:

A. Bonding Jumper
B. Bonding Jumper, Equipment
C. Bonding Jumper, Service-Side
D. Bonding Jumper, Supply-Side

2. Where used outside and not within equipment enclosures, aluminum or copper-clad aluminum grounding conductors must not be terminated within what distance of the earth?

A. 450 mm (18 in.)
B. 600 mm (24 in.)
C. 750 mm (30 in.)
D. 900 mm (36 in.)

3. The minimum size for a copper grounding electrode conductor is?

A. 12 AWG
B. 10 AWG
C. 8 AWG
D. 4 AWG

4. The size of the copper grounding electrode conductor for a 120/240 volt, 3-wire, 200-ampere service supplied by 2/0 AWG copper service-entrance conductors is required to be:

A. 10 AWG copper or 8 AWG aluminum or copper-clad aluminum
B. 8 AWG copper or 6 AWG aluminum or copper-clad aluminum
C. 6 AWG copper or 4 AWG aluminum or copper-clad aluminum
D. 4 AWG copper or 2 AWG aluminum or copper-clad aluminum

5. The definition for grounded (grounding) is "connected (connecting) to ground."

True _____ False _____

6. The grounded circuit conductor on the load side of a service disconnecting means is generally not permitted to be used for grounding non-current-carrying metal parts of equipment such as meter housings and panelboards.

True _____ False _____

7. A length of ____________ or more of a 4 AWG bare copper conductor can be used to qualify as a concrete-encased grounding electrode where installed horizontally or vertically and within that portion of a concrete footing that is in direct contact with the earth.

A. 3.0 m (10 ft)
B. 3.7 m (12 ft)
C. 6.0 m (20 ft)
D. 7.5 m (25 ft)

8. A two-family dwelling contains a 200-ampere, single-phase main service panel supplied with 2/0 AWG Type THW copper service-entrance conductors in rigid metal conduit. The minimum size of the wire-type main bonding jumper must be ________ AWG.

A. 6 AWG aluminum
B. 6 AWG copper
C. 4 AWG copper-clad aluminum
D. 4 AWG copper

9. An equipment bonding jumper is permitted to be installed outside a raceway from a dwelling A/C unit to a disconnecting means where it is routed with the raceway and not longer than:

A. 1.5 m (5 ft)
B. 1.8 m (6 ft)
C. 2.5 m (8 ft)
D. 3.0 m (10 ft)

10. Under what conditions can the earth serve as an effective ground-fault current path?

A. where salted
B. not permitted
C. with ground enhancement material
D. where approved

11. A metal underground water piping system must be used as a grounding electrode where less than 3.0 m (10 ft) of the metallic piping system is buried in the earth.

True _____ False _____

12. Motor-operated water pumps, including submersible types, require their non-current-carrying metal parts to be connected to an equipment grounding conductor at:

A. any voltage or location
B. swimming pool locations
C. commercial buildings
D. residential locations

13. The intersystem bonding termination device or other means for bonding of limited energy systems such as communications, radio and TV, and CATV circuits is required to have at least ______ intersystem bonding termination points.

A. two
B. three
C. four
D. five

14. The intersystem bonding termination device should be securely mounted and electrically connected to which of the following:

A. an enclosure for the service equipment
B. the meter enclosure
C. an exposed nonflexible metallic service raceway
D. any of the above

15. Where the required main bonding jumper is installed in service equipment enclosures, it must be identified with a green finish where it is a:

A. screw
B. strap
C. busbar
D. conductor

16. A single rod, pipe, or plate electrode is required to be supplemented by an additional electrode, such as another rod, pipe, or plate electrode.

True _____ False _____

17. Under what conditions can a metal underground gas piping system be used as a grounding electrode?

A. when approved by the AHJ and gas supplier
B. under no circumstances
C. where AFCI protection is provided
D. where GFCI protection is provided

18. For new installations, separate outbuildings supplied by a feeder or branch circuit must employ an equipment grounding conductor system.

True _____ False _____

19. Which of the following means are permitted to be used to connect equipment grounding conductors or bonding jumpers to metal enclosures?

A. Machine screw-type fasteners or thread-forming machine screws that engage not less than two threads
B. Pressure connectors listed as grounding and bonding equipment
C. Connection devices or fittings that depend solely on solder
D. Both A. and B.

20. A rebar extension from a concrete-encased electrode can be used as a conductor to interconnect other electrodes of the grounding electrode systems.

True _____ False _____

21. A single rod, pipe, or plate electrode can be installed without augmentation (supplemented) if the single electrode has a resistance to earth of ______________ or less.

A. 80 ohms
B. 45 ohms
C. 35 ohms
D. 25 ohms

22. A grounding electrode conductor that is exposed to physical damage must be installed in rigid metal conduit, intermediate metal conduit, Schedule 80 rigid PVC conduit, reinforced thermosetting resin conduit Type XW, electrical metallic tubing, or cable armor when it is sized at _____ AWG or larger.

A. 6 AWG
B. 4 AWG
C. 2 AWG
D. 1 AWG

23. A main bonding jumper must be any of the following EXCEPT a:

A. wire
B. bus
C. screw
D. rivet

24. In order to be an effective grounding electrode, a ground rod must be driven into the earth at a depth of not less than:

A. 1.22 m (4 ft)
B. 1.83 m (6 ft)
C. 2.44 m (8 ft)
D. 3.05 m (10 ft)

25. Aluminum or copper-clad aluminum terminations made within outdoor enclosures that are listed and identified for the environment are permitted to be terminated within 450 mm (18 in.) of the bottom of those listed enclosures.

True _____ False _____

26. Which of the following is not recognized as a qualifying component in a grounding electrode system?

A. ground ring
B. metal in-ground support structure
C. concrete-encased electrode
D. metric designator 16 (½ in.) galvanized pipe or conduit

27. A bonding jumper used to bond other piping systems, such as gas piping, shall be sized in accordance with_______________________, using the rating of the circuit that is likely to energize the piping system(s).

A. *NEC* Table 250.66 (IRC Table E3603.4)
B. *NEC* Table 250.102(C)(1) (IRC Table E3603.4)
C. *NEC* Table 250.122 (IRC Table E3908.13)
D. *NEC* Table 310.12 (IRC Table E3603.1.1)

28. A concrete-encased electrode is required to be not less than what portion of a concrete foundation or footing that is in direct contact with the earth?

A. 2.5 m (8 ft)
B. 3.0 m (10 ft)
C. 6.0 m (20 ft)
D. 7.5 m (25 ft)

29. Grounding electrode conductors are not permitted to be spliced under any circumstances.

True _____ False _____

30. Electrical continuity at service equipment, service raceways, and service conductor enclosures shall be ensured by any of the following methods EXCEPT:

A. Bonding equipment to the grounded service conductor
B. Connections using threaded couplings or listed threaded hubs on enclosures if made up of wrenchtight
C. Standard locknuts and bushings made up of wrenchtight
D. Other listed devices, such as bonding-type locknuts, bushings, or bushings with bonding jumpers

31. It is permitted to run a tap from a grounding electrode conductor to the inside of a panelboard where two to six service enclosures from the same service are grouped at the same location.

True _____ False _____

32. A Type UF single-branch circuit consisting of 12-3 with ground supplies and a detached garage. A grounding electrode is required at the detached garage.

True _____ False _____

33. Which of the following statements are true concerning a piece of rebar extension installed from a concrete-encased electrode to an accessible location above the slab?

A. The additional rebar section must be continuous with the grounding electrode rebar.
B. Can be connected to the grounding electrode rebar by the usual steel tie wires.
C. The rebar extension can be exposed to contact with the earth without corrosion protection.
D. Both A. and B.

34. The metal in-ground support structure of a building is an effective electrode only where __________ or more of a single structural metal member is in direct contact with the earth.

A. 1.8 m (6 ft)
B. 2.5 m (8 ft)
C. 3.0 m (10 ft)
D. 6.0 m (20 ft)

35. "Conductive material that could become energized because of the failure of electrical insulation or electrical spacing" is the definition of which of the following:

A. Energized Material
B. Likely to Become Energized
C. Conductive Energized Material
D. None of the Above

6 Requirements for Feeders

DEFINITIONS

Thefollowing definitions are relevant to this chapter and are found in *NEC* Article 100 and IRC Chapter 35).

- Feeder
- Concealed
- Exposed (as applied to wiring methods)

WHAT ARE FEEDERS?

Feeders, as typically used in one- and two-family dwellings, are the conductors between the service equipment and a feeder or sub-panelboard. A feeder is sometimes referred to as a subfeeder, and the feeder panelboard as a subpanel. The Code does not use these terms, but they are useful for understanding the difference between a service and a feeder. A feeder is always supplied by or subordinate to a service. A feeder may supply another feeder, but ultimately can be traced back to the load side of the service equipment. Feeders that supply a separate building, such as a detached garage, are often mistakenly referred to as a service to that outbuilding. If these conductors are supplied from the load side of the service equipment, they are feeder conductors by definition. Typically, at dwelling units, only utilities supply the service equipment for a given premise. The service equipment is ultimately the source for feeders on those same premises.

In some installations, the service equipment is located outside the building, with the feeder conductors running to a panelboard located inside (see photo 6.1). In other cases, the service equipment and the feeder conductors and equipment are installed outdoors. Where properly located, both the service equipment and feeder conductors and equipment can be located inside the building. However, the length of the service conductors located inside the building is limited to the minimum length needed to terminate in the service disconnect. For this reason, the service equipment must be located nearest the point where the service conductors enter

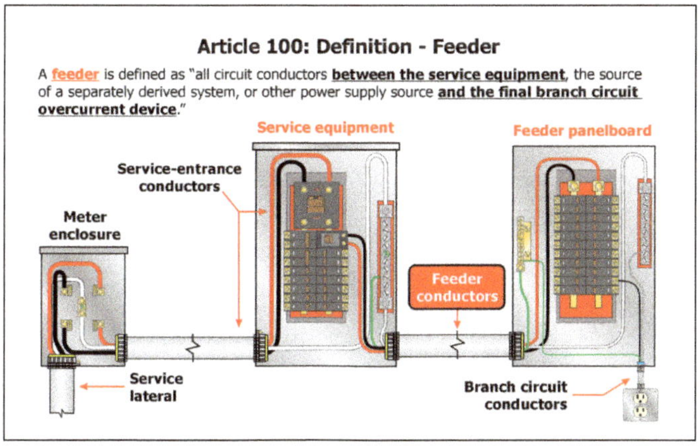

Figure 6.1. The feeder is defined as the conductor(s) between the service and the final branch-circuit overcurrent devices(s).

Photo 6.1. The split photo shows a feeder originating outside (right) and terminating on the inside at a feeder panelboard (left). Feeders must be sized to carry the load served.

the building [*NEC* 230.70(A)(1) or IRC E3601.6.2]. No such length restriction applies to feeder conductors that supply feeder panelboards located within the same building. The difference in installation requirements is due to feeder conductors having overcurrent protection at their supply end. Service conductors have only overload protection at their load end.

GENERAL REQUIREMENTS

As can be seen in the definition of a feeder, a major difference exists between feeder conductors and service conductors. Feeders usually have overcurrent protection in the form of a circuit breaker or fuse where the conductors originate or at their supply end. Service conductors have no overcurrent protection at their line or source end but are protected from overload at their load end.

Generally, feeders have overcurrent protection at the line or source end of the circuit. However, conductors that originate at a separately derived system, such as a transformer or generator, are also considered to be feeder conductors. Since transformer installations for power are not too common in one- and two-family dwellings, we will not cover them in this text (see Chapter 11 of IAEI's *Soares Book on Grounding and Bonding* for more information about this subject).

The *Code* permits feeders to be "tapped" under certain conditions. Several of the tap rules from *NEC* 240.21 allow the overcurrent protective device to be at the load end of the conductors. We will cover the 3 m (10 ft) and 7.5 m (25 ft) tap rules from the *NEC* later in this chapter. Tap rules are beyond the scope of the IRC, so installations for one- and two-family dwellings utilizing feeder taps will fall under the requirements of the *NEC*.

SIZING MAIN POWER FEEDERS AND OTHER FEEDER CONDUCTORS

A feeder must be sized to have an ampacity not less than the larger of *NEC* 215.2(A)(1) [IRC E3701.2.1] or *NEC* 215.2(A)(2) [IRC E3701.2.2]. *NEC* 215.2(A)(1) or IRC E3701.2.1 requires the feeder supplying continuous loads or any combination of continuous and noncontinuous loads to have an ampacity not less than the noncontinuous load plus 125 percent of the continuous loads. *NEC* 215.2(A)(2) or IRC E3701.2.2 requires a feeder to have an ampacity not less than the maximum load to be served after the application of any adjustment or correction factors. Feeders are also required to comply with *NEC* 110.14(C) or IRC E3705.4. These sections require temperature ratings associated with the ampacity of a conductor to be selected and coordinated so as not to exceed the lowest temperature rating of any connected terminations, conductors, or devices.

A load calculation typically must be performed to determine the minimum size feeder permitted. This calculation is required to meet the provisions of *NEC* Articles 215 and 220 or IRC E3704. See Chapter 2 for a detailed review of load calculation methods based on either the *NEC* or IRC requirements.

A load calculation for a feeder is done similarly to that for a service, except only the loads supplied by the individual feeder are considered. In some cases, this affects the demand factors allowed. For example, NEC 220.53 permits a demand factor of 75% to be applied to the nameplate rating load of four or more appliances rated ¼ horsepower or greater or 500 watts or greater (other than dryers, ranges, air-conditioning, space-heating equipment, and electric vehicle supply equipment), fastened in place and served by the same feeder. If branch circuits are installed so that fewer than the same feeder supplies four appliances, no demand factor may be applied when doing the load calculation for that feeder. IRC Table E3704.2(1) allows a 75% demand factor for all fastened-in-place appliances with a capacity of ¼ horsepower or greater, or 500 watts or greater, that are fastened in place, except for dryers, ranges, air conditioning, space heating equipment, and electric vehicle supply equipment.

Feeder conductors do not have to have an allowable ampacity that is larger than the service-entrance conductors that supply them. This means that for a given conductor material and insulation type, feeder conductors never have to be larger than their service conductors. The *Code* treats a feeder that supplies the entire load associated with a dwelling unit basically in the same manner as services. This type of feeder is commonly referred to as a "main power feeder" by users of the *Code*. Although the *Code* no longer defines a "main power feeder," for the purpose of this chapter, it can be defined as a feeder between the main service disconnect and a lighting and appliance feeder panelboard that supplies the *entire load* associated with a single-family dwelling or an individual dwelling unit in a two-family or multifamily dwelling.

NEC 310.12 or, Table 310.12, or IRC E3603.1 can be used to size conductors that qualify as the main power feeder or service conductors. It should be clear that only one main power feeder can exist in a dwelling unit. In order to qualify as a main power feeder, this main power feeder must supply *all* loads associated with the dwelling.

The number of main power feeders in an individual dwelling is limited to one for the purpose of applying *NEC* 310.12 or IRC E3603.1. *NEC* 310.12 applies where the circuits operate at 240/120-volt or a 208Y/120-volt, single-phase, 3-wire, and supplies an individual dwelling unit. IRC E3603.1 still applies only to a 120/240-volt, single-phase, 3-wire system but would need to default to the *NEC* for a 208Y/120-volt system due to IRC E3401.2. Feeder conductors that do not qualify as a main power feeder would be sized using the ampacity values from *NEC* Table 310.16 or IRC Table E3705.1. If no adjustment or correction factors are required, the ampacities of Table 310.12 are permitted to be applied.

For service and feeder conductors rated 100 through 400 amperes, the conductors supplying the entire load associated with a one-family dwelling or the feeder conductors supplying the entire load associated with an individual dwelling unit in a two-family or multifamily dwelling are permitted to have an ampacity not less than 83 percent of the service or feeder rating. The ampacity values for this service or feeder rating would be derived from the ampacity values from *NEC* Table 310.16 or IRC Table E3705.1. The grounded conductor is permitted to be sized smaller than the ungrounded conductors, provided the grounded conductor is sized to be not less than the maximum imbalance of the load and not smaller than the required minimum equipment grounding conductor size specified in *NEC* Table 250.122 or IRC Table E3908.13. [See *NEC* 310.12 or IRC E3603.1].

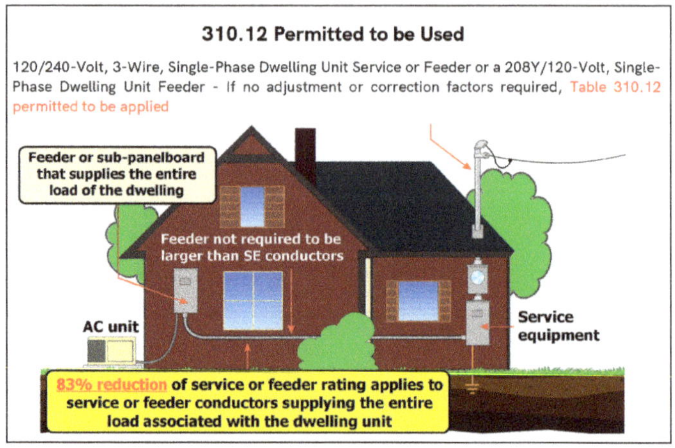

Figure 6.2. The main power feeder from service equipment to the feeder panelboard

As previously indicated, a feeder that supplies a specific load in the dwelling unit, other than the entire load, is not permitted to utilize the 83 percent ampacity rating adjustment factor specified in NEC 310.12 or IRC E3603.1. One example is a feeder from the service equipment that supplies only utilization equipment, such as several air conditioners or heat pumps. Feeders that leave the dwelling unit are not the main power feeder for the purpose of applying NEC 310.12 or IRC E3603.1. Examples of this are a feeder run to a detached garage or a feeder run to a remote panelboard that supplies swimming pool equipment. In these cases, the ampacity values in NEC Table 310.16 or IRC Table E3705.1 must be used to size these feeder conductors, as they do not supply the entire load of the dwelling unit.

FEEDER OVERCURRENT PROTECTION

Feeder conductors are required to be protected against overcurrent, which includes overload, ground faults, and short circuits. General references to protecting conductors against overcurrent in accordance with their ampacities are specified in *NEC* 310.15 and 240.4 or IRC E3705.5.

After the load calculation is performed, generally, the minimum size feeder conductor to carry this calculated load is chosen, and the appropriate overcurrent device to protect the conductor is selected. Any applicable ampacity correction factors, such as temperature correction factors, must be used to adjust the final ampacity rating of the conductor. This may necessitate increasing the size of the conductors. (Refer to

Photo 6.2. Two main-lug only (MLO) feeder panelboards supplied with two feeders with overcurrent protection to be located at the point where the feeder conductor receives their supply at the service disconnecting means (outside overcurrent devices shown in the insert at left).

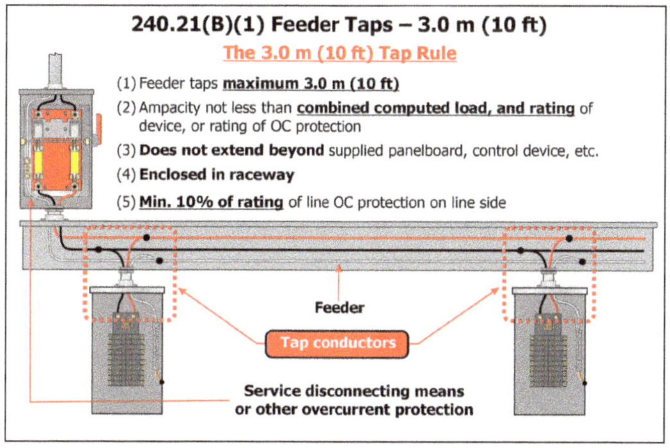

Figure 6.3. Diagram showing the *Code* requirements for a 3.0 m (10 ft) tap to a feeder

Chapter 1 for additional information on ampacity adjustment factor requirements; see also NEC 310.15(B)(1) and 310.15(C)(1) or IRC E3705.2 and E3705.3 for ampacity correction factors.)

LOCATION OF OVERCURRENT PROTECTION

Ungrounded feeder and branch-circuit conductors are normally required to be protected by overcurrent protective devices connected at the point where the conductors receive their supply (*NEC* 240.21 or IRC E3705.5).

There are installations and situations where an overcurrent protective device is not placed at the point where the conductor receives its supply. These installations require the connection of a smaller conductor to a larger one before an overcurrent device is installed to protect the conductors. These types of connections are referred to as tap conductors. Code requirements for feeder tap conductors are located in NEC 240.21(B). Feeder taps are permitted but not common in new one- and two-family dwellings. They are more common in existing dwellings. On occasion, a feeder tap is created from existing service equipment to install a panelboard or feeder to a new load, such as an air conditioner. If improperly installed, a tap can be a potential fire hazard, so a clear understanding of its limitations and conditions is crucial.

As a rule, taps do not have overcurrent protection within their ampacity on their line side or where they begin. Tap conductors are normally smaller than the feeder from which they are being tapped. Short-circuit protection for the tapped, usually smaller, conductors is provided by the overcurrent device on the line side of the tap. Overload protection is provided either by the calculated load on the feeder or by the load-side overcurrent device. For these discussions, we will consider only the 3 m (10 ft) tap rule of *NEC* 240.21(B)(1) and the 7.5 m (25 ft) tap rule of *NEC* 240.21(B)(2) as these have applications in one- and two-family dwellings and for connection of mobile homes. *NEC* 240.21 indicates it is not permissible to tap a tap conductor. In other words, this means that it is not permitted to supply a tap conductor from a conductor that is itself a tap conductor.

3 M (10 FT) TAP RULE

Where the length of the conductors is not greater than 3.0 m (10 ft), conductors are permitted to be tapped to a feeder without overcurrent protection at the tap if the following conditions are met:

1. The ampacity of the tap conductors is not less than the (a) combined calculated loads on the circuits supplied by the tap conductors, and (b) rating of the device supplied by the tap conductors or not less than the rating of the overcurrent-protective device at the termination of the tap conductors.
2. The tap conductors do not extend beyond the panelboard, disconnecting means or control devices they supply.
3. Except where the tap conductors connect to the feeder, the tap conductors are enclosed in a raceway that must extend from the tap to the panelboard or control device.
4. For field installations where the tap conductors leave the enclosure where the tap is made, the ampacity of the tap conductor must not be less than 10% of the rating of the overcurrent device on the line side of the tap.

As can be seen, the 3.0 m (10 ft) tap conductors are not necessarily required to terminate in an overcurrent device. Still, they may terminate at a main lug only (MLO) panelboard under certain more restrictive conditions [NEC 240.21(B)(1)]. Correct application of these rules may involve protecting the panelboard with a "main" overcurrent protective device. Chapter 3 addresses this subject in greater detail.

7.5 M (25 FT) TAP RULE

The requirements for installing a 7.5 m (25 ft) tap are similar in several ways to the 3.0 m (10 ft) tap rule previously discussed. In some cases, the requirements are more stringent as the conductors have a greater length without overcurrent protection within their ampacity. The requirements are contained in *NEC* 240.21(B)(2). Conductors that are not longer than 7.5 m (25 ft) are permitted to be tapped to a feeder without overcurrent protection at the tap where all the following conditions are met:

1. The ampacity of the tap conductors is not less than 1/3 of the rating of the overcurrent device protecting the feeder conductors.
2. The tap conductors must terminate in a single circuit breaker or a single set of fuses that will limit the load to the ampacity of the tap conductors. Any number of devices can be supplied on the load side of this overcurrent device.
3. The tap conductors must be suitably protected from physical damage or be enclosed in a raceway.

An example of where the 7.5 m (25 ft) tap rule may be used properly is in the connection of mobile homes. Mobile home parks often have feeders with 200-ampere overcurrent protection. The feeder will run to a disconnecting means located at each of the lots. Where a mobile home only requires a 100-ampere feeder, conductors having an ampacity or permitted rating of 100 amperes can be supplied directly from the disconnecting means that is adjacent to the mobile home. The feeder tap cannot be more than 7.5 m (25 ft) in length and must terminate in a 100-ampere overcurrent device in the mobile home panelboard. A similar procedure can be followed where the service adjacent to the mobile home is rated at 200 amperes.

FEEDER EQUIPMENT GROUNDING CONDUCTORS

The *Code* requires that where a feeder supplies branch circuits in which equipment grounding conductors are required, the feeder must include or provide a grounding means in accordance with the provisions of *NEC* 250.134 or IRC E3908.10, to which the equipment grounding conductors of the branch circuits will be connected (*NEC* 215.6 or IRC E3704.6). Non-current-carrying metal parts of equipment, raceways, and other enclosures are required to be grounded by any of the equipment grounding conductor types permitted by *NEC* 250.118 or IRC E3908.9 or by an equipment grounding conductor contained within the same raceway, cable, or cord or otherwise run with the circuit conductors. An equipment bonding jumper not longer than 1.8 m (6 ft) is permitted to be installed outside of a raceway [*NEC* 250.102(E)(2) or IRC E3908.21.2].

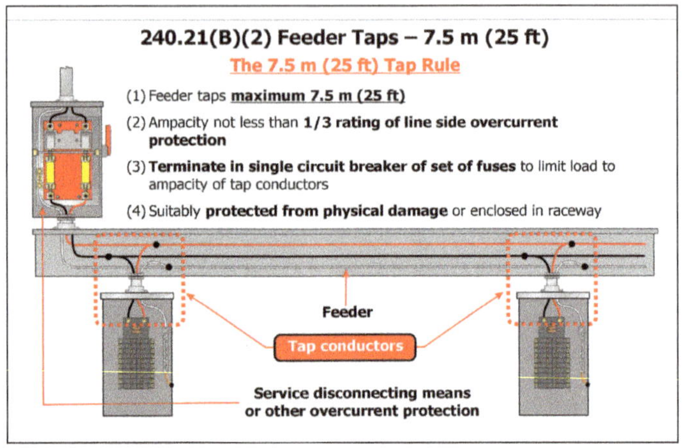

Figure 6.4. Diagram showing the *Code* requirements for a 7.5 m (25 ft) tap to a feeder

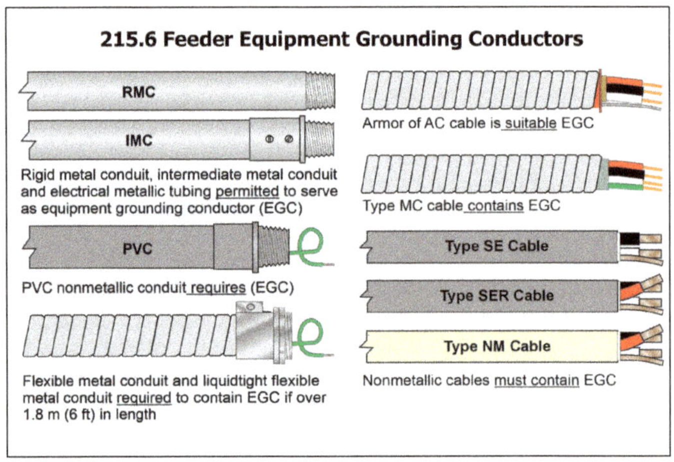

Figure 6.5. Feeders are generally required to include an equipment grounding conductor per 215.6

EQUIPMENT GROUNDING CONDUCTOR TYPES

The types of equipment grounding conductors used as part of a feeder circuit, run with or enclosing the circuit conductors, must be one or more or a combination of the following:

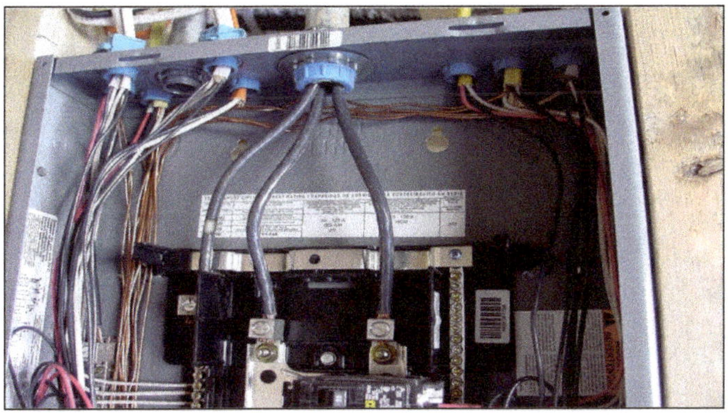

Photo 6.3. Small feeder to panelboard is to be sized based on the connected load of the branch circuits.

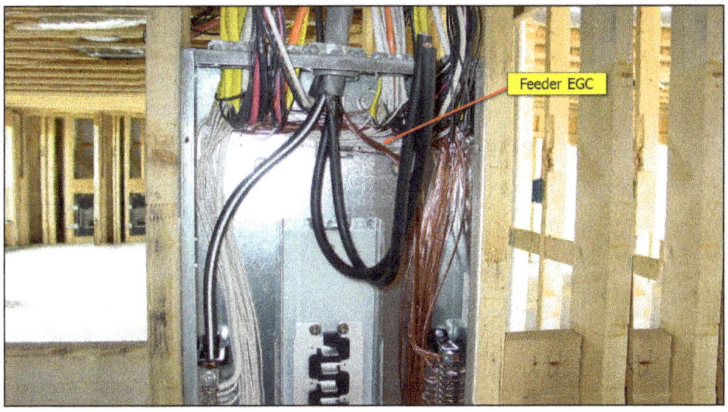

Photo 6.4. The feeder to sub-panelboard is required to include an equipment grounding conductor, as shown in this photo

1. a copper or other corrosion-resistant conductor that is solid or stranded; insulated, covered, or bare; and in the form of a wire or a busbar of any shape;
2. rigid metal conduit; intermediate metal conduit; or electrical metallic tubing;
3. flexible metal conduit or liquidtight flexible metal conduit where both the conduit and fittings are terminated in listed fittings;
4. armor of Type AC cable; surface metal raceway;
5. metal-clad cables where both the cable and the fittings are listed for grounding or
6. other electrically continuous metal raceways are listed for grounding.

Suitable equipment grounding conductors for feeders and branch circuits are provided in *NEC* 250.118 or IRC E3908.9.

IDENTIFICATION OF EQUIPMENT GROUNDING CONDUCTORS

Bare, covered, or insulated equipment grounding conductors are permitted. An individually covered or insulated equipment grounding conductor must have a continuous outer finish that is green or green with one or more yellow stripes (*NEC* 250.119 or IRC E3407.2).

An insulated or covered 4 AWG or larger copper or aluminum conductor, at the time of installation, is permitted to be permanently re-identified as an equipment grounding conductor. This identification must be present at each end and at the point where the conductor is accessible, such as at a junction box. This identification must be accomplished by one of the following methods:

1. removing the insulation or covering from the entire exposed length,
2. coloring the exposed insulation or covering green, or
3. marking the exposed insulation or covering it with green-colored tape or green-colored adhesive labels at the terminations

The *Code* also permits equipment grounding conductors to be part of a flexible cord assembly when supplying fixed equipment. The equipment grounding conductor of the cord assembly is required to be insulated. The insulated equipment grounding conductor must have a continuous outer finish of either green or green with one or more yellow stripes [*NEC* 250.119(D)].

EQUIPMENT GROUNDING CONDUCTOR SIZES

Typically, metal raceways can serve as their own equipment grounding conductor. However, some raceway installations require a separate equipment grounding conductor to be run with the circuit conductors and used to ground equipment. Non-flexible metal raceways, such as rigid nonmetallic conduits, can serve as their equipment grounding conductor. Liquidtight flexible nonmetallic conduit and flexible metal conduits where flexibility is required after installation may require a wire-type equipment grounding conductor (depending on the length) and, in most cases, can serve as their own equipment grounding conductor.

Many of the cable assemblies used in wiring one- and two-family dwellings provide an equipment grounding conductor as part of the listed cable assembly. The cable manufacturer determines the size of the equipment grounding conductor based on the largest overcurrent device that would be required to protect the largest-size ungrounded conductor contained within the cable assembly. Where raceways are installed and wire-type equipment grounding conductors are required, a suitable equipment grounding conductor must be provided. This equipment grounding conductor is also sized based on the rating of the overcurrent device ahead of the circuit. NEC Table 250.122 and IRC Table 3908.13 are the tables to be used to establish the minimum size of equipment for the grounding conductor for the branch circuit or feeder. The *Code* does not require that circuit conductors be sized to minimize voltage drop. However, suppose the circuit conductors are increased in size for any reason (including voltage drop). In that case, this change to the circuit conductors may impact the minimum size of the wire-type equipment grounding conductor as well. For example, if the size of the circuit conductors is doubled for voltage drop, the *Code* requires a proportional adjustment in the size of the equipment grounding conductor as well. Voltage drop is not usually an issue in one- and two-family dwellings but could be for a longer feeder circuit [*NEC* 250.122(B) or IRC E3908.13, *NEC* Chapter 9, Table 8].

ISOLATED GROUNDED (NEUTRAL) CONDUCTOR IN FEEDER PANEL OR SUBPANEL

In a feeder panelboard (load side or past the first means of disconnect), equipment grounding conductors must not be connected to the grounded (neutral) conductor bar or vice versa [*NEC* 250.24(B) or IRC E3607.2 and E3908.7]. In a 120-volt, single-phase, two-wire system (typical at a dwelling unit), on the load side of service equipment or past the first means of disconnect, the intentionally grounded (neutral) conductor is a current-carrying conductor. In other words, whatever current is produced by a piece of utilization equipment or light bulb, etc., carried on the ungrounded (hot) conductor, that same amount of current is flowing back to the source on the grounded (neutral) conductor (which makes it a current-carrying conductor).

Beyond the first disconnect point, connecting the equipment grounding conductor(s) to the grounded (neutral) conductor(s) would introduce unwanted or stray current onto the grounding conductor(s). This phenomenon is known as circulating currents and can cause a tingling sensation when touching a panelboard, cabinet, or lid. Such a connection between equipment grounding conductors and the grounded (neutral) conductor terminal bar is required at the service disconnecting means. The Code specifies a terminal bar or bus, specifically designated for this purpose, located where the interconnection between equipment grounding conductors and grounded circuit conductors is permitted or required by NEC Article 250.

Grounding connections are generally not allowed to be connected to any grounded (neutral) circuit conductor on the load side of the service disconnecting means. In other words,

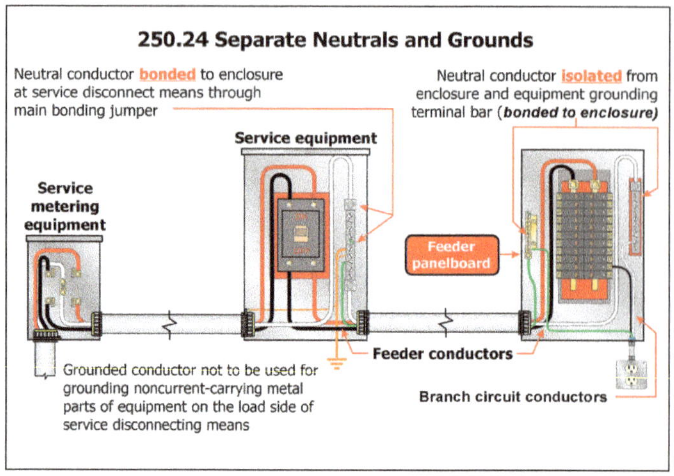

Figure 6.6. Grounded (neutral) conductors are required to be isolated from the enclosure, and EGC terminals are at feeder panelboards.

the Code does not permit the grounded (neutral) conductor to be reconnected to equipment enclosures or equipment grounding conductors unless specifically allowed under certain conditions [NEC 250.24(B), 250.142(B) or IRC E3908.7, E3908.8]. To meet this requirement, the neutral bar in a feeder panelboard is electrically and mechanically isolated from the panelboard cabinet. The manufacturer usually achieves this by mounting the neutral bar on insulators so it does not contact the enclosure; this practice is often called "floating the neutral."

Some feeder panelboards are also suitable for use as service equipment. This type of panelboard will come with a removable screw, strap, or bar that is intended to bond the neutral conductor bar to the equipment enclosure when the panelboard is used as service equipment and removed when used as a feeder panelboard. This neutral conductor bonding means should not be installed where the neutral conductor is required to be isolated past the first means of disconnect, such as at a feeder panelboard.

The *Code* also requires that the neutral conductors installed past the service be insulated. One of the reasons for this requirement is that a neutral-to-enclosure connection is not made unintentionally in a feeder panelboard. At this point in the electrical installation (on the load side, past the first means of disconnect), the neutral conductor is a current-carrying conductor.

Metal panelboard cabinets and metal panelboard frames are required to be in physical contact with each other and to be properly grounded by connection to the equipment grounding conductor system. Each panelboard used as service equipment is required to be provided with a properly sized main bonding jumper. The main bonding jumper must be placed inside the panelboard and used to connect the

Photo 6.5. Isolate the neutrals (white insulated) from the equipment grounding conductors (bare copper conductors) of feeder panelboards on the load side of the service equipment.

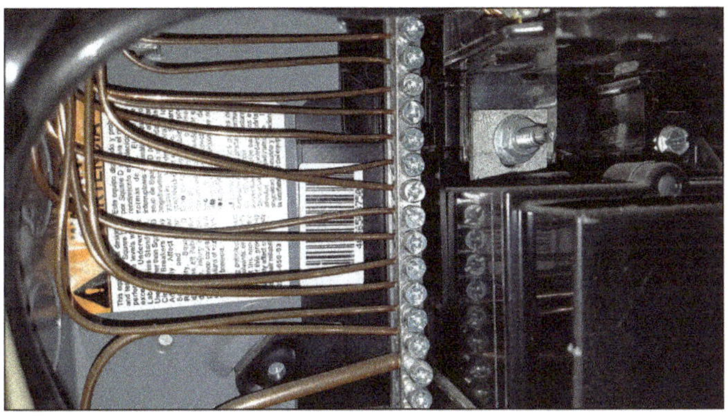

Photo 6.6. Equipment grounding terminal bar installed in panelboard for attachment of all feeders and branch-circuit equipment grounding conductors.

neutral service conductor (on its line side) to the frame of the panelboard.

When the panelboard is used with a feeder installed in a nonmetallic raceway or cable assembly, or when separate equipment grounding conductors are provided, an identified terminal bar for all the equipment grounding conductors must be secured inside the panelboard to ensure contact with the inside of the metal cabinet (see photo 6.5; NEC 408.40).

This terminal bar must be bonded to the metal cabinet and panelboard frame. It will also connect to the equipment grounding conductor run by the feeder conductors supplying the feeder panelboard. Equipment grounding terminal bars are readily available from the panelboard manufacturer as an accessory and are usually secured to the enclosure with self-tapping screws. The recommended practice is to remove any insulating paint between the terminal bar and the enclosure. This helps ensure a good electrical connection and establishes an effective fault-current path.

TYPES OF FEEDER CONDUCTORS

A wide range of wiring methods for feeder conductors is allowed in dwelling units, including both cable and raceways.

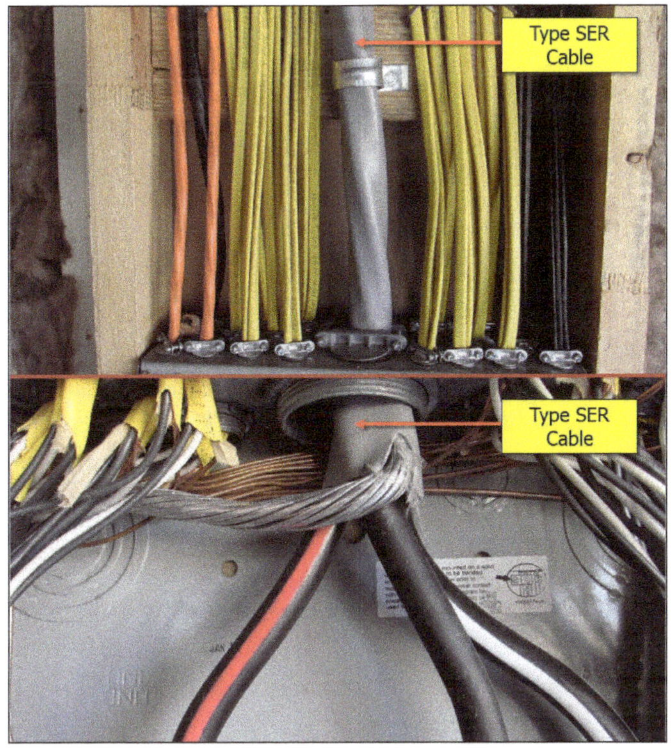

Photo 6.7(a) and (b). Typical Type SER cable installed to a feeder panelboard at the rough-in stage

When a feeder is installed in a conduit or as a cable assembly, it must include a conductor with insulation recognized by either NEC Table 310.4(1) or IRC E3406.5. Common feeder cable assemblies may feature Type AC, NM, UF, or Type SER (service-entrance round). Type SER is a typically round cable assembly that contains an insulated grounded (neutral) conductor, which is part of the assembly along with a bare equipment grounding conductor and insulated phase conductors. Typical raceways include electrical nonmetallic tubing (ENT), electrical metallic tubing (EMT), rigid polyvinyl chloride conduit (PVC), intermediate metal conduit (IMC), or rigid metal conduit (RMC) that enclose feeder conductors.

INSTALLATION OF RACEWAYS AND FEEDER CABLE ASSEMBLIES

A raceway is defined as "an enclosed channel designed expressly for holding wires, cables, or busbars, with additional functions as permitted in the Code." A raceway can also serve other purposes, such as providing physical protection (NEC Article 100 and IRC Chapter 35). Some metal raceways are also recognized as an equipment grounding conductor (NEC 250.118 or IRC E3908.9).

Conduits and electrical tubing are made from steel, aluminum, or polyvinyl chloride (PVC). Conduits and tubing most commonly used in one- and two-family dwellings include rigid metal conduit (RMC), rigid polyvinyl chloride conduit (PVC), intermediate metal conduit (IMC), liquidtight flexible metal conduit (LFMC), liquidtight flexible nonmetallic conduit (LFNC), flexible metal conduit (FMC), electrical nonmetallic tubing (ENT), and electrical metallic tubing (EMT).

Raceways are often used for service entrances and feeder conductors in new dwellings. The branch-circuit wiring may be run in raceways in existing dwellings; however, cable-type wiring methods typically prevail in new construction. Our discussion will focus on the most common types of raceways installed for dwelling unit feeders and their uses.

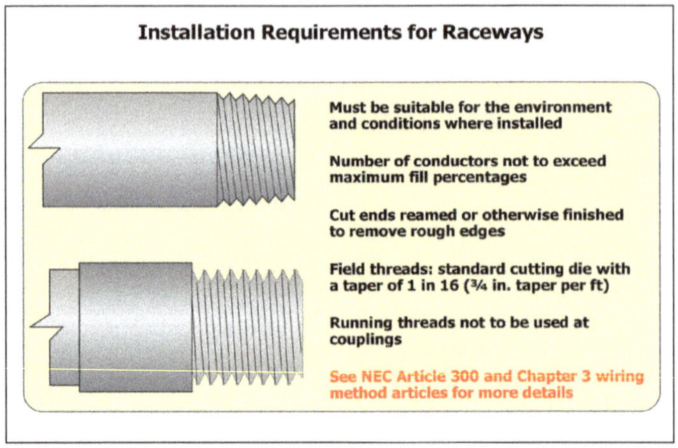

Figure 6.7. Installation requirements for raceways

NEC **Chapter 9 Tables: Table 1**

Table 1: Percent of Cross Section of Conduit and Tubing for Conductors and Cables

Number of Conductors and/or Cables	Cross Sectional Area (%)
1	53%
2	31%
Over 2	40%

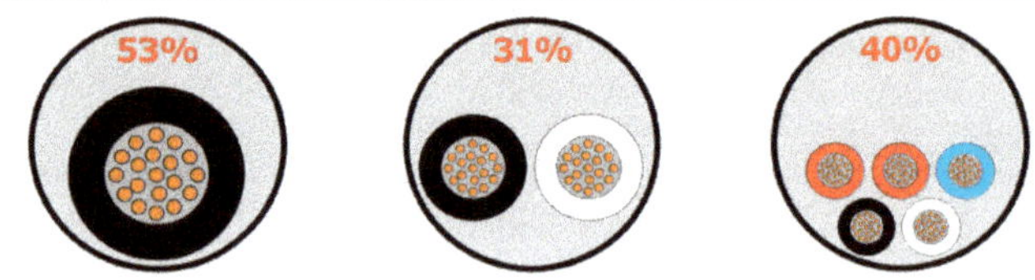

Figure 6.8. *NEC* Table 1, Chapter 9 for conduit and tubing conductor fill

MINIMUM SIZE OF RACEWAYS

The minimum size of a raceway is determined by the extent to which the conductors fill the cross-sectional area. This is commonly referred to as conduit fill, and the rules apply to conduit, tubing, and flexible raceways. Conduit fill is important for several key reasons. Keeping the quantity of conductors in a raceway limited to the values permitted by the Code (1) limits heat by allowing heat to dissipate away from the enclosed conductors properly and (2) minimizes potential damage to the conductors during the installation process. In most cases, individual conduit or raceway articles do not give a method for determining the number of conductors permitted in raceways.

In the *NEC*, "XXX.22" of the particular raceway article refers to Table 1 of Chapter 9, which is used for determining the maximum number of conductors permitted in the conduit or tubing in conjunction with Table 4 or Table 5 of Chapter 9. In the IRC, conduit or raceway fill is required to be in accordance with Table E3907.8 and the maximum number of conductors for the sizes given in Tables E3904.6(1) through E3904.6(10). The IRC is silent on conditions where combinations of conductors of different sizes are installed in the same conduit or tubing.

The percentage of conductor fill changes is based on the number of conductors that will occupy the conduit or tubing. The fill is permitted to be 53% for one conductor, 31% for two conductors, and 40% for more than two (*NEC* Table 1, Chapter 9 or IRC Table E3907.8).

Where all conductors are the same size, one can simply use the tables in *NEC* Informative Annex C or IRC Tables E3904.6(1) through E3904.6(10). These tables give the maximum number of conductors of a particular type and the same size that are permitted in a given raceway trade size. The first step in using the tables is to pick a type of raceway. The first page of *NEC* Informative Annex C lists the raceway types and the page number of the table for each raceway. The second step is to pick an insulation type. Then, follow down the left column until the type of insulation on the conductor to be used in the installation is found. The next column to the right gives the various sizes of conductors. Finally, move across the conduit or tubing size columns until the size of the raceway needed to accept the number of same-size conductors to be installed is found.

INFORMATIVE ANNEX C EXAMPLE:

As an example, let's find the minimum size EMT for a 100-ampere feeder using three 4 AWG THHN conductors.

Step 1. EMT is the raceway; therefore, Table C.1 applies.

Step 2. THHN is the insulation type. Follow down the left column until the insulation-type THHN is found. Stop at conductor size 4 AWG.

Step 3. Follow across the columns until the smallest size EMT for 3 conductors is found. Metric designator 21 (¾

Table 6.1 - 100-Ampere Feeder		
Copper Conductors[1]	Area	Total
Two, 4 AWG THW	.0973	.1946
One, 6 AWG Neutral THW	.0726	.0726
One, 8 AWG Equip. Gnd. THW	.0556	.0556
Total Sq. In. Area		**.3228**
Table 4, Chapter 9 requires:		
EMT	0.346	1 inch (MD 27)
ENT	0.564	1¼ inch (MD 35)
FMC	0.327	1 inch (MD 27)
IMC	0.384	1 inch (MD 27)
RMC	0.355	1 inch (MD 27)
PVC	0.333	1 inch (MD 27)
Aluminum Conductors[2]	**Area**	**Total**
Two, 2 AWG THW	.1194	.2388
One, 4 AWG Neutral THW	.0881	.0881
One, 6 AWG Equip. Gnd. THW	.0660	.0660
Total Sq. In. Area		**.3929**
Table 4, Chapter 9 requires:		
EMT	0.598	1¼ inch (MD 35)
ENT	0.564	1¼ inch (MD 35)
FMC	0.511	1¼ inch (MD 35)
IMC	0.659	1¼ inch (MD 35)
RMC	0.610	1¼ inch (MD 35)
PVC	0.581	1¼ inch (MD 35)
[1] ***NEC*** **Table 5, Chapter 9**		
[2] ***NEC*** **Table 5A, Chapter 9**		
[3] ***NEC*** **Table 8, Chapter 9**		

Table 6.1. Minimum raceway sizes using a combination of conductor sizes for a 100-ampere feeder

Table 6.2 - 200 Ampere Feeder		
Copper Conductors[1]	Area	Total
Two, 2/0 AWG XHHW	.2190	.4380
One, 1 AWG Neutral XHHW	.1534	.1534
Total Area W/O Gnd		**.5914**
One, 6 AWG Equip. Gnd. Bare[3]	.0270	.0270
Total Sq. In. Area		**.6184**
Table 4, Chapter 9 requires:		
EMT (W/O Gnd)	0.598	1¼ inch (MD 35)
ENT (W/Gnd)	0.774	1½ inch (MD 41)
FMC (W/Gnd)	0.743	1½ inch (MD 41)
IMC (W/O Gnd)	0.659	1¼ inch (MD 35)
RMC (W/O Gnd)	0.610	1¼ inch (MD 35)
PVC (W/Gnd)	0.794	1½ inch (MD 41)
Aluminum Conductors[2]	**Area**	**Total**
Two, 4/0 AWG XHHW	.2733	.5466
One, 2/0 AWG Neutral XHHW	.1885	.1885
Total Area W/O Gnd		**.7351**
One, 4 AWG Equip. Gnd. Bare[3]	.0420	.0420
Total Sq. In. Area		**.7771**
Table 4, Chapter 9 requires:		
EMT (W/O Gnd)	0.814	1½ inch (MD 41)
ENT (W/Gnd)	1.282	2 inch (MD 53)
FMC (W/Gnd)	1.307	2 inch (MD 53)
IMC (W/O Gnd)	0.890	1½ inch (MD 41)
RMC (W/O Gnd)	0.829	1½ inch (MD 41)
PVC (W/Gnd)	0.794	1½ inch (MD 41)
[1] ***NEC*** **Table 5, Chapter 9**		
[2] ***NEC*** **Table 5A, Chapter 9**		
[3] ***NEC*** **Table 8, Chapter 9**		

Table 6.2. Minimum raceway sizes using a combination of conductor sizes for a 200-ampere feeder

in.) is too small. Metric designator 27 (1 in.) will accept 4 conductors.

These tables provide maximum quantities of conductors for each type of common raceways and electrical tubing. For example, there is a table for electrical metallic tubing and another for electrical nonmetallic tubing. Compact stranded conductors are often installed in today's wiring systems. Compact stranded conductors are not covered in the IRC, so the provisions in *NEC* Informative Annex C are to be used by default. If compact aluminum conductors of the same size are to be installed, Table C.12(A) in *NEC* Informative Annex C gives the minimum size of conduit or tubing needed for the number and size of conductors to be installed.

Determining the minimum size raceway for combinations of conductors of different sizes or different insulation types requires a calculation. The starting point is *NEC* Table 5, Chapter 9. Table 5 gives the area in square inches for insulated conductors. The table lists diameters and areas for all conductor insulation types and sizes. Table 5A gives similar information for compact aluminum building wire. Refer to *NEC* Table 8 of Chapter 9 for the dimensions of bare conductors.

The next step is to refer to the proper table in Table 4, Chapter 9. Table 4 gives the dimensions and cross-sectional area for various types of raceways. The table for each raceway lists the trade sizes in the left column. The columns to the right give the total internal area as well as the maximum possible area that can be filled for one, two, and more than two conductors for each trade size.

Tables 6.1 and 6.2 give a couple of examples of how to determine the minimum raceway size using a combination of conductor sizes. For the purpose of our calculations, we will use the conductors permitted by NEC Table 310.16 or IRC Table E3705.1. In some examples, we will install an insulated equipment grounding conductor, and others will have a bare equipment grounding conductor that is sized based on the rating of the overcurrent protective device. Where a metal conduit or tubing is permitted to be used as the equipment grounding conductor, obviously, the equipment grounding conductor can be eliminated. In some cases, this will permit a conduit of a smaller size to be selected. Refer to Tables 6.1 and 6.2 in this chapter for examples of conduit fill calculations.

Similar procedures can be used to determine the minimum size of conduit or tubing needed for any conductor size. As shown, a smaller raceway may sometimes be suitable when using more compact conductor insulation and a bare equipment grounding conductor.

CONDUIT AND TUBING WIRING METHODS

Conduit and tubing systems made of rigid and intermediate metal conduits, rigid polyvinyl chloride conduit (PVC), electrical metallic tubing (EMT), flexible metal conduit, liquidtight flexible metal conduit, and liquidtight flexible nonmetallic conduit cannot have more than four quarter bends (totaling 360 degrees) between pull points, such as conduit bodies and boxes [see XXX.26 of the various NEC raceway and tubing articles, e.g., 344.26 for rigid metal conduit]. The radius of the bend's centerline cannot be less than what is shown in Table 2, Chapter 9, under the column "Other Bends" [see XXX.24 of the various NEC raceway and tubing articles]. Angle connectors cannot be installed where they will be hidden [see XXX.42 of the various NEC raceway and tubing articles]. In the following text, we will take a closer look at these different types of conduit or tubing wiring methods.

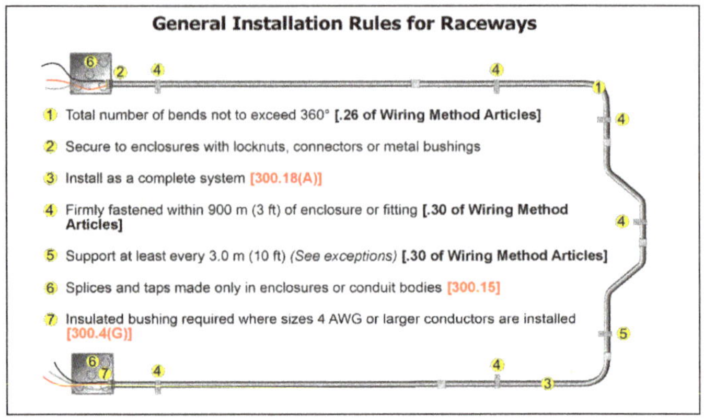

Figure 6.9. General installation rules for raceways

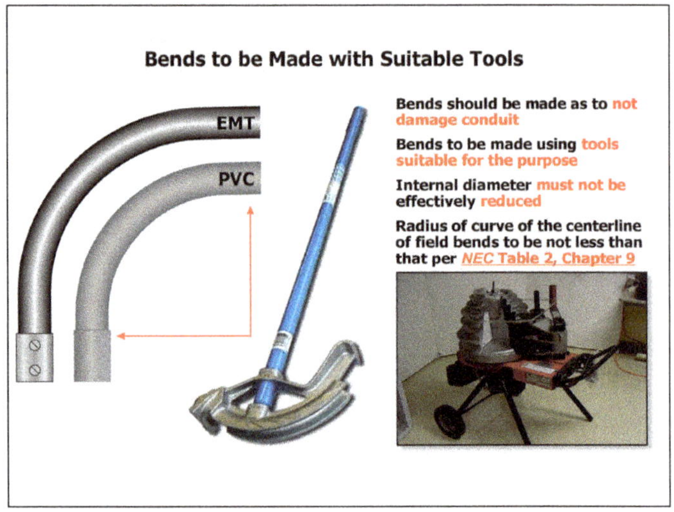

Figure 6.10. Bends are to be made with suitable tools and are not to reduce the internal diameter of the raceway.

RIGID AND INTERMEDIATE METAL CONDUIT SYSTEMS

NEC Articles 342 and 344 cover IMC and RMC, respectively. Rigid and intermediate metal conduits are metal raceways of circular cross sections with integral or associated couplings, connectors, and fittings approved for the installation of electrical conductors. These conduits are permitted to be used as a wiring method for feeders or services. Rigid and intermediate metal conduits are permitted to be used under all atmospheric conditions and occupancies unless specifically restricted. Avoid dissimilar metals in contact with each other. This will help to avoid galvanic action. Rigid and intermediate metal conduits are permitted to be used as equipment grounding conductors. Aluminum fittings and enclosures are permitted to be used with galvanized steel rigid metal and intermediate metal conduit.

Galvanized steel, stainless steel, and red brass type RMC and IMC elbows, couplings, and fittings are permitted to be installed in concrete, in direct contact with the earth, or in areas subject to severe corrosive influences where protected by corrosion protection and judged suitable for the condition [*NEC* 300.6 or IRC Table E3801.4 (footnote f)].

Galvanized steel, stainless steel, and red brass type RMC and IMC are permitted to be installed in or under cinder fill where subjected to permanent moisture. Such conduits must be protected on all sides by a layer of non-cinder concrete not less than 50 mm (2 in.) thick or where the conduit is not less than 450 mm (18 in.) under the fill or where protected by corrosion protection and judged suitable for the condition.

Where installed in wet locations, all supports, bolts, straps, screws, and so forth are required to be made of corrosion-resistant materials or protected against corrosion by corrosion-resistant materials.

The number of conductors in a single rigid or intermediate metal conduit must not exceed that permitted by the appropriate table in either *NEC* Informative Annex C or IRC Tables E3904.6(1) through E3904.6(10). For combinations of conductors, the percentage fill is specified in *NEC* Table 1, Chapter 9, using the conduit dimensions of *NEC* Table 4, Chapter 9. Refer to the previous discussion in this chapter for a description of the process and sample calculations.

All cut ends of rigid or intermediate metal conduits are required to be reamed or otherwise finished with proper tools to remove any rough edges. Where rigid or intermediate metal conduit is threaded in the field, a standard cutting die with a taper of 1 in 16 (¾-inch taper per foot) is required to be used.

Threadless couplings and connectors used with rigid or intermediate metal conduits must be made tight. Where they are used on conduits with threads, they are required to be listed for the purpose. Where buried in masonry or concrete, they are required to be the concrete-tight type. Where installed in wet locations, they must be the raintight type. Running threads are not permitted to be used on rigid or intermediate metal conduits for connection at couplings. Threaded couplings have a straight thread, not a tapered thread, as would plumbing couplings. The combination of straight couplings and tapered threads on the conduit produces a connection that will be tight and yet allow the ends of the conduit to meet in the center of the coupling. This meeting of the conduit ends that are reamed to remove sharp edges protects the contained conductors. If running threads are connected in couplings, there is no taper on the conduit or coupling, which can result in a loose connection.

Bends of rigid or intermediate metal conduit must be made so the conduit will not be damaged and the internal diameter of the conduit will not be effectively reduced. The radius of the curve of the center line of the conduit is not permitted to be less than indicated in *NEC* Table 2, Chapter 9 for bends made in the field.

Rigid or intermediate metal conduits are required to be installed as complete systems and securely fastened in place. Rigid or intermediate metal conduit is required to be firmly fastened within 900 mm (3 ft) of each outlet box, junction box, cabinet, or fitting. Rigid or intermediate metal conduit is required to be supported at least every 3 m (10 ft) (*NEC* 342.30, 344.30 or IRC Table E3802.1). Splices and taps must be made only in junction boxes, outlet boxes, or conduit bodies. Where a conduit enters a box, fitting, or other enclosure, a bushing is required to provide protection for wires from abrasion. A bushing is not required where the design of the box, fitting, or enclosure provides equivalent protection.

See *NEC* 300.4(G) or IRC E3906.1.1 for the protection of conductors 4 AWG and larger at termination points when installed in conduit or tubing. See *NEC* Article 342 (IMC) and *NEC* Article 344 (RMC) for references to the requirements previously discussed.

RIGID POLYVINYL CHLORIDE CONDUIT (PVC)

NEC Article 352 covers rigid polyvinyl chloride conduit (PVC). PVC is used frequently in all types of electrical construction. It is inherently resistant to corrosion and can provide substantial protection from physical damage where Schedule 80 PVC is specified. PVC conduit and fittings are manufactured of suitable nonmetallic material (PVC) that is resistant to moisture and chemical atmospheres. For use aboveground, it is required to be flame-retardant, resistant to distortion from heat under conditions likely to be encountered in service, and resistant to low temperature and sunlight effects. For use underground, the material is required to be acceptably resistant to moisture and corrosive agents and must be of sufficient strength to withstand abuse, such as by impact and crushing, in handling and during installation.

Where intended for direct burial without encasement in concrete, the material is also required to be capable of withstanding continued loading that is likely to be encountered after installation.

Use of Type PVC and fittings is permitted where concealed in walls, floors, and ceilings, as well as for underground installations. PVC is also permitted in locations subject to severe corrosive influences and when exposed to chemicals for which the material is specifically approved. It is permitted in wet locations when the entire conduit system, including boxes and fittings, is installed and equipped to prevent water from entering the conduit. All supports, bolts, straps, screws, and other similar components must be made of corrosion-resistant materials or protected against corrosion using approved corrosion-resistant materials. PVC is permitted in cinder fill.

Type PVC is also permitted in dry and damp locations where it is not prohibited by the Code. For exposed work, PVC cannot be installed where it is subject to physical damage unless it is identified for such use. Schedule 80 PVC is resistant to impact and crushing. It can be used in locations deemed to be at risk of potential damage.

Handling Type PVC in areas with extremely cold temperatures may render it brittle, leading to damage from physical contact.

Type PVC is not permitted to be used for the support of luminaires or other equipment. It cannot be used in a location where the ambient temperature is expected to exceed those for which the conduit is listed. PVC cannot be used in conjunction with conductors whose insulation temperature limitations would exceed those for which the conduit is listed.

Installations of Type PVC are required to comply with the applicable provisions of *NEC* Article 300 or IRC Tables E3802.1 and E3803.1. A separate equipment grounding conductor is required to be installed in Type PVC where grounding is required.

All cut ends of PVC conduit must be trimmed inside and outside to remove rough edges. All joints between lengths of PVC conduit and between conduit and couplings, fittings, and boxes must be made using an approved method.

Type PVC is required to be secured and supported as required by *NEC* Table 352.30. In addition, PVC conduit must be securely fastened within 900 mm (3 ft) of each outlet box, junction box, device box, conduit body, or other conduit termination [*NEC* 352.30(A) or IRC Table E3802.1].

Expansion fittings for PVC conduit are required to be provided to compensate for thermal expansion and contraction in a straight run between securely mounted equipment such as boxes, cabinets, elbows, or other conduit terminations. This rule applies where the calculated expansion will exceed 6 mm (¼ in.). *NEC* Tables 352.44(A) give the expansion characteristics of PVC conduit for various temperature changes. Note the table is for all sizes of conduit. As a rule of thumb, add 30°F to the expected temperature change for direct sunlight exposure. This is a result of the gray conduit having a tendency to absorb the ultraviolet (UV) rays of the sun. PVC conduit smaller than metric designator 16 (½ in. electrical trade size) is not permitted to be used. The number of conductors permitted in a single PVC conduit is not permitted to exceed the percentage fill specified in *NEC* Table 1, Chapter 9.

Where a PVC conduit enters a box or other fitting, a bushing or adapter must be provided to protect the wire from abrasion unless the design of the box or fitting is such as to provide equivalent protection. Listed nonmetallic fittings meet this requirement without the use of additional insulating bushing. Fittings are tested to ensure their edges will not damage conductors as a conductor is pulled through the fitting.

Photo 6.8. PVC conduit installed without expansion fitting can cause damage to the conduit system due to thermal expansion

Bends of PVC conduit must be made so the conduit will not be damaged and the internal diameter of the conduit will not be effectively reduced. Field bends are required to be made only with bending equipment identified for the purpose. The radius of the curve to the centerline of bends must not be less than shown in *NEC* Table 2, Chapter 9. These PVC benders are sometimes referred to as "hot boxes" due to the heat generated internally. The conduit is placed in the heater enclosure to heat it to the bending point. A wet rag is used to cool the conduit when the desired bend is obtained. Other bender types are available for this purpose, including blanket types and triethylene glycol types.

Boxes and fittings used with PVC conduit must comply with the applicable provisions of *NEC* Article 314 or IRC E3905. Splices and taps are required to be made only in junction boxes, outlet boxes, or conduit bodies.

ELECTRICAL METALLIC TUBING (EMT)

NEC Article 358 pertains to electrical metallic tubing (EMT). EMT is an unthreaded, thin-walled raceway with a circular cross-section, designed for protecting and routing conductors and cables, and for use as an equipment grounding conductor when installed with appropriate fittings. EMT is typically made of steel (ferrous) with protective coatings or aluminum (nonferrous). It may be installed for both exposed and concealed work. However, EMT must not be used where, during installation or afterward, it will be subjected to severe physical damage. EMT can be installed in concrete, in direct contact with the earth, or in areas prone to severe corrosive influences, provided galvanized steel and stainless steel EMT,

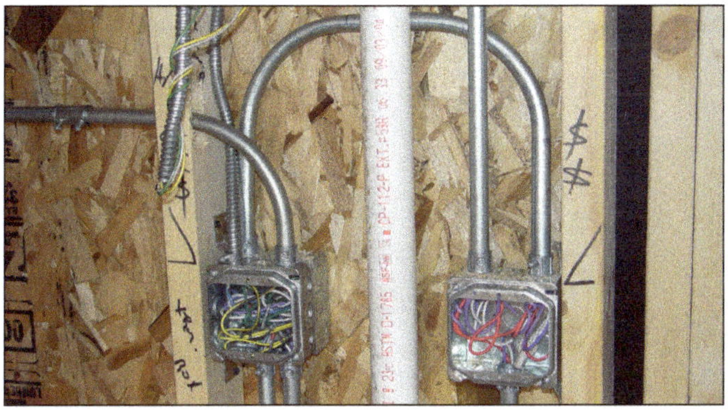

Photo 6.9. Bends in electrical metallic tubing (EMT) at a dwelling unit

elbows, and fittings are protected by corrosion-resistant coatings and are approved as suitable for such conditions.

Aluminum EMT must have approved additional corrosion protection when encased in concrete or in direct contact with the earth. Galvanized steel and stainless steel EMT can be installed in cinder concrete or cinder fill where they are exposed to permanent moisture, provided they are protected on all sides by a layer of noncinder concrete at least 50 mm (2 in.) thick or installed at least 450 mm (18 in.) below the fill [see NEC 358.10(B)].

Avoid dissimilar metals in contact with the EMT conduit to eliminate the possibility of galvanic action. Aluminum fittings and enclosures are permitted for use with galvanized steel electrical metallic tubing. Ferrous or nonferrous EMT elbows, couplings, and fittings are permitted to be installed in concrete, in direct contact with the earth, or in areas subject to severe corrosive influences where protected against corrosion and judged suitable for the condition [NEC 300.6 or IRC Table E3801.4 (footnote f)]. Where installed in wet locations, all electrical metallic tubing supports, bolts, straps, screws, connectors, couplings, and other similar components must be made of corrosion-resistant materials or protected against corrosion by corrosion-resistant materials and be listed for the specific environment.

The number of conductors in a single run of electrical metallic tubing must not exceed that permitted by the appropriate table in either *NEC* Informative Annex C or IRC Tables E3904.6(1) through E3904.6(10). For combinations of conductors, the percentage fill is specified in *NEC* Table 1, Chapter 9, using the conduit dimensions of *NEC* Table 4, Chapter 9 (see the previous discussion in this chapter).

Electrical metallic tubing is not permitted to be threaded. Where integral couplings are utilized, such couplings are permitted to be factory-threaded. Couplings and connectors used with EMT must be made up tight and are required to be identified for the application. Where buried in masonry or concrete, they must be of the concrete-tight type, and when installed in wet locations, they must be of the raintight type.

Bends in the electrical metallic tubing must be made so that the tubing is not damaged, and the internal diameter of the tubing is not reduced. The radius of the curve to the centerline of the tubing must not be less than shown in NEC Table 2, Chapter 9, for bends made in the field. All cut ends of electrical metallic tubing must be reamed or otherwise finished to remove rough edges.

Electrical metallic tubing is required to be installed as a complete system and must be securely fastened in place at least every 3.0 m (10 ft) and within 900 mm (3 ft) of each outlet box, junction box, cabinet, or fitting. Splices and taps must be made only in junction boxes, outlet boxes, or conduit bodies.

FLEXIBLE METAL CONDUIT (FMC)

NEC Article 348 covers flexible metal conduit (FMC). Flexible metal conduit is not permitted to be used in wet locations. Flexible metal conduit cannot be installed underground or be embedded in poured concrete or aggregate. Flexible metal conduit in the metric designator 12 (⅜ in.) nominal trade size is only permitted for use in lengths not longer than 1.8 m (6 ft) where it is a part of a listed assembly. It is permitted to be used for luminaire tap conductor connections to luminaires [*NEC* 410.117(C); *NEC* 348.20(A)(2)].

Securing the flexible metal conduit is required by an approved means. This must be accomplished at intervals not exceeding 1.4 m (4½ ft) in a run and not less than 300 mm (12 in.) on each side of every outlet box, junction box, cabinet, or fitting. Cable ties are permitted to secure FMC, but where used, the cable ties must be listed and identified for securement and support. Flexible metal conduits do not need to be secured where they are fished between access points through concealed spaces in finished buildings or structures, and supporting the flexible metal raceway is impractical, such as in finished gypsum-sheathed walls. Securing is not required where flexible metal conduit, metric designators 16 through 35 (trade sizes ½ through 1¼), is used in lengths not exceeding 900 mm (3 ft) at terminals where flexibility is necessary after installation and where an equipment grounding conductor is installed. For example, an equipment grounding conductor should be used where a flexible metal conduit is used to connect to a motor. The term flexibility is not defined in the *Code*. The authority having jurisdiction may require an equipment grounding conductor any time a flexible metal conduit is not supported in accordance with *NEC* 348.30. Securing is not required for lengths not exceeding 1.8 m (6 ft) from a luminaire terminal connection for luminaire tap conductor connections to flush- and recessed-type luminaires. The minimum length of flexible metal conduit used for tap connections to luminaires is 450 mm (18 in.). Securing is not required for lengths not exceeding 1.8 m (6 ft) from the last point, where the raceway is securely fastened for connections within an accessible ceiling to luminaires or other equipment.

Flexible metal conduit is permitted for equipment grounding only if it is listed, the length in any ground-fault return path does not exceed 1.8 m (6 ft), the circuit conductors contained therein are protected by overcurrent devices rated at 20 amperes or less, and the conduit is terminated in listed fittings [*NEC* 250.118(5) or IRC E3908.9.1].

Flexible metal conduits of any length are permitted where a properly sized equipment grounding conductor is installed with the circuit conductors for grounding equipment. Where an equipment bonding jumper is routed on the outside of a flexible metal conduit, it is not to exceed 1.8 m (6 ft) in length. An equipment grounding conductor is required where flexible metal conduit is used to connect equipment where flexibility is required after installation. The radius of the curve to the centerline of bends shall be not less than shown in *NEC* Table 2, Chapter 9, using the column "Other Bends." Angle-type connectors are not permitted to be used for concealed raceway installations. Fittings specifically listed for use with flexible metal conduit must be used.

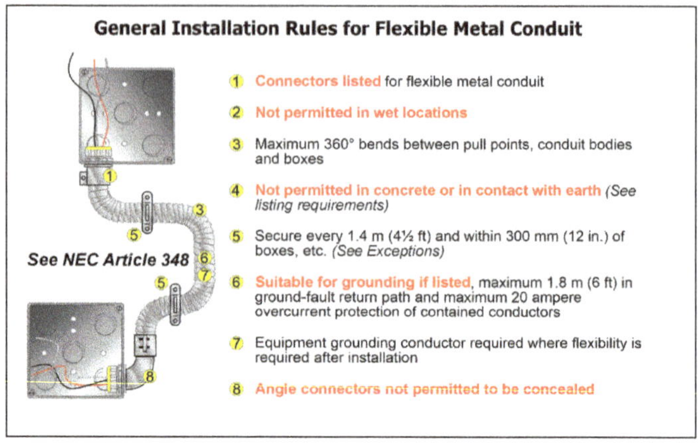

Figure 6.11. Installation rules for flexible metal conduit

LIQUIDTIGHT FLEXIBLE METAL CONDUIT (LFMC)

NEC Article 350 pertains to liquidtight flexible metal conduit (LFMC). Liquidtight flexible metal conduit is a circular raceway with an outer nonmetallic jacket that is both liquidtight and resistant to sunlight. This nonmetallic jacket surrounds an inner flexible metal core. This wiring method is used along with related couplings, connectors, and fittings. Liquidtight flexible metal conduit must be marked with durable materials. The marking must include the manufacturer's name and trademark.

LFMC can be direct buried where listed and marked for the purpose. Straight LFMC fittings are also permitted for direct burial, provided they are marked accordingly. LFMC cannot be installed where it will be subject to physical damage or where any combination of ambient and conductor temperature will produce an operating temperature in excess of that for which the material is approved.

The minimum size of LFMC allowed for use is generally not smaller than metric designation 16 (½ in. trade size). LFMC in the metric designation 12 (⅜ in.) size is available for specific applications. It is permitted for uses similar to those limitations for the metric designation 12 (⅜ in.) flexible metal conduit. Only fittings listed for use with LFMC may be used at the terminations of liquidtight flexible metal conduit.

Where LFMC is installed as a permanent raceway, it must be secured and supported. This support should be placed at intervals no more than 1.4 m (4½ ft) and secured within 300 mm (12 in.) of each side of the outlet box, junction box, cabinet, or fitting. If this wiring method is fished between access points through concealed spaces in finished buildings or structures and supporting it is impractical, securing is not required. Cable ties are allowed to secure LFMC, but when used, the ties must be listed and identified for securement and support. Securing is not needed when it is used in lengths not exceeding 900 mm (3 ft) at terminals where flexibility is necessary after installation. Additionally, securing is not required for lengths up to 1.8 m (6 ft) from the last point if the raceway is firmly fastened for connections within an accessible ceiling to luminaires or other equipment.

Listed LFMC is permitted to serve as an equipment grounding means where the total length of any ground-fault return path is 1.8 m (6 ft) or less, and the conduit is terminated in listed fittings. Where the LFMC is being used as an equipment grounding conductor, overcurrent devices are limited to 20 amperes or less for metric designators 12 and 16 (⅜ in. and ½ in. trade sizes) and 60 amperes or less for metric designators 21 through 35 (¾ in. through 1¼ in. trade sizes) and protect the circuit conductors contained therein. LFMC larger than metric designator 35 (1¼ in. trade size) is not permitted to be used as an equipment grounding conductor [*NEC* 250.118(6) or IRC E3908.9.2]. It is permitted to use LFMC of any length where a properly sized equipment grounding conductor is installed with the circuit conductors for grounding equipment. An equipment grounding conductor shall always be used where flexibility is necessary after installation. For example, an equipment grounding conductor should be used where LFMC is used to connect to a motor.

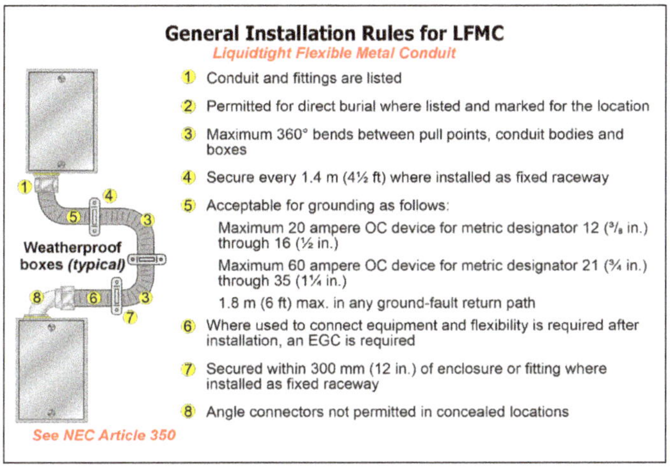

Figure 6.12. Installation rules for liquidtight flexible metal conduit

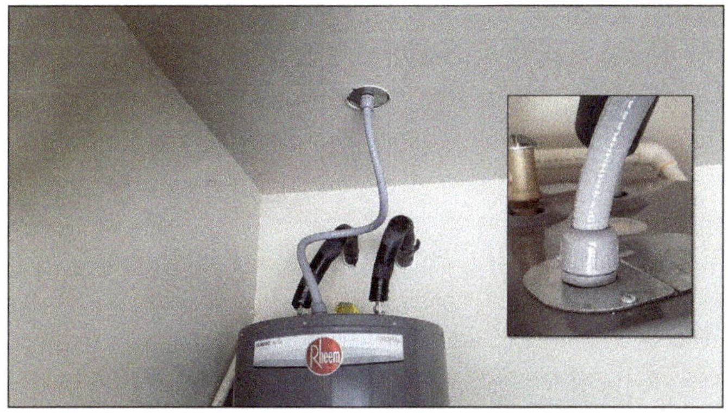

Photo 6.10. Liquidtight flexible nonmetallic conduit (LFNC) can be used in service, feeder, or branch circuit application at dwelling units

Where an equipment bonding jumper is routed on the outside of a length of LFMC, it is not to exceed 1.8 m (6 ft) in length. Longer bonding jumpers must be inside the conduit. The term flexibility is not defined in the *Code*. The authority having jurisdiction may require an equipment grounding conductor any time liquidtight flexible metal conduit is not supported in accordance with *NEC* 350.30. LFMC is recognized as an equipment grounding means in sizes up to metric designator 35 (1¼ trade size). This recognition covers only a total length in any ground return path that is not over 1.8 m (6 ft) where the conduit is terminated in fittings listed for grounding and the circuit conductors contained within it are protected by overcurrent devices rated at not over 20 amperes for the metric designator 12 (⅜ in.) and 16 (½ in.) trade sizes. Contained conductors are permitted to have overcurrent protection up to 60 amperes for metric designator 21 (¾ in.) through 35 (1¼ in.) trade sizes.

The radius of the curve to the centerline of bends shall be not less than shown in *NEC* Table 2, Chapter 9, using the column "Other Bends." Angle connectors cannot be installed where they will be concealed.

LIQUIDTIGHT FLEXIBLE NONMETALLIC CONDUIT (LFNC)

NEC Article 356 covers Liquidtight flexible nonmetallic conduit (LFNC). LFNC is a circular raceway and is available in three types. The conduit is made with a smooth, seamless inner core and cover, which are bonded together with one or more reinforcement layers between the core and cover. It is available with a smooth inner surface that is provided with integral reinforcement within the wall of the conduit. This type is marked "LFNC-A." The product marking can be found on the outer surface of the raceway. It is also available with corrugated internal and external surfaces having no integral reinforcement within the wall of the conduit. This type is marked "LFNC-B." The product marking can be found on the outer surface of the raceway.

This wiring method is recognized as flame-resistant and, with proper fittings, is approved for the installation of electrical conductors. This type is marked "LFNC-C." The product marking can be found on the outer surface of the raceway.

LFNC can be used in exposed or concealed locations where flexibility is required after installation, operation, or maintenance of equipment. It can be used where necessary to protect the enclosed conductors from vapors, liquids, or solids. It can be used for outdoor locations where it is specifically listed and marked as being suitable for the purpose. It is also permitted for direct burial or concrete encasement, provided it is listed and marked for this purpose and installed in compliance with NEC 356.42.

Where installed in lengths exceeding 1.8 m (6 ft), LFNC is required to be securely fastened at intervals not exceeding 900 mm (3 ft) and within 300 mm (12 in.) on each side of every outlet box, junction box, cabinet, or fitting. Cable ties are permitted to secure LFNC, but where used, they must be listed and identified for securement and support.Handling of LFNC in areas where the temperature is extremely cold may render it brittle and lead to damage from any physical contact. It is not permitted to be used where it will be subject to physical damage or where any combination of ambient and conductor temperatures is in excess of that for which the LFNC is identified. Type LFNC-B liquidtight flexible nonmetallic conduit is permitted in lengths longer than 1.8 m (6 ft). LFNC is required to be used only with terminal fittings listed for use with LFNC.

Where an equipment grounding conductor is required for the circuits installed in liquidtight flexible nonmetallic conduit, it is generally required to be installed on the inside of the conduit.

INSTALLATION OF FEEDER USING CABLES

Service-entrance cable is often used for feeders at dwelling units. As with raceways, the *Code* specifies securing and supporting intervals, bending radii, and restrictions on permitted uses. These rules may vary somewhat based on the type of service-entrance cable selected. A feeder installed using Type SE cable requires a thorough knowledge of *NEC* Article 338, which covers Types SE, SER, and USE service-entrance cable. This article provides descriptive information essential for the proper application of this type of wiring within its limitations of use.

Service-entrance cable is available as a single conductor or multi-conductor assembly manufactured with or without an overall covering. It is primarily used for services, but often, it is used as a feeder where the neutral conductor is insulated. Type SE cable is available in the following types:

1. Type SE cable is constructed with two insulated conductors and a bare conductor that is concentrically wound around the other two conductors. The overall covering is a flame-retardant, moisture-resistant jacket.
2. Type SER cable is constructed with three insulated conductors and a bare conductor. The overall covering is a flame-retardant, moisture-resistant jacket.
3. Type USE cable is identified for underground use. The cable has a moisture-resistant covering but is not required to have a flame-retardant covering.

Service-entrance cable is required to be marked to indicate a maximum rated voltage; proper type letters; manufacturer's name, trademark, or other marking that will include the organization responsible for the product. Type SE cable with the neutral conductor smaller than the ungrounded conductors will also be specifically marked.

Type USE conductors are usually limited to installation underground. This is due to the Type USE insulation not being of flame-retardant construction. The *Code* permits Type USE cable to rise aboveground as long as it is outside of the building. Type USE cable is permitted to be used for outside branch circuits and feeders but is generally not permitted to be used inside buildings. Where Type USE cable emerges from the ground, it must be protected from physical damage. Rigid raceways such as IMC, RMC, or Schedule 80 PVC are suitable methods for protecting the USE cable.

Type USE cable is not permitted to be used inside buildings unless the insulation is marked with the appropriate additional markings. Type USE cable is available with additional insulation designations, which include Type RHH and RHW. Some designations may include a suffix "-2" to indicate the cable is suitable for 90°C, both wet and dry. These triple-rated cables are permitted to be used at any location the individual designation permits. For example, Type RHH is a rubber-insulated wire that has insulation rated at 90°C and is permitted to be used aboveground, in or on buildings or structures, in a raceway, or in a dry location. Type RHW is a rubber-insulated wire that has a 75°C rating and is suitable for installation aboveground, in or on buildings or structures, in a wet location, or in a raceway. They are permitted to be used underground due to their USE designation. These triple-rated cables offer a great deal of flexibility as to locations where they are permitted to be installed.

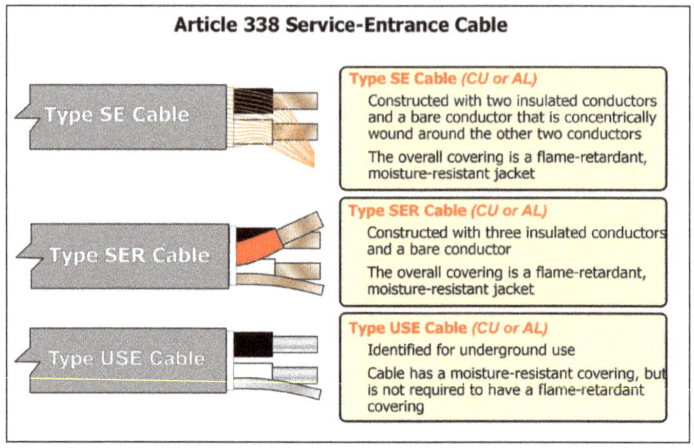

Figure 6.13. Typical service-entrance cable (Types SEU and SER)

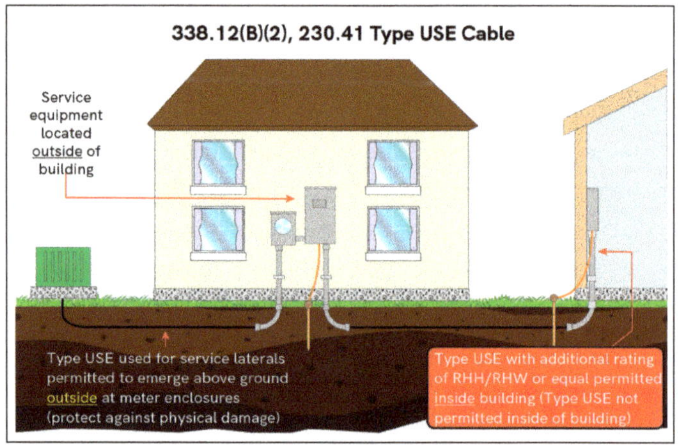

Figure 6.14. Installation of underground service-entrance conductor (USE)

GROUNDED (NEUTRAL) CONDUCTOR

A Type SE service-entrance cable is permitted to be used for branch circuits or feeders in interior wiring systems where all the circuit conductors of the cable are insulated and of the thermoset or thermoplastic type.

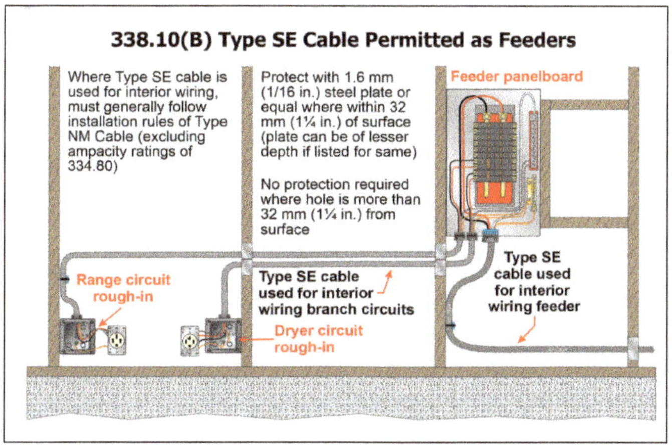

Figure 6.15. Type SE and SER used for interior wiring must comply with the installation requirements for nonmetallic-sheathed cable (excluding the ampacity ratings of 334.80).

A Type SE service-entrance cable assembly without individual insulation on the neutral circuit conductor is generally not permitted to be used for a feeder within a building. However, a Type SE cable that has a final nonmetallic outer covering and is supplied by alternating current at not over 150 volts-to-ground can be used as a feeder to supply only other buildings on the same premises where the uninsulated grounded conductor of the cable originates in service equipment.

Type SE service-entrance cable is permitted for use where the fully insulated conductors are used for circuit wiring and the uninsulated conductor is used for equipment grounding purposes only. A branch circuit supplying a 240-volt water heater is a typical example of this use.

TEMPERATURE LIMITATIONS

Where a Type SE service-entrance cable is used to supply appliances, it must not be subjected to conductor temperatures in excess of the temperature specified for the type of insulation involved. Where Type SE service-entrance cable is used for interior wiring, it must be installed in the same manner as nonmetallic-sheathed cable in *NEC* Article 334, Part II (excluding 334.80) [*NEC* 338.10(B)(4)(a)(1)]. SE cable is permitted by the IRC as a wiring method for feeders and branch circuits in accordance with IRC Table E3801.4 (footnotes b and c).

Type SER cable includes an insulated grounded (neutral) conductor as well as an equipment grounding conductor and is suitable for use as feeders where properly sized and installed in accordance with the requirements for interior nonmetallic cable wiring methods.

Requirements covering wiring methods, such as support provisions and acceptable locations, must be applied here as well. For example, the concerns for protection against physical damage for feeder cables passing through bored holes in joists, rafters, or wood members; laid-in notches cut in studs; or installed where run parallel to joists, rafters, or studs are most important here.

PROTECTION FROM PHYSICAL DAMAGE

To provide proper protection for a feeder cable and prevent it from being damaged, the installer and inspection authority must be aware of the type of construction. For example, a building constructed with a wood frame will consist of framing members, including furring strips, strapping, studs, joists, rafters, and trusses. The building code will regulate how and where these framing members can be altered in order to install the cable. The NEC and Chapter Eight of the IRC have specific requirements to permit the cable to be installed in, on, and through the framing members as long as physical protection for the cable is provided.

For example, *NEC* 300.4(A) or IRC Table E3802.1 covers protection against physical damage for cables and raceways required where they are run through bored holes in wooden framing members. Cable or raceway wiring methods installed through bored holes, in both exposed and concealed locations in joists, rafters, or wood members, are required to have all holes bored so the edge of the hole is not less than 32 mm (1¼ in.) from the nearest edge of the wood framing member.

If this distance cannot be maintained, the cable or raceway must be protected from penetration by nails or screws by a steel plate (nail plate) or suitable bushing at least 1.6 mm (1/16 in.) thick and of appropriate length and width to cover the complete area of the wiring system to be protected. Steel plates that are thinner than 1.6 mm (1/16 in.), such as heat-treated metals, are permitted to protect cables and raceways, but only if the plates are specifically listed and marked appropriately [*NEC* 300.4(A) or IRC Table E3802.1].

Raceways of intermediate metal conduit, rigid metal conduit, rigid nonmetallic conduit, and electrical metallic tubing are not required to conform to this physical protection rule for cable assemblies. Other raceways, such as electrical nonmetallic tubing and flexible metal conduit, must have the same protection from damage as that required for cable-type wiring methods such as Type NM cable.

NOTCHES IN WOOD

Where there is no objection because of weakening the building structure (check local building codes), in both exposed and concealed locations, cables or raceways are permitted to be laid in notches in wood studs, joists, rafters, or other wood members. The cable or raceway at those notched points must be protected against penetration by nails or screws by a steel plate at least 1.6 mm (1/16 in.) thick installed before the building finish is applied. Thinner steel plates are again permitted here if listed and marked appropriately. See Chapter 11 for additional information concerning the application of the building and electrical code requirements as they apply to bored holes or notches in wooden members. The *NEC* references building structure protection from drilling and notching at *NEC* 300.4(A) through (H), as does the IRC at IRC E3402.1. The specific sections in the IRC relating to requirements for notching framing members are as follows:

Steel Joists	R505.3.5
Steel Studs	R603.3.4
Wood Joists	R502.8
Wood Studs	R602.6
Wood Top Plates	R602.6.1

This requirement to protect raceways and cables does not apply to intermediate metal conduit, rigid metal conduit, rigid nonmetallic conduit, or electrical metallic tubing.

CABLES AND ELECTRICAL NONMETALLIC TUBING THROUGH METAL FRAMING MEMBERS

Where nonmetallic-sheathed feeder cables are run exposed or are installed where concealed and pass through either factory- or field-punched, cut, or drilled slots or holes in metal framing members, the cable assembly must be protected by bushings or grommets securely fastened in the opening before the cable is installed.

The bushing or grommet must cover all metal edges and be listed. Manufacturers have created various products that meet this requirement. Using materials like cardboard, garden hose, electrical tape, or other non-approved items is not allowed, as it could damage the cable.

Where driven nails or screws can penetrate feeders of nonmetallic-sheathed cable or electrical nonmetallic tubing installed through metal studs, a steel sleeve, steel nail plate, or steel clip not less than 1.6 mm (1/16 in.) in thickness is required to be used to protect the cable or tubing from possible penetration. Thinner steel plates are again permitted here if listed and marked appropriately. This requirement is not applicable to Type AC feeder cable assemblies [*NEC* 300.4(B) or IRC Table E3802.1].

Photo 6.11. Metal framing members with cable wiring methods installed

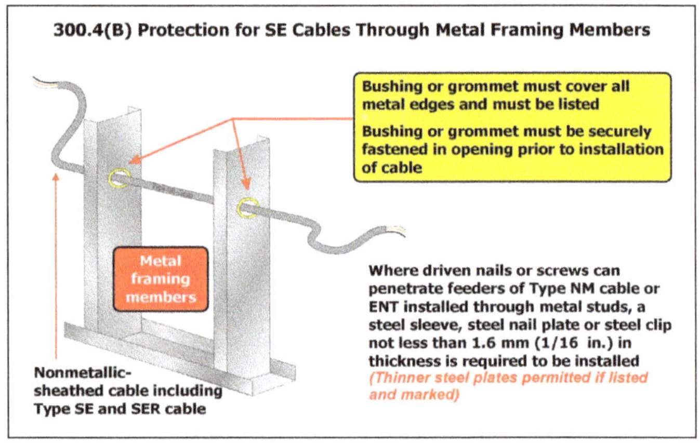

Figure 6.16. Protection of Type SE cables installed through metal framing members

FEEDER CABLES AND RACEWAYS INSTALLED PARALLEL TO FRAMING MEMBERS

Where run exposed or in concealed locations, a feeder cable or raceway-type wiring method installed parallel to wooden framing members, such as joists, rafters, or studs, must be installed and supported so the nearest outside surface of the cable or raceway is not less than 32 mm (1¼ in.) from the nearest edge of the framing member. Without this additional protection, driven nails or screws are likely to penetrate and damage the cable.

If a 32 mm (1¼ in.) distance cannot be maintained, the cable or raceway must be protected from penetration by nails or screws by a steel nail plate, sleeve, or equivalent at least 1.6 mm (1/16 in.) thick. Thinner steel plates are again permitted here if listed and marked appropriately.

This rule for the protection of cables or raceways is not applicable to feeders installed in intermediate metal conduit, rigid metal conduit, rigid nonmetallic conduit, and electrical metallic tubing. This requirement also applies to concealed work in finished buildings or finished panels for prefabricated (manufactured) buildings, wired in accordance with

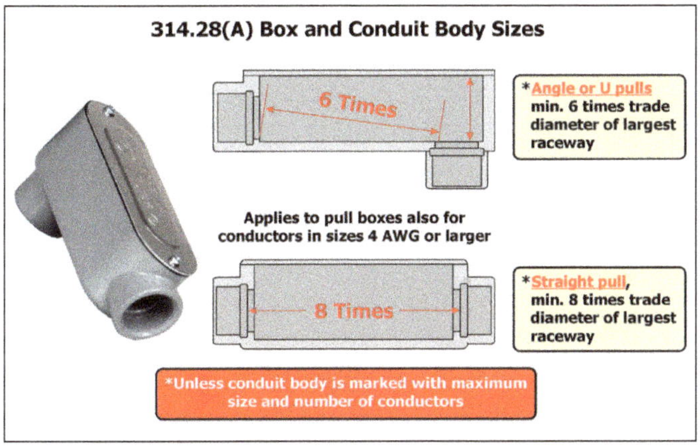

Figure 6.17. Minimum size required for conduit bodies (LB, C, LR, LL fittings, etc.)

NEC Article 545, Manufactured Buildings and Relocatable Structures.

CONDUIT BODIES

A conduit body is a separate portion of a conduit or tubing system that provides access through a removable cover(s) to the interior of the system at a junction of two or more sections of the conduit or tubing system or at a terminal point of the system. Boxes such as FS (shallow) and FD (deep) cast device boxes or larger cast or sheet metal types are not classified as conduit bodies but fall under the category of boxes.

Requirements for installing conduit bodies are found in *NEC* Article 314 or IRC E3905. Conduit bodies generally are not permitted to contain splices unless they are durably and legibly marked with their cubic inch capacity. If they are marked, conductor fill is determined using the same procedure for other than standard boxes (*NEC* 314.16(C)(2) or IRC E3905.12.3.1). Chapter 12 of this text includes additional detailed information on calculating the conductor fill of boxes.

Conduit bodies that are used as pull or junction boxes are required to comply with *NEC* 314.28. Conduit bodies used for conductors 4 AWG or larger must comply with *NEC* 314.28(A)(1) through (A)(3). For straight-through pulls, the length of the conduit body normally must not be less than *eight times* the diameter of the largest raceway. Where used with angle or U pulls, the distance between raceway entries and the opposite wall is not to be less than *six times* the diameter of the largest conduit. This dimension can be reduced where the conduit entry is opposite a removable cover, and that distance is not less than in *NEC* Table 312.6(A) for one conductor per terminal. This table deals with wire bending space at terminations of conductors. Conduit bodies of other dimensions are permitted where they are marked with both the maximum size and number of conductors.

Conduit bodies must be installed so the wiring in them can be rendered accessible without removing any part of the building.

PRODUCT SAFETY STANDARD REQUIREMENTS

It is recommended to consult the following electrical product safety guide card information and product safety standards for additional guidance on the proper installation, operation, and use of electrical equipment covered in this chapter. The four-letter code in parentheses refers to the product category in the UL Guide Information for Electrical Equipment Directory (UL Product Spec material).

- Equipment for Use in Ordinary Locations . (AALZ)
- Armored Cable (AWEZ)
- Armored Cable Type AC, Connectors (AWSX)
- Underground Feeder and Branch Cable (YDUX)
- Service-Entrance Cable (TYLZ)
- Rigid Ferrous Metal Conduit (DYIX)
- Intermediate Ferrous Metal Conduit (DYBY)
- Electrical Metallic Tubing (FJMX)
- Flexible Metal Conduit (DXUZ)
- Flexible Metal Conduit, Liquidtight (DXHR)
- Flexible Nonmetallic Conduit, Liquidtight (DXOQ)
- Rigid Nonmetallic PVC Conduit (DZYR)
- Panelboards (QEUY)
- Conduit Fittings (Bodies) (DWTT)

REFERENCES

1. *National Electrical Code* and *NEC* are registered trademarks of the National Fire Protection Association, Inc., Quincy, MA 02169. This reprinted material is not the official position of the National Fire Protection Association, which is represented only by the standard in its entirety.

2. *International Residential Code for One- and Two-Family Dwellings* (2024). Copyright © International Code Council, Inc., June 2023.

3. UL, Northbrook, IL, *UL Guide Information for Electrical Equipment Directory* (UL Product iQ® material) unless indicated otherwise.

REVIEW QUESTIONS

1. A "Feeder" is defined as "all circuit conductors between the service equipment, the source of a separately derived system, or other power supply source and the final branch-circuit overcurrent device."

True _____ False _____

2. Feeders must be protected from overcurrent in accordance with Part I of *NEC* Article 225 and IRC Chapter 36.

True _____ False _____

3. In a long run, in order to limit the voltage drop to 3 percent, the installed ungrounded conductors for a 100-ampere feeder are increased to 2 AWG THHN from 3 AWG THHN. In this case, the minimum size of the wire-type copper equipment grounding conductors is:

A. 8 AWG
B. 6 AWG
C. 4 AWG
D. 2 AWG

4. Feeder conductors for a dwelling unit are not required to have an ampacity greater (larger) than the service-entrance conductors.

True _____ False _____

5. Where a feeder supplies branch circuits in which equipment grounding conductors are required, the feeder must include or provide an equipment grounding conductor to which the equipment grounding conductor of the branch circuit must be connected.

True _____ False _____

6. A feeder supplying continuous loads or any combination of continuous and noncontinuous loads is required to have an ampacity not less than the noncontinuous load plus __________ of the continuous loads.

A. 25 percent
B. 100 percent
C. 125 percent
D. 300 percent

7. Type SER cable is a cable assembly that is round in configuration and includes an insulated grounded (neutral) conductor that is part of an assembly along with a bare equipment grounding conductor and insulated phase conductors.

True _____ False _____

8. Type ______ cable is permitted to be used as a directly buried underground feeder where run to supply a second building on the property.

A. SE
B. NM
C. USE
D. SER

9. A Type SE service-entrance cable is not permitted to be used for a feeder in interior wiring systems where all the circuit conductors of the cable are of the thermoset or thermoplastic type.

True _____ False _____

10. A typical feeder cable assembly may include Types AC, NM, UF, or SER.

True _____ False _____

11. Typical raceways employed for feeders include:

A. Electrical nonmetallic tubing and electrical metallic tubing
B. Rigid PVC conduit
C. Intermediate metal conduit or rigid metal conduit
D. All of the above

12. For a feeder, an overcurrent device is generally required to be located ______________________________.

A. where the conductor terminates
B. at the point where the conductors receive their supply
C. overcurrent protection not required
D. either A. or B.

13. The 3 m (10 ft) tap rule permits the tap conductors to be installed under specific conditions without overcurrent protection at the tap conductor origination.

True _____ False _____

14. For new installations, the grounded (neutral) conductor must be isolated from the enclosure of a feeder panelboard supplied from service equipment.

True _____ False _____

15. Conductors are permitted to fill a conduit to ________ of the area where three or more conductors are installed.

A. 31%
B. 40%
C. 53%
D. 75%

16. PVC conduit is not permitted to support luminaires.

True _____ False _____

17. An expansion fitting is required where the expansion or contraction of rigid PVC conduit is more than _______________.

A. 6 mm (¼ in.)
B. 13 mm (½ in.)
C. 19 mm (¾ in.)
D. 25 mm (1 in.)

18. Angle connectors for flexible metal and liquid-tight flexible metal conduit are not permitted to be _____________.

A. exposed
B. installed
C. installed without GFCI protection
D. concealed

19. Unless also marked with additional markings such as RHH or RHW, Type USE conductors are not permitted inside buildings.

True _____ False _____

20. Cables installed as feeders are to comply with requirements for protection from physical damage under which of the following conditions?

A. Where installed in bored holes within 25 mm (1 in.) of the nearest edge of the wood member.
B. Where installed parallel to a framing member within 25 mm (1 in.) of the nearest edge of the framing member.
C. Where installed in a shallow notch in framing members and protected with an aluminum plate at least 1.6 mm (1/16 in.) thick.
D. Where installed in a bored hole not less than 32 mm (1¼ in.) from the nearest edge of the wood member.

21. A conduit body for a straight pull must be not less than _________ times the metric designator (trade size) of the largest raceway.

A. four
B. six
C. eight
D. twelve

22. The minimum size Rigid PVC Conduit, Schedule 80, containing three 3 AWG, THHW insulated conductors is:

A. Metric designator 16 (trade size ½ in.)
B. Metric designator 27 (trade size 1 in.)
C. Metric designator 35 (trade size 1¼ in.)
D. Metric designator 41 (trade size 1½ in.)

23. Where installed in lengths exceeding 1.8 m (6 ft), LFNC is required to be securely fastened at intervals not exceeding __________ on each side of every outlet box, junction box, cabinet, or fitting.

A. 900 mm (3 ft) and within 300 mm (12 in.)
B. 1.2 m (4 ft) and within 300 mm (12 in.)
C. 1.8 (6 ft) and within 250 mm (10 in.)
D. 2.5 m (8 ft) and within 300 mm (12 in.)

24. In a feeder panelboard (load side or past the first means of disconnect), equipment grounding conductors can be connected to the grounded (neutral) conductor bar.

True ______ False ______

7 Requirements for Major Household Appliances

Photo 7.1. Gas-fired furnace (appliance) in dwelling attic supplied by an individual branch circuit.

GENERAL REQUIREMENTS

Household appliances have played a major role in the average home for more than a century. These appliances, usually gas or electrically powered, help homeowners with tasks like cooking, cleaning, and food storage.

An "appliance" is a piece of utilization equipment, generally not industrial, that is fixed in place, stationary, or portable. It is usually made in standard sizes or types and is installed or connected as a unit to perform functions like clothes washing, air conditioning, food mixing, deep frying, and more. "Utilization equipment" refers to equipment that uses electric energy for purposes such as electronic, electromechanical, chemical, heating, lighting, or similar functions.

BRANCH-CIRCUIT REQUIREMENTS

Branch-circuit conductors that supply an appliance must be sized to be capable of carrying the appliance's current without the conductor overheating under the specified conditions. Conductors that form an integral part of an appliance or factory-installed wiring are not considered when applying this requirement (NEC 422.10).

An individual appliance's branch-circuit rating or size cannot be less than the rating marked on the appliance nameplate or the rating marked on an appliance that has combined loads [*NEC* 422.10(A)(1)].

Some motor-operated appliances do not have a marked nameplate rating. In this case, the size of the branch-circuit conductors is required to be in accordance with Part II of *NEC* Article 430 [*NEC* 422.10(A)(2)]. The conductors are sized based on the full load current rating of the motors together with any other associated loads [*NEC* 430.6(A) and 430.22].

For continuously loaded appliances other than those that are motor-operated, the branch-circuit rating cannot be less than 125% of the marked nameplate rating. *NEC* 210.20(A) contains similar rules for branch-circuit overcurrent devices and conductors that serve continuous loads. Most loads in dwellings are not considered continuous loads [*NEC* 422.10(A)(3)].

Branch circuits used to supply household cooking appliances are sized according to the load calculated in accordance with NEC Table 220.55 or IRC E3702.10 and are sized in accordance with NEC 210.19(C). Refer to Chapter 2 for further information on this topic.

Where branch circuits supply an appliance and other loads, the rating of the branch circuit is determined in accordance with *NEC* 210.23 or IRC E3702. This provision is found at *NEC* 422.10(B) with reference to *NEC* 210.23, where we find permissible loads for 15- and 20-ampere, 30-ampere, 40- and 50-ampere, and larger than 50-ampere branch circuits. The installation requirements of some listed appliances may demand an individual branch circuit that meets the manufacturer's specifications. This type of instruction must be followed through *NEC* 110.3(B) or IRC E3403.3 of the *Code*.

Appliance branch-circuit conductors are required to be provided with overcurrent protection in accordance with NEC 422.11 or IRC E4101.4. Generally, branch-circuit conductors must be protected from overcurrent in accordance with NEC 240.4 or IRC E3705.5. By reference to NEC 310.14 or IRC E3705.1, NEC 240.4, or IRC E3705.5 incorporates Table 310.16 or IRC Table E3705.1 for the allowable ampacity of conductors. Specific rules for small conductors are contained in NEC 240.4(D) or IRC E3705.5.3. These sections provide that unless specifically permitted elsewhere in NEC 240.4 or IRC E3705.5.4, the overcurrent protection cannot exceed 15 amperes for 14 AWG copper, 20 amperes for 12 AWG copper, 30 amperes for 10 AWG copper, 15 amperes for 12 AWG aluminum, and 25 amperes for 10 AWG aluminum. The rating of the overcurrent protection device is determined after any correction factors for ambient temperature or number of conductors are applied to the conductors.

Nameplate information may include the maximum or minimum rating of the branch-circuit overcurrent device. If marked on an appliance nameplate, the branch-circuit overcurrent protective device rating is restricted to the marked values [*NEC* 422.11(A) or IRC E4101.4.1].

Ground-fault circuit interrupters are now required in all residences that have 125-volt through 250-volt receptacles supplying single-phase branch circuits up to 50 amps within kitchens. This code change also applies to appliances meeting the requirements of NEC 210. 8, IRC E3902.6.

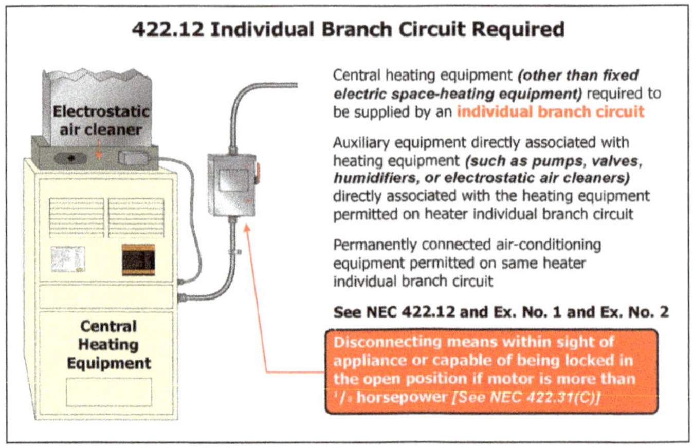

Figure 7.1. Supply to and disconnecting means for central heating equipment and electrostatic air cleaner.

Photo 7.2. Gas furnace with electrostatic air cleaner

INSTALLATION OF APPLIANCES

All appliances must be installed in an approved manner (acceptable to the authority having jurisdiction). Appliance installation methods that differ from *NEC* Article 422 are allowed only through special permission. Special permission means the written permission of the authority having jurisdiction (*NEC* 90.2(C), 110.3(B) or IRC E4101.2).

Central heating equipment, such as an oil- or gas-fired furnace, must be supplied by an individual branch circuit (NEC 422.12 or IRC E3703.1). Typically, an individual branch circuit supplies only the specific piece of utilization equipment (see NEC Article 100 or IRC Chapter 35 definitions and photos 7.1 and 7.2). In this case, this branch circuit may also supply auxiliary equipment directly associated with the central heating system.

Examples of auxiliary equipment include pump motors or solenoid valves used with a boiler, or a humidifier or electrostatic air cleaner used with a furnace. A receptacle outlet for auxiliary equipment does not include the receptacle required to be within 7.5 m (25 ft) of the equipment as mandated by NEC 210.63(A) or IRC E3901.12. The 125-volt, single-phase, 15- or 20-ampere-rated convenience receptacle

Table 7.1 Flexible Cord Connection Lengths for Specific Appliances		
Specific Appliance	**Minimum Length**	**Maximum Length**
In-Sink Kitchen Waste Disposers	450 mm (18 in.)	900 mm (36 in.)
Built-in Dishwashers	0.9 m (3 ft)	2.0 m (6.5 ft)
Trash Compactors	0.9 m (3 ft)	1.2 m (4 ft)
Range Hoods	450 mm (18 in.)	1.2 m (4 ft)

Table 7.1 Flexible Cord Lengths for Specific Appliances

422.16(A) Use of Flexible Cords Limited

Cord-and-Plug Connection Not Permitted

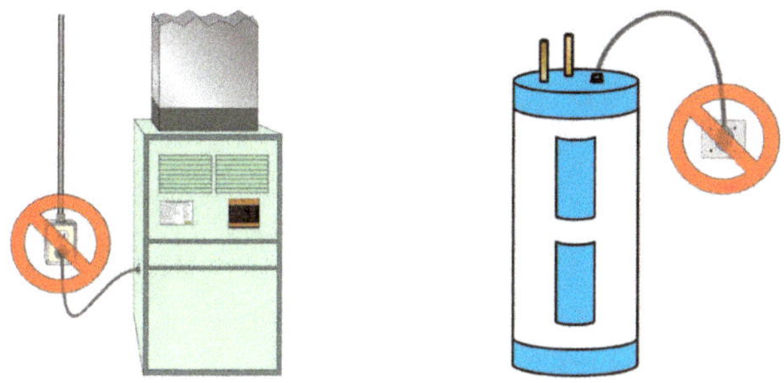

Cord-and-plug connection prohibited unless fastening means and mechanical connections are specifically designed to permit ready removal for maintenance or repair, and the appliance is intended or identified for flexible cord connection

Figure 7.2. A flexible cord connection generally is not permitted for these appliances.

outlet designated for servicing the equipment must be accessible to personnel.

Auxiliary equipment, such as an electrostatic air cleaner, is not permitted to be supplied from internal circuits of the central heating equipment unless the manufacturer of the heating equipment has made provisions for the auxiliary equipment. Check the manufacturer's instructions for such provisions. In some cases, the manufacturer's instructions will require a separate circuit. This requirement for an individual branch circuit enhances safety. The more equipment connected to a branch circuit, the more likely that power to the furnace, boiler, or heat pump could be interrupted. Additional precautions are warranted in colder climates. An overcurrent situation in the auxiliary equipment could cause the branch circuit overcurrent protective device to open and thus disrupt power to the entire heating unit.

Permanently connected air-conditioning equipment is permitted to be connected to the same branch circuit as central heating equipment. The air-conditioning load is considered non-coincident, meaning it is unlikely to operate simultaneously with the furnace, boiler, or other central heating appliance. This permission does not extend to cord- and plug-connected air conditioning units designed for installation in windows. Typical conditions where this exception applies are limited due to differences in voltage ratings, circuit ratings, and a need to comply with all applicable Code requirements (NEC 422.12 Exception No. 2 or IRC E3703.1).

Fixed electric space-heating equipment, such as baseboard or wall-mounted fan units, does not fall under the heading of central heating equipment. *NEC* Article 424 covers circuit requirements for fixed electric heating equipment. See Chapter 8 for additional. information on fixed electric space-heating equipment.

APPLIANCES CONNECTED WITH FLEXIBLE CORD

A flexible cord assembly is permitted for use in connecting certain appliances. This type of connection is permitted to facilitate frequent or occasional interchange or to prevent the transmission of noise or vibration to an appliance. A cord-and-plug connection to a sump or stormwater pump is a typical example of preventing vibration transmission through the electrical connection. A flexible cord is also permitted for use to facilitate the removal or disconnection of appliances that are fastened in place, provided the fastening means and mechanical connections are specifically designed to permit ready removal for maintenance or repair, and the appliance is specifically intended or identified for a flexible cord connection. A cord- and plug-connected household range is a typical example of an appliance specifically intended or identified for a flexible cord connection [NEC 400.10(A)(8) and 422.16(A) and (B); IRC E4101.3 and Table E4101.3].

These Code rules clearly indicate that central heating appliances and electric water heaters are not allowed to be cord- and plug-connected. The typical dryer cord or flexible cord used for a waste disposer is not "intended or identified" for use with a heating appliance. History shows that when an electric water heater was cord- and plug-connected (usually with a dryer cord), the cord deteriorated over time, leading to serious problems when someone handled these damaged cords. Electric water heaters may also include specific instructions requiring a high-temperature supply conductor at the connection point. It should be noted that flexible cords are generally rated for only 60°C use.

As shown in the following text, some appliances are specifically permitted to be cord- and plug-connected even though they are not portable. The attachment plugs must be grounding type and dead-front construction. A dead-front attachment plug has no exposed live parts on the blade side of the plug.

KITCHEN IN-SINK WASTE DISPOSER

A 120-volt kitchen in-sink food waste disposer used in a dwelling unit is permitted to be cord- and plug-connected with a flexible cord identified for the purpose. Flexible cords supplied with listed appliances are considered to be identified for the purpose. Before installing a flexible cord in the field, consult the appliance manufacturer's instructions for information on the type of cord to use and any specific installation requirements. In addition, NEC Table 400.4, Flexible Cords and Cables, specifically lists Types SP-3, SPE-3, and SPT-3 as permitted for supply to kitchen waste disposers, dishwashers, and trash compactors, as well as wall-mounted and counter-mounted cooking units, and range hoods, with references to NEC 422.16(B).

Table 7.2 Wiring for In-Sink Kitchen Waste Disposer Section 422.16(B)(1)
Permitted to be cord-and-plug connected
Flexible cord must have an equipment grounding conductor and be terminated with a grounding-type attachment plug
Length: 450 mm (18 in.) and not over 900 mm (36 in.)
Flexible Cord and Receptacle – not subject to physical damage
Receptacle must be accessible

Table 7.2. Wiring for Kitchen Waste Disposer

Photo 7.3. In-sink kitchen waste disposer is permitted to be cord-connected

The cord that connects the appliance is terminated with a grounding-type attachment plug, and all the following conditions are met:

1. supply cord length cannot be shorter than 450 mm (18 in.) nor longer than 900 mm (36 in.);
2. receptacles are required to be located to protect against physical damage to the flexible cord and attachment plug;
3. receptacles must be accessible; and
4. The flexible cord must have an equipment grounding conductor and be terminated with a grounding-type attachment plug.

Receptacles for supply to in-sink kitchen waste disposers are not required to be readily accessible. Such outlets are required to be accessible. For example, the receptacle may be located within the cabinet that supports the kitchen sink and contains the in-sink waste disposer. This cabinet may also be used to store kitchen cleaning supplies. Access to the receptacle and attachment plug may require removing these cleaning supplies. However, the receptacle is still considered accessible. Receptacle outlets installed under a sink *(such as a receptacle outlet installed under a kitchen sink for a waste disposer)* cannot be installed in the face-up position *[NEC 406.5(G)(2)]*.

BUILT-IN DISHWASHERS

A built-in dishwasher intended for dwelling unit use is permitted to be hard-wired or cord- and plug-connected with a flexible cord identified for the purpose and terminated with a grounding-type attachment plug cap [*NEC* 422.16(B)(2) or IRC E4101.3 and Table E4101.3]. If a dishwasher is cord-and-plug connected, the receptacle outlet for a built-in dishwasher is required to be located in the space adjacent to the space occupied by the dishwasher. Section 8.3.3(a) of UL 749 (Standard for Safety for Household Dishwashers) states in part, "the power-supply receptacle for the appliance (dishwasher) shall be installed in a cabinet or on a wall adjacent to the under-counter space in which the appliance is to be installed." In other words, UL 749 prohibits the installation of the power-supply receptacle outlet for the dishwasher being installed in the same cabinet space as the dishwasher. A change occurred in the 2017 *NEC* and the 2018 IRC to align *NEC* 422.16(B)(2) and IRC E4101.3 and Table E4101.3 with the product standard for household dishwashers (UL 749). Previous editions of the *Code* allowed the power supply receptacle outlet for the dishwasher to be located in the same space (behind the dishwasher).

Additional conditions for such uses are:

1. The length of the flexible cord is not less than 0.9 m (3 ft) and not to exceed 2.0 m (6.5 ft). To accommodate this potential extended length of the cord

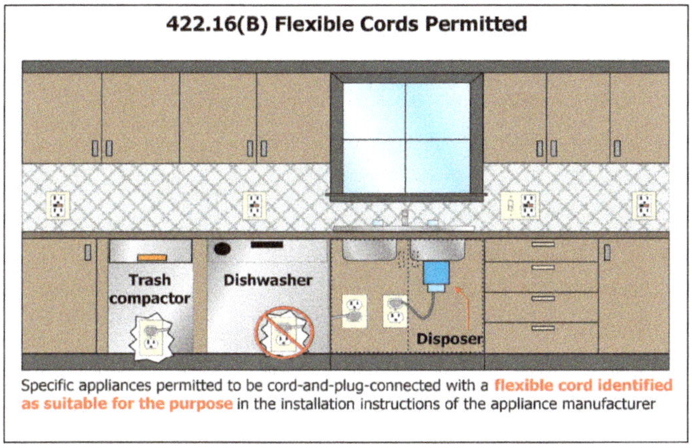

Figure 7.3. Flexible cord connection for specific appliances is permitted.

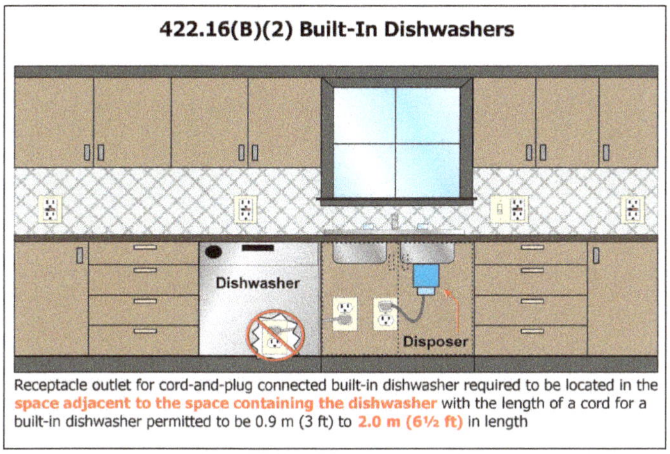

Figure 7.4. Built-in dishwashers are permitted to be cord-and-plug connected as long as the receptacle outlet is located in the space "adjacent to" the dishwasher.

for connection to a receptacle outlet in the adjacent cabinet space, the maximum length of a cord for a built-in dishwasher was extended from the previous maximum length of 1.2 m (4 ft) to the current 2.0 m (6.5 ft) measured from the face of the attachment plug to the plane created by the back of the appliance.

2. Receptacles must be located to protect against physical damage to the flexible cord and attachment plug.
3. As stated previously, receptacle outlets intended for the built-in dishwasher must be located in the space adjacent to the appliance.
4. Where a flexible cord passes through an opening, it must be protected against damage by a bushing, grommet, smoothed edge, or other approved means.
5. The receptacle must be accessible.
6. Flexible cord must have an equipment grounding conductor and be terminated with a grounding-type attachment plug.
7. A listed appliance distinctly marked to identify it as protected by a system of double insulation is not required to be terminated with a grounding-type attachment plug.

Table 7.3 Wiring for Built-in Dishwashers Section 422.16(B)(2)
Permitted to be cord-and-plug connected
Flexible cord must have an equipment grounding conductor and be terminated with a grounding-type attachment plug
A listed appliance distinctly marked to identify it as protected by a system of double insulation is not required to be terminated with a grounding-type attachment plug
Length: 0.9 m (3 ft) and not over 2.0 m (6.5 ft)
Flexible Cord and Receptacle – not subject to physical damage
Receptacle - located in space adjacent to appliance
Where a flexible cord passes through an opening, it must be protected against damage by a bushing, grommet, or other approved means
Receptacle must be accessible
GFCI protection for all outlets that supply dishwashers installed in dwelling units (receptacle outlet or a hard-wired outlet) *[See NEC 422.5(A)(7)]*

Table 7.3 Wiring for Built-in Dishwasher

Photo 7.4. Dishwashers and trash compactors are permitted to be cord-connected.

If the appliance is cord-and-plug connected, as indicated in the *Code*, the receptacle supplying power to a dishwasher is required to be located in the space adjacent to the appliance. No maximum distance away from the appliance is given for the receptacle, but the length permitted for the flexible cord limits the distance the receptacle could be away from the appliance. A cord-and-plug connection allows the dishwasher to be removed for maintenance and repair, along with the appliance being disconnected from the other utilities, such as the water and drain, before the dishwasher can be slid out and the cord cap removed from the receptacle.

NEC 422.5(A)(7) or IRC E3902.11 require GFCI protection for all outlets that supply dishwashers installed in dwelling units. This would include a receptacle outlet or a hard-wired outlet for a dishwasher.

TRASH COMPACTOR

Trash compactors intended for dwelling unit use are also permitted to be cord- and plug-connected with a flexible cord identified for the purpose and terminated with a grounding-type attachment plug cap [*NEC* 422.16(B)(2) or IRC E4101.3 and Table E4101.3]. Trash compactors typically come with a flexible cord for cord-and-plug connection from the manufacturer. Additional conditions for such uses are:

1. The length of the flexible cord is not less than 0.9 m (3 ft) and not to exceed 1.2 m (4 ft). The length is measured from the face of the attachment plug to the plane created by the back of the appliance.
2. Receptacles must be located to prevent physical damage to the flexible cord and attachment plug.
3. The receptacle can be located in the space occupied by the appliance or the space adjacent to it.
4. Where a flexible cord passes through an opening to an adjacent space, it must be protected against damage by a bushing, grommet, smoothed edge, or other approved means.
5. The receptacle must be accessible.
6. Flexible cord must have an equipment grounding conductor and be terminated with a grounding-type attachment plug.
7. A listed appliance distinctly marked to identify it as protected by a system of double insulation is not required to be terminated with a grounding-type attachment plug.

Unlike with the dishwasher, the receptacle supplying power to a trash compactor is permitted to be in the space occupied by the appliance or the space adjacent to the appliance.

Table 7.4 Wiring for Trash Compactors Section 422.16(B)(2)
Permitted to be cord-and-plug connected
Flexible cord must have an equipment grounding conductor and be terminated with a grounding-type attachment plug
Length: 0.9 m (3 ft) and not over 1.2 m (4 ft)
Flexible Cord and Receptacle – not subject to physical damage
Receptacle - located in same space or adjacent thereto
Where a flexible cord passes through an opening, it must be protected against damage by a bushing, grommet, or other approved means
Receptacle must be accessible

Table 7.4 Wiring for Trash Compactor

Photo 7.5. Cord-and-plug connected dishwasher with receptacle outlet located in adjacent cabinet space *(split receptacle with two branch circuits feeding receptacle)*

RANGE HOODS AND MICROWAVE OVEN/RANGE HOOD COMBINATIONS

Range hoods and over-the-range microwave ovens with integral range hoods are permitted to be cord- and plug-connected. This includes appliances that are a combination microwave oven and range hood. The installation instructions will identify the types of flexible cords that are suitable for this application. Flexible cords supplied with listed appliances are considered to be suitable. The cord must terminate in a grounding-type attachment plug. All of the following additional conditions must also be met:

1. Supply cord length cannot be shorter than 450 mm (18 in.) nor longer than 1.2 m (4 ft).
2. Receptacles are required to be located to prevent physical damage to the flexible cord and attachment plug.
3. Receptacles must be accessible.
4. The receptacle must be supplied by an individual branch circuit.
5. Flexible cord must have an equipment grounding conductor and be terminated with a grounding-type

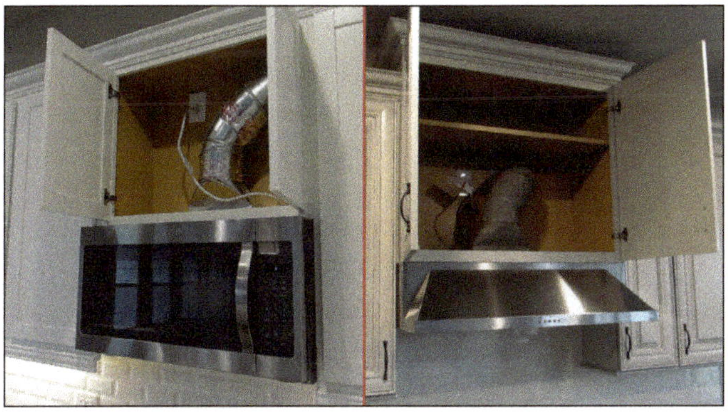

Photo 7.6. Two photos: (L) combination microwave-oven/range-hood (R) range hood, both using a suitable cord-and-plug connection identified for such use

Photo 7.7. Range receptacle for counter-mounted cooking unit

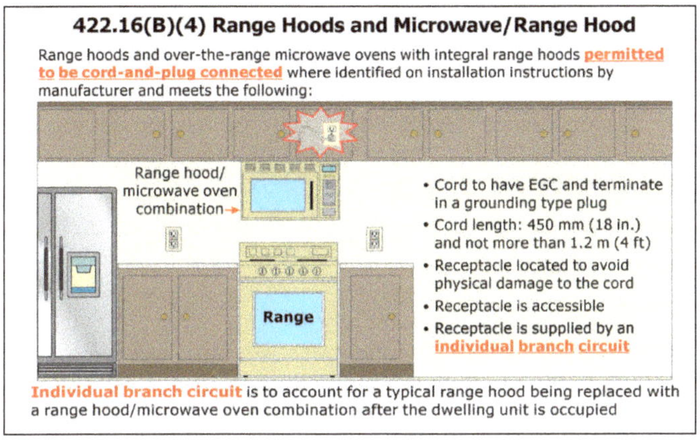

Figure 7.5. Range hoods and microwave ovens with built-in range hoods both call for an individual branch circuit when cord-and-plug connected

attachment plug.

6. A listed appliance distinctly marked to identify it as protected by a system of double insulation is not required to be terminated with a grounding-type attachment plug.
[*NEC* 422.16(B)(4) or IRC E4101.3].

Even though a typical range hood (without a microwave oven) draws very little current (2-3 amperes), one of the requirements to permit a range hood to be cord-and-plug connected is for the outlet to be supplied by an individual branch circuit (typically 15- or 20-amperes). This requirement for an individual branch circuit is to account for an ordinary range hood (without a microwave oven) being replaced with a range hood/microwave oven combination, which has become a common practice at dwelling units after the dwelling unit is occupied. These range hood/microwave oven combos can demand up to 8 to 12 amperes of current. An individual branch circuit (supplying only one utilization equipment) is necessary to accommodate such a load.

WALL-MOUNTED OVENS AND COUNTER-MOUNTED COOKING UNITS

Wall-mounted ovens and counter-mounted cooking units are permitted to be either permanently connected or, for ease in servicing or installing the unit, cord-and-plug connected. The *Code* does not specify a minimum or maximum length of the cord for this type of appliance.

The cord, the separable connector, or plug and receptacle combination in the supply circuit to the oven or cooking unit is required to be suitable for the maximum temperature of the space in which it is located. Consult the manufacturer's installation instructions for the cooking appliance for additional information on this topic.

The *Code* prohibits the use of flexible cords where they will run through holes in walls, structural ceilings, suspended ceilings, dropped ceilings, or floors. However, the flexible cord is permitted to pass through a hole in the side of a kitchen cabinet in order to reach a receptacle outlet located in an adjacent cabinet. Such an installation does not violate the rules of the *Code* simply because the side of a kitchen cabinet is not considered a wall. The requirement for protecting the flexible cord from physical damage must be complied with for cord installations *[See 422.16(B)(3)]*.

While the general requirement calls for a grounding-type attachment plug, an exception allows listed kitchen in-sink food waste disposers, dishwashers, trash compactors, and range hoods that are protected by a double insulation system or its equivalent to be ungrounded and supplied by a two-wire attachment plug cap. When such a system is used, the equipment must be clearly marked [See Exception to NEC 422.16(B)(1), (B)(2), and (B)(4) or IRC E4101.3].

SUPPORT OF CEILING-SUSPENDED (PADDLE) FANS

Ceiling-suspended (paddle) fans are considered appliances by the Code. They must be supported independently of an outlet box by means of a listed outlet box or a listed outlet box system identified for use with a ceiling-suspended (paddle) fan. The outlet box or system must be installed to meet the requirements of NEC 314.27(C) or IRC E3905.8. An outlet box system consists of a box and a special mounting bracket or bar. A listed outlet box system with a listed weight-supporting ceiling receptacle (WSCR) and a compatible weight-supporting attachment fitting (WSAF) identified and designed for support, identified for the use, and installed in accordance with NEC 314.27(E) or IRC E3905.6.3 can also be used to support a ceiling-suspended (paddle) fan. Refer to Chapter 12 for further information on outlet box installations for ceiling-suspended (paddle) fans. [NEC 422.18 or IRC E4101.6]

Outlet boxes mounted in the ceilings of habitable rooms of dwelling units in a location acceptable for the installation of a ceiling-suspended (paddle) fan are required to comply with one of the following:

(1) Listed for the sole support of ceiling-suspended (paddle) fans.

(2) Installed so as to allow direct access through the box to structural framing capable of supporting a ceiling-suspended (paddle) fan without removing the box *[See NEC 314.27(C) or IRC E3905.8]*.

ELECTRIC WATER HEATERS

Where a branch circuit supplies a fixed storage-type water heater with a capacity of 450 L (120 gallons) or less, it must be rated at not less than 125 percent of the ampere rating of the water heater as shown on the nameplate. This includes both the overcurrent device and conductors. A typical nameplate on an electric water heater will include the information in Figure 7.6.

Branch circuit sizing requirements for water heaters with a capacity of 450 L (120 gallons) or less are provided in NEC 422.13 or IRC E3702.10.

Table 7.6
Electric Water Heater Branch-Circuit Ampacity Example:

Nameplate = 4500 watts

4500 watts x 125% (for continuous load) = 5625 watts

5625 watts ÷ 240 volts = 23.4 amperes

23.4 amperes = Minimum ampacity rating for branch-circuit conductors for water heater

Table 7.6. Electric Water Heater, Branch-Circuit Ampacity

Table 7.7
Electric Water Heater Overcurrent Protection Example:

Nameplate = 4500 watts

4500 watts ÷ 240 volts = 18.8 amperes

18.8 amperes x *150 % = 28.2 amperes

**NEC* 422.11(E)(3) requires the overcurrent protection for a single non-motor-operated *appliance (such as a water heater)* to not exceed 150% of the appliance rated current.

NEC 422.11(E)(3) also permits the overcurrent protection to be "rounded up" to the next corresponding standard overcurrent device.

Minimum overcurrent device = 30 amperes

Table 7.7. Electric Water Heater, Overcurrent Protection

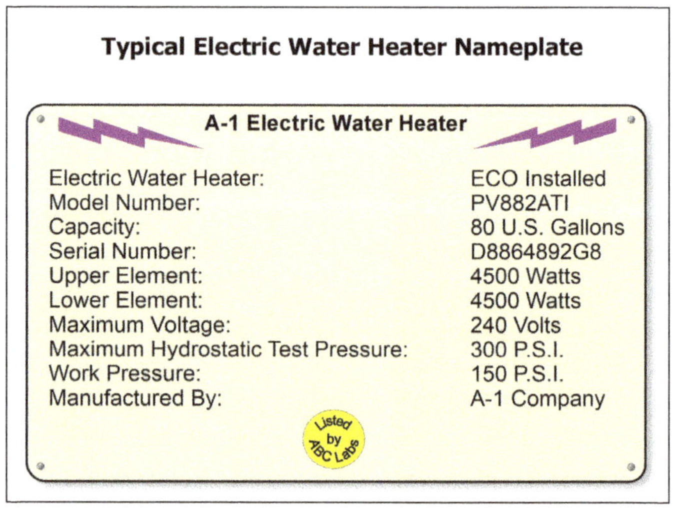

Figure 7.6. Example of Water Heater Nameplate Information

DISCONNECTING MEANS FOR APPLIANCES

A means of disconnect is generally required for permanently connected appliances (*NEC* 422.31 or IRC E4101.5). There are also provisions for the disconnection of cord- and plug-connected appliances, as previously covered in this chapter (*NEC* 422.16 or IRC E3909.1).

A disconnecting means must be provided to simultaneously disconnect each appliance from all ungrounded (hot) conductors. Where an appliance is supplied by more than one branch circuit or feeder, the disconnecting means must be grouped and properly identified as the appliance disconnect (*NEC* 110.22 and 422.30 or IRC E3404.13 and E4101.5).

Photo 7.8. Ceiling-suspended (paddle) fans (appliance) must be generally supported independent of an outlet box by a listed outlet box system identified for use with a ceiling-suspended (paddle) fan

Photo 7.9. Circuit breaker locking accessory device installed at panelboard

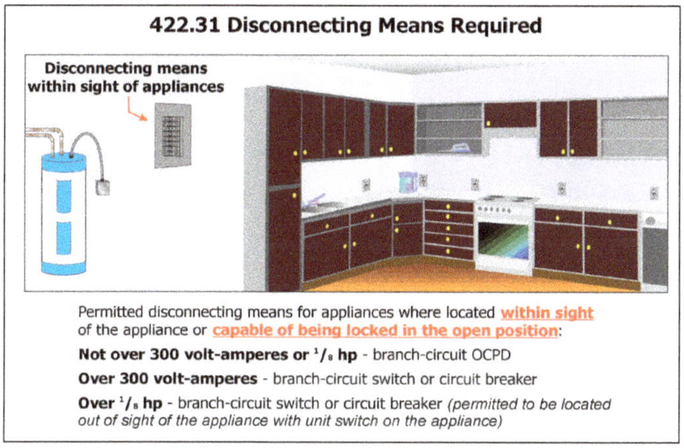

Figure 7.7. Disconnecting means are required for appliances

TYPES OF DISCONNECTING MEANS FOR PERMANENTLY CONNECTED APPLIANCES

Appliances Rated at Not Over 300 VA or ⅛ HP

For permanently connected appliances rated not over 300 volt-amperes (watts) or not over ⅛ horsepower, the branch-circuit overcurrent device is permitted to serve as the required disconnecting means where the switch or circuit breaker is within sight from the appliance or is capable of being locked in the open (off) position in accordance with *NEC* 110.25 or IRC E4101.8. The locking device must remain in place with or without the lock installed [*NEC* 422.31(A) and 110.25 or IRC Table 4101.5 and E4101.8].

Appliances Rated Over 300 VA

For permanently connected appliances rated greater than 300 volt-amperes (watts), the branch-circuit switch or circuit breaker is permitted to serve as the disconnecting means where the switch or circuit breaker is within sight from the appliance or is capable of being locked in the open (off) position in accordance with *NEC* 110.25 or IRC E4101.8. The provision for locking or adding a lock to the disconnecting means must remain in place with or without the lock installed [*NEC* 422.31(B), 110.25 or IRC Table E4101.5, E4101.8]

MOTOR-OPERATED APPLIANCES RATED OVER ⅛ HP

For permanently connected motor-operated appliances with motors rated over ⅛ horsepower, the branch circuit switch or circuit breaker is permitted to serve as the disconnecting means where the switch or circuit breaker is within sight of the appliance, or the disconnecting means must be capable of being locked in the open (off) position in accordance with *NEC* 110.25 or IRC E4101.8. This disconnecting means is generally required to be one of the following types: a listed motor-circuit switch rated in horsepower, a listed molded case circuit breaker, a listed molded case switch, an instantaneous trip circuit breaker that is part of a listed combination motor controller, a listed self-protected combination controller, a listed manual motor controller additionally marked "Suitable as Motor Disconnect" where installed between the final motor branch-circuit short-circuit protective device and the motor. For stationary motors rated at 2 hp or less and 300 volts or less, the disconnecting means is permitted to be one of the following devices:

1. A general-use switch having an ampere rating not less than twice the full-load current rating of the motor.
2. On AC circuits, a general-use snap switch is suitable only for use on AC, not general-use AC–DC snap switches, where the motor's full-load current rating is not more than 80 percent of the switch's ampere rating.

3. A listed manual motor controller with a horsepower rating not less than the motor rating is marked "Suitable as Motor Disconnect."

The disconnecting means for motor-operated appliances rated over ⅛ horsepower with circuits rated 1000 volts, nominal, or less must have an ampere rating not less than 115 percent of the full-load current rating of the motor. A listed unfused motor-circuit switch having a horsepower rating not less than the motor horsepower is permitted to have an ampere rating less than 115 percent of the full-load current rating of the motor [*NEC* 422.31(C), 430.109, 430.110, or IRC Table 4101.5].

If a motor-operated appliance rated over ⅛ horsepower is supplied with a unit switch, the unit switch with a marked-off position is permitted to serve as the disconnecting means where other means for disconnection are provided. The service disconnecting means is permitted to be the other disconnecting means. The branch-circuit switch or circuit breaker serving as the other disconnecting means is permitted to be out of sight from the appliance. The unit switch must have a marked-off position that is a part of the appliance and disconnects all ungrounded conductors [*NEC* 422.34 or IRC Table E4101.5].

The term "within sight from the appliance" means the disconnecting means must be clearly visible from the appliance and located no more than 15 m (50 ft) away. One option is to install a suitable disconnecting means at an electric water heater, built-in electric range, cooktop, or oven, as the branch-circuit switch or circuit breaker is not typically located near the appliances. Providing lock-off capability at the supply circuit breaker is an acceptable alternative for many non-motor-operated appliances. Disconnect switches are typically designed by the manufacturer to be capable of being locked open.

Most manufacturers of circuit breakers produce an accessory device that mounts on the breaker, allowing the circuit breaker to be locked in the open (off) position. Either the switch or circuit breaker with the accessory kit installed, as described here, will satisfy this Code requirement. This disconnecting means provides an additional level of safety for the person who services, maintains, or repairs the appliance since that person is typically the one in control of the locking device.

Disconnect switches are usually designed by the manufacturer so they are capable of being locked open. Most manufacturers of circuit breakers make an accessory device that mounts on the breaker and will permit the circuit breaker to be locked in the open (off) position. Either the switch or circuit breaker with the accessory kit installed described here will satisfy this *Code* requirement. This disconnecting means provides an additional level of safety for the person that services, maintains, or repairs the appliance since that person is typically the one in control of the locking device.

Photo 7.10. Unit switch on electric water heater

ATTACHMENT PLUG AND RECEPTACLE OR ATTACHMENT FITTINGS USED AS DISCONNECTING MEANS

For cord- and plug-connected appliances or attachment fitting-connected appliances, an accessible separable connector or an accessible attachment plug and receptacle is permitted to serve as the required disconnecting means. This plug and receptacle connection must meet the definition of accessible. The cord and plug or attachment fitting must be able to be removed without damaging the building structure or finish. The cord, plug, attachment fitting, and receptacle outlet must not be permanently closed in by the structure or finish of the building. Attachment fittings must be a factory-installed part of the appliance and must be suitable for disconnection. Attachment fittings are devices intended for the direct connection of electrical utilization equipment designed to mate with a corresponding contact device [*NEC* Article 100 and 422.33(A) or IRC Chapter 35 and Table E4101.5].

For a cord- and plug-connected household electric range, an attachment plug and receptacle connection located at the rear base of the range is permitted as the disconnecting means, provided it is accessible from the front by removing a bottom utensil drawer [NEC 422.33(B) or IRC E4101.5 and Table E4101.5].

The rating of the receptacle, or a separable connector, cannot be less than the rating of any appliance connected to it unless demand factors authorized elsewhere in the *Code* are applied. They must also be capable of interrupting their rated current without hazard to the operator.

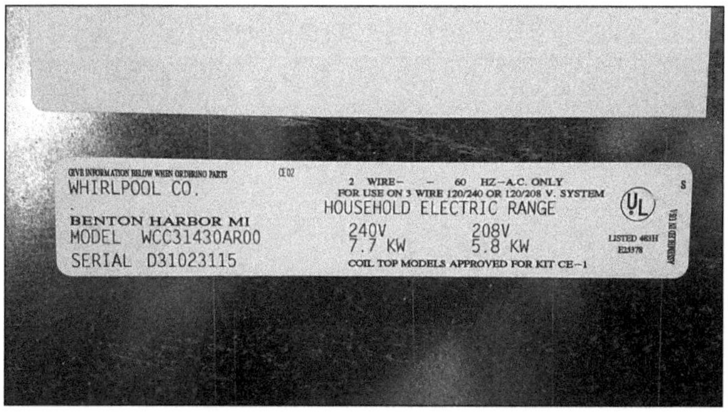

Photo 7.11. A nameplate (typical) for a household range

APPLIANCE UNIT SWITCH(ES) USED AS DISCONNECTING MEANS

A unit switch with a marked-off position, which is part of an appliance and disconnects all ungrounded conductors, is permitted to serve as the disconnecting means for multifamily, two-family, and single-family dwellings where other means for disconnection are provided. The service disconnecting means is permitted to be the same as the other disconnecting means. The branch-circuit switch or circuit breaker serves as the other disconnecting means, and is permitted to be out of sight from the appliance. [NEC 422.34 or IRC Table E4101.5].

In multifamily dwellings, the other disconnecting means must be within the dwelling unit or on the same floor as the dwelling unit in which the appliance is installed, and the control of lamps and other appliances must be permitted.

In two-family dwellings, the other disconnecting means can be the feeder or service disconnecting means, located either inside or outside the dwelling unit where the appliance is installed. In this case, an individual switch or circuit breaker for the dwelling unit is permitted and may also be used to control lamps and other appliances. This recognizes that more than one appliance, receptacle outlet, or luminaire may be on the same circuit.

In one-family dwellings, the service disconnecting means is permitted to be the other disconnecting means [*NEC* 422.34(C)].

Other types of switches on the appliance, such as thermostats and control switches, may turn the appliance on and off but may not be the required disconnecting means. Such switches may not have a marked "off" position and do not disconnect all power to the appliance. A typical example is a dishwasher, where the door switch, control buttons, or dial are switches that control a relay or controller within the appliance. This type of switch does not qualify as the appliance disconnecting means. A disconnecting means located within sight of the dishwasher or capable of being locked in the open position must be provided if the appliance is permanently connected. If the dishwasher is cord- and plug-connected, the attachment plug can serve as the disconnecting means.

Where switches and circuit breakers are used as disconnecting means, they must be of the indicating type. A properly rated general-use snap switch — AC only — with marked "ON" and "OFF" positions would be an acceptable within-sight disconnecting means. The motor full-load current shall not be greater than 80% of the ampere rating of the switch (*NEC* 422.35; 404.14 or IRC E4101.5, E4001.1).

OVERCURRENT PROTECTION OF APPLIANCES

Appliances are required to be protected against overcurrent. The limits on the maximum size of the protective device are based on the type of appliance and the information provided on the nameplate. If a protective device rating is marked on the appliance, the branch-circuit overcurrent device rating cannot exceed the protective device rating marked on the appliance.

A household-type appliance provided with surface heating elements that have a maximum demand of more than 60 amperes is required to have its power supply subdivided into two or more circuits, each of which is provided with overcurrent protection rated at not over 50 amperes. This rarely occurs in household-type dwelling unit appliances [*NEC* 422.11(B)].

Where the branch circuit supplies a single non-motor-operated appliance, the rating of overcurrent protection must not be in excess of that marked on the appliance nameplate.

If the overcurrent protection rating is not marked on the nameplate and the appliance is rated at over 13.3 amperes, overcurrent protection cannot exceed 150% of the appliance-rated current. Where 150% of the appliance rating does not correspond to a standard overcurrent device ampere rating, the next higher standard rating is permitted. The standard ratings of fuses and circuit breakers are found in *NEC* 240.6(A) or IRC E3705.6. If the overcurrent protection rating is not marked and the appliance is rated 13.3 amperes or less, the maximum rating of the protective device is 20 amperes. Lower values are permitted [*NEC* 422.11(E) or IRC E4101.4.1].

Electric heating appliances employing resistance-type heating elements that are rated at more than 48 amperes must have their heating elements subdivided. Each subdivided load cannot exceed 48 amperes and each subdivided load is required to be protected at not more than 60 amperes. Supplementary-type overcurrent protective devices are permitted for use in this application.

These supplementary overcurrent protective devices must be factory-installed within or on the heater enclosure or supplied as a separate assembly by the heater manufacturer. They must be accessible but not necessarily readily accessible, and suitable for branch-circuit protection. The main conductors feeding these protective devices are considered branch-circuit conductors unless they are household-type heating appliances with surface heating elements [NEC 422.11(F)(1)].

Resistance-type immersion electric heating elements are permitted to be subdivided into circuits not exceeding 120 amperes and protected at not more than 150 amperes. This would apply where the heating elements (1) are contained in ASME-rated and stamped vessels, (2) are included in listed instantaneous water heaters, and (3) are installed in low-pressure water heater tanks or open-outlet water heater vessels [*NEC* 422.11(F)(3)].

Overload protection for the motors of motor-operated appliances must be provided. In many cases, the manufacturer builds this overload protection into the appliance.

Overload protection must be provided for hermetic refrigerant motor compressors located within air-conditioning or refrigerating equipment. The marking on the nameplate gives the maximum size and type of the permitted protective device. See Chapter 9 for additional information on this topic.

Where motor-operated appliance overcurrent protective devices separate from the appliance are required, information for the selection of these devices must be marked on the appliance nameplate.

GROUNDING OF APPLIANCES

Appliances often present an increased risk of electric shock due in part to the fact that many appliances are manufactured with exposed conductive materials. In kitchens and laundry rooms, it can be quite common for two or more appliances to be located within the range of a person's reach. Without being properly grounded, these separate appliances could have different voltage potentials. In addition, one or more appliances may be installed within reach of a conductive sink, such as laundry, utility, and kitchen sinks, that may be both wet and grounded.

If one of the appliances is not properly grounded and a line-to-ground fault was to occur, any exposed metal parts present a shock or electrocution hazard to a person who may inadvertently provide the electrical path between the ungrounded appliance and a grounded surface such as a metal sink.

All exposed, normally non-current-carrying metal parts of appliances that are fastened in place or connected by a permanent wiring method must be properly grounded by

250.110 Grounding of Appliances

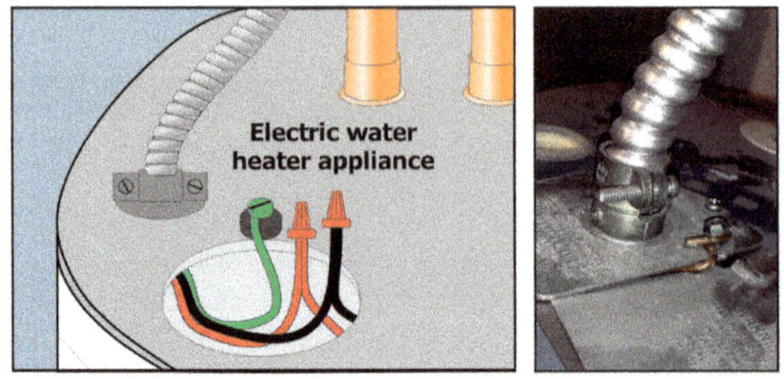

Exposed, normally non-current-carrying metal parts of appliances **(fixed equipment)** that are likely to become energized **must be properly grounded** by connection an equipment grounding conductor

Figure 7.8. Grounding of appliances required

connection to an equipment grounding conductor where their metal parts are likely to become energized (*NEC* 250.134 or IRC E3908.10). Becoming energized could be the result of an internal failure of insulated parts or from contact with an external source of electrical current, such as portable appliances. Dwelling unit appliances must be properly grounded by connection to an equipment grounding conductor under the following conditions:

1. Where they are located within 2.5 m (8 ft) vertically or 1.5 m (5 ft) horizontally of ground or grounded metal objects and subject to contact by persons.
2. Where they are located in a wet or damp location and not isolated.
3. Where they are in electrical contact with metal.
4. Where they are supplied by a metal-clad, metal-sheathed, metal-raceway, or other wiring method that provides an equipment grounding conductor [*NEC* 250.110 or IRC E3908.2].

Exposed, normally non-current-carrying metal parts of cord- and plug-connected equipment shown in the following list must also be grounded by connection to an equipment grounding conductor:

1. air conditioners, freezers, icemakers, and refrigerators;
2. clothes-washing, clothes-drying, dishwashing machines, kitchen waste disposers, range hoods, computers, sump pumps, electrical aquarium equipment;
3. hand-held motor-operated tools; stationary and fixed motor-operated tools;

Photo 7.12. Dryer rough-in wiring, including insulated grounded conductor and bare equipment grounding conductor (4-wire circuit)

Photo 7.13. Typical electric dryer cord and plug connection in a dwelling unit

4. motor-operated appliances of the following types: hedge clippers, lawn mowers, snow blowers and wet scrubbers;

5. portable handlamps

6. accessible non-current-carrying metal parts of appliances, such as central vacuum outlet assemblies, are likely to become energized [see *NEC* 250.114].

The type of equipment used to ground the conductor and appliances will depend on the wiring method employed. For the most commonly used wiring method at dwelling units, cable wiring systems, an equipment grounding conductor is typically contained in the cable. For metal-clad cables, the metal cable jacket may qualify as the equipment grounding conductor. For metal raceways, the raceway itself is permitted as the equipment grounding conductor. For nonmetallic raceways, a wire-type equipment grounding conductor must be installed inside the raceway. Cord- and plug-connected equipment may be grounded by means of an equipment grounding conductor consisting of a bare conductor or

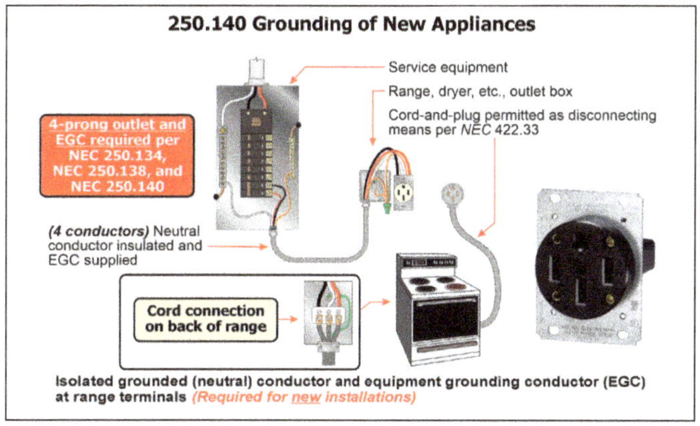

Figure 7.9. Grounding of frames of new ranges and dryers requires a separate insulated neutral conductor and an equipment grounding conductor for grounding purposes (Note the 4-wire cable and the 4-wire receptacle)

insulated conductor that is green or green with one or more yellow stripes that is part of a cable assembly or flexible cord. The flexible cord or cable assembly must terminate within a grounding-type attachment plug with one fixed grounding contact (NEC 250.118 or IRC E3908.9)

GROUNDING FRAMES OF RANGES AND CLOTHES DRYERS

The frames of electric ranges, wall-mounted ovens, counter-mounted cooking units, clothes dryers, and outlet or junction boxes that are part of the circuit supplying the normally non-current-carrying metal parts of appliances must always be properly grounded by connection to an equipment grounding conductor. The grounding requirements are separated into two categories (new and existing). Grounding requirements for new installations are different from those for existing installations that were grounded under older editions of the Code.

GROUNDING COOKING APPLIANCES AND CLOTHES DRYERS ON NEW BRANCH CIRCUITS

In new installations, the *Code* requires the frames of ranges, wall-mounted ovens, counter-mounted cooking units, and clothes dryers to be grounded by an equipment grounding conductor. Outlet or junction boxes that are part of the circuit must be grounded in a similar manner. In a typical 120/240-volt, single-phase dwelling unit system, this means it is necessary to run four conductors to these appliances: two ungrounded (hot) conductors, an insulated grounded (neutral) conductor, and an equipment grounding conductor. The equipment grounding conductor is permitted to be any of the methods described in *NEC* 250.118 or IRC E3908.9 and can be an individual conductor of the cable assembly or an acceptable metal raceway. Cord- and plug-connected equipment must be grounded as specified in *NEC* 250.138. Typically, the equipment grounding conductor will terminate in a grounding-type attachment plug with a fixed grounding contact (see Figure 7.9)

Receptacle outlets for these new branch-circuit installations are required to be the 3-pole with ground type. Four-wire, grounding-type flexible cords are required for pigtails for new ranges and dryers. This has not always been the case, as prior to the 1996 *NEC*, three-wire cords and attachment plugs had been permitted for many years, using the grounded (neutral) conductor for grounding the frames of ranges and clothes dryers. Where a new cooking appliance or new clothes dryer has a factory-installed supply cord, verify that the equipment grounding conductor and the neutral conductor are not bonded together by the manufacturer.

Check the junction box of the appliance prior to installation. For new installations, ensure that the bonding strap or screw is not installed between the neutral conductor and the appliance frame. Proper bonding and grounding connections must be verified where existing appliances that might use a different grounding method are moved between dwellings. In existing dwellings where three wires are run, the bonding jumper between the equipment grounding terminal and the neutral conductor should be installed on the appliance. In new installations, this bonding jumper cannot be installed.

The IRC does not contain allowances for the grounding of frames of existing ranges and dryers by means of connection to the grounded conductor. The restriction on this method of grounding for these appliances was placed in the 1996 NEC. The NEC continues to permit the grounded (neutral) conductor for grounding dryers and ranges only in existing installations in existing dwellings. New installations, in both new and existing dwellings, must be provided with a separate, insulated, grounded (neutral) conductor, as well as an equipment grounding conductor, for grounding purposes only. The IRC does not permit grounding connections to a grounded conductor on the load side of the service disconnect at locations other than at existing separate buildings under certain restrictive conditions [IRC E3908.7; E3908.8 and NEC 250.140, 250.134, and 250.138]. See also footnote (c) to IRC Table E3801.4, which indicates the grounded conductor must be insulated. Installations of existing dryers and ranges falling under the scope of the IRC would have to default to the provisions in NEC 250.140(B) in order to use the grounded conductor for the grounding of frames of existing ranges and dryers.

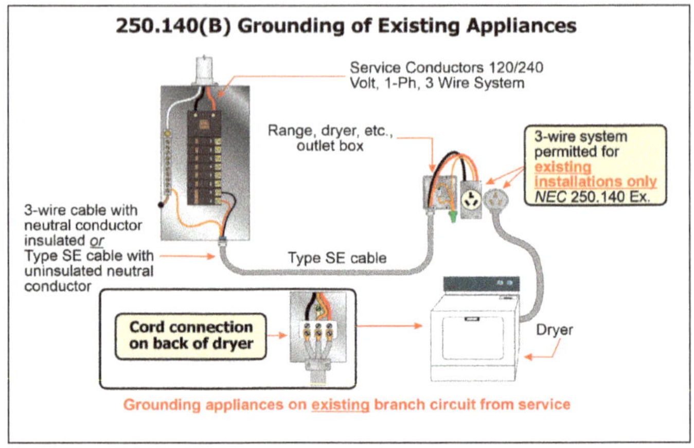

Figure 7.10. Grounding the frame of a range or dryer by using the grounded (neutral) conductor of an existing branch circuit that originates at the service equipment

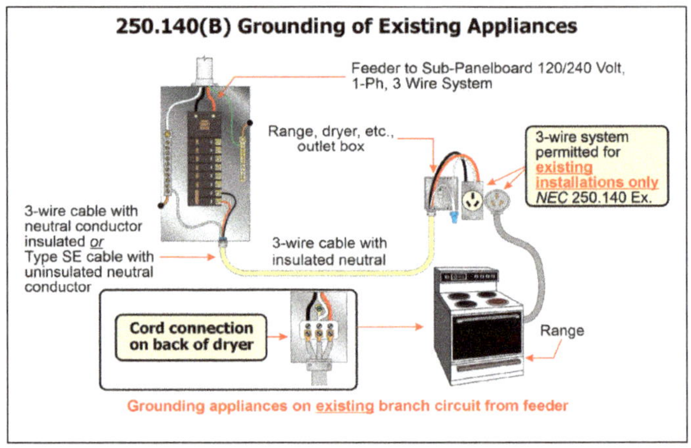

Figure 7.11. Grounding the frame of a range or dryer by the grounded (neutral) conductor of an existing branch circuit that originates in a sub-panelboard supplied by a feeder

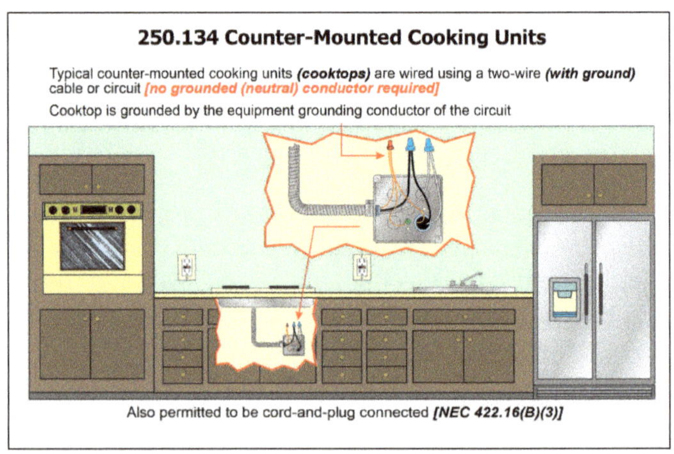

Figure 7.12. Cooktops can be wired with a two-wire with ground cable, as no neutral is required (follow the manufacturer's instructions)

Photo 7.14. Counter-mounted cooking units are typically wired with a two-wire circuit

WIRING AND GROUNDING COOKTOPS

Many cooktops (counter-mounted cooking units) can be wired properly with a two-wire (with ground) cable or branch circuit, as no grounded (neutral) conductor is typically required for these cooking appliances. This will be the case for almost all one- and two-family dwelling cooking appliances. These counter-mounted range tops must be grounded by connection to the equipment grounding conductor provided with the supplying branch circuit [*NEC* 250.134 or IRC E3908.10]. Counter-mounted cooking units and wall-mounted ovens are permitted to be permanently connected (hardwired) or, for ease in servicing or for installation, cord- and plug-connected. These units need to be complete with provisions for mounting and for making electrical connections in this manner. The manufacturer of counter-mounted cooktops will typically supply a wiring method or "pigtail" connector consisting of flexible metal conduit from the cooking unit. Either a separable wiring method or a plug and receptacle combination in the supply line to the cooking unit must be identified for the temperature of the space in which it is located [*NEC* 422.16(B)(3) or IRC Table 4101.5].

POLARITY IN CORD- AND PLUG-CONNECTED APPLIANCES

The *Code* requires that certain types of appliances be provided with a polarized attachment plug. Where an appliance is provided with a manually operated, line-connected, single-pole switch used for appliance on-off operation, or an Edison-base lampholder, or either a 15- or 20-ampere rated receptacle, the attachment plug that connects the appliance to the supply circuit is required to be the polarized or grounding-type (*NEC* 422.40).

Cord- and plug-connected portable freestanding hydromassage units and hand-held hair dryers that are subject to immersion are required to be constructed to provide protection for personnel against electrocution when immersed while in the "ON" or "OFF" position. Typically, the manufacturer of the appliance incorporates the protective device into the construction of the product (*NEC* 422.41). See Chapter 14, for information concerning grounding and bonding requirements for hydromassage bathtubs.

PRODUCT SAFETY STANDARD REQUIREMENTS

It is recommended to consult the following electrical product safety guide card information and product safety standards for additional guidance on the proper installation, operation, and use of electrical equipment covered in this chapter. The four-letter code in parentheses indicates the product category in the UL Guide Information for Electrical Equipment Directory (UL Product Spec material).

- Electrical Equipment for Use in Ordinary Locations (AALZ)
- Fans, Ceiling Suspended (GPRT)
- Waste Disposers (ZDHR)
- Household Trash Compactors (XUUM)
- Dishwashers, Household (DMIY)
- Attachment Plugs (AXGV)
- Attachment Plugs, Fuseless (AXUT)

REFERENCES

1. *National Electrical Code* and *NEC* are registered trademarks of the National Fire Protection Association, Inc., Quincy, MA 02169. This reprinted material is not the official position of the National Fire Protection Association, which is represented only by the standard in its entirety.

2. *International Residential Code for One- and Two-Family Dwellings* (2024). Copyright © International Code Council, Inc., June 2023.

3. UL, Northbrook, IL, *UL Guide Information for Electrical Equipment Directory* (UL Product iQ® material) unless indicated otherwise.

REVIEW QUESTIONS

1. For permanently connected appliances rated at not over 300-volt amperes or ⅛ hp, the branch-circuit overcurrent device can be used as the disconnecting means ________ ______________________________.

A. where the switch or circuit breaker is within sight from the appliance
B. is capable of being locked in the open position
C. is capable of being locked in the closed position
D. Either A. or B.

2. A disconnecting means must be provided to ____________ disconnect each appliance from all ungrounded (hot) conductors. Where an appliance is supplied by more than one branch-circuit or feeder, the disconnecting means must be ____________ and properly identified as the appliance disconnect.

A. completely, accessible
B. properly, readily accessible
C. simultaneously, grouped
D. simultaneously, readily accessible

3. The maximum length of a suitable flexible cord for a cord- and plug-connected built-in dishwasher must not be longer than what distance?

A. 450 mm (18 in.)
B. 600 mm (24 in.)
C. 1.2 m (4 ft)
D. 2.0 m (6.5 ft)

4. A fixed storage-type water heater of 300 L (80 gallons) is required to be supplied by a branch circuit sized at not less than _______ percent of the nameplate rating of the water heater.

A. 100
B. 110
C. 115
D. 125

5. Which of the following is the maximum allowable rating of a permanently connected appliance where the branch-circuit overcurrent device is used as the appliance disconnecting means where the switch or circuit breaker is within sight from the appliance or is lockable in accordance with *NEC* 110.25?

A. ⅛ horsepower
B. ¼ horsepower
C. ½ horsepower
D. 1 horsepower

6. Receptacle outlets installed under a sink (such as a *receptacle outlet* installed under a kitchen sink for a *waste disposer*) cannot be installed in the face-up position.

True _____ False _____

7. A flexible cord is not permitted to be used for the connection of a kitchen in-sink waste disposal.

True _____ False _____

8. A wall-mounted oven is permitted to be cord- and plug-connected.

True _____ False _____

9. A built-in trash compactor, intended for dwelling unit use, is permitted to be cord- and plug-connected provided the length of the cord is at least what distance?

A. 450 mm (18 in.)
B. 600 mm (24 in.)
C. 900 mm (36 in.)
D. 1.2 m (48 in.)

10. Permanently connected appliances rated over 300 volt-amperes may be disconnected from their supply by:

A. an accessible overcurrent protective device (regardless of its location).
B. the circuit breaker protecting the branch circuit (regardless of its location).
C. the branch-circuit switch or circuit breaker that is within sight from the appliance or is capable of being locked in accordance with *NEC* 110.25.
D. any of the above

11. The disconnecting means for motor-operated appliances rated over ⅛ horsepower with circuits rated 1000 volts, nominal, or less must have an ampere rating not less than ________ percent of the full-load current rating of the motor.

A. 100
B. 115
C. 125
D. 150

12. Outlet boxes mounted in the ceilings of habitable rooms of dwelling units in a location acceptable for the installation of a ceiling-suspended (paddle) fan is required to comply with which of the following:

A. Listed for the sole support of ceiling-suspended (paddle) fans.
B. A ceiling-mounted outlet box providing direct access through the box to structural framing capable of supporting of a ceiling-suspended (paddle) fan without removing the box.
C. Located in the center of the room.
D. Either A. or B.

13. Each electrically heated appliance that is intended by size, weight, and service to be located in a fixed position shall be placed so as to provide ample protection between the appliance and adjacent combustible material.

True _____ False _____

14. Central heating equipment (such as an oil- or gas-fired burner) is required to be supplied by a/an____________ branch circuit that supplies only one piece of utilization equipment (other than auxiliary equipment or permanently connected air-conditioning equipment).

A. general purpose
B. multiwire
C. individual
D. appliance

15. An equipment grounding conductor shall be used and connected where a central vacuum outlet assembly has accessible non-current-carrying metal parts that are likely to become energized.

True _____ False _____

16. A flexible cord assembly is permitted to be used for the connection of an appliance to facilitate frequent interchange, or to prevent the transmission of noise or vibration to an appliance.

True _____ False _____

17. For cord- and plug-connected household electric ranges, an attachment plug and receptacle connection at the rear base of a range, accessible from the front by removal of a drawer, shall be considered as meeting the disconnecting means requirements of the appliance.

True ______ False ______

18. What is the minimum permitted branch-circuit overcurrent protection for a 240-volt, 4500-watt water heater on an individual branch circuit?

A. 20 A
B. 25 A
C. 30 A
D. 40 A

19. If a protective device rating is marked on the appliance, the branch-circuit overcurrent device rating cannot exceed the protective device rating marked on the appliance.

True ______ False ______

20. A listed appliance distinctly marked to identify it as protected by a system of double insulation is not required to be terminated with a grounding-type attachment plug.

True ______ False ______

8 Fixed Electric Space-Heating Equipment

Photo 8.1. Fixed electric baseboard heater

All fixed electric space-heating equipment must be properly installed in accordance with installation instructions and listing or labeling requirements before the authority having jurisdiction (AHJ) will grant approval. All exposed non-current-carrying metal parts of fixed electric space-heating equipment that are subject to becoming energized must be properly grounded by connection to an equipment grounding conductor [*NEC* 110.2, 110.3(B); 250.110; and 250.134; IRC E3403.3 and E3908.2].

The heating equipment discussed in this chapter includes heating cables, unit heaters, boilers, central systems employing electric heating elements, and other fixed electric space-heating equipment. NEC Article 424 does not cover processing, heating, or room air conditioning. The requirements of Article 440 apply to fixed electric heating equipment incorporating a hermetic refrigerant motor compressor. Refer to Chapter 9 of this book for information on air-conditioning and heat pump equipment.

Receptacle outlets are generally prohibited from being installed above baseboard heating units. This prohibition is not directly addressed by the *Code*. The restriction is found in the typical installation instructions that accompany listed units. Such restrictions effectively become *Code* requirements through the application of *NEC* 110.3(B) or IRC E3403.3.

Basic requirements for appliances and fixed electric heating equipment are provided in IRC Chapter 41. Many specific requirements for the various types of electric heat used in one- and two-family dwellings are beyond the scope and requirements included in the IRC, so by default, the rules in the *NEC* must be followed (IRC E3401.2).

The product safety standard listing for electric baseboard heaters covers space heaters of the portable and permanently mounted types intended to be positioned or installed on, or in, the wall at the baseboard level or on the floor. The UL Guide Information for Electrical Equipment Directory

(UL Product Spec material) on Baseboard Heaters (KLDR) states, "Baseboard heaters have been investigated and found to incorporate suitable safeguards against the establishment of fire hazards that might result from contact with draperies, furniture, carpeting, bedding, and the like; however, discoloration or scorching (but no glowing embers or flaming) may result on adjacent materials.

> "Heaters, other than those marked to indicate that they are not for residential use, have been investigated to determine that the accessible surface temperatures are low enough to reduce the likelihood of burns from accidental contact.

> "Electrical cords, drapes, and other furnishings should be kept away from baseboard heaters. To reduce the likelihood of cords contacting the heater, the heater is not to be located beneath electrical receptacles. Listings of receptacle accessories for use with an individual manufacturer's baseboard heaters are included under the category Baseboard Heater Accessories (KLQZ)."

A receptacle outlet incorporated into a listed baseboard heater is recognized by the *Code* as being permitted to serve as the required receptacle outlet for the wall space. These receptacles are either factory-installed or furnished as accessories by the manufacturer. Such receptacles cannot be connected to the same branch circuits that supply the heaters. They are to be connected to general-purpose, or if appropriate, small-appliance branch circuits [NEC 110.3(B); 210.52; 210.52 (Informational Note); IRC E4101.2; E3403.3, E3901.1].

FIXED ELECTRIC SPACE HEATING — SUPPLY CONDUCTOR INSULATION RATING

Some fixed electric space-heating equipment includes marking that requires supply conductor insulation to be rated greater than 60°C. This marking must be clear and permanent so it will be plainly visible after the equipment is installed. This marking is permitted to be adjacent to the field-connection box. Where fixed electric space-heating equipment is not so marked, it may be wired with 60°C-rated conductors (*NEC* 424.11).

For example, the nameplate on the heater may require a conductor with 75°C insulation. This is to help ensure that the termination within the junction box of the heating equipment is adequately protected against heat damage. The installer must consider both ends of the conductor to be certain it complies with all rules.

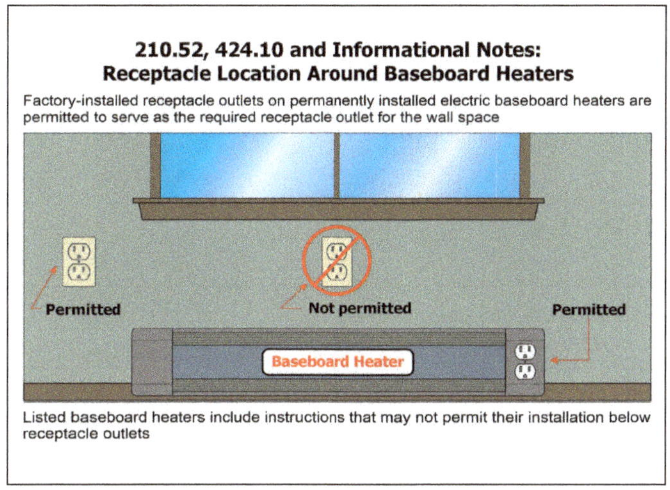

Figure 8.1. Baseboard heaters may be prohibited from being installed below receptacle outlets by their manufacturer's specifications

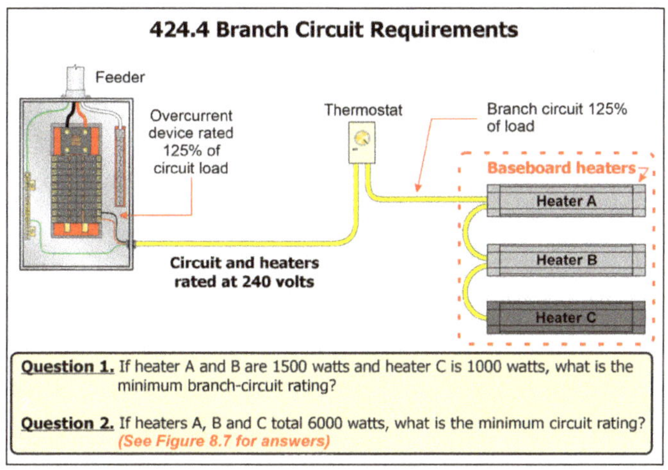

Figure 8.2. A branch circuit is required to be rated at 125% of the load *(see Figure 8.7 for the answer to questions)*

The *Code* indicates that unless marked otherwise, terminals of circuit breakers and switches rated 100 amperes or less are rated for 60°C conductor insulation. However, it is quite common to find that modern circuit breakers are marked with a dual 60/75°C rating. To determine the proper size conductor where the terminals are only rated for 60°C, refer to *NEC* Table 310.16 or IRC Table E3705.1 and use the 60°C column. There, choose a conductor with an allowable ampacity that equals or exceeds that required for the circuit. This will be the minimum size conductor permitted. To comply with the higher termination requirements marked on the heater, choose the same size conductor with 75°C or 90°C insulation. This conductor satisfies both the ampacity and temperature requirements.

PHYSICAL DAMAGE

Fixed electric space-heating equipment may not be installed

Photo 8.2. Fixed electrical space-heating equipment (electric furnace) with supplementary overcurrent protection for subdivision loads.

where exposed to physical damage unless the equipment is adequately protected in an approved manner. It is difficult to provide physical protection for baseboard units [*NEC* 424.12(A)].

Fixed electric space heaters and related equipment installed in damp or wet locations must be listed for such locations and must be constructed and installed so water cannot enter or accumulate in or on any wired sections, electrical components, or ductwork [*NEC* 424.12(B)].

For unit heaters, permanently wired radiant heaters, and radiant heat cables used for heating swimming pool deck areas, see Chapter 14 for additional information (*NEC* 680.27(C) or IRC E4206.13).

REQUIRED SPACING FROM COMBUSTIBLE MATERIAL

Fixed electric space-heating equipment must be installed to provide the required spacing between the equipment and any adjacent combustible material. Some common examples of combustible materials found in dwelling units are drapes, curtains, towels, furniture, doors, and boxes. The required spacing is marked on the equipment or furnished by the manufacturer. This spacing may not be required for equipment that has been identified and listed for installation in direct contact with combustible material (*NEC* 424.13).

BRANCH-CIRCUIT REQUIREMENTS FOR FIXED ELECTRIC SPACE-HEATING EQUIPMENT

The *Code* provides the minimum requirements for determining the proper size of the branch circuit.

Individual branch circuits are permitted to supply any volt-ampere (VA) or wattage rating of fixed electric space-heating equipment for which the branch circuit is rated. For example, an electric furnace may require a single 60-ampere branch circuit. Branch circuits supplying two or more outlets for fixed electric space-heating equipment must be rated at 15, 20, 25, or 30 amperes. For example, if a branch circuit supplies several baseboard heaters, the maximum size of the branch circuit is limited to 30 amperes. Fixed electric space heating equipment consisting of resistance elements and any associated motors is considered a continuous load. The minimum allowable ampacity of the branch-circuit conductors must be not less than 125% of the total load. If the heating equipment contains motors, the motor loads must be added to the load of the resistance elements. Similarly, the minimum rating of the branch-circuit fuse or circuit breaker is 125% of the total load (NEC 424.4 or IRC E3702.11).

The rating or setting of overcurrent protective device(s) is permitted to be the next higher standard overcurrent device to be used to protect the conductor where the allowable ampacity of the conductor does not match the standard rating of the circuit breaker or fuse. For example, if the total load is calculated to be 42 amperes. The minimum required circuit size is 52.5 amperes (based on 125% of the total load). A 6 AWG copper conductor with an ampacity of 55 amperes is selected from the allowable ampacity tables. A 60-ampere fuse or circuit breaker can be used to protect the 6 AWG conductors.

For electrode-type boilers, where listed for continuous operation at 100% of its rating, a contactor, thermostat, relay, or similar device is permitted to supply its full-rated load as provided in *NEC* 210.19(A)(1), Exception to (1). This type of equipment would not commonly be used in dwellings.

The size of the branch-circuit conductors and overcurrent protective devices used for the supply to fixed electric space-heating equipment, including a hermetic refrigerant motor-compressor with or without resistance units, is computed in accordance with *NEC* 440.34 (conductors supplying motor-compressor loads in addition to other loads) and 440.35 (conductors supplying multimotor and combination-load equipment).

DISCONNECTING MEANS FOR FIXED SPACE-HEATING EQUIPMENT

A disconnecting means is required for the heater(s), motor controller(s), and supplementary overcurrent protective device(s) of all fixed electric space-heating equipment. The disconnecting means must be capable of simultaneously opening all ungrounded (hot) conductors. Multiple disconnects are permitted as a disconnecting means. Where heating equipment is supplied by more than one source, feeder, or branch circuit, the individual disconnects that comprise the disconnecting means must be grouped and identified as having multiple disconnecting means. Each disconnecting means is required to simultaneously disconnect all ungrounded conductors that it controls. Each disconnecting means shall be marked to indicate its purpose unless located such that the purpose is evident. The disconnecting means it is required to have an ampere rating of not less than 125 percent of the total load of the motors and the heaters. The provision for locking or adding a lock to the disconnecting means is required to be installed on or at the switch or circuit breaker used as the disconnecting means and must remain in place with or without the lock installed (*NEC* 424.19, 110.25 or IRC E4101.5 and Table E4101.5).

The disconnecting requirements can be broken down into three broad categories: (1) for heating equipment that has supplementary overcurrent protection, (2) for heating equipment that does not have supplementary overcurrent protection, and (3) for heating equipment that has unit switches as the disconnecting means. The rules are further broken down depending upon whether the heating equipment has a motor that is ⅛ horsepower or more.

DISCONNECTING MEANS FOR FIXED ELECTRIC SPACE-HEATING EQUIPMENT PROVIDED WITH SUPPLEMENTARY OVERCURRENT PROTECTION

The disconnecting means for fixed electric space-heating equipment provided with supplementary overcurrent protection must be within sight of the supplementary overcurrent protective device(s). If the devices are fuses, the disconnecting means must be located on the supply side, and the disconnecting means must comply with either of the following conditions:

1. Heater not containing a motor rated over ⅛ horsepower. The disconnecting means, or unit switches with a marked "off" position, are permitted to serve as the required disconnecting means for both the motor controller(s) and heater under one of the following:
   a) The disconnecting means provided is within sight from the motor controller(s) and the heater.
   b) The disconnecting means provided is capable of being locked in the open "off" position.
   c) The disconnecting means is a unit switch with a marked "off" position

2. Heater containing a motor rated over ⅛ horsepower. The disconnecting means is permitted to serve as the required disconnecting means for both the motor controller(s) and heater by one of the means specified below:

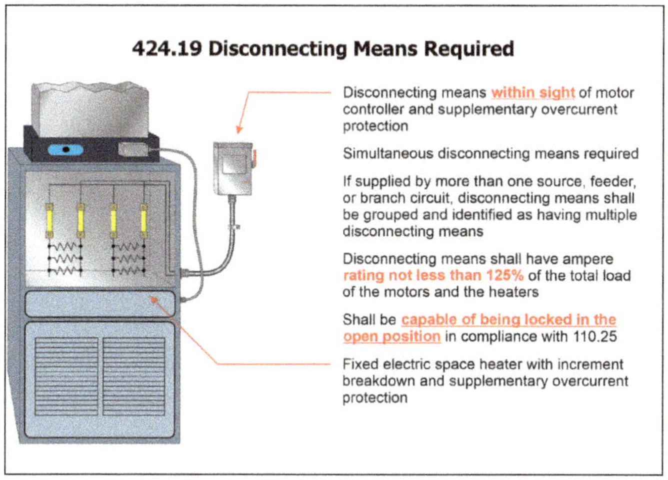

Figure 8.3. Disconnecting means within sight of supplementary overcurrent protection

   a) Where the disconnecting means is in sight from the motor controller(s) and the heater and complies with Part IX of Article 430 (Disconnecting Means for Motors).
   b) Where a motor(s) of more than ⅛ hp and the heater are provided with a single unit switch that complies with the unit switch equipment of *NEC* 422.34, the disconnecting means is permitted to be out of sight from the motor controller.[*NEC* 424.19(A) or IRC E4101.5, Table IRC E4101.5].

DISCONNECTING MEANS FOR FIXED ELECTRIC SPACE-HEATING EQUIPMENT WITHOUT SUPPLEMENTARY OVERCURRENT PROTECTION

Fixed electric space-heating equipment without a motor (such as baseboard heaters) or with a motor rated not over ⅛ horsepower is permitted to be disconnected by the branch-circuit switch or circuit breaker where it is within sight from the heater. Since the branch-circuit switch or circuit breaker is rarely installed to be within sight of the heater, it must be capable of being locked in the open position if intended to be used as the required disconnecting means.

This requirement for heating equipment makes the disconnecting means requirements for these heaters identical to those for water heaters or appliances. For motor-driven electric space-heating equipment provided with a motor rated over ⅛ horsepower, the disconnecting means for the heating equipment must be located within sight from the motor controller. Where the disconnecting means is not within sight from the heater, a separate disconnecting means must be provided within sight from the controller,

Photo 8.3. Fixed electric heating equipment equipped with a unit switch.

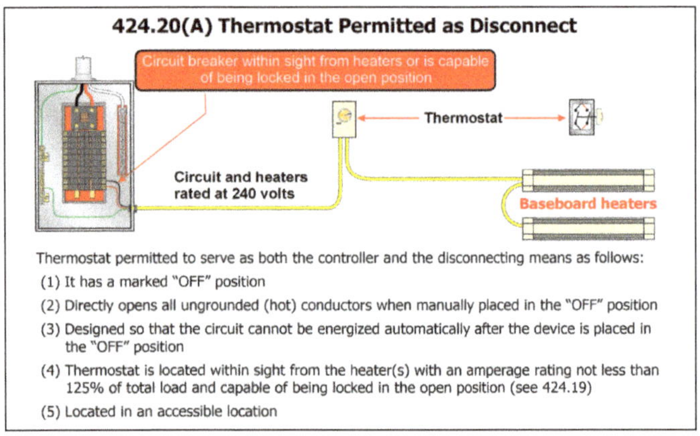

Figure 8.4. The thermostat is permitted as the disconnecting means.

or the disconnecting means that is located out of sight must be capable of being locked in the open position [*NEC* 424.19(B), 110.25 or IRC E4101.5, Table IRC E4101.5].

UNIT SWITCHES AS DISCONNECTING MEANS

Unit switches with a marked "off" position that are part of a fixed heater and that disconnect all ungrounded (hot) conductors are permitted as the disconnecting means required by *NEC* Article 424 where *other disconnecting means* are provided [*NEC* 424.19(C)].

In two-family dwellings, the *other disconnecting means* is permitted either inside or outside of the dwelling unit in which the fixed heater is installed. In this case, an individual switch or circuit breaker for the unit is permitted. It is also permitted to control lamps and appliances, such as the main service disconnecting means [*NEC* 424.19].

In one-family dwellings, the service disconnecting means is permitted to be the *other disconnecting means* along with the unit switch [*NEC* 424.19(B)].

Unit switches are sometimes installed on wall-insert-type heaters and portable heaters, but are seldom seen on fixed baseboard heaters.

THERMOSTATICALLY CONTROLLED SWITCHING DEVICES

Thermostatically controlled switching devices, and combination thermostats and manually controlled switches are permitted to serve as both controllers and the required disconnecting means for heaters only when all the following conditions are met:

1. They are provided with a marked "off" (open) position
2. They directly open all ungrounded (hot) conductors when manually placed in the "off" (open) position
3. They are designed so the circuit cannot be energized automatically after the device has been manually placed in the "off" open position
4. They are located within sight of the heater(s) they control
5. Located in an accessible location

Where the thermostat is not within sight from all of the heaters that it controls, the branch-circuit disconnect or circuit breaker supplying that circuit must be capable of being locked in the "off" position. Otherwise, a disconnecting means is required within sight of the heater(s) [*NEC* 424.20(A) or IRC E4101.5 and Table 4101.5].

Thermostats that do not directly interrupt all ungrounded (hot) conductors and thermostats that operate remote control circuits are not required to meet the requirements included above. These devices cannot be used as the required disconnecting means.

Where used as disconnecting means, all switches and circuit breakers are required to be of the indicating type (*NEC* 424.21 or IRC E4101.5 and Table 4101.5). An *indicating type* switch or circuit breaker is one that indicates that it has tripped, such as one whose handle positions to the middle after a trip or has a red indicator for the tripped position.

OVERCURRENT PROTECTION FOR ELECTRIC SPACE-HEATING EQUIPMENT

Overcurrent protection for fixed electric space-heating equipment generally is the branch-circuit fuse or circuit breaker. Overcurrent protection for motor-operated equipment is required to have additional overcurrent protection by Parts III and IV of *NEC* Article 430 or Parts III and IV of

NEC Article 440 as applicable [*NEC* 424.22(A)].

Resistance-type heating elements in electric space-heating equipment are required to be protected at not more than 60 amperes. Equipment rated more than 48 amperes and employing such elements must have its heating elements subdivided, and each subdivided load cannot exceed 48 amperes. Where a subdivided load is less than 48 amperes, the rating of the supplementary overcurrent protective device is required to be not less than 125% of the total load [*NEC* 424.22(B)].

In accordance with *NEC* 424.22(C), supplementary overcurrent protective devices for the subdivided loads specified above are required to be:

1. factory-installed within or on the heater enclosure or supplied for use with the heater as a separate assembly by the heater manufacturer
2. accessible *(but not required to be readily accessible)*
3. suitable for branch-circuit protection

Where supplementary overcurrent protection is used for the equipment covered in this chapter, it is not required to be readily accessible. Supplementary overcurrent protection cannot be used as a substitute for branch-circuit overcurrent devices or in place of the protection required for branch circuits [*NEC* 240.10].

Where cartridge fuses are used to provide this supplementary overcurrent protection, a single disconnecting means is permitted to be used as the disconnecting means for all of the subdivided loads [*NEC* 424.22(C)].

ELECTRIC SPACE-HEATING CABLES

Heating cables are supplied as a complete factory assembly with nonheating leads not less than 2.1 m (7 ft) in length. Heating cables are required to be marked with the manufacturer's identifying name or symbol, catalog number, and rating in volts and watts or volts and amperes. When encountering older, previously installed heating cables, keep in mind that prior to the 2017 *NEC*, each unit length of the heating cable was to be marked on the nonheating lead within 75 mm (3 in.) of the terminal end. In the 2014 edition of the *NEC* and prior, *NEC* 424.35 provided a color identification for heating cable leads.

The color of the marking was to correspond to the circuit voltage the heating cable was designed to use. The color yellow was used for 120-volt heating cables and red was used for 240-volt units. These marking requirements were deemed unnecessary as heating cable products are already required per *NEC* 424.28(A) to be permanently marked with the rated voltage on the nameplate marking, which will identify the voltage rating of the product (*NEC* 424.34 and 424.35).

INSTALLATION OF OTHER CONDUCTORS LOCATED ABOVE HEATED CEILINGS

Branch-circuit and feeder wiring located above heated ceilings must be located at least 50 mm (2 in.) above the heated ceiling. Branch-circuit and feeder wiring installed in this manner is considered to be operating at an ambient temperature of not less than 50°C (122°F). The ampacity of these branch-circuit and feeder conductors must be computed on the basis of the ampacity adjustment factors in accordance with NEC 310.15(B)(1) for the allowable ampacity tables of the Code. If the branch-circuit or feeder conductors are located above thermal insulation that is at least 50 mm (2 in.) thick, the conductor ampacity is subject to ambient correction in accordance with NEC 310.15(B)(1).

Adjusting the ampacity of the branch-circuit conductors due to the higher temperatures in an attic or similar location may still be necessary. Chapter 1 contains additional information on adjusting conductor ampacity for high ambient temperatures (*NEC* 424.36).

ELECTRIC SPACE-HEATING CABLES, RESTRICTED LOCATIONS

Electric space-heating cables are permitted to extend beyond the room or area in which they originate unless they are prohibited by the following provisions of *NEC* 424.38.

Electric space-heating cables cannot be installed in closets [other than noted at *NEC* 424.38(C)], over walls, or over partitions that extend to the ceiling. Heating cables cannot be installed under or through walls or in tub or shower walls. Space-heating cables are also prohibited over cabinets where the clearance from the ceiling is less than the minimum horizontal dimension of the cabinet to the nearest cabinet edge that is open to the room or area.

For example, if the cabinet is 300 mm (12 in.) deep, the minimum space between the cabinet and the ceiling would need to be 300 mm (12 in.) if heating cables are installed in the ceiling above the cabinet. Isolated single runs of embedded electric space-heating cables are permitted to pass over partitions that extend to the ceiling.

Use of electric space-heating cable in closet ceilings as a low-temperature heat source to control relative humidity is permitted where it is used only in those portions of the ceiling that are unobstructed to the floor by shelves or other permanent luminaires [*NEC* 410.16 and 424.38(C)].

CLEARANCE OF ELECTRIC SPACE-HEATING CABLES

Electric heating elements of cables installed in ceilings must be positioned at least 200 mm (8 in.) away from the edge of outlet boxes and junction boxes used for mounting surface-type luminaires. Recessed luminaires and their trims, ventilating openings, and any other openings on room surfaces require a clearance of at least 50 mm (2 in.). Proper planning is essential for the installation of surface-mounted luminaires. Sufficient area must be provided around the proposed location to ensure that any ceiling-mounted equipment does not cover any electric space-heating cable (NEC 424.39).

SPLICING OF SPACE-HEATING CABLES

Embedded space-heating cables are permitted to be spliced only where necessary and only where using splice kits identified in the manufacturer's instructions. Where splicing is necessary, use a properly sized compression sleeve applied with a companion crimp tool. The length of the space-heating cable cannot be altered under any circumstances. Reducing the length of the cable reduces the cable resistance and increases current flow. Increased current flow may cause the cable to overheat. Any alterations to these leads, such as loose splices, improper insulation, or improperly sized conductors, create a potential fire hazard (NEC 424.40).

INSTALLATION OF SPACE-HEATING CABLES ON DRY BOARD, IN PLASTER, AND ON CONCRETE CEILINGS

Space-heating cables identified to be used only for use in ceiling installations are not permitted to be installed in walls, unless they are installed as isolated single runs that are run down on a vertical surface to reach a dropped or suspended-type ceiling. Space-heating cables have to be installed in accordance with the manufacturer's instructions.

Adjacent runs of space-heating cable must be installed in accordance with the manufacturer's instructions. Prior to the 2017 *NEC*, that meant adjacent runs of space-heating cables, not over 9 watts/m (2 and ¾ watts/foot), had to be installed at least 38 mm (1½ in.) on centers.

Space-heating cables are permitted to be installed only on gypsum board, plaster lath, or other fire-resistant material. Where metal lath or other electrically conductive surfaces are encountered, a coat of plaster must be applied to the metal lath prior to installing the cables. This will effectively separate the metal lath or conductive surface from the space-heating cable.

Splices between all heating cables and nonheating leads, together with at least 75 mm (3 in.) of the nonheating leads at the splices, are required to be embedded in plaster or dry board in the same manner as the heating cable.

The entire ceiling surface is required to have a finish of thermally non-insulating sand plaster having a nominal thickness of 13 mm (½ in.). Other non-insulating material specifically identified as being suitable for this use is permitted where properly applied in accordance with the material manufacturer's specified thickness and directions.

Electric space-heating cables must be secured at intervals not greater than 400 mm (16 in.). Permitted means to secure the cable include stapling, tape, plaster, nonmetallic spreaders, or other means acceptable to the AHJ. This distance may be increased to a maximum of 1.8 m (6 ft) if the cable is identified for the greater distance.

Staples or metal fasteners that straddle the cable cannot be used in conjunction with metal lath or other electrically conductive surfaces [*NEC* 424.41(A) through (H)].

DRY BOARD INSTALLATIONS

In dry board installations, the entire ceiling below the space-heating cable is required to be covered with gypsum board not over 13 mm (½ in.) in thickness. The heating cables are installed between the upper and lower layers of the gypsum board. This may be described as "double dry board or sandwich type" construction. The void between the upper layer of the gypsum board, plaster lath, or other fire-resistant material and the surface layer of the gypsum board must be completely filled with thermally conductive, non-shrinking plaster or other approved material of equivalent thermal conductivity.

Electric space-heating cables must be installed so they will not contact metal or other electrically conductive surfaces.

Electric space-heating cables installed in dry board applications must be installed so they are parallel to joists, leaving a clear space, centered under the joists, of 65 mm (2½ in.) in width between centers of adjacent runs of cable. The surface layer of the gypsum board must be mounted to the ceiling-framing members in such a fashion that supporting nails, screws, or other fasteners, will not pierce the heating cable.

Electric space-heating cables are generally only permitted to cross joists at the ends of the room. The space-heating cable is permitted to cross joists elsewhere so the installer can avoid placing the space-heating cable too close to ceiling penetrations and luminaires. Specific installation details are found in the instructions from the manufacturer of the heating cables [*NEC* 424.41(I), (J)].

Dry board ceilings containing heating cables are not permitted to be covered with decorative panels or beams constructed of materials that have thermal insulating properties. Common finish materials such as wood, fiber, or plastic have thermal insulating properties and are not permitted to be used. Materials such as paint, wallpaper, or other approved surface finishes are permitted [*NEC* 424.42].

INSTALLATION OF NONHEATING LEADS OF SPACE-HEATING CABLES

Nonheating leads of electric space-heating cables are required to be installed with *NEC* Chapter 3 wiring methods or other listed means from the junction box to the heating cable located within the ceiling. The nonheating lead is permitted to be in the form of single conductors installed in approved raceways, single or multiconductor Type UF, Type NMC, or other approved conductors. Not less than 150 mm (6 in.) of free nonheating leads are required to be accessible for splicing within the junction box. The marking of these nonheating leads must be visible within the junction box [*NEC* 424.43(A) and (B)].

Excess leads of space-heating cables cannot be cut under any circumstances but are required to be secured to the underside of the ceiling and embedded in plaster or other approved material, leaving only a length sufficient to reach the junction box with not less than 150 mm (6 in.) of free nonheating lead to be left within the junction box [*NEC* 424.43(C)].

INSTALLATION OF SPACE-HEATING CABLES IN CONCRETE OR POURED MASONRY FLOORS

Space-heating cables installed in concrete or poured masonry floors must be installed ac-cording to the manufacturer's instructions. These instructions can differ from one manu-facturer to another. A typical manufacturer's instructions usually do not allow the space-heating cables to exceed 54 watts per linear meter (16½ watts per linear foot) [NEC 424.44(A)]. The spacing between adjacent runs of space-heating cables must also follow the manufacturer's instructions [NEC 424.44(A)].

Space-heating cables are required to be secured in place by nonmetallic frames, spreaders, or other approved means while the concrete or other finish is being applied [*NEC* 424.44(B)]. Space-heating cables cannot bridge expansion joints unless the cables are protected from damage due to

Photo 8.4. Space-heating cables are required to be secured in place by nonmetallic frames, spreaders, or other approved means *(Photo Courtesy of WarmlyYours US)*

expansion and contraction. Product instructions may give additional guidance on proper protection techniques. Generally, space-heating cables are required to be spaced away from metal objects embedded in the floor. Consult the product instructions for specific distances. The grounded metal-clad heating cable is permitted to be in contact with metal embedded in the floor.

Nonheating leads must be protected against physical damage from the point where they leave the floor. Acceptable means of protection include rigid metal conduit, intermediate metal conduit, Schedule 80 rigid PVC conduit, electrical metallic tubing, or other means acceptable to the AHJ. Bushings or approved fittings are required to be used with these protective sleeves. A protective bushing or fitting is required on the end of the sleeve at the point where the leads emerge within the floor slab. Protection on the other end may be required if the insulation of nonheating lead is subject to damage. For example, if EMT is used as the sleeve, the cut ends of the tubing should be first reamed out, and then a bushing or approved fitting should be attached to both ends [*NEC* 424.44(C) and (D)].

The *Code* contains specific ground-fault circuit interrupter (GFCI) provisions for the installation of space-heating cables located in concrete or poured masonry bathroom and kitchen floors and in hydromassage bathtub locations. GFCI protection for personnel is required to be provided regardless of the type of electrically heated floor system [*NEC* 424.44(E), 424.45(E), or IRC E3902.14].

Extreme care must be exercised for the installation of space-heating cables to prevent physical damage to the heating cable assembly. Heating cables are required to be inspected, tested, and approved by the AHJ before they are permitted to be covered or concealed. The test performed on this type of installation may be accomplished by the use of

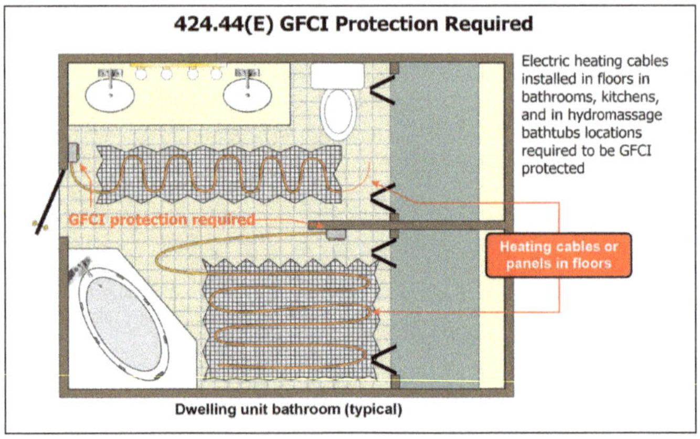

Figure 8.5. Installations of heating cables or panels in floors; GFCI protection is required

Photo 8.5. Ceiling-mounted radiant heating panels *(Courtesy of Sun-Ray Comfort Heating Pty Ltd)*

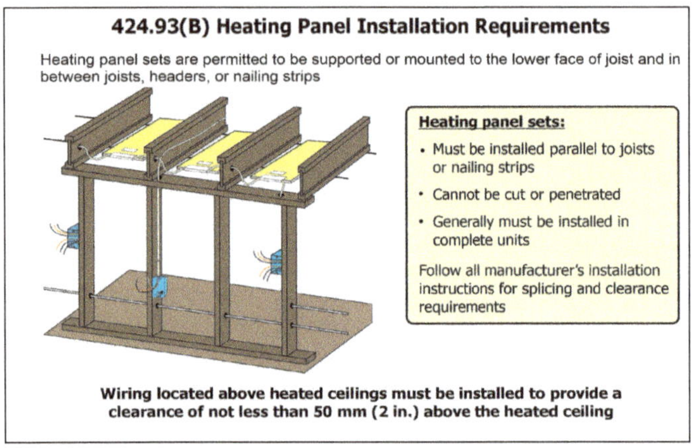

Figure 8.6. Installation of radiant heating panel sets

a suitable continuity tester or mega-ohm meter (Megger) to make sure no damaged or "open" cables are present.

Beginning with the 2023 edition of the *NEC*, with a future effective date of January 1, 2026 *[see 424.48]*, heating cables and heating cable sets will be permitted to be installed in, on, or behind walls provided all of the following conditions are met concerning the wall-installed heating cables or heating cable set(s):

1. Identified as suitable for installation in, on, or behind walls
2. GFCI protected
3. Grounding means *(such as copper braid, metal sheath, or other approved means)* is provided
4. AFCI protected
5. Installed no more than 1.2 m (4 ft) above the floor

While concerns exist regarding heating cables installed in walls, the need for such installations with proper protection is evident. According to the substation for this revision, wall-mounted heating cable installations were already being considered in Canada to permit these types of wall-mounted installations with proper protection. With an understanding that installation codes, product standards, and certification must be correlated to determine the heating cable products that will operate with the required stated protection, the future effective date of January 1, 2026, was established. This revision for heating cables correlates with added requirements in NEC 424.93(C) for heating panels.

ELECTRIC RADIANT-HEATING PANELS AND HEATING PANEL SETS

A heating panel is a complete assembly equipped with a junction box or a length of flexible conduit for connection to a branch circuit. A heating panel set is defined as a rigid or non-rigid assembly that includes nonheating leads or a terminal junction assembly suitable for connection to a wiring system. The manufacturer will mark each unit with an identifying name or symbol. Additional required markings include the catalog number and the rating in volts and watts or volts and amperes. These markings are placed so that they are visible before the panel trim cover is installed.

Manufacturers of heating panels or heating panel sets provide marking labels. These labels will indicate that the space-heating installation incorporates heating panels or heating panel sets. The instructions will state that these labels are to be affixed to the panelboards that supply the panels or panel sets. The label is intended to identify the branch-circuit fuse or circuit breaker with the corresponding panel or panel set that it supplies. This helps ensure that the branch circuit that supplies the heating panel or heating panel set is properly identified.

The requirement to install the manufacturer's marking labels is more stringent than the general requirements for the identification of disconnecting means. Heating panels and heating panel set installations that are visible and distinguishable after installation are not required to have labels affixed to the panelboard. However, the general requirements for identification of the disconnecting means still apply [*NEC* 424.92(A), (B), and (C)].

INSTALLATION REQUIREMENTS FOR RADIANT-HEATING PANELS AND HEATING PANEL SETS

Heating panels and heating panel sets must be installed in accordance with the manufacturer's instructions [*NEC* 424.93(A)(1)]. These instructions indicate that the heating portion cannot be installed on or behind surfaces that are subject to physical damage. They are not permitted to be run through or above walls, partitions, cabinets, cupboards, or similar portions of structures that extend to the ceiling or run in or through thermal insulation. The heating panel is permitted to be in contact with the surface of the thermal insulation. They are not permitted to be run in walls except as permitted by *NEC* 424.93(C) [*NEC* 424.93(A)(2)].

The edges of electric radiant-heating panels and heating panel sets must be installed so that they are at least 200 mm (8 in.) away from the edges of any outlet boxes and junction boxes used for mounting surface-type luminaires. A clearance of at least 50 mm (2 in.) is required from recessed luminaires and their trims, ventilating openings, and other such openings in room surfaces. Where specifically listed and marked for lesser clearances, electric radiant-heating panels and heating panel sets can be installed at their marked clearances. A sufficient area must be provided around the location of the panel in order to ensure that any surface-mounted units will not cover any portion of the panel [NEC 424.93(A)(3)].

After the heating panels or heating panel sets are installed and inspected, a surface such as gypsum drywall (sheetrock) is installed over the heating panels or heating panel sets. The type of surface material is identified in the instructions provided by the panel manufacturer. The drywall (sheetrock) surface should be secured so nails, screws, or other fasteners will not pierce the heating panels [*NEC* 424.93(A)(4)].

Surfaces permitted by *NEC* 424.93(A)(4) installed over the heating panels or heating panel sets may be covered with paint, wallpaper, or other approved finish materials as identified in the manufacturer's instructions.

Heating panel sets are permitted to be secured to the lower face of joists or mounted in between joists, headers, or nailing strips. Heating panel sets are required to be installed parallel to joists or nailing strips.

Nailing or stapling of heating panel sets must be accomplished only through the unheated portions provided for this purpose. Heating panel sets cannot be cut or nailed through any point closer than 6 mm (¼ in.) to the element. Nails, staples, or other fasteners cannot be used where they penetrate current-carrying parts. Heating panel sets are required to be installed as complete units unless they are specifically identified as being suitable for field cutting in an approved manner [*NEC* 424.93(B)(1), (2), (3), and (4)].

Beginning with the 2023 edition of the *NEC*, with a future effective date of January 1, 2026 *[see 424.93(C)]*, heating panels will be permitted to be installed in, on, or behind walls provided all of the following conditions are met concerning the wall-installed heating panel(s):

1. Identified as suitable for installation in, on, or behind walls
2. Installed per the manufacturer's instructions and in accordance with the product listing
3. GFCI protected *(low-voltage heating panels not required to be GFCI protected)*
4. Grounding means *(such as copper braid, metal sheath, or other approved means)* is provided
5. AFCI protected
6. Installed no more than 1.2 m (4 ft) above the floor

While concerns exist regarding heating panels and heating panel sets installed in walls, the need for installing heating panels in walls with proper protection is evident. With an understanding that installation codes, product standards, and certification will all need correlation to determine the heating panel products that will operate with the required stated protection, the future effective date of January 1, 2026, was established. This revision for heating panels correlates with added requirements in NEC 424.48 for heating cables.

Branch-circuit and feeder wiring located above heated ceilings must be installed to provide a clearance of not less than 50 mm (2 in.) above the heated ceiling. Wiring installed in this manner is considered as operating at an ambient temperature of not less than 50°C (122°F). The ampacity of these conductors would have to be adjusted utilizing the temperature correction factors found in the allowable ampacity of *NEC* 310.15(B)(1). However, if the branch circuit or feeder conductors are located above thermal insulation at least 50 mm (2 in.) thick, it would be subject to ambient correction to account for the presence of heating panels. Adjusting the ampacity of the branch-circuit conductors due to the higher ambient temperatures in an attic or similar location may still be necessary (*NEC* 424.94).

424.4 Branch Circuit Requirements

Answers to Figure 8.2 Questions

Answer to Question 1:
If heater A and B are 1500 watts and heater C is 1000 watts, what is the minimum branch-circuit rating?

1500 + 1500 + 1000 watts = 4000 watts
4000 watts x 125% = 5000 watts
5000 watts ÷ by 240 volts = 20.8 amperes
Minimum branch circuit rating is 25 amperes

Answer to Question 2:
If heaters A, B and C total 6000 watts, what is the minimum circuit rating?

6000 watts x 125% = 7500 watts
7500 watts ÷ by 240 volts = 31.25 amperes
Minimum branch circuit rating is 31.25 amperes

Note: Two circuits must be used since the minimum rating of the circuit exceeds the maximum circuit size of 30 amperes ***(two or more FESHE outlets)***

Figure 8.7. Answers to Figure 8.2 questions

Any branch circuit or feeder wiring installed behind heating panels or heating panel sets that are located in interior walls or partitions is considered to be operating at an ambient temperature of 40°C (104°F). The reduced ampacity of this wiring is required to be computed based on the ampacity correction factors for high ambient temperatures.

CONNECTION TO BRANCH-CIRCUIT CONDUCTORS

Electric radiant heating panels or heating panel sets are often assembled together in the field to form a heating installation in one room or area. This type of heating assembly must be connected together in accordance with the manufacturer's instructions [*NEC* 110.3(B) and *NEC* 424.96(A) or IRC E3403.3].

Heating panel sets must be connected to a branch circuit using a wiring method specifically identified as being suitable for the purpose. Instructions for heating panels and heating panel sets intended for single conductor supply indicate the type of cable to be used. Types NM and NMC are not suitable for installation in these types of radiant heating panels or panel sets. Some manufacturers provide a terminal junction assembly. This type of heating panel set is permitted to have its nonheating leads attached at the time of installation in accordance with the manufacturer's instructions. These connections must be made in junction boxes. Open splices to the branch-circuit conductors are not permitted [*NEC* 424.96(C)].

Excessive lengths of nonheating leads of electric radiant-heating panels or heating panel sets can be cut to the required length as indicated in the manufacturer's installation instructions. The nonheating leads are required to meet the installation requirements for the type of wiring method used. Nonheating leads that are an integral part of an electric radiant-heating panel or a heating panel set and are not subject to the ampacity requirements or correction factors [*NEC* 424.97].

Installation in concrete or poured Masonry Electric radiant-heating panels or heating panel sets should be secured by the means specified in the manufacturer's instructions. The securing means must be identified as being suitable for the installation [*NEC* 424.98(A)].

These panels or panel sets cannot be installed where they bridge expansion joints unless provision is made for expansion and contraction. Generally, spacing is required between electric radiant-heating panels and heating panel sets and metal embedded in the floor. No specific distance is stipulated by the *Code*. Consult the manufacturer's instructions for specific installation details. Grounded metal-clad electric radiant-heating panels are permitted to be in contact with metal embedded in the floor [*NEC* 424.98(B) and (C)].

Nonheating leads must be protected against physical damage from the point where they leave the floor. Acceptable means of protection include rigid metal conduit, intermediate metal conduit, Schedule 80 rigid PVC conduit, electrical metallic tubing, or other means acceptable to the AHJ. Bushings or approved fittings are required to be used with these protective sleeves. A protective bushing or fitting is required on the end of the sleeve at the point where the leads emerge within the floor slab. Protection on the other end may be required if the insulation of nonheating lead is subject to damage. For example, if EMT is used as the sleeve, the cut ends of the tubing should be first reamed out, and then a bushing or approved fitting attached to both ends (*NEC* 424.98(D) and (E), and *NEC* 300.4).

INSTALLATION OF RADIANT-HEATING PANELS OR PANEL SETS UNDER FLOOR COVERINGS

Listed heating panels or panel sets shall be installed on floor surfaces that are smooth and flat if the panel is to be installed under floor coverings. The panels shall be firmly anchored to the floor using an adhesive or anchoring system identified for the use. The panel or panel set must be protected from expansion and contraction if installed across expansion joints. After installation, the panels are to be inspected prior to covering. The floor covering should be secured to the heating panel with a release-type adhesive or other approved attachment means. The floor covering must be of a type identified by the manufacturer of the panel as being suitable for the use [*NEC* 424.99(A), (B)(1) through (4)].

The branch circuit(s) supplying heating panels and panel sets are required to have ground-fault circuit-interrupter (GFCI) protection for personnel. The manufacturer of the panel will provide an internal device that opens the ungrounded conductors when it detects a low- or high-resistance fault. The device must detect line-to-line, line-to-ground, and line-to-grounded conductor faults. One way such a fault occurs is from a sharp object penetrating a heating element. An internal grounding shield is one method used to provide line-to-ground protection [*NEC* 424.99(B)(5) and (6)].

PRODUCT SAFETY STANDARD REQUIREMENTS

It is recommended to consult the following electrical product safety guide card information and product safety standards for additional guidance on the proper installation, operation, and use of electrical equipment covered in this chapter. The four-letter code in parentheses indicates the product category in the Underwriters Laboratories Guide Information for Electrical Equipment (UL Product Spec material).

- Electrical Equipment for Use in Ordinary Locations (AALZ)
- Baseboard Heaters (KLDR)
- Circuit Breakers, Molded-Case, and Circuit Breaker Enclosures (DIVQ)
- Heating and Cooling Equipment (LZFE)
- Radiant Heating Equipment (KQYZ)
- Switches, Enclosed (WIAX)
- Switches, Knife (WIOV)
- Switches, Molded Case (WJAZ)
- Temperature Indicating and Regulating Equipment (XAPX)

REFERENCES

1. *National Electrical Code* and *NEC* are registered trademarks of the National Fire Protection Association, Inc., Quincy, MA 02169. This reprinted material is not the official position of the National Fire Protection Association, which is represented only by the standard in its entirety.

2. *International Residential Code for One- and Two-Family Dwellings* (2024). Copyright © International Code Council, Inc., June 2023.

3. UL, Northbrook, IL, *UL Guide Information for Electrical Equipment Directory* (UL Product iQ® material) unless indicated otherwise.

REVIEW QUESTIONS

1. Combined thermostats and manually controlled switches can serve as the disconnecting means for electric heaters without a marked "off" position.

True _____ False _____

2. Wiring located above ceilings heated by electric radiant-heating panels are required to be spaced not less than _____ above the ceiling.

A. 13 mm (½ in.)
B. 19 mm (¾ in.)
C. 25 mm (1 in.)
D. 50 mm (2 in.)

3. A branch circuit supplying two or more outlets for fixed electric space heating equipment in a residential occupancy cannot be rated over _____ amperes.

A. 20
B. 25
C. 30
D. 50

4. Embedded heating cable installations must be inspected and approved before being covered or concealed.

True _____ False _____

5. An installation of heating cable identified only for use in ceiling installations cannot be installed in walls, EXCEPT under what condition?

A. To reach the floor for the heating cable in the floor.
B. An isolated single run of cable is to be installed down a vertical surface to reach a dropped ceiling.
C. To conceal the heating cable.
D. To install the heating cable in a neat and workmanlike manner.

6. Heating cables installed by plaster must be supported at intervals not exceeding what distance unless the cables are identified for a greater spacing.

A. 400 mm (16 in.)
B. 425 mm (17 in.)
C. 450 mm (18 in.)
D. 500 mm (20 in.)

7. What is the minimum required clearance from an electric space heating cable to the edge of an outlet box supplying a recessed luminaire and its trim?

A. 25 mm (1 in.)
B. 50 mm (2 in.)
C. 100 mm (4 in.)
D. 150 mm (6 in.)

8. Fixed electric space-heating equipment requiring supply conductors with an insulation rating greater than _____ must be clearly and permanently marked.

A. 25°C
B. 30°C
C. 45°C
D. 60°C

9. A heating panel set is a rigid or non-rigid assembly that is provided with nonheating leads or a terminal junction assembly identified as being suitable for the connection to a wiring system.

True _____ False _____

10. In accordance with product instructions provided with the listed electric baseboard heaters, their installation below receptacle outlets may be prohibited.

True _____ False _____

11. Where a motor(s) of more than ⅛ hp and the heater are provided with a ___________, the disconnecting means shall be permitted to be out of sight from the motor controller.

A. single unit switch
B. double unit switch
C. a locking means
D. knife switch

12. Branch-circuit and feeder wiring are required to be spaced so that they are not less than what distance above the heated ceiling and are considered as operating at an ambient temperature of 50°C (122°F)?

A. 16 mm (⅝ in.)
B. 19 mm (¾ in.)
C. 25 mm (1 in.)
D. 50 mm (2 in.)

13. With a future effective date of January 1, 2026, heating cables and heating cable sets will be permitted to be installed in, on, or behind walls provided which of the following conditions are met concerning the wall-installed heating cables or heating cable set(s):

A. Identified as suitable for installation in, on, or behind walls
B. GFCI protected
C. AFCI protected
D. All of the above

14. Heating cables are prohibited in all of the following locations EXCEPT:

A. In closets (under any condition)
B. Over walls (under any condition)
C. Over kitchen cabinets that are fastened to the ceiling
D. Embedded in concrete or poured masonry floors

15. GFCI protection is required for cables installed in electrically heated floors in which of the following locations?

A. bathrooms
B. kitchens
C. hydromassage bathtub locations
D. All of the above

16. Heaters installed in an air duct shall be ___________ as suitable for the installation.

A. labeled
B. listed
C. approved
D. identified

17. For low-voltage fixed electric space-heating equipment, the power unit shall be an isolating type with a rated output not exceeding the following:

A. 25 amperes, 30 volts
B. 60 volts dc
C. Both A and B.
D. Neither A and B.

18. A receptacle outlet incorporated into a listed baseboard heater is permitted to serve as the required receptacle outlet for the wall spacing.

True _____ False _____

19. Fixed electric space-heating equipment is required to be installed to provide the required spacing between the equipment and adjacent combustible material unless it is _____________ to be installed in direct contact with combustible material.

A. identified
B. listed
C. approved
D. certified

9 Air-Conditioning Equipment Installations

AIR-CONDITIONER AND HEAT PUMP COMPRESSOR NAMEPLATE INFORMATION

Reviewing and understanding the information on air conditioner or heat pump equipment nameplates is a crucial step in accurately evaluating an air conditioner or heat pump installation. It simplifies the selection of the correct branch-circuit conductor size, the type and rating of the overcurrent protection device, and the size of the disconnecting means. The installer or inspector can find most of the necessary information by referring to the data plate located on the unit. Figure 9.1 and photo 9.1 are examples of the type of information found on the nameplate for a residential central air-conditioning unit.

Air-conditioning and heat pump equipment have a hermetic motor compressor and are not treated the same as a conventional electric motor. The hermetic motor compressor operates in a refrigerant environment. A hermetic motor-compressor does not have a horsepower or full-load current rating, such as a standard motor. Special terms are used to provide the necessary information to install wiring properly for this equipment.

One of the terms noted on the nameplate in Figure 9.1 is Compressor RLA, which is the rated load in amperes or rated-load current for the motor compressor. The value for a given motor-compressor unit is established by the manufacturer. The rated load current is the current resulting when the motor compressor is operated at the rated load, rated voltage, and rated frequency of the equipment it serves. This value is used to calculate the minimum circuit ampacity and overcurrent protective device ratings specified on the nameplate of this equipment. The RLA for the compressor in Figure 9.1 is 18.0 amperes.

Another term noted on the air conditioner nameplate is Compressor LRA, which represents locked-rotor amperes or locked-rotor current. It is the maximum current flowing to the motor when it is in a locked or non-turning condition. This value is necessary to ensure that the air conditioner disconnecting means and controller have proper interrupting capacities. The LRA for the compressor in Figure 9.1 is 96 amperes.

Minimum supply circuit ampacity, or similar wording, is the minimum circuit ampacity required to select conductor size and switch rating for the unit. This value is determined by the following formula:

Minimum Circuit Ampacity =
(RLA x 1.25) + Other Loads

Maximum fuse or circuit breaker size indicates the maximum fuse or circuit breaker size permitted. The manufacturer has selected a rating that will permit the motor to start as well as provide overcurrent protection. The minimum fuse or circuit breaker size indicates the smallest rating that will allow the motor to start. The mark indicating the minimum rating of the fuse or circuit breaker is an optional marking and is not found on all nameplates.

Rated load amperes (RLA) are used to calculate the maximum overcurrent protection permitted for the unit. Later in this chapter, we will use the information from the nameplate example in Figure 9.1 to do a sample calculation of an AC unit.

In some cases, the equipment is also marked with a branch-circuit selection current (BCSC). This is a value in amperes that must be used instead of the rated-load current for determining ratings of motor branch-circuit conductors, disconnecting means, controllers, and branch-circuit short-circuit and ground-fault protective devices whenever the running overload protective device permits a sustained current greater than the rated-load current. The value of the branch-circuit selection current will always be equal to or greater than the marked rated-load current.

Where required, the value of the branch-circuit selection current will be found on the nameplate where the value for the rated-load current appears. Since the sample air-conditioning nameplate data included in this chapter does not include a branch-circuit selection current, the rated-load current (RLA) is used for the calculations [*NEC* Articles 100 and 440 or IRC E3702.11].

Circuit breakers manufactured today are no longer required to be HACR-rated. A HACR circuit breaker is a heating, air-conditioning, and refrigeration-rated circuit breaker. The listed inverse time circuit breakers are acceptable for the protection of HVAC equipment without any additional marking. Remember that the *Code* requires compliance with nameplate markings. If the nameplate specifies fuses only, the substitution of even a listed circuit breaker is a violation of *NEC* 110.3(B) or IRC E3403.3. Generally, the manufacturer offers a choice of either fuses or circuit breakers. Check the nameplate information to make sure.

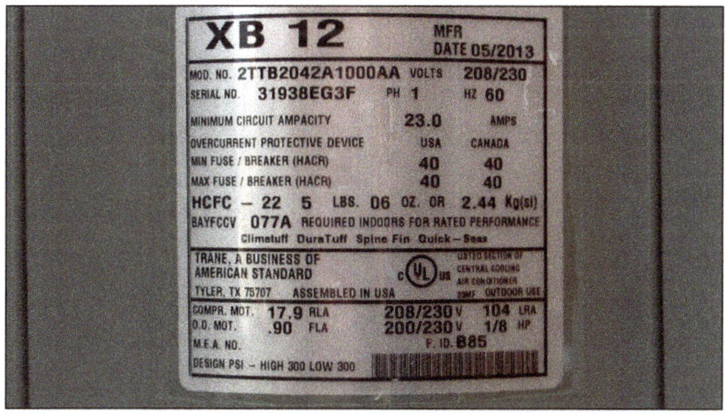

Photo 9.1. Air-conditioning equipment nameplate showing maximum and minimum fuse or circuit breaker rating as specified by the manufacturer.

440.4 Nameplate Required

Breezy Cool Air-Conditioning Equipment

Model Number RAKA-36JAZ	MFD 03/99
Serial Number 5340 F0994 8995	Outdoor Use
Volts - 208/230 Phase 1	Hertz 60
Compressor RLA - 18.0/18.0	LRA 96
Outdoor Fan Motor FLA - 1.3	HP 1/5
Minimum supply circuit ampacity	24/24 amp
Max. fuse or circuit breaker size*	40/40 amp
Min. fuse or circuit breaker size*	30/30 amp
Design pressure high	300 PSIG
Design pressure low	150 PSIG
Outdoor units factor charge	77 OZ. R22
Total system charge	OZ. R22

***HACR type breaker for USA**
See instructions inside access panel

Listed by ABC Labs

Figure 9.1. Air-conditioning equipment nameplate for dwelling units

440.4(B) Minimum Circuit Ampacity

Using the information from the **AC nameplate in Figure 9.1**, determine the maximum circuit ampacity for the branch circuit conductors to the AC unit

Minimum Circuit Ampacity =
(RLA of Highest Rated Motor x 1.25) + Other Loads
[NEC 440.4(B), 440.6, 440.7, 430.24]

Minimum Circuit Ampacity =
Compressor *(18.0 x 1.25 = 22.5)* + Fan Motor FLA *(1.3)* = **23.8 amperes**

Rounded up to 24 amperes, is the minimum supply circuit ampacity stated on the nameplate and is the required minimum ampacity of the branch-circuit conductors

Figure 9.2. Minimum circuit ampacity for Figure 9.1 AC unit

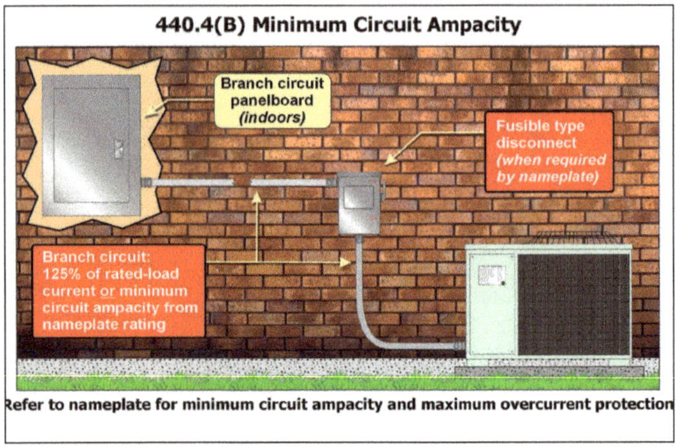

Figure 9.3. The branch circuit size must meet the minimum circuit amperes required by the nameplate

BRANCH CIRCUIT REQUIREMENTS

The value of the rated-load current is the starting point for calculating the required ampacity of branch-circuit conductors and rating of electrical equipment for a hermetic refrigerant motor compressor. The rated load current is determined by the manufacturer and is marked on the nameplate of the unit. This value is necessary to determine the rating or ampacity of the disconnecting means, the branch-circuit conductors, the controller, branch circuit short-circuit and ground-fault protection, and the separate motor overload protection [*NEC* 440.4(B), 440.35 or IRC E3702.11].

BRANCH-CIRCUIT CONDUCTOR SIZE FOR AIR-CONDITIONING EQUIPMENT

For a typical air-conditioning unit or heat pump outdoor unit that has a motor compressor and additional load(s), such as a fan motor, the conductors supplying this equipment must have an ampacity not less than 125% of either the rated load or branch circuit selection current, whichever is larger, plus the full-load current of the fan motor. The values in Figure 9.2 are calculated from the example nameplate in Figure 9.1.

This value, rounded up to 24 amperes, is the minimum supply circuit ampacity stated on the nameplate and is the required minimum ampacity of the branch-circuit conductors selected from *NEC* Table 310.16 or IRC Table E3705.1. Since the equipment nameplate includes this value, the installer or inspector doesn't need to perform this calculation.

The allowable ampacity of a 12 AWG Type TW, THW, or THWN copper conductor is 25 amperes when operated in an ambient temperature not exceeding 86°F. Where installed in temperatures in excess of 86°F, the ampacity correction factors listed in the conductor allowable ampacity tables must be applied [*NEC* Table 310.15(B)(1)(1) or IRC E3705.2 and Table E3705.2].

The Code generally limits the overcurrent protection of a 12 AWG conductor to 20 amperes unless otherwise specifically permitted. NEC 240.4(G) and Table 240.4(G) permit air-conditioning circuit conductors to be protected in accordance with Parts III, IV, and VI of Article 440. NEC 440.21 specifically states that the provisions of Part III add to or amend the general requirements for overcurrent protection found elsewhere in the Code, such as in Article 240. For example, the air conditioner nameplate marking indicates a maximum fuse or circuit breaker size of 40 amperes. The minimum supply circuit ampacity is 24 amperes. Consequently, if no adjustment for ambient temperature higher than 86°F is required, all 12 AWG conductors with an allowable ampacity of 25 amperes are acceptable as the branch-circuit conductors supplying this unit.

However, this does not include 12 AWG copper conductors that are part of a cable assembly, such as Type NM or UF, which have an allowable ampacity of 20 amperes at 60°C. The conductor ampacity requirements for Type NM and UF cable at *NEC* 334.80 and IRC E3705.4.4 limit the ampacity of these conductors to the 60°C column of *NEC* Table 310.16 or IRC Table E3705.1. In the IRC, Section E3705.5.4 refers back to Section E3702.12 for overcurrent protection of air-conditioning and heat pump equipment.

The concept for protection where a 12 AWG copper wire with an ampacity of 20 amperes is permitted to have overcurrent protection of 40 amperes is as follows:

1. The 40-ampere fuse or circuit breaker at the origination of the circuit will protect the conductors from a short circuit and a ground fault. A short-circuit or line-to-line fault is where the ungrounded (hot) conductors fault together. A ground fault is where the ungrounded (hot) conductor(s) fault to the equipment grounding conductor or grounded equipment.
2. The conductor is protected from overload by the running overcurrent device. The overload device may be a separate device. A thermal element built into a motor controller is one type of overload. The overload for a hermetic motor compressor is usually contained in the motor housing and directly senses motor temperature.

The combination of the two elements provides the overcurrent protection necessary for the safe and proper operation of motor-operated equipment.

BRANCH CIRCUIT MAXIMUM OVERCURRENT PROTECTION RATING

The Code requires that the branch-circuit overcurrent protection be capable of carrying the starting current of the motor. The maximum rating is 175% of the rated load current of the hermetic refrigerant compressor motor. If this value is not sufficient to allow the motor to start, the rating or setting can be increased to allow the motor to start, but cannot exceed 225% of the rated-load current [NEC 440.22(A) and 440.22(A), Ex. No. 2].

A typical AC unit or heat pump will contain a hermetic refrigerant compressor motor plus one or more fan motors. The branch-circuit protective device provides short-circuit and ground-fault protection for these motors. The rated-load current(s) for these motors are added to the adjusted value of the hermetic refrigerant motor-compressor RLA in order to determine the maximum rating or setting of the branch-circuit protective device. Using the values for RLA for all motors listed on the nameplate in Figure 9.1, the maximum rating or setting of the fuse or circuit breaker is based on calculations in Figure 9.4.

440.22(A) Branch-Circuit Short-Circuit and Ground-Fault Protection

Using the information from the AC **nameplate in Figure 9.1**, determine the maximum branch-circuit short-circuit and ground-fault protection for the AC unit

Maximum rating = 175% of the rated-load current of the hermetic refrigerant compressor motor + other loads

If this value is not sufficient to allow the motor to start, the rating or setting can be increased to allow the motor to start but cannot exceed 225% of the rated-load current

(Compressor RLA x 2.25) + Other Loads *(Fan Motor FLA)*
(18.0 x 2.25 = 40.5 amperes) + 1.3 = 41.8 amperes

Maximum rating or setting of fuse or circuit breaker = 40 amperes

Figure 9.4. Maximum branch-circuit short-circuit and ground-fault protection for Figure 9.1 AC unit

Photo 9.2. Disconnecting means is required to be within sight from and readily accessible from the equipment.

If this calculated value is not a standard current rating of an overcurrent protective device, the *Code* requires the use of the next lower standard rating. In this example, the next lower standard device is 40 amperes, as stated on the sample nameplate. It is not necessary to actually perform this calculation in the field. The manufacturers of the AC unit or heat pump calculate the value and determine the maximum rating. They include this rating on the unit's nameplate. A lower-rating protective device is permitted as long as the fuse or circuit breaker is capable of carrying the unit's starting current.

If the nameplate includes a "MINIMUM FUSE OR CIRCUIT BREAKER SIZE," then the *Code* would also require compliance with that marking, in addition to the marking of the "MAXIMUM FUSE OR CIRCUIT BREAKER SIZE." A fuse or circuit breaker rated lower than the minimum value marked on the nameplate will not be capable of carrying the starting current of the unit.

DISCONNECTING MEANS RATING

Where the air-conditioning or heat pump compressor unit consists of a hermetic refrigerant motor-compressor(s) in combination with other loads, such as the fan motor, the horsepower rating of the disconnecting means is based on the summation of all currents at rated-load condition and also at locked-rotor condition.

Using the nameplate information from Figure 9.1., the 18.0-ampere RLA of the compressor motor is added to the 1.3-ampere FLA of the fan motor. The total of 19.3 amperes is then considered to be the equivalent full-load current for the combined load. According to *NEC* Table 430.248, the full-load current rating of a 230-volt, single-phase, 3-horsepower motor is 17 amperes, while the full-load current rating of a 230-volt, single-phase, 5-horsepower motor is 28 amperes. Consequently, since the equivalent full-load current of the example A/C unit is 19.3 amperes, we must use the next higher rating, and the disconnect switch must have a minimum of a 5-horsepower, 230-volt, single-phase rating.

The ampere rating of the disconnecting means must also be at least 115% of the sum of all currents at rated-load condition [*NEC* 440.12(B)(2)]. This minimum rating would then be 115% x 19.3 amperes = 22.3 amperes. If the disconnecting means includes or serves as the branch-circuit overcurrent protection for the unit, the rating required for the overcurrent device, rather than this minimum rating, would generally be the determining factor in sizing the disconnecting means. A fused disconnect switch containing either the maximum or

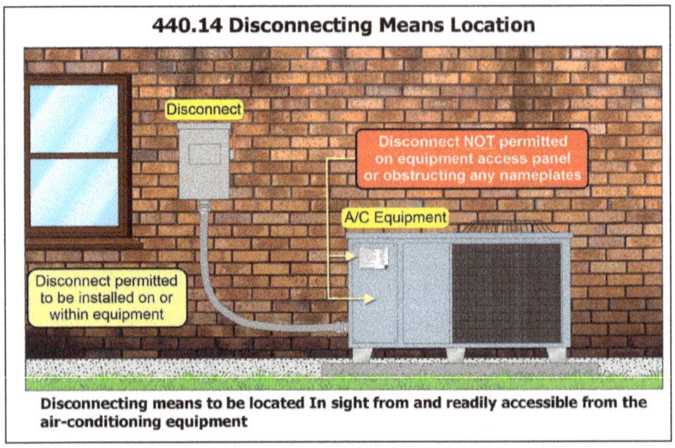

Figure 9.5. Disconnect is to be located within sight from and readily accessible from the air-conditioning equipment

Photo 9.3. A receptacle is required within 7.5 m (25 ft) and on the same level as the air-conditioning equipment

minimum sizes of fuses listed on the nameplate would exceed this 115% minimum requirement. However, if an unfused disconnect switch is used as the disconnecting means, then the larger of the two values obtained by the 115% calculation and the equivalent horsepower rating would establish the minimum switch rating.

There is another consideration in determining the correct size of the disconnecting means serving the air conditioning unit. The disconnecting means rating must also be based on currents at locked-rotor condition [refer to NEC Table 430.251(A)], which lists various values of locked-rotor current and an equivalent horsepower rating.

In our example, the nameplate indicates that the motor-compressor LRA is 96 amperes. The nameplate does not give an LRA for the fan motor. A typical assumption for AC motors is to suppose the LRA is six times the FLA. For the fan motor in the example, this gives an approximate value for LRC of 6 x 1.3 amperes = 7.8 amperes. Adding this to the motor-compressor LRA of 96 amperes gives us an equivalent LRA for the combined load of 103.8 amperes. Referring to *NEC* Table 430.251(A), we find that for a single-phase, 230-volt motor with a 103.8 amperes motor locked-rotor current, the equivalent horsepower rating falls between 3 and 5 horsepower. A 3-horsepower disconnect switch would be slightly undersized. The disconnecting means for the unit should be based on a 5-horsepower rating.

DISCONNECTING MEANS LOCATION FOR AIR-CONDITIONING EQUIPMENT

A disconnecting means is required to be located "within sight from and readily accessible" from the air-conditioning equipment [NEC 440.14 or IRC Table E4101.5]. This disconnecting means it can be installed on or within the air-conditioning equipment. This disconnecting means must meet the working space requirements of NEC 110.26(A) or IRC E3405.1].

"In Sight From (Within Sight From, Within Sight)" indicates that one piece of equipment must be visible from, within, or within sight of another, with a maximum distance of 15 meters (50 feet). "Accessible, Readily (Readily Accessible)" means that it can be reached quickly for operation, maintenance, or inspections without needing tools (other than keys), climbing over or under obstacles, removing barriers, or using portable ladders.

This disconnecting means it can be installed on or within the air-conditioning equipment. This disconnecting means must meet the working space requirements of *NEC* 110.26(A) or IRC E3405.1].

A disconnecting means that the equipment is field mounted directly on the equipment, which must be mounted in a manner that will not restrict access through panels designed to allow service and repair operations and access to the components inside the unit. This field-mounted disconnecting means also must not obscure the equipment nameplate(s). An air conditioner or heat pump compressor is generally located on a concrete pad outside the dwelling. It may also be located under the house in a properly ventilated crawl space or on the roof. It is important to remember that the definitions of within sight and readily accessible have a significant meaning in this application.

The disconnecting means is usually located next to the air-conditioning unit and is, therefore, considered as being readily accessible. Access to the disconnecting means may be restricted if it is located above or behind the unit. If the disconnecting means is likely to require examination, adjustment, servicing, or maintenance while energized (and it will), sufficient working space should be provided. The proper location is subject to the judgment of the authority having jurisdiction after applying the working space provisions of

NEC 110.26(A) or IRC E3405.1.

The purpose of the required disconnecting means is to provide a ready and visible means of disconnection for the person who will service or repair the equipment. At dwelling units, a locking type of disconnecting means located out of sight from the unit is not an acceptable alternative for air-conditioning equipment (*NEC* 440.14 or IRC E4101.5 and Table E4101.5).

The attachment plug and receptacle combination is considered the disconnecting means for most cord- and plug-connected equipment, such as room air conditioners.

WORKING SPACE

The working space around electrical equipment, such as an AC unit disconnect that is likely to "require examination, adjustment, servicing, or maintenance while energized,' must be provided according to NEC 440.14 and Table 110.26(A) or IRC E3405.2. The clearance must be in the direction of access to the equipment or the part of the equipment that is likely to be worked on while live, exposed parts are present. Service personnel often examine or test this equipment while it is energized. Safe access must be available to the control equipment inside the unit. For a typical dwelling with a 120/240-volt, 3-wire, single-phase service, the minimum dimensions of this workspace are 762 mm (30 in.) wide—either the width of the equipment or, if greater, this measurement—and 900 mm (36 in.) deep. The height of this space should be clear and extend from the ground to 2.0 m (6½ ft) or the height of the equipment, whichever is greater. Meeting this working space requirement involves ensuring safe access during installation. Clear working space is needed in front of removable panels that provide access to live parts for testing or inspection, without requiring the worker to stretch across or lie on the AC unit itself.

In addition to adequate working space, service personnel often require power for portable electrical tools and equipment associated with servicing air-conditioning equipment. The *Code* recognizes the need for a service receptacle to be installed near this equipment for this purpose. A 125-volt, single-phase, 15- or 20-ampere receptacle outlet for servicing air-conditioning and refrigeration equipment is required to be located not less than 7.5 m (25 ft) from and on the same level as the equipment itself. The receptacle must be in a readily accessible location from the equipment and must be GFCI-protected as applicable to outdoor receptacle outlets at dwelling occupancies (*NEC* 210.63 and 210.63(A) or IRC E3901.12); see figure 9.6 and photos 9.3 and 9.4).

Some electrical equipment manufacturers have combined the required disconnecting means and service

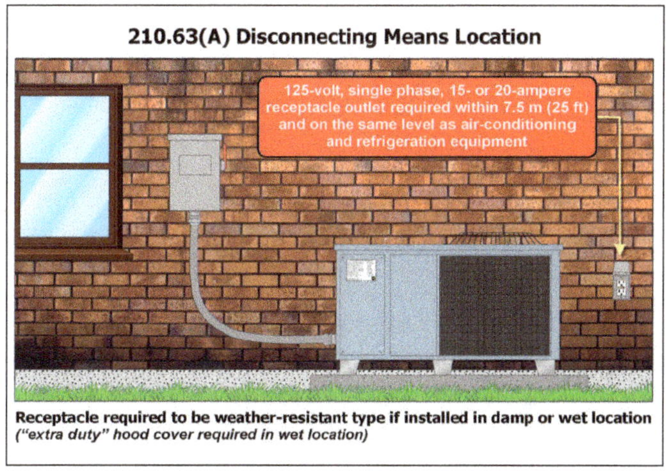

Figure 9.6. A 125-volt, single-phase, 15- or 20-ampere receptacle is required within 7.5 m (25 ft) of air-conditioning equipment

Photo 9.4. Required receptacle outlet provided within the A/C disconnecting means enclosure

receptacle outlet into one product (see photo 9.4). Separate circuits for the receptacle and the compressor unit are required. The service receptacle outlet should be a weather-resistant, GFCI-protected device and the extra duty-rated in-use cover for an outlet box hood complying with the requirements of *NEC* 406.9(B)(1) or IRC E4002.9 for use in wet locations. Other listed products, enclosures, or assemblies providing weatherproof protection that do not utilize an outlet box hood are not required to be marked "extra duty." Per the manufacturer's specifications, this product typically has a minimum mounting height to prevent snow or water from entering the opening at the bottom of the hinged cover.

INDOOR HEAT PUMP EQUIPMENT

A heat pump transfers thermal energy from one location, typically called the "source," which is at a lower temperature, to another location called the "sink" or "heat sink," which is at a higher temperature. A heat pump always moves thermal

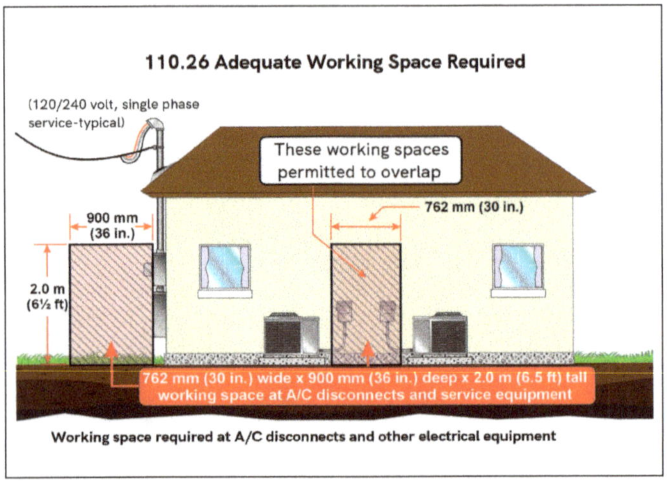

Figure 9.7. Proper and adequate working space is required around A/C equipment as well as service equipment

Photo 9.5. Working space is required for electrical equipment. The larger A/C unit appears to be violating the working space required for the two disconnects.

energy in the opposite direction from temperature, but a heat pump that maintains a thermally conditioned-space can be used to provide either heating or cooling, depending upon whether the environment is cooler or warmer than the conditioned-space. In heating mode, the outdoor coil becomes the evaporator, while the indoor coil becomes the condenser, which absorbs the heat from the refrigerant and dissipates it into the air flowing through it. In cooling mode, the outdoor coil is now the condenser. This makes the indoor coil now the evaporator. The indoor coil is now the evaporator in the sense that it is going to be used to absorb the heat from inside the enclosed space. The evaporator absorbs the heat from the inside and takes it to the condenser, where it is rejected into the outside air.

With heat pump split systems, an indoor air-handling unit is typically installed. Refrigerant lines are run from the outdoor unit to the indoor unit, with the indoor unit typically being located in the crawl space, attic, utility closet, or room. The indoor unit includes a fan for circulating air through the dwelling as well as the refrigeration coil. Resistance heating elements are sometimes added to the indoor unit to serve as backup heat in case of compressor failure. The resistance heating elements will provide extra heating capacity and promote faster heating recovery.

Many of the same rules apply to indoor units as apply to electric furnaces. A disconnecting means rated for the load to be served is required within sight of the indoor unit. In some cases, this disconnecting means consists of one or more circuit breakers that are located in the unit and are operational from outside the unit.

The minimum size of the branch circuit to the indoor unit is required to be not less than 125% of the resistance heat and motor load.

ROOM AIR CONDITIONERS

A room air conditioner is considered an alternating-current appliance that is installed in the conditioned room and incorporates a hermetic motor compressor. The room air conditioner may be of the air-cooled window, freestanding console, or in-wall type. The following requirements cover equipment rated not over 250 volts, single-phase, and such equipment may be cord- and plug-connected.

When determining branch-circuit requirements for a room air conditioner, a cord- and attachment-plug-connected unit is considered a single motor unit if its rating is not more than 40 amperes, 250 volts, single-phase. Additionally, the total rated-load current is displayed on the air conditioner's nameplate. The rating of the branch-circuit protective device must not exceed the ampacity of the branch-circuit conductors or the rating of the receptacle, whichever is less.

The total marked rating of a cord-and-attachment-plug-connected room air conditioner shall not exceed 80% of the branch-circuit rating where no other loads are supplied. Conditioned space can be used to provide either heating or cooling, depending on whether the environment is cooler or warmer than the conditioned space (IRC E3702.13).

An attachment plug and properly rated receptacle or cord connector is permitted to serve as the required disconnecting means for a single-phase room air conditioner rated 250 volts or less if the following conditions are met: (a) the manual controls on the room air conditioner are readily accessible and located within 1.8 m (6 ft) from the floor, or (b) an approved manually operable disconnecting means is installed in a readily accessible location that is within sight from the room air conditioner (*NEC* 440.63).

Where a flexible cord is used to supply a room air conditioner, the length of such cord cannot exceed (a) 1.3 m (10 ft) for a nominal 120-volt rating or (b) 1.8 m (6 ft) for a nominal 208- or 240-volt rating (*NEC* 440.64).

Single-phase room air conditioners are required to be provided with either leakage-current detector-interrupter (LCDI), heat-detecting circuit interrupter (HDCI), or arc-fault circuit interruption (AFCI) protection. This protection is required only for cord- and plug-connected units. The protective device must be factory installed and either be an integral part of the attachment plug or be located in the power supply cord within 300 mm (12 in.) of the attachment plug (*NEC* 440.65).

This requirement is intended to protect against hazards caused by damaged or broken cord insulation. Seasonal use of room air conditioners promotes cord damage unless the unit is stored properly.

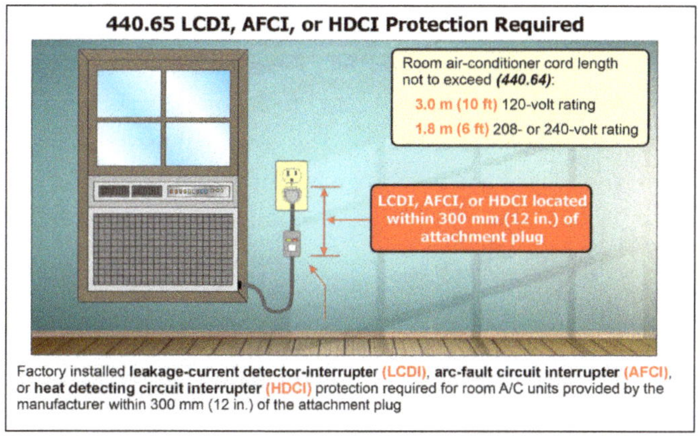

Figure 9.8. Room air conditioner cord length maximums are based on voltage; LCDI or AFCI protection is required for room air conditioners.

Photo 9.6. Typical window room air-conditioner installed in a dwelling unit

SAFETY STANDARD INFORMATION FOR OVERCURRENT PROTECTION REQUIREMENTS FOR AIR-CONDITIONING EQUIPMENT

The *Code* requires that listed or labeled equipment be installed, used, or both in accordance with any instructions included in the listing or labeling [*NEC* 110.3(B) or IRC E3403.3].

Qualified laboratories typically list air-conditioning and heating equipment in accordance with the following standards:

UL 1995, Heating and Cooling Equipment
This standard covers central heating, central air-conditioning, and heat pumps.

UL 484, ROOM AIR CONDITIONERS

These product safety standards detail the necessary safety tests and determine the required nameplate markings and instructions that are included by the manufacturer of the equipment. For example, the *Code* would require that the branch-circuit overcurrent protective device correspond with the type and size specified on the air conditioner nameplate.

For room air conditioners, the nameplate will also be marked to show the type and maximum size of overcurrent protection permitted for the unit. Previous comments relative to the use of fuses or HACR circuit breakers are applicable to room or window units as well as to central air conditioners and heat pumps. Where a room air conditioner is added to an existing dwelling and supplied from an existing panelboard or load center, it is important to verify the branch-circuit overcurrent device for the circuit supplying the air conditioner complies with the type and size overcurrent protection stated on the unit's nameplate.

PRODUCT SAFETY STANDARD REQUIREMENTS

It is recommended to consult the electrical product safety guide card information and product safety standards for further guidance on the proper installation, operation, and use of electrical equipment covered in this chapter. The four-letter code in parentheses indicates the product category in the Underwriters Laboratories Guide Information for Electrical Equipment Directory (UL Product Spec material).

- Electrical Equipment for Use in Ordinary Locations (AALZ)
- Heating and Cooling Equipment: Air Conditioners, Central Cooling (LZFE)
- Air Conditioners, Packaged Terminal (ACKZ)
- Air Conditioners, Room (ACOT)
- Heating and Cooling Equipment: Heat Pumps (LZFE)

REFERENCES

1. *National Electrical Code* and *NEC* are registered trademarks of the National Fire Protection Association, Inc., Quincy, MA 02169. This reprinted material is not the official position of the National Fire Protection Association, which is represented only by the standard in its entirety.

2. *International Residential Code for One- and Two-Family Dwellings* (2024). Copyright © International Code Council, Inc., June 2023.

3. UL, Northbrook, IL, *UL Guide Information for Electrical Equipment Directory* (UL Product iQ® material) unless indicated otherwise.

REVIEW QUESTIONS

1. Air-conditioning equipment disconnecting means must be in a readily accessible location and never more than what distance away from the air conditioner?

A. 15 m (50 ft)
B. 18 m (60 ft)
C. 23 m (75 ft)
D. 30 m (100 ft)

2. A 240-volt single-phase room air conditioner may be disconnected from its source of supply by an attachment plug and receptacle if its manual controls are readily accessible and not more than what distance from the floor?

A. 900 mm (3 ft)
B. 1.2 m (4 ft)
C. 1.7 m (5½ ft)
D. 1.8 m (6 ft)

3. For a single air conditioner motor compressor, the branch-circuit conductor ampacity must be calculated at ___ % of either the motor-compressor rated-load current or the branch-circuit selection current, whichever is greater.

A. 110
B. 125
C. 150
D. 175

4. An air conditioner disconnecting means is permitted to be installed on a removable access panel of an air-conditioning unit.

True _____ False _____

5. In general, an additional disconnecting means is not required for a single-phase, 240-volt, cord- and plug-connected room air-conditioning unit.

True _____ False _____

6. The required ampacity of branch-circuit conductors and the rating of electrical equipment for a hermetic refrigerant motor compressor is based upon the markings found on the unit nameplate, giving the rated load current.

True _____ False _____

7. The AC unit disconnecting means shall not be located on the unit itself, where the disconnecting means will obscure the equipment nameplate(s).

True _____ False _____

8. The total marked rating of a cord- and attachment plug-connected room air conditioner shall not exceed _____ of the rating of a branch circuit where lighting outlets or other appliances are also supplied.

A. 20%
B. 50%
C. 80%
D. 100%

9. A 125-volt, 15- and 20-ampere receptacle outlet for servicing air-conditioning and refrigeration equipment is required to be located not less than __________ from, and on the same level as, the equipment itself.

A. 1.8 m (6 ft)
B. 3.0 m (10 ft)
C. 7.5 m (25 ft)
D. 15 m (50 ft)

10. The motor-compressor branch-circuit short-circuit and ground-fault protective device shall be capable of carrying the starting current of the motor. The maximum rating of this device is _____ percent of the rated load current of the hermetic refrigerant compressor motor. If this value is not sufficient to allow the motor to start, the rating or setting can be increased to allow the motor to start but cannot exceed _____ percent of the rated-load current.

A. 125 / 175
B. 150 / 225
C. 175 / 225
D. 175 / 300

11. Where a flexible cord is used to supply a 120-volt-rated room air conditioner, the length of such cord shall not exceed __________.

A. 900 mm (3 ft)
B. 1.2 m (4 ft)
C. 1.8 m (6 ft)
D. 3.0 m (10 ft)

12. Single-phase cord- and plug connected room air conditioners shall be provided with which of the following factory-installed devices as an integral part of the attachment plug?

A. Leakage-current detector-interrupter (LCDI)
B. Arc-fault circuit interrupter (AFCI)
C. Heat-detecting circuit interrupter (HDCI)
D. Any of the above

13. Where an air-conditioning or heat pump compressor unit consists of a hermetic refrigerant motor-compressor(s) in combination with other loads (such as a fan motor), the ampere rating of the disconnecting means must also be at least _____ of the sum of all currents at rated-load condition.

A. 100%
B. 115%
C. 125%
D. 175%

10 Personnel Protection

The Code addresses various provisions related to protecting personnel at one- and two-family homes. These protection devices include ground-fault circuit-interrupter (GFCI) and arc-fault circuit-interrupter (AFCI) devices. Additionally, other equipment such as smoke alarms and burglar alarm systems, as regulated by building codes, also provide safety for occupants of dwelling units.

This chapter will focus on these personnel protection devices that, when performing their jobs, provide a certain level of security and safety that we often take for granted.

DEFINITIONS

The following definitions are provided in *NEC* Article 100 and IRC Chapter 35. If a review is needed, please refer back to Article 100 in the *NEC*.

- ***Ground-Fault Circuit Interrupter (GFCI)***
- ***Arc-Fault Circuit Interrupter (AFCI)***
- ***Multiple-Station Smoke Alarm***
- ***Single-Station Smoke Alarm***

GROUND-FAULT CIRCUIT INTERRUPTER (GFCI) PRINCIPLES OF GFCI OPERATION

The primary function of a GFCI device is to protect persons from hazards relating to an electrical shock hazard. The GFCI sensing system continuously monitors the current balance in the ungrounded (hot) conductor and the grounded (neutral) conductor. If the current in the grounded conductor becomes less than the current in the ungrounded conductor, a ground fault or loss of current in the field would exist. With this fault, a portion of the current returns to the supply source by some path other than the grounded conductor. With a current imbalance as low as 4 to 6 mA, the Class A GFCI will interrupt the circuit, and this will be shown by a trip or off indicator on the GFCI device (see Figure 10.1).

For the purposes of determining if a receptacle outlet is going to need GFCI protection and a measurable distance is involved [such as within 1.8 m (6 ft) of a sink], the distance is required to be measured as the "shortest path" the power supply cord of an appliance connected to the receptacle would follow without piercing a floor, wall, ceiling, or fixed barrier. At one point, the *Code* would not allow this GFCI measurement to pass through a door, doorway, or window. The prohibition of passing through a door or doorway was removed in the 2020 *NEC* revision process primarily to ensure GFCI protection for a receptacle outlet located in a cabinet below a kitchen sink. The prohibition of a measurement passing through a window was removed during the 2023 *NEC* revision cycle. Removing the reference to a window for measurements ensures receptacle outlets within the measured distance of 1.8 m (6 ft) will have GFCI protection even if the measurement has to be taken through an interior pass-through, such as at a bar area (which could be perceived as a window) (see the parent text at *NEC* 210.8 or IRC E3902.9).

GROUND-FAULT CIRCUIT-INTERRUPTER PROTECTION, WHERE REQUIRED

A ground-fault circuit interrupter (GFCI) is a device that carries current without using electric energy. A Class A GFCI is intended to minimize electric shock hazards for the protection of people that will open a circuit with a loss of current within a current range of 4 to 6 milliamperes (mA) as required by UL 943, Standard for Ground-Fault Circuit Interrupters. It will function to de-energize a circuit or portion of a circuit within an established period of time where a ground-fault current exceeds some predetermined value that is less than that required to operate the overcurrent protective device protecting the supply circuit (see photo 10.1).

For the 2023 NEC, GFCI protection for personnel at the dwelling unit was expanded to include all 125-volt through 250-volt receptacles supplied by single-phase branch circuits rated 150 volts or less to ground in the specific locations specified at NEC 210.8(A) or IRC Section E3902. Prior to this 2020 NEC revision, GFCI protection at dwelling units was limited to 125-volt, single-phase, 15- and 20-ampere receptacles. The necessity for GFCI protection for areas such as kitchens and laundry areas has been proven for these receptacles over several Code cycles. GFCI expansion to 250-volt-rated receptacles was warranted, as they present similar shock hazards. The substantiation submitted for this change demonstrated the need for GFCI protection on receptacles rated greater than 125 volts. Including these higher-rated receptacles for GFCI protection at dwelling

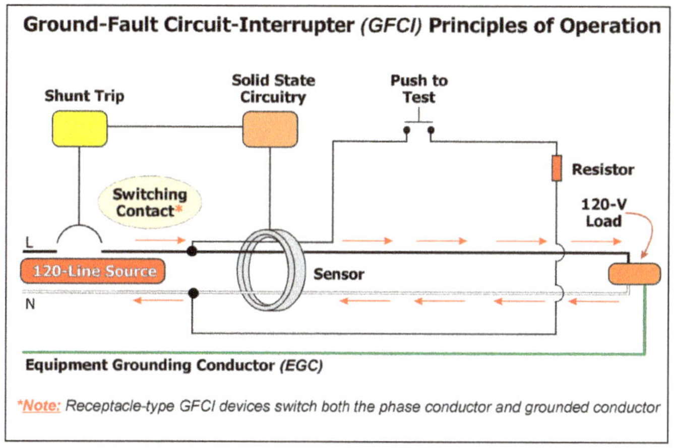

Figure 10.1. GFCI principles of operation

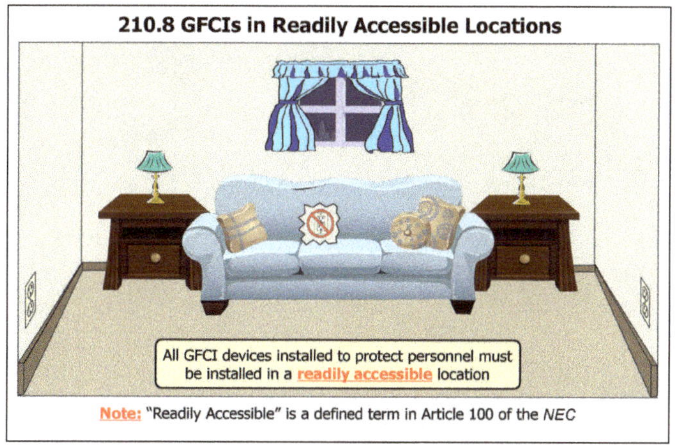

Figure 10.2. GFCI devices are required to be installed in a readily accessible location

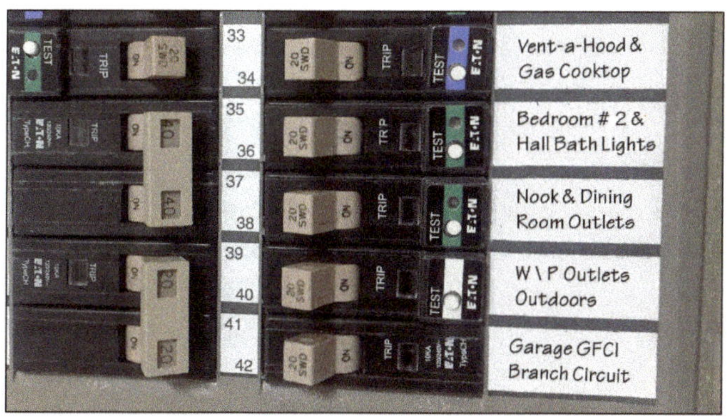

Photo 10.1. The GFCI circuit breaker type installed in the panelboard protects the entire branch circuit

units is compatible with the GFCI protection provisions that occurred for other than dwelling units at 210.8(B) during the 2017 NEC revision cycle.

While many features of protective devices primarily protect the electrical system from overcurrent, ground faults, and short circuits, GFCIs are primarily installed to protect people from shock hazards. Where GFCI protection is required at dwelling units, the *Code* is not specific as to the type of GFCI protec-

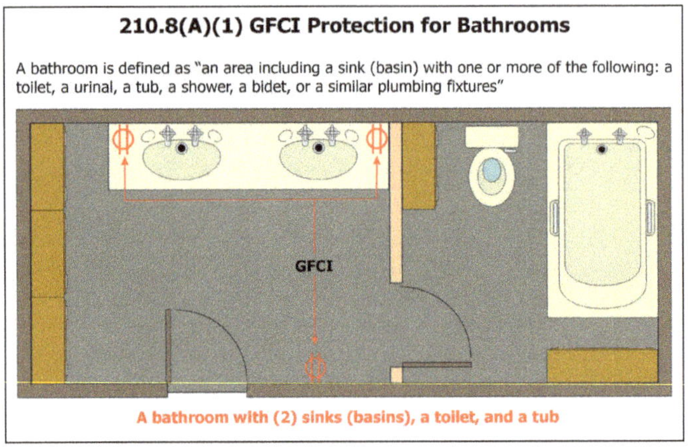

Figure 10.3. GFCI protection is required for all 125-volt through 250-volt receptacles installed in dwelling unit bathrooms

Photo 10.2. Bathroom receptacles are required to be GFCI-protected

tion being delivered. This can be achieved by either a GFCI device (receptacle or blank-face) or by GFCI overcurrent devices. This is solely a design or economic decision. All GFCI devices required by *NEC* 210.8 or IRC E3902 are required to be installed in a readily accessible location.

This requirement would prevent installing a GFCI device under a hydromassage bathtub skirt if access requires tools, for example. Listed weight-supporting ceiling receptacles (WSCR) used with compatible weight-supporting attachment fittings (WSAF) to support a ceiling luminaire or ceiling-suspended fan are not required to have GFCI protection in the areas described at NEC 210.8(A) or IRC E3902, where a ceiling-suspended fan may be installed (such as in bathrooms, basements, kitchens, etc.) [see NEC 210.8(A), Ex. No. 3].

GFCI protection is required for all 125-volt through 250-volt receptacles supplied by single-phase branch circuits rated at 150 volts or less to ground installed in dwelling units at the following locations.

GFCI-BATHROOMS

All 125-volt through 250-volt receptacles supplied by single-phase branch circuits rated 150 volts or less to ground, installed in dwelling unit bathrooms, must be protected by a ground-fault circuit interrupter (GFCI) (see Photo 10.2). An exception applies to bathroom exhaust fans. Typically, a residential exhaust fan has a factory-installed receptacle inside the housing that is an integral part of the unit. These factory-installed receptacles are usually not easily accessible and are mounted inside the bathroom exhaust fan assembly. Starting with the 2023 NEC (2021 IRC), these internal exhaust fan receptacles do not need to be GFCI protected unless required by the manufacturer's instructions or listing of the exhaust fan [see NEC 210.8(A), Ex. No. 4 or IRC E3902.1, Exception].

The bathroom GFCI requirements apply to all other receptacles of this voltage when they are located in a bathroom. This includes receptacles installed in luminaires or intended to supply appliances in the bathroom. Laundry equipment that happens to be located in a bathroom must be GFCI-protected if supplied by a 125-volt through 250-volt receptacle outlet [NEC 210.8(A)(1) or IRC E3902.1].

According to NEC Article 100 or IRC Chapter 35, a bathroom is defined as an area that contains a sink (basin) and one or more of the following: a toilet, a urinal, a tub, a shower, a bidet, or a similar plumbing fixture. A key word in this definition is "area," which clarifies where GFCI protection is needed and indicates that this protection is not limited to just one room within the bathroom area (refer to the NEC or IRC for the bathroom definition).

GFCI-GARAGES AND ACCESSORY BUILDINGS

All 125-volt through 250-volt receptacles supplied by single-phase branch circuits rated 150 volts or less to ground installed in attached or detached dwelling unit garages must be protected by a ground-fault circuit interrupter. Earlier editions of the *Code* included a few exceptions for GFCI protection at select dwelling unit garage receptacles. These exceptions were removed in the 2008 *NEC* and the 2009 IRC. These previous exceptions, one of which permitted non-GFCI-protected receptacles where they were not readily accessible (such as those used for a garage door opener) as an alternative to GFCI protection, were removed due to the inconsistency of what is considered "readily accessible" depending on one's height and the fact that there is no substantiated evidence of a conflict between garage door openers and GFCI protection.

The other previous exception, which permitted a single non-GFCI receptacle that was used for an appliance located in a dedicated space (such as a single receptacle used for the supply to a freezer) or a non-GFCI duplex receptacle that supplied two appliances that occupied a dedicated space (such as a duplex receptacle used for the supply to a freezer and a refrigerator), was removed as well (see figure 10.4). These previous exceptions were removed to make certain that all 125-volt through 250-volt receptacles supplied by single-phase branch circuits rated 150 volts or less to ground installed in garages are provided with GFCI protection for any user of appliances or other equipment located in garages, regardless of the receptacle location. The tolerated appliance leakage currents permitted by today's appliance product standards are far less than the operational threshold of a GFCI. If a GFCI trips while supplying a modern appliance such as a refrigerator while utilizing today's technology regarding GFCIs, this is typically an indication of a voltage leak out onto the frame of that appliance in excess of 4–6 milliampere.

With the increase in voltage rating from 125-volt only to 125-volt through 250-volt receptacle outlets requiring GFCI protection, the GFCI requirements will now include receptacle outlets that were previously not required to have GFCI protection. This would include receptacle outlets, such as a 240-volt, 30-ampere dryer receptacle, installed in a dwelling unit's garage. Generally, any receptacle rated up to 250 volts supplied by single-phase branch circuits rated 150 volts or less to the ground and installed in a dwelling unit kitchen, bathroom, laundry area, garage, or any other dwelling unit location addressed at 210.8(A)(1) through (A)(12) will now require GFCI protection for personnel.

The requirement for GFCI protection for garages also applies to accessory buildings that have a floor at or below grade level. Such buildings have areas that are not typically used as habitable rooms. Receptacles in storage areas, work areas, and areas of similar use within such accessory structures require GFCI protection the same as their counterparts in garages [*NEC* 210.8(A)(2) or IRC E3902.2]

GFCI-OUTDOOR RECEPTACLE OUTLETS AND OUTDOOR OUTLETS

At a dwelling unit, the *Code* requires all 125-volt through 250-volt receptacles supplied by single-phase branch circuits rated 150 volts or less to ground installed outdoors to be protected by a ground-fault circuit interrupter, with one exception. This GFCI requirement applies regardless of location or elevation on the building, including those installed on the roof or under a roof overhang.

An exception excludes GFCI requirements for those receptacles that are not readily accessible and are installed on a dedicated branch circuit for electric snow-melting or deicing

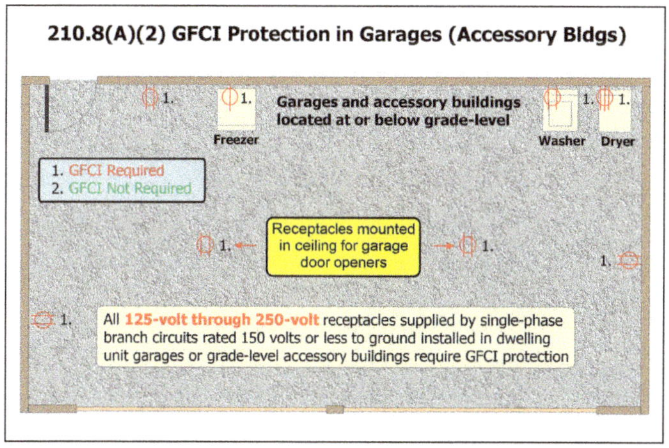

Figure 10.4. Garage and accessory building receptacle GFCI requirements

Photo 10.3. GFCI protection is required for all 125-volt through 250-volt receptacles installed in dwelling unit garages

equipment in compliance with *NEC* 426.28. Ground-fault protection of equipment for fixed outdoor electric deicing and snow-melting equipment, except for equipment that employs mineral-insulated, metal-sheathed cable embedded in a noncombustible medium. Ground-fault protection (GFP) of equipment cannot be used as a substitute for required GFCI protection. GFP protection does not provide the same level of Class A (4 to 6 mA) personnel protection against electrocution [*NEC* 210.8(A)(3) and 210.8(A), Ex. No. 1 or IRC E3902.3, E4101.7].

For the 2020 *NEC*, a new requirement was added at *NEC* 210.8(F) calling for GFCI protection for outdoor outlets *(not just receptacle outlets)* that are supplied by single-phase branch circuits rated 150 volts to ground or less, 50 amperes or less. Receptacles that are not readily accessible and are supplied by a branch circuit dedicated to electric snow-melting, deicing, or pipeline and vessel heating equipment are exempt from this GFCI protection. Outdoor lighting outlets are also exempt.

The most dramatic effect this 2020 *NEC* requirement had was requiring GFCI protection for dwelling units, outdoor-in-

Figure 10.5. GFCI requirements for outdoor receptacles

Photo 10.4. Close-up of an outdoor weather-resistant, tamper-resistant, GFCI-protected receptacle outlet

stalled heat pumps, and air-conditioning units. With this requirement applying to "all outdoor outlets," this would include outdoor hard-wired AC units. This new section requiring GFCI protection on outdoor outlets for dwellings is related to the submitted substantiation detailing a couple of facilities associated with outdoor outlet-connected equipment, such as an outdoor HVAC condensing unit. A Tentative Interim Amendment (TIA) was issued for 210.8(F) under the 2020 *NEC* exempting listed HVAC equipment. This TIA had a future effective date of September 1, 2026 (see TIA 1593). TIAs are amendments to an NFPA standard such as the *NEC*. TIAs are enforceable as part of the *Code* but have not been published in a First Draft Report and Second Draft Report for review and comment. An issued TIA automatically becomes a public input for the next edition of the *Code* and is then subject to all of the standards development process procedures.

Changes occurred at 210.8(F) for the 2023 *NEC* as the text was added to make it clear that outdoor outlets that are supplied by single-phase branch circuits rated 150 volts to ground or less, 50 amperes or less at garages that have floors located at or below grade level, accessory buildings, and boathouses require GFCI protection as well. The 2023 *NEC* also requires existing outdoor outlets that are replaced to be supplied with GFCI protection as well.

A new Exception No. 2 was also added to 210.8(F) for the 2023 *NEC*. This exception exempts GFCI protection for listed HVAC equipment. This exception is in effect until September 1, 2026. One would be hard-pressed to find any dwelling unit HVAC equipment installed at dwelling units that is *NOT* listed equipment.

This exception basically exempts all HVAC equipment installed outdoors from GFCI protection until at least September 1, 2026. The *NEC* Correlating Committee formed a Task Group to look at any incompatibility issue involving GFCI protection and the most common AC equipment installed at dwelling units. This Task Group pointed out in TIA 1654 that UL 943 (*Standard for Ground-Fault Circuit-Interrupters*) requires that Class A ground-fault circuit-interrupters are set to trip with a loss of current of 6 to 4 mA. UL 60335-2 (*Standard for Household and Similar Electrical Appliances - Safety - Part 2-40: Particular Requirements for Electrical Heat Pumps, Air Conditioners, and Dehumidifiers*) allows a maximum leakage current value onto the frame of a typical AC unit of 10 mA for these appliances accessible to the general public.

Data was also submitted to the Task Group showing that listed HVAC equipment typically can have a leakage current higher than what would trip a Class A GFCI, but the touch current is well below levels that would injure or harm an individual. The data showed that the number of potential deaths from electrocution involving HVAC equipment may be as high as four per year. However, the number of fatalities (no cooling during a heat wave period) due to nuisance trips associated with GFCI protection of HVAC equipment where no hazard exists may be as high as 750 per year. Hopefully, by the time September 1, 2026, rolls around, all of these incompatibility issues will be resolved.

GFCI-CRAWL SPACE RECEPTACLES AND CRAWL SPACE LIGHTING OUTLETS

All 125-volt through 250-volt receptacles supplied by single-phase branch circuits rated 150 volts or less to ground installed in crawl spaces (where the crawl space is at or below grade level) are required to have GFCI protection for personnel. These receptacles are often installed for sump pumps or for servicing heating or air-conditioning equipment installed in these crawl spaces [*NEC* 210.8(A)(4) or IRC E3902.4]. It should be noted that all sump pumps require GFCI protection (regardless of their location), as stated in *NEC* 422.5(A)(6). This would include cord-and-plug connected as well as hard-wired sump pumps.

Lighting outlets that do not exceed 120 volts installed in crawl spaces located at or below grade level also require GFCI protection. Crawl space lighting outlets, particularly incandescent light bulbs of keyless lampholders, can easily be accidentally broken, resulting in possible electrocution upon unintentional contact with the live, exposed parts of the broken light bulb and the earth. The number of open-bulb keyless or pull-chain lampholders installed in crawl spaces is countless and poses a shock hazard when not GFCI-protected.

GFCI-BASEMENTS

The *Code* requires all 125-volt through 250-volt receptacles supplied by single-phase branch circuits rated 150 volts or less to ground installed in dwelling unit basements to have GFCI protection for personnel, with one exception.

The areas of concern here are unfinished portions or areas of the basement not intended as habitable rooms and finished basements [*NEC* 210.8(A)(5) or IRC E3902.5]. *NEC* Section 210.52(G)(3) and IRC E3901.9 make it clear that at least one 125-volt, single-phase, 15- or 20-ampere receptacle outlet is required to be installed in each unfinished portion of a basement. GFCI protection must be provided for each general-use receptacle in each portion of the unfinished basement. The definition of a "Habitable Room" in *NEC* Article 100 and IRC Chapter 35, along with the local building code, should be consulted for determining whether rooms in basements are permitted to be classified as a habitable room or space.

The *NEC* and IRC define a "Habitable Room" as "a room in a building for living, sleeping, eating, or cooking, but excluding bathrooms, toilet rooms, closets, hallways, storage or utility spaces, and similar areas" [*NEC* Article 100 and IRC Chapter 35]. "Habitable Space" is defined in the IRC as "any space in a building for living, sleeping, eating or cooking. Bathrooms, toilet rooms, closets, halls, storage or utility spaces and similar areas are not considered habitable spaces" [IRC R202 Definitions].

Master controls for permanently installed burglar and fire alarm systems are often located in the unfinished parts of a basement. An exception exists regarding the requirement for GFCI protection of all receptacles in unfinished basement areas and in all locations mentioned at 210.8(A). The language in the 2023 NEC refers not to burglar or fire alarm systems but to "premises security system." GFCI protection of branch circuits for fire alarm systems is prohibited by 760.41(B) and 760.121(B). Currently, the NEC does not prohibit a receptacle for a fire alarm system from having GFCI protection.

The exception in Section 210.8(A)(5) used to prohibit it in

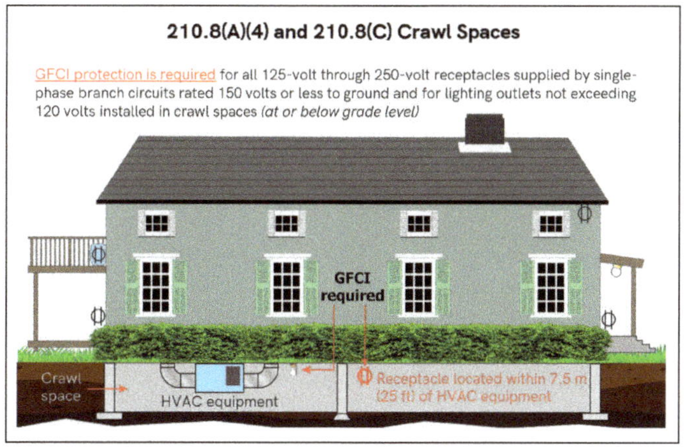

Figure 10.6. GFCI requirements for crawl space receptacles

Photo 10.5. GFCI protection is required for crawl space lighting outlets

the 2020 NEC. The 2023 language no longer bans it because the words "security system" replaced "fire alarm or burglar alarm" in the exception.

Two former exceptions similar to the former exceptions for receptacles located in garages and accessory structures were deleted from the 2008 *NEC* and the 2009 IRC. These were exceptions that applied to receptacles that were not readily accessible and receptacles that supplied appliances that occupied dedicated spaces. Just as in the garage areas, these former exceptions no longer apply to the unfinished basement areas.

For the 2020 *NEC* (and still in effect for the 2023 *NEC*), changes to the *Code* affected the GFCI requirements for receptacles in both an unfinished basement and a finished basement intended as a habitable room or space. The voltage and amperage thresholds that limited GFCI protection at dwelling units to 125-volt, 15- and 20-ampere receptacles were revised to include all 125-volt through 250-volt receptacles supplied by a single-phase branch circuit rated 150 volts or less to ground *[see NEC 210.8(A) and IRC E3902]*. The second change occurred at the GFCI provision of *NEC* 210.8(A)(5) and IRC E3902.5. While this GFCI requirement was historically reserved for "unfinished" basements, changes to the 2020

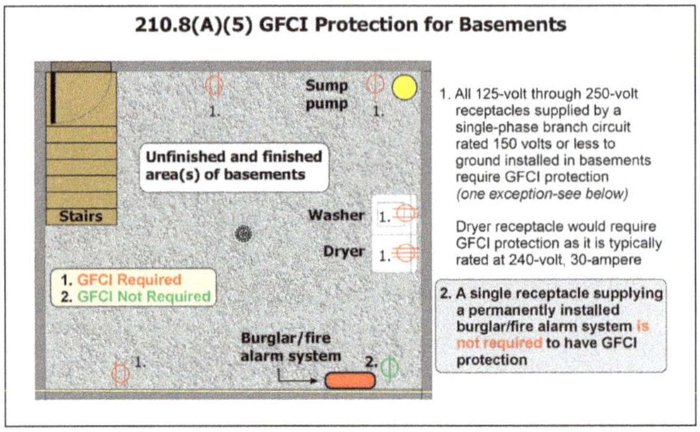

Figure 10.7. GFCI requirements for receptacles in unfinished portions of basements

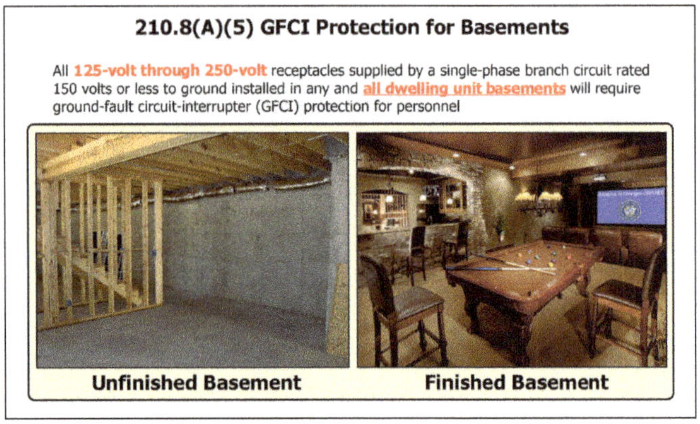

Photo 10.6. GFCI protection is required for both finished and unfinished basements

NEC threw this GFCI provision open to ***ALL*** dwelling unit basements *(not just unfinished basements)*, including basements that are finished out to be a habitable room or space such as a bedroom, exercise room, or game room.

Conductive floor surfaces may exist in finished and unfinished basements, and those basements (whether finished or unfinished) are prone to moisture, including flooding, thus making GFCI protection a requirement for all basements of a dwelling unit.

GFCI-KITCHEN(S)

All 125-volt through 250-volt receptacles supplied by single-phase branch circuits rated 150 volts or less to ground installed in kitchens of dwellings are now required to be provided with ground-fault circuit interrupter (GFCI) protection. In previous editions of the *Code*, GFCI protection was only required for kitchen receptacles that were installed to serve a kitchen countertop surface. In other words, if a 125-volt, 20-ampere receptacle outlet were installed behind a refrigerator (to serve the refrigerator), and the receptacle outlet was not located within 1.8 m (6 ft) of a sink, that receptacle outlet would not have required GFCI protection under previous editions of the *Code*. That is no longer the case, as the requirement that the kitchen receptacle be intended to serve a kitchen countertop or work surface was removed for the 2023 *NEC*.

Under the 2023 *NEC*, all 125-volt through 250-volt receptacles supplied by single-phase branch circuits rated 150 volts or less to ground installed in an area defined as a "*kitchen*" must be GFCI protected. This will make knowledge of the definition of a "*kitchen*" important. According to *NEC* Article 100 and IRC Chapter 35, a "*kitchen*" is defined as "an area with a sink *and* permanent provisions for food preparation *and* cooking."

Changes to the 2014 *NEC* (2015 IRC) at *NEC* 210.8(A)(7) and IRC E3902.7 required all 125-volt, single-phase, 15- and 20-ampere receptacles located within 1.8 m (6 ft) of the top inside edge of any dwelling unit sink (including the kitchen sink) to be GFCI protected. This voltage rating was expanded to all 125-volt through 250-volt receptacle outlets supplied by single-phase branch circuits rated 150 volts or less to ground for the 2020 *NEC* and the 2021 IRC. With changes to subsequent editions of the *Code*, there is no specific distance from the kitchen sink for a receptacle outlet located in a kitchen to be placed in order to not require GFCI protection. For example, if a 125-volt, single-phase, 20-ampere receptacle serves (or does not serve) a kitchen countertop and is located across the room 4.5 m (15 ft) from the kitchen sink, that receptacle would still require GFCI protection.

Furthermore, GFCI protection would also be required for receptacle outlets located within dwelling unit kitchen cabinets, such as for a range hood, microwave oven, or waste disposal if these receptacle outlets are located in an area defined as a kitchen even though they do not serve a kitchen countertop. [*NEC* 210.8(A)(6) or IRC E3902.6; see also figure 10.8].

Attention must also be given to a new 2023 *NEC* provision found at *NEC* 210.8(D). This section calls for GFCI protection for branch circuits or outlets supplying specific appliances rated 150 volts or less to the ground and 60 amperes or less, single- or 3-phase. Included in this list of appliances now demanding GFCI protection (regardless of their location) are drinking water coolers and bottle fill stations, sump pumps, dishwashers, electric ranges, wall-mounted ovens, counter-mounted cooking units, clothes dryers, and microwave ovens. All of these appliances can easily be located in a dwelling unit or kitchen area.

This section uses the terms "branch circuits or outlets," indicating that GFCI protection would be required regardless of whether the appliance in question was cord-and-plug connected or hard-wired. Previous editions of the *Code* might call for GFCI protection on some of these appliances when located within 1.8 m (6 ft) of the top inside edge of any dwelling unit sink (including the kitchen sink). With the changes at *NEC* 210.8(A)(6), IRC E3902.6, and *NEC* 210.8(D), this measurement and distance requirement is no longer a consideration when it comes to GFCI protection.

GFCI AREAS WITH SINKS, FOOD PREPARATION, BEVERAGE PREPARATION, OR COOKING

Beginning with the 2023 edition of the *Code*, all 125-volt through 250-volt receptacles supplied by single-phase branch circuits rated 150 volts or less to ground installed in areas with a sink(s) and permanent provisions for food preparation, beverage preparation, *or* cooking will require GFCI protection. First and foremost, notice that this area is different and does not qualify as a "kitchen," as the definition of a kitchen states that the area in question would have "a sink *and* permanent provisions for food preparation *and* cooking."

Some areas in dwelling units have a sink and perform many of the functions of a kitchen but lack permanent facilities for both food preparation and cooking (as defined in a kitchen). Whether or not they have permanent cooking or food prep facilities, the appliances and equipment used are similar. These spaces typically have countertops and cabinets and may include refrigeration units, dishwashers, toasters, blenders, coffee makers, and ovens, all of which place similar demands on the electrical system. Electrical hazards in these non-kitchen spaces are comparable to those in a kitchen and depend on the activities being performed, moisture presence, and nearby equipment [NEC 210.8(A)(7) or IRC E3902.7].

GFCI-DWELLING UNIT SINKS

All 125-volt through 250-volt receptacles supplied by single-phase branch circuits rated 150 volts or less to ground installed within 1.8 m (6 ft) from the top inside edge of the bowl of a dwelling unit sink are required to have ground-fault circuit-interrupter (GFCI) protection for personnel. This would include any type of sink, including such things as laundry, utility, mud room, kitchen, or wet bar sinks located in a dwelling unit. This requirement applies to a laundry or utility sink regardless of the location, such as in a basement.

If the receptacle outlet supplying the washer, gas dryer, refrigerator, or any other 125-volt through 250-volt receptacle outlet is located within 1.8 m (6 ft) of the laundry or kitchen sink, the receptacle(s) must be GFCI-protected. There are no exceptions to this requirement. When it comes to a laundry area, keep in mind that the 1.8 m (6 ft) distance from the laundry sink is no longer relevant with the addition of *NEC* 210.8(A)(11) or IRC E3902.9 that requires GFCI protection for receptacles in *laundry areas* (regardless of the distance from a sink).

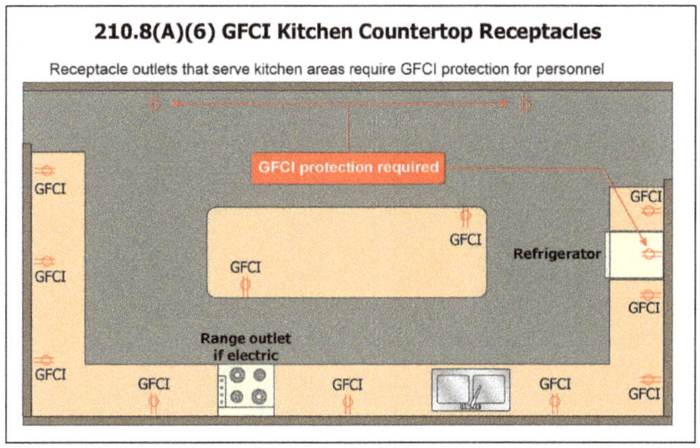

Figure 10.8. GFCI requirements for counter receptacle outlets in kitchens

Photo 10.7. Wet bar sink showing GFCI receptacle within 1.8 m (6 ft) of sink

A receptacle installed in a kitchen requires GFCI protection, regardless of the distance from the sink. The rule in 210.8(A)(8) applies to any other sink in the house, except for laundry rooms, where the entire area must have GFCI protection for all receptacles. It also applies to a receptacle in an adjoining area within 6 feet of the kitchen sink, but not inside the kitchen. Refer to photos 10.7 and 10.8 and figures 10.9 and 10.10 for examples of this requirement.

In previous editions of the Code, this specific GFCI requirement for dwelling unit sinks did not apply to kitchen sinks because the language at NEC 210.8(A)(7) or IRC E3902.7 explicitly stated that the requirement was for sinks "located in areas other than kitchens." This language and the exclusion of kitchen sinks were removed in the 2014 NEC and 2015 IRC. [See NEC 210.8(A)(7) and IRC E3902.7]. Now, all receptacles within a kitchen must be GFCI protected for 120-volt through 250-volt, single-phase branch circuits rated 150 volts or less to ground.

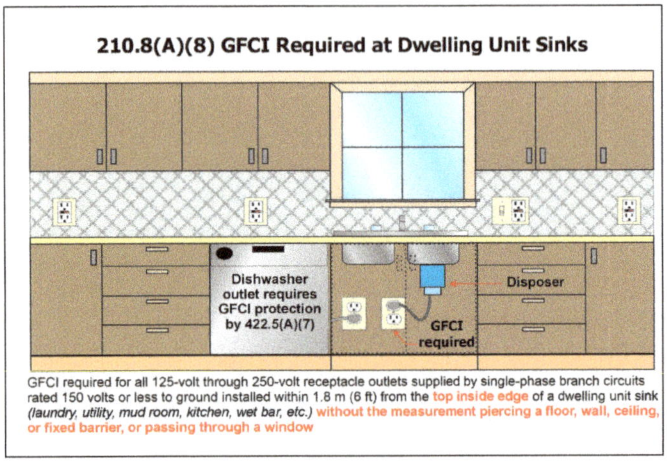

Figure 10.9. Receptacles within 1.8 m (6 ft) of sinks require GFCI protection.

Photo 10.8. GFCI protection is required for all 125-volt through 250-volt receptacles within 1.8 m (6 ft) of the top inside edge of sinks (including kitchen sinks)

Figure 10.10. GFCI requirements at dwelling unit sink, such as a laundry room or mud room sink

GFCI-BATHTUBS OR SHOWER STALLS

Receptacle outlets installed near bathtubs or shower stalls must have GFCI protection for all 125-volt through 250-volt receptacles supplied by single-phase branch circuits rated 150 volts or less to ground. This applies when installed within 1.8 m (6 ft) of the outside edge of a dwelling unit's bathtub or shower stall, even if these fixtures are not in a designated bathroom. Bathtubs or shower stalls do not always meet the definitions of a bathroom in NEC Article 100 or IRC Chapter 35. Without this rule, receptacle outlets in such areas might not need GFCI protection if not classified as bathrooms. An example could be a room or area connected to a bedroom with only a bathtub or shower stall as the plumbing fixture, while a basin sink and toilet are in another common area off the bedroom (see figure 10.11). According to NEC Article 100 or IRC Chapter 35, a bathroom is "an area including a sink (basin) with one or more of the following: a toilet, a urinal, a tub, a shower, a bidet, or similar plumbing fixtures." If there is no sink or basin along with any of these fixtures in one area, it would not qualify as a bathroom and would not be covered by the GFCI requirements of NEC 210.8(A)(1) or IRC E3902.1. [NEC 210.8(A)(10) or IRC E3902.9].

In an area with a bathtub or shower (such as the space on the left in Figure 10.11), it's very common to find a vanity or makeup table. It's also common to find a 125-volt receptacle installed to serve that vanity area. However, this receptacle is not required by the Code. The Code specifies that a receptacle must be installed within 900 mm (3 ft) of the outside edge of a sink, but note that there is no sink in the bathtub or shower stall area shown in Figure 10.11. Before the 2014 NEC (2015 IRC) update requiring GFCI protection in bathtub and shower areas, there was no Code requirement for GFCI protection for a receptacle placed by choice at the vanity in Figure 10.11.

GFCI-LAUNDRY AREAS

GFCI protection is required for all 125-volt through 250-volt receptacles supplied by single-phase branch circuits rated 150 volts or less to ground installed in a laundry room or area. As with previous editions of the *Code*, the presence of a laundry room sink is not the driving factor in whether GFCI protection is required or not. As described in the definition of a "laundry area," these areas typically involve electrical appliances and the presence of water, resulting in an increased risk of electric shock hazards. A "*laundry area*" is defined in *NEC* Article 100 or IRC Chapter 35 as "an area containing or designed to contain a laundry tray, clothes washer, or clothes dryer."

This GFCI requirement for laundry areas addresses the increased shock hazard risk. It is consistent with other *Code* requirements for GFCI protection of receptacles in areas in close proximity to water. With the increase in the voltage rating from 125-volt only to 125-volt through 250-volt receptacle outlets requiring GFCI protection, the GFCI requirements now include receptacle outlets such as a 240-volt, 30-ampere dryer receptacle installed in a laundry or utility room. Any receptacle rated up to 250 volts supplied by single-phase branch circuits rated 150 volts or less to the ground and installed in a dwelling unit laundry area requires GFCI protection for personnel.

As the requirements for GFCI protection have been expanded throughout the NEC over the last 18 Code cycles, the number of electrical shock incidents related to consumer products has continued to decline over that time. Increased usage of GFCI protection for personnel at receptacles of residential homes is a highly effective means of further reducing the potential for electrical shock hazards. [*NEC* 210.8(A)(11) or IRC E3902.10].

210.8(A)(10) GFCI Bathtub or Shower Stalls

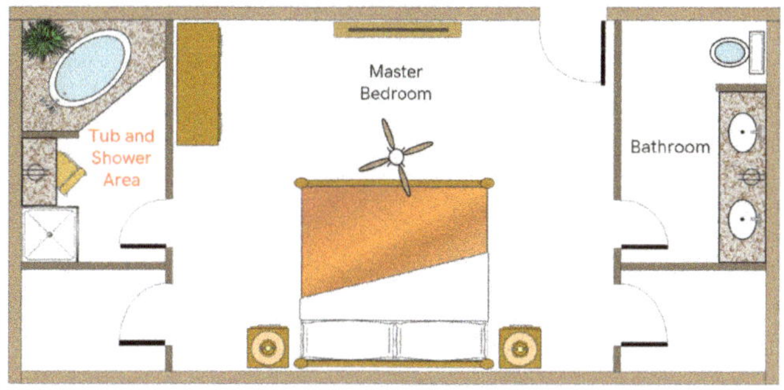

All 125-volt through 250-volt receptacle outlets supplied by single-phase branch circuits rated 150 volts or less to ground installed within 1.8 m (6 ft) of the outside edge of a dwelling unit **bathtub or shower stall** requires GFCI protection

Figure 10.11. GFCI requirements at dwelling unit bathtub or shower stall areas (area on the left side of illustration does not meet the definition of a bathroom)

210.8(A)(11) GFCI Dwelling Laundry Areas

GFCI protection required for all 125-volt through 250-volt receptacle outlets supplied by single-phase branch circuits rated 150 volts or less to ground located in a laundry room areas ***(with or without a sink present)***

Figure 10.12. GFCI requirements at dwelling unit laundry areas

GFCI-INDOOR DAMP AND WET LOCATIONS

Usually, if someone finds a damp or wet spot inside a dwelling, those areas likely need GFCI protection, especially if they are in a bathroom, laundry room, or near a bathtub or shower. Receptacle outlets in these zones are generally within 1.8 m (6 ft) of sinks, which require GFCI protection according to the rules at NEC 210.8(A) or IRC E3902.

What if one were to encounter an area of the dwelling unit that could be considered a damp or wet location, and that location was not within 1.8 m (6 ft) of a sink, bathtub, or shower area? What if this potential damp or wet location was not located in one of the previously GFCI-protected areas, such as a kitchen or laundry area?

To answer these questions, a GFCI requirement for dwelling unit indoor damp and wet locations was added to the 2020 *NEC* and the 2021 IRC that requires GFCI protection for all 125-volt through 250-volt receptacle outlets supplied by a single-phase branch circuit rated 150 volts or less to ground installed in indoor damp or wet locations that GFCI requirements might not cover for laundry areas, areas around sinks, etc. An example of an area that comes to mind that this will affect are areas like a mud room with no sink or a mud room with a sink, but receptacles in that area are located greater than 1.8 m (6 ft) from said sink. These areas are typically used to wash down pets (or sometimes kids) before they are permitted to re-enter the main dwelling unit [*NEC* 210.8(A)(12) or IRC E3902.11].

When considering if an area is a damp or wet location (be it indoors or outdoors), who is responsible for making this interpretation? Of course, the answer to this question is the authority having jurisdiction (AHJ). What is an indoor damp or wet location? Hopefully, the AHJ will depend upon the definitions for a damp, wet, or dry location found in *NEC* Article 100 or IRC Chapter 35 in making the determination as to an area's location being considered a damp, wet, or dry location.

OTHER DWELLING UNIT AREAS UNDER CONSIDERATION FOR GFCI PROTECTION

GFCI-BOATHOUSES AND BOAT HOIST

All 125-volt through 250-volt receptacle outlets supplied by a single-phase branch circuit rated 150 volts or less to ground installed in dwelling unit boathouses used for storage or work areas are required to have ground-fault circuit-interrupter protection (GFCI) for personnel [*NEC* 210.8(A)(9) or IRC E3902.15].

The *Code* also requires GFCI protection for outlets that

Photo 10.9. GFCI protection is required for all 125-volt through 250-volt receptacles in laundry areas, regardless of the presence of a sink.

Photo 10.10. GFCI is required for laundry area receptacles regardless of the presence of a sink

Photo 10.11. GFCI protection is required for all outlets not exceeding 240 volts that supply dwelling unit boat hoist

supply boat hoists, including those associated with dwelling units. This requirement applies to all boat hoist outlets that do not exceed 240 volts branch circuits. Note that the requirement applies to receptacle outlets as well as permanently connected or hard-wired boat hoists, as the term used here is "outlets" (not just receptacle outlets) [*NEC* 555.35(C) or IRC E3902.16].

GFCI-DWELLING UNIT KITCHEN DISHWASHER BRANCH CIRCUIT

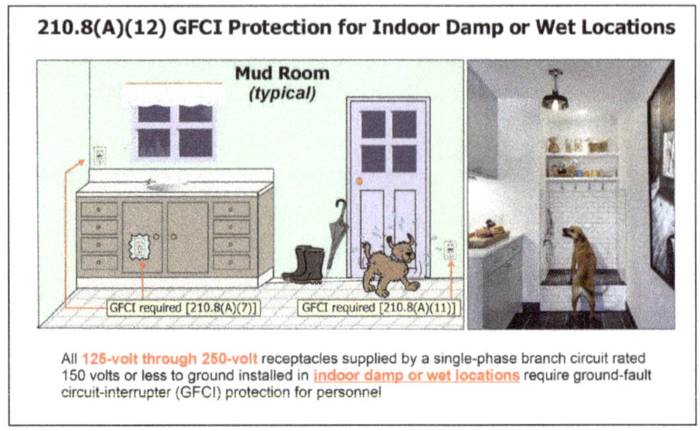

Figure 10.13. GFCI protection is required for dwelling unit indoor damp and wet locations

Class A GFCI protection is required for all outlets that supply dishwashers installed in dwelling units. This would include a receptacle outlet or a hard-wired outlet for a dishwasher. This GFCI requirement applies to appliances supplied by branch circuits rated 150 volts or less to ground and 60 amperes or less, single- or 3-phase. Modern-day electronically controlled dishwashers have different failure modes than their electromechanical ancestors. The electronic components of today's modern residential dishwashers can result in a greater "end of life" arc flash of ground fault, which can result in an increased risk of electrical shock. This can be mitigated by providing GFCI protection for outlets supplying dishwashers. [*NEC* 422.5(A)(7) and *NEC* 422.5(B) or IRC E3902.11]. On a related topic, keep in mind that if a dishwasher is cord-and-plug connected, the receptacle outlet for a built-in residential dishwasher is required to be located in the space adjacent to the space occupied by the dishwasher [*NEC* 422.16(B)(2)(5) and IRC E4101.3 and Table E4101.3]. See Chapter 7 for specific details on the wiring methods for dwelling unit dishwashers.

GFCI-UNDERGROUND-RESIDENTIAL BRANCH CIRCUITS

Sometimes, the Code provides additional benefits to GFCI-protected branch circuits beyond electrical safety. One of those requirements is the burial depth. 120-volt branch circuits installed under residential driveways and lawn areas are typically required to be installed at least 450 mm (18 in.) to 600 mm (24 in.) below grade. However, if the branch circuit for these underground conductors is protected by an overcurrent device rated at not more than 20 amperes and is provided with ground-fault circuit-interrupter (GFCI) protection for personnel, the burial depth may be reduced to 300 mm (12 in.) [see NEC Table 300.5(A), Column 4 or IRC

Table E3803.1, Column 4]. The GFCI protection can be in the form of an overcurrent device or a receptacle device.

GFCI-ELECTRICALLY HEATED FLOORS

Ground-fault circuit-interrupter (GFCI) protection for personnel is required to be provided for cables installed in electrically heated floors of bathrooms, kitchens, and hydromassage bathtub locations.

This applies where heating cables are installed in concrete or poured masonry floors to reduce potential shock hazards to persons, usually with bare feet, in these areas. This requirement applies regardless of the type of floor covering over the concrete or poured masonry, such as wood flooring, ceramic tile, etc. (see Figure 10.15). [*NEC* 424.44(E) or IRC E3902.17]

GFCI-SWIMMING POOLS AND SIMILAR INSTALLATIONS

There are several ground-fault circuit-interrupter (GFCI) requirements in *NEC* Article 680 and Chapter 42 of the IRC pertaining to swimming pools, spas, hot tubs, and the like. These GFCI requirements are discussed in greater detail in Chapter 14.

GFCI-HYDROMASSAGE BATHTUBS

Hydromassage bathtubs, which are very popular in dwelling units, require additional GFCI protection above that required in dwelling unit bathrooms. Additionally, all 125-volt, single-phase receptacles not exceeding 30 amperes installed within 1.83 m (6 ft) of the inside wall of the hydromassage bathtub are required to be GFCI protected.

The GFCI device protecting the hydromassage bathtub must be located on an individual branch circuit and be readily accessible. If the access door is held on with magnets or Velcro, the receptacle under the tub could still meet the definition of "readily accessible." If the access door requires a tool to remove, such as a screwdriver, then the receptacle is not readily accessible. According to the definition of an individual branch circuit, this must be a branch circuit with no other outlets. To comply with both of these requirements, it is common to see a "faceless" GFCI device installed on a wall in the bathroom containing the hydromassage bathtub. "Faceless," in this context, refers to the absence of receptacle outlet openings on the front of the device (see Figure 10.16). This accomplishes the readily accessible provisions for the GFCI devices as well as the individual branch circuit requirement, as no other type of utilization equipment (such as a hair dryer) can be plugged into this branch circuit.

Photo 10.12. GFCI protection is required for residential dishwashers

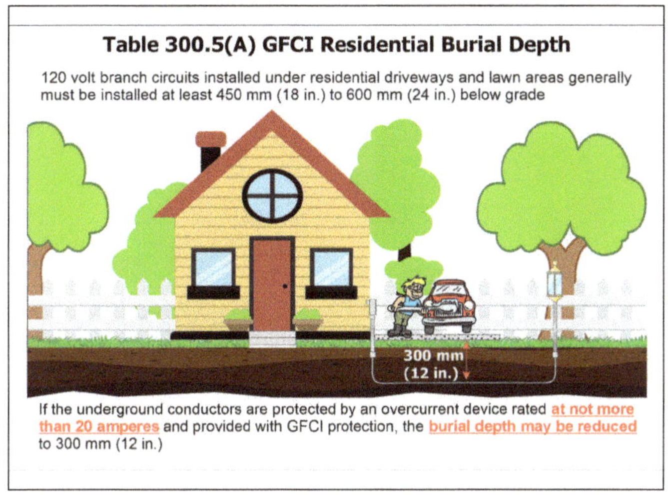

Figure 10.14. Burial depth allowances for underground wiring at dwelling units with GFCI protection

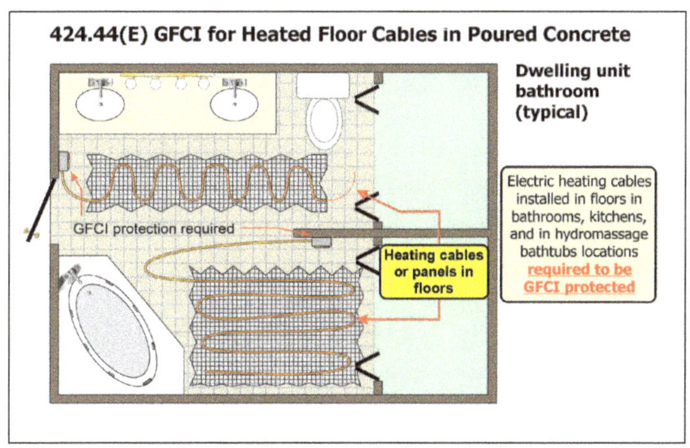

Figure 10.15. GFCI protection is required for heating cables on the floors of bathrooms, kitchens, and hydromassage tub areas.

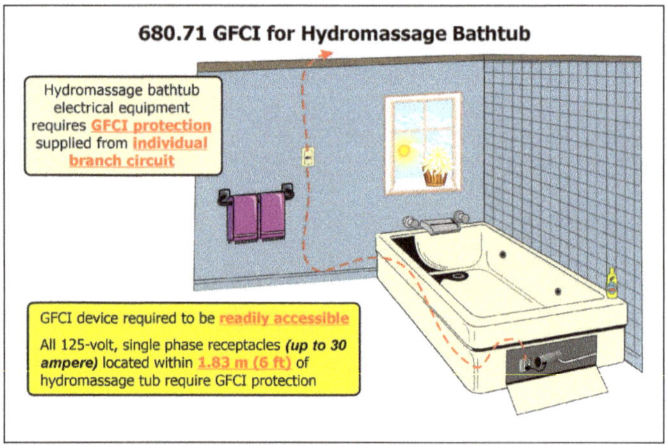

Figure 10.16. GFCI requirements for hydromassage tubs

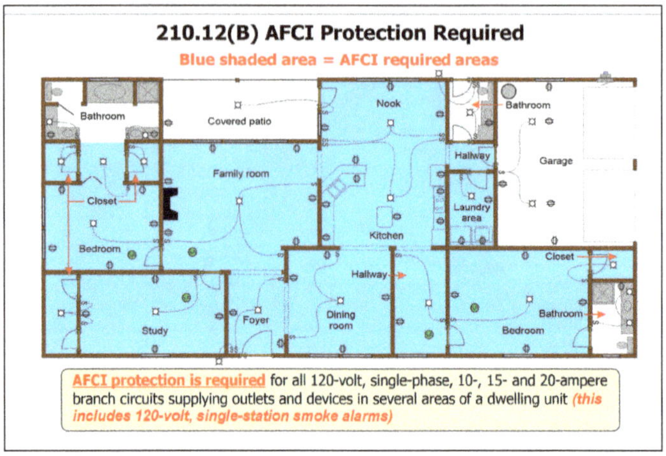

Figure 10.17. Arc-fault circuit-interrupter protection is required for all branch circuits supplying outlets or devices in many areas of the dwelling unit

Luminaires, switches, and other electrical equipment associated with or installed around hydromassage bathtubs are treated the same as regular bathtubs in regard to their installation requirements [See *NEC* 680.71 and *NEC* 680.72 or IRC E4209.2].

GFCI-REPLACEMENT RECEPTACLES

In existing dwellings, GFCI protection is required for replacement receptacles in any location where the receptacle would now require GFCI protection under today's *Code* requirements [*NEC* 406.4(D)(3)]. An example of this would be an older, existing, non-GFCI-protected kitchen countertop receptacle or a bathroom receptacle being replaced. See Chapter 16 for further information on the replacement of existing receptacle outlets.

ARC-FAULT CIRCUIT-INTERRUPTER (AFCI) PROTECTION

An arc-fault circuit interrupter (AFCI) is a device intended to protect against the effects of arcing-type faults. The device recognizes the characteristics unique to arcing and will quickly de-energize the affected circuit when it detects a sustained arcing-type fault (see definition in NEC Article 100 or IRC Chapter 35, along with photos 10.13 and 10.14).

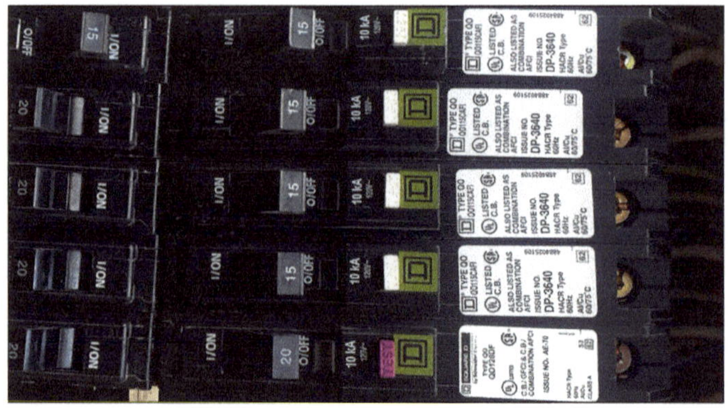

Photo 10.13. Combination-type AFCI circuit breakers protecting the entire branch circuit originating from the panelboard

The arc-fault circuit interrupter (AFCI) does not protect people from electric shock in the manner of ground-fault protection for personnel. It is also not intended to replace the overcurrent protective device located at the circuit's source. An AFCI device will potentially reduce the number of residential fires by mitigating arcing effects in damaged electrical wiring. The damaged conductors may be part of the building's permanent wiring or any electrical cords and equipment connected to the protected circuit.

The *Code* requires the entire branch circuit supplying outlets or devices in certain areas of the dwelling unit to be AFCI-protected with conditions. Specifically, the *Code* requires that "all 120-volt, single-phase, 10-, 15- and 20-ampere branch circuits supplying outlets or devices installed in dwelling unit kitchens, family rooms, dining rooms, living rooms, parlors, libraries, dens, bedrooms, sunrooms, recreation rooms, closets, hallways, laundry areas, or similar rooms or areas to be protected by any of the means described in *NEC* 210.12(A)(1) through (A)(6) [or IRC E3902.21]." Beginning with the 2014 edition of the *NEC* (2015 IRC), the list of areas in a dwelling unit that require AFCI protection includes kitchens and laundry areas. The expansion into the kitchens and laundry areas was another step in the incremental approach for AFCI protection at dwelling units. AFCI protection is required for all 120-volt, single-phase, 10-, 15- and 20-ampere branch circuits supplying not just outlets but devices installed in the list of rooms or areas requiring AFCI protection.

With *outlets* or *devices* included in the requirements for AFCI protection, it is important to understand the definition of these two terms. In *NEC* Article 100 and IRC Chapter 35, the definition of an *outlet* is "a point on the wiring system

Photo 10.14. Anatomy of an AFCI overcurrent device

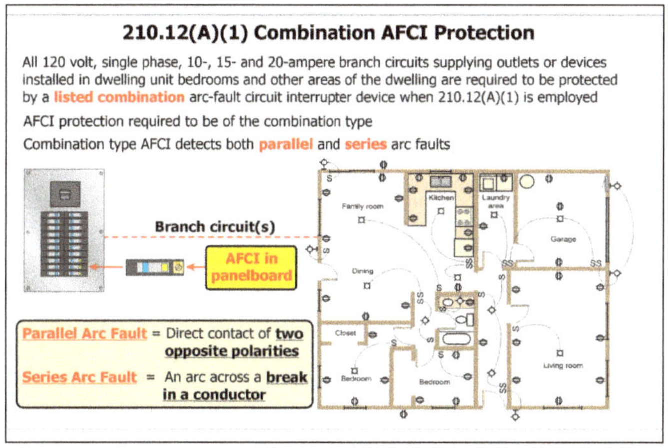

Figure 10.18. Combination AFCI protection protects against both parallel and series arcing events

at which current is taken to supply utilization equipment." This would include receptacle outlets, lighting outlets, an outlet for a single station, a 120-volt smoke alarm, etc. This definition does not include a device such as a switch. At a switch location, the current is not taken to supply utilization equipment; the switch allows current to "pass-through" the switch to supply utilization equipment. A device is defined in *NEC* Article 100 and IRC Chapter 35 as "a unit of an electrical system, other than a conductor, that carries or controls electric energy as its principal function." This definition would describe such things as a switch.

The parent text of NEC 210.12 or IRC E3902.21 requires all AFCI devices to be listed and installed in a readily accessible location. This accessible feature meets the requirement for AFCI devices and will help facilitate the ability to reset the AFCI device if it detects an arcing event and interrupts the current to the arc-fault.

The *Code* describes AFCI protection methods in a list format. Provisions for outlet branch circuit (OBC) AFCI devices, as well as provisions for AFCI overcurrent devices, are permitted with conditions. The *Code* permits six (6) different options on how one can deliver the required AFCI protection at dwelling units (see Figure 10.22).

The ***first AFCI option*** is the most popular method, involving a listed combination-type AFCI overcurrent device at the origin of the branch circuit (at the panelboard) to protect the entire branch circuit. Combination AFCI devices provide protection from both parallel (two opposite polarities) and series (break in a conductor) arcing events. Manufacturers have integrated the technology into inverse-time circuit breakers as well as other types of devices, such as outlet branch circuit (OBC) AFCI receptacle-type devices.

The ***second AFCI option*** involves a listed branch circuit/feeder-type AFCI overcurrent device installed at the origin of the branch circuit (at the panelboard) in combination with a listed outlet branch-circuit (OBC) type AFCI device installed at the first outlet box on the branch circuit being protected. The first outlet box in the branch circuit containing the OBC AFCI device must be marked to indicate that it is the first outlet of the circuit. The branch circuit/feeder-type AFCI devices were pretty much rendered obsolete with the requirement for "combination" AFCI protection introduced in the 2005 *NEC* (2006 IRC) with a future enforcement date of January 1, 2008. Branch circuit/feeder-type AFCI devices provide "series" protection at about 80-85 amperes, whereas a combination AFCI device will provide series protection at around 5 amperes of current.

The ***third AFCI option*** involves a listed "supplemental" AFCI overcurrent device. This method would require a listed supplemental arc protection circuit breaker installed at the origin of the branch circuit (at the panelboard) in combination with a listed outlet branch circuit (OBC) type AFCI device installed at the first outlet box on the branch circuit. The branch circuit wiring must be continuous from the overcurrent device to the (OBC) type AFCI device located at the first outlet box in the branch circuit being protected. This first outlet would need to be marked or identified to indicate that it is the first outlet of the circuit. Using this "supplemental" AFCI option, the maximum length of the branch circuit wiring from the branch circuit overcurrent device to the first outlet cannot exceed 15.2 m (50 ft) for a 14 AWG or 21.3 m (70 ft) for a 12 AWG conductor. A wise recommendation here would be to contact your overcurrent device manufacturer to inquire about the availability of a "supplemental" AFCI overcurrent device.

The ***fourth option*** for delivering AFCI protection involves what is referred to as the listed "system combination" type AFCI protection. The system combination-type AFCI protection would involve a listed outlet branch circuit (OBC) type AFCI device installed at the first outlet on the branch circuit in combination with a standard listed branch circuit

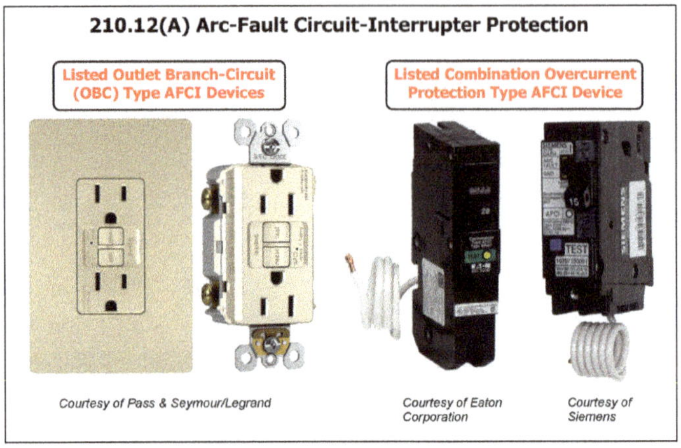

Figure 10.19. AFCI protection can be delivered in the outlet branch circuit (OBC) receptacle type or the circuit breaker type

overcurrent protective device. As with the supplemental AFCI protection, the branch circuit wiring must be continuous from the branch circuit overcurrent device to the (OBC) type AFCI device, and the branch circuit wiring is subject to the same length limitations.

The first outlet box in the branch circuit would again need to be marked to indicate that it is the first outlet of the branch circuit. With this type of system, the manufacturers of the OBC-type AFCI device would need to provide the installer and the inspection community information in their manufacturer's specifications as to the specific brand and type of standard overcurrent protective device that can be used in conjunction with their OBC-type AFCI device that would constitute a "system combination" type AFCI protection technique.

The ***fifth AFCI option*** permits an outlet branch-circuit (OBC) type AFCI device to be installed at the first outlet of the branch circuit. The branch circuit conductors between the branch-circuit overcurrent device and the (OBC) type AFCI device at the first outlet would have to be installed in RMC, IMC, EMT, Type MC, or steel armored cable Type AC, meeting the requirements of NEC 250.118 or IRC E3908.9 using a metal outlet and/or junction boxes. This option was an exception to NEC 210.12(A) and IRC E3902.12 in the 2011 NEC and 2012 IRC, as well as in previous editions of the Code.

The ***sixth option*** allows an outlet branch-circuit (OBC) type AFCI device to be installed at the first outlet to provide protection for the remaining portion of the branch circuit, where the portion of the branch circuit between the branch-circuit overcurrent device and the first outlet is installed with metal or nonmetallic conduit or tubing that is encased in not less than 50 mm (2 in.) of concrete. This option was Exception No. 2 to *NEC* 210.12(A) and IRC E3902.12 in the 2011 *NEC* and 2012 IRC and previous editions of the *Code*.

Exception No. 1 to *NEC* 210.12(B) or IRC E3902.17 permits the omission of AFCI protection for an individual branch circuit supplying a permanently installed burglar/fire alarm system installed in accordance with *NEC* 760.41(B) or 760.121(B), where the individual branch circuit is installed in RMC, IMC, EMT, Type MC, or steel armored cable Type AC, meeting the requirements of *NEC* 250.118 or IRC E3908.9, with metal outlet and junction boxes. This exception does not apply to single- or multiple-station smoke alarms, which require AFCI protection as an outlet located in the areas or rooms mentioned in *NEC* 210.12(B) or IRC E3902.17 of a dwelling.

Single and multiple-station smoke alarms are supplied by a branch circuit covered by the requirements of NEC Article 210 and IRC Chapter 37, not through a non-power-limited fire alarm circuit powered by a fire alarm control panel.

New for the 2023 *NEC*, Exception No. 2 was added that omits AFCI protection for an individual branch circuit supplying an outlet for arc welding equipment at a dwelling unit. This AFCI exception has an expiration date of January 1, 2025. This exception provides a window of opportunity to continue the research currently underway between AFCI manufacturers and manufacturers of arc welding equipment and address any compatibility concerns should any be identified.

It should be noted that several manufacturers have developed and made available a "dual function" circuit breaker that delivers both GFCI protection and AFCI protection on the same device. This can serve as a convenient solution when branch circuits and receptacle outlets require both GFCI and AFCI protection, such as in a kitchen or laundry area.

AFCI PROTECTION FOR EXISTING BRANCH CIRCUIT MODIFICATIONS

NEC 210.12(E) or IRC E3902.22 addresses AFCI protection for branch circuit wiring extensions, modifications, or replacements at existing dwelling units. This AFCI requirement for existing dwelling units was introduced in the 2011 *NEC* and the 2012 IRC. At an existing dwelling unit where branch-circuit wiring is modified, replaced, or extended in any of the rooms or areas specified in *NEC* 210.12(B) or IRC E3902.17 (bedrooms, family rooms, kitchens, etc.), the existing or new branch circuits supplying these rooms or areas must be AFCI-protected.

This provision allows the required AFCI protection to be in the form of a listed outlet branch-circuit (OBC) AFCI device located at the first receptacle outlet of the existing branch circuit to be extended or modified or a listed combination AFCI overcurrent device at the existing service or panelboard equipment. For outlet branch-circuit (OBC) type AFCI devices, the first receptacle outlet is specified as the location in an effort to limit confusion if the first outlet on the existing branch circuit happens to be a switch, luminaire, smoke alarm outlet, or something other than a receptacle outlet.

An exception to AFCI "Branch Circuit Wiring Extensions, Modifications, or Replacements" was added to the 2014 *NEC* and the 2015 IRC to indicate what is considered a "dwelling unit branch circuit extension" and clarify that branch circuit conductors can be extended up to 1.8 m (6 ft.) without AFCI protection where no additional outlets or devices are installed *(other than splicing devices)*. This measurement shall not include the conductors inside an enclosure, cabinet, or junction box.

There are times when a service upgrade occurs at an existing dwelling unit. This action can sometimes involve relocation of a panelboard location, even a short distance, from the original location of the existing panelboard location, such as when the existing panelboard location is in a clothes closet, which is prohibited under current *Code* requirements at *NEC* 240.24(D) or IRC E3705.7(3) *(prohibited locations of overcurrent devices)*. This situation can easily result in the existing branch circuit conductors being too short to reach their new location at the new panelboard. Often, these branch circuits are spliced and extended using the old panelboard cabinet as a junction box with the busbars removed. This is a good example of an extension or modification that is exempted from AFCI protection, with the exception of *NEC* 210.12(E) or IRC E3902.18.

AFCI PROTECTION FOR REPLACEMENT RECEPTACLES AT EXISTING DWELLING UNITS

NEC 406.4(D)(4) requires AFCI protection for replacement receptacle outlets supplied by branch circuits that require AFCI protection elsewhere in the *Code*. The language at *NEC* 406.4(D)(4) allows AFCI protection to be provided to the replacement receptacle by either an AFCI circuit breaker upstream on the supplying branch circuit or by an outlet branch-circuit type AFCI device located at the replaced outlet. Both the receptacle and circuit breaker AFCI will provide protection for all receptacles and other outlets that are downstream on the branch circuit. This AFCI requirement for replacement receptacles is not addressed in the IRC.

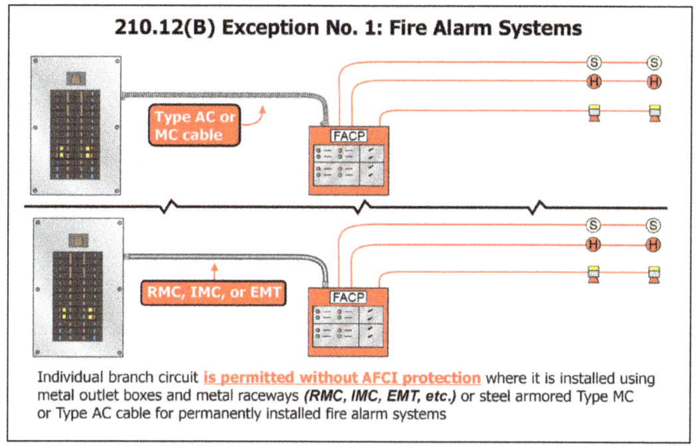

Figure 10.20. Permanently installed burglar/fire alarm system branch circuit exempted from AFCI protection

Photo 10.15. A 125-volt, single-station smoke alarm installed in dwelling unit bedrooms, hallways, living rooms, etc., requires AFCI protection per NEC 210.12(B) or IRC E3902.16

Photo 10.16. The individual branch circuit for a power source of non–power-limited (NPLFA) and power-limited fire alarm (PLFA) circuits are prohibited from AFCI or GFCI protection per NEC 760.41(B) and 760.121(B), respectively

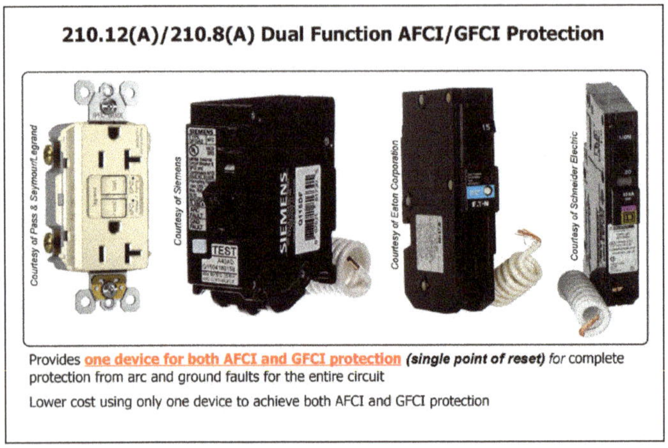

Figure 10.21. Dual-function devices offer both AFCI and GFCI protection in one device

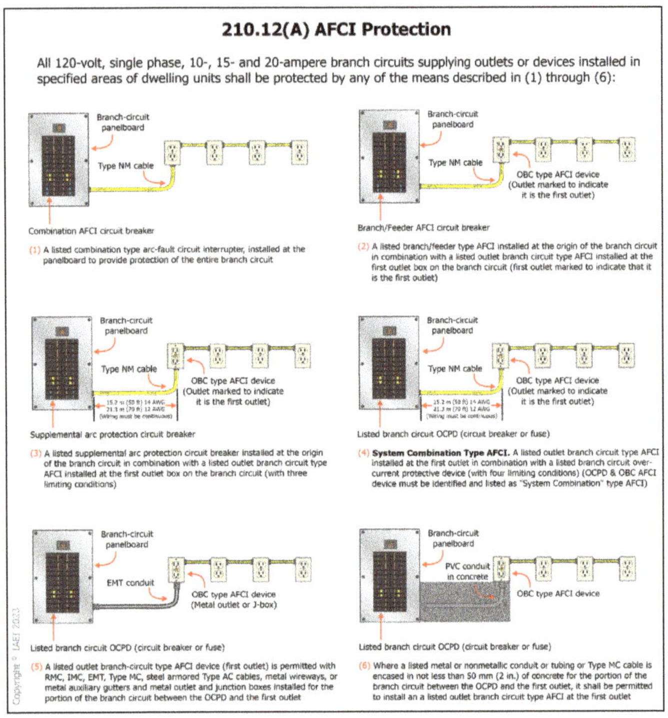

Figure 10.22. Six options are available for providing AFCI protection

TAMPER-RESISTANT RECEPTACLES

Tamper-resistant receptacles are another safety feature that protects personnel in dwelling units. These receptacles have a built-in shutter system that prevents foreign objects like keys, paper clips, and hairpins from touching live electrical components if inserted into the slots, which is especially important as young children, age two and under, may try to insert objects. However, these shutters do not hinder normal cord cap insertion because both the ungrounded and grounded conductor slots must be engaged simultaneously before the shutters open. All 15- and 20-amp, 125- and 250-volt nonlocking receptacles in all areas of dwelling units, including boathouses, mobile homes, manufactured homes, and

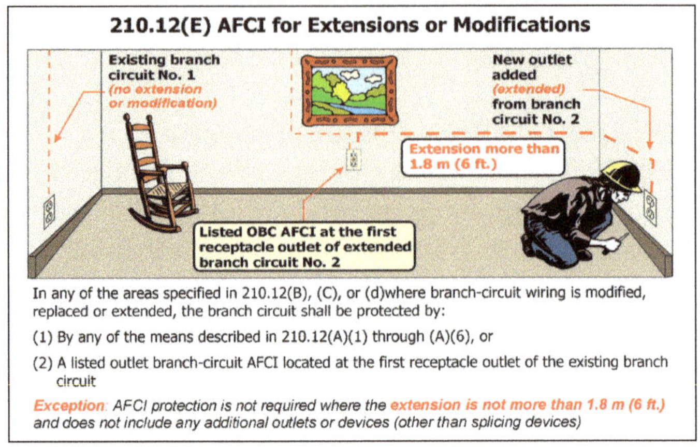

Figure 10.23. AFCI requirements for extended or modified branch circuits

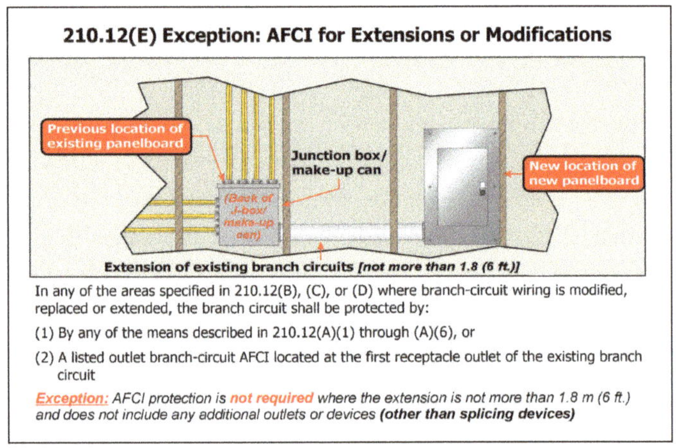

Figure 10.24. Existing branch circuit conductors can be extended up to 1.8 m (6 ft.) without AFCI protection where no additional outlets or devices are installed

their attached and detached garages, accessory buildings, and common areas, are now required to be listed tamper-resistant receptacles. This is a change for the 2023 NEC (2024 IRC).

As in previous editions, only those receptacles referenced in NEC 210.52 and 550.13 or IRC E3901 needed to be tamper-resistant. Since these references cover most rooms and areas in dwelling units, the previous requirement applied to most receptacles in these areas. Now, the rule expands to require ALL 15- and 20-amp, 125- and 250-volt nonlocking receptacles in dwelling units to be tamper-resistant. [NEC 406.12 or IRC E4002.14].

An exception to *NEC* 406.12 or IRC E4002.14 omits tamper-resistant receptacles in four specific conditions or locations of the dwelling. One of the areas addressed by this exception deals with receptacles located more than 1.7 m (5½ ft) above the floor. *NEC* 210.52 and 550.13 or IRC E3901 deal with dwelling unit receptacle outlets and, in particular, spacing requirements for these dwelling unit receptacles. Required spacing of dwelling unit receptacles must be installed at a height below 1.7 m (5½ ft.) to be considered as meeting the requirements for wall spacing. Receptacles installed above 1.7 m (5½ ft.) are not typically accessible to small children and are well out of their reach and, therefore, do not require tamper-resistant receptacles.

The second part of the exception exempts receptacles that are part of a luminaire or appliance from tamper-resistant receptacle requirements, as these receptacles are typically not accessible to small children as well.

The third part of the exception addresses a single receptacle or duplex receptacle that is not readily accessible and located within a dedicated space for appliances. This will eliminate the need for tamper-resistant receptacles to be installed behind dishwashers, refrigerators, washing machines, etc. Again, these receptacles are typically not accessible to small children, and the need for tamper-resistant receptacles is rendered moot.

The fourth location identified by the exception is non-grounding receptacles used for replacements of another non-grounding-type receptacle as permitted in NEC 406.4(D)(2)(a), as there are no known 125-volt, 15- or 20-ampere non-grounding receptacles listed as tamper-resistant receptacles. This fourth part of the exception addressing non-grounding receptacles is not part of the IRC, as the IRC does not address non-grounding receptacles.

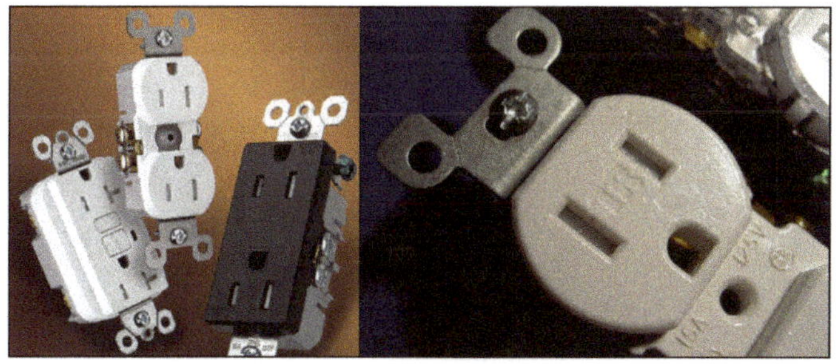

Photo 10.17. Tamper-resistant receptacles are available in a wide variety of models for various applications *(Photos courtesy of Pass & Seymour/Legrand)*

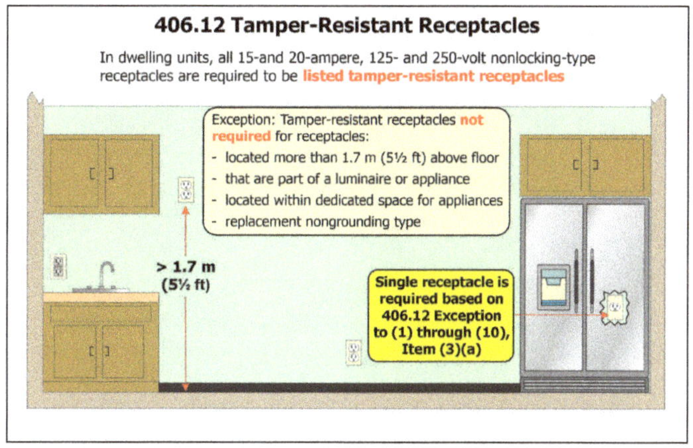

Figure 10.25. Exceptions to tamper-resistant receptacles

SMOKE ALARM OUTLETS AND WIRING

Smoke alarms provide a great deal of protection for the occupants of one- and two-family dwellings. They have become mandatory building code requirements, and the number of lives and properties that have been saved due to their expanded use is evidence of their worth. Unlike an arc-fault circuit interrupter that detects an arcing event before a source of a fire can occur, the primary purpose of the smoke alarm is early warning for the occupant to evacuate the building. Building code(s) typically require smoke alarms (not smoke detectors) in one- and two-family dwellings and these requirements are consistent with the provisions of NFPA 101, Life Safety Load, and NFPA 72, National Fire Alarm *Code*. Most building codes adopt NFPA 72 by reference. Be sure to refer to the local building code official to determine which building code is applicable. The IRC requires smoke alarms to be listed and installed in accordance with the provisions of Section R310 and the household fire warning equipment provisions of NFPA 72. Combination smoke and carbon monoxide alarms are permitted to be used in lieu of smoke alarms.

LOCATION(S)

Smoke alarms are required to be installed in multiple locations of one- and two-family dwelling occupancies. There are three specific locations where the detection/alarm device is required.

A smoke alarm is required to be installed in each sleeping room of the dwelling and outside of each sleeping room (see figure 10.26) and on each additional story of the dwelling, including basements and cellars, but excluding crawl spaces. Providing smoke alarms in these spaces and spaces other than the required space provides additional levels of protection. In split-level types of construction, smoke alarms are required on each level where the levels are closed off by an intervening door between the levels. Where there is no door separating the levels, a smoke alarm installed on the upper level is acceptable for both levels in cases where the lower level is less than one story below the upper level (see Figure 10.27).

WIRING AND INTERCONNECTIONS

Where more than one smoke alarm is installed within an individual dwelling, the smoke alarms are required to be interconnected in a manner that triggers all alarms to announce

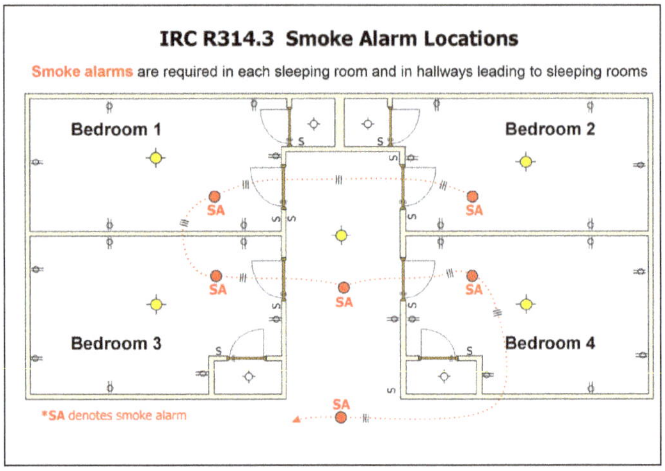

Figure 10.26. Smoke alarms are required in each sleeping room and each hallway leading to sleeping rooms and must be interlocked

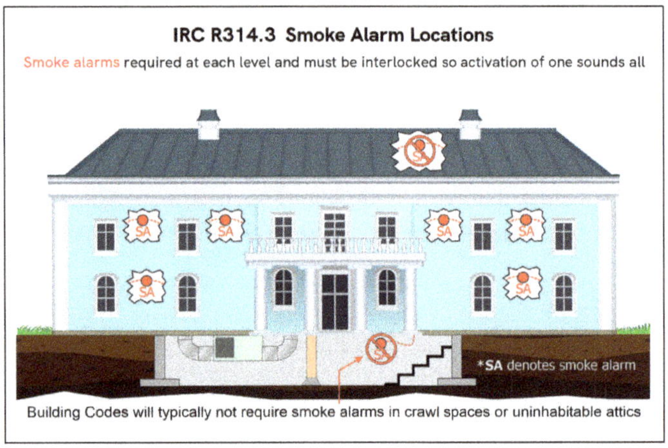

Figure 10.27. Smoke alarms are required at each level and must be interlocked

if one detects smoke (see Figure 10.28). The audibility of the alarm must be adequate to be heard in all bedrooms of the dwelling and must be heard over any background noise with all intervening doors closed. Physical interconnection of smoke alarms is not required where listed wireless alarms are installed and all alarms sound upon activation of one alarm. Physical interconnection of smoke alarms can be omitted where listed wireless alarms are installed, and all alarms sound upon activation of one alarm.

Smoke alarms are listed devices and must be installed in accordance with the manufacturer's installation instructions. Smoke alarms are typically installed in dwellings, both inside and outside each bedroom, and on every level, including basement locations. Sometimes, central fire alarm systems are installed in dwellings, which can then be connected to local fire departments and provide annunciation directly to the fire department in the event of an alarm. This is an added benefit of the central alarm that is monitored.

POWER SOURCE

Smoke alarms are connected to either the local area branch circuit wiring or sometimes to a branch circuit dedicated to just smoke alarms. A dedicated branch circuit for smoke alarms is not a mandatory requirement but an option. The IRC requires that smoke detection devices receive their power from the building wiring system when the building wiring is supplied from a commercial source and requires a battery backup when power is interrupted. Wiring circuitry is to be permanent, and no switch other than an overcurrent protective device is permitted in the circuit. Smoke alarms are permitted to be of the battery-only type where installed in existing buildings that are remodeled or renovated and fall under the exceptions to IRC Section R314.6 or where installed in buildings without commercial power. It is important that the battery be changed regularly to ensure proper protection (see photo 10.16).

PRODUCT SAFETY STANDARD REQUIREMENTS

It is recommended to consult the following electrical product safety guide card information and product safety standards for additional guidance on the proper installation, operation, and use of electrical equipment covered in this chapter. The four-letter code in parentheses indicates the product category in the UL Product iQ® at www.ul.com/PiQ.

- Equipment for Use in Ordinary Locations (AALZ)
- Armored Cable (AWEZ)
- Armored Cable Type AC, Connectors (AWSX)
- Flexible Conduit, Liquidtight (DWWY)
- Flexible Metal Conduit (DXUZ)
- Grounding and Bonding Equipment (KDER)
- Nonmetallic-Sheathed Cable (PWVX)
- Nonmetallic-Sheathed Cable Connectors (PXJV)
- Receptacles for Attachment Plugs and Plugs (RTRT)
- Service Cable (TXKT)
- Underground Feeder and Branch-Circuit Cable (YDUX)

REFERENCES

1. *National Electrical Code* and *NEC* are registered trademarks of the National Fire Protection Association, Inc., Quincy, MA 02169. This reprinted material is not the official position of the National Fire Protection Association, which is represented only by the standard in its entirety.

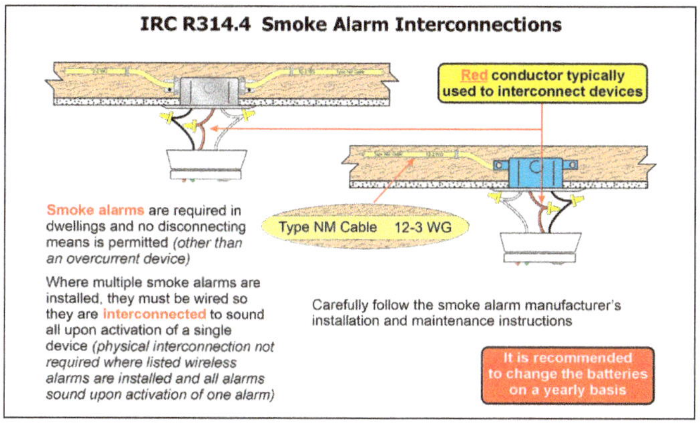

Figure 10.28. Smoke alarms are interlocked so all sound upon activation of one alarm

Photo 10.18. Smoke alarm showing power and interconnection wiring (3-wire connection) and battery backup

2. *International Residential Code for One- and Two-Family Dwellings* (2024). Copyright © International Code Council, Inc., June 2023.

3. UL, Northbrook, IL, *UL Guide Information for Electrical Equipment Directory* (UL Product iQ® material) unless indicated otherwise.

REVIEW QUESTIONS

1. For dwelling units, where ground-fault circuit-interrupter (GFCI) protection for personnel is required, GFCI protection is required for all ______________________ receptacles supplied by single-phase branch circuits rated 150 volts or less.

A. 125-volt, 15- and 20-ampere
B. 240-volt, 30-ampere
C. 125-volt through 250-volt
D. All of the above

2. Which of the following locations do not require ground-fault circuit-interrupter protection for 125-volt through 250-volt grounded receptacle outlets?

A. bathrooms in dwellings
B. kitchens in dwellings
C. laundry area at dwellings (*with no laundry room sink present*)
D. bedrooms [*not within 1.8 m (6 ft) of a sink]*

3. The receptacle outlet for a garbage disposer or refrigeration equipment in a dwelling unit kitchen is permitted to be supplied from an individual branch circuit, but the branch circuit must be rated at 20 amperes or greater.

True _____ False _____

4. Class A ground-fault circuit interrupter devices are set to trip when the current to ground is between ___________ or higher.

A. 3 to 4 mA
B. 4 to 6 mA
C. 6 to 8 mA
D. to 6 amps

5. All GFCI devices required by *NEC* 210.8 or IRC E3902 are required to be installed in a __________ location.

A. accessible
B. readily accessible
C. indoor
D. convenient

6. All 125-volt through 250-volt receptacles installed at which of the following dwelling unit locations must be provided with ground-fault circuit interrupter (GFCI)?

A. 125-volt, 15-ampere receptacles installed in the attic for servicing a gas furnace
B. 240-volt outlet installed in a laundry area for a clothes dryer
C. 125-volt lighting outlet installed outdoors
D. Both A. and B.

7. Which of the following 125-volt through 250-volt receptacles installed in a dwelling unit bathroom must be provided with ground-fault circuit interrupter (GFCI) protection?

A. 125-volt receptacle installed to serve the bathroom vanity/sink area
B. 240-volt receptacle installed to serve laundry equipment located in the bathroom area
C. A factory-installed receptacle that is not readily accessible and mounted internally to a bathroom exhaust fan assembly
D. Both A and B

8. A 125-volt, 15-ampere duplex receptacle installed in a dwelling unit garage ceiling for a garage door opener requires GFCI protection.

True _____ False _____

9. An individual branch circuit to a permanently installed burglar/fire alarm system installed in a dwelling unit bedroom closet where the individual branch circuit is installed in steel armored cable meeting the requirements of *NEC* 250.118, with ______________________ does not require AFCI protection.

A. fiberglass insulation
B. metal boxes and enclosures
C. plastic outlet and junction boxes
D. a sprinkler system

10. All 125-volt through 250-volt receptacles installed outdoors at dwelling units are required to be protected by a ground-fault circuit interrupter except for those outdoor receptacles that are ________________ and are installed on a dedicated branch circuit for electric snow-melting or deicing equipment.

A. located on a balcony
B. not accessible
C. not readily accessible
D. located above direct grade access

11. Listed weight-supporting ceiling receptacles (WSCR) utilized in combination with compatible weight-supporting attachment fittings (WSAF) installed for the purpose of supporting a ceiling luminaire or ceiling-suspended fan are required to be GFCI-protected when installed in a dwelling unit basement

True _____ False _____

12. A single non-GFCI receptacle that is used for and installed behind a freezer located in a dedicated space in a dwelling unit garage is exempted from GFCI protection.

True _____ False _____

13. All 125-volt through 250-volt receptacles installed within ____________ of the outside edge of a dwelling unit sink (including kitchen, laundry, utility, or wet bar sinks) are required to have ground-fault circuit-interrupter protection for personnel.

A. 900 mm (36 in.)
B. 1.2 m (4 ft)
C. 1.8 m (6 ft)
D. GFCI protection is not required in these locations.

14. The measurements from a dwelling unit sink to a potential GFCI-protected receptacle are to be measured from

A. the bottom of the bowl of the sink
B. the top inside edge of the bowl of the sink
C. the top outside edge of the bowl of the sink
D. any of the above

15. All 120-volt, single-phase, 15- and 20-ampere branch circuits supplying outlets or devices installed in which of the following locations require protection by a listed, combination-type, arc-fault circuit interrupter(s) (AFCI)?

A. dining rooms
B. closets
C. bathrooms
D. Both A and B

16. Generally, a smoke alarm is required to be installed in which of the following locations in a dwelling unit?

A. outside of each sleeping room
B. garages
C. sleeping rooms
D. A. and C.

17. All 125-volt through 250-volt receptacle outlets supplied by a single-phase branch circuit rated 150 volts or less to ground installed at dwelling unit indoor damp or wet locations require GFCI protection for personnel.

True _____ False _____

18. Tamper-resistant receptacles are required at dwelling units at which of the following locations?

A. Mobile homes
B. Only in all areas specified in *NEC* 210.52 and 550.13 or IRC E3901.1 (before any exceptions are applied)
C. All 15- and 20-ampere, 125- and 250-volt nonlocking-type receptacles installed at dwelling units
D. Both A and C

19. The GFCI device protecting a hydromassage bathtub is required to be ________________.

A. accessible
B. readily accessible
C. the receptacle-type GFCI device only
D. listed for a wet location

20. Which of the following 15- and 20-ampere, 125- and 250-volt nonlocking-type receptacles installed at a dwelling unit are NOT required to be listed as tamper-resistant type receptacles?

A. A receptacle located more than 1.7 m (5½ ft) above the floor in a living room
B. A receptacle that is part of a luminaire or appliance in a kitchen
C. A receptacle installed 300 mm (12 in.) above the floor in a bedroom
D. Both A. and B.

21. GFCI protection is required for all 125-volt through 250-volt receptacles installed within ____________ of the outside edge of a dwelling unit bathtub or shower stall, even if these bathtub or shower stalls are not located in a defined bathroom.

A. 900 mm (3 ft)
B. 1.5 m (5 ft)
C. 1.8 m (6 ft)
D. 3.0 m (10 ft)

22. When located in a laundry area, GFCI protection is not required for a 250-volt, single-phase, 30-ampere receptacle when located more than 1.8 m (6 ft) from a laundry room sink.

True ______ False ______

23. An existing branch circuit can be modified or extended up to ______________ without AFCI protection where no additional outlets or devices are installed.

A. 600 mm (24 in.)
B. 900 mm (3 ft)
C. 1.2 m (4 ft)
D. 1.8 m (6 ft)

24. GFCI protection is required for all outlets supplied by branch circuits rated 150 volts or less to ground that supply dishwashers installed in dwelling units.

True ______ False ______

25. For the purposes of determining if a receptacle outlet in the vicinity of a sink is going to need GFCI protection, the distance is required to be measured as the shortest path the power supply cord of an appliance connected to the receptacle would follow without piercing a floor, wall, ceiling, or fixed barrier. The power supply cord is permitted to pass through which of the following?

A. door or doorway
B. cabinet door
C. window
D. all of the above

26. Lighting outlets that do not exceed __________volts installed in crawl spaces located at or below grade level require GFCI protection.

A. 30
B. 110
C. 120
D. 40

27. All 125-volt through 250-volt receptacles installed in dwelling unit boathouses require ground-fault circuit-interrupter (GFCI) protection for personnel.

True ______ False ______

28. "A room in a building for living, sleeping, eating, or cooking, but excluding bathrooms, toilet rooms, closets, hallways, storage or utility spaces, and similar areas" defines what specific term in a dwelling unit?

A. Living Room
B. Habitable Room
C. Kitchen
D. Bedroom

29. All 125-volt through 250-volt receptacles installed in the following locations supplied by single-phase branch circuits rated 150 volts or less to ground require ground-fault circuit-interrupter (GFCI) protection for personnel (*with an exception for a receptacle supplying only a permanently installed fire alarm or burglar alarm system*).

A. Basements
B. Unfinished basements
C. Neither A. or B.
D. Both A. and B.

30. All of the 125-volt through 250-volt receptacles installed in the following kitchen locations supplied by single-phase branch circuits rated 150 volts or less to ground require ground-fault circuit-interrupter (GFCI) protection for personnel.

A. Only the receptacles installed to serve a kitchen countertop or work surface
B. All of the receptacles installed in an area defined as a kitchen
C. Only those receptacles installed within 1.8 m (6 ft) of the kitchen sink
D. Only those receptacles installed within 1.8 m (6 ft) of the kitchen sink and intended to serve a kitchen countertop or work surface

31. GFCI protection is required for branch circuits or outlets supplying which of the following appliances rated 150 volts or less to ground and 60 amperes or less, single- or 3-phase.

A. Electric ranges
B. Wall-mounted ovens
C. Microwave ovens
D. All of the above

11 Power and Lighting Distribution

GENERAL REQUIREMENTS

Numerous branch circuits are required to achieve the necessary power and lighting distribution in a typical one- and two-family dwelling. The *Code* provides minimum requirements for the branch circuits that supply receptacles, lighting, and appliance outlets used in these buildings and related structures.

Two of the primary requirements for dwelling unit branch circuits are the minimum ratings and where such circuits are required. This chapter focuses on the distribution requirements for power and lighting branch circuits based on the minimum requirements of the *Code*.

DEFINITIONS

A ***branch circuit*** is defined as "the circuit conductors between the final overcurrent device protecting the circuit and the outlet(s)" (*NEC* Article 100 or IRC Chapter 35).

Branch circuits are rated by the ampere rating or setting of the overcurrent protective device. Typical branch-circuit ratings for dwelling units are 15, 20, 30, 40, and 50 amperes. Provisions to allow a 10-ampere branch circuit were introduced to the *Code* during the 2023 *NEC* (2024 IRC) revision cycle. If the installer or building owner chooses to install these smaller branch circuits at a dwelling unit, it should be noted that these 10-ampere branch circuits have some limitations.

A 10-ampere branch circuit *can* supply loads for lighting outlets, exhaust fans for bathroom and laundry areas within dwelling units, and a gas fireplace unit served by an individual branch circuit. A 10-ampere branch circuit *cannot* supply loads for receptacle outlets, fixed appliances *(except as permitted for individual branch circuits)*, garage door openers, or laundry equipment. The minimum size branch circuit conductors for a 10-ampere branch circuit is 14 AWG copper or 12 AWG aluminum or copper-clad aluminum

Photo 11.1. Branch circuits connected to circuit breakers in a panelboard.

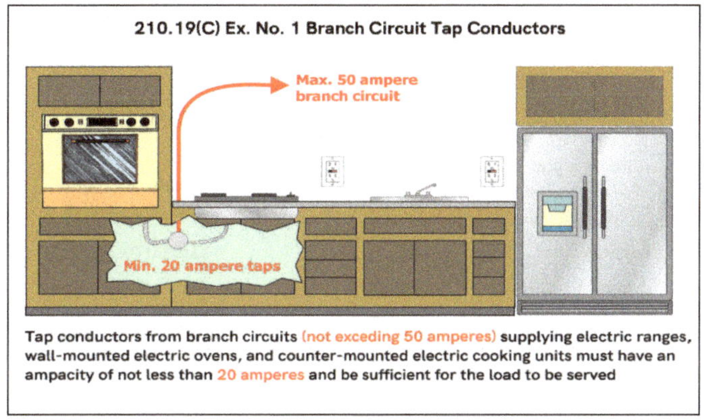

Figure 11.1. Taps permitted to kitchen appliances

conductors as described in *NEC* Table 210.24(1) and *NEC* Table 210.24(2). See *NEC* 210.23(A)(1) and (2) for specific information on permitted 10-ampere branch circuits.

Where conductors of higher ampacity are installed for any reason, such as to compensate for voltage drop, the ampere rating or setting of the specified overcurrent protective device is required to be used to determine the actual rating of the branch circuit (*NEC* 210.18 or IRC E3702.2).

For example, this means a 10 AWG copper Type THW circuit conductor that has an allowable ampacity of 30 amperes is considered a 20-ampere branch circuit where it is protected by a 20-ampere overcurrent device.

BRANCH-CIRCUIT VOLTAGE LIMITATIONS

In dwelling units, voltages are not permitted to exceed 120 volts, nominal, between conductors where they supply the terminals of luminaires or cord- and plug-connected loads operating at 1440 volt-amperes, nominal or less, or less than ¼ horsepower.

Branch circuits energized at up to 120 volts, nominal, between conductors, are permitted to supply the following: terminals of lampholders applied within their voltage ratings, auxiliary equipment of electric-discharge lamps (see *NEC* 410.137 for auxiliary equipment limitations), and cord- and plug-connected or permanently connected utilization equipment (*NEC* 210.6(A) and (B) or IRC E3702.1).

MINIMUM SIZES AND TYPES OF CONDUCTORS

The minimum size of conductors used for branch circuits, whether solid or stranded, is 14 AWG copper or 12 AWG aluminum or copper-clad aluminum, except as permitted elsewhere in the *Code* (*NEC* 310.3(A) or IRC E3406.3). The minimum size of conductors used in Class 1, Class 2, and Class 3 circuits are permitted to be in accordance with *NEC* 724.49(A) and 725.127 and Exception. Similarly, the minimum size of conductors used in fire protective signaling circuits is determined in accordance with *NEC* 760.49(A), 760.127 and Exception, and 760.142. The IRC covers limited energy systems and circuits; however, it is limited to Class 2 remote-control, signaling, and power-limited circuits that are not an integral part of a device or an appliance. Other classes of these types of circuits or systems are required to follow the provisions of *NEC* Article 725 (IRC E4301.1). See Chapter 17 for additional information on limited energy circuits.

The ampacity of branch-circuit conductors cannot be less than the maximum load that will be served. Where a branch circuit supplies continuous loads or any combination of continuous and noncontinuous loads, the minimum branch-circuit conductor size, before application of any adjustment or correction factors, must have an allowable ampacity equal to or greater than the noncontinuous load plus 125% of the continuous load [*NEC* 210.19(A) and 210.20(A)]. Few loads in a dwelling occupancy are considered to be continuous in nature.

In addition, the conductors of a branch circuit supplying more than one receptacle for cord- and plug-connected portable loads must have an ampacity of not less than the rating of the branch circuit [*NEC* 210.21(B)(2)].

BRANCH-CIRCUIT CONDUCTORS SUPPLYING SPECIFIC APPLIANCES

The ampacity of branch-circuit conductors used to supply household ranges, wall-mounted ovens, counter-mounted cooking units, and other household cooking appliances cannot be less than the branch-circuit rating. It must not be less than the maximum load to be served. Where a range of 8¾ kW or more in rating is to be supplied, a branch-circuit rating of not less than 40 amperes is required.

Under certain conditions, branch-circuit tap conductors

Branch Circuits Required

Branch Circuits Required for Dwelling Units		
Type of Circuit	Number of Circuits	Reference
General Lighting	?	220.41 and Calculation*
Small Appliance	2	210.11(C)(1)
Laundry	1	210.11(C)(2)
Bathroom	1	210.11(C)(3)
Garage	1	210.11(C)(4)

* Section 210.11(A) requires the minimum number of branch circuits to be determined from the total computed load and the size and rating of the branch circuits used

Table 11.1. Branch circuits required for dwelling units

Specific Branch Circuits References

References for Branch Circuit Installations	
Branch Circuit Supplying	NEC Reference
Electric Range(s)	210.19(A)(3), 220.55
Electric Dryer(s)	220.14(B), 220.54
Appliances *(General)*	422.10
Central Heating	422.12
Water Heater	422.13
Fixed Electric Space Heating	424.4(A) and (B)
Motors *(Single)*	430.22
Air-Conditioners	440.32, 440.33
Window A/C Units	440.62

Table. 11.2. References for branch-circuit installations

used to supply electric ranges, wall-mounted electric ovens, and counter-mounted electric cooking units are permitted from a 50-ampere branch circuit. These branch-circuit conductors are permitted to have an ampacity of not less than 20 amperes and must be sufficient for the load to be served. These taps cannot be longer than necessary for servicing the appliance [see figure 11.1 and *NEC* 210.19(C) and Ex. No. 1].

The neutral conductor of a 3-wire branch circuit used to supply a household electric range, a wall-mounted oven, or a counter-mounted cooking unit is also permitted to be smaller than the ungrounded (hot) conductors. The maximum demand of the range has been computed in accordance with *NEC* Table 220.55 or IRC Table E3704.2(2). In no case can the ampacity of the neutral conductor be less than 70% of the rating of the branch circuit. The grounded conductor (neutral conductor) cannot be smaller than 10 AWG [*NEC* 210.19(C), Ex. No. 2 and 220.61(B)(1) or IRC E3704.3].

NEC Table 310.16 or IRC Table E3705.1 covers the allowable ampacity ratings of conductors before any correction factors are applied. Ampacity values contained in these tables are not the conductors' true ampacity as defined in the *Code*; rather, the table values indicate the amount of current various size conductors having the insulation type shown are permitted to carry. See Chapter 1 for a more complete discussion of the subject, including details for proper application of ampacity correction factors.

Branch-circuit conductors supplying loads other than cooking appliances are required to have an ampacity sufficient for the loads served and cannot be smaller than 14 AWG copper or 12 AWG aluminum or copper-clad aluminum [*NEC* 210.19(A)(2) and 310.3(A) or IRC E3701.2 and E3406.3].

BRANCH CIRCUITS REQUIRED

The *Code* contains specific requirements for providing branch circuits for general lighting circuits, small appliances, laundry, and bathroom branch circuits. A summary of the branch

Photo 11.2. Residential service panelboard showing branch circuits, including a multiwire branch circuit *(lower right)*

circuits required for dwelling units and the applicable *Code* sections are provided in Table 11.1.

General-purpose receptacles in rooms such as living rooms and bedrooms in dwelling units are permitted to be on the general lighting circuits. The *Code* does not require that receptacles be on different branch circuits from the lighting outlets on these general lighting circuits in these areas. Electrical designers often specify this separation of circuits in commercial and industrial occupancies. See Chapter 2 for the method for determining the minimum number of general-purpose lighting circuits required at a dwelling unit.

Other references for branch-circuit installations depending on specific equipment are included in Table 11.2.

MULTIWIRE BRANCH CIRCUITS

A ***Multiwire Branch Circuit*** is defined as "a branch circuit that consists of two or more ungrounded conductors that have a voltage between them, and a neutral conductor that has equal voltage between it and each ungrounded conductor of the circuit and that is connected to the neutral or grounded conductor of the system" (*NEC* Article 100 or IRC Chapter 35).

The *Code* requires that each multiwire branch circuit be provided with a means to disconnect simultaneously all

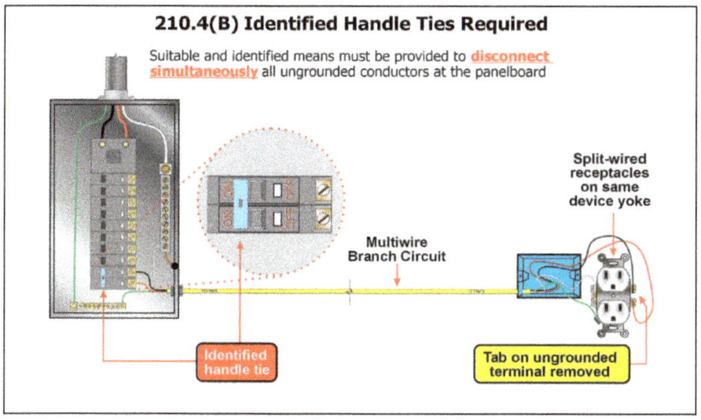

Figure 11.2. Identified handle ties are required for multiwire branch circuits

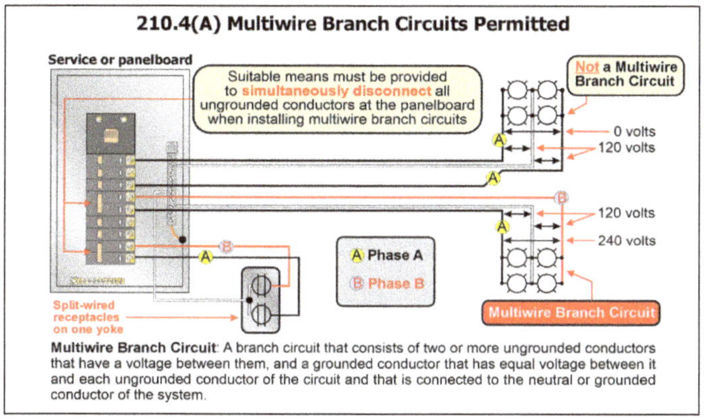

Figure 11.3. Multiwire branch circuits share a common neutral conductor

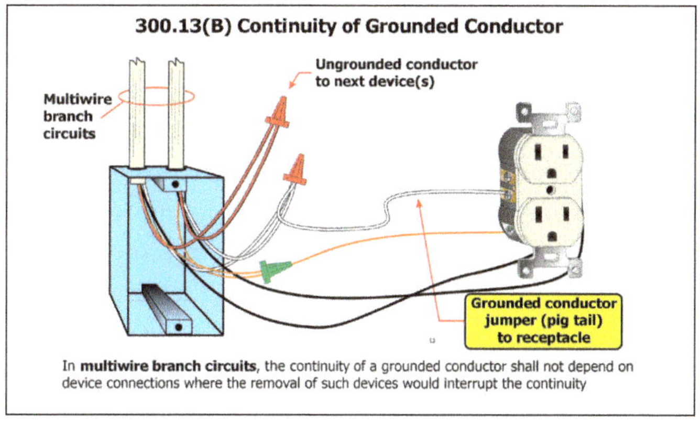

Figure 11.4. Continuity is required for grounded (neutral) conductors where multiwire branch circuits pass through outlet boxes

ungrounded conductors at the panelboard where the circuit originates [*NEC* 210.4(B) or IRC E3701.5]. Identified handle ties listed and manufactured for the particular brand of circuit breaker installed should be used. The use of a screw, wire, or a nail tying two single-pole breakers together is not acceptable and does not meet Code requirements (see Figure 11.2).

All conductors of a multiwire branch circuit (including the grounded circuit conductor) are required to be grouped together in at least one location within the panelboard by cable ties or similar means to identify these circuits as multiwire circuits to future users or installers of this electrical system. This grouping requirement is not required if the circuit enters from a cable or raceway unique to the circuit that makes the grouping obvious or if the conductors are identified at their terminations with numbered wire markers corresponding to the appropriate circuit number [*NEC* 210.4(D), *NEC* 200.4(B) and Ex. No. 1 or IRC E3701.5.2 and Exception].

A multiwire branch circuit will often supply one common receptacle or other equipment on the same yoke or strap. Often, the break-off tab of the receptacle is removed so that the top and bottom half of a receptacle can be energized by different circuits from opposite phases. One example is where a properly located receptacle outlet supplies both a dishwasher and a garbage disposal. The load of each appliance often necessitates that each be supplied by a separate circuit. Supplying the receptacle with a 3-wire cable and "split-wiring" the receptacle is an efficient way of providing power to both appliances utilizing one receptacle outlet and device.

Disconnecting both ungrounded (hot) conductors is necessary to help avoid the hazards associated with the installation of a multiwire branch circuit. One hazard is associated with an open neutral conductor. A multiwire circuit typically has 240 volts between the ungrounded (hot) conductors and 120 volts from each of the ungrounded conductors to the grounded or neutral conductor.

Both hot (ungrounded) conductors of the multiwire circuit rely on the neutral conductor to complete their circuit. This ensures that 120-volts-to-ground loads are maintained at the required voltage by the common neutral conductor. If the neutral conductor is broken or disconnected from a device used to complete the neutral circuit, the loads beyond that point operate in series at 240 volts. This can result in an unbalanced voltage across some loads. Higher voltages on electronic equipment such as computers, televisions, and audio equipment will damage the equipment.

An open neutral conductor on a multiwire branch circuit will produce a potential difference (voltage) between the open conductors. This voltage is a hazard to an individual working on the circuit and can result in a severe electrical shock.

It is also equally important that each ungrounded (hot) conductor contained within a 3-wire cable assembly be connected on opposite phases. A properly connected neutral conductor in a 3-wire cable carries only the unbalanced load. This is the difference between the loads on each phase conductor. If both ungrounded conductors are connected to the same phase, the neutral will carry the sum of the currents. This situation has been the cause of conductor failures, some of which have caused fires. See Figure 11.3 for examples of 3-wire circuits that illustrate both "multiwire branch circuits" and "not multiwire branch circuits."

The general rule at *NEC* 210.4(C) permits multiwire branch circuits to be used only for the supply of line-to-neutral loads. The *Code* allows a multiwire branch circuit to supply line-to-line loads where only one piece of utilization equipment that utilizes electric energy for mechanical, chemical, heating, lighting, or similar purposes is supplied. The *Code* also permits a multiwire branch circuit to supply line-to-line loads if all ungrounded conductors of the multiwire branch circuit are opened simultaneously by the branch-circuit overcurrent device [*NEC* 210.4(C), Ex. No. 1 and Ex. No. 2].

Ungrounded (hot) conductors of a multiwire branch circuit to supply 240-volt equipment, such as an electric range or dryer, are also permitted to be supplied by a two-pole circuit breaker. Two-pole circuit breakers are required where loads are connected line-to-line. Only a two-pole circuit breaker will ensure that both poles open simultaneously when the circuit breaker opens due to an overcurrent.

All conductors of a multiwire branch circuit are required to originate from the same panelboard or similar distribution equipment [*NEC* 210.4(A) or IRC E3701.5]. In multiwire branch circuits, the continuity of a grounded (neutral) conductor cannot depend on device connections such as lampholders, receptacles, and so forth, where the removal of such devices would interrupt the continuity of the grounded (neutral) conductor [see *NEC* 300.13(B) or IRC E3406.11.2].

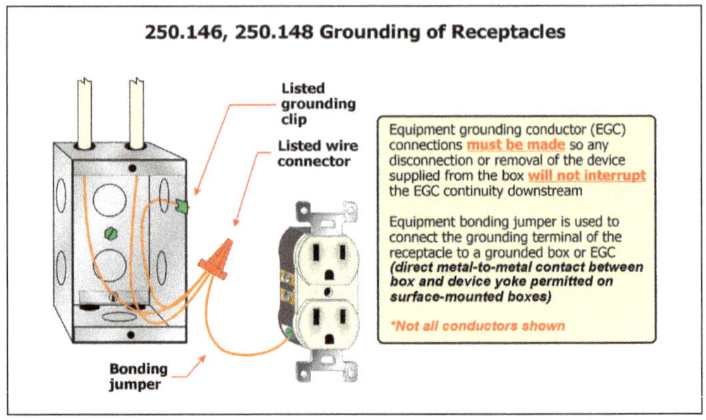

Figure 11.5. Equipment grounding conductor connections at receptacle outlet boxes

Photo 11.3. Equipment grounding conductor connected to metal box using a grounding screw

EQUIPMENT GROUNDING CONDUCTOR (EGC), GENERAL

The equipment grounding conductor is the key part of the electrical safety circuit that enables the fuse or circuit breaker to function in the event of a ground fault. Care must be taken when installing branch circuits and feeders to ensure the integrity and functionality of the equipment grounding conductor. The equipment grounding conductor of a branch circuit can be identified by a continuous green color or a continuous green color with one or more yellow stripes. As with the most popular dwelling unit wiring method (Type NM cable), it is also permitted to be bare. Ungrounded or grounded conductors are not permitted to be identified using the color green or green with yellow stripes.

The Code requires equipment grounding conductors to be installed in the same raceway, cable, or cord as the circuit conductors. This helps keep the circuit impedance as low as possible during normal operation and when fault current flows. Maintaining low impedance allows high levels of current to return to the overcurrent device, aiding its operation. In older homes where the cable lacks an equipment grounding conductor, NEC 250.130(C) must be followed. Replacing receptacles and extending branch circuits in older installations without a grounding conductor are detailed further in Chapter 16.

Bare, covered, or insulated equipment grounding conductors are permitted. An individually covered or insulated grounding conductor has a continuous outer finish (insulation) that is green or green with one or more yellow stripes. The Code allows an insulated or covered conductor 4 AWG or larger to be permanently identified as a grounding conductor at the time of installation. Identification must be made using approved methods at each end and wherever the conductor is accessible. These methods include stripping the insulation or cover from the entire exposed length, marking the insulation or cover with green color, or attaching green-colored tape or labels (NEC 250.119 or IRC E3407.2).

The *Code* also includes specific requirements for ungrounded (hot) conductor identification. Single or multiconductor cables are required to be finished so as to be "clearly distinguishable" from the neutral conductor (white or gray) and equipment grounding conductors (green, green with a yellow stripe, or bare) [*NEC* 310.6(C) or IRC E3407.3].

Photo 11.4. Equipment grounding conductor connected to metal box using a listed grounding clip

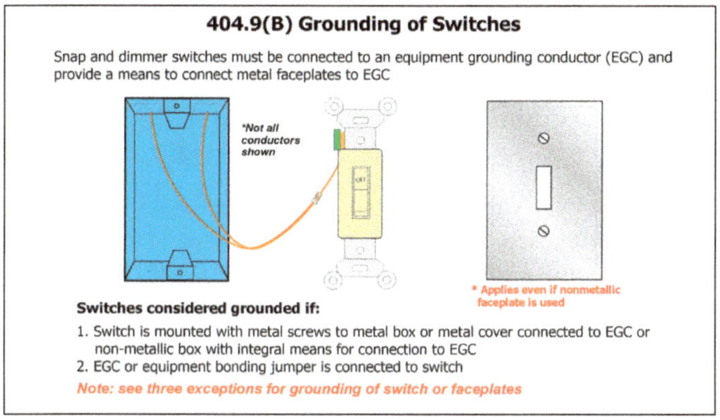

Figure 11.6. Grounding of switches and metal faceplates is required

EQUIPMENT GROUNDING CONDUCTORS

The *Code* includes a list of acceptable equipment grounding conductors for use in wiring systems, including wiring methods that are also recognized as suitable equipment grounding conductors. These vary from actual wire conductors to conduit and tubing types (see *NEC* 250.118 or IRC E3908.9 for equipment grounding conductors suitable for use as an EGC). The minimum size for separately installed equipment grounding conductors must not be less than *NEC* Table 250.122 or IRC Table E3908.13. Equipment grounding conductor sizing is based on the rating of the overcurrent protective device (circuit breaker or fuse) protecting the branch-circuit conductors.

EQUIPMENT GROUNDING CONDUCTOR CONNECTIONS

The *Code* requires that where circuit conductors are spliced together or supply equipment or devices in a box, all equipment grounding conductors are to be either joined together or all joined to the enclosure. The equipment grounding conductors are joined or spliced together in an enclosure with an identified wire connector. An alternative is to use listed devices, such as ground clips or ground screws, for securing each individual equipment grounding conductor to the metal enclosure or box. Insulation is not required on the equipment grounding conductor(s) splice(s) [*NEC* 250.148(A) or IRC E3908.14].

An equipment grounding conductor connection must be made so any disconnection or removal of a receptacle, luminaire, or other device supplied from the box will not interfere with or interrupt the equipment grounding conductor continuity downstream. This is usually accomplished by splicing the equipment grounding conductors together with short "pigtails" or bonding jumpers used to connect to the grounding terminal on each switch or receptacle. That way, the receptacle or switch can be removed without disturbing the continuity of the equipment grounding conductor (*NEC* 250.146 and 250.148 or IRC E3908.15 and E3908.15; see photos 11.3 and 11.4).

Where a metal enclosure is used, a connection between one or more equipment grounding conductors is required to be made from the splice mentioned above to the enclosure by means of a grounding screw or listed grounding clip. Identified grounding screws typically employ 8-32 or 10-32 machine threads to maximize the metal-to-metal contact. Sheet metal screws or other types of screws that do not engage a minimum of two threads are not permitted to be used for connecting grounding conductors or bonding jumpers to enclosures (*NEC* 250.8 or IRC E3406.14.).

This requirement for grounding a metal box by a "grounding screw that shall be used for no other purpose" precludes the use of a cable clamp screw for connecting the equipment grounding conductor. Equipment grounding conductors wrapped around the outer jacket of an older existing nonmetallic-sheathed cable assembly and secured under a cable clamp where they are back wrapped are not, and have never been, permitted to be used as the termination to a metal box or other enclosure [*NEC* 250.148(C) or IRC E3908.15].

The typical metal enclosure (box) is required to be provided with a tapped hole engraved with the marking "GRND" or "GR" that is to be used for this purpose.

Where a nonmetallic outlet or device box is used, the equipment grounding conductors are required to be spliced together and arranged inside the nonmetallic enclosure so that a connection can be made to any fitting or device that requires grounding. This is accomplished by splicing the equipment grounding conductors together with short "pigtails" or bonding jumpers used to connect to the grounding terminal on each switch, receptacle, or mounting bracket for a luminaire.

METHODS OF GROUNDING SWITCHES

The *Code* requires that snap switches be effectively grounded by means of connection to an equipment grounding conductor. In addition to grounding the switch, a means must be provided to bond metal faceplates (present or future) to the equipment grounding conductor. This requirement applies even if a nonmetallic faceplate is used with the switch. Snap switches are considered effectively grounded if either of the following conditions is met.

1. The switch is mounted with metal screws to a metal box or metal cover that is connected to an equipment grounding conductor, or the switch is mounted to a nonmetallic box that contains an integral means for connecting to an equipment grounding conductor. The *Code* does not require the use of "self-grounding" snap switches.
2. An equipment grounding conductor or equipment bonding jumper is connected to an equipment grounding termination of the snap switch. If multiple switches are ganged together in a single box, each switch must be bonded.

In older homes, where an equipment grounding means does not exist in the box or where the wiring method does not contain an equipment grounding conductor, a snap switch having a metal yoke is permitted to be installed without being grounded; this applies to the replacement of switches only. If the switch is located within 2.5 m (8 ft) vertically or 1.5 m (5 ft) horizontally of earth, grade, concrete floors, or other conducting surfaces, a nonmetallic cover plate must be installed. A permitted alternative is to protect the switch with a ground-fault circuit interrupter. Listed switch kits and listed switch assemblies constructed with nonmetallic yokes and equipped with nonmetallic faceplates are not required to be grounded as well. These switches are not designed for and have no provisions for attaching a metallic faceplate to them. Snap switches with integral nonmetallic enclosures do not need a means of providing a connection to a metal faceplate to an equipment grounding conductor as these switches incorporate integral nonconductive (nonmetallic) faceplates and inherently cannot interchange faceplates, metal or otherwise. [*NEC* 404.9(B) and Exceptions or IRC E4001.11.1 and Exceptions]. See Chapter 16 for additional information on the replacement of switches in existing dwellings.

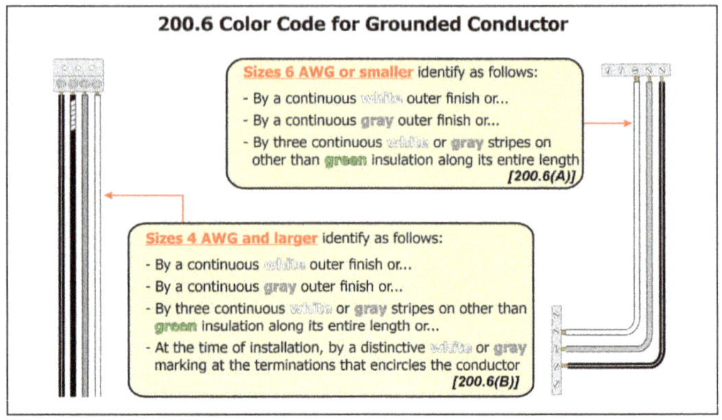

Figure 11.7. Identification of grounded (neutral) conductors

RECEPTACLES AND CORD CONNECTORS

In new construction, all receptacles installed in 15- and 20-ampere branch circuits are required to be of the grounding type [*NEC* 406.4(A) or IRC E4002.2]. A grounding-type receptacle is generally permitted to be installed only on circuits of the voltage class and current for which they are rated. Receptacles and cord connectors that have grounding contacts shall have the contacts effectively grounded [*NEC* 406.4(B)]. This is accomplished by connecting the grounding contact(s) to the equipment grounding conductor of the branch circuit supplying the receptacle. Where an effective grounding means exists, grounding-type receptacles must be used.

In older, existing dwellings employing non-grounding-type receptacles, some existing circuits may not contain an effective grounding means. In this case, non-grounding-type receptacles may continue to be used. The *Code* also permits a non-grounding-type receptacle to be replaced with a properly marked GFCI receptacle or receptacle protected by a GFCI [*NEC* 406.4(D)(2)]. However, if the *Code* requires a particular appliance or other cord- and plug-connected equipment to be grounded, GFCI protection cannot be used as a substitute. Equipment such as refrigerators, freezers, air-conditioners, washing machines, dishwashers, waste disposers, computers, sump pumps, and aquarium pumps and filters are required to be grounded via an equipment grounding conductor (*NEC* 250.114 or IRC E4101.3). Chapter 17 contains additional information on this topic.

METHODS OF GROUNDING RECEPTACLES

The branch-circuit cable assemblies or branch-circuit raceways must include or provide an equipment grounding conductor to which the grounding contacts of the grounding-type receptacle or cord connector will be connected. The

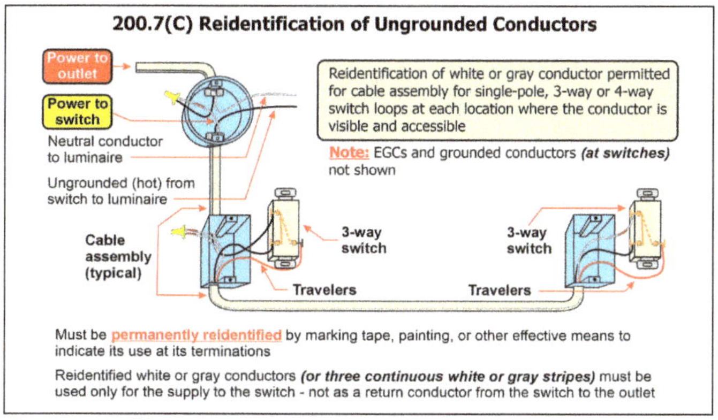

Figure 11.8. White conductor (re-identified) in cable is power conductor (line side) from lighting outlet to 3-way switches

equipment grounding conductor terminal on a receptacle or switch is required to be identified by a:

1. green-colored terminal screw provided with a hexagonal head that is not readily removable;
2. green-colored hexagonal terminal nut that is not readily removable or
3. green-colored pressure wire connector.

Where the terminal for the equipment grounding conductor is not visible, the conductor entrance hole is required to be marked with the word "green" or "ground" or be otherwise identified as required by *Code* and product standard (*NEC* 250.126).

An equipment bonding jumper is used to connect the grounding terminal of the receptacle to a grounded box or equipment grounding conductor. Where the box is surface-mounted, direct metal-to-metal contact between the box and the device yoke is permitted as the bonding means. At least one of the insulating washers must be removed from the receptacle in order to accomplish the metal-to-metal contact. An alternative is a contact yoke that is self-grounding. This type of device is sometimes called a self- or auto-grounding receptacle (*NEC* 250.146 or IRC E3908.15).

COLOR CODING FOR BRANCH-CIRCUIT CONDUCTORS

The grounded (neutral) branch-circuit conductor is a system or circuit conductor that is intentionally grounded. This conductor is required to be specifically identified by a continuous white or gray outer finish or by three continuous white or gray stripes on any color other than green insulation along its entire length [*NEC* 200.6(A) and (B) or IRC E3407.1]. The *Code* permits a cable containing an insulated conductor with a white or gray outer finish or a marking of three continuous white or gray stripes to be used for the supply to single-pole, three-way, or four-way switches. If used for single-pole, 3-way, or 4-way switch loops, the re-identified conductor can only be used for the supply to the switch but not as a return conductor from the switch to the switched outlet, such as a luminaire. This is illustrated in Figure 11.8.

In this type of circuit, permanent re-identification of the white or gray conductor insulation is required. Painting, marking tape, or other effective means may be used to meet this requirement. Re-identification is required where the conductor is visible and accessible, including terminations such as at switches and junction boxes. [NEC 200.7(C)(1) or IRC E3407.3 Exceptions].

Remember that permission to re-identify conductors with white or gray or white- or gray-striped conductor insulation applies only to cable assemblies. The conductor with white, gray, or other identified insulation in a raceway is not permitted to be used as an ungrounded (hot) conductor. The *Code* does not permit painting, taping, or otherwise re-identifying individual grounded conductors run through raceways and using them as ungrounded phase or equipment grounding conductors.

RATING OF OVERCURRENT DEVICES

Branch-circuit conductors (not exceeding 1000 volts ac) and equipment are required to be protected by properly rated and sized overcurrent protective devices (*NEC* 210.20 or IRC E3705.5). Where a branch circuit supplies continuous loads (three hours or more of continuous operation) or any combination of continuous and noncontinuous loads, the rating of the overcurrent device cannot be less than the noncontinuous loads plus 125% of the continuous loads. This has the effect of limiting the circuit and overcurrent device on these loads to not more than 80% of their ratings. The majority of loads in dwelling units are considered to be noncontinuous loads. An example of a continuous load at a dwelling unit would be an electric swimming pool pump motor, and/or electric heater can be expected to operate for three hours or more continuously and would thus meet the definition of a continuous load.

Conductor protection must be in accordance with the allowable ampacity tables for the particular conductor (*NEC* Tables 310.16 through 310.19 or IRC Table E3705.1). The rating or setting of the overcurrent protective device is also not permitted to exceed the values specified for the equipment.

The *Code* does permit the rating of a branch circuit supplying a receptacle outlet to exceed the ampere rating of the device if two or more receptacles or outlets are supplied. Proper application of this rule requires a brief review of the definition of a receptacle.

A *receptacle* is defined as "a contact device installed at the outlet for the connection of an attachment plug or for the direct connection of electrical utilization equipment designed to mate with the corresponding contact device. A single receptacle is a single contact device with no other contact device on the same yoke or strap. A multiple receptacle is two or more contact devices on the same yoke or strap."

Outlet devices are always required to have an ampere rating that is not less than the load to be served. A single receptacle is required to be rated at not less than that of the branch circuit where the receptacle is installed on an individual branch circuit. For example, a single receptacle is installed on a 20-ampere branch circuit. This receptacle must have a rating of 20 amperes.

The rules are different when two or more receptacles are connected to the same branch circuit. First, a receptacle can only be used to supply a cord- and plug-connected load equal to or less than the value specified in either NEC Table 210.21(B)(2) or IRC Table E3702.14. For example, a 15-ampere receptacle cannot supply more than a 12-ampere load.

Second, the receptacle ratings are required to conform to the values included in *NEC* Table 210.21(B)(3) or IRC Table E4002.1.2. This permits a 15-ampere receptacle on a 20-ampere branch circuit and a 40- or 50-ampere receptacle on a 40-ampere branch circuit. A receptacle rated at more than 50 amperes cannot be less than the rating of the branch circuit. A 60-ampere receptacle must be installed on a branch circuit rated 60 amperes or less.

MAXIMUM LOAD ON BRANCH CIRCUITS

The total load on any branch circuit cannot exceed the actual rating of the branch circuit. Also, the branch circuit cannot exceed the maximum loads specified as follows indicated in *NEC* 220.11 or IRC E3701.2.

Motor Operated and Combination Loads. *NEC* Article 430 is required to be applied where a branch circuit supplies only motor-operated loads. Article 440 is required to be applied where a branch circuit supplies only air-conditioning and/or refrigerating equipment. For branch circuits that supply loads consisting of motor-operated utilization equipment that is fastened in place and has a motor rated larger than ⅛ horsepower in combination with other loads, the total computed load must be based on 125% of the largest motor load plus the sum of all the other loads [*NEC* 220.11(A) or IRC E3702.7].

Inductive and LED Lighting Loads. For circuits supplying lighting units having ballasts, such as those found in fluorescent luminaires, transformers, autotransformers, or LED drivers, the total computed load is required to be based upon the total ampere ratings of such units and not on the total watts of the lamps [*NEC* 220.11(B) or IRC E3702.8].

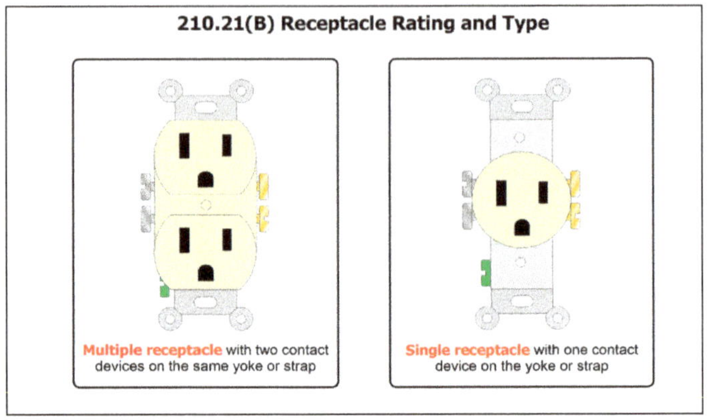

Figure 11.9. Typical 125-volt, 15-ampere receptacles (single and duplex)

Electric Cooking Appliances. Demand factors given in *NEC* Table 220.55 or IRC Table E3704.2(2) are permitted to be applied to electric cooking appliances when calculating the rating of a range, wall-mounted oven, counter-mounted cooking unit, and other household cooking appliance load in excess of 1¾ kW. Note 4 to *NEC* Table 220.55 is permitted to be used to calculate the branch-circuit rating of one range.

NEC Table 220.55, Note 4, applies to the minimum branch-circuit load for one electric range. Note 5 indicates that the branch-circuit load for one wall-mounted oven or one counter-mounted cooking unit must be the nameplate rating of the appliance. Note 6 stipulates that the branch-circuit load for a counter-mounted cooking unit and not more than two wall-mounted ovens, all supplied from a single branch circuit and located in the same room, is computed by adding the nameplate rating of the individual appliances and treating this total as equivalent to one range. After doing this, the demand factors of *NEC* Table 220.55 can be applied to the total.

PERMISSIBLE LOADS ON BRANCH CIRCUITS

The load cannot exceed the branch-circuit ampere rating under any circumstance. An individual branch circuit is permitted to supply any load for which it is rated (*NEC* 210.22).

The 2023 *NEC* has introduced provisions to allow a 10-ampere branch circuit under certain conditions in dwelling units. A 10-ampere branch circuit *is permitted* to supply lighting outlets, dwelling unit exhaust fans on bathroom or laundry

210.21(B) Receptacle Ratings
[NEC Tables 210.21(B)(2) and 210.21(B)(3)]

Maximum Cord- and Plug-Connected Load to Receptacle

Circuit rating (amperes)	Receptacle rating (amperes)	Maximum load
15 or 20	15	12
20	20	16
30	30	24

Receptacle Ratings for Various Size Circuits

Circuit rating (amperes)	Receptacle rating (amperes)
15	Not over 15
20	15 or 20
30	30
40	40 or 50
50	50

Figure 11.10. Receptacle ratings for individual branch circuits and general-purpose branch circuits

room lighting circuits, and/or a gas fireplace unit supplied by an individual branch circuit. A 10-ampere branch circuit *is not permitted* to supply receptacle outlets, fixed appliances (except as permitted for individual branch circuits), garage door openers, or laundry equipment [*NEC* 210.23(A)].

A 15- or 20-ampere branch circuit is allowed to supply lighting outlets, lighting units, or other utilization equipment, or a combination of these. Small appliances, laundry, bathroom, and garage branch circuits are limited to supplying receptacle outlets only, as specified in NEC 210.11(C)(1), (2), (3), and (4) or IRC E3703.2, E3702.3, E3702.4, and E3702.5, respectively. Refer to NEC 210.23(B) or IRC E3702.3.

The rating of any one cord- and plug-connected piece of utilization equipment (not fastened in place) cannot exceed 80% of the ampere rating of the branch circuit [*NEC* 210.23(B)(1) or IRC E3702.3]. The total rating of utilization equipment that is fastened in place (other than luminaires) cannot exceed 50% of the branch-circuit ampere rating where lighting units, cord- and plug-connected utilization equipment not fastened in place, or both, are also supplied by the branch circuit (*NEC* 210.23(B)(2) or IRC E3702.3).

A 30-ampere branch circuit can be used to supply utilization equipment such as electric dryers or water heaters. The rating of any one cord-and-plug-connected utilization equipment cannot exceed 80 percent of the branch-circuit ampere rating. A 40- or 50-ampere branch circuit is permitted to supply cooking appliances that are fastened in place in one- and two-family dwelling unit(s) [*NEC* 210.23(C) and (D) or IRC E3702.4].

BRANCH-CIRCUIT REQUIREMENTS - SUMMARY

The requirements for circuits having two or more outlets, other than the receptacle circuits of *NEC* 210.11(C)(1), (C)(2), and (C)(3) or IRC E3703.2, E3702.3, and E3702.4 are summarized in *NEC* Tables 210.24(1) and (2) and IRC Table E3702.14 [*NEC* 210.24 or IRC E3702.14]. An outlet or outlets installed for the purpose of supplying electric vehicle supply equipment (EVSE) greater than 16 amperes or 120 volts are to be supplied by an individual branch circuit for electric vehicle charging only. Per the definition of an *individual branch circuit*, this circuit shall have no other outlets. This EVSE branch circuit is permitted to supply multiple EVSE stations for the purpose of load management or supplying EVSEs with adjustable settings [*NEC* 625.40 or IRC E3702.13].

COMMON AREA BRANCH CIRCUITS

Branch circuits installed in dwelling units are permitted to supply only loads within that dwelling unit or loads associated only with that dwelling unit. Branch circuits required for the purpose of lighting, central alarm, signal, communications, or other needs for common or public areas of a two-family dwelling, a multifamily dwelling, or a multi-occupancy building are not permitted to be supplied from an individual dwelling unit panelboard (*NEC* 210.25(A) and (B) or IRC E3702.14).

Loads associated with an individual dwelling unit are permitted to be supplied by the dwelling unit panelboard. Examples are the tenant's storage space, laundry area, or garage stall. These spaces are not required to be contiguous to the tenant's dwelling unit.

Branch circuits for lighting, central alarm, signal, communications, and other needs for public or common areas are not permitted to be supplied from the panelboard located in a resident manager's dwelling unit. An additional panelboard must supply such loads, often called a "house panel." This will reduce the likelihood of tenants turning off these essential and safety circuits or the loss of power to common area loads due to the utility disconnecting power to an individual dwelling unit. The house panel is permitted in addition to the service equipment that supplies each dwelling unit. *NEC* 230.40, Exception No. 4 permits an additional set of service conductors to supply the house panel for a two-family or a multifamily dwelling.

DWELLING UNIT RECEPTACLE OUTLETS

Required receptacle outlets specified in *NEC* 210.52 and IRC E3901 are those convenience receptacle outlets for branch circuits rated 125-volt, 15- and 20-amperes. This makes it clear that any receptacle outlets beyond those ampacity and voltage configurations are not permitted to be counted as the required receptacles for wall spacing, etc. For

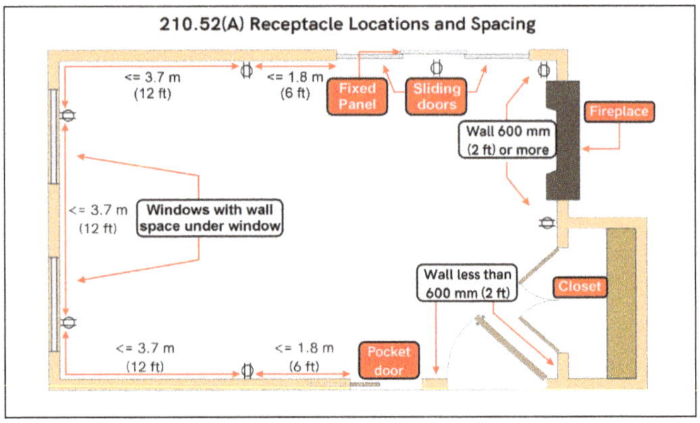

Figure 11.11. Spacing requirements for general-use receptacle placement

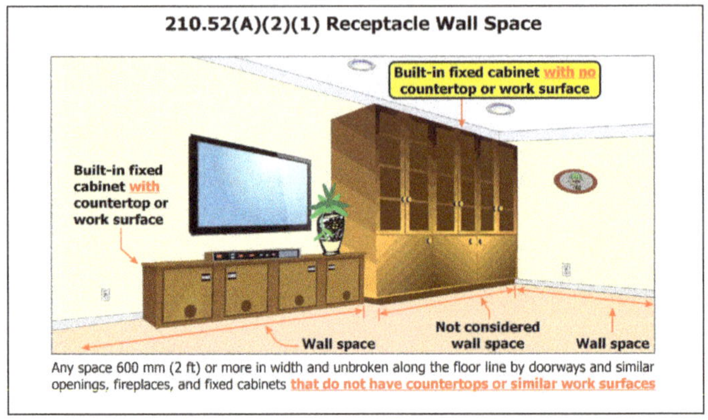

Figure 11.12. Fixed cabinets that do not have a countertop or similar work surface can break up wall space for receptacle outlets

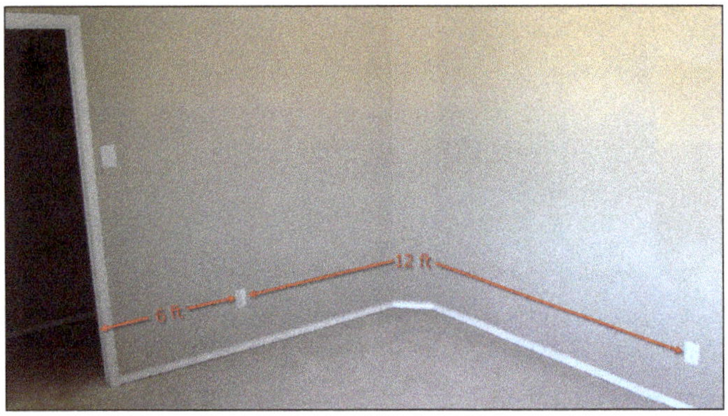

Photo 11.5. Maximum receptacle outlet placement requirements in a typical bedroom

Photo 11.6. Fixed-in-place railings and banisters are considered wall space of receptacle outplacement

example, a 30-ampere, 240-volt receptacle outlet at a kitchen countertop location is considered over and above the required 125-volt, 20-ampere receptacle outlet required to serve the countertop space.

The *Code* requires receptacle outlets for general use in every dwelling unit: kitchen, family room, dining room, living room, parlor, library, den, sunroom, bedroom, recreation room, or similar room or area. The receptacle outlets are required to be installed so that no point along the floor line in any wall space is greater than 1.8 m (6 ft), measured horizontally, from an outlet in that space.

The *Code* describes wall space as: "Any space 600 mm (2 ft) or more in width (including space measured around corners) and unbroken along the floor line by doorways and similar openings, fireplaces, stationary appliances, and fixed cabinets that do not have countertops or similar work surfaces." Only "fixed cabinets that do not have countertops or similar work surfaces" are considered as an item (along with doorways, stationary appliances, and fireplaces) that would not be counted as "wall space" concerning receptacle spacing and location requirements. Fixed cabinets such as kitchen pantry-type cabinets (but not limited to kitchen cabinets) and built-in bookcases that do not have countertops or similar work surfaces are considered to break up the wall space for the spacing of receptacle outlets. Fixed cabinets with countertops and similar work surfaces, such as the back side of a peninsular cabinet/countertop that separates the kitchen countertops from the nook or living room area, do require receptacle outlets as this back side of the cabinet would be considered wall space.

Every wall space of 600 mm (2 ft) or more in width is required to be treated individually and separately from all other wall spaces within the room. A wall space includes two or more walls of a room (around corners) where they are unbroken at the floor line. Spaces occupied by fixed panels in exterior or interior walls are considered wall space for required outlet spacing. Sliding patio doors or deck panels in exterior walls are exempt from this requirement [*NEC* 210.52(A)(1) and (A)(2) or IRC E3901.2.1 and E3901.2.2]. See Figure 11.11 for the receptacle layout in a typical room. Space created by fixed room dividers, such as freestanding bar-type counters or stairway railings, is considered wall space for the purpose of applying these receptacle outlet requirements. Railings create wall spaces where electrical appliances such as televisions, cable boxes, DVRs, or lamp-type luminaires are commonly located. Floor receptacles are a common solution for fixed railings and

210.52(A) Receptacle Locations and Spacing *(Typical)*

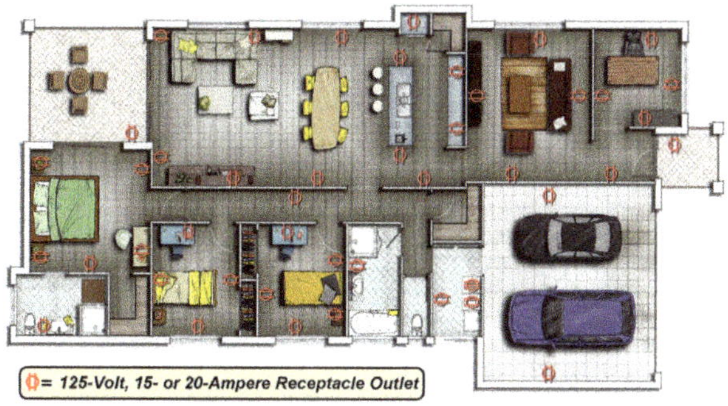

Figure 11.13. Typical receptacle placement requirements at dwelling units

210.52(A)(3) Floor Receptacles

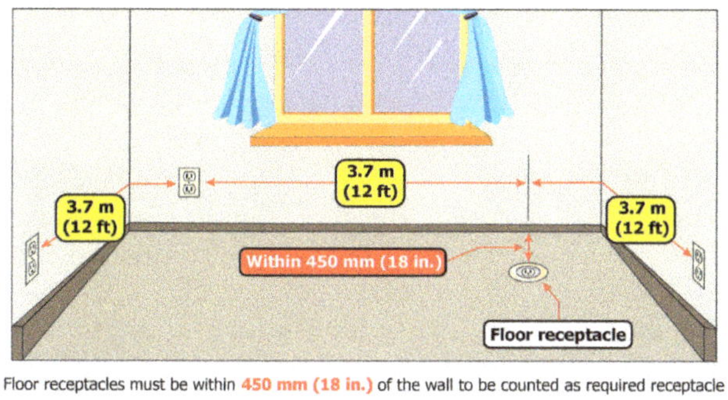

Figure 11.14. Receptacles installed in the floor to meet placement requirements

fixed banister locations where receptacle outlets are required to be placed [*NEC* 210.52(A)(3) or IRC E3901.2.3].

Where floor receptacles are installed to meet the receptacle outlet spacing provisions, they must be located within 450 mm (18 in.) of the wall they are intended to serve. Without receptacles located for these spaces, occupants often resort to the unsafe practice of using extension cords across walkways or through doorways, or they place these extension cords under carpeting or other floor covering, creating a possible fire hazard [*NEC* 210.52(A)(3) or IRC E3901.2.3].

Although not an enforceable requirement, a good rule of thumb for receptacle outlet spacing is to space receptacle outlets equal distances apart. Required wall receptacle outlets are in addition to any receptacle that is part of any luminaire or appliance that may be located within cabinets or cupboards or located over 1.7 m (5½ ft) above the floor (*NEC* 210.52 or IRC E3901.1).

Switched receptacles for lighting as permitted by *NEC* 210.70(A)(1), Ex. No. 1 or IRC E3903.2, Ex. No. 1 are not acceptable for required wall receptacle outlet spacing requirements if the entire receptacle is switched. The receptacles required for wall spacing requirements must meet the objective of providing convenience receptacles in those

210.52 Dwelling Unit Receptacle Outlets

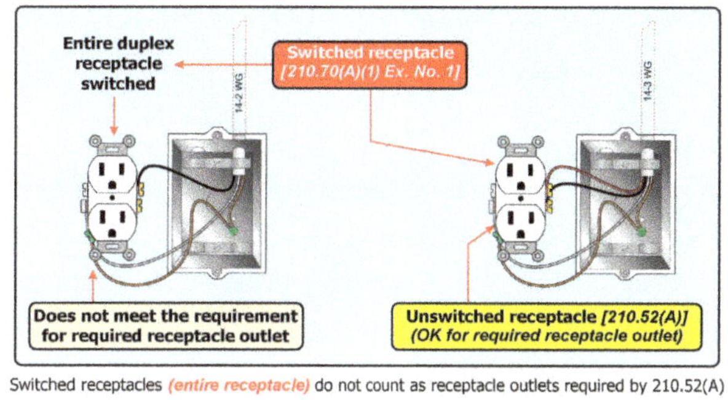

Figure 11.15. Switched receptacles that switch the entire receptacle outlet do not count as receptacle outlets required by *NEC* 210.52 or IRC E3901.1

prescribed locations for users. Switching both receptacles of a duplex-type receptacle could leave an area of a wall without a receptacle that provides continuous service to meet these minimum requirements. Leaving only one of the receptacles on a duplex receptacle switched and the other receptacle on that yoke continuously energized would satisfy the required wall receptacle outlet spacing requirements [*NEC* 210.52(2) or IRC E3901.1].

Electric baseboard heating units (fixed electric space heating) create a potential hazard. A flexible cord that is draped in front of or on the hot fins of the baseboard heater can create a fire and shock hazard. Listed baseboard heaters may contain instructions that prohibit the installation of the unit below receptacle outlets. There is no specific vertical distance or separation above which receptacle outlets are then permitted. Installing a baseboard heater below a fixed receptacle outlet is not a good practice as, again, this could subject any flexible cord plugged into this receptacle outlet to physical damage from the heat of the baseboard heater. The *Code* permits factory-installed receptacle outlets or outlets provided as a separate assembly by the manufacturer in permanently installed electric baseboard heaters to serve as the required outlet or outlets for the wall space utilized by such permanently installed heaters. These receptacle outlets are not permitted to be connected to the electric baseboard heater circuits. The branch circuit that serves that area must be used to supply the receptacle located in the baseboard heater (*NEC* 210.52, 424.10, or IRC E3901.1).

SMALL APPLIANCE CIRCUITS (KITCHEN)

The *Code* requires that two or more 20-ampere small-appliance branch circuits serve all receptacle outlets in the kitchen, pantry, breakfast room, dining room, or similar area of a dwelling unit. One of the 20-ampere small-appliance circuits is required to supply refrigeration equipment located in these kitchen areas as well unless the exception for a specific appliance is employed. [*NEC* 210.52(B)(1) or IRC E3901.3 (also see exceptions)].

Other than in kitchens and bathrooms, switched receptacle(s) sometimes used for general illumination are permitted, as stated earlier. These switched receptacles are in addition to those required to serve the wall spaces in the most habitable rooms. Switched receptacles in lieu of lighting outlets are also not permitted in hallways and stairways.

As mentioned above, the *Code* also permits an individual branch circuit for a specific appliance (such as the refrigerator) per *NEC* 210.52(B)(1) Ex. No. 2 or IRC E3901.3 Ex. No. 2. Where installed, the specific appliance branch circuit must not be rated smaller than 15 amperes. As of the 2023 *NEC*, the receptacle supplying refrigeration equipment in the kitchen does require GFCI protection as all 125-volt through 250-volt receptacles supplied by single-phase branch circuits rated 150 volts or less to ground installed in an area defined as a "*kitchen*" must be GFCI protected. In previous editions of the *Code*, this receptacle outlet was located behind the appliance and not readily accessible to serve the kitchen countertop and not located within 1.8 m (6 ft) of a sink such as a kitchen sink would not have required GFCI protection. This would not hold true for the latest edition of the *Code*. See Chapter 10 in this book for more information on GFCI requirements for dwelling unit kitchens.

The two or more small-appliance branch circuits cannot be connected to receptacle outlets outside of the kitchen, pantry, breakfast room, dining room, or similar area of a dwelling unit or *other outlets*, with two exceptions. In this case, the term "other outlets" refers to lighting, appliance, and receptacle outlets that do not serve countertops or wall spaces in the kitchen, pantry, breakfast room, dining room, or similar areas. These rules make it plain that it is a *Code* violation to put a lighting outlet on one of the small-appliance branch circuits. This means that additional branch circuits (typically general lighting branch circuits) must be provided to supply lighting outlets for kitchen hood fans, built-in microwave ovens, dishwashers, garbage disposals, etc. The exceptions that would allow "other outlets" on one of the small-appliance branch circuits include a receptacle installed solely for an electrical clock in kitchens, dining rooms, or related spaces permitted to be supplied by one of the small-appliance branch circuits. The other exception permits receptacles to be installed to provide power for supplemental equipment and lighting on gas-fired appliances such as ranges, ovens, or counter-mounted cooking units, which are to be supplied from one of the

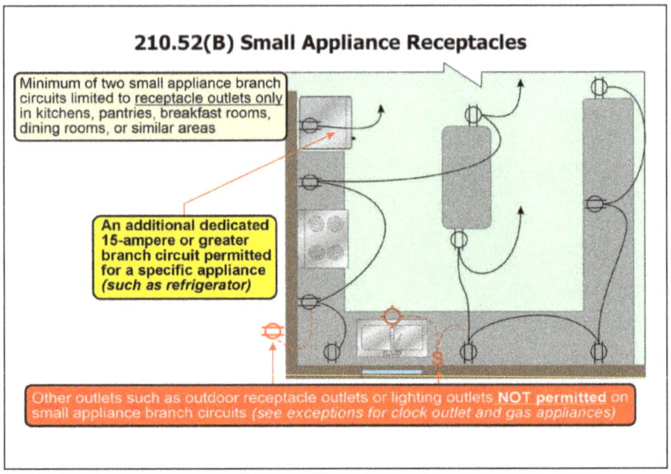

Figure 11.16. Small-appliance branch circuits for kitchen, pantry, breakfast room, and dining room receptacle outlets

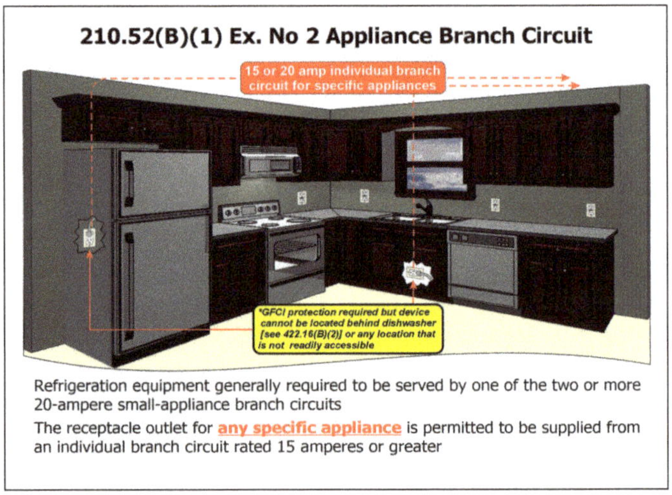

Figure 11.17. Individual specific appliance branch circuits permitted in the kitchen area

small-appliance branch circuits. Outdoor receptacle outlets, kitchen lighting outlets, and other outlets not specified above are not permitted to be connected to the small-appliance branch circuit(s) [*NEC* 210.52(B)(2) and exceptions or IRC E3901.3.1 and exceptions].

At least two of the 20-ampere small-appliance branch circuits are required to serve the countertop surfaces in the kitchen. Either or both of these two circuits can also supply other receptacle outlets in the same room or adjacent areas such as a dining room or pantry receptacle outlets. No small-appliance branch circuit is allowed to serve more than one kitchen. Some dwellings have a kitchen on the patio or in a "mother-in-law" type apartment in addition to the main dwelling kitchen. Each kitchen must be supplied by not less than two 20-ampere, small-appliance branch circuits [*NEC* 210.52(B)(3) or IRC E3901.3.2].

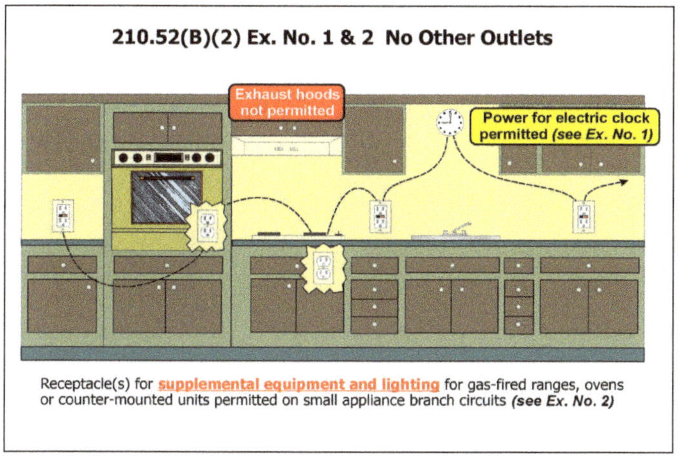

Figure 11.18. Small-appliance branch circuits are permitted to supply electric clock outlets, gas-fired ranges, ovens, or counter-mounted units

Photo 11.7. Kitchen receptacle placement at wall countertop locations

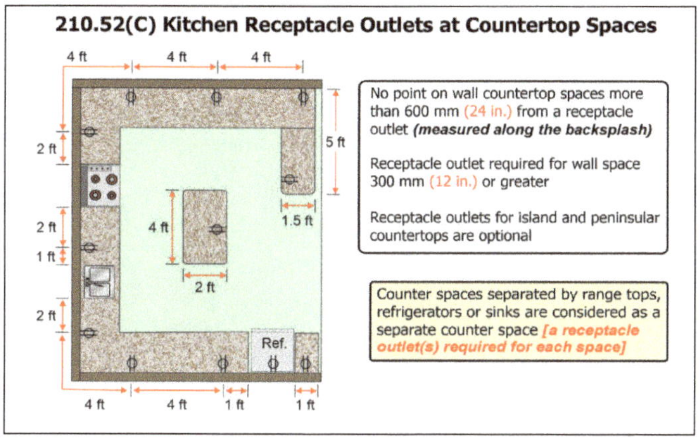

Figure 11.19. Kitchen receptacle spacing requirements at countertop surface *(typical layout)*

A kitchen is defined in the *Code* as "an area with a sink and permanent provisions for food preparation and cooking." Permanent provisions for food preparation and cooking are not defined; therefore, is a microwave oven "permanent provisions for food preparation and cooking?" The enforcing agency often has to use good judgment and discretion relative to making this determination to ensure that adequate circuitry and receptacle outlets are provided for the anticipated use. It is not uncommon for the AHJ to consider a built-in microwave oven a "permanent provision," while a microwave oven setting on the countertop is a "not permanent provision."

KITCHEN BASE WALL COUNTERTOP RECEPTACLE SPACING

There are typically three types of countertops that can be encountered in dwelling unit kitchens. These would be base or wall countertops, peninsular countertops, and island countertops. The *Code* has different rules for required receptacle outlets at these different countertops. At kitchen and dining area countertop spaces of dwelling units, a receptacle outlet must be installed at each wall counter space that is 300 mm (12 in.) or wider. In addition, receptacle outlets must be installed so that no point along the wall line is more than 600 mm (24 in.) measured horizontally from a receptacle outlet in that space.

For applicable wall counter spaces, this means the first receptacle outlet must be located not more than 600 mm (24 in.) from the edge of the counter space, and the next receptacle outlet can be no more than 1.2 m (48 in.) from the first receptacle outlet, and so on along the wall counter space. Though not specifically mentioned here, the provision of measuring around the corner of the floor for wall receptacles is commonly accepted for receptacle outlet spacing at kitchen wall countertop spaces as well. This 300 mm (12 in.) and 600 mm (24 in.) rule is only applicable to wall or base countertops.

The term "appliance garage" is referenced in *NEC* 210.52(C)(3) and IRC E3901.4.3. An appliance garage is often an integral part of the dwelling unit kitchen cabinetry. Although it is permitted to install a receptacle outlet within the appliance garage, that receptacle outlet is not permitted to serve as the required receptacle outlet for the countertop space. Also, any receptacle outlet installed within an appliance garage located on the countertop must be GFCI-protected.

Receptacle outlets may be required to serve wall or countertop space located behind sinks or ranges. Two figures are included in the *Code* to illustrate this condition (see *NEC* Figure 210.52(C)(1) or IRC Figure E3901.4.1). One illustration is for a sink or range located on a straight countertop. If the space between the sink or range and the wall or backsplash is 300 mm (12 in.) or greater, a receptacle outlet is required to serve that space behind the sink or range top. The other illustration is of a sink or range mounted in a corner of a countertop. If the space between the sink or range and

Photo 11.8. Receptacles located in an appliance garage cannot count as the required receptacle for kitchen countertop spacing

Photo 11.9. Kitchen countertop receptacles at the corner-mounted sink area

the corner is 450 mm (18 in.) or greater, a receptacle outlet is required to serve that space in the corner of the countertop. The third function of both illustrations is to show where the measurement to the first required receptacle outlet starts at the edge of the sink or range.

KITCHEN ISLAND AND PENINSULAR COUNTERTOP SPACES

In this current edition of the *Code* and the previous edition of the *Code*, the receptacle outlet requirements for the island and peninsular kitchen countertops and work surfaces have been dramatically revised. See *NEC* 210.52(C)(2) or IRC E3901.4.2 for these island and peninsular requirements. In the 2017 *NEC* (2018 IRC) and previous editions of the *Code*, a linear measurement was required across the countertop with at least one receptacle outlet required to be installed at each island countertop space or peninsular countertop space with a long dimension of 600 mm (24 in.) or greater and a short dimension of 300 mm (12 in.) or greater. These island or peninsular countertop measurements were in place from the 1993 *NEC* through the 2017 *NEC*. With this linear measurement, only one receptacle outlet was required at an island or peninsular countertop, regardless of its size. A 600

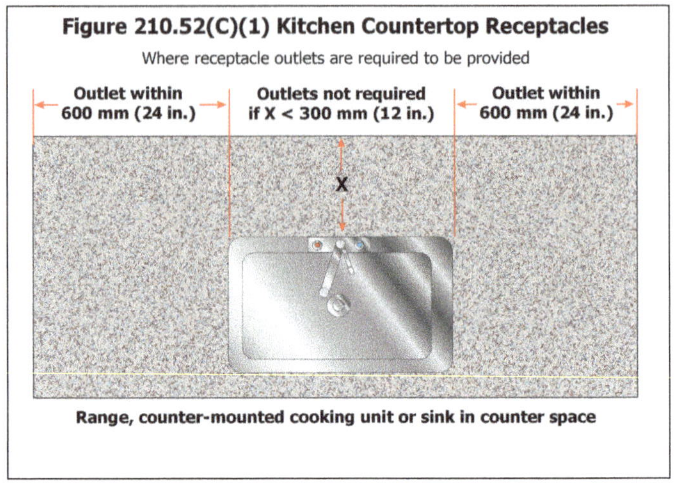

Figure 11.20. Kitchen countertop receptacle outlet spacing requirements are required behind range or sink mounted on the straight countertop if "x" is equal to or greater than 300 mm (12 in.)

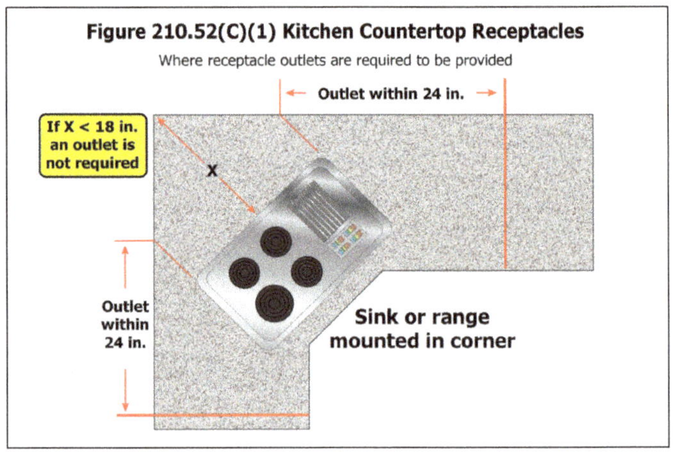

Figure 11.21. Kitchen countertop receptacle outlet spacing requirements are required behind the corner-mounted range or sink if "x" is equal to or greater than 450 mm (18 in.) than 300 mm (12 in.)

mm (24 in.) X 300 mm (12 in.) island or peninsular countertop required the same number of receptacle outlets (one) as a 1.2 m (4 ft) X 1.8 m (6 ft) island or peninsular. This "at least one receptacle shall be installed" requirement for island and peninsular countertops has been the standard for the *Code* since the 1990 edition of the *NEC*. It is not uncommon to find large to very large islands and peninsular countertops in today's modern dwelling unit kitchens.

For the 2023 *NEC* (2024 IRC), revisions were put in place to make receptacle outlets installed at dwelling unit kitchen islands and peninsular countertops ***optional***. In the event a receptacle outlet is not provided, the electrical contractor must provide a wiring method to the island or peninsula for the future addition of a receptacle outlet should the homeowner desire such a receptacle outlet at the island or peninsular location. A Type UF cable installed under the slab

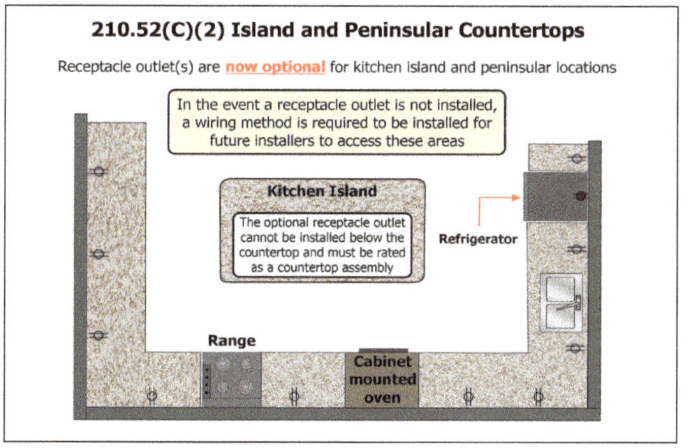

Figure 11.22. Kitchen Island countertop receptacle requirements

Photo 11.10. The receptacle is located on the kitchen island

to the island or peninsular location secured in a box with a cover would be an example of a wiring method installed for future use. This will allow for the future addition of a receptacle outlet at the island or peninsular location without extensive damage to the floor and surrounding area in the vicinity of the island or peninsular [*NEC* 210.52(C)(2) or IRC E3901.4.2].

If (by choice) a receptacle outlet *is* installed to serve a kitchen island and/or peninsular countertop, the receptacle outlet *cannot* be installed below the countertop or work surface. If installed, the receptacle outlet(s) would be required to be installed *on or above*, but not more than 500 mm (20 in.) above the kitchen island or peninsular countertop or work surface. In a dwelling unit kitchen counter, the receptacle must be listed for a countertop. A receptacle listed for a work surface *does not* meet the requirements for a countertop receptacle. If the kitchen has a *work surface* that is not a countertop, then a receptacle listed for work surfaces is acceptable. However, if the area is a *countertop*, only a receptacle listed for countertop use is permitted. See 210.52(C)(2). A work surface can use either product, but a counter cannot. See Photo 11.11 for an example of a listed receptacle outlet assembly installed in an island countertop or work surface.

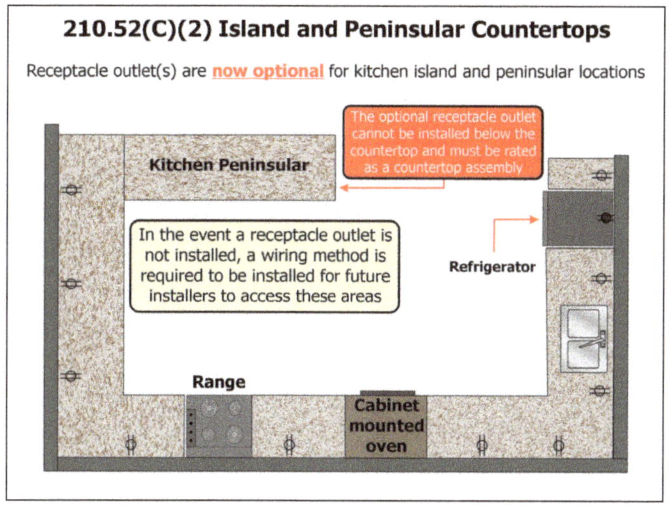

Figure 11.23. Kitchen peninsular countertop receptacle requirements

With such dramatic changes to the kitchen island and peninsular countertop receptacle outlet requirements over a relatively short period of time, it will be of the utmost importance for both the installer and the inspector to know what edition of the *Code* is being enforced. As an example, let's say we have a dwelling unit kitchen island that measures 2.1 m (7 ft) long by 1.2 m (4 ft) wide [2.60 m^2 (28 ft^2)]. If this 8.4 m^2 (28 ft^2) island were installed under the 2017 *NEC* (2018 IRC), it would require *one* receptacle outlet. If this same 2.60 m^2 (28 ft^2) island were installed under the 2020 *NEC* (2021 IRC), it would require *three* receptacle outlets. Installed under the 2023 *NEC* (2024 IRC), this same 28 sq. ft. island would require *zero* receptacle outlets, but a wiring method to the island would be required.

SEPARATE KITCHEN COUNTERTOP SPACES FOR WALL-CONNECTED COUNTERTOPS

Kitchen countertop spaces separated by range tops, refrigerators, or sinks are generally considered separate countertop spaces, applying the same rules as wall-countertop spaces. This was addressed in the Code at one time [see 2017 NEC 210.52(C)(4) and 2018 IRC E3901.4.4], but is no longer specifically addressed in the Code, as this is considered general knowledge. This means each side of one of these separators is considered an individual countertop space, and measurements must be made to determine if a receptacle is required for each space. An example of this would be a range top installed at a dwelling unit's kitchen wall, separating two separate countertop spaces.

KITCHEN COUNTERTOP RECEPTACLE OUTLET LOCATIONS

Receptacle outlets that serve countertops or work surfaces in kitchens, pantries, etc. are required to be located on or above, but not more than 500 mm (20 in.) above the countertop or work surface in order to serve that countertop or work surface [*NEC* 210.52(C)(3); IRC E3901.4.3]. Beginning with the 2023 *NEC* (2024 IRC), receptacle outlets are not permitted to be installed below a countertop or work service and serve that countertop or work surface. Keep in mind that it is very common for the backside of a kitchen island or peninsular to be considered wall space for the nook, living room, dining room, etc. In that case, the backside of that island or peninsular would require a receptacle outlet to serve that wall space (see Photos 11.10 and 11.11). This creates a receptacle outlet installed "below" the countertop, but that receptacle outlet is not intended to serve that countertop; it is intended to serve the nook, living room, dining room, etc.

In addition, receptacles are not permitted to be installed in a face-up position on the work surfaces or countertops [*NEC* 406.5(G)(1) or IRC E4002.16]. A receptacle outlet that is not readily accessible by an appliance garage or an appliance that is fastened in place or otherwise restricts ready access to the receptacle cannot be considered one of the required receptacles. Receptacle outlet assemblies listed for the application are also permitted to be installed in kitchen countertops and work surfaces. New technology and design concepts in recent years have produced innovative receptacle construction and housing designs for receptacles to be safely installed in countertops. These listed outlet assemblies will allow for receptacles to be installed flush in kitchen and bathroom countertops. Some of these new design concepts consist of a hydraulic cylinder that actuates with pressure being placed on top of the cover. Upon release, the hydraulic cylinder raises providing a standard duplex receptacle mounted in the vertical position above the countertop. These listed assemblies have passed the water saturation test while in the closed position. Additionally, when the receptacle of these listed assemblies is closed below the countertop, they are de-energized (see Figure 11.26 and photo 11.11).

Previous editions of the *Code* limited the installation of kitchen countertop receptacle outlets mounted below the countertop to only locations constructed for the physically impaired or to countertops that are flat across their entire surface with no backsplash. The below-counter restrictions were eliminated for the 2020 NEC and the 2021 IRC and are no longer relevant to the current edition of the Code.

DWELLING UNIT BATHROOMS

At least one receptacle outlet is required to be installed in bathrooms within 900 mm (3 ft) of the outside edge of each basin (sink). This bathroom receptacle outlet must be located

Photo 11.11. Listed receptacle outlet assembly for installation in a countertop.

Photo 11.12. Kitchen island countertop space with a receptacle outlet is installed "below" the countertop to serve the adjacent living room (not to serve the kitchen countertop) extends more than 150 mm (6 in.) beyond its support base (cabinet)

on a wall or a partition adjacent to the sink or sink countertop. Typically, this means a wall or a partition that is in contact with the countertop in which the basin is installed (see photo 11.13). Receptacle outlet assemblies listed for the application are also permitted to be installed on or in bathroom countertops (see photo 11.11). A receptacle outlet is also permitted to be mounted in the side or face of the sink cabinet that supports the sink, but not more than 300 mm (12 in.) below the top of the sink or the sink countertop (see photo 11.14) [*NEC* 210.52(D) or IRC E3901.6].

Additional receptacle outlets are permitted in the bathroom area. The bathroom receptacle outlets required by *NEC* 210.52(D) or IRC E3901.6 are required to be supplied by one or more 120-volt, 20-ampere circuit(s) supplying only the bathroom receptacle outlets. Any countertop or similar work surface receptacle outlets with an exception that will be discussed in the next paragraph [*NEC* 210.11(C)(3) or IRC E3703.4]. All receptacles located in the bathroom must have ground-fault protection for personnel [*NEC* 210.8(A)(1) or IRC E3902.1]. Receptacle outlets are not permitted to be installed in a face-up position in the work surfaces or countertops in a bathroom

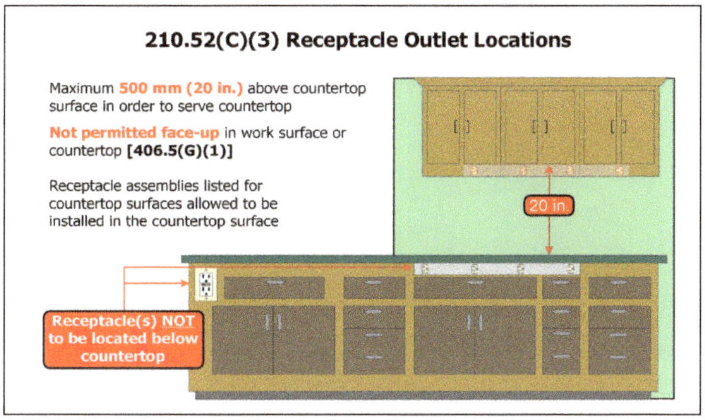

Figure 11.24. Counter receptacle outlet location requirements

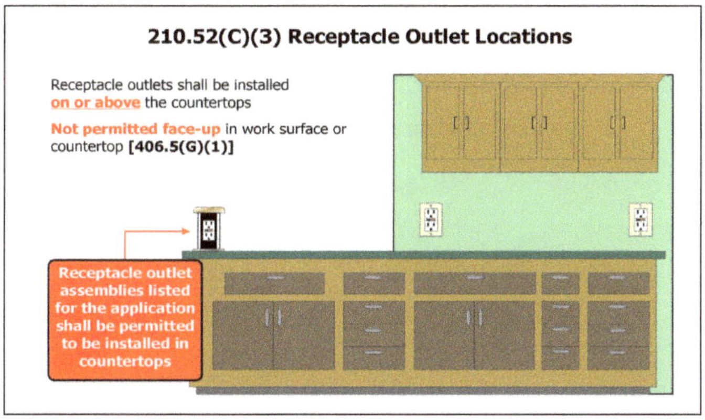

Figure 11.25. Receptacle outlet assemblies listed for the application are permitted to be installed in kitchen countertop

[*NEC* 406.5(G)(1) or IRC E4002.16].

Branch circuits for bathrooms are permitted to be installed in one of two configurations. These bathroom branch circuits can be installed (1) with one or more 20-ampere bathroom branch circuit(s) supplying only the receptacle outlets in all the bathrooms or (2) (by exception) with a 20-ampere bathroom branch circuit for *each* bathroom providing power to those bathroom receptacle outlets along with the luminaires and other equipment such as an exhaust fan in that same bathroom. At a minimum, the *Code* requires at least one 120-volt, 20-ampere branch circuit for the supply of all bathroom receptacle outlet(s) required by *NEC* 210.52(D) or IRC E3901.6 in the dwelling unit. This general rule requires this bathroom circuit(s) to have no other outlets. This is in response to the increased use of electrical grooming appliances (hair dryers, etc.), which is placing a greater demand on the bathroom branch circuits.

As discussed above, an exception to this rule permits a 120-volt, 20-ampere circuit to supply outlets for other equipment, such as lighting and exhaust fans, as well as the receptacle outlets, in a single bathroom. In this case, the 20-ampere circuit can only supply loads in that bathroom it supplies [see figure 11.28; *NEC* 210.11(C)(3) and Exception

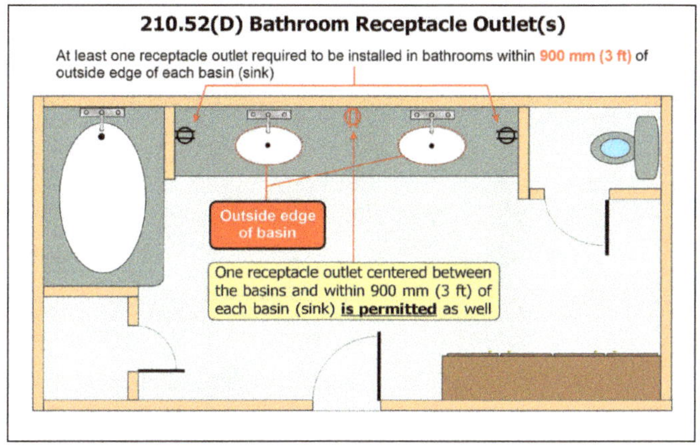

Figure 11.26. Receptacle placement in dwelling unit bathrooms relative to sink locations

or IRC E3703.4 and Exception].

OUTDOOR RECEPTACLE OUTLETS

The *Code* requires at least two 125-volt, 15- or 20-ampere outdoor receptacle outlets to be readily accessible from grade level and not more than 2.0 m (6½ ft) above grade level for a one-family dwelling and for each unit of a two-family dwelling unit that is at grade level. A receptacle outlet is required at both the front and back of the dwelling. Of course, these receptacle outlets must have GFCI protection per *NEC* 210.8(A)(3) or IRC E3902.3 for outdoor receptacle outlets [*NEC* 210.52(E)(1) or IRC E3901.7].

The *Code* also requires at least one 125-volt, 15- or 20-ampere receptacle outlet to be installed at all balconies, decks, and porches that are within 102 mm (4 in.) horizontally of the dwelling unit. This receptacle outlet(s) must be accessible from the balcony, deck, or porch and located not more than 2.0 m (6½ ft) above the balcony, deck, or porch walking surface. The IRC calls for these outdoor receptacle outlets at balconies, decks, or porches when the balconies, decks, or porches are accessible from inside of the dwelling unit.

Previous editions of the *NEC* required this same specific condition: the balcony, deck, or porch had to be "accessible from inside the dwelling unit." This was a key condition that was tied to the history of this requirement. When this requirement first appeared in the 2008 *NEC*, there was an exception for a balcony with "less than 1.86 m^2 (20 ft^2)." This exception was, in effect, to eliminate the need for a receptacle outlet at a small balcony that was attached to the building simply for aesthetics (looks) and had no access from the living quarters. Later, this exception was dropped, but the condition that the balcony be "accessible from inside the dwelling unit" remained *(until the 2020 NEC)* to still elim-

Photo 11.13. Receptacle outlet installed on the roughed-in centered between the two future basins of a dwelling unit bathroom

Photo 11.14. The Receptacle outlet located in the knee space above is permitted as the required receptacle outlet to serve the basin per *NEC* 210.52(D)

inate the need for an outdoor receptacle outlet at a balcony that no one had access to unless one resorted to climbing a ladder from the outside. Beginning with the 2020 edition of the *NEC*, a 125-volt, 15- or 20-ampere outdoor receptacle outlet is required at every balcony, deck, or porch regardless of its size or accessibility from the living quarters as long as the balcony, deck, or porch is located within 102 mm (4 in.) horizontally of the dwelling unit.

The *Code* requires a 125-volt, 15- or 20-ampere receptacle for servicing air-conditioning and refrigeration equipment. This receptacle is required to be not less than 7.5 m (25 ft) from the equipment and on the same level as the equipment and cannot be connected to the load side of the equipment's branch-circuit disconnecting means. The receptacle must be in an accessible location. A receptacle outlet installed to meet *NEC* 210.52(E) or IRC E3901.7 (outdoor outlets) can be used to meet this service receptacle outlet requirement if properly located. This receptacle outlet is not required at one- and two-family dwellings for the service of evaporative coolers. See Chapter 9 for additional information about this required receptacle outlet for use when servicing heating, ventilating, and air-conditioning equipment (*NEC* 210.63

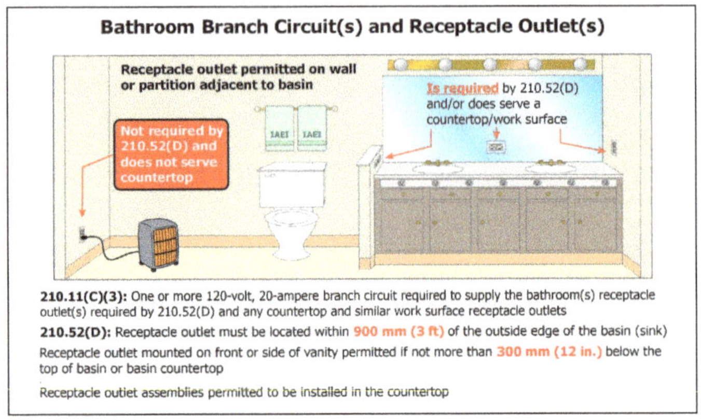

Figure 11.27. Bathroom receptacle outlets are permitted to be supplied by one or more 120-volt, 20-ampere circuit(s) and mounted on the front or side of the vanity

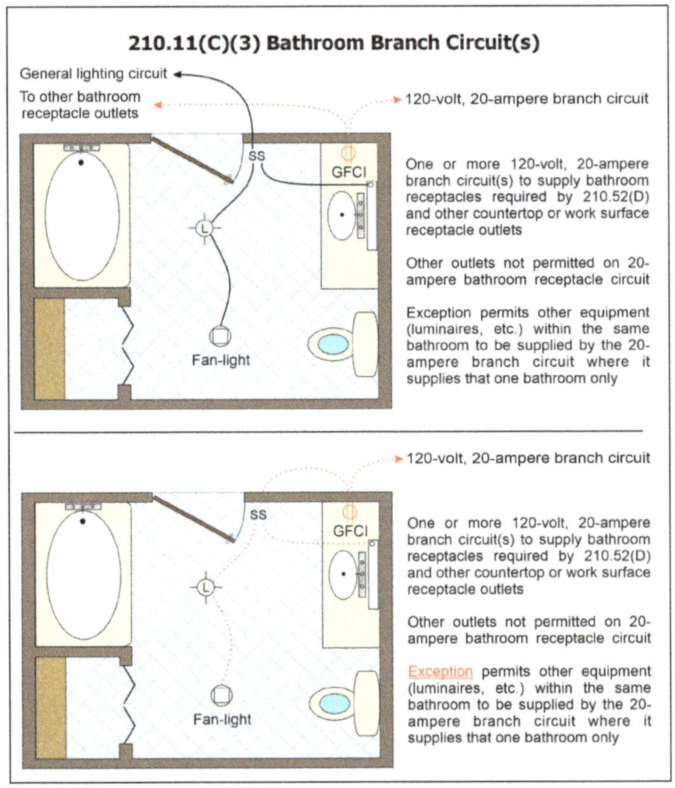

Figure 11.28. At least one or more 20-ampere branch circuits are required for dwelling bathroom(s)

and 210.63(A) or IRC E3901.12).

This A/C service receptacle outlet is required to be GFCI-protected by *NEC* 210.8(E). If this receptacle outlet is located outside, it would demand GFCI protection from *NEC* 210.8(A)(3) or IRC E3902.3 as well.

All 15- or 20-ampere, 125-volt receptacle outlets installed in a wet location, such as outdoors and exposed to the weather, are required to be installed in a weatherproof enclosure. The enclosure must be weatherproof regardless of whether or not an attachment plug is inserted. An outlet box hood cover installed for this purpose

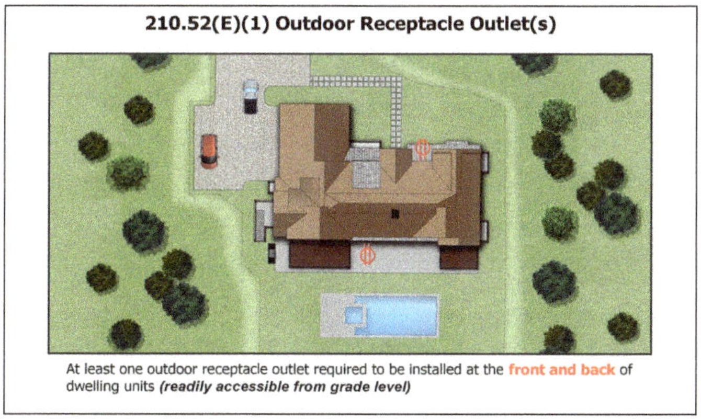

Figure 11.29. Outdoor receptacle outlets are required at both front and back of dwellings

Photo 11.15. Receptacle outlets are required at all balconies, decks, and porches located within 102 mm (4 in.) horizontally of the dwelling unit

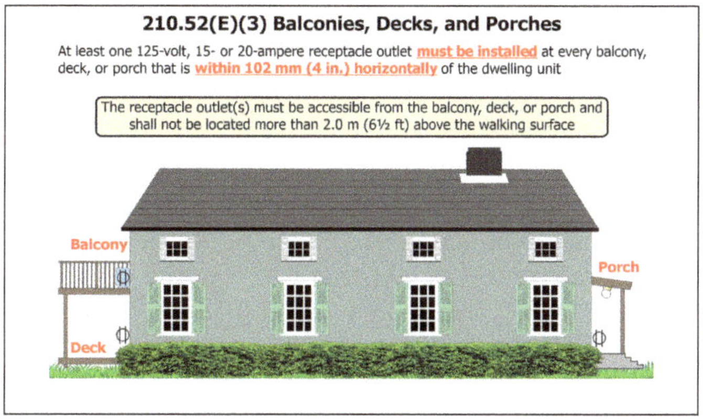

Figure 11.30. Receptacles at balconies, decks, and porches

Photo 11.16. Receptacle outlet required to be accessible from the balcony, deck, or porch where the balcony, deck, or porch is located within 102 mm (4 in.) horizontally of the dwelling unit

must be listed and identified as an "extra duty" type hood cover. All 15- and 20-ampere, 125- and 250-volt non-locking receptacles installed in damp or wet locations are required to be of the listed weather-resistant type [*NEC* 406.9(A) and (B)(1); IRC E4002.8 and E4002.9].

Determining if a location is a damp, wet, or dry location is the responsibility of the authority having jurisdiction (AHJ) utilizing the damp, wet, or dry location definitions located in *NEC* Article 100 or IRC Chapter 35. See Chapter 13 for additional information on receptacle outlets installed in damp and wet locations.

LAUNDRY AREAS

At least one receptacle outlet is required to be installed for the laundry equipment in dwelling units [*NEC* 210.52(F) or IRC E3901.8]. Laundry outlet(s) must be supplied by a 20-ampere branch circuit. This branch circuit is not permitted to have other outlets other than receptacle outlets in the laundry area [*NEC* 210.11(C)(2) or IRC E3703.3]. More than one receptacle outlet is permitted to be supplied by the 20-ampere laundry circuit as long as they are located in the same laundry area. Other receptacle outlets connected to a different circuit are permitted to be located in the laundry room. In an apartment or multifamily building dwelling unit where laundry facilities are provided on the premises and are available to all building occupants, a laundry receptacle outlet is not required in the unit itself [*NEC* 210.52(F) Ex. No. 1].

Where laundry outlet(s) are installed in a dwelling unit to be used for specific appliances, such as laundry equipment, the laundry receptacle outlet(s) must be located not further than 1.8 m (6 ft) from the intended location of the appliance. Laundry equipment includes appliances such as a washing machine, a gas clothes dryer, or a clothes iron [*NEC* 210.50(C) or IRC E3901.5].

BASEMENTS, GARAGES, AND ACCESSORY BUILDINGS

At least one 125-volt, 15- or 20-ampere receptacle outlet is required to be installed in each separate unfinished portion of a basement, in each attached garage and each detached garage with electric power, and each accessory building supplied with electric power. This receptacle outlet is in addition to those that are provided for specific equipment in these areas, such as laundry equipment or a sump pump in a one-family dwelling. Receptacle outlets provided for the purpose of supplying a permanently installed premises security system (burglar or fire alarm system) cannot be considered as meeting this requirement [*NEC* 210.52(G) or IRC E3901.9].

In each attached garage and each detached garage with electric power, at least one receptacle is required to be installed in each vehicle bay and located not more than 1.7 m (5½ ft) above the floor. This would call for three receptacles for a 3-car garage, etc. This reduces the use of extension cords that might otherwise be used to extend the branch circuit wiring and provides a safer environment for the homeowner. The garages encountered today at most dwelling units have gone from a simple place to park vehicles out of the elements in a bygone era to "do-it-all" locations where homeowners service their vehicles and/or convert a portion of the garage space to serve as a workshop. It is not uncommon in these situations for the homeowner to resort to running an extension cord from a receptacle outlet behind an appliance or the garage door receptacle outlet, stapling the cord to the ceiling and down the wall to have an additional outlet for convenience use for such things as hand drills, car vacuum cleaners, etc.

At least one 120-volt, 20-ampere branch circuit is required to be provided in these attached garages and in detached garages with electric power to supply the garage receptacle outlets, including the receptacle outlets required by *NEC* 210.52(G)(1) or IRC E3901.9. Additional branch circuits rated at 15-ampere or greater are permitted to serve other receptacle outlets in these garage areas other than the receptacle outlets required by *NEC* 210.52(G)(1) or IRC E3901.9. An exception permits the required 120-volt, 20-ampere garage receptacle outlet branch circuit to also supply outdoor receptacle outlets.

A lot of the appliances and tools used in today's dwelling unit garages are rated at 12 to 16 amperes or even higher and demand at least a 20-ampere-rated branch circuit. A 15-ampere rated branch circuit in the modern dwelling unit garage is typically not sufficient for these loads. A small, portable, 120-volt, 1-horsepower air compressor, having an ampacity of 16 amperes, would be an example of such a tool resulting in this 20-ampere rated branch circuit requirement. Having these branch circuits generally limited to just the garage will greatly assist in the ability of these garage receptacle outlets to possess enough capacity to handle a load such as a 120-volt electric vehicle (EV) charging station or electric vehicle supply station (EVSE). With this EV charging concept in mind, it makes

Photo 11.17. Outlet box hood covers are required to be identified as the "extra duty" type (metallic or nonmetallic)

Photo 11.18. The laundry circuit installed at the rough-in wiring stage was installed within 1.8 m (6 ft) of the intended laundry equipment location (receptacle for dryer also installed at right)

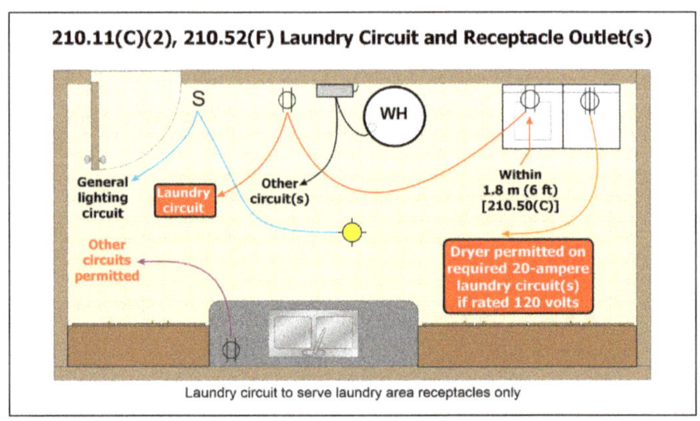

Figure 11.31. Receptacle is required within 1.8 m (6 ft) of the intended laundry equipment location

sense to expand the requirements for receptacle outlets in dwelling unit garages to provide at least one receptacle outlet in each vehicle bay space. [*NEC* 210.11(C)(4) and 210.52(G)(1) or IRC E3703.5 and E3901.9]. When it comes to electric vehicles and an EVSE, keep in mind that an outlet installed for the purpose of supplying an EVSE greater than 16 amperes or greater than 120 volts must be supplied by an *individual* branch circuit with no other outlets (*NEC* 625.40).

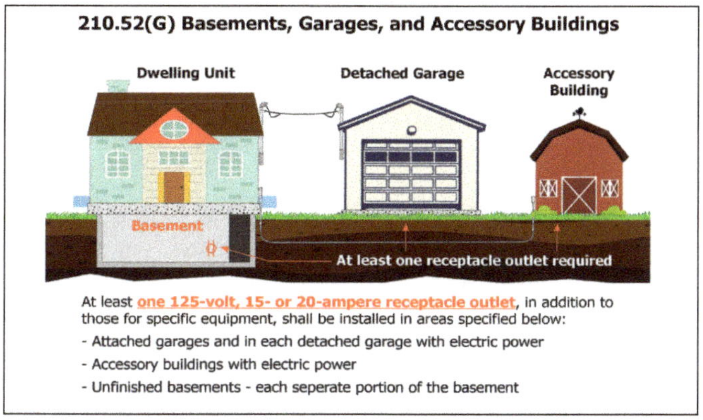

Figure 11.32. Receptacle outlet(s) required for basements, garages, and accessory buildings (with power)

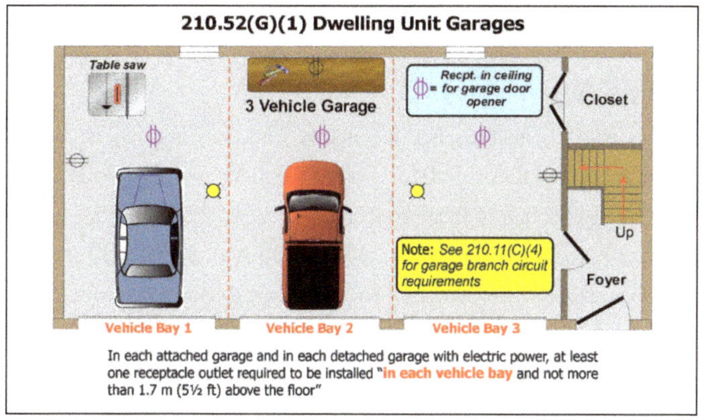

Figure 11.33. Dwelling unit garage receptacle requirements

Photo 11.19. The dwelling unit with a three-vehicle bay attached garage would require a minimum of (3) receptacle outlets in the garage area

A new exception was added for the 2023 *NEC* that allows the 20-ampere garage branch circuit that supplies a single-vehicle bay garage to supply the required receptacle outlet(s) and all other outlets for other equipment within the same garage. Note that the 2023 *NEC* references *NEC* 210.23(A)(1) and (A)(2) at this new exception. This reference should be *NEC* 210.23(B)(1) and (B)(2).

At least one 125-volt, 15- or 20-ampere receptacle outlet is required to be installed in each accessory building that is provided with electric power. It is not uncommon to provide electric power to an accessory building for a minimum amount of lighting in these outbuildings. If electrical power is provided for any reason, at least one 125-volt, 15- or 20-ampere receptacle outlet is also required. Often, accessory buildings are used and treated as such things as workshops, home office workspaces, or art studios. When an accessory building is used in such a manner, the necessity of receptacle outlet(s) is warranted [*NEC* 210.52(G)(2) or IRC E3901.9].

At least one 125-volt, 15- or 20-ampere receptacle outlet is also required to be installed in each separate unfinished portion of a basement. Where a portion of the basement is finished into a habitable room(s), a receptacle outlet must be located in the unfinished portion of the basement or each separate unfinished portion of the basement [*NEC* 210.52(G)(3) or IRC E3901.9].

GFCI protection is required for these receptacle outlets (s) in these basement, garages, and accessory building areas as well [*NEC* 210.8(A)(2) and (A)(5) or IRC E3902.2 and E3902.5]. An exception to this GFCI requirement applies to a single receptacle that provides power to a permanently installed premises security system (burglar or fire alarm system). This exception to GFCI protection is consistent with provisions at *NEC* 760.41(B) and 760.121(B) as well as with NFPA 72, The National Fire Alarm *Code*.

The *Code* also requires a 125-volt, 15- or 20-ampere receptacle for heating, air-conditioning, and refrigeration equipment. This receptacle is required to be not less than 7.5 m (25 ft) from the equipment and on the same level as the equipment. This receptacle must be in an accessible location. If this heating, air-conditioning, or refrigeration equipment happens to be located in a basement, garage, and accessory building, a receptacle outlet installed to meet *NEC* 210.52(G) or IRC E3901.9, *(Basements, Garages, and Accessory Buildings)* could be used to meet this HVAC receptacle outlet requirement as well if properly located. See Chapter 9 for additional information about this required receptacle outlet for use when servicing heating, ventilating, and air-conditioning equipment (*NEC* 210.63 and 210.63(A) or IRC E3901.12).

HALLWAYS

At least one 125-volt, 15- or 20-ampere receptacle outlet is to be installed in a hallway when the hallway is of 3.0 m (10 ft) or more in length. This rule will help to avoid having to pass an appliance cord through a doorway for connection to a receptacle outlet in a bedroom for such things as a vacuum cleaner. The length of the hallway is considered to be the length along the centerline of the hallway without passing through a doorway [*NEC* 210.52(H) or

IRC E3901.10].

FOYERS

Foyers that are not part of a hallway in accordance with *NEC* 210.52(H) or IRC E3901.10 and that have an area greater than 5.6 m^2 (60 ft^2) are required to have a 125-volt, 15- or 20-ampere receptacle(s) located in each wall space 900 mm (3 ft) or more in width and unbroken by doorways, floor-to-ceiling windows, and similar openings [*NEC* 210.52(I) or IRC E3901.11].

This foyer requirement is *not* required to follow the requirements found in *NEC* 210.52(A) or IRC E3901.2 for receptacle wall spacing requirements *[no point measured horizontally along the floor line in any wall space is more than 1.8 m (6 ft) from a receptacle outlet]*. Under this foyer requirement, it does not matter if a foyer wall space is 900 mm (3 ft) wide or 4.3 m (14 ft) wide; only one receptacle outlet would be required. Foyers or entryways with a floor area larger than 5.6 m^2 (60 ft^2) is the minimum area requirement for this section to exempt smaller foyers that typically do not have the space to facilitate home furnishings and electrical equipment being located in these areas of the dwelling. Some of today's larger dwelling units have foyers the size of bedrooms. Before this foyer receptacle outlet requirement became part of the *Code*, these large foyers may have had only one receptacle outlet and, in some cases, no receptacle outlets at all. This often leads the homeowner to resort to the use of extension cords placed under rugs or across doorways as a substitute for the fixed wiring. Properly placed receptacle outlets should be available for the connection of table lamps and other cord-connected products typically found in larger foyers.

LIGHTING OUTLETS REQUIRED

At least one lighting outlet controlled by a listed wall-mounted control device *(such as a snap-switch or occupancy sensor)* is required to be installed in every habitable room, kitchen, laundry area, and in every bathroom. The wall-mounted control device is required to be located near an entrance to the room on a wall. Some luminaires and ceiling-mounted paddle fans are now controlled by a remote-control device that does not require a physical wire from the switching device to the luminaires and/or ceiling-mounted paddle fans. In these cases, the switch or wall-mounted control device cannot rely solely on a battery for power unless a means is provided for automatically energizing the lighting outlet upon the battery failure [*NEC* 210.70 and 210.70(A)(1) or IRC E3903.2].

In other than kitchens, laundry areas, and bathrooms,

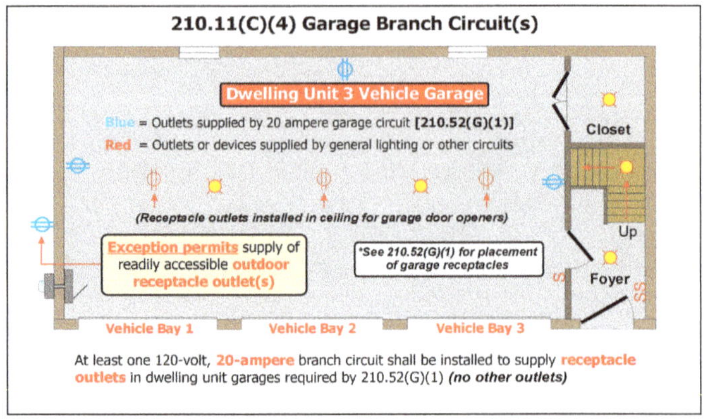

Figure 11.34. At least one 20-ampere branch circuit required for garage receptacle outlets required by *NEC* 210.52(G)(1)

Photo 11.20. Dwelling unit accessory building with power

Photo 11.21. Receptacle outlet required in each portion of a dwelling unit unfinished basement

one or more receptacle(s) that are controlled by a listed wall-mounted control device are permitted in lieu of the required lighting outlets [*NEC* 210.70(A)((1) Ex. No. 1 or IRC E3903.2 Ex. No. 1].

The *NEC* and the IRC define a "Habitable Room" as "a room in a building for living, sleeping, eating, or cooking, but excluding bathrooms, toilet rooms, closets, hallways, storage or utility spaces, and similar areas (see *NEC* Article 100 or IRC Chapter 35). The IRC similarly defines a "Habitable Space" as "a space in a building for living, sleeping, eating or cooking. Bathrooms, toilet rooms, closets, halls, storage or

Photo 11.22. Hallway receptacle outlet at rough wiring stages

Photo 11.23. Foyer area at dwelling unit front door area

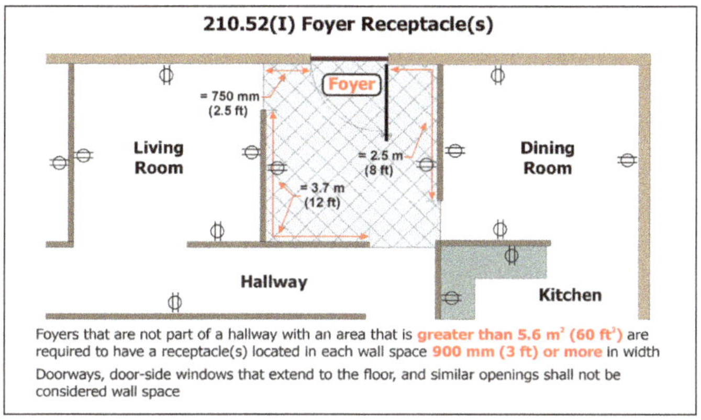

Figure 11.35. Receptacle outlet requirements for foyers

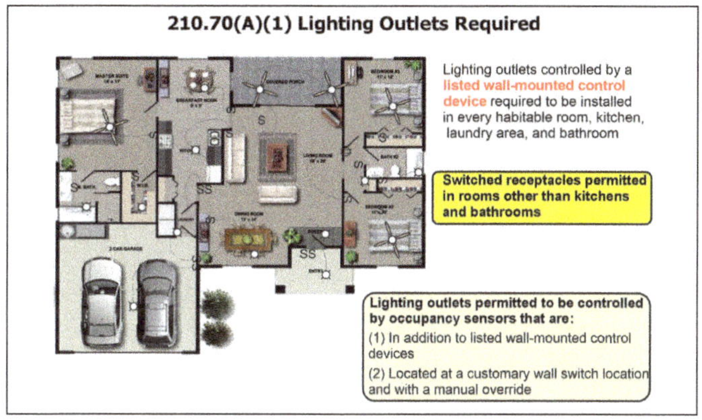

Figure 11.36. Lighting outlets required (controlled by a listed wall-mounted control device)

utility spaces and similar areas are not considered habitable spaces."

Exception No. 2 to *NEC* 210.70(A)(1) or IRC E3903.2 permits control of the required lighting outlet with occupancy sensors where the occupancy sensor(s) is:

1. in addition to listed wall-mounted control devices, or
2. located at a customary wall switch location and equipped with a manual override that will allow the sensor to function as a wall switch.

ADDITIONAL LIGHTING OUTLETS REQUIRED

Additional locations where wall switch-controlled lighting outlets are required are identified in the *Code* at *NEC* 210.70(A)(2) or IRC E3903.3. At least one lighting outlet controlled by a listed wall-mounted control device must be installed in hallways, stairways, attached garages, detached garages, and accessory buildings with electrical power.

For dwelling units, attached garages, and detached garages with electric power, the **exterior side of outdoor entrances or exits** with grade-level access must be illuminated with at least one exterior lighting outlet controlled by a listed wall-mounted control device. A remote source of illumination is permitted. No minimum amount of illumination level (foot candles) is specified. The location of the switch for these outdoor luminaires is not specified by the *Code* other than the statement made in the main rule of *NEC* 210.70(A)(1) or IRC E3903.2, which says wall-mounted control devices are to be located near an entrance to the room on a wall (these "additional locations" are not located at *NEC* 210.70(A)(1) or IRC E3903.2). Good work practice would suggest that these listed wall-mounted control devices be located near the outdoor entrance or exit doorways closest to the outdoor luminaire. A switch is not required at a vehicle door in a garage. A vehicle door (such as an overhead garage door) is not considered an outdoor entrance for the purpose of determining the necessity of an outdoor lighting outlet. However, a switch can be installed at a vehicle door as an option [*NEC* 210.70(A)(2) or IRC E3903.3].

A new exception related to the lighting outlet for outdoor entrances or exits with grade-level access having to be located on the exterior of the building was added for the 2023 edition of the *NEC*. For an outdoor, grade-level, bulkhead door with stairway access to a sub-grade-level basement, the required lighting outlet that provides illumination on the stairway steps is permitted to be located in the basement *interior* within 1.5 m (5 ft) horizontally of the bottommost stairway riser. This interior lighting outlet is permitted to be controlled by a listed wall-mounted control device or by a unit switch of the interior luminaire or interior lampholder [see *NEC* 210.70(A)(2)(2), Exception to (2)].

Lighting outlets are required for **interior stairways**. No minimum illumination level (foot candles) is specified by the *Code*. The lighting outlet for interior stairs must have a listed wall-mounted control device at each floor level where the difference in floor levels is six or more risers (steps). Where an intermediate landing is in the stairway, and a stairway entry is provided at the landing location, a listed wall-mounted control device is required to control the stairway lighting at the landing location as well [*NEC* 210.70(A)(2)(3) or IRC E3903.3.1].

In hallways, stairways, or at outdoor entrances, remote, central, or automatic control of lighting is permitted [*NEC* 210.70(A)(2), Exception to (1), (2), and (3) or IRC E3903.3.1, Exception]. Dimmer control of lighting outlets installed in interior stairways with six or more risers (steps) between floor levels (with at least two different switching devices, one at each level) must have the capacity to provide the full range of dimming control at *each location*. This will allow each dimmer control device the ability to achieve maximum brightness at each control location for the interior stairway illumination [*NEC* 210.70(A)(2)(4)].

At least one lighting outlet containing a switch or listed wall-mounted control device is required for spaces such as **attics**, **underfloor spaces**, **utility rooms**, and **basements**, where these spaces are used for storage or have equipment that requires servicing. No minimum illumination levels are required. A point of control for lighting in these storage or equipment spaces must be located at each entry that permits access to the attic or underfloor space, utility room, or basement. A lighting outlet containing a switch (such as a pull chain) may be used to meet this requirement, provided a means to operate the switch is located at the point of entry to the space [*NEC* 210.70(C) or IRC E3903.4].

Crawl spaces often contain equipment requiring servicing. Where a lighting outlet is installed for equipment requiring service, the lighting outlet is required to be installed at or near the equipment requiring servicing. Ground-fault circuit-interrupter (GFCI) protection is required for lighting outlets (not exceeding 120 volts) installed in a crawl space [*NEC* 210.8(C) and IRC E3902.4]. See Chapter 10 in this book for additional information on GFCI requirements.

Equipment requiring servicing includes heating and air-conditioning equipment, water heaters, central vacuum system power units, laundry equipment, and similar equipment. It is not the intent of this rule to require a separate lighting outlet for equipment if other listed wall-mounted control devices or switch-controlled lighting outlets, such as room or space lighting outlets, provide adequate illumination.

Photo 11.24. A switch is required at each level of an interior stairway with six risers or more

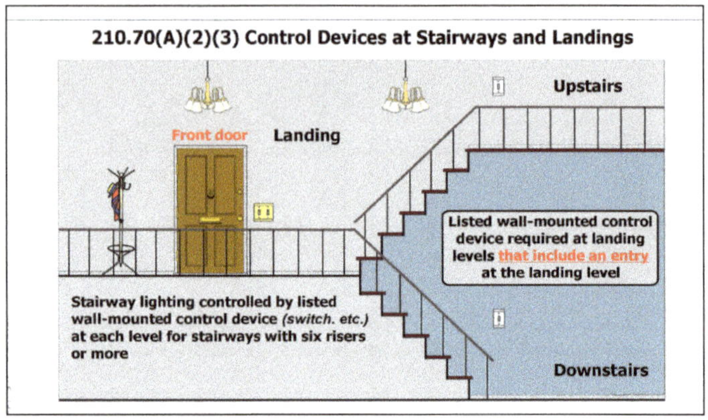

Figure 11.37. Stairway lighting outlet(s) and control device (switch, etc.) requirements

Similarly, illumination is required for all working spaces for service equipment and panelboards installed indoors, and this light source cannot be controlled by automatic means only, such as a motion sensor, which controls all illumination within the workspace. Additional lighting outlets are not required where the workspace is illuminated by an adjacent light source, such as the light source for the room or garage containing the panelboard [*NEC* 110.26(D) or IRC E3405.7].

WIRING METHODS

Chapter 3 of the *NEC* covers wiring methods and materials. *NEC* Article 300 *(General Requirements for Wiring Methods and Materials)* includes general requirements for wiring and electrical installations with emphasis placed on protection from physical damage, general securing, and support requirements, rules for conductors, underground and aboveground wiring provisions, and so forth. *NEC* Chapter 3 includes requirements for boxes and cabinets and a list of wiring

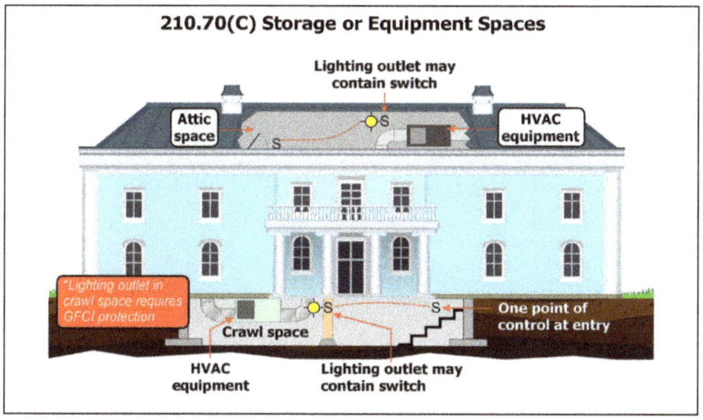

Figure 11.38. A switch-controlled lighting outlet or lighting outlet controlled by a listed wall-mounted control device is required near mechanical equipment

Photo 11.25. Switched lighting outlet at the entrance to attic space where equipment is located

methods for use in electrical systems. Chapter 3 of the *NEC* also provides articles dedicated to each wiring method. For example, nonmetallic-sheathed cable (Type NM) is located in *NEC* Article 334, and rigid metal conduit is found in *NEC* Article 344, and so on. The most common type of wiring method employed in modern dwelling unit construction is by far in the form of nonmetallic-sheathed cable (Type NM cable), which consists of two or more individually insulated conductors contained inside a protective polyvinyl chloride (PVC), nylon sheathing. Type NM cable usually contains one or more "hot" (current-carrying) insulated conductors, an insulated grounded (neutral wire), and an uninsulated or bare equipment grounding conductor wrapped in a paper insulator.

Within each wiring method article are requirements regarding installation and uses specific to that wiring method. Additionally, each wiring method article provides some key information, such as uses permitted and uses not permitted. Key definitions related to each article and a definition of each wiring method can be found in *NEC* Article 100. For example, Cable, Nonmetallic-Sheathed (Type NM) is defined as "insulated conductors enclosed within an overall nonmetallic jacket." Additionally, other requirements, such as securing and support, listing requirements, conductor fill, bending radius, construction specifications, etc., are provided within each wiring method article. This information becomes an important resource when deciding which type of wiring method is to be used for a one- or two-family dwelling electrical system. Some of the specifics about each method are provided in the IRC in a tabular format. The *NEC* would need to be referenced by default for all the specific information concerning each particular wiring method.

The IRC handles wiring methods for one- and two-family dwellings a bit differently in the way the information is laid out in the *Code* itself. The allowable wiring methods are listed for installers in a tabular format (see Table 11.3 of this publication for a reproduction of IRC Table E3801.2).

Moving further into IRC Chapter 38, the uses permitted and not permitted are also included in a table, with the wiring methods listed horizontally across the top of the table and the application listed in a vertical column on the left (see IRC Table E3801.4). Important footnotes relative to use requirements are also provided at the bottom of this IRC table. Just as notes to tables in the *NEC*, these are applicable as *Code* requirements, not just information. IRC Table E3802.1 provides general support and installation requirements for each method allowed for one- and two-family dwellings. This table is set up similarly with the wiring method listed horizontally across the top and the installation requirements, such as support, number of bends, etc., listed vertically in the column on the left. As previously noted, one must refer to the *NEC* for more specific information relative to each wiring method, as this is not provided within the IRC.

A table for underground wiring methods that include minimum burial depths and specific location criteria is also provided in IRC Chapter 38 (see IRC Table E3803.1). This table also includes important footnotes that relate directly to the provisions in the table. This table is derived from the *NEC* and is identical to *NEC* Table 300.5.

As stated earlier, nonmetallic-sheathed cable (Type NM) is the most popular wiring method of choice for one- and two-family dwelling electrical systems and their attached or detached garages and storage buildings. Conduit and tubing are also used for wiring in dwellings but are not as common. Where conduit or tubing is used, the specific requirements for the conduit or tubing wiring method must be complied with. For example, where Type AC or Type MC cable is selected and installed as the wiring method for a dwelling unit, the specific rules for each of these cable methods must be applied. The *NEC* provides the specifics for Type AC cable in *NEC* Article 320, and Type MC cable requirements are found in *NEC* Article 330. Construction characteristics, securing and support provisions, protection from physical damage, and limitations of use are provided within these articles. Some areas and jurisdictions place local limitations and restrictions on the types of wiring permitted for residential wiring. Others will not permit cable types of wiring but require conduit or tubing. It is important for contractors, designers, etc., to be familiar with local codes and regulations when working within these jurisdictions. Being proactive in these circumstances is worth the effort. Always check with the enforcing agency when in doubt as to which *Code* edition has been adopted and is being used and if there are any local amendments or regulations that overlay the minimum requirements of *NEC* or IRC.

Nonmetallic-sheathed cables, such as Type NM, UF, SE, and SER, are the most commonly used wiring methods in one- and two-family dwelling electrical wiring. The following paragraphs will focus on the requirements for these types of cables.

TYPE NM CABLE

One of the important and often discounted characteristics of Type NM cable is the need to protect it from damage. Nonmetallic jacketed cables can be damaged during or after initial installation. The *Code* includes a plethora of rules and restrictions for installations of nonmetallic-sheathed cable wiring methods. The requirements discussed in the following paragraphs concerning Type NM cable can be found in *NEC* Article 334 and IRC Chapter 38.

Type NM and Type NMC cables are required to be listed. Type NM cable is permitted for both exposed and concealed work in normally dry locations. It also can be fished through air voids in masonry blocks or tile walls. Type NM cable cannot be embedded in masonry, adobe, concrete, fill, or plaster. It cannot be run in a shallow chase and covered with these materials. Type NMC can be run in a shallow chase and covered with adobe or plaster if properly protected from damage.

Type NM cable cannot be used where exposed or subject to excessive amounts of moisture or dampness. It can be used in locations that are normally dry. The cable may be exposed to incidental amounts of moisture during construction without permanent damage. The outer sheath is moisture-resistant. However, if the ends of the cable are submerged in water, damage to the cable from weeping may occur, and the cable should be replaced.

Manufacturers produce Type NM cables with distinguishing colors for the jackets or sheaths. For example, it is common to find the outer sheathing of 14-2 Type NM cable to be white in color, 12-2 may be yellow, and 10-2 may be orange. This is common and convenient for installers and inspectors but is *not* a requirement of the *Code*. Color-coding of the outer sheathing is handy for installers for things such as keeping 15-ampere circuits separate from 20-ampere circuits. It is also handy for quick identification of the load-side wiring of a GFCI device. Keep in mind that this is not a requirement of the *Code* and that all manufacturers may not produce their Type NM cable with this same color option on the jacket of their product. Those cables produced with colored jackets or sheaths can vary in color from manufacturer to manufacturer. Currently, there is no Code requirement or industry standard regarding this type of outer sheathing identification. It is important for installers and inspectors to always verify the identification markings on the cable when determining suitability for use and not rely just on the colors of the outer sheathings.

IRC Table E3801.2 Allowable Wiring Methods

Allowable Wiring Methods	Designated Abbreviation
Armored cable	AC
Electrical metallic tubing	EMT
Electrical nonmetallic tubing	ENT
Flexible metal conduit	FMC
Intermediate metal conduit	IMC
Liquidtight flexible conduit	LFC
Metal-clad cable	MC
Nonmetallic sheathed cable	NM
Rigid metallic conduit	RMC
Rigid polyvinyl chloride conduit (Type PVC)	RNC
Reinforced thermosetting resin conduit	RTRC
Service entrance cable	SE
Surface raceways	SR
Underground feeder cable	UF
Underground service cable	USE

Table 11.3. IRC Table E3801.2. Allowable Wiring Methods

PROTECTION OF NM CABLES

In exposed work, the cable must closely follow the building surface of the building finish. If this is not possible, the cable can be secured to the running boards. The cable is required to be protected from physical damage. Routing the cables through rigid raceways such as EMT or Schedule 80 PVC is a method of protecting cables exposed to physical damage.

Where Type NMC cable is installed in shallow chases in masonry, concrete, or adobe, the cable must be protected by a steel plate at least 1.59 mm (1⁄16 in.) thick and covered with plaster or adobe or a similar finish [*NEC* 334.10(B)(3)]. Several conditions apply to apply to NM and NMC cables running in unfinished basements and crawl spaces. If the cable is run at angles to the joists, the cable can be secured to the bottom of the joists as long as the conductors are not smaller than two (2) 6 AWG or three (3) 8 AWG. Smaller cables at angles to the joists must be run through bored holes or secured to running boards. Cables installed on the wall of an unfinished basement are permitted in listed conduit or tubing or be protected in accordance with *NEC* 300.4 or

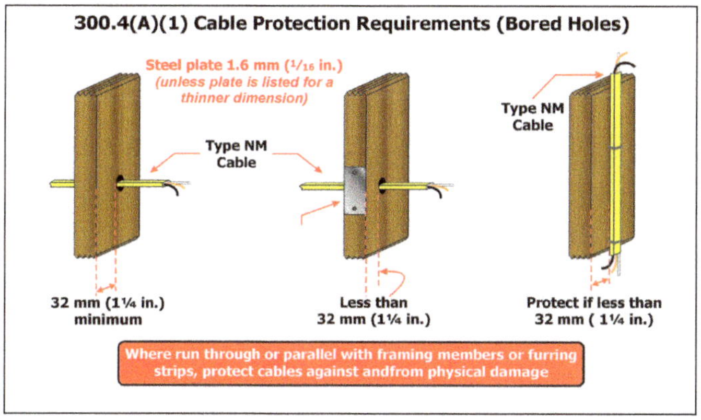

Figure 11.39. Installation and protection of Type NM cables

IRC Table E3802.1. Electrical metallic tubing (EMT) is a type of tubing identified for this type of physical protection. A nonmetallic fitting or bushing must be installed at the point where the cable enters the conduit or tubing. The fitting is intended to protect the cable from abrasion or damage as it is pulled through the raceway. The outer sheathing of the Type NM cable is required to extend through the conduit or tubing and into the outlet, device, or junction box not less than 6 mm (¼ in.). The cable also needs to be secured within 300 mm (12 in.) of the point where the cable enters the conduit or tubing (*NEC* 334.15(C) or IRC E3802.4).

The requirements of *NEC* 300.4 and IRC Table E3802.1 apply where cables are installed through or parallel to framing members. These rules require that the cables be set back from the face or edge of the framing member at least 32 mm (1¼ in.) or be protected by a steel plate at least 1.6 mm (1/16 in.) thick and of the appropriate length and width to cover the entire wiring area. Listed and marked steel plates less than the required 1.6 mm (1/16 in.) thick are permitted that afford equal or better protection against nail or screw penetration. This would include heat-treated type metals thinner than 1.6 mm (1/16 in.) but stronger than steel. Boxes, cable, or raceway-type wiring methods are also prohibited from being installed directly under metal-corrugated sheet roof decking unless the nearest outside surface of the cable or raceway is not less than 38 mm (1½ in.) from the nearest surface of the roof decking [see *NEC* 300.4(E)].

The *Code* requires listed bushings or grommets to be installed in metal framing members through which Type NM cable will be installed. The listed bushing or grommet is required to cover all metal edges and to be securely fastened in the opening prior to installation of the cable [*NEC* 300.4(B)(1) or IRC Table E3802.1].

Bends in nonmetallic cables have to be made so that the cable and conductors will not be damaged. The radius of the curve of the inner edge of the cable shall be at least five times the diameter of the cable. For flat cables, the major diameter dimension of the cable is used to determine the bending radius [*NEC* 334.24 or IRC E3802.5].

AMPACITY OF TYPE NM CABLE

The insulated conductors shall be rated for 600 volts and range from sizes 14 AWG through 2 AWG copper conductors and 12 AWG through 2 AWG aluminum or copper-clad aluminum conductors [*NEC* 334.104]. In addition to the insulated conductors, the cable must include an insulated or bare conductor to be used as an equipment grounding conductor only [*NEC* 334.108 or IRC E3908.9.3]. The insulation has to be rated for 90°C (194°F) [*NEC* 334.112 or IRC E3705.4.4]. The ampacity of Type NM cable is that of 60°C (140°F) conductors in accordance with *NEC* Table 310.16 or IRC Table E3705.1 [NEC 334.80 or IRC E3705.4.4. The overall covering of Type NM cable is required to be flame-retardant and moisture-resistant [*NEC* 334.116(A)].

Precaution needs to be exercised when installing more than two Type NM cables through the same opening in wood framing members. When more than two Type NM cables containing two or more current-carrying conductors are installed through the same hole (without maintaining spacing). That hole is fire- or draft-stopped using thermal insulation, caulk, or sealing foam, the allowable ampacity of each conductor is required to be adjusted in accordance with *NEC* Table 310.15(C)(1) *(Adjustment Factors for More Than Three Current-Carrying Conductors in a Raceway or Cable)*, or IRC Table E3705.3 [*NEC* 334.80 or IRC E3705.4.4]. The *Code* does not identify what distance is required between cables in order to "maintain spacing" between said cables. This is going to be an AHJ call.

As an example, if six (6) 12-2 (with ground) Type NM cables were installed through a single fire-caulked drilled hole in a top plate, that would equal twelve (12) current-carrying conductors. According to *NEC* Table 310.15(C)(1) or IRC Table E3705.3, those twelve (12) current-carrying conductors could only be loaded to *50 percent* of their ampacity values from *NEC* Table 310.16 or IRC Table E3705.1. The provisions of *NEC* 310.14(A)(2), Exception or IRC E3701.3, Exception, (two different ampacities) permit the higher ampacity to be used if the higher ampacity distance is less than or equal to 3.0 m (10 ft) or 10 percent of the total circuit length. However, for adjustment factors for Type NM cable as described above, NEC 334.80 or IRC E3705.4.4 indicate that the exception to NEC 310.14(A)(2) or IRC E3701.3 does not apply in this application.

Also note that where more than two Type NM cables con-

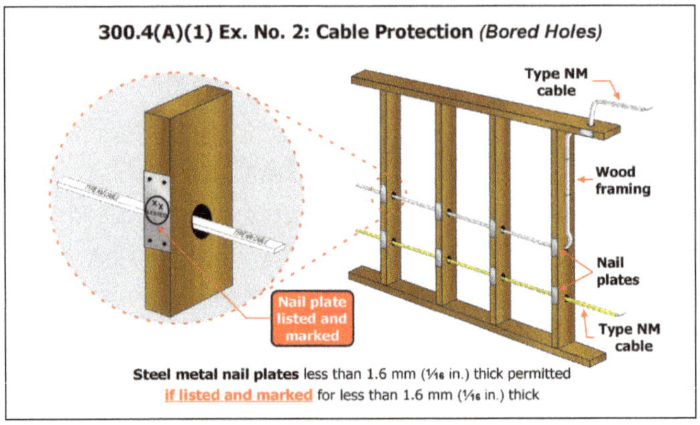

Figure 11.40. Listed and marked steel plates less than the required 1.6 mm (1/16 in.) thick are permitted for protection of Type NM cable

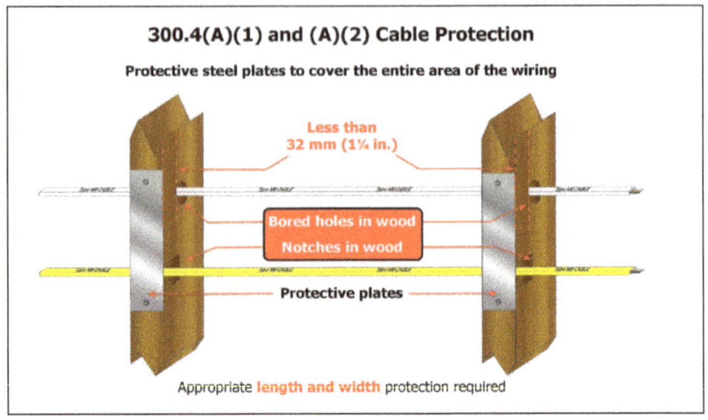

Figure 11.41. Metal nail protectors of the appropriate length and width

Photo 11.26. Listed metal nail protector minimum 1.6 mm (1/16 in.) thick

Photo 11.27. Steel plate(s) at least 1.6 mm (1/16 in.) thick needed where 32 mm (1¼ in.) distance cannot be maintained

taining two or more current-carrying conductors are installed in contact with thermal insulation without maintaining spacing between cables, the allowable ampacity of each conductor is required to be adjusted in accordance with *NEC* Table 310.15(C)(1) or IRC Table E3705.3 as well [*NEC* 334.80 or IRC E3705.4.4].

SECURING AND SUPPORTING TYPE NM CABLES

The nonmetallic-sheathed cable must be properly secured by staples, cable ties (listed and identified for securement and support), straps, hangers, or similar fittings. Two conductor (flat) cables are not permitted to be stapled on the edge. Cables run through holes in wood or metal joists, rafters, or studs are considered to be supported [*NEC* 334.30 or IRC E3802.6]. Securing methods are required to be designed and installed so as not to damage the cable. Care must be observed by the installer to not drive staples too deep or too strong as to damage the cable and enclosed conductors, which can result in potential fires at dwelling units.

When installing the cable, securing the cable with things like a staple is the desired result. This does not have to result in driving the staple too tight and damaging the cable. Movement could be caused by the building settling, framing members twisting, or just normal deflection of the ceiling and floor. A properly installed insulated staple provides the necessary support and decreases the possibility that the cable jacket is tightly pinched. While the *Code* does not require the use of insulated staples, some local inspection agencies require their use. Insulated staples should be considered for the extra protection they provide [*NEC* 334.30 and 300.4 or IRC E3802.6 or Table E3802.1].

Type NM cables are required to be secured in place at intervals not exceeding 1.4 m (4½ ft) and within 300 mm (12 in.) from every cabinet, box, or fitting. Where Type NM cable or multiconductor Type UF cable is used with a nonmetallic single gang box not larger than a nominal size 57 mm x 100 mm (2¼ in. x 4 in.) mounted in walls or ceilings, and where the cable is fastened within 200 mm (8 in.) of the box measured along the sheath and where the sheath extends through a cable knockout not less than 6 mm (¼ in.), securing the cable to the box is not be required [*NEC* 314.17(B)(2), Exception or IRC E3905.3.1, Exception]. For concealed work in finished buildings, it is permissible to fish the cable in hollow spaces in the structure between access points without supports.

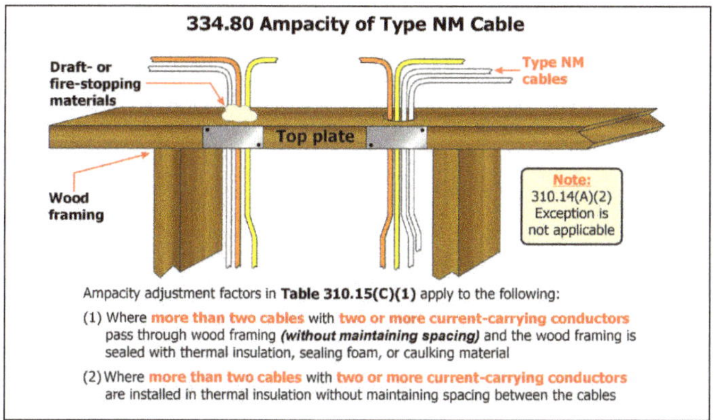

Figure 11.42. Ampacity adjustment factors for Type NM cable

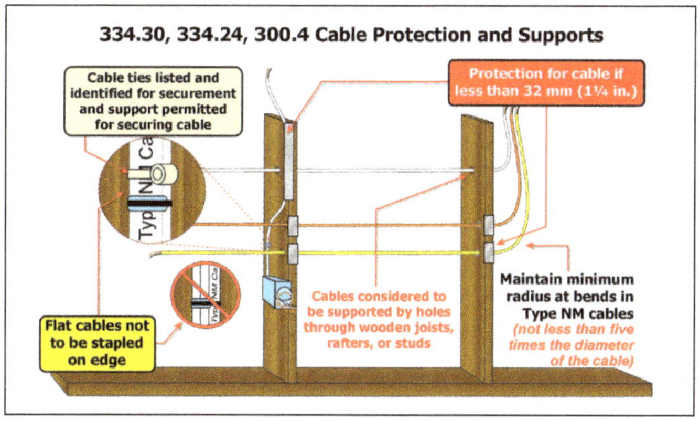

Figure 11.43. Types of support for Type NM Cable

Photo 11.28. Ampacity adjustment factors must be applied when more than two Type NM cables containing two or more current-carrying conductors are installed through the same sealing foam-filled hole

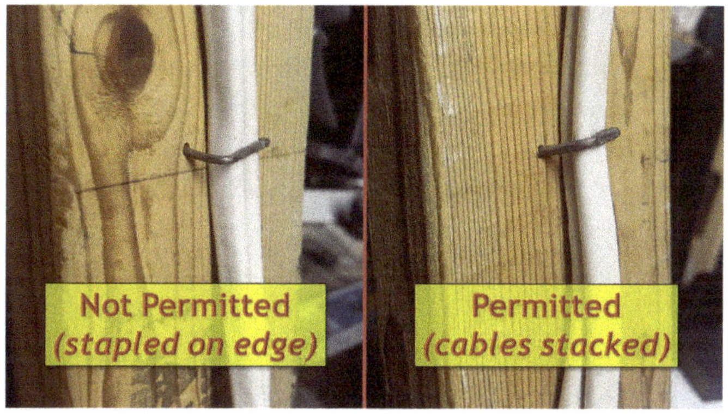

Photo 11.29. Cables cannot be stapled on edge

Table 310.15(C)(1) Adjustment Factors for More Than Three Current-Carrying Conductors

Number of Conductors[1]	Percent of Values in Tables 310.16 through 310.19 as Adjusted for Ambient Temperature if Necessary
4-6	80
7-9	70
10-20	50
21-30	45
31-40	40
41 and above	35

[1]Number of conductors is the total number of conductors in the raceway or cable, including spare conductors. The count shall be adjusted in accordance with 310.15(E) and (F), and shall not include conductors that are connected to electrical components but that cannot be simultaneously energized.

Table 11.4. Adjustment Factors for More Than Three Current-Carrying Conductors *[Reproduction of NEC Table 310.15(C)(1)]*

Cables such as Type NM, SE, and SER are required to be secured to the outlet box, cabinet, cutout box, or meter socket enclosure, with the exception mentioned in the previous paragraph for a single-gang nonmetallic box with the cable secured within 200 mm (8 in.) of the box. A variety of connectors and fittings are available, including metal and nonmetallic, clamp-type, and squeeze-type connectors for this purpose. It is important that the proper connector be installed for the size and configuration of the Type NM cable. It should be assumed that each connector is designed for the connection of a single cable unless the carton containing the connector is marked for more than one cable. In addition, the carton will include the size and number of cables permitted. For example, the connector may be suitable for two 12-2 cables but only a single 12-3 cable.

The cable connector not only secures the cable to the cabinet or outlet box but also closes the hole through which the cable enters the box or cabinet. The practice of installing a large EMT, PVC, or cable connector at the top of a panelboard cabinet and routing several cables through the connector is a *Code* violation and should not be permitted. This practice leaves an opening through which debris or rodents can enter the cabinet and also provides a chimney for hot gases or flames to pass through should there be a problem of that nature in the cabinet.

Cables are required to be individually secured to a cabinet or other enclosure [*NEC* 312.5(C) or IRC E3907.8]. An exception to this requirement permits multiple Type NM cables to be fished through a protective sleeve such as conduit or tubing into an enclosure without securing the cables to the enclosure. All of the following conditions must be met before this alternate method of cable entry is permitted in cabinets:

1. The cables must enter the enclosure through the top of the enclosure.
2. The enclosure must be surface-mounted.

3. The nonflexible raceway must be used as the protective sleeve (not less than 450 mm (18 in.) and not more than 3.0 m (10 ft) in length.
4. Each Type NM cable is fastened within 300 mm (12 in.) of the outer end of the raceway.
5. The nonflexible raceway extends directly above the enclosure and does not penetrate the structural ceiling. (Keep in mind that a suspended ceiling is not a structural ceiling.)
6. A suitable fitting or bushing is provided on each end of the raceway (chase) to protect the cable(s) from abrasion, and the fittings remain accessible after installation.
7. The raceway is sealed or plugged at the outer end using an approved means so as to prevent access to the enclosure through the raceway. (Duct seal, caulk, or fiberglass insulation are examples of materials used to comply with this requirement.)
8. The cable sheath is continuous through the raceway and extends into the cabinet or other enclosure beyond the fitting not less than 6 mm (¼ in.).
9. The raceway is fastened at its outer end and the other points in accordance with the applicable article.
10. The allowable cable fill does not exceed that permitted for complete conduit or tubing systems by *NEC* Table 1 of *NEC* Chapter 9 and all applicable notes. Note 2 to the tables in *NEC* Chapter 9 does not apply to this condition, as Note 2 states that the percent of cross-sectional areas of Table 1 of *NEC* Chapter 9 applies only to complete conduit or tubing systems and is not intended to apply to sections of conduit or tubing used to protect exposed wiring from physical damage.

SERVICE-ENTRANCE CABLE

The *Code* permits service-entrance cables to be used in interior wiring systems with restrictions. All the circuit conductors are required to be covered with thermoset or thermoplastic insulation. Type SE cables without individual insulation on the branch-circuit conductor are not permitted as a branch circuit or as a feeder within a building. Type SE service-entrance cable is permitted for branch-circuit wiring where the fully insulated conductors are used for circuit wiring and the uninsulated conductor is used for equipment grounding. An example of this type of permitted use is branch-circuit conductors to a 240-volt, single-phase (no neutral) appliance such as a water heater or electric furnace. Type SE cable conductors must not be subject to conductor temperatures in excess of the temperature rating of the contained conductor insulation. Typically, Type SE cables are manufactured with

Photo 11.30. Type NM cable stapled to secure and support

conductors that have 90°C insulation.

Type SE cable contains Type RHW, XHHW, or THWN conductors. Type USE contains conductors that have insulation equivalent to RHW or XHHW. The type designation of the conductors may be marked on the cable. When used, the marking indicates the equivalent temperature rating of the insulation. When the marking does not appear, the temperature rating of the cable is 75°C. Type USE-2 contains insulation that is equivalent to RHW-2 or XHHW-2 and is rated 90°C wet or dry.

Type SE cable used for interior wiring must comply with the same installation requirements for Type NM cable in Part II of *NEC* Article 334 (excluding 334.80). For Type SE cable with ungrounded conductor sizes 10 AWG and smaller, where installed in thermal insulation, the ampacity of the Type SE conductors must be in accordance with the 60°C (140°F) conductor temperature rating. The maximum conductor temperature rating is permitted to be used for ampacity adjustment and correction purposes if the final derated ampacity does not exceed that for a 60°C (140°F) rated conductor.

Where more than two Type SE cables containing two or more current-carrying conductors in each cable are installed in contact with thermal insulation, caulk, or sealing foam without maintaining spacing between cables, the ampacity of each conductor shall be adjusted in accordance with Table 310.15(C)(1) *(Adjustment Factors for More Than Three Current-Carrying Conductors in a Raceway or Cable)*, or IRC Table E3705.3.

The *NEC* continues to allow the grounded conductor of Type SE cable to be uninsulated where it supplies existing ranges and dryers in *existing installations only*. An uninsulated grounded conductor is also permitted for a feeder to be placed in an existing separate building or structure. New installations at ranges, dryers, and separate buildings or structures require a separate insulated grounded conductor

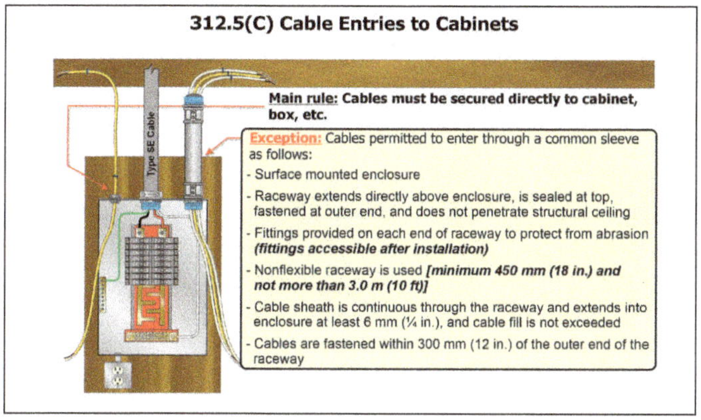

Figure 11.44. Cables with nonmetallic sheaths are permitted to enter enclosures without being secured to the panelboard cabinet under specific conditions

Photo 11.31. Type SER cable installed in the interior of a dwelling unit for a range circuit (4-wires)

(neutral) and an equipment grounding conductor.

Where installed as exterior wiring, the installation must comply with the applicable *Code* provisions for outside feeders and branch circuits and be installed in accordance with Part I of Article 225. Type SE and SER cables are required to be supported and secured in accordance with NEC 334.30 at intervals not exceeding 1.4 m (4½ ft) and within 300 mm (12 in.) from every cabinet, box, or fitting. Outside installations are permitted to be supported by a messenger wire in accordance with *NEC* Articles 225 and 396. Type USE cable installed as underground feeder and branch circuit cable must comply with Part II of Article 340 with the exception of single-conductor Type USE and multi-rated USE conductors.

See *NEC* Article 338, in particular, *NEC* 338.10 (Uses Permitted) and *NEC* 338.12 (Uses Not Permitted) for uses of SE cable as branch circuits or feeders. Also see IRC Table E3802.1.

UNDERGROUND FEEDER AND BRANCH-CIRCUIT CABLE (TYPE UF)

Underground feeder and branch-circuit cable (Type UF) is another wiring method suitable for indoor and outdoor use. This cable is available in sizes 14 AWG copper or, 12 AWG aluminum or copper-clad aluminum through 4/0 AWG.

The conductors of Type UF are required to be one of the moisture-resistant types listed in *NEC* Table 310.4(A) that are suitable for branch-circuit wiring or one that is identified for such use. In addition to the insulated conductors, Type UF cable is permitted to have an insulated or bare conductor to be used for equipment grounding conductor purposes only. Type UF cable is addressed in detail in Article 340 of the *NEC*.

The overall covering of Type UF is flame-retardant and moisture-, fungus- and corrosion-resistant. It is also suitable for direct burial in the earth. It is important to regulate rough or rocky backfill when using Type UF cable and to protect it from physical damage. Type UF cable is required to have a distinctive marking specifying the cable type on the exterior for its entire length.

Type UF cable is also permitted for use as feeder or branch-circuit cable and, like other conductors, must be provided with properly rated overcurrent protection.

Where single-conductor Type UF cables are installed, all conductors of the circuit, including any neutral conductor and equipment grounding conductor, are required to be run together in the same trench or raceway.

Where Type UF cables emerge from underground runs that are secured to the side of a building and are subjected to physical damage, they must be protected by a sleeve of rigid metal conduit, intermediate metal conduit, or Schedule 80 PVC conduit. The ends of this raceway are required to be protected by a bushing or terminal fitting. Short metallic sleeves are not required to be grounded. The recommended practice is to properly ground the metal sleeve to avoid the possibility of energizing it when a raceway is damaged by some powerful force, such as a car or truck bumper. Type UF cable assemblies are permitted to be used as overhead branch circuits supported by a messenger wire.

Type UF cable is permitted for interior wiring in wet, dry, or corrosive locations. Where Type UF cable is installed indoors as nonmetallic-sheathed cable, the installation requirements are required to follow all the provisions in Article 334, Parts II and III for Type NM cable. Type UF cable is not permitted to be used as service-entrance cable or embedded in poured cement, concrete, or aggregate, except where embedded in plaster as non-heating leads. This type of cable is also restricted for use where exposed to direct rays of the sun unless specifically identified as sunlight resistant. The ampacity of Type UF cable is that of 60°C (140°F) conductors in accordance with *NEC* Table 310.16 or IRC Table E3705.1.

UNDERGROUND INSTALLATIONS AND COVER REQUIREMENTS (IN GENERAL)

Minimum cover requirements for cable assemblies, conduits, or other raceways that are directly buried are included in NEC Table 300.5(A) or IRC Table E3803.1. "Cover" refers to the shortest distance in millimeters (or inches) from a point on the top surface of any directly buried conductor, cable, conduit, or raceway to the top surface of the finished grade, concrete, or similar cover. (see Note 1 to NEC Table 300.5 or footnote "e" to IRC Table E3803.1). The term finished grade makes it necessary for the installer and inspection authority to stay alert, especially when the final grading of the earth covering the underground run has not been completed.

NEC Table 300.5(A) and IRC Table E3803.1 provide information for the location of the following types of wiring methods or circuits:

1. direct burial of cables or conductors;
2. rigid metal or intermediate metal conduit;
3. electrical metallic tubing (EMT), rigid nonmetallic (PVC) conduit listed for direct burial without concrete encasement or other approved raceways;
4. residential branch circuits rated 120 volts or less provided with GFCI protection and overcurrent protection not over 20 amperes;
5. and circuits for control of irrigation and landscape lighting limited to not more than 30 volts and installed with Type UF or in other identified cable or raceway.

All underground installations using metallic methods must be properly grounded and bonded. Where an underground cable is installed under a building, it must be in a raceway [see exceptions for Type MI cable and Type MC cable listed for direct burial at *NEC* 300.5(C) or IRC E3803.11].

Where direct-buried conductors and cables emerge from grade and are specified in columns 1 and 4 of *NEC* Table 300.5(A) or IRC Table E3803.1, they are required to be protected from physical damage by enclosures or raceways that extend from the minimum cover distance below grade to a point at least 2.5 m (8 ft) above the finished grade. In no case is this protection required to exceed 450 mm (18 in.) below the finished grade [*NEC* 300.5(D)(1) or IRC E3803.3]. Conductors entering a building are required to be protected to the point of entrance.

Where the enclosure or raceway is subject to physical damage, the conductors are required to be installed in rigid

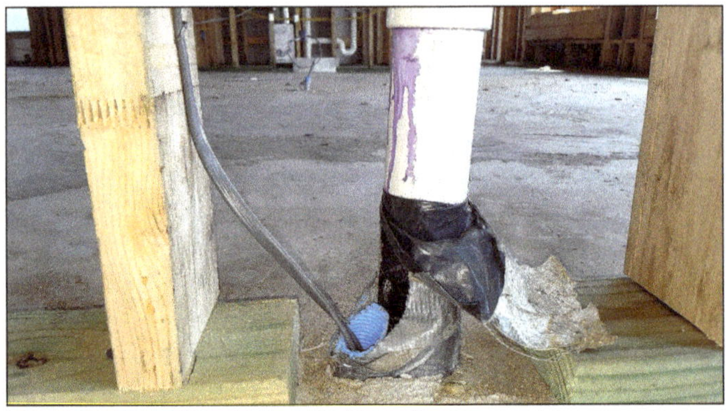

Photo 32. Type UF cable installed in and under the slab to supply power to a kitchen island (background)

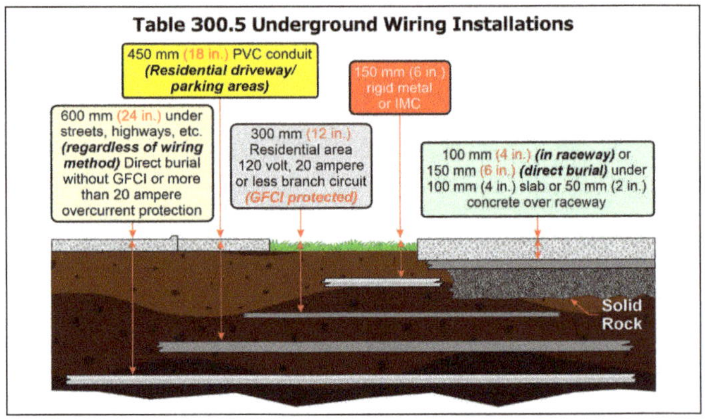

Figure 11.45. Minimum cover requirements for various wiring methods

metal conduit, intermediate metal conduit, Schedule 80 PVC conduit, or equivalent. Where direct-buried conductors, raceways, or cables are subject to damage by ground movement due to frost or settlement, wiring must be arranged to prevent damage to the conductors or connected equipment. A couple of common methods can be used to provide this protection. Slack in the direct burial cables or conductors that create an "S" loop prior to the transition to the conduit sleeve is acceptable. Conduit expansion fittings and flexible wiring methods installed in risers from the ground to connected equipment are other popular methods of providing protection [see photo 11.33; *NEC* 230.32 and 300.5(J) or IRC E3803.9].

SPLICES AND TAPS IN DIRECT-BURIED CONDUCTORS

Direct-buried conductors or cables are permitted to be spliced or tapped without the use of splice boxes. Splices or taps are required to be made using methods identified for the use of listed materials in accordance with *NEC* 110.14(B). See Chapter 1 for additional information on requirements for splicing underground conductors. Backfill containing large rocks, paving materials, cinders, large or sharply angular substances, or corrosive materials cannot be placed in an

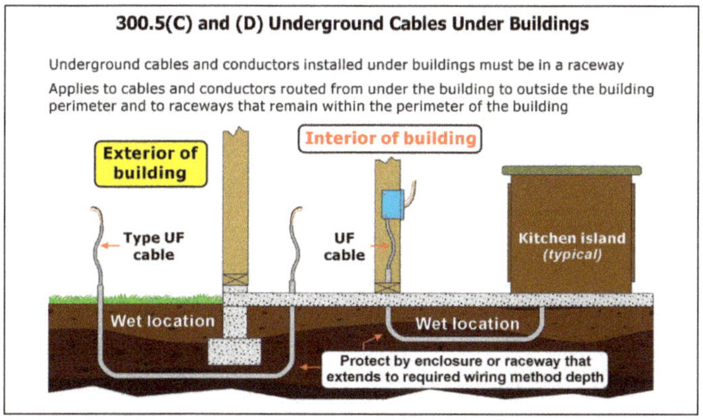

Figure 11.46. Protecting cables from damage with sleeves is required where installed under buildings

Photo 11.33. Expansion coupling in service lateral conduit (see *NEC* 300.5(J) or IRC E3803.9)

excavation where they may damage raceways, cables, or other substructures or prevent adequate compacting of fill or contribute to corrosion of raceways, cables or other substructures. Where necessary to prevent physical damage to the raceway or cable, protection must be provided in the form of granular or selected backfill material, suitable running boards, suitable sleeves, or other approved means. Conduits or raceways through which moisture may contact energized live parts are required to be sealed or plugged at either or both ends.

The presence of hazardous gases or vapors (such as methane) may also necessitate the sealing of underground conduits or raceways entering buildings.

A bushing or terminal fitting provided with an integral bushed opening is required to be used at the end of a conduit or other raceway that terminates underground, where the conductors or cables emerge as a direct-burial wiring method. A seal incorporating the physical protection characteristics of a bushing is permitted to be used in lieu of a bushing. [*NEC* 300.5 or IRC Section E3803]

FLEXIBLE CORDS

Flexible cords cannot be used as a substitute for fixed wiring methods such as cables or raceways. Cords are permitted for the connection of portable lamps or appliances. Cords can also be used for the connection of luminaires. Some additional conditions apply.

Flexible cords can be used for a cord- and plug-connection to stationary equipment for the following purposes:

1. to facilitate their frequent interchange;
2. to prevent transmission of noise or vibration; and
3. where the fastening means and mechanical connections are specifically designed to permit ready removal for maintenance and repair, and the appliance is intended or specifically identified to be supplied by a flexible cord connection.

Flexible cords are not permitted to be run through holes in walls, structural ceilings, suspended ceilings, dropped ceilings, or floors. They are not permitted to be concealed by walls, floors, or ceilings or located above suspended ceilings. Flexible cords cannot be attached to the building surfaces. Chapter 7 of Book II as well as Chapter 13 in this book offer examples of some of the commonly permitted uses of cords.

See *NEC* Article 400 and, in particular, NEC 400.10 (Uses Permitted) and *NEC* 400.12 (Uses Not Permitted) for the use of flexible cords and flexible cables.

PRODUCT SAFETY STANDARD REQUIREMENTS

It is recommended to consult the following electrical product safety guide card information and product safety standards for additional guidance on properly installing, operating, and using electrical equipment covered in this chapter. The four-letter code in parentheses indicates the product category in the UL Product iQ®.

- Equipment for Use in Ordinary Locations (AALZ)
- Armored Cable (AWEZ)
- Armored Cable Type AC, Connectors (AWSX)
- Flexible Conduit, Liquidtight (DWWY)
- Flexible Metal Conduit (DXUZ)
- Grounding and Bonding Equipment (KDER)
- Nonmetallic-Sheathed Cable (PWVX)
- Nonmetallic-Sheathed Cable Connectors (PXJV)
- Receptacles for Attachment Plugs and Plugs (RTRT)
- Service Cable (TXKT)
- Underground Feeder and Branch-Circuit Cable (YDUX)

REFERENCES

1. *National Electrical Code* and *NEC* are registered trademarks of the National Fire Protection Association, Inc., Quincy, MA 02169. This reprinted material is not the official position of the National Fire Protection Association, which is represented only by the standard in its entirety.

2. *International Residential Code for One- and Two-Family Dwellings* (2021). Copyright © International Code Council, Inc., June 2023.

3. UL, Northbrook, IL, *UL Product iQ*® unless indicated otherwise.

REVIEW QUESTIONS

1. A wall receptacle outlet is required to serve kitchen base counter spaces when the space is wider than what dimension?

A. 150 mm (6 in.)
B. 225 mm (9 in.)
C. 250 mm (10 in.)
D. 300 mm (12 in.)

2. A wall space located between two closet doors in a bedroom requires a receptacle outlet in that space where the wall measures ______ or more in width.

A. 600 mm (2 ft)
B. 1.2 m (4 ft)
C. 1.8 m (6 ft)
D. 3.7 m (12 ft)

3. A receptacle outlet installed for a washing machine and gas dryer must be located within what distance of the intended location of the appliance?

A. 1.8 m (6 ft)
B. 2.1 m (7 ft)
C. 2.5 m (8 ft)
D. 2.7 m (9 ft)

4. A lampholder with a pull chain (*no wall switch*) is permitted as the required illumination within a detached garage with electric power.

True _____ False _____

5. Where connected to a branch circuit supplying two or more receptacles or outlets, the maximum cord- and plug-connected load permitted on a 15-ampere rated receptacle is:

A. 15 amperes
B. 12 amperes
C. 10 amperes
D. 7.5 amperes

6. A cord- and plug-connected household electric range supplied by an individual 40-ampere circuit is permitted to be connected to a single receptacle rated at:

A. 40 amperes
B. 50 amperes
C. 40 or 50 amperes
D. None of the above

7. A 10-ampere branch circuit installed at a dwelling unit is permitted to supply which of the following loads:

A. Receptacle outlets
B. Laundry equipment
C. Lighting outlets
D. All of the above

8. If a point along the baseboard of a wall of a living room is 2.1 m (7 ft) from the nearest receptacle outlet, an additional receptacle outlet must be installed.

True _____ False _____

9. Where Type NM cable is used with a nonmetallic single gang box not larger than a nominal size 57 mm x 100 mm (2¼ in. x 4 in.) mounted in a wall, and where the cable is fastened within ____________ of the box measured along the sheath and where the sheath extends through a cable knockout not less than 6 mm (¼ in.), securing the cable to the box is not be required.

A. 150 mm (6 in.)
B. 200 mm (8 in.)
C. 300 mm (12 in.)
D. 450 mm (18 in.)

10. The rating of any one portable cord- and plug-connected appliance used on a 30-ampere branch circuit cannot exceed:

A. 30 amperes
B. 27 amperes
C. 24 amperes
D. 21 amperes

11. Listed switch kits and listed switch assemblies constructed with nonmetallic yokes and equipped with nonmetallic faceplates are not required to be bonded to the equipment grounding conductor.

True _____ False _____

12. Outdoor lighting outlets installed where located at exterior outdoor entrances or exits of dwelling unit(s) are generally required to be controlled by:

A. a pull-chain lampholder
B. a wall-mounted control device
C. a circuit breaker only
D. a wall switch only

13. An electric range rated at 8¾ kW or more must have a minimum branch-circuit rating of how many amperes?

A. 30
B. 40
C. 50
D. 60

14. Receptacle outlets that are located on or above the countertop or work surface must not be located more than ________________ above the countertop or work surface in order to serve that countertop or work surface.

A. 400 mm (16 in.)
B. 450 mm (18 in.)
C. 500 mm (20 in.)
D. 600 mm (24 in.)

15. At dwelling unit island and peninsular kitchen countertops and work surfaces, at least one receptacle outlet is required to be provided to serve the countertop or work surface.

True _____ False _____

16. Receptacle outlets are permitted to be installed below a dwelling unit kitchen countertop or work surface in order to serve that countertop or work surface.

True _____ False _____

17. Each multiwire branch circuit shall be provided with a means that will ____________________ all ungrounded conductors at the point where the branch circuit originates.

A. provide arc-fault circuit-interrupter protection for
B. disconnect
C. simultaneously disconnect
D. provide ground-fault circuit protection for

18. The maximum voltage permitted between branch-circuit conductors in dwelling units that supply luminaires is:

A. 250 volts
B. 240 volts
C. 150 volts
D. 120 volts

19. A 240-volt branch circuit installed in rigid nonmetallic (PVC) conduit that is installed below a driveway between a house and a garage must be buried at least what distance?

A. 250 mm (10 in.)
B. 300 mm (12 in.)
C. 400 mm (16 in.)
D. 450 mm (18 in.)

20. At least ________ of the free conductor must be left at each outlet and switch point for splices or the connection of luminaires or devices.

A. 150 mm (6 in.)
B. 100 mm (4 in.)
C. 75 mm (3 in.)
D. 50 mm (2 in.)

21. Protection is required for exposed or concealed cables and raceways that are installed parallel to framing members, such as joists, rafters, or studs, where the nearest outside surface of the cable or raceway is less than what distance from the nearest edge of the framing member where a nail or screw is likely to penetrate?

A. 13 mm (½ in.)
B. 19 mm (¾ in.)
C. 25 mm (1 in.)
D. 32 mm (1-¼ in.)

22. All 15- and 20-ampere, 125- and 250-volt nonlocking receptacles installed in damp or wet locations are required to be of the listed __________ type receptacles.

A. weather-proof
B. wet location
C. weather-resistant
D. any of the above

23. The minimum cover permitted for Type UF cable buried under a driveway to a post light at a one-family dwelling is not less than what distance where a 20-ampere, GFCI-protected branch circuit rated at 120 volts or less is installed?

A. 150 mm (6 in.)
B. 300 mm (12 in.)
C. 450 mm (18 in.)
D. 600 mm (24 in.)

24. Sheet metal screws or other types of screws that do not engage a minimum of two threads are not permitted to be used for connecting equipment grounding conductors or bonding jumpers to enclosures.

True ______ False ______

25. Where more than ________ Type NM cables containing two or more current-carrying conductors are installed in contact with thermal insulation without maintaining spacing between cables, the allowable ampacity of each conductor is required to be adjusted in accordance with *NEC* Table 310.15(C)(1) or IRC Table E3705.3

A. one
B. two
C. three
D. four

26. Conductors emerging from the ground require protection from physical damage when installed on a utility pole. What is the minimum height above the finished grade for this protection?

A. 3.7 m (12 ft)
B. 3.4 m (11 ft)
C. 3.0 m (10 ft)
D. 2.5 m (8 ft)

27. For wall spaces in dwelling unit kitchens, receptacle outlets are required to be installed so that no point along the wall line is more than ______________ measured horizontally from a receptacle outlet in that space.

A. 450 mm (18 in.)
B. 600 mm (24 in.)
C. 900 mm (36 in.)
D. 1.2 m (48 in.)

28. Type UF cable must not be used when exposed to direct rays of the sun unless it is identified as being sunlight resistant.

True ______ False ______

29. Foyers that are not part of a hallway in accordance with *NEC* 210.52(H) or IRC E3901.10 and that have an area that is greater than 5.6 m^2 (60 ft^2) shall have a 125-volt, 15- or 20-ampere receptacle(s) located in each wall space __________ or more in width and unbroken by doorways, floor to ceiling windows, and similar openings.

A. 1.8 m (6 ft)
B. 1.2 m (4 ft)
C. 900 mm (3 ft)
D. 600 mm (2 ft)

30. The minimum size of conductors used for branch circuits at dwelling units, whether solid or stranded, is generally 14 AWG copper or 12 AWG aluminum or copper-clad aluminum.

True _____ False _____

31. At dwelling units, at least one receptacle outlet is required to be installed in hallways where that hallway reaches a length of 3.0 m (10 ft) or more.

True _____ False _____

32. Fixture wires cannot be sized smaller than 16 AWG.

True _____ False _____

33. Each conductor entering a nominal 295 cm3 (18 cu. in.) device box shall extend a minimum of what distance past the opening of the box?

A. 65 mm (2-½ in.)
B. 75 mm (3 in.)
C. 90 mm (3-½ in.)
D. 150 mm (6 in.)

34. Branch circuit(s) supplying garage receptacle(s) are also permitted to supply which of the following outlets as well.

A. only readily accessible outdoor receptacle outlets located outside the garage
B. outdoor receptacle outlets
C. bathroom receptacle outlets
D. lighting outlets of a two-vehicle bay garage

35. Crawl spaces with a lighting outlet not exceeding 120 volts shall be protected by an arc-fault circuit interrupter.

True _____ False _____

36. Receptacles located in an appliance garage cannot count as the required receptacle for kitchen countertop spacing.

True _____ False _____

37. In each attached garage and in each detached garage with electric power, at least one receptacle is required to be installed in each vehicle bay and located not more than ____________ above the floor.

A. 1.5 m (5 ft)
B. 1.7 m (5½ ft)
C. 1.8 m (6 ft)
D. 2.5 m (8 ft)

38. If a receptacle outlet is not provided to serve a dwelling unit kitchen island or peninsular countertop or work surface, provisions such as a wiring method must be provided to the island or peninsula for the future addition of a receptacle outlet to serve the island or peninsular location.

True _____ False _____

39. At least _______ 125-volt, 15- or 20-ampere receptacle outlet is required to be installed in each accessory building that is provided with electric power.

A. one
B. two
C. three
D. no receptacle out required

40. At least one receptacle outlet is required to be installed at all balconies, decks, and porches that are _______________ _______________ the dwelling unit. This receptacle outlet(s) must be accessible from the balcony, deck, or porch and located not more than ____________ above the balcony, deck, or porch walking surface.

A. accessible from inside the dwelling unit and attached to / 1.5 m (5 ft)
B. within 102 mm (4 in.) horizontally of / 2.0 m (6½ ft)
C. attached to / 1.8 m (6 ft)
D. within 102 mm (4 in.) horizontally of / 2.5 m (8 ft)

12 Outlet Boxes and Other Enclosures

One of the most common code violations encountered at any location, including one- or two-family dwelling units, is an overloaded outlet box with too many conductors, devices, etc. All boxes (enclosures) must be large enough to provide sufficient free space for all conductors and devices contained within them to prevent overcrowding and potential physical damage when installing or finishing the devices or splices. The code includes general rules for boxes, such as sizing and support requirements. It also outlines specific situations where a box or conduit body is necessary in the wiring system. Several general requirements exist, along with additional sections that address particular wiring methods, enclosures, or conditions. In this chapter, we will review the general requirements and those specific conditions commonly found in a one- or two-family dwelling.

DEFINITIONS

The following definitions are outlined in NEC Article 100 and IRC Chapter 35. If a review is necessary, please consult Article 100 in the NEC for further details.

- ***Cabinet***
- ***Conduit Body.*** [Boxes such as FS and FD or larger cast or sheet metal boxes are not classified as conduit bodies.]
- ***Device***
- ***Enclosure.*** [See Table 110.28 for examples of enclosure types.]
- **Fitting**
- ***Handhole Enclosure***
- ***Outlet***
- ***Outlet Box Hood***

See *NEC* Article 100 or IRC Chapter 35.

TYPES OF NONMETALLIC AND METAL BOXES

An outlet or switch box is a type of enclosure used for supporting and enclosing wiring devices such as switches or receptacles. This type of box typically has threaded mounting holes for securing the yoke or strap of the device to the box. For an outlet or switch box, the device mounting holes accept a 6-32 machine screw. The boxes are usually square or rectangular and are available in a variety of depths. These enclosures may be used in the one, two, three, or more gang types.

Boxes for supporting luminaires and ceiling paddle fans are available in round, pancake, and octagonal types. The mounting holes on this type of box accept an 8-32 machine screw. Where a square metal box is used, it may be used for the support of a luminaire or device, depending on the type of raised cover used. These covers are often referred to as *plaster* or *mud rings*.

A box used to enclose flush devices is required to be of such design that the devices will be completely enclosed on the back and sides, and substantial support for the devices will be provided.

Photo 12.1. Typical 4-gang nonmetallic switch box installed at a dwelling unit front door area

Photo 12.2. Nonmetallic receptacle outlet box (single-gang) [Note that cable is secured within 200 mm (8 in.) of the box in accordance with *NEC* 314.17(B)(2), Ex.]

Photo 12.3. Pancake-style luminaire box

CONDUIT BODIES

A conduit body is a conduit fitting. It is often used with a conduit or tubing system instead of a pull box. A conduit body has a removable cover. The conduit body is often referred to by the orientation of the conduit entry with respect to the cover. For example, an "LB-type" conduit body has a conduit entry opposite the removable cover.

BOXES OR CONDUIT BODIES, WHERE REQUIRED

The *Code* contains specific requirements for where a box or conduit body is required in the wiring system. There are several general requirements with additional sections that cover specific wiring methods, enclosures, or conditions. We will review the general requirements and those specific conditions that are usually encountered in a one- or two-family dwelling.

The general requirement applies to wiring methods such as conduit, electrical metallic tubing, Type MC cable, Type AC cable, and nonmetallic-sheathed cable. Where the conductors are spliced or terminated at outlet, switch, junction, pull, or splice points, a box or conduit body generally must be installed. In addition, a box is required to be installed at outlet and switch points for concealed knob-and-tube wiring (*NEC* 300.15 or IRC E3905.1).

A box is not required where cables enter or exit from conduit or tubing that is used to provide cable support or protection against physical damage. A fitting is required on the end(s) of the conduit or tubing to protect the wires or cables from abrasion. This type of fitting should have a nonmetallic protective bushing or adaptor or metallic fittings with smooth throats at the point where the cable enters the raceway. This installation method is often used in exposed locations such as the basement or garage. Typically, a length of electrical metallic tubing (EMT) or rigid polyvinyl chloride conduit (PVC) is secured to the wall. The conduit terminates with one end in a conventional device box. Nonmetallic cable is pulled through a bushing or adaptor on the other

Photo 12.4. Conduit body (typical)

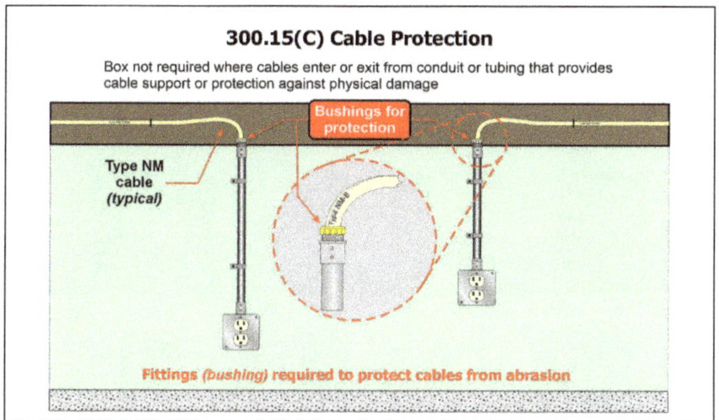

Figure 12.1. The box is not required where the cable enters or exits a conduit or tubing, providing physical protection

end of the conduit. Enough of the nonmetallic cable jacket is stripped back to permit the conductors to be terminated on the switch or receptacle within the box [*NEC* 300.15(C) or IRC E3905.1.2].

A box is not required if the wiring device has an integral enclosure. This type of device is required to be identified for use and has brackets that securely fasten the device to walls or ceilings. Switches and receptacles of this type are identified for use with nonmetallic-sheathed cables. The termination compartment is an integral part of the device, and no separate enclosure is required. While typically installed in manufactured housing such as manufactured homes, manufactured dwellings, and recreational vehicles, this type of device is permitted in conventional stick-built wood frame construction. In addition to *NEC* 300.15(E), *NEC* sections 334.30(C); 545.10; 550.15(I); 551.47(E), Exception No. 1; and 552.48(E), Exception No. 1 or IRC E3905.1.3 contain specific provisions for wiring devices with integral enclosures used in applications covered by these articles.

Direct-buried conductors or cables are permitted to be spliced or tapped without the use of a splice or junction box. A handhole enclosure is often used where access to the splices after installation is desired. Wire connectors or splicing devices used with direct burial conductors and cables must

314.24 Depth of Boxes

Boxes that enclose devices or utilization equipment are to be sized based on the supply conductors as follows:

Conductor Size Supplying the Device or Utilization Equipment	Minimum Depth of Box Containing Device or Utilization Equipment
14 AWG or Smaller	23.8 mm (15/16 in.)
12 or 10 AWG*	30.2 mm (1 3/16 in.)* ^
8, 6, or 4 AWG	52.4 mm (2 1/16 in.)
Larger than 4 AWG	Identified for their specific function *(see exception)*
Large Equipment *[Projects more than 48 mm (1 7/8 in.) rearward]*: Not less than depth of equipment plus 6 mm (¼ in.)	
Devices or utilization equipment that is listed to be installed with specified boxes shall be permitted	
**Where the equipment projects rearward from the mounting plane of the box by more than 25 mm (1 in.), the box shall have a depth not less than that of the equipment plus 6 mm (¼ in.)*	
^Where wiring enters the center portion of the rear of a box opposite to the equipment, the minimum clearance shall be increased to 13 mm (½ in.)	

Table 12.1. Minimum depth of boxes for outlets, devices, or utilization equipment

be listed or identified for such use [*NEC* 110.14(B) or IRC E3406.11]. Specific installation details are often found in the manufacturer's instructions. It is important to follow these instructions in order to make the splice or joint watertight. Separately installed pressure connectors must be used with conductors at ampacities that do not exceed the ampacity at the listed and identified rating of the connector. See Chapter 1 for additional information on this subject. In addition, the depth of direct-buried cables must comply with the burial depth requirements [*NEC* 300.5(E); 300.15(G); or IRC E3803.4].

GENERAL RULES FOR BOXES

The *Code* includes general rules for boxes, such as sizing and support provisions. All boxes (enclosures) must be large enough to provide sufficient free space for all conductors and devices that will be enclosed within them to prevent overcrowding and possible physical damage when the devices or splices are installed or completed. Outlet boxes that do not contain devices or utilization equipment are not permitted. See Section 314.24(A). Boxes that enclose large equipment are required to have a depth that is not less than the depth of the equipment plus 6 mm (¼ in.). Large equipment is defined as devices or utilization equipment that projects more than 48 mm (1 ⅞ in.) rearward from the mounting plane of the box [*NEC* 314.24 or IRC E3905.4].

Other boxes that enclose devices or utilization equipment are to be sized based on the supply conductors that supply the devices or utilization equipment (see Table 12.1).

In completed installations, each box is required to have a cover, faceplate, or luminaire canopy. One reason behind this rule is that if a failure occurs, they are likely to occur at a joint, splice, or termination. This type of failure often starts with a loose connection and may result in a great deal of heat and be accompanied by arcing. A box or conduit body reduces the likelihood of adjacent combustible materials being ignited by heat or arcing (*NEC* 314.25; 410.22 and IRC E3906.9).

Screws used for the purpose of attaching covers, or other equipment, to the box are required to be either machine screws matching the thread gauge or size that is integral to the box or must be in accordance with the manufacturer's instructions. This means the use of screws like drywall screws, wood screws, or sheet-metal screws for attaching covers, luminaires, or other equipment to boxes is unacceptable. This crude practice can result in damage to the box and inadequate support of the attached cover, luminaire, or equipment itself [*NEC* 314.25, 404.10(B), 406.5 or IRC E3906.9, E4001.10, E4002.6].

Metal boxes, covers, or cover plates are required to be grounded. *NEC* 250.110 or IRC E3908.2 specifically requires that exposed non-current-carrying metal parts of fixed equipment be grounded by connection to the equipment grounding conductor. Specifically, 250.110(5) requires that a metal box or enclosure be grounded if supplied by a metal raceway, metal-clad or metal-sheathed cable, or other wiring method that includes an equipment grounding conductor. The attachment of a metal cover or plate to an effectively grounded enclosure or device usually fulfills this requirement [*NEC* 250.110(5); 314.4; 314.28(C) or IRC E3908.2, E3906.10].

Metallic box covers cannot be used on nonmetallic boxes unless the cover is attached to the EGC as per 314.25(A).

Metal boxes have a means for connecting the box to an equipment grounding conductor; typically, this consists of a 10-32 tapped hole. Some nonmetallic boxes contain an integral means for bonding switches and metal faceplates to the equipment grounding conductor. The bonding means may be a metal strap or band that links the device mounting screw to an equipment grounding terminal [*NEC* 314.30(D); and IRC E3905.2 and E3905.3].

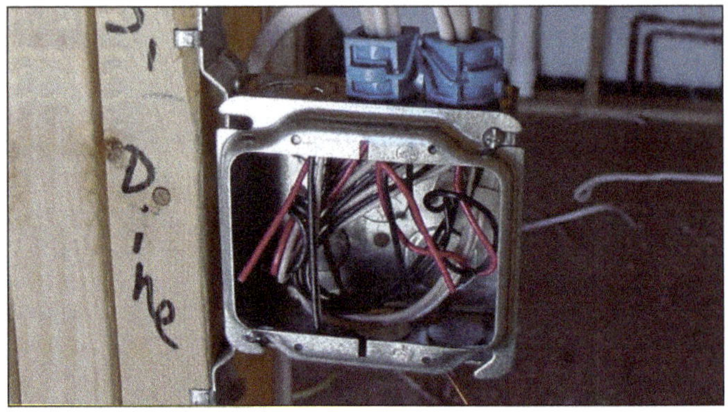

Photo 12.5. The volume of metal boxes is based on the total volume of assembled sections, including plaster rings or covers with marked volumes

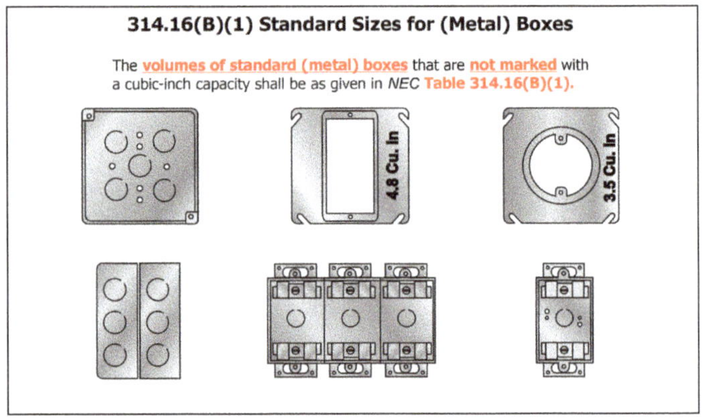

Figure 12.2. The volume of standard metal boxes is given in *NEC* Table 314.16(B)(1)

BOX VOLUME CALCULATIONS

The volume of a box is the total volume in cubic millimeters or cubic inches of the assembled sections. Where used, the space provided by plaster rings, domed covers, and extension rings, which are marked with their volume in cubic inches, is included in the total volume of the box and assembled sections. If the extension ring is made of metal and corresponds to a size listed in *NEC* Table 314.16(A) or IRC Table E3905.12.1, the appropriate volume for that size box is used as the volume of the extension ring. As stated, the total volume is the volume of the box plus the volume of any attached plaster rings, extension rings, or domed cover.

This total volume (space) determines the number and size of conductors and wiring devices that the *Code* permits to be contained in the box. Conductors, internal clamps, barriers, support fittings, and devices such as switches and receptacles take up space within the box. So, the *Code* assigns an associated volume allowance to each conductor, clamp, barrier, support fitting, device, and equipment grounding conductor.

This volume allowance is listed in cubic inches or cubic centimeters. *NEC* Table 314.16(B)(1) and IRC Table E3905.12.2.1 list the volume allowance as a function of conductor size. The volume allowance for each conductor, clamp, support fitting, device, and equipment grounding conductor(s) is added together. The box must have a volume that equals or exceeds the total volume required for the contained items and conductors.

Sometimes, more conductors end up in boxes than were originally intended. For example, the conduit may be of sufficient size for a new circuit; however, the pull or outlet box may not have sufficient volume. Where practicable, an extension ring that is the same shape as the box can be installed; this will add the required additional volume (space) so the original box does not have to be replaced. Sometimes, it is necessary to replace the original box with a deeper metallic or nonmetallic box to achieve the required volume of space.

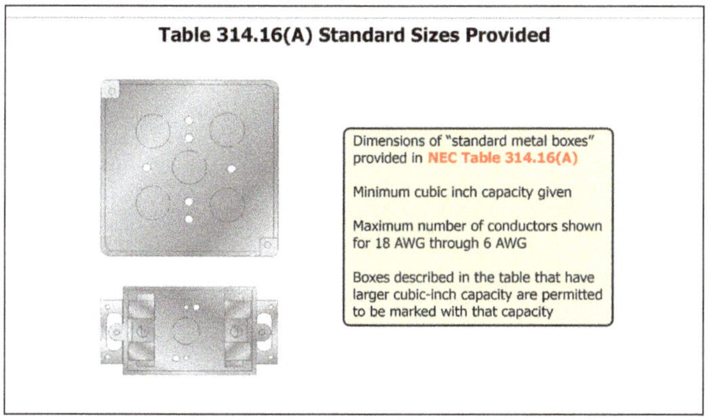

Figure 12.3. Box volume for standard boxes

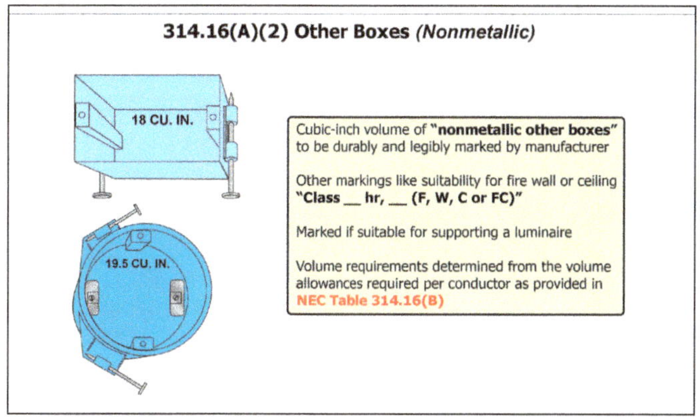

Figure 12.4. Conductor fill for nonmetallic other boxes

Photo 12.6. Nonmetallic box with plaster extension ring

STANDARD METAL BOXES

The *Code* covers the maximum number of conductors permitted within a standard metal box. These boxes may have their cubic millimeter (cubic inch) capacity marked on the box where their capacity is larger than given in the volume allowance table, but that is not required.

Table 314.16(A) in the *NEC* and Table E3905.12.1 in the IRC provide box dimensions and trade sizes in inches for standard metal boxes. The minimum cubic millimeter (cubic inch) capacity for each standard-size metal box is given along with the maximum number of conductors of sizes 18 AWG through 6 AWG permitted in the box. As shown in the table, the number of conductors permitted applies only where all conductors are the same size. Where a box contains conductors of different sizes, the required volume of the box must be calculated.

A calculation is also required if the box contains devices, clamps, barriers, or support fittings. No volume allowances are required for small fittings such as locknuts and bushings [*NEC* 314.16(B) or IRC E3905.12.2].

OTHER BOXES (NONMETALLIC BOXES)

Metal boxes may be manufactured in sizes other than those listed in *NEC* Table 314.16(A) or IRC Table E3805.12.1. This type of box is called a non-standard box. Nonmetallic boxes made in custom sizes are called non-standard as well if their dimensions are different from those in *NEC* Table 314.16(A) or IRC Table E3905.12.1. All non-standard boxes must have their cubic millimeter (cubic inch) capacity durably and legibly marked by the manufacturer on the inside of the box. Additional markings on nonmetallic boxes may include the type of fire-resistive wall, floor, or ceiling assembly for which the box is suitable [*NEC* 314.16(A)(2) or IRC E3905.12.1.2].

A box listed for the support of a luminaire or lampholder in a ceiling must be designed for the purpose and is required to support a luminaire weighing a minimum of 23 kg (50 lb.). Boxes used at luminaire or lampholder outlets in a wall are required to be designed for the purpose. They are required to be marked on the interior of the box itself to indicate the maximum weight of the luminaire that is permitted to be supported by that box in the wall [if other than 23 kg (50 lb.)]. It is important to note that "old work" round boxes are not listed for luminaire support in a ceiling. They are permitted for junctions and for utilization equipment such as smoke detectors, but not for luminaires. Vertical surface or wall-mounted device boxes are permitted to support luminaires that weigh not more than 3 kg (6 lb.). These device boxes may not be marked as being suitable for luminaire support since that is not their primary purpose. The luminaire or its supporting yoke must be secured to the box with at least two 6-32 screws [*NEC* 314.27(A) or IRC E3905.6].

Requirements for determining the maximum number of conductors permitted in non-standard or nonmetallic boxes are given in *NEC* 314.16(B) or IRC E3905.12.2. The individual volume of each conductor, clamp, and fitting is provided in the volume allowance table [*NEC* Table 314.16(B)(1) and IRC Table E3905.12.2.1]. The minimum volume of the box is the total of the individual volumes required for each conductor, device, or fitting.

Table 12.2 Volume Allowance Required per Conductor *[NEC Table 314.16(B)(1)]*		
Conductor Size *(AWG)*	Free Space Within Box for Each Conductor	
	cm³	in.³
18	24.6	1.50
16	28.7	1.75
14	32.8	2.00
12	36.9	2.25
10	41.0	2.50
8	49.2	3.00
6	81.9	5.00

Table 12.2. Volume allowance required per conductor [NEC Table 314.16(B)(1)]

A nonmetallic, single-gang device box typically used for new construction does not have internal clamps. The openings on a single-gang nonmetallic box are considered knockouts. As a result, no volume allowance for clamps is required on a single-gang nonmetallic box. Multiple-gang devices, such as nonmetallic boxes and round or octagonal boxes, have internal clamps. These clamps may be field-installed or incorporated into the box by the manufacturer. A single volume allowance is made for one or more internal clamps that are present in these nonmetallic multi-gang boxes. The volume allowance for clamps that are an integral part of the box construction is included in the marked volume of the box.

BOX FILL CALCULATIONS

The *Code* provides the method for determining the minimum cubic millimeter (cubic inch) volume required in non-standard or nonmetallic boxes. This same method is used to determine the minimum volume of standard boxes that have more than one size of conductor or any devices installed in them. This method requires the addition of all the allowances required for various items, and this sum then becomes the minimum volume required for the box. At that point, the size of a standard (metal) box or nonmetallic box with a cubic inch volume that equals or exceeds the volume required can be determined. As stated previously, no allowance is required for small fittings in boxes, such as for locknuts, bushings, and parts of cable connectors that are inside single-gang nonmetallic boxes.

The cubic millimeter (cubic inch) volume for each item that must be considered is given in *NEC* Table 314.16(B)(1) or IRC Table E3905.12.2.1. This table gives the cubic millimeter (cubic inch) volume allowance required for conductors from 18 AWG through 6 AWG.

Photo 12.7. Single-gang nonmetallic box-fill calculated by the volume required per conductor, devices, etc. (conductors have been pulled out of the box for ease of counting the total number of conductors)

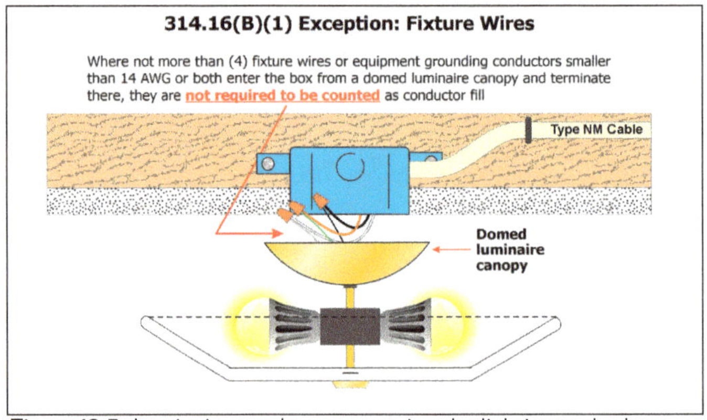

Figure 12.5. Luminaire conductors entering the lighting outlet box

CONDUCTOR FILL

The *Code* requires each conductor that originates outside the box and terminates or is spliced within the box to be counted once. Each conductor that passes through the box without splice or termination is also required to be counted once. A looped, unbroken conductor 300 mm (12 in.) or longer is counted as two conductors. A conductor that originates in and of which no part leaves the box is not required to be counted. An example of this would be a jumper or "pigtail" to a wiring device such as a switch.

An exception to this rule applies to the conductors that are part of the luminaire wiring. It permits equipment grounding conductor(s) or not more than four fixture wires smaller than 14 AWG, or both, to be omitted from the fill calculations. The conductors must enter a box from a domed luminaire or similar canopy and are required to terminate within the outlet box.

CLAMP FILL

The *Code* provides information on the volume required for clamp fill. Where one or more internal cable clamps, whether factory or field-supplied, are present in the box, a single volume allowance in accordance with *NEC* Table 314.16(B)(1) or IRC Table E3905.12.2.1 is required to be made based

314.16(B) Box Fill Calculations

Conductor, Device or Type of Fitting	Conductor Volume Allowance
Each conductor originating outside and terminating inside the box	1
Each conductor passing through the box unbroken	1
Looped or coiled conductor longer than 300 mm (12 in.)	2
Conductors that do not leave box	0
Max. of (4) fixture wires smaller than 14 AWG plus EGC *(domed luminaire canopy)*	0
Cable clamps *(all)* (LB)	1
Securely installed barriers *(if not marked with its volume)*	½ in^3 metal or 1 in^3 nonmetallic
Support fittings *(luminaire stud or hickey)(per fitting)* (LB)	1
Device or equipment *(receptacle, switch, etc.)(per yoke)* (LC)	2
Large device wider than 50 mm (2 in) with double mounting yokes (LC)	4
Equipment grounding conductors *(first 4)* (LB)	1
Additional equipment grounding conductors *(after the first 4)* (LB)	¼

LB = Largest conductor in box
LC = Largest conductor connected to device
****No allowance is required for small fittings like locknuts and bushings***

Table 12.3. Box volume calculations for standard and nonstandard boxes

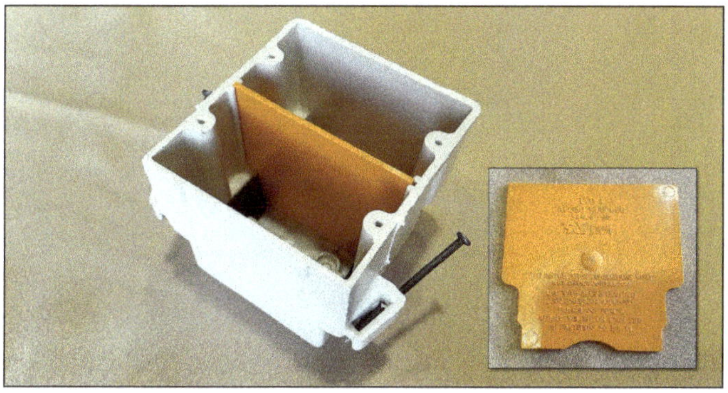

Photo 12.8. A nonmetallic barrier designed to be installed in a nonmetallic box

on the largest conductor present in the box. This section requires a single volume allowance for all the clamps that are internal to the box, regardless of the number of clamps. Volume allowance is required for nonmetallic boxes (other than single-gang nonmetallic boxes) with integral cable clamps. Before the total box volume can be determined on nonmetallic boxes (other than single-gang boxes), the manufacturer is required to remove the internal clamping mechanism; then, the total box volume is determined by a nationally recognized testing laboratory (NRTL). Single-gang boxes are determined to have knockouts rather than cable clamps, so no deduction is required for cable clamps for single-gang nonmetallic boxes when securing nonmetallic-sheathed cable within 200 mm (8 in.) of the box as allowed by *NEC* 314.17(B)(2) Exception or IRC E3905.3.1 Exception. No allowance is required to be made for a cable connector with its clamping mechanism outside the box (see Photo 12.5). A clamp assembly listed and marked for use with specific nonmetallic boxes that incorporates a cable termination for the cable conductors is a standard in the electrical industry. Conductors that originate within the clamp assembly are required to be included in conductor fill calculations as though they entered from outside the box [*NEC* 314.16(B)(2) or IRC E3905.12.2.2].

BARRIERS INSTALLED IN BOXES

Where a box is provided with one or more securely installed interior barriers, the *Code* requires the volume of that barrier to be allocated to each of the resulting spaces. Each barrier, if not marked with its volume, will need to be considered as taking up 8.2 cm^3 (½ $in.^3$) of space if metal and 16.4 cm^3 (1 $in.^3$) if nonmetallic. This volume allowance is based on a simple volume calculation of 100 mm x 50 mm x 1.6 mm (4 in. x 2 in. x 1/16 in.) for metal and double that volume for nonmetallic barriers. Each space within a box installed with an interior barrier will need to be calculated separately as two separate boxes or spaces [*NEC* 314.16(A) and (B) or IRC E3905.12.1 and E3905.12.2].

SUPPORT FITTINGS FILL

Where one or more fixture studs or hickeys are present in the box, a single volume allowance is required to be made for each type of fitting in the box.

Where both a fixture stud and a hickey are present in the box, two-volume allowances must be made. One volume allowance is made for the fixture stud, and the other is for the hickey. Each volume allowance is based on the largest conductor present in the box [*NEC* 314.16(B)(3) and IRC E3905.12.2.3].

A fixture stud is a fitting that mounts to the top of the box, usually inserts through the knockout of a metal box, and is threaded to accommodate the fixture stem. A hickey is a fitting that can be described as a coupling that has threads the same size as the fixture stem and has an oval-shaped hole on one or more sides for the fixture wires to exit inside the box. The hickey is usually threaded onto the fixture stud with the fixture stem threaded into it. These fittings, while used in many existing installations, have fallen out of popular use. Nowadays, a fixture strap or hanger strap is typically secured to the box with two 8-32 screws, and the luminaire is mounted to the strap.

DEVICE OR EQUIPMENT FILL

Minimum requirements for device or equipment fill are also provided in the *Code*. A double volume allowance is required for each yoke or strap containing one or more devices or equipment. Table 314.16(B)(1) or IRC Table E3905.12.2.1 gives the volume allowance. This volume allowance is based on the largest conductor connected to a device(s) or equipment supported by that yoke or strap. This requirement applies to devices such as switches, timers, dimmers, and receptacles. The word equipment is intended to include items like pilot lights.

Each device or equipment is considered individually, where more than one item is contained in the box. For example, if a switch has 14 AWG wires connected to it, a volume allowance of 2 x 2.0 cubic inches or 4 cubic inches total is required for that switch. If a receptacle has 12 AWG wires connected to it, a volume allowance of 2 x 2.25 or 4.5 cubic inches must be made for that particular device.

Any device or utilization equipment wider than a single 50 mm (2 in.) device box is required to have a double volume allowance provided for each gang required for mounting. This would apply to a device such as a dryer receptacle that would require two yokes for mounting. A device such as this would be required to be counted as four conductors based on the largest conductor connected to such a device. If this dryer receptacle were supplied with 10 AWG conductors, the total volume allowance just for this dryer receptacle would be 10 cubic inches (4 x 2.5 = 10 cubic inches) [*NEC* 314.16(B)(4) or IRC E3905.12.2.4].

EQUIPMENT GROUNDING CONDUCTOR FILL

A single volume allowance has been required for all equipment grounding conductors within a box since the 1971 *NEC*. This single volume deduction was based on the largest equipment grounding conductor or equipment bonding jumpers present in the box. This was for all equipment grounding conductors or equipment bonding jumpers regardless of the number of equipment grounding conductors or equipment bonding jumpers installed. If six 12 AWG equipment grounding conductors were present in a box, a single volume allowance or deduction of 36.9 cm^3 (2.25 $in.^3$) would be required. This single volume allowance was based in part on the fact that these equipment grounding conductors or equipment bonding jumpers are not current-carrying conductors under normal conditions.

Beginning with the previous *Code* cycle, the single volume allowance or deduction applies to the first four equipment grounding conductors or equipment bonding jumpers. Any equipment grounding conductor or equipment bonding jumper installed above four requires an additional ¼ volume allowance for each conductor to the existing volume allowance of a single conductor volume allowance based on

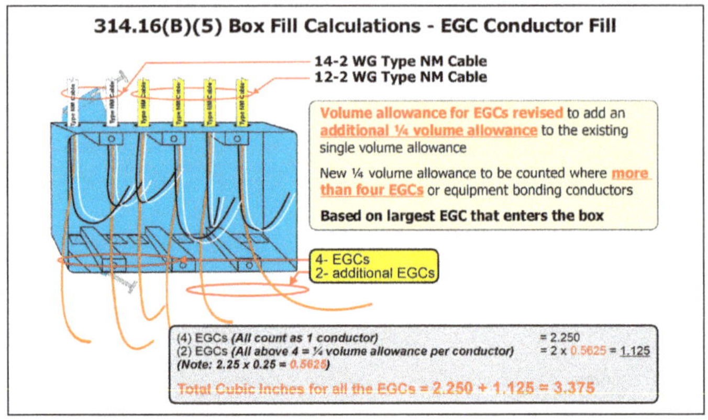

Figure 12.6. Additional volume allowance is required for more than four equipment grounding conductors

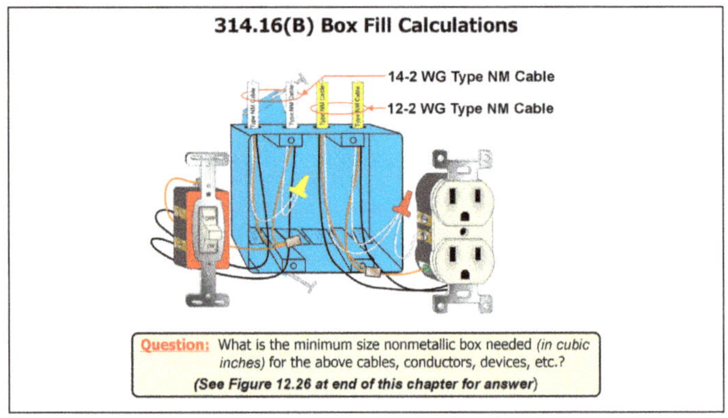

Figure 12.7. Minimum cubic inches needed for the box? *(See Figure 12.26 at the end of Chapter 12)*

the largest equipment grounding conductor or equipment bonding jumper entering the box. In multiple gang boxes, taking only one volume allowance based on the largest EGC is not adequate in many cases, resulting in significant undue crowding of conductors and not enough free space to allow heat to dissipate from the contained conductors. This 2020 *NEC* (2021 IRC) change also eliminated the need to specifically address the equipment grounding conductors for isolated ground receptacles permitted by *NEC* 250.146(D) [*NEC* 314.16(B)(5) or IRC E3905.12.2.5].

TERMINAL BLOCK FILL

Where a terminal block is present in a box, a single volume allowance from *NEC* Table 314.16(B)(1) or IRC Table E3905.12.2.1 must be made for each terminal block assembly based on the largest conductor(s) terminated to the assembly. Terminal blocks in boxes are not common in dwelling unit construction but could exist and will need to be addressed accordingly.

BOX FILL CALCULATION

Based on the information previously covered and the requirements of *NEC* 314.16(B)(1) through (B)(6) or IRC

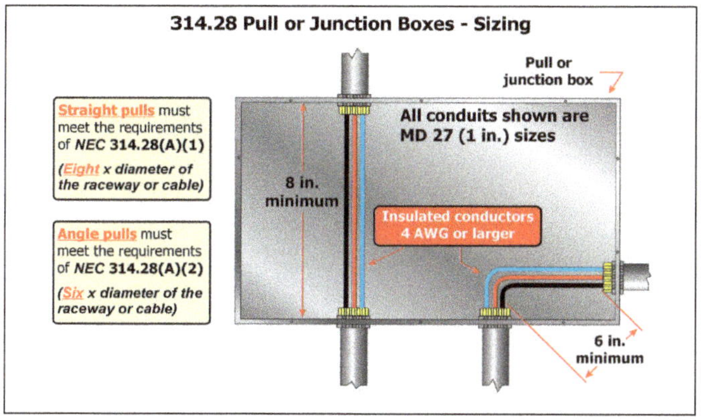

Figure 12.8. Dimensions of pull or junction boxes containing conductors of sizes 4 AWG or larger

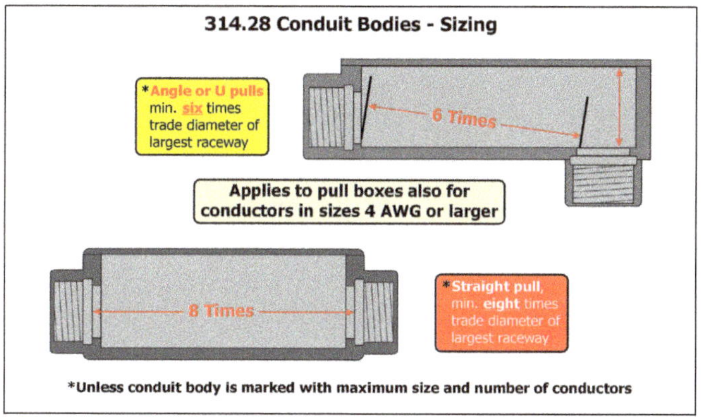

Figure 12.9. Dimensions of conduit bodies containing conductors of sizes 4 AWG or larger

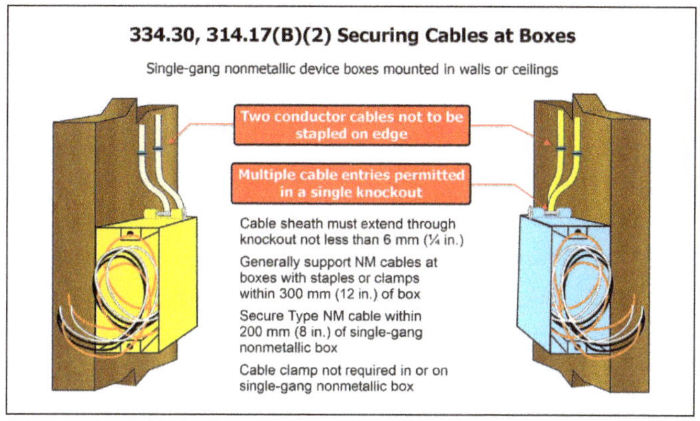

Figure 12.10. Multiple cables are permitted through a single opening on the box

E3905.12.2.1 through 6, what would be the minimum cubic inches needed inside the two-gang box illustrated in Figure 12.7 for all of the contained conductors, devices, cable clamps, etc.? See the answer at the end of the chapter in Figure 12.26.

PULL AND JUNCTION BOXES AND CONDUIT BODIES

Pull and junction boxes used with conductors 4 AWG and larger must be sized to meet the requirements of *NEC* 314.28. If the conductors are pulled straight through the box, the length of the box is eight times the trade size of the largest raceway. When transposing cable size into raceway size, the minimum size raceway required for the number and size of the conductors in the cable is used in the calculation.

Where splices are made in the box or where the conductors are formed into a U-pull or angle-pull, the calculation is slightly more complex. The dimension between the wall of the box where the conduit enters, and the opposite wall of the box is calculated by multiplying the trade size of the largest raceway size by six. This distance is increased if there are any other raceway entries in the same row on the same wall of the box. The trade sizes of the additional raceways are added up and added to the initial result.

Where conduit bodies contain conductors in sizes 4 AWG or larger, dimension requirements must be met. For straight pulls, the length of the conduit body is to be not less than eight times the trade diameter of the largest raceway. Where an angle or U-pull is being made, the distance between the raceway entry and the opposite wall of the conduit body must be not less than six times the largest raceway.

In addition, the distance between the conduit entries (containing the same conductor) in the conduit body must be not less than six times the trade diameter of the larger raceway. Short radius conduit bodies, such as capped elbows and service-entrance elbows, enclosing 6 AWG or smaller conductors are not permitted to contain splices, taps, or devices. These conduit bodies must be of adequate size to provide free space for all conductors enclosed in the fitting [*NEC* 314.28, 314.16(C)(3) or IRC E3905.12.3.2]. See Chapter 6 for additional information on conduit bodies.

MULTIPLE CABLE ENTRIES

In general, the *Code* requires that cables be clamped or otherwise secured to the box. The *Code* permits (by exception) multiple cable entries in a single knockout opening in nonmetallic boxes.

The exception permits Type NM cable or Type UF cable to enter a nonmetallic box without being clamped to the enclosure. This exception applies only to single-gang nonmetallic boxes with nominal dimensions of 57 mm (2¼ in.) x 100 mm (4 in.).

The box must be mounted on a wall or ceiling. Cable entries into the box are not required to be secured to the box if the cable is secured or fastened within 200 mm (8 in.) measured along the sheath of the box [*NEC* 314.17(C) Exception or IRC E3905.3.2 Exception].

BOXES IN WET OR DAMP LOCATIONS

Boxes, conduit bodies, and fittings installed in damp or wet locations must be positioned or equipped so as to prevent moisture from entering or accumulating within the box, conduit body, or fitting. Boxes, conduit bodies, outlet box hoods, and fittings installed in wet locations must be specifically listed for use in wet locations. Drainage openings that are not smaller than 3 mm (⅛ in.) and not larger than 6 mm (¼ in.) are permitted to be installed in the field in boxes or conduit bodies listed for use in damp or wet locations. Larger openings are permitted to be installed in the field in accordance with the manufacturer's instructions [*NEC* 314.15 or IRC E3905.11].

Wet locations include installations underground or in concrete slabs or masonry in direct contact with the earth, locations subject to saturation with water or other liquids, such as the area outdoors where the A/C unit is located, and locations exposed to weather and unprotected. The interior of enclosures and raceways installed underground and raceways installed in wet locations above grade are considered to be a wet location [*NEC* 300.5(B), 300.9 or IRC E3803.10, E3802.8]. See Figure 12.11.

Photo 12.9. Multiple cables through a single clamp where the box is identified for same. The cable sheath must extend at least 6 mm (¼ in.) inside the box.

Photo 12.10. Box installed in combustible finish material is required to be flush with the finished surface

Photo 12.11. Boxes installed in noncombustible material (gypsum board) are allowed to be installed flush with the finished surface or have up to a 6 mm (¼ in.) setback

INSTALLATION OF, LOCATION OF, AND EXTENSION FROM FLUSH BOXES

A flush-mounted box is required to be mounted within a wall, floor, or ceiling such that the front edge of the box is essentially even with the finished surface. Boxes employing a flush-type cover or faceplate that are installed in walls or ceilings of concrete, tile, or other non-combustible material are required to be installed so that the front edge of the box, plaster ring, extension ring, or listed box extender will not be set back of the finished surface more than 6 mm (¼ in.). Gypsum wallboard (sheetrock) is considered a noncombustible finish for the purpose of meeting the requirements of this section. Outlet boxes, plaster rings, extension rings, or listed box extenders installed in walls and ceilings constructed of wood or other combustible materials are required to be flush with or project out from the finished surface (*NEC* 314.20 or IRC E3906.5).

Noncombustible surfaces of plaster, drywall, or plasterboard that are broken or incomplete are required to be repaired so there will be no gaps or open spaces greater than 3 mm (⅛ in.) at the edge of the box. This rule applies to boxes that employ a flush-type cover or faceplate. The rule covers both new installations and boxes installed in an existing dwelling unit. Sealing or repairing oversize or uneven holes around boxes helps maintain the integrity of any fire rating the outlet or device box has in combination with the wall covering or wall/ceiling assembly (*NEC* 314.21 or IRC E3906.6; see also Figure 12.13).

An exposed surface extension made from a flush-mounted outlet box is required to be mounted and mechanically secured by the use of a box or extension ring installed over the concealed box. A surface extension from the cover of a concealed box is permitted by the exception to be made when the cover is designed so that it cannot fall off or be removed if its security means becomes loose (see Figure 12.14). The wiring method for this so-called "whip" is required to be of the flexible type

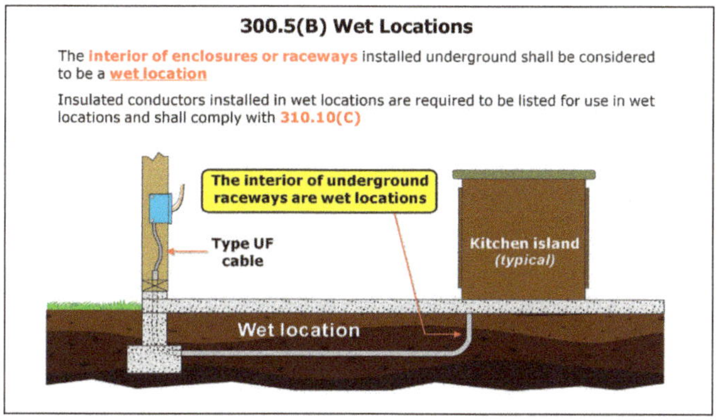

Figure 12.11. The interior of enclosures or raceways installed underground are considered wet locations

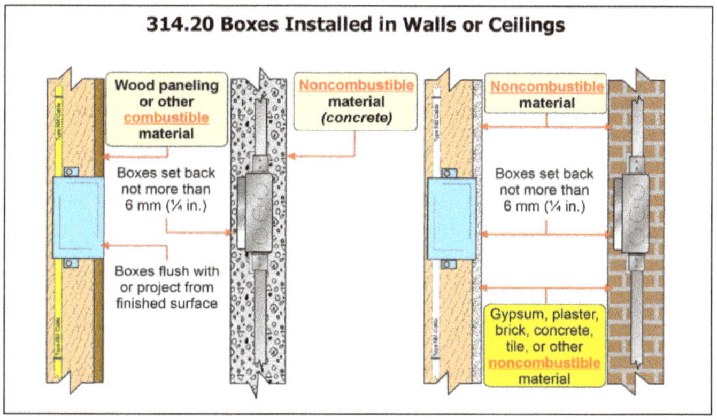

Figure 12.12. Installation requirements for boxes relative to both combustible and noncombustible finished surfaces

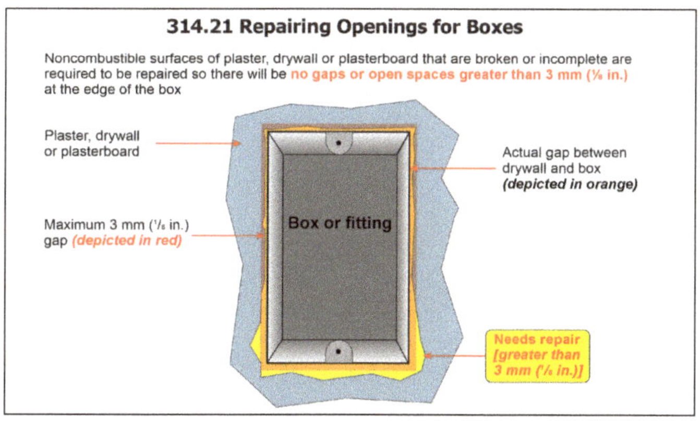

Figure 12.13. Repairing spaces or gaps around flush boxes in plaster, drywalls, or plasterboard

for an approved length and arranged to include the required equipment grounding conductor within it. This will provide access and ensure electrical continuity between the box and metal cover as well as equipment supplied by the branch circuit should there be a need to remove the cover to inspect or check the splices. This method reduces the possibility of an electric shock hazard (*NEC* 314.22 or IRC E3906.7 and exceptions).

SECURING AND SUPPORTING BOXES

Enclosures are required to be rigidly and securely fastened in

Photo 12.12. Openings around boxes need to be sealed (repaired) tight to box edges *[gap no greater than 3 mm (⅛ in.)]*

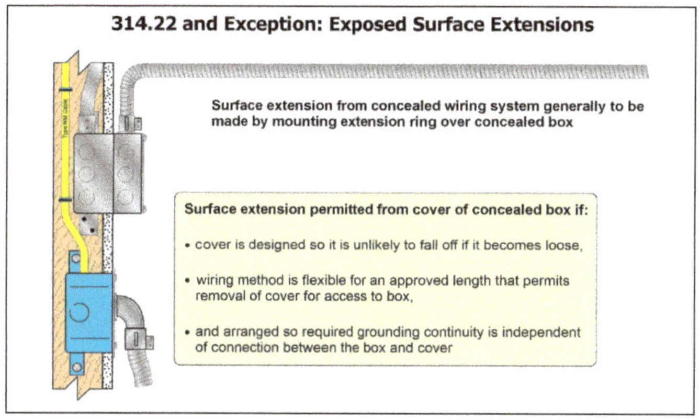

Figure 12.14. Surface extensions permitted from flush-mounted boxes

Photo 12.13. Surface extension from flush-mounted box

place. Surface-mounted enclosures are required to be securely fastened to the building surface. If the mounting surface does not provide adequate support, the enclosure must be rigidly supported by a structural member of the building. The enclosure may be mounted directly to the structure or mounted indirectly by using a metal, wooden, or polymeric brace of sufficient strength.

Where a nail is used as a fastening means for the support of a box, it is permitted to pass through the interior of the enclosure only where it is located within 6 mm (¼ in.) of the back or ends of the inside of the enclosure. Screws are not permitted to pass through the box unless the exposed threads are protected by an approved means to avoid abrasion of the enclosed conductor's insulation. Some types of box kits, such as those intended for support of ceiling-suspended paddle fans, will have mounting screws that pass through the box. Often the box manufacturer will provide a protective tube or sleeve to be used with these screws.

A metal brace used to secure an enclosure to a structural member is required to be protected against corrosion and formed from metal not less than 0.51 mm (0.020 in.) thick, uncoated. A wooden brace used to secure an enclosure to a structural member is required to have a cross-section not less than a nominal 25 mm x 50 mm (1 in. by 2 in.). Wood braces in wet locations are required to be treated for the conditions (*NEC* 314.23 or IRC E3906.8.2.2).

Screws used for supporting boxes for flush-mounted devices are not permitted to be used for the attachment of the device contained within the box (*NEC* 314.19 or IRC E3905.5).

Although not common, a suspended drop tile ceiling can be found it dwelling units from time to time. Support wires used for enclosure or luminaire support must provide adequate support. The *Code* permits independent support wires to be added to a suspended ceiling assembly. Independent support wires are those that were added in addition to the wires that support the suspended ceiling for the support of electrical enclosures or luminaires. Identifying these independent support wires is required for both fire-rated and non-fire-rated ceiling assemblies. Identification by distinguishable color, tagging, or other effective means will clarify the intended purpose for all personnel involved with the construction.

Only the independent wires are permitted to be used for raceway, cable, luminaire, or enclosure support. In addition, the enclosure must be secured to the support wires using methods identified for the purpose. Support wires used for enclosure support must be fastened at each end so as to be taut within the ceiling cavity [*NEC* 314.23(D)(2); 300.11(A)].

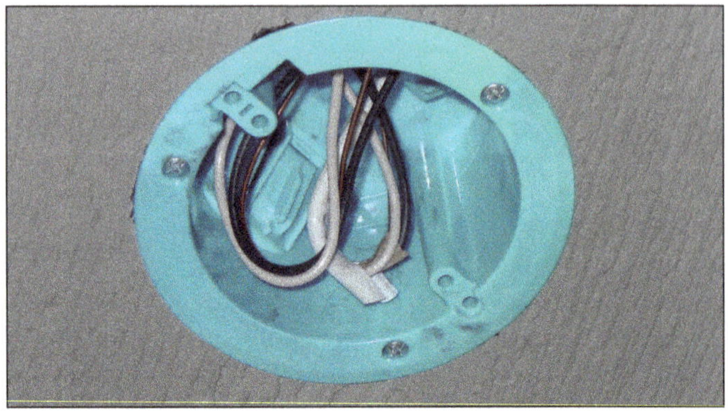

Photo 12.14. Old-work style round nonmetallic box cut into existing building surface

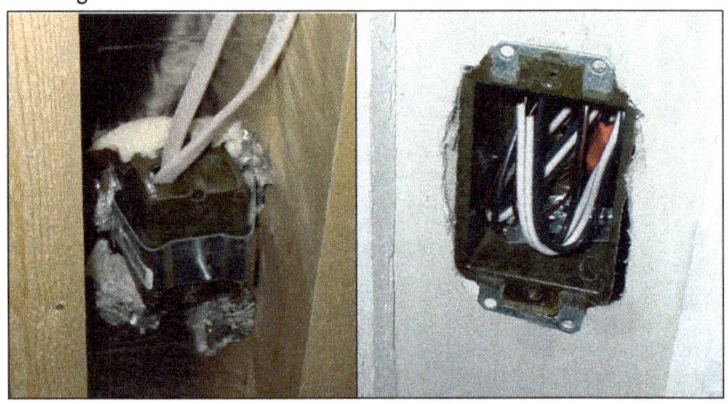

Photo 12.15. Old-work style outlet box with front and back view

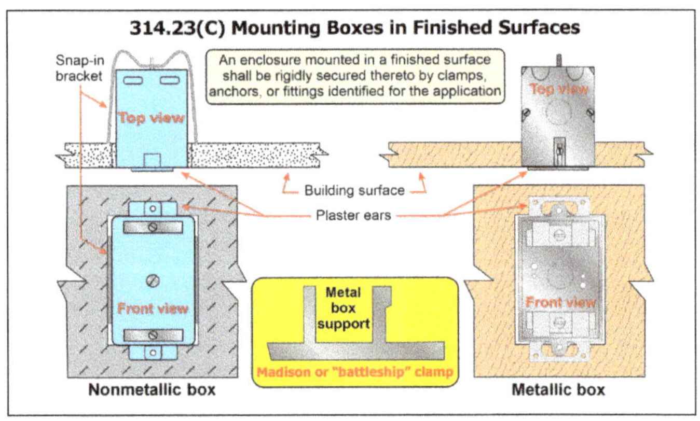

FIGURE 12.15 Nonstructural securing or mounting of boxes.

RACEWAY-SUPPORTED ENCLOSURES AND LUMINAIRES

It is not uncommon to see landscape lighting installed at one- or two-family dwellings where the junction boxes and/or luminaires are supported by the conduit system.

There are two sets of rules that address this installation. The first condition is where the conduits support a box or other enclosure. The second condition involves an installation where the conduits support a box, and the box supports a luminaire or contains a device. Devices that could be used in this context include switches and receptacles. They do not include splicing devices such as twist-on wire connectors. An enclosure that is supported by the connected conduits and does not contain a device(s) or does not support a luminaire must not exceed 1650 cm^3 (100 in.^{3}) in size. The enclosure must have threaded entries or identified hubs. A minimum of two conduits threaded wrenchtight into the enclosure is required for support. Each supporting conduit must be secured within 900 mm (3 ft) of the enclosure if the conduits are on at least two sides. If both conduits are on one side, the support must be within 450 mm (18 in.) [*NEC* 314.23(E) and (F) or IRC E3906.8.4 and E3906.8.5].

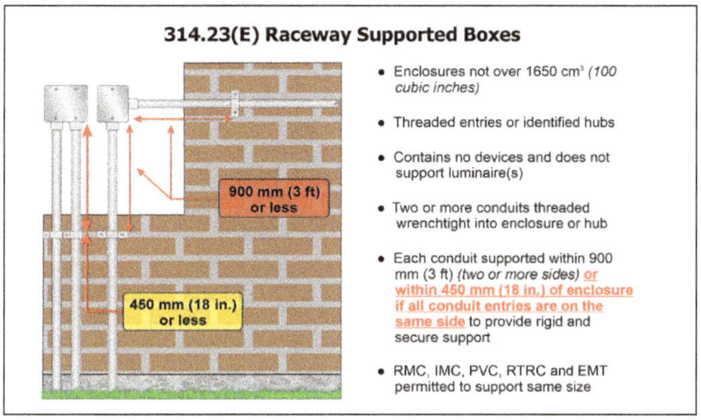

Figure 12.16. Raceway used for support of boxes that do not contain devices or support luminaires

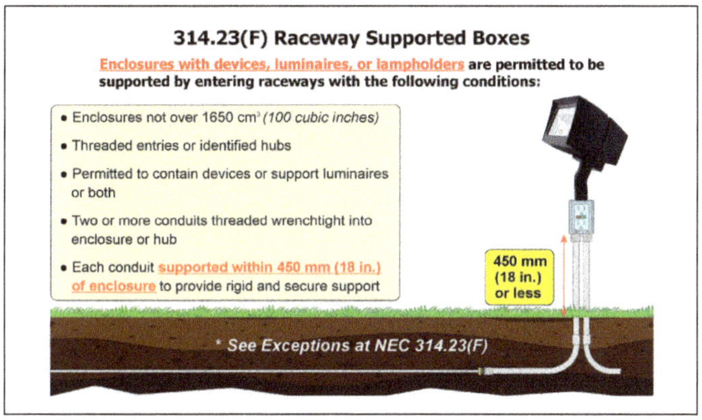

Figure 12.17. Raceway used for supporting boxes that contain devices or support luminaires

Rigid metal (RMC), intermediate metal (IMC), rigid polyvinyl chloride (PVC), reinforced thermosetting resin (RTRC) conduit or electrical metallic tubing (EMT) are permitted to support conduit bodies where the conduit bodies are not larger than the largest trade size of the conduit or electrical metallic tubing [*NEC* 314.23(E) Exception or IRC E3906.8.4 Exception]. Schedule 40 rigid PVC conduit may not be used where subject to physical damage. Schedule 80 rigid PVC conduit is permitted in these locations [*NEC* 352.12(C) and 352.10(K)].

Raceways are also permitted to support enclosures that contain devices or support luminaires or lampholders. However, rigid metal conduits (RMC) or intermediate metal conduits (IMC) are the only conduits permitted to support these boxes or boxes with luminaires. A rigid polyvinyl chloride PVC conduit is not permitted to be used for the support of luminaires or other equipment [*NEC* 352.12(B)]. Likewise, electrical metallic tubing is not permitted for box and luminaire support. Electrical metallic tubing is not suitable for installation in the ground or in concrete slabs on grade, in most cases, unless it is protected from corrosion.

Just like with an enclosure that is supported by the connected conduits and does not contain a device(s) or does not support a luminaire, an enclosure that is supported by the connected conduits and does contain a device(s) or supports a luminaire must not exceed 1650 cm³ (100 in.³) in size. The enclosure must have threaded entries or identified hubs. A minimum of two conduits threaded wrenchtight into the enclosure is required for support. Each supporting conduit must be secured within 900 mm (3 ft) of the enclosure if the conduits are on at least two sides [*NEC* 314.23(F) or IRC E3906.8.5].

Enclosure(s) supported by embedment in concrete or masonry must be identified as suitably protected from corrosion. All metal boxes, except for aluminum alloy boxes, are provided with corrosion protection suitable for installation in concrete or masonry. A metal or nonmetallic box designated concrete tight is permitted to be supported solely by embedment. The enclosure is required to be securely embedded in the concrete or masonry unless additional support is provided [*NEC* 314.23(G) or IRC E3906.8.6].

UNUSED OPENINGS

Unused openings, other than those intended for the operation of equipment, those intended for mounting purposes, or those permitted as part of the design for listed equipment, shall be closed to afford protection substantially equivalent to the wall of the equipment. Where metallic plugs or plates are used with nonmetallic enclosures, they shall be recessed at least 6 mm (1⁄4 in.) from the outer surface of the enclosure. An enclosure is defined as the case or housing of apparatus, or the fence or walls surrounding an installation to prevent personnel from accidentally contacting energized parts or to protect the equipment from physical damage. A device box is not an enclosure, therefore the rule regarding the 1/4" setback of a plug or plate used in a nonmetallic enclosure does not apply to a nonmetallic device box. See Chapter 1 for additional information on unused openings.

BOXES AT LUMINAIRE OUTLETS

Ceiling-Mounted Luminaires

Boxes used at ceiling-mounted luminaire or lampholder outlets must be designed specifically for that purpose. They are also required to support a luminaire weighing up to 23 kg (50 lb.). A luminaire that weighs more than 23 kg (50 lb.) must be supported independently of the outlet box unless the outlet box is listed for not less than the designated luminaire weight to be supported. At every ceiling-mounted outlet used exclusively for lighting, the box has to be designed or installed so that a luminaire or lampholder can be attached. The interior of the box is required to be marked by the manufacturer to indicate the maximum weight the box is permitted to support *[if other than 23 kg (50 lb.)]* [*NEC* 314.27(A)(2) or IRC E3905.6.2].

Vertical Surface-Mounted Luminaires

Boxes used at luminaire or lampholder outlets in or on a vertical surface such as a wall are also required to be identified and marked on the interior of the box to indicate the maximum weight of the luminaire that is permitted to be supported by said box in or on the vertical surface *[if other than 23 kg (50 lb.)]*.

An exception permits vertical surface-mounted luminaires or lampholders to be supported on other boxes, such as device boxes. The conditions limit the weight of the luminaire to 3 kg (6 lb.). The luminaire or its support yoke, or lampholder, must be secured to the box with at least two No. 6 or larger screws [*NEC* 314.27(A)(1), or IRC E3905.6.1].

Photo 12.16. Ceiling-mounted outlet box for luminaire

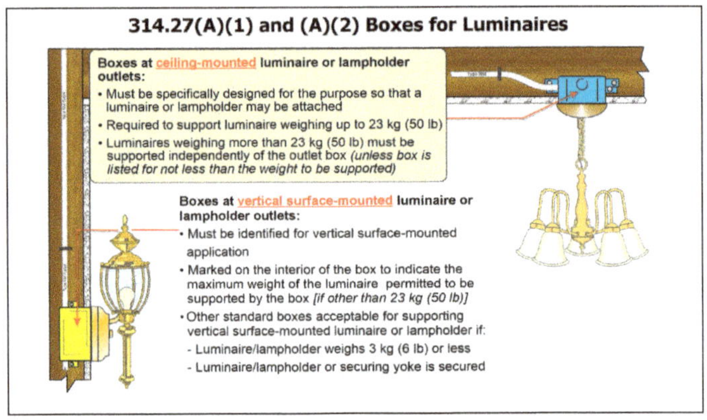

Figure 12.18. Boxes at luminaire or lampholder outlets

Utilization Equipment

Utilization equipment such as smoke alarms and carbon monoxide detectors weighing not more than 3 kg (6 lb.) are permitted to be supported on nonmetallic single-gang boxes or plaster rings that are secured to boxes, provided the equipment or its supporting yoke is secured to the box with no fewer than two No. 6 or larger screws [*NEC*.314.27(D) or IRC E3905.9].

Weight-Supporting Ceiling Receptacles (WSCR) and Weight-Supporting Attachment Fittings (WSAF)

In recent *Code* cycles, listed power supply devices and listed locking support and mounting receptacles and supporting means for luminaires, lampholders, and ceiling suspended (paddle) fans have been incorporated into the *Code* as an option for supporting and mounting these devices. These listed weight-supporting attachment fittings (WSAF) and matching weight-supporting ceiling receptacles (WSCR) are designed to be installed in boxes designed for supporting luminaires, lampholders, and ceiling-suspended (paddle) fans. These optional support and mounting fittings can be used to support and power the luminaire or ceiling suspended (paddle) fan directly thus facilitating replacement of the luminaire or ceiling suspended (paddle) fan when attached in or to the lighting outlet boxes. These listed WSAFs and WSCRs incorporate newly listed technology designed to power and support luminaires and or ceiling-suspended (paddle) fans from a receptacle and mounting means located in the box or mounted directly to the box. This listed device will provide a secure mounting mechanism and will facilitate the interchange of luminaires and ceiling-suspended (paddle) fans in a safe and efficient manner [*NEC* 314.27(E) or IRC E3905.6.3].

These provisions for listed WSAFs and WSCRs for luminaires coincide with the definition of a "receptacle" in *NEC* Article 100 and IRC Chapter 35. A receptacle is defined as "a contact device installed at the outlet for the connection of an attachment plug or for the direct connection of electrical utilization equipment designed to mate with the corresponding contact device." In conjunction with these requirements, similar *Code* language can be found at *NEC* 314.27(C) and 422.18 or IRC E3905.8 and E4101.6 for the support of a ceiling-suspended (paddle) fan.

FLOOR BOXES

A box listed specifically for installation in a floor is required to be used where receptacles are located in a floor. Floor boxes and fittings must be installed in accordance with the installation instructions provided with the product.

Boxes designated for floor installation, as specified in the Code, are equipped with covers and gaskets to exclude surface water and sweeping compounds that may be present during floor cleaning operations. Those boxes are marked on the carton to identify boxes of this type, such as "floor box" or "floor box, concrete tight," as appropriate. Additionally, floor

Photo 12.17. A device box installed in the ceiling for support of a device such as a smoke alarm or receptacle

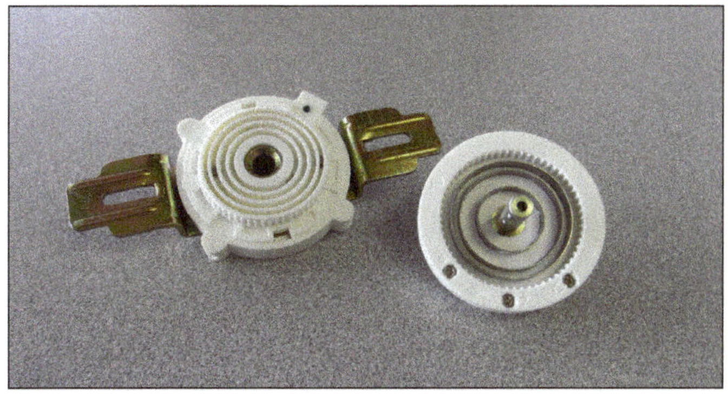
Photo 12.18. Listed locking support and mounting receptacle for luminaires and ceiling (paddle) fans

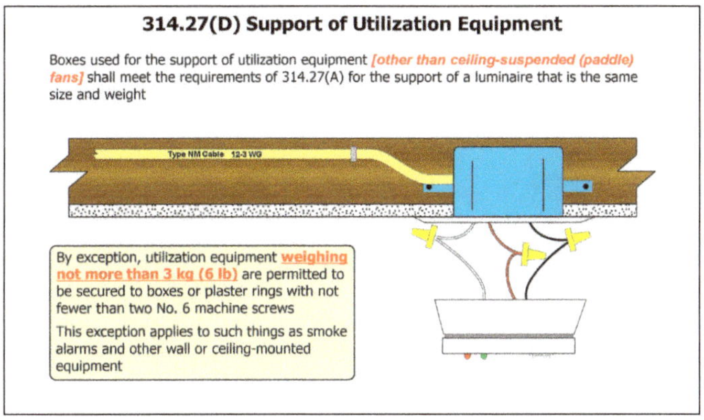

Figure 12.19. Outlet box for support of utilization equipment

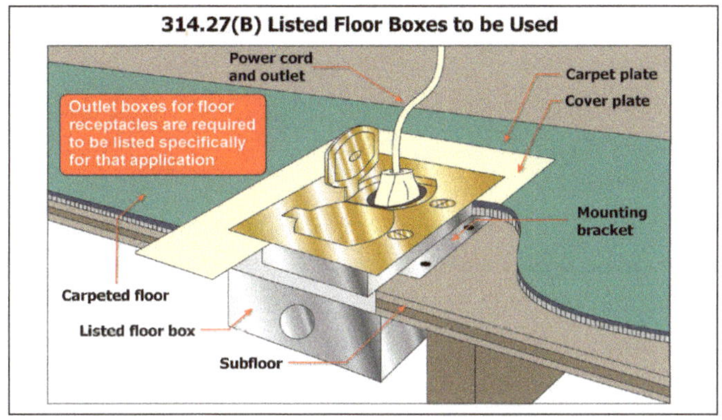

Figure 12.20. Boxes for floor receptacles are required to be listed for that purpose

boxes designed and listed for use in wood-frame construction are also available. Frequently, the floor box is provided as an assembly. The manufacturer provides a box, gasket, and cover plate to be used together. Some inspectors permit cast or formed metal boxes with a weatherproof plate to be used in dwelling units located outside heavily traveled areas as a reasonable alternative [NEC 314.27(B) or IRC E3905.7].

BOXES AT CEILING-SUSPENDED (PADDLE) FAN OUTLETS

A box used at ceiling fan outlets is not permitted to be used as the sole support for ceiling-suspended (paddle) fans unless it is specifically listed for the application and for the weight of the fan to be supported. This requirement applies to both metal and nonmetallic boxes. Outlet boxes or outlet box systems used as the sole support for a ceiling-suspended (paddle) fan are required to be listed and marked on the interior of the box by the manufacturer as suitable for ceiling-suspended paddle fan support. These listed boxes cannot support ceiling-suspended (paddle) fans that weigh more than 32 kg (70 lb.). For outlet boxes or outlet box systems designed to support ceiling-suspended (paddle) fans that weigh more than 16 kg (35 lb.), the required marking must also include the maximum weight the box or box system is listed to support [*NEC* 314.27(C) or IRC E3905.8].

This box or box system must be rigidly supported by a structural member of the building. The outlet box system often has a bracket that is directly secured to a structural member. The outlet box and the ceiling-suspended (paddle) fan are then typically supported by the bracket using the hardware provided in the kit.

Manufacturers of listed ceiling-suspended (paddle) fan boxes or outlet box systems typically supply braces that can be installed for secure box support. Different types of braces are available. Some types are suitable for both new and retrofit applications. As stated earlier, where the weight of the listed ceiling-suspended (paddle) fan, with or without accessories, exceeds 16 kg (35 lb.), the *Code* requires that the box or outlet box system be marked with the maximum weight that can be supported. All manufacturer instructions must be followed for the proper installation of the box intended to support a ceiling-suspended (paddle) fan. Hardware and fasteners supplied by the manufacturer must be completely installed to ensure a safe installation. Incidents have been confirmed concerning ceiling-suspended (paddle) fans that have fallen from the ceiling and caused injury to persons or property. Often, the ceiling-suspended (paddle) fans were installed improperly by untrained personnel, using boxes for support that were not listed for ceiling-suspended (paddle) fan use.

All outlet boxes must be supported on a surface or to a structural member or brace that provides rigid and secure support. This is especially important with outlet boxes used for ceiling-suspended (paddle) fans. The vibrations of the fan will impose additional forces on the outlet box, and the support means. The manufacturer's instructions will often specify the type of fasteners or specific mounting means suitable for use with the listed outlet box.

For example, the instruction on the box may indicate that the listed box must be mounted directly to a ceiling joist for secure support. Or the markings on the box may specify that a minimum of two (2) 2-inch-long, No. 10 or 12, Type A self-tapping screws must be used to attach the box to the joist or truss.

The listed box is equipped with two threaded holes accepting standard 8-32 screws for luminaire support. However, instructions may indicate that all ceiling-paddle fans or heavy luminaires are not to be mounted to the outlet box ears. The manufacturer will often provide a bracket with larger screws for a ceiling-suspended (paddle) fan or heavy luminaire support.

Careful attention must be paid to this requirement. As previously stated, there have been many reported instances of property damage as well as injury to persons caused by a ceiling-suspended (paddle) fan that broke loose and fell due to improper mounting. It is critical that all the manufacturer's instructions are carefully followed and that all the correct hardware items be used, in addition to ensuring the proper outlet box is used (*NEC* 422.18 or IRC E4101.6).

This is an especially severe problem in existing occupancies where an existing luminaire is to be replaced by a ceiling-suspended (paddle) fan. One solution is a type of ceiling-suspended (paddle) fan outlet box with a retrofit-type bracket. This type of assembly is readily available. These boxes and brackets can be installed from below the ceiling through the existing luminaire box hole in the ceiling or a hole that is cut for the outlet box without having access to the attic or joist space above the box. The bracket is installed through this hole and then adjusted to make contact with the ceiling joists. After securely tightening the bracket to the framing, the listed ceiling-suspended (paddle) fan support box is then attached to the bracket. Where the space above the box is accessible, proper structural bracing can be installed for listed outlet boxes identified for support of ceiling-suspended (paddle) fans.

Photo 12.19. Ceiling-suspended paddle fans require suitable support

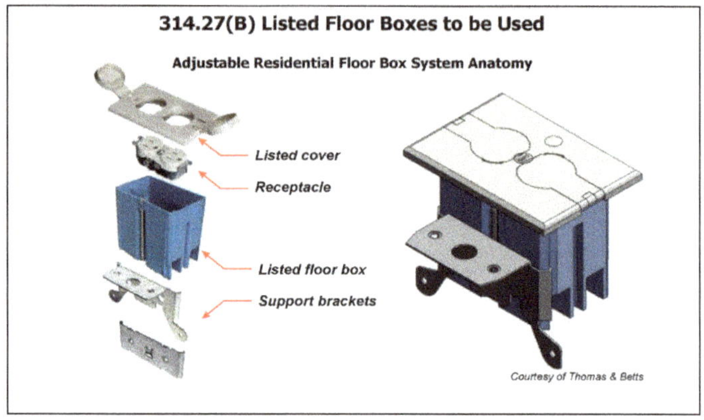

Figure 12.21. "Adjust-a-Box" listed specifically for installation in a floor

Prior to the 2020 *NEC* (2021 IRC), where a "spare," separately switched, ungrounded conductor was provided to a ceiling-mounted outlet box in a location acceptable for a ceiling-suspended (paddle) fan in one-family, two-family, or multifamily dwellings, the outlet box or outlet box system was required to be listed for sole support of a ceiling-suspended (paddle) fan or the ceiling-suspended (paddle) fan had to be mounted directly to a structural framing member. A spare switched conductor is oftentimes installed at single-family, two-family, and multi-family dwelling units in locations such as bedroom switch locations, which supply ceiling-mounted luminaire or lighting outlets. This spare switched conductor is typically installed in the event a ceiling fan will be installed (at a later date) by the homeowner. The problem arises when these ceiling-mounted outlet boxes are served with a standard luminaire outlet box at the initial installation. The homeowner will typically not know the proper application would be to replace the standard luminaire box with a ceiling box listed for support of a ceiling-suspended (paddle) fan.

During the 2020 *NEC* and the 2021 IRC revision cycle, the requirement outlet box or outlet box system required to be listed for sole support of a ceiling-suspended (paddle) fan or the ceiling-suspended (paddle) fan having to be mounted directly to a structural framing member was expanded to all outlet boxes mounted in the ceilings of habitable rooms of

Photo 12.20 Outlet box listed for support of a ceiling-suspended (paddle) fan.

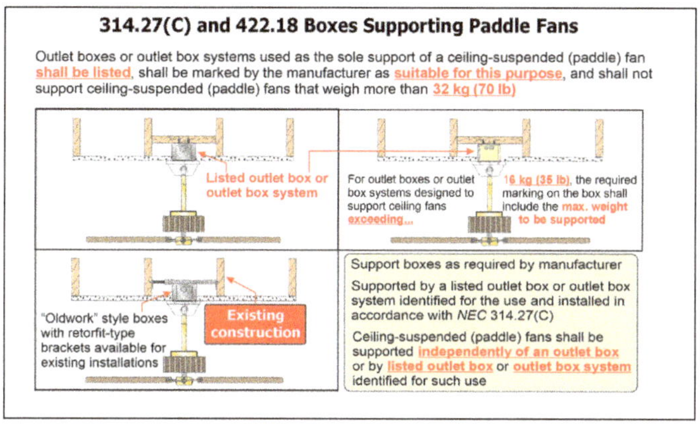

Figure 12.22. Support of ceiling-suspended (paddle) fans with boxes

Figure 12.23. All outlet boxes mounted in the ceilings of habitable rooms of dwelling units are generally required to be listed ceiling fan outlet boxes or systems

dwelling units in locations acceptable for the installation of a ceiling-suspended (paddle) fan. These requirements, being predicated upon a spare separately switched conductor being present in the box in previous editions of the Code, become an outdated installation technique as many ceiling-suspended (paddle) fans are now remote-controlled, requiring only a two-wire installation.

This requirement is applicable only in locations acceptable for the installation of a ceiling-suspended (paddle) fan. As an example, a ceiling-mounted lighting outlet box located 300 mm (12 in.) from a wall for something like track lighting would not require a lighting outlet box listed for the sole support of a ceiling-suspended (paddle) fan as this box would not be able to accommodate the installation of a ceiling-suspended (paddle) fan [*NEC* 314.23; 314.27(C); or IRC E3906.8, E3905.8].

BOXES TO BE ACCESSIBLE

Boxes, conduit bodies, and handhole enclosures must be installed so that the wiring and devices contained are accessible. No part of the building or structure can restrict access to the wiring within the box. For underground installations, access to the wiring is required without having to excavate sidewalks, paving, earth, concrete, or other substances used to establish the finish grade. An exception permits listed underground boxes or handhole enclosures to be covered with sand or gravel. The location of the enclosure must be effectively identified and accessible for excavation to those authorized to access, maintain, or inspect the wiring or devices (*NEC* 314.29 or IRC E3905.10).

ELECTRICAL PENETRATIONS OF FIRE-RATED STRUCTURAL MEMBERS

In most single-family dwelling units, the wall and ceiling between the attached garage and the remainder of the dwelling are the only walls or ceilings that are required to have a fire rating. Requirements for the ratings of firewalls, ceilings, area separations, and building separations are found in the applicable building code, not in the *NEC*. Building separation firewalls may be required where single-family dwellings, such as condominiums, are a part of an overall larger structure. It is common for these building separation firewalls to have a two-hour fire rating (*NEC* 300.21 or IRC E3402.2 and R302.4).

These sections additionally require that openings around electrical penetrations through fire-resistant-rated walls, partitions, floors, or ceilings be firestopped using approved methods. The approved firestopping methods must maintain the fire-resistant rating of the assembly. The IRC provides detailed information relative to electrical penetrations in walls, partitions, floors, or ceilings that are fire-rated at IRC Section R302.

FIRESTOPPING PENETRATIONS

Power, lighting, remote-control, signaling, and power-limited circuits and equipment must be installed not substantially to increase the spread of fire or products of combustion. The

Code requires suitable firestopping be installed at any openings around electrical penetrations where the wiring methods for such circuits penetrate into or through fire-resistance-rated walls, partitions, floors, or ceilings (*NEC* 300.21 or IRC E3402.2 and E3402.3). An approved firestop method must be used to maintain the fire-resistant rating. Many caulking or sealing products have been specifically designed for sealing penetrations in fire-rated assemblies. Simply stuffing the bored holes or similar openings with pieces of insulation, caulk, foam, newspaper, or steel wool is not acceptable. The common practice of installing a device ring without a box is not permitted in a fire-rated assembly. For example, a device-ring rough-in for a telephone receptacle outlet would not be permitted in a fire-rated assembly. A suitable "back-box" listed for use in the fire-rated assembly is required.

When installing Type NM cable through firestopping penetrations, keep in mind the requirements of *NEC* 334.80 or IRC E3705.4.4, which state that where more than two NM cables containing two or more current-carrying conductors are installed, without maintaining spacing between the cables, through the same opening in wood framing that is to be sealed with thermal insulation, caulk, or sealing foam, the allowable ampacity of each conductor must be adjusted for bundling of conductors. See Chapter 11 in this book for additional information on the requirements of *NEC* 334.80 or IRC E3705.4.4.

Photo 12.21. Typical building area separation construction at two-family dwelling units

Photo 12.22. A 3-gang switch box and wiring method installed in the environmental air-handling space *(environmental air-handling is shaded in red for photo clarification)*

WIRING IN AIR-HANDLING SPACES

It is best to avoid placing any electrical wiring in ducts or other spaces for environmental air (plenums) due to the restrictive wiring methods required for these locations. As used in dwellings, return air ducts and plenums generally consist of prefabricated round or rectangular sheet metal ducts used to transport environmental air. The term plenum is defined in Article 100 of the *NEC* and in Chapter 35 of the IRC. Plenums in dwellings are usually fabricated chambers built for the sole purpose of transmitting environmental air. For example, a plenum chamber is used to connect the ducts to a forced-air furnace.

Electrical wiring in ducts or plenums is covered in *NEC* 300.22(B). Wiring methods, such as electrical metallic tubing, flexible metallic tubing, and intermediate or rigid metal conduit, are required in ducts or plenums used for environmental air. The circuits and equipment are restricted to those directly associated with the operation of the heating, ventilation, or air-conditioning system. For example, a humidistat may be installed in the plenum to monitor the relative humidity of the conditioned air.

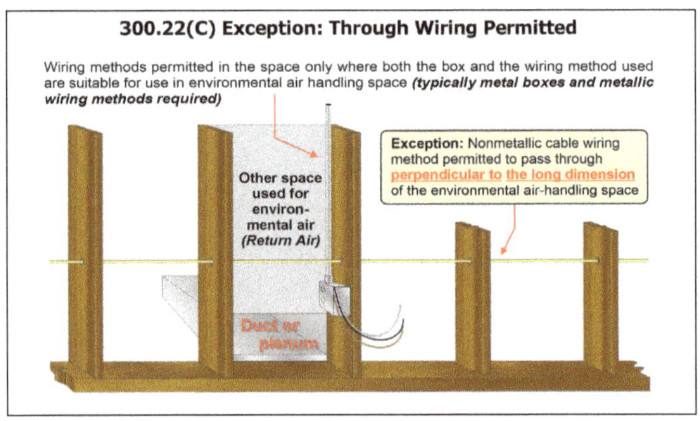

Figure 12.25. Electrical wiring in "other spaces used for environmental air (plenum)" is permitted using specific installation criteria

OTHER SPACES USED FOR ENVIRONMENTAL AIR

NEC 300.22(C) provides the requirements for wiring in "Other Spaces Used for Environmental Air (Plenums)." It is common in dwellings to use stud or joist spaces that are closed in with gypsum board or sheet metal to provide return

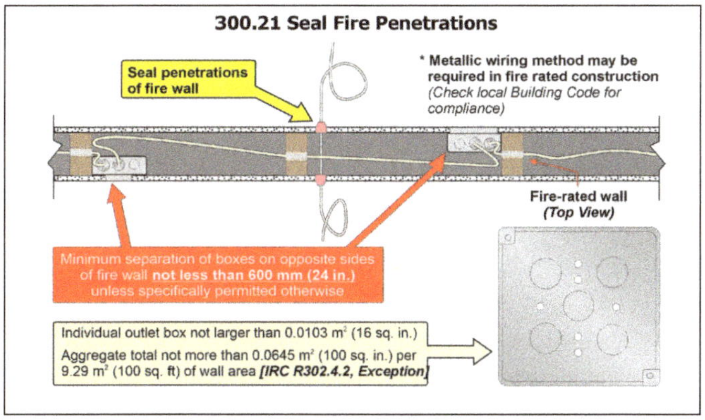

Figure 12.24. Electrical outlet boxes in firewalls

Photo 12.23. Type NM cable installed perpendicular to the long dimension of a stud space being used as a return air plenum

Photo 12.24. Typical handhole enclosure *(Photo courtesy of Strongwell)*

air space used for environmental air-handling purposes. These spaces are examples of "other spaces" to which *NEC* 300.22(C), 722.179(A)(1), and 725.3(B) or IRC E4303.2.1 would apply. The requirements that limit the types of wiring methods and equipment in these spaces are less restrictive than those regulating ducts and plenums. However, if possible, avoid placing wiring and equipment in these spaces. The mechanical codes would also limit the number of foreign systems in the space in order to avoid unduly restricting air movement.

NEC 300.22(C) Exception or IRC E3904.7 permits nonmetallic wiring methods in the "joist or stud spaces" of

Photo 12.25. Identification marking on a handhole cover

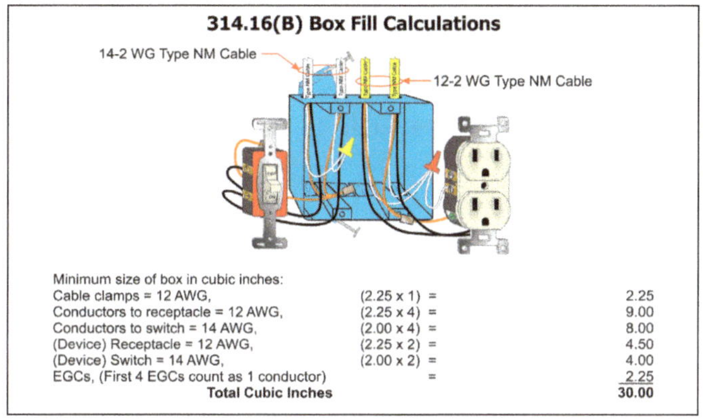

Figure 12.26. Answer key to Figure 12.7

dwelling units where the wiring passes through such spaces "*perpendicular to the long dimension of such spaces.*" This permits the use of a wiring method, such as Type NM nonmetallic-sheathed cable, to be installed only where it crosses the narrow axis of the space that is used to transport environmental air (see Figure 12.25 and Photo 12.23). It is not permitted to pass through the space in the long axis, even if secured to the framing member. The intent of this rule is to limit the amount of this nonmetallic (flammable) wiring material in the air-handling space. Type NM cable is not permitted to supply equipment, such as device boxes, in these air-handling spaces. If boxes are required in this space, the wiring method must be one of the types specified in *NEC* 300.22(C)(1) for environmental air spaces, such as a metal raceway or metal-clad cable. Remember that the building and mechanical codes may prohibit electrical boxes in this space in order to avoid unduly restricting air movement. Listed Class 2 cables, such as CMX and CL2X, are also permitted to pass through the short dimension of these "other spaces" without being installed in a metal raceway or other wiring method required by *NEC* 300.22(C)(1).

HANDHOLE ENCLOSURES

Handhole enclosures are typically used with underground circuits. Handhole enclosures are addressed by the *NEC* at

314.30(A) through (D). The removable cover gives access to wiring or equipment. The enclosure must be designed and installed to withstand safely all loads that are likely to be imposed on them. They are also required to be identified for use in underground systems. The minimum size of the enclosure is determined based on *NEC* 314.30(A). Underground raceways and cables are required to be extended into the handhole enclosure but are not required to be mechanically connected to the enclosure.

A typical handhole enclosure is not watertight. Splices or equipment inside it must be listed as suitable for wet locations. The cover of a handhole should have a clear mark or logo indicating its function. Identification marks, such as "Electric," "Communication," or "Landscape Lighting," clearly indicate the purpose. The cover must be designed so that it requires tools to open or weigh over 45 kg (100 lb..). Metal covers or other exposed conductive surfaces need to be properly bonded to ensure electrical continuity and safely carry any fault current. A field-installed equipment-bonding jumper might be required to achieve this [NEC 314.30(D]. The IRC does not specifically address handhole enclosures.

CONDUCTOR ROUTING AND INDUCTION

Where conductors carrying alternating current are installed in ferrous metal enclosures or ferrous metal raceways, they are required to be arranged so as to avoid heating the surrounding metal by induction. To accomplish this, all ungrounded phase conductors and, where used, the grounded (neutral) conductor and all equipment grounding conductors are generally required to be grouped together by installing them in the same raceway, cable, or trench. Switch loops are not required to include a grounded (neutral) conductor because of the canceling effect due to counter electromotive force (emf) (*NEC* 300.20).

PRODUCT SAFETY STANDARD REQUIREMENTS

It is recommended to refer to the following electrical product safety guide card information and product safety standards for additional guidance on the proper installation, operation, and use of electrical equipment covered in this chapter. The four-letter code in parentheses indicates the product category in the Underwriters Laboratories Guide Information for Electrical Equipment Directory (Product Spec).

- Electrical Equipment for Use in Ordinary Locations (AALZ)
- Outlet Boxes and Fittings (QBPZ)
- Metallic Outlet Boxes (QCIT)
- Nonmetallic Outlet Boxes (QCMZ)
- Conduit Fittings (Bodies) (DWTT)
- Fans, Ceiling Suspended (GPRT)

REFERENCES

1. *National Electrical Code* and *NEC* are registered trademarks of the National Fire Protection Association, Inc., Quincy, MA 02169. This reprinted material is not the official position of the National Fire Protection Association, which is represented only by the standard in its entirety.

2. *International Residential Code for One- and Two-Family Dwellings* (2024). Copyright © International Code Council, Inc., June 2023.

3. UL, Northbrook, IL, UL Product iQ® material unless indicated otherwise.

REVIEW QUESTIONS

1. How many 14 AWG conductors are permitted in a 100 mm (4 in.) square metal box sized at 38 mm (1-½ in.) deep?

A. 14
B. 12
C. 10
D. 9

2. All cable clamps added together represent ______ conductor allowance(s) when calculating the required size of a junction or device box.

A. 0
B. 1
C. 2
D. 3

3. Fittings such as capped elbows and service-entrance elbows are classified as conduit bodies.

True _____ False _____

4. What is the minimum volume of space required for four (4) 14 AWG and four (4) 12 AWG conductors installed within a metal junction box?

A. 230 cm^3 (14 cu. in.)
B. 246 cm^3 (15 cu. in.)
C. 263 cm^3 (16 cu. in.)
D. 279 cm^3 (17 cu. in.)

5. The minimum volume of space required within a nonmetallic junction box for four (4) 10 AWG, and four (4) 12 AWG conductors is:

A. 197 cm^3 (12 cu. in.)
B. 246 cm^3 (15 cu. in.)
C. 312 cm^3 (19 cu. in.)
D. 344 cm^3 (21 cu. in.)

6. Plates or metal plugs used to close up unused openings in a nonmetallic box must be recessed at least:

A. 19 mm (¾ in.)
B. 13 mm (½ in.)
C. 6 mm (¼ in.)
D. 3 mm (⅛ in.)

7. A metal brace used to secure a box to a structural member is required to be protected against corrosion and formed from metal not less than 1.6 mm (1⁄16 in.) thick, uncoated.

True _____ False _____

8. A ceiling-suspended (paddle) fan that does not exceed _______ pounds in weight, with or without its accessories, is permitted to be supported by a listed outlet box or outlet box system identified for such use.

A. 6
B. 20
C. 35
D. 70

9. Where one (1) cable clamp and one (1) device are included, how many 12-2 W/G Type NM cables are permitted in a 3" x 2" x 3-½" metal device box?

A. 1
B. 2
C. 3
D. 4

10. The same screw used to support the box to a framing member is permitted to be used to secure a device to the box.

True _____ False _____

11. What is the minimum volume of space required for a ceiling box containing a luminaire (fixture) stud and two (2) cable clamps with one (1) 14 AWG two-wire W/G Type NM cable feed in, and one (1) 14 AWG two-wire W/G Type NM cable feed out?

A. 98 cm^3 (6 cu. in.)
B. 115 cm^3 (7 cu. in.)
C. 230 cm^3 (14 cu. in.)
D. 295 cm^3 (18 cu. in.)

12. Outlet boxes mounted in the ceilings of habitable rooms of dwelling units in a location acceptable for the installation of a ceiling-suspended (paddle) fan is required to comply with which of the following:

A. Listed for the sole support of ceiling-suspended (paddle) fans.
B. Installed so as to allow direct access through the box to structural framing capable of supporting of a ceiling-suspended (paddle) fan without removing the box.
C. Located in the center of the room.
D. Either A. or B.

13. Of the boxes listed below, what is the minimum size box required in the volume of space for a nonmetallic junction box containing (8) 12 AWG conductors and (4) equipment grounding conductors that are brought into the box for splicing?

A. 295 cm^3 (18 cu. in.)
B. 328 cm^3 (20 cu. in.)
C. 344 cm^3 (21 cu. in.)
D. 395 cm^3 (24 cu. in.)

14. A metal device box contains eight (6) 12 AWG conductors, no clamps, hickeys, or devices. This device box also contains (6) equipment grounding conductors. Which of the following is the minimum allowable box permitted to contain these conductors?

A. 100 mm x 32 mm (4 in. x 1-¼ in.) (round/octagonal)
B. 100 mm x 38 mm (4 in. x 1-½ in.) (round/octagonal)
C. 100 mm x 54 mm (4 in. x 2-⅛ in.) (round/octagonal)
D. 100 mm x 32 mm (4 in. x 1-¼ in.) (square)

15. A two-gang nonmetallic box with internal cable clamps contains (2) 14/3 AWG with ground nonmetallic-sheathed cables and (2) 14/2 AWG with ground nonmetallic-sheathed cables. This box will also contain (2) 3-way switches. Which of the following is the minimum allowable volume of space capacity of the box?

A. 426 cm^3 (26 cu. in.)
B. 459 cm^3 (28 cu. in.)
C. 525 cm^3 (32 cu. in.)
D. 590 cm^3 (36 cu. in.)

16. The maximum weight of a luminaire which can be directly supported from a ceiling-mounted outlet box designed for the support of luminaires is:

A. 35 lbs.
B. 40 lbs.
C. 50 lbs.
D. 60 lbs.

17. Boxes, conduit bodies, outlet box hoods, and fittings installed in wet locations must be specifically listed for use in wet locations.

True ______ False ______

18. The required box fill volume allowance for a switch or receptacle on a single yoke or strap in a typical nonmetallic box is:

A. single volume allowance
B. double volume allowance
C. triple volume allowance
D. no volume allowance required

19. Outlet boxes that employ a flush-type cover and that are installed in walls and ceilings constructed of wood or other combustible material are required to be installed so that the front edge of the box is flush with the finished surface or project therefrom.

True ______ False ______

20. For a standard metal box containing (1) luminaire stud and (2) cable clamps, the total conductor volume allowance required for these two items combined is:

A. one
B. two
C. three
D. four

21. An extension ring is permitted to be used when extending a branch circuit from a concealed octagon box.

True ______ False ______

22. No box (that does not enclose devices or utilization equipment) is permitted to have an internal depth of less than 12.7 mm (½ in.).

True ______ False ______

23. Before applying any exceptions, a box that is intended to enclose devices or utilization equipment supplied by 14 AWG conductors is required to have an internal depth of not less than _________.

A. 12.7 mm (½ in.)
B. 15.8 mm (⅝ in.)
C. 19.0 mm (¾ in.)
D. 23.8 mm (15⁄16 in.)

24. A nonmetallic box is required to be suitable for the lowest temperature-rated conductor entering the box.

True ______ False ______

25. Boxes employing a plaster ring and installed in walls or ceilings of concrete, tile, or other noncombustible material are required to be installed so that the front edge of the plaster ring will not set back of the finished surface more than what distance?

A. 6 mm (¼ in.)
B. 13 mm (½ in.)
C. 16 mm (⅝ in.)
D. 19 mm (¾ in.)

26. Where a nail is used as a fastening means for the support of a box, it is permitted to pass through the interior of the enclosure only where it is located within what distance of the back or ends of the inside of the enclosure?

A. 6 mm (¼ in.)
B. 13 mm (½ in.)
C. 16 mm (⅝ in.)
D. 19 mm (¾ in.)

27. A vertically mounted luminaire such as a wall-mounted luminaire weighing not more than ___________ is permitted to be supported on other boxes not designed for the purpose or by plaster rings that are secured to other boxes, provided the luminaire or its supporting yoke is secured to the box with no fewer than two (2) No. 6 or larger screws.

A. 1.4 kg (3 lb.)
B. 2.3 kg (5 lb.)
C. 3 kg (6 lb.)
D. 4 kg (8 lb.)

28. Any device or utilization equipment wider than a single 50 mm (2 in.) device box is required to have a _________ volume allowances provided for each gang required for mounting.

A. single
B. double
C. triple
D. no volume alliance required

29. Openings around electrical penetrations through fire-resistant-rated walls, partitions, floors, or ceilings must be firestopped using __________ methods.

A. approved
B. listed
C. identified
D. fire-proof

30. Nonmetallic wiring methods are permitted in the joist or stud spaces used for environment air (plenum) in dwelling units where wiring passes through __________ to the long dimension of such spaces.

A. vertical
B. perpendicular
C. horizontal
D. sideways

31. Screws used for the purpose of attaching covers, or other equipment, to the box are required to be either machine screws matching the thread gauge or size that is integral to the box or must be in accordance with the ___________ ___________.

A. NEMA specifications
B. manufacturer's instructions
C. UL Product Standard 790
D. UL Product iQ™ (*the old UL White Book*)

32. Drainage openings that are not larger than ___________ are permitted to be installed in the field in boxes or conduit bodies listed for use in damp or wet locations. Larger openings are permitted to be installed in the field in accordance with manufacturer's instructions.

A. 6 mm (¼ in.)
B. 13 mm (½ in.)
C. 16 mm (⅝ in.)
D. 19 mm (¾ in.)

33. Where a metal device box is provided with a securely installed metal interior barrier, the barrier is considered as take up ___________ of space.

A. 1.0 cm^3 ($\frac{1}{16}$ $in.^3$)
B. 2.1 cm^3 (⅛ $in.^3$)
C. 4.1 cm^3 (¼ $in.^3$)
D. 8.2 cm^3 (½ $in.^3$)

34. A ___________ is a listed product that is an optional support and mounting fitting that can be used to support and power a luminaire or ceiling suspended (paddle) fan directly thus facilitating replacement of the luminaire or ceiling suspended (paddle) fan when attached in or to the lighting outlet boxes.

A. weight-supporting attachment fittings (WSAF)
B. weight-supporting ceiling receptacles (WSCR)
C. separable attachment fitting
D. Both A. and B.

35. For box fill calculations, where up to _______ equipment grounding conductors enter a box, a single volume allowance shall be made based on the largest equipment grounding conductor entering the box. An additional ______ volume allowance shall be made for each additional equipment grounding conductor that enters the box, based on the largest equipment grounding conductor.

A. four / ¼
B. four / ½
C. six / ¾
D. four / single

13 Requirements for Luminaires, Receptacles, and Switches

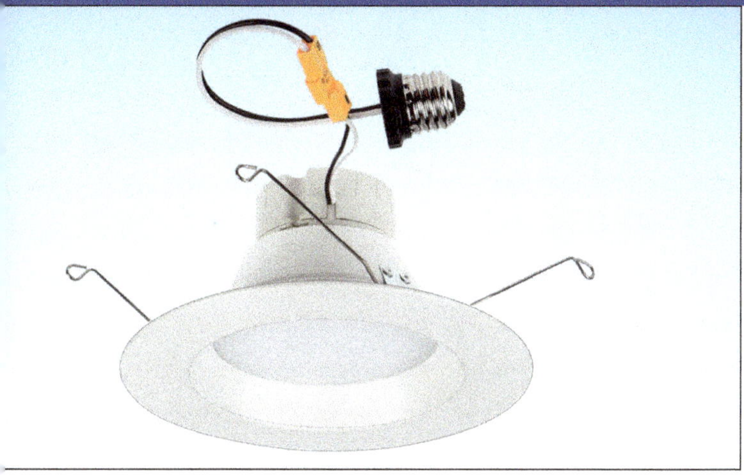

Photo 13.1. LED retrofit kit for incandescent recessed luminaire

Luminaires, receptacles, and switches are the "nuts and bolts" of the dwelling unit. These items and devices make life much easier for the occupant. They have become so commonplace that we tend to take them and their functions for granted. The Code has a lot to say about these essential items. This chapter will provide a closer examination of their working functions and requirements.

LUMINAIRES

The term luminaire is used in international standards for what has been commonly identified in the United States and past editions of the *Code* as a lighting fixture. Including a definition for luminaire in the *Code* is important. The term "fixture" is still used in other parts of the *Code*, such as Article 402, *Fixture Wire*, which covers the requirements for and construction specifications of fixture wire, and it still limits the use of fixture wire for installation in and connection to luminaires.

DEFINITION

A "luminaire" is a complete lighting unit that includes a light source such as a lamp or multiple lamps, along with the components designed to position the light source and connect it to the power supply. It may also contain parts to protect the light source, the ballast, or to distribute the light. A lampholder itself is not considered a luminaire.

As a rule, luminaires must be installed in compliance with three sets of requirements:

1. Article 410 of the *National Electrical Code* or E4003, E4004, and E4005 of the IRC
2. Product safety standard and guide card information of UL Product iQ®
3. The manufacturer's instructions or markings on the luminaire, carton, or other materials provided with the luminaire

NEC Article 410 and the latter part of Chapter 40 of the IRC contain many requirements for the safe installation of luminaires. General requirements are provided that cover all luminaires. Limitations are placed upon the type of luminaires that can be installed in specific locations such as clothes closets, shower and bath areas, and wet, damp, dry, and corrosive locations. Rules for supporting, grounding, and wiring luminaires are contained in *NEC* Article 410 and Chapter 40 of the IRC as well. Some general construction requirements are included in *NEC* Article 410 Part VII. Most of the specific requirements that govern luminaire construction are left to the product safety standard.

Many requirements regarding the proper construction of electric luminaires are contained in product safety standards produced by Underwriters Laboratories (UL). These standards spell out in precise detail every facet of the construction, testing, marking, and performance requirements for luminaires. The standard most applicable to luminaires used in one- and two-family dwellings is *UL Standard for Safety for Luminaires*, UL 1598. This standard applies to luminaires for use in non-hazardous locations and that are intended for installation on branch circuits of 600 volts nominal or less between conductors in accordance with the NEC. UL standards are updated on a regular basis and are available from Underwriters Laboratories or other sources.

Many marking requirements are contained in the product safety standards. In addition, the product standards require the luminaire manufacturer to include installation instructions or use information in sheets that accompany the luminaire. Such installation instructions or restrictions on use are requirements of the *Code*. Some markings are permitted or required to be on the luminaire itself, while other markings are permitted or required to be in or on the shipping carton [*NEC* 110.3(B) or IRC E3403.3].

INSTRUCTIONS OR MARKINGS ON LUMINAIRES BY THE MANUFACTURER

All luminaires, lampholders, and retrofit kits for luminaires are required to be listed (see *NEC* 410.6). It is critical to the safe installation and operation of luminaires that they be installed and used in full compliance with all instructions, markings, and specifications provided by the luminaire manufacturer. Some of these instructions limit the luminaire to installation in certain locations, such as on a wall or ceiling or in a dry or damp location, or they require certain minimum clearances from combustible materials. Other instructions or markings specify certain trims on recessed luminaires or perhaps indicate requirements on the insulation temperature rating of branch-circuit supply conductors to the luminaire.

RETROFIT KITS FOR LUMINAIRES

The definition of "Retrofit Kit" in Article 100 of the NEC states that a retrofit kit is "a complete subassembly of parts and devices for field conversion of utilization equipment." Extensive upgrades are underway in the lighting industry to achieve greater energy efficiency in luminaires by replacing existing in-place illumination systems with newer, energy-efficient lighting methods, such as light-emitting diode (LED) technology. In the case of NEC Article 410, retrofit kits for luminaires will incorporate field modifications of existing luminaires. Testing laboratories, such as Underwriters Laboratories (UL), have developed protocols for these field conversions, ensuring that when performed within the testing laboratory's parameters, the field conversion "retrofit kits" do not compromise the safety profile of the listed luminaire. For example, to ensure compatibility with field modifications, UL requires that all parts for luminaire conversions be assembled into a kit labeled as Classified by UL. Adding "retrofit kits" to the required listing requirements of NEC Article 410 will drive the use of these "classified" kits and the use of conversion subassemblies. The change in illumination systems in luminaires presents a hazard for installers performing maintenance after the conversion if not installed using the listed subassembly retrofit kits and installed properly according to the manufacturer's instructions.

Luminaires that are improperly installed or have a lamp that is larger than the luminaire is designed for have been the cause of countless fires. These fires have resulted in the loss of a significant number of lives and the loss of millions of dollars in the value of buildings or structures.

Many marking or warning requirements are contained in the safety standard for luminaires, UL 1598. Markings that are required vary to some extent based on the type of luminaire involved and the type of lamp required to be used. A typical marking will state "CAUTION—RISK OF FIRE. USE MAX ___ WATTS TYPE ___ONLY." A symbol for the type of lamp to be installed includes types A, B, C, R, T, or PAR. Lamp-type names include Reflector, Elliptical Reflector, Parabolic Reflector, and Globe (see Table 13.1).

There are many lamp types available for use in luminaires. Symbols are often on the luminaire markings as installation use guides and instructions for installers and consumers. It is important that the appropriate lamp be inserted in luminaires and that maximum wattages are not exceeded. Please refer to Table 13.1 for a list of the various lamp symbols as derived from UL Standard 1598 Table D.1.

The common thread in these warning statements is the risk of fire from incandescent luminaires. Overlamping of luminaires presents a major potential threat. Overlamping means

Figure D.1
Pictogram format
(See Clause D.2.)

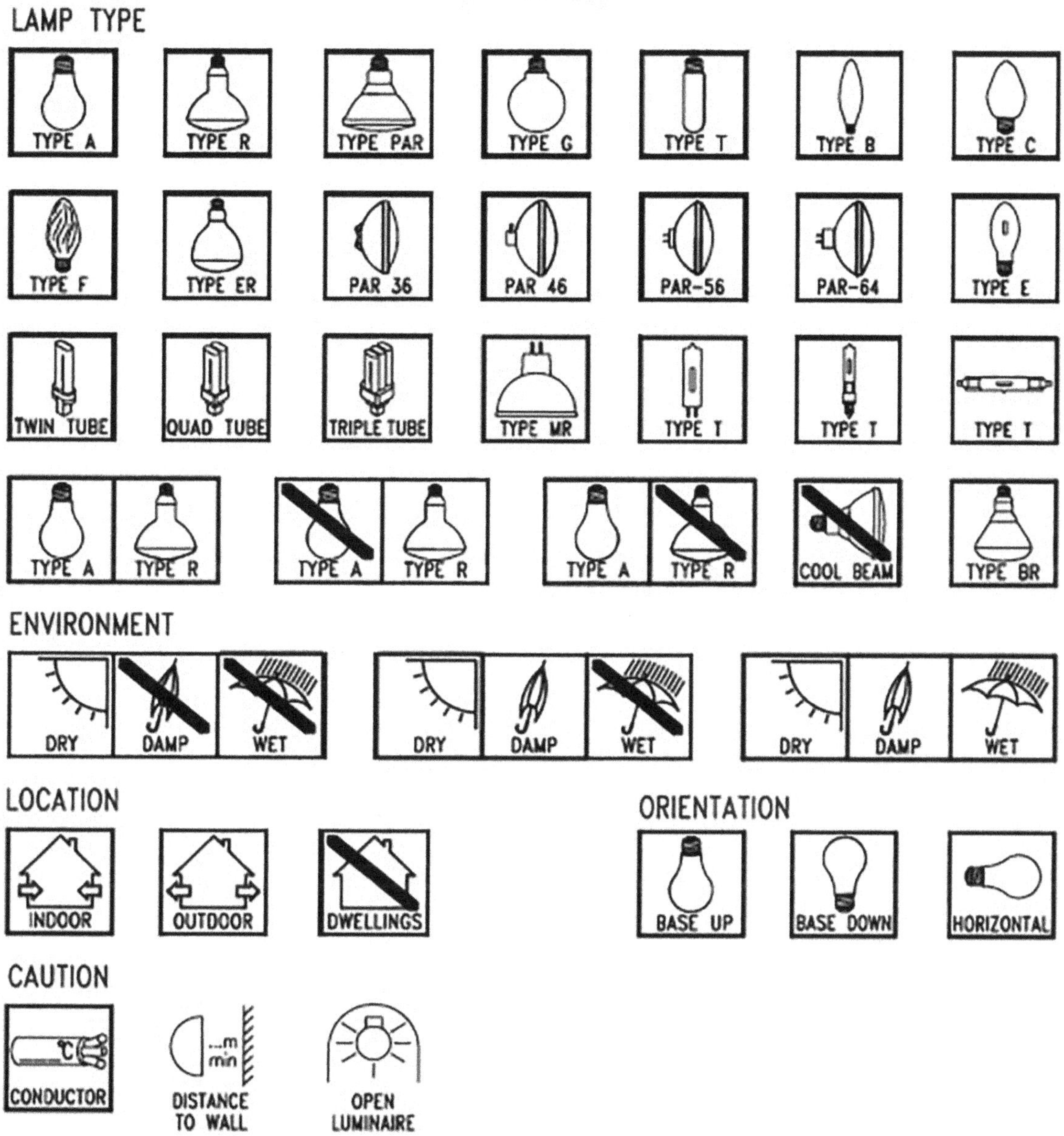

MAX is a symbol that is a minimum of 4.8 mm 0.188 in) high and 15 mm (0.8 in) wide with a character stroke width of 1 mm (0.40 in). Uppercase typestyle Univers Bold 19 provides the correct size, character spacing, and character stroke width.

NOTE Pictogram format frame shown is a 16 mm (0.625 in) square.

S3846B

Table 13.1. UL 1598 Table D.1 (in part) *Courtesy of Underwriters Laboratories*

Photo 13.2. Luminaires are designed in a wide variety of styles and designs

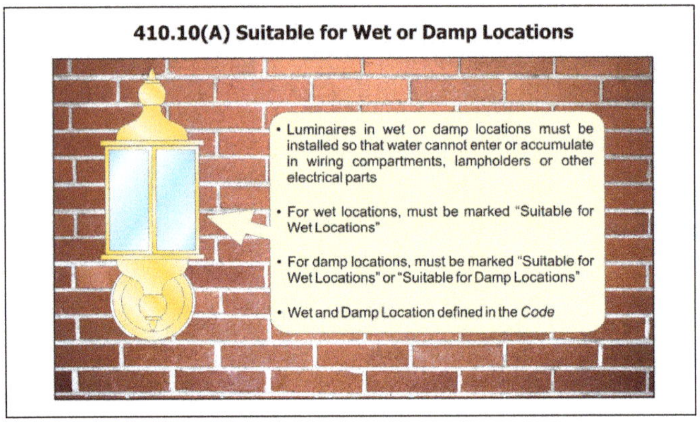

Figure 13.1. Luminaires installed in wet locations

using a lamp with a wattage greater than the maximum wattage for which a luminaire is designed. Incandescent lamps are rated in watts, which is a measure of the energy used to produce light. A higher-wattage lamp generally produces a brighter light. The higher-wattage lamp also radiates more energy in the form of heat.

The size of the lamp does not always correspond to the wattage. For a given type and physical size lamp, several wattage ratings are often available. This increases the chance of overlamping. The Edison base lampholder, also referred to as a medium base, has been the standard lamp base for many years. No scheme is available for limiting lamp wattage in a manner similar to a Type S fuse, which is a fuse of a given ampere rating that is not interchangeable with a fuse of a different ampere rating.

Keep in mind that the U.S. Department of Energy banned the sale of energy-efficient light bulbs (with a few exceptions) beginning in August of 2023. Given that incandescent and halogen light bulbs are far less energy efficient than LED light bulbs, many experts are treating this as a ban on incandescent and halogen bulbs.

LUMINAIRE DESIGNS

Luminaires come in a wide variety of sizes, shapes, styles, and designs to supply illumination for an equally wide variety of locations. Lamps are usually incandescent or electric discharge such as fluorescent, metal halide, or high-pressure sodium. Quartz, tungsten, and halogen, with LED (light emitting diode), are becoming the most popular form of lamps being sold and used today. Many supplemental or decorative applications are now employing low-voltage luminaires. Some luminaires are designed to be installed in specific locations, such as surfaces, recessed, or pole-mounted. Other types of luminaires are designed for and intended to be chain-, cable-, or cord-hung.

LUMINAIRES IN DRY, DAMP, AND WET LOCATIONS

Dry Locations

Luminaires are permitted to be installed only in dry locations unless they are marked "Suitable for Damp Locations" or "Suitable for Wet Locations." The term *dry location* is defined as "A location not normally subject to dampness or wetness. A location classified as dry may be temporarily subject to dampness or wetness, as in the case of a building under construction." The definitions of *damp location* and *wet location* are also found in Article 100 of the *NEC* or Chapter 35 of the IRC.

Wet Locations

Luminaires installed in wet or damp locations must be installed so that water will not enter or accumulate within the luminaire wiring compartment, lampholder, or other parts associated with the construction of the luminaire [*NEC* 410.10(A) or IRC E4003.9].

Only luminaires marked "Suitable for Wet Locations" or "Suitable for Damp/Wet Locations" are permitted to be installed in wet locations. A wet location includes any installation underground, in concrete slabs, in masonry that is in direct contact with the earth, and outside locations that are exposed to rain, ice, and snow. Luminaires installed on the outside of a dwelling unit that are exposed to the weather and unprotected are luminaires that the AHJ will consider to be in a wet location. A luminaire marked "Suitable for Wet Locations" is also permitted to be installed in a damp or dry location.

Damp Location

Only luminaires marked "Suitable for Wet Locations" or "Suitable for Damp Locations" are permitted to be installed in damp locations. A damp location includes locations such as the area under an outside open porch, patio, carport, or entryway. The location is covered and has limited protection from the weather, but may be subject to some moderate degrees of moisture. Some basements, cellars, or crawl spaces located in one- and two-family dwelling units may be subject to moderate degrees of moisture and be considered damp locations.

LUMINAIRE INSTALLATION IN WET OR DAMP LOCATIONS

A listed luminaire marked "Suitable for Wet Locations" is intended to be installed at least 1.2 m (4 ft) above the ground where it is not likely to be subjected to water from things like a lawn or garden sprinkler system. A wall- or post-mounted luminaire that may be used for landscape lighting may be installed within 1.2 m (4 ft) of ground level if it is marked "Suitable for Mounting within 1.2 m (4 ft) of the Ground." Luminaires with this marking are intended to be subjected to water from lawn or garden sprinkler systems but are not intended to be installed at or below ground level, where they may be subjected to immersion in water.

INSTALLATION REQUIREMENTS IN WET LOCATIONS

Many wet location luminaires are mounted to an outlet box that is flush with the exterior surface of the dwelling unit. The *Code* requires the luminaire to be mounted on an outlet box since the typical luminaire does not have an integral wiring compartment. The outlet box may be incorporated into a siding accessory, sometimes called a "siding block," or the outlet box may be recessed into the exterior surface of the wall. Consult the manufacturer's instructions for guidance on proper installation. These instructions identify for the installer (1) that the use of the caulking compound is required to provide a watertight seal between the luminaire and the building and (2) the generic type of caulking compound to be used.

LUMINAIRES AROUND SWIMMING POOLS

NEC Article 410 or Chapter 40 of the IRC does not include any specific rules covering luminaires installed in, around, or above swimming pools or spas and hot tubs. Although rules for these luminaires are covered in *NEC* Article 680 and IRC Chapter 42, any luminaire installed in a wet location, such as outdoors, to provide lighting for a swimming pool, spa, or hot tub must still comply with the requirements for luminaires used in wet locations. *NEC* Article 680 or IRC Chapter 42 limits the placement of these luminaires if in close proximity to the pool, spa, or hot tub. See Chapter 14 for additional information on luminaires for swimming pools and related locations.

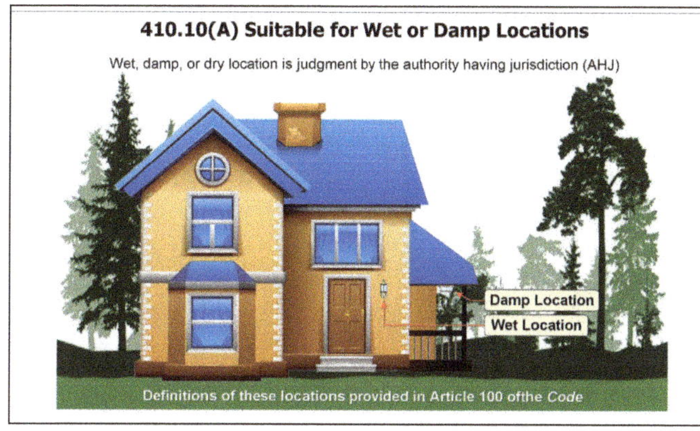

Figure 13.2. Luminaires in wet as compared to damp locations

Photos 13.3. The wall-mounted luminaire clearly shows the bead of caulking around the top and sides of the mounting plate edge to seal against the entry of water and moisture. A gap in the caulking should be left at the bottom of the luminaire mounting base to release any condensation buildup, as shown.

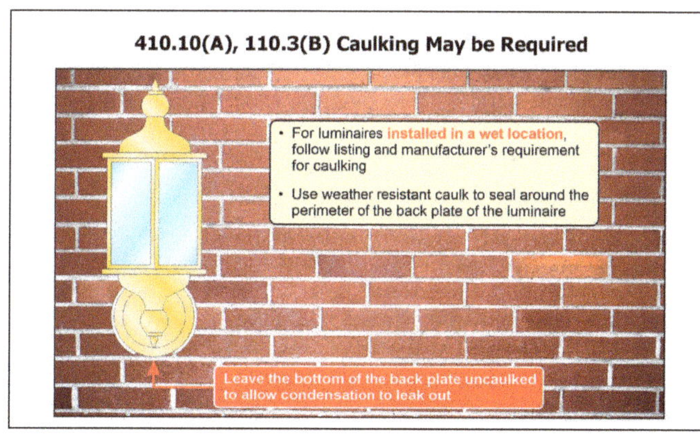

Figure 13.3 Caulking is generally required for luminaires installed in wet locations.

LUMINAIRES INSTALLED IN BATHROOMS

The *Code* limits some types of luminaires from being installed in the tub or shower space. No part of cord-connected luminaires, chain-, cable-, or cord-suspended luminaires, lighting track, pendants, or ceiling-suspended (paddle) fans can be located within a zone measured 900 mm (3 ft) horizontally and 2.5 m (8 ft) vertically from the top of a bathtub rim or

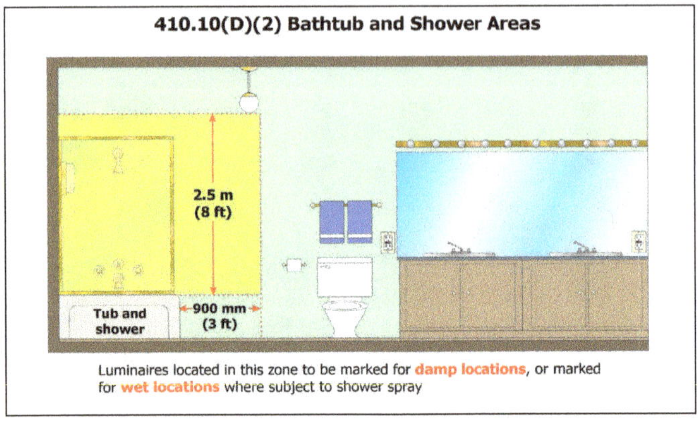

Figure 13.4. Luminaires installed in bathrooms are generally prohibited within the *Code*-identified tub or shower zone *(illustrated in yellow)*

Photo 13.4. A hanging pendant luminaire installed directly over a bathtub inside the 900 mm (3 ft) by 2.5 m (8 ft) zone is a clear *Code* violation

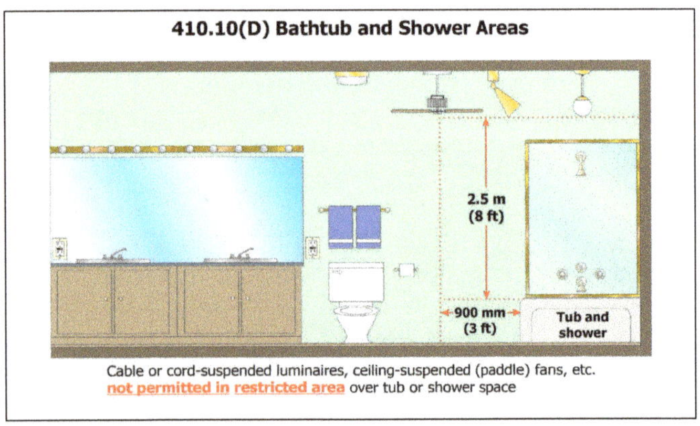

Figure 13.5. Luminaires installed in the bathtub or shower zone *(illustrated in yellow)* must be listed for the location

shower stall threshold. This zone is all-encompassing and includes the space directly over the tub or shower stall. This requirement also applies to hydromassage bathtubs.

Other types of luminaires may be located in this zone if listed for the proper location. At a minimum, the luminaires located in this zone must be listed for damp locations. If a permitted type of luminaire is subject to spray from the showerhead, the luminaire must be listed for wet locations. Properly listed recessed luminaire assemblies and wall sconce-type luminaires are examples of luminaire types permitted in the tub or shower zone if properly listed for a wet or damp location. *NEC* Article 680 or IRC Chapter 42 does not modify this rule for hydromassage bathtubs [*NEC* 410.10(D)(1) and (D)(2) or IRC E4003.11].

Other types of luminaires can be installed in bathrooms, depending on whether they are suitable for the location. These include surface and recessed types. No minimum height above the tub is given for surface and recessed luminaires. Recessed luminaires installed above shower stalls should be sealed with a gasket and marked as suitable for a damp location.

Some inspectors also consider recessed luminaires installed above bathtubs without showerheads to be in a damp location and require a luminaire suitable for damp locations, while other AHJs consider that location to be dry and permit dry location luminaires. Consult with the proper authority with jurisdiction (AHJ) prior to installing the luminaire. Luminaires above bathroom vanities containing sink(s) are typically considered to be in a dry location by the AHJ.

Some manufacturer markings or installation instructions on recessed luminaires require protection by a GFCI device if the luminaires are located in a damp or wet location. GFCI protection for luminaires in damp or wet locations is typically not required by the *Code*; however, the manufacturer is permitted to be more restrictive than the *Code*. Such instructions or markings must comply with the terms of *NEC* 110.3(B) and IRC E3403.3, which are covered in Chapter 1 of this text.

LUMINAIRES INSTALLED IN CLOTHES CLOSETS

The *Code* does not specifically require that a luminaire be installed in a clothes closet. If, by choice, a luminaire is installed in a clothes closet, the installation must meet the minimum specific requirements of the *Code*. These rules have evolved over the years and have become more stringent due to fires that have originated in areas with highly ignitable materials, such as clothes closets. Careful consideration must be given to spacing luminaires away from the storage spaces in a clothes closet. This storage space is often used to stack combustible materials such as clothing or boxes.

A fire can be started if the combustible materials tip over and contact open incandescent lamps. Broken incandescent lamps can also cause fires. Fracture of the lamp causes pieces of hot filament to drop onto nearby combustible material. If the combustible material is near the lamp, the filament will not have time to cool.

A lengthy definition of storage space is provided at *NEC*

410.16(A), Figure 410.16(A), and IRC E4003.12, and Figure E4003.12. The lower part of storage space is defined as "the volume bounded by the sides and back closet walls and planes extending from the closet floor vertically to a height of 1.8 m (6 ft) or the highest clothes-hanging rod and parallel to the walls at a horizontal distance of 600 mm (24 in.) from the sides and back of the closet walls respectively..."

Storage space above 1.8 m (6 ft) or the highest clothes-hanging rod continues to extend "vertically to the closet ceiling parallel to the walls at a horizontal distance of 300 mm (12 in.) or the width of the shelf, whichever is greater..."

For a closet that permits access to both sides of a hanging rod, the storage space "includes the volume below the highest rod extending 300 mm (12 in.) on either side of the rod on a plane horizontal to the floor extending the entire length of the rod." By definition, "storage space" can be delineated even if the closet rods or shelves are not installed. Such is the situation often encountered during the rough-in stage(s) of construction when the closet luminaire box is "roughed-in" so that its location can be inspected and determined to be *Code*-compliant or not. The *Code* contains a very useful figure that graphically illustrates the storage spaces within a typical clothes closet (*NEC* Figure 410.16(A) or IRC Figure E4003.12; (see Figure 13.7 for additional details).

Incandescent luminaires with open (exposed) lamps, partially enclosed lamps, pendant luminaires, or lampholders are not permitted in clothes closets. Use of a lampholder *(keyless or pull chain)* in a clothes closet (regardless of the type of lamp employed) is a violation of the *Code*. Keep in mind that installing a compact fluorescent lamp in a lampholder does not turn the lampholder into a fluorescent luminaire.

The term "lampholder" should be clearly understood; it is a device that is intended to support an electric lamp mechanically and to connect it electrically to a circuit (NEMA). A lampholder may have a switch with a pull chain or be of the keyless type. Lampholders are typically made of porcelain, plastic, fiberglass, or other nonmetallic materials. A lampholder is not a luminaire and should not be referred to as a luminaire (see the last sentence of the definition of "Luminaire" in *NEC* Article 100 or IRC Chapter 35).

A lighting outlet is intended for the direct connection of a lampholder, a luminaire, or a pendant cord terminating in a lampholder. A luminaire installed in a clothes closet is permitted to be one of the following:

(1) Surface-mounted or recessed incandescent or LED luminaires with completely enclosed light sources
(2) Surface-mounted or recessed fluorescent luminaires

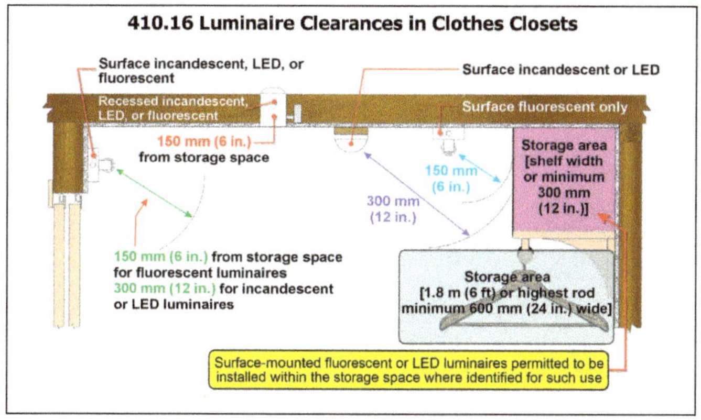

Figure 13.6. Luminaires in clothes closets must be of a certain type and comply with minimum clearance requirements

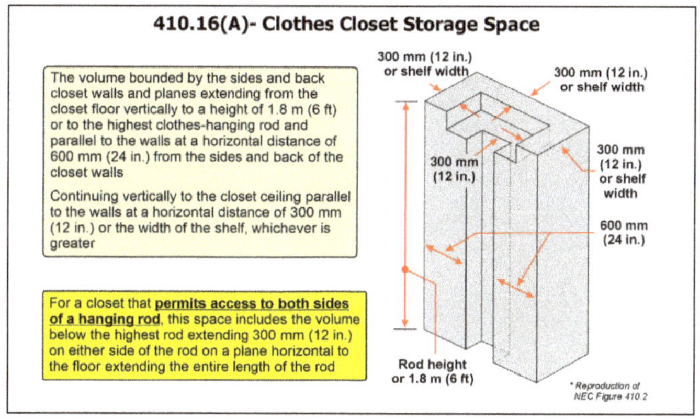

Figure 13.7. Reproduction of *NEC* Figure 410.16(A) or IRC Figure E4003.12 for closet storage spaces

Photo 13.5. Recess luminaires installed in a clothes closet with proper clearance from storage

(3) Surface-mounted fluorescent or LED luminaires identified as suitable for installation within the clothes closet storage space

LED lighting has become a growing and popular option for closet lighting, as well as for other areas, because of its relatively low heat source.

UL 1598 (Standard for Safety of Luminaires) and UL 2108 (Standard for Safety for Low Voltage Lighting Systems) specify listing requirements for luminaires used

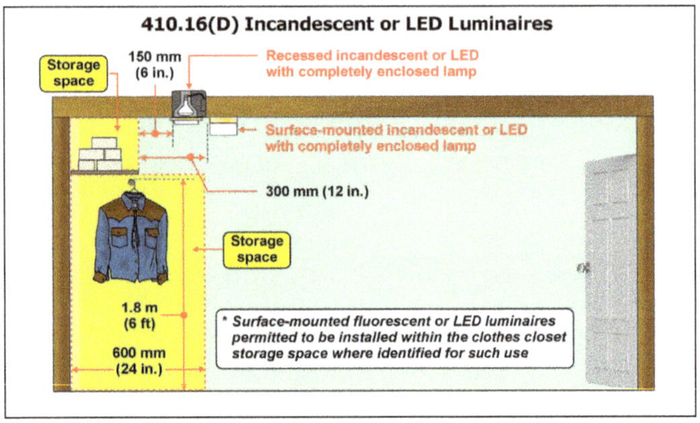

Figure 13.8. Incandescent or LED luminaires installed in clothes closet.

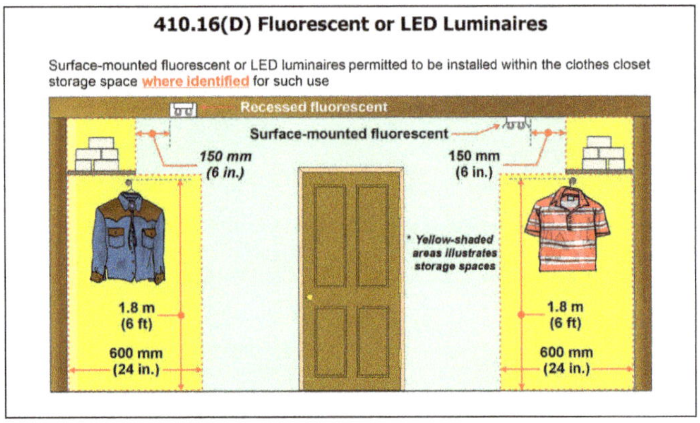

Figure 13.9. Fluorescent luminaires installed in clothes closets

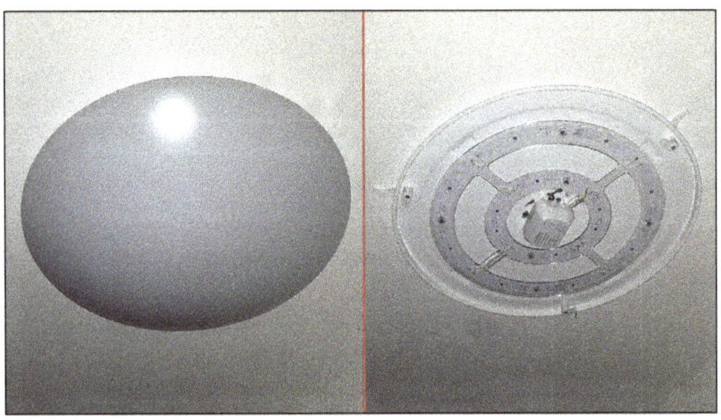

Photo 13.6. LED-type luminaires are a popular choice for installation in clothes closets

in clothes closet storage spaces. One requirement for a luminaire to be considered suitable for direct instal-lation within the closet is a temperature test, where the luminaire operates with glass fiber insulation batting positioned over and touching its exposed surface. This test simulates a scenario where the luminaire is in contact with clothing in a closet. Temperatures on the materials of the luminaire must not exceed their maximum ratings, and temperatures on the luminaire's enclosure in contact with the batt insulation must not exceed 90°C. Luminaires that pass this test are allowed to note in their installation instructions that they are suitable for installation within a clothes closet storage space. UL certifies (lists) luminaires for use in a clothes closet storage space under the product categories "IEUZ" (Surface-Mounted Fluorescent Luminaires), "IFAM" (Surface-Mounted LED Luminaires), and "IFDR" (Low-Voltage Lighting Systems Power Units, Luminaires, and Fittings). The installation instruc-tions specify which luminaires are suitable for use in a clothes closet storage area.

A surface-mounted incandescent or LED luminaire of the totally enclosed type is permitted to be installed in a clothes closet. Permitted locations are on the inside wall above the top of the door or on the ceiling. A clearance of not less than 300 mm (12 in.) must be provided between the luminaire and the closest point of the storage area. Recessed incandescent or LED luminaires with completely enclosed lamps are permitted to be installed in the wall or on the ceiling where a clearance of not less than 150 mm (6 in.) is provided between the luminaire and the nearest point of the storage area (*NEC* 410.16(D) or IRC E4003.12; see figures 13.6 – 13.10 for additional details).

A surface-mounted or recessed fluorescent or LED luminaire is permitted to be installed on the inside wall of a clothes closet, either above the door or on the ceiling. In either case, a clearance of not less than 150 mm (6 in.) is required between the luminaire and the closest point of a storage area. A surface-mounted fluorescent or LED luminaire is permitted to be installed within the clothes closet storage space where the luminaire is identified for this use [*NEC* 410.16(D), Exception or IRC E4003.12].

It is very important that the lamp wattage marked on the luminaire is not exceeded to avoid the generation of excessive heat. This is a potential source of ignition for nearby combustible material.

RECESSED INCANDESCENT LUMINAIRES

The *Code* requires recessed incandescent luminaires to be provided with thermal protection and identified as being thermally protected. Thermal protection is not required for listed recessed incandescent luminaires that provide by design, construction, and thermal performance characteristics that are the equivalent to thermally protected luminaires [*NEC* 410.115(B)]. The marking "INHERENTLY PROTECTED" identifies such luminaires. The thermal sensing device is installed by the luminaire manufacturer and may be placed in various locations. It is direct acting and interrupts power to the luminaire when it operates.

To avoid these thermal protection issues and to freshen up the lighting in a home, retrofit kits to upgrade incandescent luminaires to LED lighting have become very common in dwelling unit construction. This often can change the rating of a recessed lighting source from a non-Type IC rating to a Type IC rating. Some of the advantages to this process include significant savings in energy, a very long life of the lighting outsources, and a clean, modern look. In new construction, LED lighting has become a popular choice for residential lighting as well.

CLEARANCE FROM COMBUSTIBLE MATERIALS

There are two basic types of recessed luminaires: Type IC and Non-Type IC Rated. "IC" stands for "Insulation Contact." Non-type IC housings are prohibited from directly contacting any insulation in contact with the housing. With a non-type IC-rated recess luminaire, the insulation can trap the heat created by the bulb and either damage the insulation on the electrical conductors, creating a fire hazard or actually igniting combustible materials. Other than at the points of their support, recessed portions of non-type IC-rated luminaire enclosures must be spaced at least 13 mm (½ in.) from combustible materials. This requirement applies to incandescent, LED, and fluorescent luminaires. This spacing is not applicable where the recessed luminaire is identified as a Type IC recessed luminaire, being suitable for thermal insulation and combustible materials to be in direct contact with the luminaire [*NEC* 410.116(A)(1) and (A)(2) or IRC E4004.8].

THERMAL INSULATION

In general, thermal insulation is not permitted to be installed above or within 75 mm (3 in.) of a recessed luminaire's enclosure, wiring compartment, ballast, transformer, LED driver, or power supply. Thermal insulation cannot be installed above the luminaire, where the insulation will entrap heat and prevent the free circulation of air. Thermal insulation is permitted to make contact with a recessed luminaire enclosure, wiring compartment, ballast, transformer, LED driver, or power supply where the luminaire is identified as being suitable for insulation to be in direct contact with the luminaire. These luminaires are marked "TYPE IC" or "INHERENTLY PROTECTED." The luminaire is equipped with thermal protection or is inherently protected. The thermal protector is designed to automatically deactivate the lamp should the luminaire be overheated due to improper lamping or any other cause [*NEC* 410.116(B) or IRC E4004.9].

A typical marking on a Type IC luminaire will read

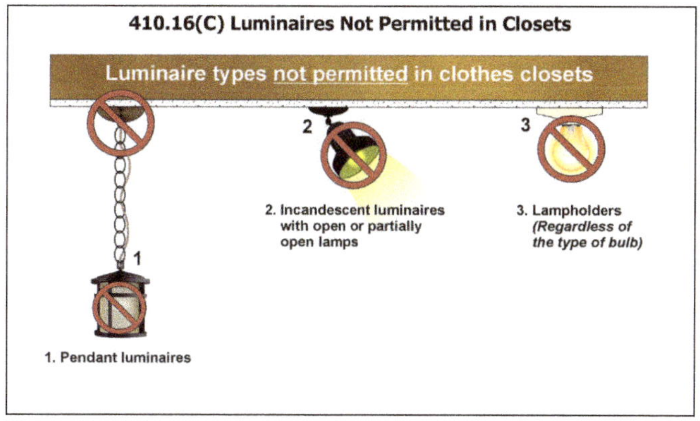

Figure 13.10. Luminaire types prohibited in clothes closets

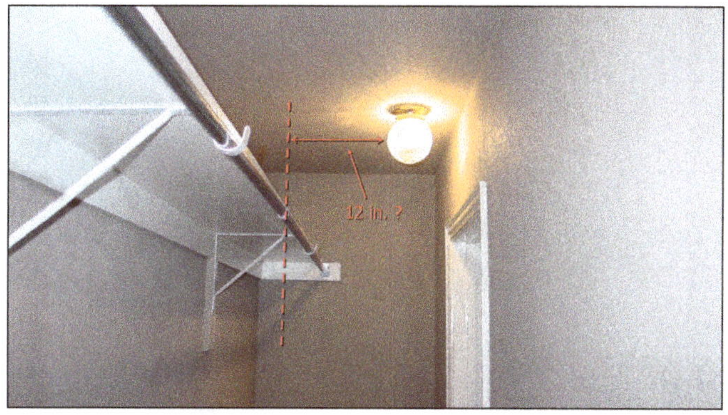

Photo 13.7. Incandescent luminaire installed in clothes closet that does not appear to have proper clearance from storage area

Photo 13.8. Recessed incandescent luminaire shown at the rough-in stage of construction

"BLINKING LIGHT OF THIS THERMALLY PROTECTED LUMINAIRE MAY INDICATE (OVERHEATING)." Conditions that could result in the light blinking may be stated in place of the phrase provided in parentheses. An example is "BLINKING LIGHT MAY INDICATE INSULATION TOO CLOSE TO LUMINAIRE."

BRANCH-CIRCUIT CONDUCTOR TEMPERATURE LIMITATIONS

Branch-circuit conductors with insulation temperature

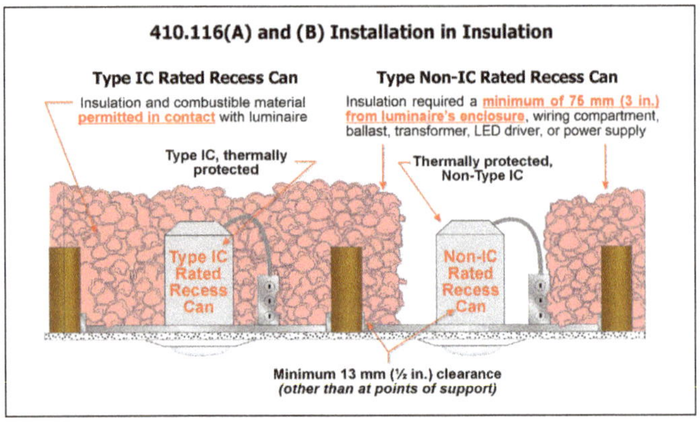

Figure 13.11. Recessed incandescent luminaires generally require clearance from insulation and combustible materials

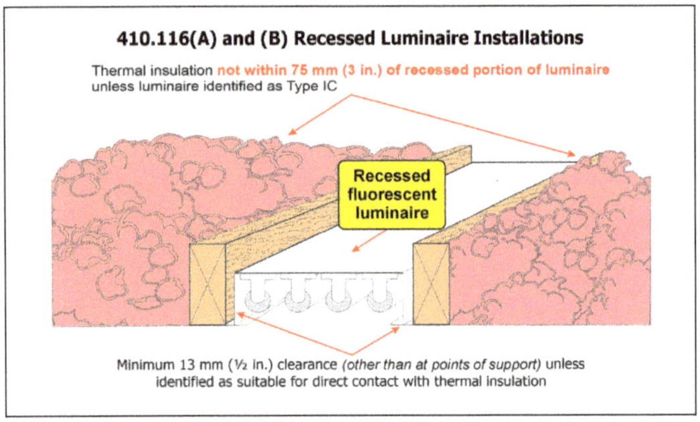

Figure 13.12. Installation of recessed fluorescent luminaires

ratings suitable for the temperature encountered at the luminaire terminations are permitted to terminate in the luminaire. The marking found on luminaires will indicate the required minimum temperature rating of the conductors that supply the luminaire where the minimum temperature exceeds 60°C.

Many recessed incandescent luminaires are marked, "For supply connections, use wire rated for at least 90°C." Where Type NM-B nonmetallic-sheathed cable is used as the branch-circuit wiring method, the cable may directly supply the luminaire outlet box or splice compartment. Although the conductors within Type NM-B cable must be sized based on its 60°C ampacity, their temperature rating is 90°C, and they could be used for direct connection of a luminaire marked for 75°C or 90°C supply conductors [*NEC* 410.117(A) and (B)].

MARKINGS ON RECESSED INCANDESCENT LUMINAIRES

Specific markings for incandescent recessed luminaires are found in the UL Safety Standard for Luminaires, UL 1598. All recessed luminaires, except for those marked "FOR USE IN POURED CONCRETE ONLY," are marked "BLINKING LIGHT OF THIS THERMALLY PROTECTED LUMINAIRE MAY INDICATE OVERHEATING." Other possible conditions that could result in the light blinking may be included in the markings. An example marking is "BLINKING LIGHT MAY INDICATE INSULATION TOO CLOSE TO LUMINAIRE."

Some incandescent recessed luminaires are convertible and can be used in either Type IC or non-Type IC applications. The same rough-in section or housing is used for both types of applications. The choice of trim and lamp type determines the suitability of the completed assembly for a given application. Details for making the proper choice of finishing section or trim are included in the installation instructions.

The convertible recessed luminaire has several types of markings that are unique to this type of product. The rough-in section is marked with the following two statements:

> A. "BLINKING LIGHT OF THIS THERMALLY PROTECTED LUMINAIRE MAY INDICATE OVERHEATING."
>
> B. "DO NOT INSTALL INSULATION WITHIN 76 MM (3 IN.) OF ANY PART OF THE LUMINAIRE."

The marking in "B" is on a peel-off label and is to be removed when the luminaire is used in a Type IC application.

The rough-in housing has additional markings to clarify the proper combination of housing and trim. For example, the rough-in housing is marked "ROUGH-IN SECTION FOR CONVERTIBLE RECESSED LUMINAIRE." Another marking gives the proper trims for both Type IC applications. For example, the rough-in housing is marked "TYPE IC TRIMS/FINISHING SECTIONS: AA, BB, CC, etc." and additionally marked "TYPE IC/NON-IC TRIMS/FINISHING SECTIONS: XX, YY, ZZ, etc." The finishing section has a corresponding marking. For example, the marking on the trim will indicate "FINISHING SECTION FOR USE WITH ROUGH-IN SECTION T-9A."

Recessed luminaires that consist of a housing (rough-in section) and various permitted trims (finishing section) are marked on each separable part with markings that correlate the permitted combinations.

Branch-circuit wiring cannot pass through an outlet box that is an integral part of an incandescent luminaire unless the luminaire is identified through wiring. This limitation does not apply to the branch-circuit conductors that are part of a 2-wire or multi-wire branch circuit that supplies power to multiple luminaires that are connected together.

Luminaires must be constructed or installed so that the conductors within outlet boxes will not be subjected to temperatures higher than that for which the conductors are rated. Manufacturers of surface-mounted luminaires often use an insulation pad for this purpose. The insulation pad is a critical part of the listed luminaire and must be installed at the time of installation (*NEC* 410.21 or IRC E4003.13).

FLUORESCENT LUMINAIRES

Fluorescent luminaires are referred to in the *Code* in many locations as "electric-discharge luminaires." These fluorescent luminaires have enjoyed an increased popularity over time due to their energy efficiency. Many models of fluorescent lamps are available that are self-contained without an external ballast.

Where these luminaires are installed indoors, they are required to have ballasts that have thermal protection that is integral to the ballast. Replacement ballasts for existing luminaires are also required to have thermal protection [*NEC* 410.130(E)(1) or IRC E4003.6]. This thermal protection is provided by a switching device integral to the ballast that interrupts the power to the lamps if the ballast temperature rises to 105°C or greater. This switching device will reset so the ballast will resume operation once the ballast cools down below 105°C.

Where a surface-mounted fluorescent luminaire is installed on a low-density fiberboard, it must be listed for this condition, or it must be installed on spacers not less than 38 mm (1½ in.) from the surface. Low-density fiberboard includes sheets, panels, and tiles that have a density of 320 kg/m^3 (20 lb/ft^3) or less and are formed of bonded plant fiber. This material does not include solid or laminated wood or fiberboard that has a density in excess of 320 kg/m^3 (20 lb/ft^3). The rules for recessed fluorescent luminaires parallel the rules for recessed incandescent luminaires. While fluorescent luminaires do not generate the same quantity of heat as incandescent luminaires do during operation, insulation must be kept clear of the luminaire to avoid cycling of the thermal protector. Fluorescent luminaires that are intended to have thermal insulation in contact with the luminaire are marked "TYPE IC," identical to the marking for recessed incandescent luminaires (*NEC* 410.136 or IRC E4004.7).

LED LUMINAIRES

Currently, the hottest trend in lighting is in the form of light-emitting diode (LED) luminaires. Dramatic advances in technology for LED luminaires broadened the applicability of this type of illumination, creating an exciting option for dwelling units and other applications. Compared to

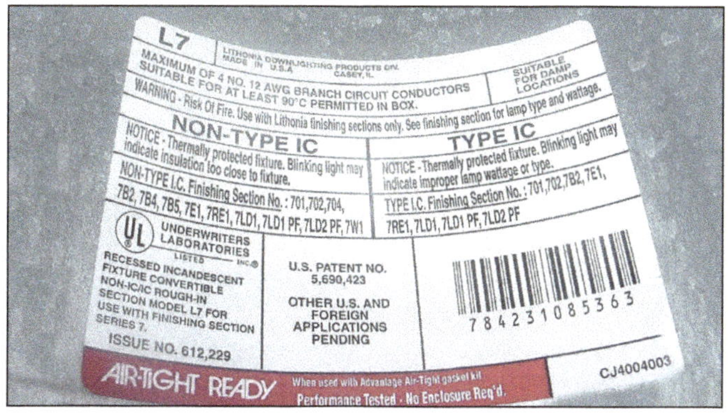

Photo 13.9. Typical markings on recessed luminaires

traditional light sources such as incandescent lighting, LED luminaires can deliver longer life in a smaller size, faster switching, lower energy consumption, lowered maintenance demands with improved physical robustness, and equal or better quality of light. They also have fewer environmental concerns linked to their disposal. A typical 13-watt LED lamp emits 450 to 650 lumens, which is equivalent to a standard 40-watt incandescent bulb. A standard 17-watt LED bulb is equivalent to a 100-watt incandescent bulb and produces roughly 1600 lumens of light.

An LED is a two-lead semiconductor light source. When a suitable current is applied to the leads, electrons are able to recombine with electron holes within the device, releasing energy in the form of photons. This effect is called electroluminescence, and the color of the light (corresponding to the energy of the photon) is determined by the energy band gap of the semiconductor. LEDs are typically small (less than 1 mm^2), and integrated optical components may be used to shape the radiation pattern. LEDs use heat sinks to absorb the heat produced by the LED and dissipate it into the surrounding environment. This keeps LEDs from overheating and burning out.

LEDs are incorporated into bulbs and sometimes the luminaire itself for general lighting applications. Small in size, LEDs provide unique design opportunities. Some LED bulb solutions may physically resemble familiar light bulbs and better match the appearance of traditional light bulbs. Some LED luminaires may have LEDs built in as a permanent light source (see photo 13.6). There are also hybrid approaches where a non-traditional "bulb" or replaceable light source format is used and specially designed for a unique luminaire. LEDs offer a tremendous opportunity for innovation in lighting form factors and fit a wider breadth of applications than traditional lighting technologies.

Typically, luminaires are required to be installed with a certain amount of clearance from the storage space within a closet. However, surface-mounted fluorescent or LED lumi-

Photo 13.10. LED luminaire *(identified for the use)* installed directly in the storage space of a clothes closet

naires are permitted to be installed within the closet storage space where the luminaire is identified for this use [*NEC* 410.16(B)(3) or IRC E4003.12].

SUPPLY CONDUCTORS

Luminaires are required to be wired with conductors having insulation suitable for the environmental conditions, current, and voltage. The conductor insulation must have a temperature rating that is suitable for the temperature to which the conductors will be subjected (*NEC* 410.52 or IRC E4003.13).

Where feeder or branch-circuit conductors pass within 75 mm (3 in.) of the ballast, transformer, LED driver, or power supply in the luminaire, the conductors are required to have not less than 90°C insulation temperature rating. The following conductor insulation types are permitted: RHH, THW, THHN, THHW, FEP, FEPB, SA, and XHHW. Type THW insulation usually has a 75°C temperature rating. The *Code* permits type THW insulation to have a 90°C rating where the conductor is used within electric-discharge lighting equipment [*NEC* 410.68 and Table 310. 4(1)].

CAUTION FOR EXISTING INSTALLATIONS

The most common wiring method for many dwelling units is nonmetallic-sheathed cable branch-circuit wiring. Prior to 1984, Type NM cable was manufactured with conductor insulation of 60°C material for many years. Starting with the 1984 edition of the *NEC*, Type NM cable was required to be produced with insulated conductors that had a 90°C rating. This product is required to be identified with the suffix "-B" so the 90°C rated product will read "Type NM-B."

In some cases, and some areas, dwelling units were wired with the 60°C rated wiring for a few years beyond 1984, depending on which edition of the *Code* was locally adopted, how it was enforced, and on the local supply of the older 60°C rated wiring material.

This change in the temperature rating of branch-circuit wiring, along with a change in the design of many incandescent luminaires, presents a situation of which installers need to be particularly aware. Many incandescent luminaires of the "drum" or enclosed type that fit against a ceiling have been designed to require branch-circuit supply conductors rated more than 60°C. Particular care should be taken when replacing existing luminaires. If the dwelling unit is wired with 60°C Type NM cable, verify prior to installation that the new luminaire does not require 75°C or 90°C supply conductors if it is not possible to find a luminaire rated for 60°C conductor insulation, additional electrical work is required (see Figure 13.13).

Installing a luminaire requiring conductors with a temperature rating higher than the supply conductors can create temperatures in the outlet boxes that exceed the rating of the insulation on the existing wire. Damage to the branch-circuit wiring can occur where this excess temperature exists for an extended period. See Chapter 16 for additional information on the subject of existing electrical installations and wiring.

The Safety Standard for Luminaires, UL 1598, requires specific markings to alert the installer to potential safety hazards. A luminaire that may be installed in a dwelling unit and is intended to be connected to or mounted over an outlet box will be marked. The marking will state "CAUTION— RISK OF FIRE" and "CONSULT A QUALIFIED ELECTRICIAN TO ENSURE CORRECT BRANCH-CIRCUIT CONDUCTOR." The minimum temperature rating of the supply conductors will also be marked on the label of the new luminaire.

CONNECTION OF ELECTRIC-DISCHARGE AND LED LUMINAIRES

Where electric-discharge or LED luminaires are supported independently of an outlet box, they are required to be supplied and connected through a metal raceway, nonmetallic raceway, Type MC cable, Type AC cable, or nonmetallic-sheathed cables [*NEC* 410.24(A)].

Listed electric-discharge or LED luminaires are permitted to be cord-connected if the luminaire is located directly below the outlet box and the cord is continuously visible for its entire length (except at terminations). It is not subject to strain or physical damage. These cord-equipped luminaires must terminate at the supply end of the cord in a grounding-type attachment. If grounding is not required, a polarized plug is permitted. The *Code* does not recognize the practice of directly connecting cords to outlet boxes, with or without strain relief. Listed cord-and-canopy assemblies are required for direct connection. The listed cord-and-canopy assembly will incorporate proper strain relief for the cord [*NEC* 410.62(C)(1)].

ACCESS TO SPLICES

All connections between luminaire conductors and branch-circuit conductors must be accessible for maintenance and inspection. Splices and taps are not permitted to be located within luminaire arms or stems. Access to splices supplying a luminaire mounted within a cove must also be accessible. Some electric-discharge or LED luminaires are listed for mounting on the surface of a vertical surface such as a wall or ceiling. Such luminaires are not supported solely by the outlet box. The branch-circuit supply conductors are within a concealed outlet, pull, or junction box behind the luminaire. The luminaire must be provided with a suitable opening on the back of the luminaire so that the conductors and splices within the outlet box are accessible without removing the luminaire. Simply using a standard chase nipple for this purpose is not permitted. One method used to meet this requirement is to remove the large knockout provided by the luminaire manufacturer in the portion of the luminaire housing intended to be mounted against the ceiling or wall. The luminaire should be mounted to the surface with the large knockout opening centered over the outlet box or mud ring. It is usually necessary to install vinyl electrical tape or other approved insulating material over the sharp edge of the knockout opening. This will help protect the conductor insulation from damage [*NEC* 410.24(B) or IRC E4004.3].

SUPPORTING OF LUMINAIRES

In most cases, only an outlet box that is specifically designed for the purpose can be used for luminaire or lampholder support. At every outlet used exclusively for lighting, the box must be so designed or installed that a luminaire can be properly attached. Using the luminaire mounting hardware provided by the manufacturer helps ensure the proper security and support of the luminaire [*NEC* 314.27(A) or IRC E3905.6].

UL Product iQ®, Metallic Outlet Boxes (QCIT) adds, "A box, with or without bracket or bar hanger, intended for support of a luminaire weighing 50 lbs or less, is provided with a marking 'FOR FIXTURE SUPPORT' on the carton." Listed boxes that are suitable for supporting ceiling-mounted luminaires weighing more than 23 kg (50 lb) are marked with the maximum weight of the luminaire to which that particular box can support. Device plaster rings are not listed for support of ceiling-mounted luminaires unless marked for

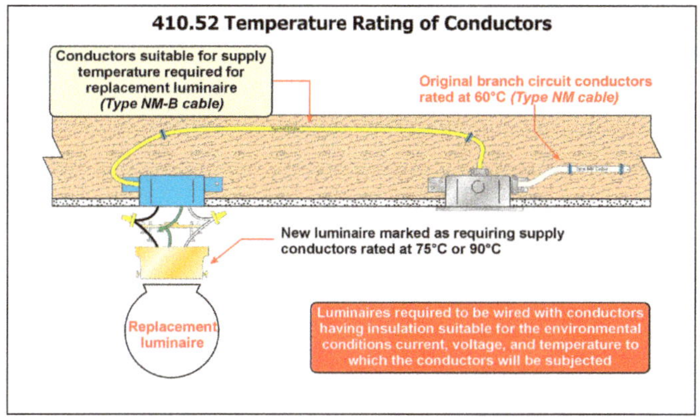

Figure 13.13. Supply conductors must be suitable for the maximum temperature

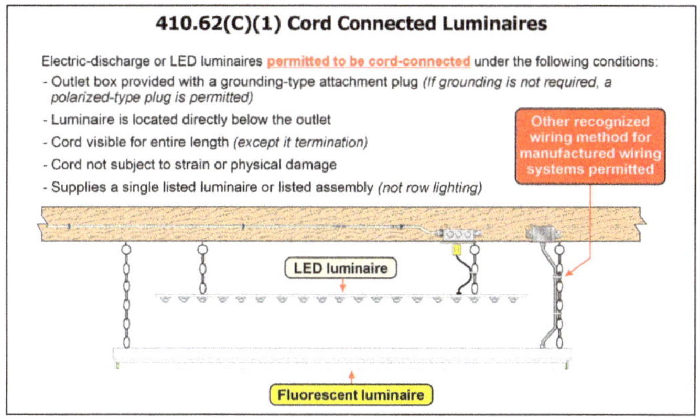

Figure 13.14. Cord-connected electric discharge or LED luminaires

use in ceilings and with the maximum weight of the product to be supported [*NEC* 314.27(A)(2) or IRC E3905.6.2]. The UL Product iQ® information for Nonmetallic Outlet Boxes (QCMZ) has similar marking requirements and restrictions.

Device boxes are generally not recognized for the support of luminaires. Such boxes are typically intended only to support switches, receptacles, and similar devices or utilization equipment such as smoke alarms. An exception to the *Code* permits device boxes to support lighter luminaires. A vertical surface-mounted luminaire that weighs not more than 3 kg (6 lb) is permitted to be supported on a device box. The luminaire or its supporting yoke must be secured to the box with at least two No. 6 or larger screws [*NEC* 314.27(A)(1) Exception or IRC E3905.6.1 Exception].

The UL directory indicates that old-work boxes are also permitted to support luminaires. Old-work boxes are also commonly referred to in the field as "snap-in," "cut-in," or "pop-in" boxes. These types of boxes are generally used where wiring is fished into existing walls or ceilings. An old-work-type box is typically secured to the building finish by brackets and not directly by a structural member. Both metallic device boxes and nonmetallic device boxes intended to be installed in an existing structure have been investigated for

Photo 13.11. Luminaires are required to be supported by boxes designed for the purpose

Photo 13.12. Wall-mounted luminaires weighing not more than 3 kg (6 lb) are permitted to be mounted by the device box

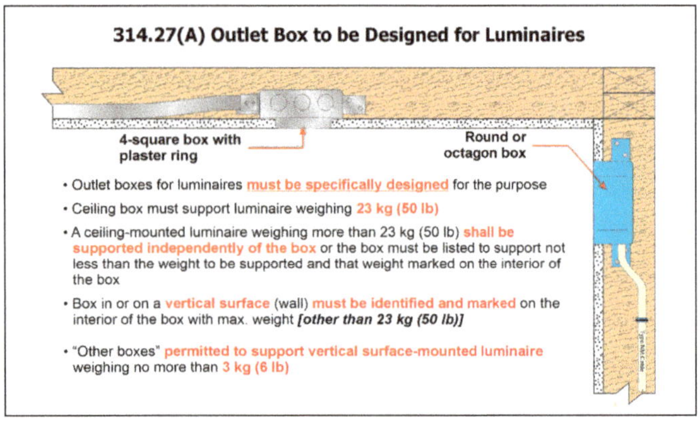

Figure 13.15. Boxes at luminaire outlets

Photo 13.13. Some old work boxes are permitted to support luminaires as well

the support of luminaires, smoke alarms, or carbon monoxide detectors. The maximum weight of the supported luminaire, smoke alarm, or carbon monoxide detector is 3 kg (6 lb). The UL product category Nonmetallic Outlet Boxes (QCMZ) contains similar information.

Manufacturers may have additional requirements that are more specific than the general requirements found in the *Code*. Keep in mind that the instructions of the manufacturer become part of the *Code* per its listing.

For example, instructions from a major luminaire manufacturer might read, "These luminaires, with the exception of the post type, are intended to be mounted to a 100 mm square x 54 mm deep (4 in. x 2-⅛ in.) metal octagon outlet box at a minimum of 1.8 m (6 ft) above ground level. The box must be directly supported by the building structure."

Another major manufacturer of luminaires includes in the instructions a requirement for a metal box to be used to support the luminaire. Where they are not a part of outlet boxes, luminaire studs and hickeys must be made of steel, malleable iron, or other material suitable for the application [see Section 410.36(C)]. Where used to support luminaires, raceway fittings must be capable of supporting the weight of the complete luminaire assembly, including the lamp(s) [see Section 410.36(E)].

LUMINAIRES SUPPORTED BY TREES

Vegetation, such as trees, is permitted to support outdoor luminaires and associated equipment [*NEC* 410.36(G)]. Vegetation is not permitted to support overhead spans for service, feeder, or branch-circuit conductors [*NEC* 225.26, 230.10, 590.4(J)]. Branch circuits are permitted to be run underground to supply the luminaires supported by trees. Conductors or cables extending from underground up the side of the tree must be protected from physical damage by an approved means such as in rigid metal conduit, intermediate metal conduit, or Schedule 80 PVC conduit or equivalent.

This added protection would prevent a weed cutter or lawn mower from damaging the Type UF cable that would otherwise be exposed at the bottom of the tree. The *Code* contains additional requirements for protecting underground conductors that transition from underground to above grade. See Chapter 11 and *NEC* 300.5 or IRC E3803 for additional details on underground wiring methods.

GROUNDING OF LUMINAIRES

Luminaires that have exposed metal parts are required to be manufactured with a means for the connection of an equipment grounding conductor for grounding purposes (*NEC* Part V, Grounding and 410.44 or IRC E4003.3). Luminaires and equipment are required to be mechanically connected to an equipment grounding conductor as specified in *NEC* 250.118 or IRC E3908.9 and sized in accordance with *NEC* 250.122 or IRC E3908.13 [*NEC* 410.44 or IRC E3908.11].

A bare or green-insulated equipment bonding jumper that is riveted to the metal canopy of the luminaire is frequently used to bond a luminaire to the metallic outlet box. With a nonmetallic box, the bonding jumper is usually directly connected to the branch-circuit equipment grounding conductor(s). Some nonmetallic boxes incorporate an equipment-bonding terminal strip. In this case, grounding is accomplished by connecting the luminaire support screw to the equipment-grounding terminal strip. The branch-circuit supply, or switch leg that is run to the luminaire outlet box, will include an equipment grounding conductor that must be connected to the same equipment-bonding terminal strip.

In addition to grounding the luminaire, the equipment grounding conductors that are part of the supply circuit must also be used to ground the metal outlet box. Only one conductor is permitted under the green bonding screw. Splicing together all the equipment grounding conductors with the bonding jumpers or "pigtail" to the luminaire and the metal outlet box will accomplish this task.

NEC 410.44 Ex. No. 1 allows replacement luminaires are permitted to connect an equipment grounding conductor in the same manner as replacement receptacles in compliance with *NEC* 250.130(C). The luminaire must then comply with *NEC* 410.42 or IRC E4003.3, which states that exposed conductive parts that are accessible to unqualified persons are required to be connected to an equipment grounding conductor or be separated from all live parts and other conducting surfaces by a listed system of double insulation.

Small, isolated parts, such as mounting screws, clips, and decorative bands on glass spaced at least 38 mm (1½ in.) from lamp terminals and portable luminaires with a polarized attachment plug, are not required to be connected to an equipment grounding conductor [*NEC* 410.42 or IRC E4003.3].

Where an equipment grounding means does not exist at a lighting outlet (as is the case for many older homes) and replacement luminaires are installed, replacement luminaires that are GFCI protected or do not have exposed conductive parts are also not required to be connected to an equipment grounding conductor [*NEC* 410.44 Ex. No. 2]. See Chapter 16 for additional information on this topic.

Photo 13.14 Trees permitted to support outdoor luminaires and associated equipment.

Photo 13.15. LED track lighting in a dwelling unit

TRACK LIGHTING

Track lighting offers an alternative approach to providing illumination in dwelling units. Lighting track is available in a variety of styles and configurations, including surface and recessed types. Requirements relative to the safe installation of lighting track are found in *NEC* Article 410 Part XIV or IRC Section E4005.

The lighting track is required to be permanently installed and supplied by a branch circuit with a rating not greater than that of the track. Track and associated fittings must be suitable for use and used in accordance with the manufacturer's installation instructions. Using the fittings in the proper fashion ensures that proper polarity and grounding for the track installation can be established in the installation (*NEC* 410.150(A) and (B) or IRC E4005.1, E4005.2).

The lighting track is required to be securely mounted so that the track will accept the weight it is intended to carry by attached luminaires (track lighting heads). Lighting track must be supported at least every 1.2 m (4 ft) unless otherwise identified. Every single section of track that is 1.2 m (4 ft) in length or shorter must be supported at each end of the track, and the track must be a single length (*NEC* 410.150(D), 410.154, and 410.155(A) or IRC E4005.5). The ends of the track must be capped.

The lighting track must be properly grounded by connection of the equipment grounding conductor and must

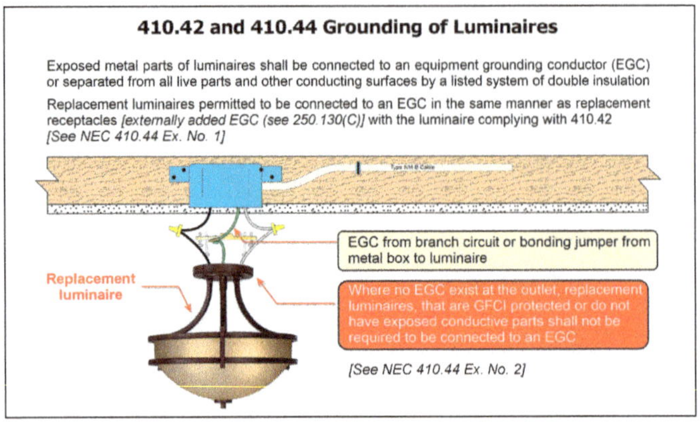

Figure 13.16. Grounding of luminaires with exposed conductive parts

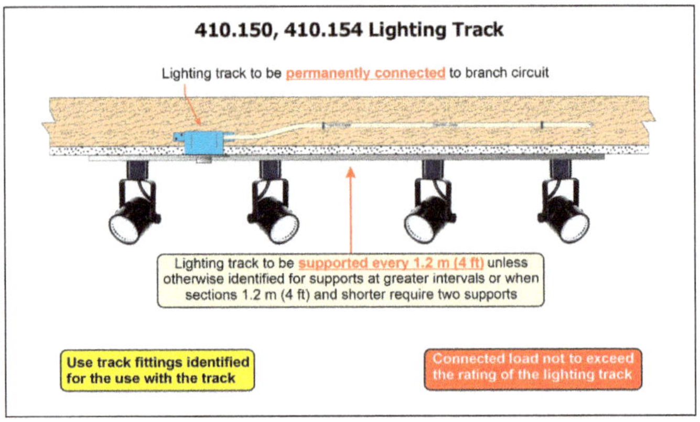

Figure 13.17. Lighting track installation

extend the grounding to any attached luminaires that are required to be grounded. Track lighting sections must be securely fastened together with suitable fittings that maintain the circuitry, polarity, and grounding connections between track sections [*NEC* 410.155(B) or IRC E4005.6].

LOCATIONS FOR TRACK LIGHTING

Track lighting is permitted to be installed in a variety of locations in dwelling units but is not permitted to be installed in any of the following locations or areas:

1. Where likely to be subject to physical damage
2. In wet or damp locations
3. Where subject to corrosion or corrosive vapors (not likely to be found in dwelling units)
4. In battery storage rooms (not likely to be found in dwelling units)
5. In hazardous locations (not likely to be found in dwelling units)
6. Where concealed
7. Extending through walls or partitions
8. Less than 1.5 m (5 ft) above the finish floor unless protected from damage, or the track operates at less than 30 volts rms open-circuit voltage
9. Within the restricted bathtub zone [900 mm (3 ft) horizontally and 2.5 m (8 ft) vertically from the top of the bathtub rim] described at *NEC* 410.10(D) or IRC E4003.11

See *NEC* 410.150(C) or IRC E4005.4.

LOW-VOLTAGE LIGHTING SYSTEMS

A low-voltage lighting system operates at 30 volts AC or 60 volts DC or less and has an isolating type power supply. The low-voltage luminaires and associated equipment are required to be identified for use with such systems. The output supply is required to be not more than 25 amperes maximum under all load conditions. These lighting systems must be listed as complete systems.

The luminaires, power supply, and luminaire fittings (including the exposed bare conductors) of an exposed bare conductor lighting system must be listed for use as part of the same identified lighting system [*NEC* 411.2(A)]. A low-voltage lighting system must be supplied from a maximum 20-ampere branch circuit. See *NEC* Article 411 for low-voltage lighting system requirements.

Lighting equipment connected to a Class 2 power source is also considered a low-voltage lighting system. Listed Class 2 lighting equipment must be rated in conformance with *NEC* Chapter 9, Table 11(A) or Table 11(B). Table 11(A) applies to alternating current sources, and Table 11(B) applies to direct-current sources. The power for Class 2 circuits is either (1) inherently limited, requiring no overcurrent protection, or (2) not inherently limited, requiring a combination of power source and overcurrent protection. Power sources designed for interconnection shall be listed for the purpose. As part of the listing, Class 2 power sources are durably marked where plainly visible to indicate the class of supply and its electrical rating. A Class 2 power source that is not suitable for wet location use must be marked.

Low-voltage lighting systems are primarily covered by Article 411 in the *NEC*. Low-voltage lighting systems are not dealt with specifically in the IRC, so by default, the requirements of *NEC* Article 411 are applicable to these systems even if the adopted code is the IRC.

Several hazards are addressed through the listing process and *Code* requirements for such systems. Some illumination sources are tungsten halogen lamps, which produce significant amounts of heat. For these types of low-voltage luminaires, the manufacturer may restrict the location of the lamps to open spaces that are reasonably well-ventilated. There may be restrictions against mounting the luminaire directly to a combustible surface as well. As with other lighting applications, LED technology is making great strides in the low-voltage lighting industry. The heat-related restrictions associated with tungsten halogen lamps are not an issue with LED technology.

The wiring method may be supplied by the manufacturer as part of the system. Some instructions may specify the type of cable as well as size and length limitations. The wiring method may not be concealed or extended through a building wall unless it is (1) one of the types permitted by Chapter 3 of the *NEC* or (2) installed using wiring supplied by a Class 2 power source and installed to meet the requirements of *NEC* 725.130. Nonmetallic-sheathed cable is an example of a wiring method that is permitted to be concealed or extended through building walls.

When installed around a swimming pool, these low-voltage lighting systems cannot be installed within 3 m (10 ft) horizontally from the nearest edge of the water unless permitted by *NEC* Article 680. *NEC* 680.22(B)(6) allows specific low-voltage luminaires to be located less than 1.5 m (5 ft) from the inside walls of the pool under certain conditions. These luminaires must be of the type that does not require connection to an equipment grounding conductor. These luminaires cannot exceed the voltage limitations defined in the definition of "Low Voltage Contact Limit" in *NEC* Article 100 or IRC E4201.2.

These luminaires must also be supplied by listed transformers or power supplies that comply with *NEC* 680.23(A)(2) or IRC E4206.1 for transformers or power supplies listed for swimming pool and spa use. See Chapter 14 for additional information on low-voltage lighting around pools.

It is important to follow all applicable rules of the *Code* for luminaire installations. The installer must become familiar with and follow the installation instructions. The installer and user must take into account any limitations on the use based on product listings.

Luminaires are a source of heat. Connecting the luminaire to branch-circuit conductors that do not have the minimum required insulation temperature rating may cause the wiring to fail over time. Luminaires that are overlamped or improperly installed too close to combustible materials are a potential fire hazard.

RECEPTACLES

A receptacle is a contact device installed at the outlet for the connection of an attachment plug or for the direct connection of electrical utilization equipment designed to mate with the corresponding contact device. A single receptacle is a single contact device with no other contact device on the same yoke or strap. A duplex receptacle has two contact devices on the same yoke or strap. Consequently, a duplex receptacle is a multiple receptacle and would not be acceptable for those applications where the *Code* requires a single receptacle.

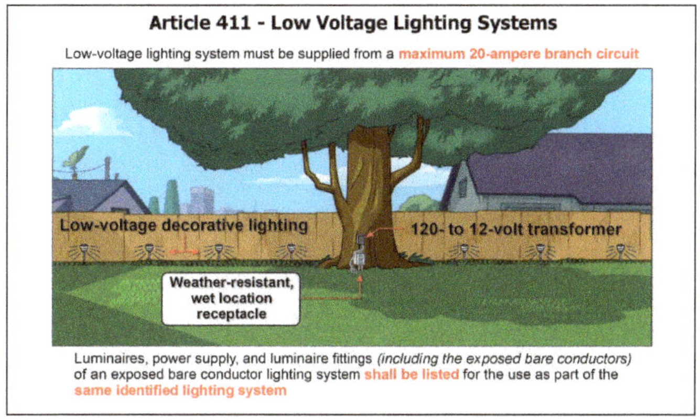

Figure 13.18. Low-voltage lighting systems must meet the requirements of *NEC* Article 411

Photo 13.16. Typical landscape lighting system with power supply and wet location receptacle (insert)

An attachment plug (cord cap or plug) is a device that, by insertion in a receptacle, establishes a connection between the conductors of the attached flexible cord and the conductors connected permanently to the receptacle.

Receptacles must be listed and marked with the manufacturer's name or identification and voltage and ampere ratings. Receptacles and cord connectors must be rated not less than 15 amperes, 125 volts, or 15 amperes, 250 volts and must be of a type that is not suitable for use as a lampholder [*NEC* 406.3(A) and (B)].

Receptacles rated 20 amperes or less and directly connected to aluminum conductors must be marked CO/ALR [*NEC* 406.3(C)]. Detailed reports on repairing aluminum-wiring connections at terminations are available from the U. S. Consumer Product Safety Commission and UL Solutions, Inc.. In addition, see Chapter 16 for additional information on "pigtailing," the practice of connecting insulated copper jumper wires from the receptacle or luminaire to aluminum branch-circuit conductors.

RECEPTACLES, WHERE REQUIRED

The *Code* contains general requirements for where recepta-

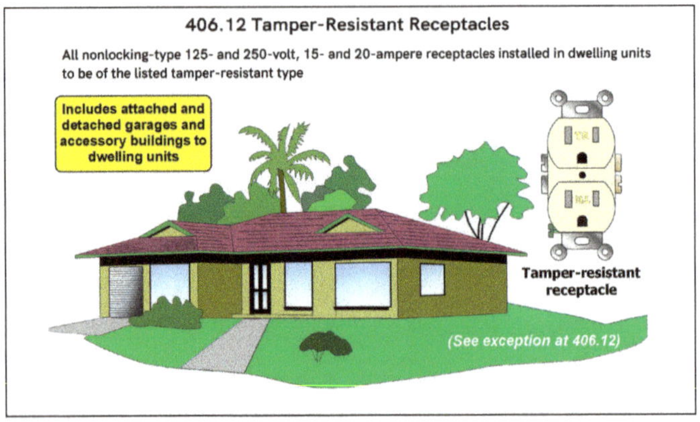

Figure 13.19. Tamper-resistant receptacles are required at all dwelling unit receptacle outlets (with exception)

cle outlets are required in dwelling units [see *NEC* 210.52 through 210.63]. Where permitted, cord pendants (a flexible cord having a receptacle-type cord cap) may be permanently installed in place of receptacle outlets. A receptacle is required wherever flexible cords with attachment plugs might be used. For appliances, receptacles are generally required to be installed within 1.8 m (6 ft) of the intended location of the appliance (*NEC* 210.50(C) or IRC E3901.5). There are many specific requirements for receptacle outlet locations in one- and two-family dwellings. See Chapter 11 of this text for additional information on required receptacle outlets and locations of such outlets.

TAMPER-RESISTANT RECEPTACLES

All nonlocking-type 125- and 250-volt, 15- and 20-ampere receptacles in all dwelling units, boathouses, mobile homes, and manufactured homes, including their attached and detached garages, accessory buildings, and their common areas, are required to be listed tamper-resistant receptacles (with an exception that we will discuss in the next paragraph). In previous editions of the *Code*, tamper-resistant receptacles were limited to only the receptacles of a dwelling unit referred to in *NEC* 210.52 and 550.13 or IRC E3901.1. This would include most of the receptacles in a dwelling unit (but not all). An example of a receptacle that was not required to be tamper-resistant under previous editions of the *Code* would be a receptacle at an air conditioning unit or a furnace in an attic. This receptacle is required at a dwelling unit by *NEC* 210.63(A) or IRC E3901.12, not *NEC* 210.52 or IRC E3901.1. However, the current edition of the Code would require this receptacle to be of the tamper-resistant type. All nonlocking-type 125- and 250-volt, 15- and 20-ampere receptacles in all dwelling units were encompassed in the 2023 *NEC* (2024 IRC). See Chapter 10 for additional information on tamper-resistant receptacles.

Tamper-resistant receptacles feature a built-in shutter system that prevents penetration of foreign objects such as keys, paper clips, and hair pins from touching electrically live components if they are inserted into the slots of these receptacles, as they often are by young children. However, these shutters do not impair normal cord cap insertion, as both the ungrounded and grounded conductor slots must be engaged simultaneously before the shutters will open. The requirement for tamper-resistant receptacles would include the majority of the receptacles in dwelling units (with exception), including attached and detached garages, accessory buildings, and common areas to dwelling units [*NEC* 406.12 or IRC E4002.14].

Dwelling unit receptacles in the following locations are not required to be of the tamper-resistant type:

1. Receptacles located more than 1.7 m (5½ ft) above the floor.
2. Receptacles that are part of a luminaire or appliance.
3. A single receptacle that is not readily accessible and supplies one appliance or a duplex receptacle that is not readily accessible and supplies two appliances installed within a space occupied by or designated for an appliance that, in normal use, is not easily moved from one place to another and is cord-and-plug-connected for the purpose of frequent interchange, prevention of noise or vibration, or the appliance is intended or identified for flexible cord connection (such as behind a refrigerator in a kitchen area).
4. Nongrounding receptacles used for replacements of existing non-grounding receptacles

See *NEC* 406.12 Exception or IRC E4002.14 Exception.

In existing dwellings, listed tamper-resistant receptacles would need to be provided where replacement receptacles are installed at receptacle outlets that are now required to be tamper-resistant receptacles under today's *Code* requirements. Tamper-resistant receptacles are not required at replacement receptacles where a non-grounding receptacle is replaced with another non-grounding receptacle or where aluminum branch-circuit conductors are directly terminated on a CO/ALR receptacle, installed as a replacement [*NEC* 406.4(D)(5)].

GFCI AND AFCI PROTECTION FOR REPLACEMENT RECEPTACLES

In existing dwellings, GFCI protection is required for replacement receptacles in any location where the receptacle would now require GFCI protection under today's *Code* requirements [*NEC* 406.4(D)(3)]. An example of this would be a non-GFCI-protected kitchen countertop receptacle or a bathroom receptacle being replaced at an existing dwelling unit. An exception to *NEC* 406.4(D)(3) permits the GFCI device protecting a replacement receptacle to be located elsewhere where the existing outlet box size will not permit the installation of the GFCI receptacle. In this case, the replaced receptacle is permitted to be replaced with a new receptacle of the existing type (non-grounding, etc.). This new GFCI-protected receptacle would need to be marked "GFCI Protected" and "No Equipment Ground."

If an existing receptacle outlet is replaced and that receptacle outlet is located in any areas specified in *NEC* 210.12(A), (B), (C), the replacement receptacle at this outlet is required to be AFCI protected. This AFCI protection can be provided by either a listed outlet branch-circuit type AFCI receptacle, a receptacle protected by a listed outlet branch-circuit type AFCI type receptacle, or a receptacle protected by a listed combination type AFCI circuit breaker [*NEC* 406.4(D)(4)]. See Chapter 16 for additional information on the replacement of existing receptacles.

MOUNTING OF RECEPTACLES

Receptacles are required to be mounted in identified boxes or assemblies. These boxes or assemblies are generally required to be securely fastened in place.

Screws used for the purpose of attaching receptacles to a box must be of the type provided with a listed receptacle or must be machine screws having 32 threads per inch or part of listed assemblies or systems, and in accordance with the manufacturer's instructions [*NEC* 406.5 or IRC E4002.6]. This will eliminate the use of drywall or sheet-metal screws for attaching receptacles, covers, luminaires, and other equipment to mounting boxes. This shoddy support method has become a concern in the electrical industry. This crude practice should be unacceptable and can result in damage to the box and inadequate support of the attached receptacle itself. Installers should always follow the manufacturer's installation instructions. This same requirement for mounting receptacles also applies to mounting switches or attaching covers and canopies to boxes [*NEC* 404.10(B) and 314.25 or IRC E4001.10 and E3906.9].

RECEPTACLE FACEPLATES

Metal receptacle faceplates are required to be of ferrous metal not less than 0.76 mm (0.030 in.) in thickness. Where faceplates are made of nonferrous metal, they cannot be less than 1.02 mm (0.040 in.) in thickness. Metal faceplates must be properly grounded to the equipment grounding conductor that is present in a metallic box. A metal faceplate is considered grounded if connected with metal screws to a grounded metal device yoke or strap. There is no requirement

Photo 13.17. Typical listed tamper-resistant receptacle *(notice the shutters behind the ungrounded and grounded blade slots)*

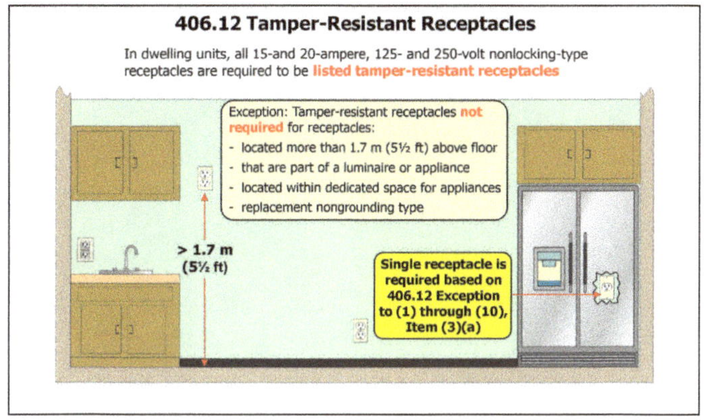

Figure 13.20. Tamper-resistant receptacles are not required in certain locations (by exception)

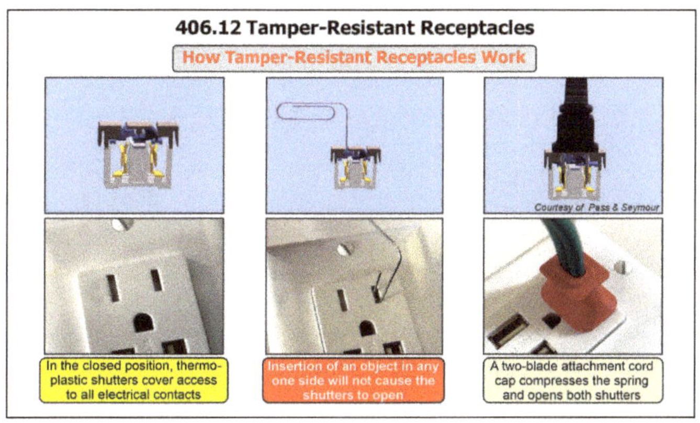

Figure 13.21. Illustration of how a tamper-resistant receptacle works

to remove the insulating washer to ensure the grounding of a metal faceplate unless the receptacle is grounded on a surface mounted box by direct contact with the box.

The device yoke is considered grounded if connected with metal screws to a metal box or a nonmetallic box with integral means for grounding devices. The box must be surface-mounted on the wall, ceiling, or floor surface to use this provision. When connected to a grounded metal box or nonmetallic box with integral grounding means, a listed self-grounding or auto-grounding type device may also be used for grounding the faceplate. Connecting the equipment grounding conductor or equipment bonding jumper to the equipment grounding termination on the device is anoth-

Photo 13.18. The use of a drywall or sheet-metal screw for mounting of a receptacle or switch is prohibited

Photo 13.19. Nonmetallic (insulating material) faceplate for receptacle

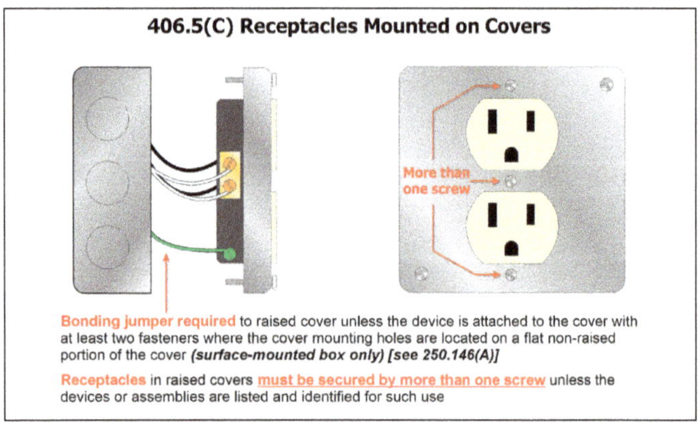

Figure 13.22. Receptacles in raised covers require a minimum of two mounting screws or means

er permitted method of grounding the yoke [*NEC* 406.6, 250.146(A) or IRC E4002.5, E3908.14].

Faceplates of insulating material must be noncombustible and not less than 2.54 mm (0.10 in.) in thickness but shall be permitted to be less than 2.54 mm (0.10 in.) in thickness if formed or reinforced to provide adequate mechanical strength.

Receptacle faces are required to be flush with or project from faceplates of insulating material and are required to project a minimum of 0.4 mm (0.015 in.) from metal faceplates. Faceplates are required to be installed so as to cover the opening and seat against the mounting surface completely. Receptacles mounted in boxes that are set back from the wall surface are required to be installed so that the mounting yoke or strap of the receptacle is held rigidly at the surface of the wall. Receptacles mounted in boxes that are flush with the wall surface or project from that place are required to be so installed that the mounting yoke or strap of the receptacle is seated against the box or raised box cover [*NEC* 406.6 or IRC E4002.5, E4002.6, E4002.7].

Receptacles installed in raised covers are not permitted to be secured solely by a single screw unless the device, assembly, or box cover is listed and identified for such use. Common practice is to provide additional holes in the raised cover to secure the receptacle to the cover through the receptacle mounting yoke holes.

All 15- and 20-ampere attachment plugs and connectors are required to be constructed with no exposed current-carrying parts except the prongs, blades, or pins. The cover for wire terminations is required to be a part that is essential for the operation of an attachment plug or connector (dead-front construction).

Attachment plugs must be installed so that their prongs, pins, or blades are not energized unless inserted into an energized receptacle. This effectively prohibits the practice of using a flexible cord with an attachment plug at both ends for the purpose of providing temporary power. A receptacle cannot be installed so as to require an energized attachment plug as its source of supply [*NEC* 406.7(B) or IRC E3909.4].

Backfeeding a permanently installed receptacle to energize the entire system during a power outage is a dangerous and unacceptable practice. Without a suitable transfer mechanism, individuals working on other parts of the system are exposed to electrical shock or electrocution. It is important to provide transfer equipment suitable for the intended purpose and designed to prevent the inadvertent interconnection of both sources.

Receptacles are generally not permitted to be controlled by dimmer switches. General-use dimmer switches are intended for the connection of only permanently installed incandescent luminaires unless listed for other loads and installed accordingly [*NEC* 404.14(F) or IRC E4001.12].

A receptacle cannot be grouped or ganged in a box with other receptacles, snap switches, or similar devices unless they are arranged so that the voltage between adjacent devices does not exceed 300 volts or unless they are installed in boxes or enclosures equipped with identified, securely installed barriers between the receptacle and the adjacent devices [*NEC* 406.5(J)].

RECEPTACLE FACEPLATES WITH AN INTEGRAL NIGHT LIGHT AND/OR

USB CHARGER

A flush device cover plate that additionally provides a night light and/or Class 2 output connector(s) is required to be listed and constructed such that the night light and/or Class 2 circuitry is integral to the flush device cover plate. Listed receptacle faceplates with integral night lights, USB chargers, or both that rely solely on spring-tensioned contacts are permitted to be connected to only brass or copper alloy receptacle terminal screws rated 1 watt or less [*NEC* 406.6(D)].

RECEPTACLES INSTALLED IN DAMP OR WET LOCATIONS

Damp Locations

Precaution needs to be taken when receptacles are installed in damp or wet locations. All 15- and 20-ampere, 125- and 250-volt nonlocking type receptacles installed in damp locations are required to be of the listed weather-resistant type. Weather-resistant receptacles installed outdoors in wet and damp locations are subject to more demanding temperature variations, direct sunlight, and possible mechanical abuse than those installed indoors in a typical dry environment. Weather-resistant receptacles are specifically designed to withstand these environments and conditions. Several manufacturers of weather-resistant receptacles are producing these receptacles with a "WR" designation on the face of the receptacle to identify the receptacle as the weather-resistant type readily. This "WR" marking is not a *Code* requirement.

Examples of outdoor damp locations are partially protected locations such as under canopies, open-roofed porches, and the like. A receptacle installed in such a location will not be directly subjected to a beating rain or runoff. A receptacle installed outdoors in a damp location is required to have an enclosure for the receptacle that is weatherproof when the receptacle is covered (the attachment plug cap may not be inserted, and the receptacle's cover is closed). In both a damp or a wet location, a hinged cover of an outlet box hood cover is required to be able to open at least 90 degrees or fully open if the cover is not designed to open 90 degrees from the closed to open position after installation. An installation suitable for wet locations is also suitable for damp locations [*NEC* 406.9(A) or IRC E4002.8].

Weatherproof is defined as being so constructed or protected that exposure to the weather will not interfere with successful operation.

Rainproof, raintight, or watertight equipment can fulfill the requirements for weatherproofing where varying weather conditions other than wetness, such as snow, ice, dust, or temperature extremes, are not a factor.

Photo 13.20. Weather-resistant receptacle with the "WR" marking *(receptacle is also tamper-resistant)*

Photo 13.21. Receptacle cover listed and identified to be used in damp location *(not listed and identified for a wet location as the cover does not provide weatherproof protection when attachment plug cap is installed)*

Watertight is defined as being so constructed that moisture will not enter the enclosure under specified test conditions (see *NEC* Article 100 or IRC Chapter 35).

Wet Locations

A 15- or 20-ampere, 125- and 250-volt receptacle installed in a wet location such as outdoors and exposed to the weather or in other wet locations is required to be installed in a weatherproof enclosure. The enclosure must be weatherproof regardless of whether or not the attachment plug is inserted. An outlet box hood installed for this purpose must be listed and shall be identified as "extra duty." Other listed products, enclosures, or assemblies providing weatherproof protection that do not utilize an outlet box hood need not be marked "extra duty." [*NEC* 406.9(B)(1); IRC E4002.9]. An outlet box hood is defined as "a housing shield intended to fit over a faceplate for flush-mounted wiring devices, or an integral component of an outlet box or of a faceplate for flush-mounted wiring devices." The outlet box hood does not

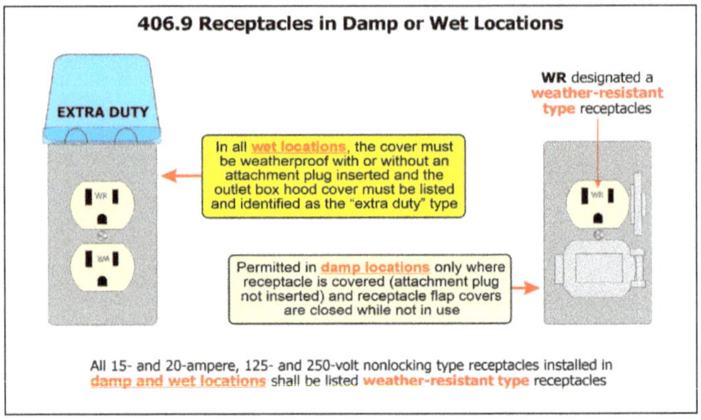

Figure 13.23. Receptacles installed in damp and wet locations

Photo 13.22. Outdoor receptacle outlets with in-use "Extra Duty" type covers *(extra duty covers come in cast metal or nonmetallic styles)*

serve to complete the electrical enclosure; it does reduce the risk of water coming in contact with electrical components within the hood, such as attachment plugs or wiring devices [*NEC* Article 100].

Several receptacle hoods or covers currently on the market comply with this requirement. They usually have a domed cover that will allow a flexible cord to enter the bottom of the cover and close with the cord cap inserted in the receptacle. Both cast metal and nonmetallic types of covers are available.

For receptacles other than 15- or 20-ampere, 125- and 250-volt, the requirements for the type of cover vary. If the product is attended while being used, the receptacle enclosure is only required to be weatherproof when the attachment plug is removed [*NEC* 406.9(B)(2) or IRC E4002.10]. Receptacle outlets other than 15- or 20-ampere, 125- and 250-volt receptacles are not very common at dwelling units.

The enclosure for a receptacle installed in an outlet box that is flush-mounted on a wall surface is required to be weatherproofed by means of a weatherproof faceplate assembly that provides a watertight connection between the plate and the surface of the wall. A receptacle outlet installed outdoors is required to be located so that water accumulation is not likely to touch the outlet cover or plate [*NEC* 406.9(D) or IRC E4002.12].

It should be understood that the authority having jurisdiction (AHJ) has the responsibility to determine whether a location is considered a wet, damp, or dry location. Definitions contained in the *Code* provide excellent guidance for the AHJ on which to base this determination. Where the location is questionable, such as a damp, wet, or dry location, the installer should always check with the local AHJ.

Careful attention must be paid to the proper installation and support of receptacles or luminaires where they may be installed in an enclosure supplied by an underground conduit type wiring method. Attention to landscaping details helps ensure the receptacle outlet will not end up buried [*NEC* 314.23(F) or IRC E3906.8.5].

Just like with damp locations, all 15- and 20-ampere, 125- and 250-volt nonlocking type receptacles installed in wet locations are required to be of the listed weather-resistant type.

In existing dwelling units, weather-resistant receptacles would need to be provided where replacement receptacles are installed at receptacle outlets that are now required to be weather-resistant receptacles under today's *Code* requirements [*NEC* 406.4(D)(6)].

BATHTUB AND SHOWER SPACES

Historically, the *Code* has prohibited the installation of receptacle outlets within or directly over a bathtub or shower space. This bathtub or shower "space" was not defined by the *Code* and was subject to the interpretation of the AHJ. A bathtub or shower space could reasonably be interpreted to include the walls that form the bathtub or shower walls or the walls that are immediately adjacent to the bathtub or shower. Periodically, some installers and AHJs would attempt to enforce the 900 mm (3 ft) by 2.5 m (8 ft) "zone" requirements that apply to luminaires from *NEC* Article 410 to receptacle outlet placements in bathrooms as well. In previous editions of the *Code*, this was going above the minimum requirements of the *Code*.

For the 2020 *NEC Code* cycle (2021 IRC), rules were implemented to prohibit receptacle outlets from being installed within a zone measured 900 mm (3 ft) horizontally and 2.5 m (8 ft) vertically from the top of the bathtub rim or shower stall threshold with this identified zone being all-encompassing and including the space directly over the tub or shower stall. This 900 mm (3 ft) by 2.5 m (8 ft) "zone" for bathroom receptacles had been rejected in past editions of the *Code* as *NEC* 210.52(D) and IRC E3901.6 requires a wall receptacle outlet within 900 mm (3 ft) of the outside edge of each basin (sink) in bathrooms. This requirement is to ensure that a wall receptacle is present in the bathroom to eliminate the need to serve the bathroom with an extension cord. These two separate *Code* rules *[NEC 406.9(C) or IRC E4002.11 for receptacles prohibited in a "zone" around a tub or shower and the requirement demanding a receptacle within 900 mm (3 ft) of bathroom sinks at NEC 210.52(D) and IRC E3901.6]* could actually be in conflict with each other in smaller sized bathrooms.

To remedy that possible *Code* conflict where the bathroom is dimension-wise, simply too small to accommodate this 900 mm (3 ft) horizontal measurement, an exception was added stating that in bathrooms with less than the required zone, the receptacle(s) are permitted to be installed opposite the bathtub rim or shower stall threshold on the farthest wall within the room [*NEC* 406.9(C), Ex. No. 2 or IRC E4002.11]

For the 2023 *NEC* (2024 IRC), language was added to include the space below the 900 mm (3 ft) by 2.5 m (8 ft) "zone" outside the bathtub or shower stall in the area prohibited from housing a receptacle outlet. This "zone" now includes the space measured vertically from the floor to 2.5 m (8 ft) above the top of the bathtub rim or shower stall threshold. The identified zone is all-encompassing and includes the space directly over the bathtub or shower stall and the space below this zone, but not the space separated by a floor, wall, ceiling, room door, window, or fixed barrier. A receptacle installed for a hydromassage bathtub motor under the skirt of the tub is exempt [see *NEC* 406.9(C), Ex. No. 1 and 680.73 or IRC E4002.11 and E4209.4].

A new Ex. No. 4 to *NEC* 406.9(C) (IRC E4002.11) was added to address electronic toilets or personal hygiene devices such as an electronic bidet seat. This new exception will allow a single receptacle for these devices in the prohibited zone, with the receptacle required to be readily accessible and not located in the space between the toilet and the bathtub or shower.

Another new exception exists pertaining to weight-supporting ceiling receptacles (WSCR). These WSCRs are permitted to be installed for listed luminaires that employ a weight-supporting attachment fitting (WSAF) in damp locations. These listed luminaires would have to comply with *NEC* 410.10(D) or IRC E4003.11 for luminaires installed in and around bathtub or shower areas [*NEC* 406.9(C), Ex. No. 3 or IRC E4002.11].

SWITCHES

Many different types of switches are installed in one- and two-family dwellings. Switches such as the single-pole, 3-way, 4-way, double-pole, and enclosed knife switches in the fused and unfused types are just a few examples of switches installed in dwelling units.

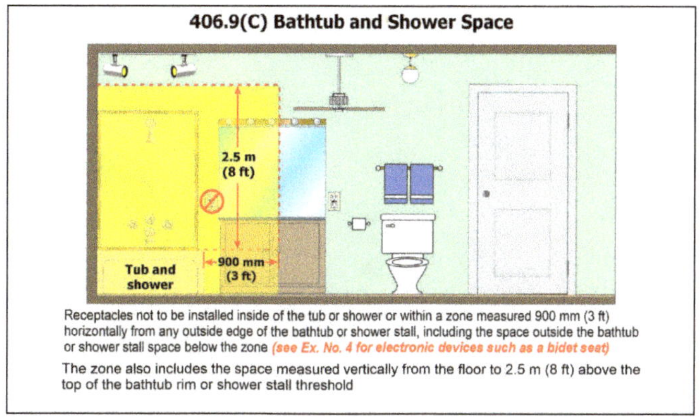

Figure 13.24. Bathroom receptacles are not permitted in a 900 mm (3 ft) by 2.5 m (8 ft) "zone" around bathtub or shower spaces

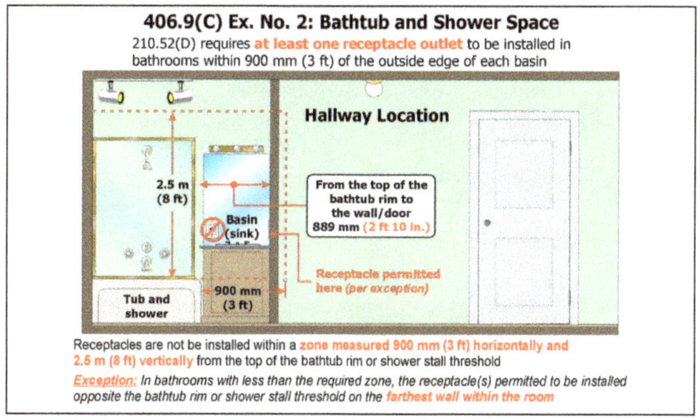

Figure 13.25. Bathroom receptacle(s) permitted within 900 mm (3 ft) by 2.5 m (8 ft) "zone" around the bathtub or shower space, with the exception of small bathrooms

Where surface-mounted switches are installed in any damp or wet location, such as outside of a dwelling, that switch or circuit breaker is required to be enclosed in a weatherproof enclosure or cabinet. A flush-mounted switch or circuit breaker in a damp or wet location must be equipped with a weatherproof cover. Time switches, flashers, and similar devices are required to be of the enclosed type or must be mounted in cabinet boxes or equipment enclosures. Energized parts are required to be protected by a barrier to prevent operator exposure when making manual adjustments or switching [*NEC* 404.4 or IRC E4001.7].

Switches are not permitted to be installed within tubs or shower spaces unless installed as part of a listed tub or shower assembly [*NEC* 404.4(C) or IRC E4001.7].

A single-throw knife switch must be connected so the switch blades are not energized when the switch is in the open or "off" position. Single-throw knife switches are to be placed so that gravity will not tend to close them. Single-throw knife switches (approved for use in the inverted position) are required to be provided with an integral mechanical means that ensures that the blades remain in the open (off) position when set to that position [*NEC*

Photo 13.23. Switches are permitted to be installed outside of a tub or shower space *(as shown)* but cannot be installed within the tub or shower space

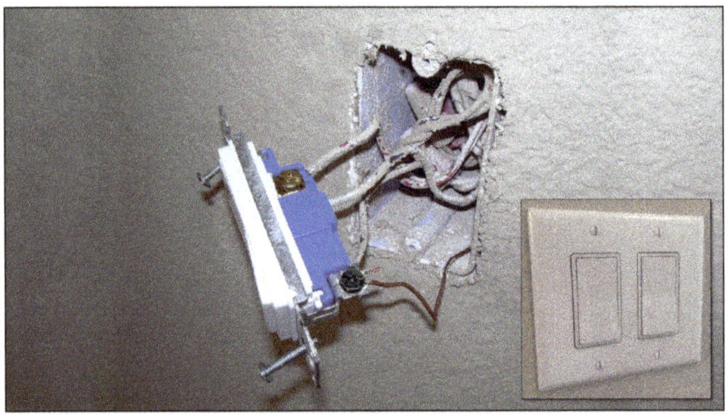

Photo 13.24. Typical single-pole switch (rocker-style) installed in a dwelling unit with means for grounding the switch and a metal switch plate

404.6(A)].

General-use and motor-circuit switches and circuit breakers, where mounted in an enclosure, must clearly indicate *(in a location that is visible when accessing the external operating means)* whether they are in the open "off" or closed "on" position. Where these switch or circuit breaker handles are operated vertically rather than rotationally or horizontally, the "up" position of the handle must be in the "on" position (*NEC* 404.7 or IRC E4001.3).

All switches and circuit breakers that are used as switches must be located so they can be operated from a readily accessible place. All switches and circuit breakers must be installed so the center of the grip of their operating handles when in their highest position, will not be more than 2.0 m (6 ft 7 in.) above the surface of the floor or other working platform [*NEC* 404.8(A) or IRC E4001.6].

Multipole snap switches are not permitted to control more than a single circuit unless they are listed and marked for the use, or unless their voltage ratings are not less than the applied voltage of the connected circuits [*NEC* 404.8(C) or IRC E4001.13].

Snap switches are generally required to be effectively grounded by connection to an equipment grounding conductor. In addition, the snap switch must provide a means to ground a metal faceplate. This means of grounding metal faceplates is required even if a nonmetallic faceplate is installed. This requirement intends to ensure proper grounding of a metal faceplate if, at any time in the future, one is installed. This requirement also applies to dimmer and control switches.

By the requirements of *NEC* 404.9(B) or IRC E4001.11.1, snap switches, dimmers, control devices, and metal faceplates are considered to be effectively grounded by connection to an equipment grounding conductor if either of the two following conditions are met:

1. The switch is mounted with metal screws to a metal box or metal cover that is connected to an equipment grounding conductor or to a nonmetallic box that has an integral means for connecting to an equipment grounding conductor. This section does not mandate a self-grounding or auto-grounding type switch yoke. Certainly, such a device could be used, if available, to meet the requirements of this section.
2. An equipment-grounding conductor or equipment bonding jumper is connected to an equipment-grounding termination of the snap switch or device.

Listed switch kits and listed switch assemblies constructed with nonmetallic yokes and equipped with nonmetallic faceplates are not required to be grounded. These switches are not designed for and have no provisions for attaching a metallic faceplate to them. Snap switches with integral nonmetallic enclosures are also exempt from connecting to an equipment grounding conductor. These switches incorporate integral nonconductive (nonmetallic) faceplates and inherently cannot interchange faceplates, metal or otherwise [*NEC* 404.9(B), Ex. No. 2 or IRC E4001.11.1, Ex. No. 2].

All snap switches mounted in boxes shall have faceplates installed so as to cover the opening and seat against the finished surface completely. Where mounted in boxes that are set back of a wall surface a maximum of 6 mm (¼ in.) in noncombustible walls or ceilings, a flush-type snap switch is required to be installed so the extension plaster ears are seated against the wall surface. A flush-type snap switch mounted in a box that is flush with the wall surface, or projects therefrom, is required to be installed so the mounting yoke or strap of the switch is seated against the box [*NEC* 314.20 and 404.10(B) or IRC E3906.5 and E4001.10].

As stated earlier with the mounting of receptacles, screws

used for the purpose of attaching a snap switch to a box must be of the type provided with a listed snap switch, or they must be machine screws having 32 threads per inch or part of listed assemblies or systems, in accordance with the manufacturer's instructions. Again, this will eliminate the use of nonconforming screws such as drywall or wood screws from being used to mount snap switches to a box.

Metal enclosures used for switches or circuit breakers must be properly grounded by connection to an equipment grounding conductor. Where nonmetallic enclosures are used with metal-armored cables or metallic conduits, provision is required to be made for grounding continuity [*NEC* 404.12 or IRC E4001.5].

A snap switch that is rated at 20 amperes or less and directly connected to aluminum conductors must be listed and marked "CO/ALR," which stands for "Copper/Aluminum Revised." Devices marked CO/ALR are specifically designed and identified to ensure a good connection through the use of a larger contact area and compatible aluminum materials [*NEC* 404.14(C) or IRC E4001.2].

For termination of conductors on snap switches, the *Code* has some specific requirements located at *NEC* 404.14(D). For 15-ampere and 20-ampere snap switches not marked CO/ALR, only copper and copper-clad aluminum conductors can terminate on said switches. Terminals on snap switches marked CO/ALR are permitted to be used with copper, aluminum, and copper-clad aluminum conductors.

Snap switches that employ screwless terminals of the conductor push-in type construction (also known as conductor push-in terminals) can only be installed on 15-ampere or less branch circuits using only 14 AWG solid copper wire unless the switch is listed and marked for other types and sizes of conductors.

GROUNDED CIRCUIT CONDUCTORS AT SWITCH LOCATIONS

The grounded circuit conductor (typically the neutral conductor) for the controlling lighting circuit is generally required to be installed at the switching device location where switches control lighting loads such as luminaires, ceiling paddle fans, etc., and are supplied by a general-purpose branch circuit that is grounded. These lighting loads serve bathrooms, hallways, stairways, or rooms suitable for human habitation or occupancy as defined in the applicable building code. This would eliminate the need for the grounded conductor at such locations as a switch that serves a closet, as the closet is not considered a habitable room.

This grounded (neutral) conductor is typically needed for the successful operation of such things as electronic lighting

Photo 13.25. Switch locations generally require the presence of a grounded circuit conductor for such devices as an occupancy sensor, lighted dimmer, etc.

control devices, such as occupancy sensors. The grounded conductor needs to be extended to any switch location as necessary and connected to the switching devices that require line-to-neutral voltage to operate the electronics of the switch in standby mode.

Where multiple switch locations control the same lighting load such that the entire floor area of the room or space is visible from the single or combined switch locations, the grounded conductor would only be required at one switching location. An example of this would be a room such as a dining room with two 3-way switches in the same room. The grounded conductor would only be required at one of these 3-way switch locations as it is unlikely that an occupancy sensor would ever be installed at both of these 3-way switch locations.

This grounded conductor requirement goes on to indicate that where any of the following conditions exist, the grounded (neutral) conductor is permitted to be omitted from the switch enclosure:

(1) Conductors for switches controlling lighting loads enter the box through a raceway with that raceway large enough for all contained conductors, including a grounded conductor

(2) Where snap switches with integral enclosures identified for the use, having brackets that securely fasten the device to walls or ceilings of conventional onsite frame construction, for use with nonmetallic-sheathed cable

(3) Where lighting in the area is controlled by automatic means, such as a motion sensor located in the ceiling or the corner of a wall

(4) Where a switch controls a receptacle load

Many electronic lighting control devices, such as occupancy sensors, require a standby current to maintain a ready state of detection for the function of these devices. This also allows

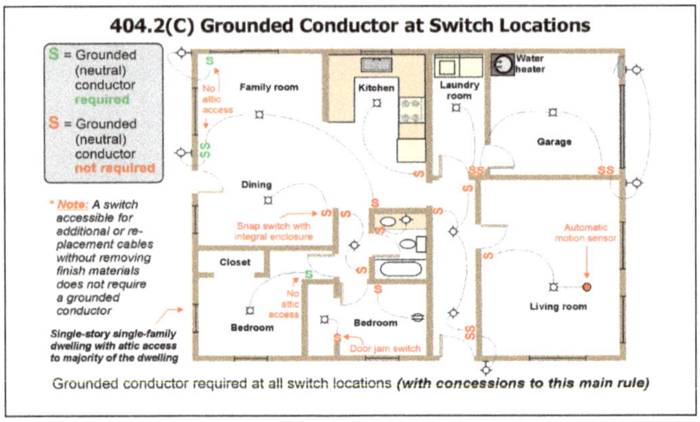

Figure 13.26. A grounded conductor is generally required at all switch locations *(with conditions)*

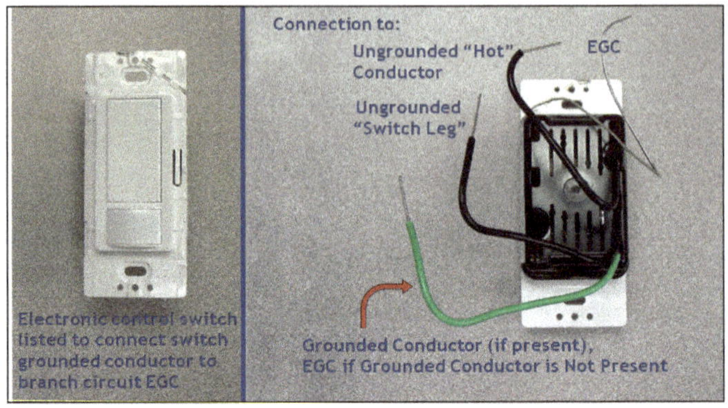

Photo 13.26. Electronic control switches are no longer allowed to introduce current on the EGC

immediate switching of the load to the "on" condition. These devices require a standby current when they are in the "off" state as well. These devices typically utilize the grounded conductor for the standby current flow. In some situations, such as where the grounded conductor is provided at the lighting outlet location, a grounded conductor is typically not provided in the switch box for switches controlling lighting loads.

When a snap switch is replaced with an occupancy sensor at a later date, installers have typically employed the equipment grounding conductor for the standby current of these control devices. This is not a good practice due to the introduction of circulating current onto the equipment grounding conductor. UL 773A permits occupancy sensors to have a current of up to 0.5 mA on the equipment grounding conductor. This current flow at the load, when the device is presumed to be off, could create a potential risk to installers. [*NEC* 404.2(C) or IRC E4001.15].

An exception to *NEC* 404.2(C) states that the grounded conductor connection requirements do not apply to replacement or retrofit switches installed in locations prior to local adaption of *NEC* 404.2(C) and where the grounded conductor cannot be extended without removing finish materials.

Because of the permitted current of up to 0.5 mA on the equipment grounding conductor for occupancy sensors, the number of electronic control switches on any branch circuit cannot exceed five. The number connected to any feeder on the load side of a system or main bonding jumper cannot exceed 25 [*NEC* 404.2(C), Exception or IRC E4001.15, Exception].

Electronic control switches are required to be listed. Electronic control switches are no longer permitted to introduce current on the equipment grounding conductor during normal operation. There are listed electronic control devices (occupancy sensors) currently in the marketplace that are allowed to connect the device-grounded conductor to the branch circuit equipment grounding conductor (if the branch circuit grounded conductor is not present in the switch box). This practice obviously introduces current onto the equipment grounding conductor. These listed devices will be phased out as the manufacturer's inventory is depleted. Currently, electronic control switches that introduce current on the equipment grounding conductor are required to be listed and marked for use in replacement or retrofit applications only [*NEC* 404.22].

PRODUCT SAFETY STANDARD REQUIREMENTS

It is recommended to consult the following electrical product safety guide card information and safety standards for additional guidance on the proper installation, operation, and use of electrical equipment covered in this chapter. The four-letter code in parentheses indicates the product category in the UL Product iQ® at www.ul.com/PiQ.

- Equipment for Use in Ordinary Locations (AALZ)
- Luminaires and Fittings (HYXT)
- Fluorescent Lamp Type Luminaires (IEUT)
- Fluorescent Recessed Luminaires (IEVV)
- Incandescent Lamp Type Luminaires (IEYV)
- Incandescent Recessed Luminaires (IEZX)
- Metallic Outlet Boxes (QCIT)
- Nonmetallic Outlet Boxes (QCMZ)
- Receptacles (RTDV)
- Receptacles for Plugs and Attachment Plugs (RTRT)
- Dimmers, General Use Switch (EOYX)

REFERENCES

1. *National Electrical Code* and *NEC* are registered trademarks of the National Fire Protection Association, Inc., Quincy, MA 02169. This reprinted material is not the official position of the National Fire Protection Association, which is represented only by the standard in its entirety.

2. *International Residential Code for One- and Two-Family Dwellings* (2024). Copyright © International Code Council, Inc., June 2023.

3. UL, Northbrook, IL, *UL Guide Information for Electrical Equipment Directory* (UL Product iQ® material) unless indicated otherwise.

REVIEW QUESTIONS

1. Conductors with an insulation rating of 90°C are required for branch-circuit wiring within the ballast compartment of fluorescent luminaires where located within what distance of a ballast?

A. 38 mm (1-½ in.)
B. 50 mm (2 in.)
C. 65 mm (2-½ in.)
D. 75 mm (3 in.)

2. A splice or tap is permitted to be located within an arm or stem of a luminaire.

True _____ False _____

3. Luminaires installed outdoors under canopies, marquees, and roofed open porches are typically considered to be located in a ____________ location.

A. wet
B. damp
C. dry
D. corrosive

4. No parts of a pendant, lighting track, ceiling-suspended paddle fan, or hanging-type luminaire can be located within a zone ____________ horizontally and ____________ vertically from the top of a bathtub rim.

A. 600 mm (2 ft) / 1.5 m (5 ft)
B. 900 mm (3 ft) / 2.5 m (8 ft)
C. 900 mm (3 ft) / 3.0 m (10 ft)
D. 1.2 m (4 ft) / 2.5 m (8 ft)

5. Before applying any exceptions, receptacle outlets are not permitted to be located within 900 mm (3 ft) of the bathtub or shower space.

True _____ False _____

6. Indoor installations of electric-discharge fluorescent luminaires of 1000 volts or less, other than those with a simple reactance type ballast, must or may:

A. contain thermal protection integral within the ballasts
B. contain ballasts that are not integrally protected
C. be cord-connected by Type TP cord
D. have a 60°C branch-circuit conductor within 75 mm (3 in.) of a ballast

7. A flush-type receptacle outlet installed outdoors in a wet or damp location on a wall surface is required to be installed:

A. at least 300 mm (12 in.) from grade level
B. so a watertight connection is made between the wall and the cover plate
C. with a raintight fitting connected to a raceway
D. with a nonmetallic cover plate

8. Receptacle outlets are generally prohibited from being installed within a zone measured ____________ horizontally and ____________ vertically from the top of the bathtub rim or shower stall threshold, with this identified zone being all-encompassing and including the space directly over the tub or shower stall.

A. 600 mm (2 ft) / 1.5 m (5 ft)
B. 900 mm (3 ft) / 2.5 m (8 ft)
C. 900 mm (3 ft) / 3.0 m (10 ft)
D. 1.2 m (4 ft) / 2.5 m (8 ft)

9. This all-encompassing zone for receptacles around a bathtub or shower stall also includes the space directly below this zone.

True _____ False _____

10. Before applying any exceptions, all 125-volt, 15- and 20-ampere receptacles in dwelling units are required to be listed as tamper-resistant receptacles.

True _____ False _____

11. A single receptacle for an electronic bidet is allowed to be installed in the prohibited zone for receptacles around a bathtub or shower space with the receptacle required to be _______________ and _________________________ ___________________________.

A. readily accessible
B. not located in the space between the toilet and the bathtub or shower
C. located in the space between the toilet and the bathtub or shower
D. Both A. and B.

12. A clearance of what distance is required for the recessed portion of a recessed incandescent luminaire not marked Type IC from thermal insulation?

A. 75 mm (3 in.)
B. 100 mm (4 in.)
C. 125 mm (5 in.)
D. 150 mm (6 in.)

13. Electronic control switches that introduce current on the equipment grounding conductor are required to be listed and marked for use in replacement or retrofit applications only.

True _____ False _____

14. The grounded (neutral) conductor must be connected to the screw shell of a lampholder.

True _____ False _____

15. A luminaire that weighs more than _____ lbs. must be supported independently of the outlet box unless the outlet box is listed and marked for not less than the weight to be supported.

A. 6
B. 35
C. 50
D. 70

16. A receptacle outlet grounding terminal is to be used for equipment grounding conductor connection purposes only.

True _____ False _____

17. Where a bathroom is, dimension-wise, too small to accommodate the required zone for receptacle outlet placement around a bathtub or shower stall, the receptacle(s) are permitted to be installed opposite the bathtub rim or shower stall threshold _________________________.

A. outside the bathroom
B. on the wall closest to the bathtub rim or shower stall threshold
C. on the farthest wall within the room
D. in a location determined by the designer, homeowner, or AHJ

18. All 15- and 20-ampere, 125- and 250-volt nonlocking receptacles installed in damp or wet locations are required to be of the listed _______________ type.

A. weatherproof
B. weather-resistant
C. tamperproof
D. rain-resistant

19. Unless identified as being suitable for insulation to be in contact with the luminaire, a recessed luminaire must have a clearance from the luminaire to combustible material by at least what distance?

A. 6 mm (¼ in.)
B. 13 mm (½ in.)
C. 16 mm (⅝ in.)
D. 25 mm (1 in.)

20. A surface-mounted or recessed incandescent type luminaire that is installed in a clothes closet must have all lamp(s) of the completely enclosed type.

True ______ False ______

21. Where switches control lighting loads such as luminaires, ceiling paddle fans, etc., and are supplied by a general-purpose branch circuit that is grounded, the grounded circuit conductor for the controlled lighting circuit is not required to be provided at which of the following switch locations:

A. Where conductors enter the box enclosing the switch through a raceway, provided that the raceway is large enough for all contained conductors, including a grounded conductor
B. Where lighting in the area is controlled by an automatic means
C. Both A and B
D. All switch locations other than those described in A.

22. In existing dwellings, weather-resistant receptacles are required, whereas replacement receptacles are installed at receptacle outlets that are now required to be weather-resistant receptacles under today's *Code* requirements.

True ______ False ______

23. Luminaires that are not permitted to be installed in a clothes closet include:

A. incandescent luminaire with open (exposed) or partially enclosed lamps
B. pendant luminaires
C. lampholders
D. All of the above

24. Which of the following luminaires or parts of a luminaire are not required to be connected to an equipment grounding conductor?

A. Small, isolated parts, such as mounting screws, clips, and decorative bands on glass spaced at least 38 mm (1½ in.) from lamp terminals
B. Portable luminaires with a polarized attachment plug
C. Replacement luminaires that are GFCI protected where an equipment grounding means does not exist at a lighting outlet
D. All of the above

25. Listed receptacle faceplates with integral night light, USB charger, or both, that rely solely on spring-tensioned contacts are permitted to be connected to only brass or copper alloy receptacle terminal screws rated _________ or less.

A. 1 watt
B. 2 watts
C. 3 watts
D. 10 watts

26. Which of the following statements concerning track lighting is NOT true?

A. Lighting track shall be permanently installed and connected to a branch circuit
B. Only lighting track fittings shall be installed on the lighting track
C. Lighting track can be installed in a wet or damp location
D. Lighting track fittings shall not be equipped with general-purpose receptacles

27. Lighting track may be installed where the track is extended through a wall or partition.

True ______ False ______

28. Screws used for the purpose of attaching a snap switch to a box must be of the type provided with a listed snap switch, or they must be machine screws having __________ threads per inch or part of listed assemblies or systems, in accordance with the manufacturer's instructions.

A. 24
B. 28
C. 16
D. 32

29. A general term for a complete subassembly of parts and devices for field conversion of utilization equipment best describes:

A. a listed luminaire
B. a retrofit kit
C. a listed equipment power-panel
D. a luminaire dimming panel

30. The two basic types of recessed luminaires are Type IC and Type Non-IC rated.

True ______ False ______

31. Light-emitting diode (LED) luminaires use heat sinks to absorb the heat produced by the LED and dissipate it into the surrounding environment.

True ______ False ______

32. For a 15 ampere rated receptacle installed in a wet location that utilizes other listed products, enclosures, or assemblies providing weatherproof protection that do not utilize an outlet box hood need not be identified as __________.

A. damp duty
B. extra duty
C. wet duty
D. weather duty

33. Switches are not permitted to be installed within bathtub or shower spaces unless installed as part of a listed tub or shower assembly.

True _____ False _____

14 Swimming Pools and Other Similar Installations

NEC Article 680 addresses the construction of electrical wiring and equipment related to installations such as swimming pools, hot tubs, spas, fountains, splash pads, hydromassage bathtubs, and electrically powered pool lifts. This section will concentrate on these aquatic environments associated with residential units. The NEC generally considers these types of installations as bodies of water. Because this term is also used elsewhere in the Code, it is important to recognize that NEC Article 682 also covers bodies of water, but of a different kind, such as aeration ponds, fish farm ponds, storm retention basins, sewage treatment ponds, irrigation channels or ponds, and similar structures. Review the scope of each article to determine which installations are included. The IRC covers permanently installed pools, storable pools, spas, hot tubs, and hydromassage tubs. The IRC does not address fountains, therapeutic tubs, or electrically powered pool lifts that are occasionally found in residential settings.

NEC Article 680, Part I, General Requirements, applies to all bodies of water covered by Parts I through VI. Part VII (Hydromassage Bathtubs) is a stand-alone part of *NEC* Article 680. Part VIII (Electrically Powered Pool Lifts) is also a stand-alone part except where the requirements of other parts of Article 680 are specifically referenced. Other parts of Article 680 cover installations such as permanently installed pools, spas, hot tubs, and fountains, and yet refer to the requirements in Part I. The rules of Part I must be applied and are not limited to just permanently installed swimming pools.

It is important to understand the organization and arrangement of the *NEC* and Article 680 to apply proper requirements to pools, spas, etc. An overview of the organization of Article 680 in the *NEC* is provided as follows.

NEC ARTICLE 680 ARRANGEMENT

Article 680 is divided into several parts, namely Part I through Part VIII. Table 14.1 identifies the specific bodies of water covered by each part of NEC Article 680.

Part I contains ***general requirements*** and applies to all equipment and installations covered in the article (except for Part VII).

Part II provides rules that are specific to ***permanently installed pools***. Other parts of the article cover the equipment or installations identified in the title of the part and do not necessarily apply to all equipment included within the article. For example, the rules in Part II for equipotential bonding for permanently installed pools do not apply to the wiring associated with a storable pool (Part III) or hydromassage bathtubs, covered in Part VII. Specific bonding requirements are provided for those particular bodies of water. *NEC* 680.70 clearly indicates that hydromassage bathtubs are to comply with Part VII only. As stated earlier, Part VII is a stand-alone part of *NEC* Article 680.

Part III provides rules that are specific to ***storable pools, storable spas, and storable hot tubs***. Storable pools, spas, and hot tubs are subject to Part I and Part III of Article 680 (see *NEC* 680.30). As an example, *NEC* 680.22(A)(1) or IRC E4203.1.3 requires at least one 125-volt, 15- or 20-ampere receptacle to be installed and located not less than 1.83 m (6 ft) from and not more than 6.0 m (20 ft) from, the inside wall of a permanently installed pool. Notice that this requirement is located in Part II of *NEC* Article 680. This required receptacle outlet would not be required for a storable pool that must comply with Parts I and III of *NEC* Article 680. IRC Section E4203.1.3 states that this required receptacle outlet must be located not less than 6 feet (1829 mm) from and not more than 20 feet (6096 mm) from the inside wall of *permanently installed pools* and *outdoor spas and hot tubs (does not apply to a storable pool)*.

The title of **Part IV** is ***Spas and Hot Tubs***. *NEC* Section 680.40 indicates that spas or hot tubs are required to comply with the rules of Parts I and IV. Part IV is subdivided to provide requirements for outdoor installations in 680.42 and indoor installations in 680.43. Outdoor installations must comply with Parts I and II (Permanently Installed Pools) except as permitted in 680.42(A) or 680.42(B). Indoor installations of spas and hot tubs must follow Parts I and II, except where Part IV modifies the more general requirements. Section 680.43 also generally permits indoor units to be connected with *NEC* Chapter 3 wiring methods. As we will see, the requirements of Part IV are slightly different from the rules for permanently installed swimming pools in Part II and storable pools in Part III. Keep this differentiation in mind as we review several of the grounding and bonding requirements that apply to each of these particular situations and equipment. Basically, the requirements in Part II apply equally to permanently installed pools, spas, and hot tubs that are installed outdoors.

NEC Article 680 Arrangement	
Part I	General
Part II	Permanently Installed Pools
Part III	Storable Pools, Storable Spas, Storable Hot Tubs, and Storable Immersion Pools
Part IV	Spas, Hot Tubs, and Permanently Installed Immersion Pools
Part V	Fountains
Part VI	Pools and Tubs for Therapeutic Use
Part VII	Hydromassage Bathtubs
Part VIII	Electrically Powered Pool Lifts

Table 14.1 Arrangement of *NEC* Article 680

IRC Chapter 42 Arrangement	
E4201	General
E4202	Wiring Methods for Pools, Spas, Hot Tubs, and Hydromassage Bathtubs
E4203	Equipment Location and Clearances
E4204	Equipotential Bonding
E4205	Bonding and Grounding
E4206	Equipment Installation
E4207	Storable Swimming Pools, Storable Spas, and Storable Hot Tubs
E4208	Spas and Hot Tubs
E4209	Hydromassage Bathtubs

Table 14.2. Specific Parts and Arrangement of IRC Chapter 42

Part V of *NEC* Article 680 includes the rules for ***fountains***. Permanent fountains installed at dwelling units must meet the requirements of both Parts I and V. Part V does not cover self-contained portable fountains. Portable fountains must be installed to meet the requirements of Parts II and III of Article 422, Appliances. See Chapter 7 for additional information on appliances.

Part VI of *NEC* Article 680 relates to ***therapeutic tubs and pools***. These types of tubs and pools must comply with the provisions of Part I and Part VI of *NEC* Article 680. These types of tubs and pools are typically used for therapeutic use in health care facilities, gymnasiums, athletic training rooms, and similar areas. They are not typically found in dwelling units. Portable therapeutic appliances have to comply with Parts II and III of *NEC* Article 422 (Appliances).

Part VII of *NEC* Article 680 covers the requirements of ***hydromassage bathtubs***. *NEC* 680.70 gives direction for hydromassage bathtubs to comply with Part VII of *NEC* Article 680. As stated previously, these tubs do not have to comply with other parts of *NEC* Article 680.

Part VIII of *NEC* Article 680 establishes *Code* rules for ***electrically powered pool lifts***. Electrically powered pool lifts are electrically powered lifts that provide accessibility to and from a pool or spa for people with disabilities. The electrical requirements of these devices must comply with Part VIII of *NEC* Article 680 and are not subject to the requirements of other parts of Article 680 except where the requirements are specifically referenced. Part VIII and electrically powered pool lifts were introduced to the *NEC* during the 2017 revision of the *NEC*.

IRC CHAPTER 42 ARRANGEMENT

Requirements contained in the IRC for swimming pools and similar installations are typical of those contained in the *NEC* with, once again, a difference in numbering. Rules for these installations and this type of equipment are provided in IRC Chapter 42, Swimming Pools. The scope of Chapter 42 indicates the installations and equipment covered; this scope is similar to that of *NEC* Article 680, from which the rules are derived. Keep in mind that electrical systems, equipment, or components not specifically addressed in the chapters of the IRC must default to the applicable requirements of the *NEC* as indicated in IRC E3401.2.

Chapter 42 is divided into several sections, namely Sections E4201 through E4209. The following table identifies the general requirements and the specific body of water covered by each section.

APPLICATION OF OTHER ARTICLES

Wiring or equipment in or adjacent to pools and fountains should comply with other applicable provisions of the *Code*, except where modified by *NEC* Article 680 or IRC Chapter 42. An example of this is another requirement in *NEC* 411.4(B) that applies to low-voltage lighting systems installed around pools. Such systems are prohibited from being installed within 3.0 m (10 ft) of pools, spas, hot tubs, and fountains unless specifically permitted by Article 680. Article 680 permits certain types of listed low-voltage contact limit lighting to be used in the walls of the pool, spa, or hot tub or installed in fountains. *NEC* 680.22(B)(6) gives permission to use typical low-voltage yard lighting located less than 1.5 m (5 ft) from the inside walls of the pool, spa, or hot tub.

APPLICABLE DEFINITIONS

Several definitions unique to swimming pools, spas, hot tubs, and other equipment are found in *NEC* Article 100 and IRC Chapter 42. These definitions, which are unique to the pool and spa environment historically, were found in NEC Article 100 and 680.2; however, all definitions were moved to *NEC* Article 100 for the 2023 edition of the *NEC*. There are twenty-eight (28) definitions found in NEC Article 100 and seventeen (17) definitions located at IRC E4201.2 used exclusively for these aquatic environments. The definitions contained in the *Code* that precede the actual *Code* requirements can assist users with the proper application of the appropriate rule to these types of installations and equipment. For example, a storable swimming pool is defined as a pool that is "installed entirely on or above the ground that is intended to be stored when not in use and is designed for ease of relocation, regardless of water depth." Those storable pools that meet the criteria specified in the definition would be required to comply with Part III of *NEC* Article 680 or IRC Section E4207. If the pool in question were constructed in the ground, it would no longer qualify for applying the requirements of Part III of *NEC* Article 680 or IRC Section E4207, and the general requirements of Part II of *NEC* Article 680 and IRC E4201 through E4206 for a permanently installed swimming pool would be applicable.

For the complete listing of definitions of words and terms used in the *Code*, refer to *NEC* Article 100 or IRC Section E4201.2. The following definitions are provided in *NEC* Article 100 and IRC Chapter 35. If a review is needed, please refer back to Article 100 in the *NEC* for further review.

- ***Corrosive Environment***
- ***Fixed (as applied to equipment)***
- ***Portable (as applied to equipment)***
- ***Stationary (as applied to equipment)***
- ***Forming Shell***
- ***Fountain***
- ***Hydromassage Bathtub***
- ***Lighting Assembly, Cord-and-Plug-Connected (Cord-and-PlugConnected Lighting Assembly)***
- ***Lighting Assembly, Through-Wall (Through-Wall Lighting Assembly)***
- ***Low Voltage Contact Limit***
- ***Luminaire, Dry-Niche (Dry-Niche Luminaire)***
- ***Luminaire, No-Niche (No-Niche Luminaire)***
- ***Luminaire, Wet-Niche (Wet-Niche Luminaire)***
- ***Maximum Water Level***
- ***Permanently Installed Decorative Fountains and Reflection Pools***
- ***Pool***
- ***Pool, Immersion (Immersion Pool)***
- ***Pool, Permanently Installed Swimming, Wading, Immersion, and Therapeutic (Permanently Installed Swimming, Wading, Immersion, and Therapeutic Pools).*** *Note to Reader: In previous editions of the Code, any pool capable of holding water at a depth greater than 1.0 m (42 in.) was considered (by definition) a permanently installed pool.*

- *Pool, Storable; used for Swimming, Wading, or Immersion (Storable Immersion Pool) or (Storable Pool). Note to Reader: In previous editions of the Code, storable pools were limited to pools that were capable of holding water to a maximum depth of 1.0 m (42 in.).*
- *Pool Cover, Electrically Operated (Electrically Operated Pool Cover)*
- *Pool Lift, Electrically Powered. (Electrically Powered Pool Lift).*
- *Spa or Hot Tub*
- *Spa or Hot Tub, Packaged Equipment Assembly. (Packaged Spa or Hot Tub Equipment Assembly).*
- *Spa or Hot Tub, Self-Contained (Self-Contained Spa or Hot Tub)*
- *Spa or Hot Tub, Storable. (Storable Spa or Hot Tub).*
- *Splash Pad.*

Figure 14.1. Corrosive environments are addressed in *NEC* Article 680 and IRC Chapter 42

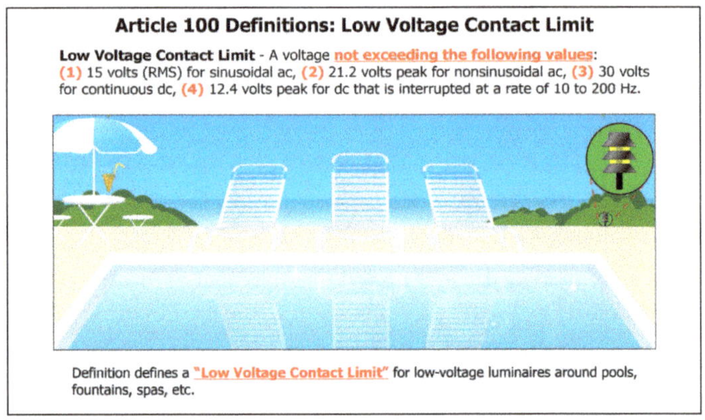

Figure 14.2. Low voltage contact limits are used throughout *NEC* Article 680 and IRC Chapter 42

Photo 14.1. Examples of a storable/portable spa or hot tub

GROUND-FAULT CIRCUIT-INTERRUPTER TYPES

Ground-fault circuit interrupter (GFCI) protection is required for several different types of circuits and equipment associated with bodies of water, such as pools, spas, hot tubs, and hydromassage bathtubs (see Table 14.3). The very nature of the conditions and environment in and around swimming pools and other similar locations where we intentionally mix people, water, and electricity produces a recipe for the need for GFCI protection.

A GFCI device may be a self-contained unit, circuit breaker or receptacle type, or other listed type. For a dwelling unit, a ground-fault circuit interrupter is a device whose function is to interrupt the electric circuit to the load when a fault current to ground exceeds those values established for a Class A GFCI device. There are several classes of ground-fault circuit interrupters (Class A, C, D, and E), and each class has a distinct interruption function. The most common class of GFCI devices is a Class A GFCI device. When referring to a dwelling unit, the type of GFCI device will always be a Class A (4 to 6 mA) GFCI device. There is also a class of GFCI protection known as a Special Purpose Ground-Fault Circuit-Interrupter (SPGFCI). SPGFCIs would not typically have an application in the dwelling unit setting (see definition of SPGFCI below).

Class A GFCI Device

A ground-fault circuit interrupter is to be used only in a circuit that has a solidly grounded (neutral) conductor. A Class A ground-fault circuit interrupter trips when the current to ground has a value in the range of 4 through 6 milliamperes and is suitable for use in branch circuits and feeder circuits, including swimming pool circuits. Swimming pool circuits that were installed before the local adoption of *NEC*-1965 may include sufficient leakage current to cause a Class A ground-fault circuit interrupter to trip. As a result, product safety standards require a Class A device to be marked: "To minimize false tripping, do not connect to swimming-pool equipment installed before the adoption of the 1965 *National Electrical Code*." All GFCI devices associated with dwelling units, swimming pools, and similar installations are required

PERMANENT VS. STORABLE POOL

For a number of years in the *Code*, what primarily distinguished a storable pool from a permanently installed pool was its water depth capacity. In past *Code* cycles *(including the 2020 NEC and the 2021 IRC)*, a ***Storable Pool*** was defined as a pool intended to be stored when not in use that was constructed on or above the ground and capable of holding water to ***a maximum depth of 1.0 m (42 in.)***. On the other hand, a ***permanently installed Swimming Pool*** has historically been defined as a pool constructed in the ground or partially in the ground or any other pool capable of holding water at a depth ***greater than 1.0 m (42 in.)***.

As clearly indicated in these previous definitions, there seemed to be a dividing line and something magic about this 1.0 m (42 in.) depth of water when it comes to a storable pool versus a permanently installed pool. Up to the 1975 edition of the *NEC*, the definition of a storable or a permanently installed pool did not contain a prescriptive wall height or water depth. For the 1978 *NEC*, a proposed wall height of ***900 mm (3 ft)*** was added to the definition of a storable pool. The substantiation for this change indicated that the availability of increasingly larger sizes of above-ground pools made it necessary to clarify the intent of the definition by including maximum dimensions.

During the 1981 *NEC* revision cycle, a proposal was accepted to revise the definition of a storable pool by increasing the maximum ***900 mm (3 ft)*** wall height limitation for a storable pool to a pool capable of holding water to a maximum depth of ***1.0 m (42 in.)***. The submitter of this change stated that the current 1978 *NEC* definition of storable pool "inaccurately and inadequately describes the current products on the marketplace today." The submitter went on to say that approximately 85 percent of the above-ground pools sold today (1981) are greater than 900 mm (3 ft) in depth *[usually 1.0 m (42 in.) to 1.2 m (48 in.)]* and come in a variety of shapes and may have a maximum dimension of greater than 4.5 m (15 ft). In looking at these historical changes, there was not a whole lot of substantiation as to why the maximum water depth of 1.0 m (42 in.) *[and not 1.2 m (48 in.) or some other dimension]* was chosen by the technical committee. There is some indication that the 1.0 m (42 in.) maximum water depth was tied to a storable pool wall height of 1.1 m (45 in.). This 1.0 m (42 in.) maximum water depth has lived on in the definition of a storable pool from the 1981 *NEC* until the 2020 edition of the *NEC*.

Storable pool manufacturers have been producing storable pools with wall heights of 1.2 m (48 in.) for over 40 years without any negative impact on electrical safety. Today, it is very common to see storable pools that exceed the previously defined 1.0 m (42 in.) water depth but qualify in every other way as a storable pool. From an electrical safety standpoint, these storable pools capable of holding water at a depth greater than 1.0 m (42 in.) *(not meeting the previous definition of a storable pool)* are no different than a pool that has a maximum 1.0 m (42 in.) water depth *(meeting the previous definition of a storable pool)*. Providing requirements for electrical safety around a storable pool is not impacted if the maximum water depth is greater than 1.0 m (42 in.). However, while enforcing the 2020 *NEC* (2021 IRC), the Authority Having Jurisdiction (AHJ) has no choice but to treat this seemly storable pool as a permanently installed pool if the maximum water depth is greater than 1.0 m (42 in.).

Once this defined "42-inch" limitation is broken, the AHJ knows that Part II of *NEC* Article 680 must somehow be applied. The difficulty comes as the equipotential bonding grid cannot be bonded to this "storable pool" at four points [*NEC* 680.26(B)(2)], the pool water cannot be bonded to the equipotential bonding grid [*NEC* 680.26(C)], the pool pump for this "storable pool" is double insulated and cannot be bonded [*NEC* 680.26(B)(6) and Exception], and a solid 8 AWG copper equipotential bonding conductor cannot be connected to the equipment grounding conductor of the branch circuit supplying the pump motor [*NEC* 680.26(B)(6)(a)], just to name a few.

For the 2023 *NEC* (2024 IRC), the definitions for a *storable pool* and a *permanently installed pool* have been revised to address the water depth issues of the previous editions of the *Code*. The definition of a "Storable Pool" now indicates that a storable pool would be installed entirely on or above the ground, intended to be stored when not in use, and designed for ease of relocation, ***regardless of water depth***. A "Permanently-Installed Pool" is now defined as a pool that is constructed or installed in the ground or partially in the ground, and all pools are installed inside of a building. Notice that both of these revised definitions are **void of any qualifying depth of the pool or water level.** These definitions were revised to affirm that any electrical hazards associated with a storable or permanently installed pool are not related to the depth of said pool. These revised definitions were necessary to address multiple adverse impacts created with a prescriptive limitation of a 1.0 m (42 in.) water depth. Newer designs in storable pools allow for depths greater than 1.0 m (42 in.), and the addition of a few inches of water has absolutely no impact on electrical safety. These revised definitions will make it much easier to comply with and enforce safe and proper electrical rules for both of these different aquatic applications.

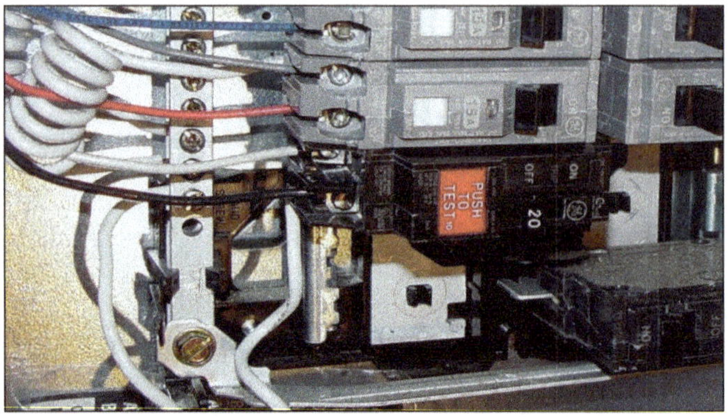

Photo 14.2 GFCI device in the form of a circuit-breaker type

Photo 14.3. GFCI device in the form of a receptacle outlet type

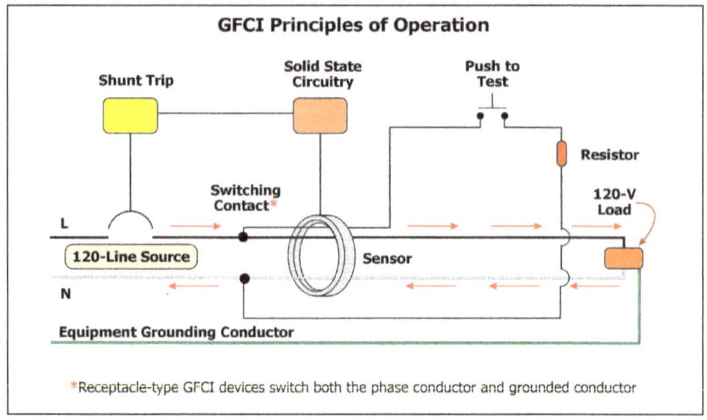

Figure 14.3. GFCI principles of operation

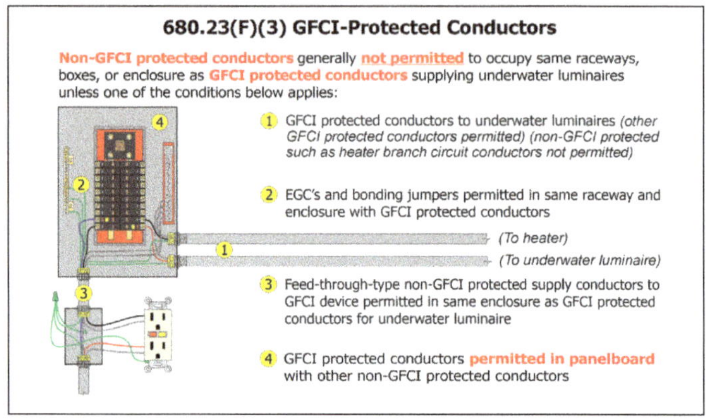

Figure 14.4. GFCI conductors in enclosures with other non-GFCI-protected conductors

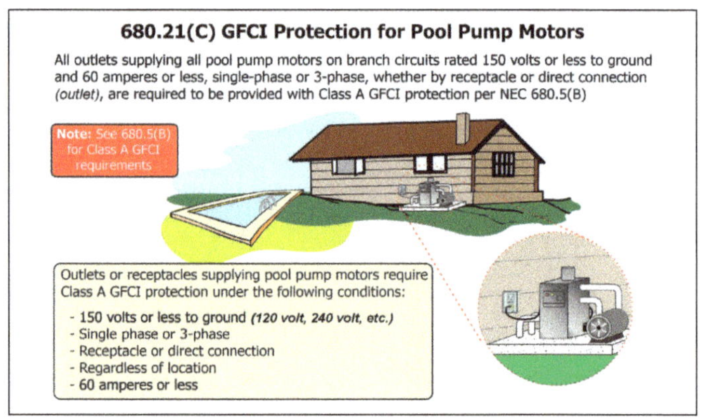

Figure 14.5. GFCI protection is required for both cord-connected and hard-wired pump motors

to be Class A type. For further information, see UL 943, Standard for Ground-Fault Circuit Interrupters."

Other Classes of GFCI Devices

Class C, D, and E GFCIs trip at 20 mA rather than the 4 to 6 mA trip current mandated for Class A GFCI device. This increase in GFCI trip level is allowed by UL 943 assuming the availability of a reliable grounding path in parallel with the body. During a fault, the equipment grounding conductor will shunt the fault current around the body and cause the device to trip. This provides the "let-go" protection, while the 20 mA threshold provides protection against fibrillation. If there is no equipment grounding conductor path, such as in two-wire household products, then the GFCI device must provide both let-go and fibrillation protection, and a Class A device is required.

SPGFCI Device

A special purpose ground-fault circuit-Interrupter (SPG-FCI) is defined as "a device intended for the detection of ground-fault currents, used in circuits with voltage to ground greater than 150 volts, that functions to de-energize a circuit or portion of a circuit within an established period of time when a ground-fault current exceeds the values established for Class C, D, or E devices." With SPGFCI protection associated with "voltage to ground greater than 150 volts," this would eliminate SPGFCI protection at dwelling units as the voltage to ground typically involved at dwelling unit is 120 volts to ground. SPGFCI protection provisions were added to *NEC* Article 680 at a handful of locations for the 2023 edition of the *NEC*.

Table 14.3 Various GFCI Requirements for Swimming Pools and Similar Locations
See *NEC* 680.5(A), (B), and (C) for General Requirements, GFCI Provisions, and SPGFCI Provisions

Equipment (Location)	*NEC* Section	IRC Section
Pool Equipment Room Receptacle(s)	680.12(B)	E4203.1.7
Pool Pump Motors (Permanently Installed Pools)	680.21(C)	E4203.1.4
Pool Pump Motor Replacements	680.21(D)	*NEC* Default
Receptacles - Circulation and Sanitation System	68.22(A)(2)	E4203.1.1
Receptacles (Permanently Installed Pools & Outdoor Spas & Hot Tubs)	680.22(A)(4)	E4203.1.4
Luminaires, Ceiling Fans - Indoor Clearances	680.22(B)(2)	E4203.4.2
Luminaires, Ceiling Fans - Existing Installations	680.22(B)(3)	E4203.4.4
Luminaires, Ceiling Fans (Permanently Installed Pools)	680.22(B)(4)	E4203.4.6
Luminaires; Lamping, Relamping, and Servicing	680.23(A)(3)	E4206.4
Underwater Luminaires (Permanently Installed Pools) *(1st GFCI requirement in the NEC)*	680.23(A)(8)	E4206.4
Conductors on Load Side of GFCI for Underwater Luminaire Conductors	680.23(F)(3)	E4206.3
Enclosures for GFCIs for Underwater Forming Shell	680.24(B)	E4206.9.2
Pool Cover Equipment (Permanently Installed Pools)	680.27(B)(2)	E4206.11
Gas-Fired Water Heater	680.28	*NEC* Default
Pool Pump Motor (Storable Pools, Storable Spas, or Storable Hot Tubs)	680.31	E4207.1
Storable Pool, Storable Spa, or Storable Hot Tubs Receptacles/Equipment	680.32	E4207.2
Luminaires (Storable Pools, Storable Spas, or Storable Hot Tubs)	680.33(B)	E4207.3.2
Self-Contained Storable and Portable Immersion Pools	680.35(A)	*NEC* Default
Storable and Portable Heaters for Storable and Portable Immersion Pools	680.35(C)	*NEC* Default
Self-Contained Storable and Portable Immersion Pools (Audio Equipment)	680.35(D)	*NEC* Default
Self-Contained Storable and Portable Immersion Pools (Receptacles)	680.35(G)	*NEC* Default
Outdoor Spas or Hot Tubs (Cord-n-Plug)	680.42(A)(2)	E4202.3.3
Outdoor Spas or Hot Tubs (Underwater Luminaires)	680.42(C)	*NEC* Default
Receptacles (Indoor Spas & Hot Tubs)	680.43(A)(2)	E4203.1.6
Outlets that Supply Indoor Spas or Hot Tubs	680.43(A)(3)	E4208.1
Luminaires, Ceiling Fans (Spa and Hot Tubs)	680.43(B)(1)	E4203.4.2
Indoor Spas or Hot Tubs (Underwater Luminaires)	680.43(B)(2)	*NEC* Default
Self-Contained or Packaged Spas or Hot Tubs	680.44(A)	E4208.1
Self-Contained or Packaged Spas or Hot Tubs (with integral GFCI protection)	680.44(B)	E4208.1
Self-Contained or Packaged Spas or Hot Tubs (Gas-Fired Water Heaters)	680.44(C)	*NEC* Default
Permanently Installed Immersion Pools (Integral Pumps and/or Heaters)	680.45(A)	*NEC* Default
Permanently Installed Immersion Pools (Storable and Portable Pumps)	680.45(B)	*NEC* Default
Permanently Installed Immersion Pools (Permanently installed Heaters)	680.45(C)(1)	*NEC* Default
Permanently Installed Immersion Pools (Storable and Portable Heaters)	680.45(C)(2)	*NEC* Default
Permanently Installed Immersion Pools (Audio Equipment)	680.45(D)	*NEC* Default
Splash Pads *(comply with Part V of Article 680 per definition of a Splash Pad)*	680.50	*NEC* Default
Fountains (Luminaires, Submersible Pumps, and Other Submersible Equipment)	680.51(A)	*NEC* Default
Fountains (Cord-n-Plug Connected Equipment, Including Power-Supply Cords)	680.56(A)	*NEC* Default
Signs (In or Adjacent to Fountains)	680.57(B)	*NEC* Default
Fountains - Adjacent Receptacles	680.58	*NEC* Default
Fountains - Permanently Installed Nonsubmersible Pumps	680.59	*NEC* Default
Therapeutic Pools and Tubs (Outlet Supplied)	680.62(A)	*NEC* Default
Therapeutic Pools and Tubs (with Integral GFCI protection)	680.62(A)(1)	*NEC* Default
Receptacles (Therapeutic Pools and Tubs)	680.62(E)	*NEC* Default
Hydromassage Bathtubs	680.71	E4209.2
Electrically Powered Pool Lifts	680.82	*NEC* Default
Electrically Powered Pool Lifts (Receptacles)	680.84	*NEC* Default

Table 14.3. A collection of the GFCI requirements for pools, spas, hot tubs, fountains, hydromassage bathtubs, and other aquatic equipment and environments found in both the IRC and the *NEC*.

GROUND-FAULT CIRCUIT INTERRUPTER (GFCI) PRINCIPLES OF GFCI OPERATION

The primary function of a GFCI device is to protect persons from hazards relating to a shock hazard. This is especially relevant in and around swimming pools, spas, hot tubs, etc. The GFCI sensing system continuously monitors the current balance in the ungrounded (hot) conductor and the grounded (neutral) conductor. If the current in the grounded conductor (supply path) becomes less than the current in the ungrounded conductor (return path), a ground fault or loss of current in the field would exist. With this fault, a portion of the current returns to the supply source by some path other than the grounded conductor. With a current imbalance as low as 4 to 6 mA, the Class A GFCI device will interrupt the circuit, and this will be shown by a trip or off indicator on the device (see Figure 14.3).

GFCI REQUIREMENTS FOR SWIMMING POOLS

The *Code* contains several specific requirements relative to ground-fault circuit interrupters for swimming pools (see Table 14.3). Any conductors installed on the load side of a ground-fault circuit interrupter used to provide protection for underwater luminaires are generally not permitted to occupy raceways, boxes, or enclosures containing other (non-GFCI protected) conductors, even if the other conductors are for pool-related equipment such as for a pool pump motor.

Four specific alternatives to this general rule are provided in the *Code*, and the conditions are as follows:

1. GFCIs are permitted in a panelboard that contains circuits protected by other than GFCIs.
2. Supply conductors to a feed-through-type GFCI are permitted in the same enclosure.
3. Conductors on the load side of a GFCI are permitted to occupy raceways, boxes, or enclosures containing only conductors protected by GFCIs.
4. Equipment grounding conductors and bonding jumpers are permitted in the same raceway, box, or enclosure.

It is clear that conductors that supply pool equipment, such as pool heaters that do not have GFCI protection, cannot be installed in the same conduits and junction and pull boxes with conductors having GFCI protection that supply underwater luminaires [*NEC* 680.23(F)(3) or IRC E4206.3].

It should be noted that the GFCI requirements for permanently installed pool pump motors have changed quite a few times over the last few *Code* cycles. Today, all outlets supplying all pool pump motors on branch circuits rated 150 volts or less to the ground and 60 amperes or less, single-phase or 3-phase, whether by receptacle or direct connection, are required to be provided with Class A ground-fault circuit-interrupter protection for personnel [*NEC* 680.21(C) or IRC E4203.1.4]. An exception allows listed low-voltage motors not requiring grounding, with ratings not exceeding the low-voltage contact limit that is supplied by listed transformers or power supplies to be installed without GFCJ protection [*NEC* 680.21(C) Exception].

As Table 14.3 indicates, there are a lot of GFCI requirements for pools, spas, etc., in *NEC* Article 680 (IRC Chapter 42). The very nature of the conditions and environment in and around swimming pools and other similar locations where we intentionally mix people, water, and electricity is a natural fit for GFCI protection.

GROUNDING REQUIREMENTS

Electrical equipment associated with bodies of water, such as pools, spas, hot tubs, and hydromassage bathtubs, must be grounded as well as bonded. Grounding is accomplished by connection to an equipment grounding conductor (EGC). Grounding and bonding serve two different functions. Bonding serves to equalize voltage potential in or around the swimming pool, outdoor hot tub, or spa to keep all metal components at the same voltage potential so a person in contact with two different metal components at possibly different potentials won't serve as the conductive path to equal out that voltage difference. Subsequently, the equipment grounding conductor serves to provide a reference to the ground for metal parts and equipment and also provides a low-impedance path for fault current to return to the source and facilitate overcurrent device operation in clearing a fault.

When it comes to swimming pools, spas, and similar aquatic environments, corrosive conditions and wet locations are very common. **Feeders and branch circuits** installed in a corrosive environment or a wet location are required to contain an equipment grounding conductor (EGC) that is an insulated copper conductor sized in accordance with *NEC* Table 250.122 or IRC Table E3908.13, but in no case, smaller than 12 AWG. This insulated EGC is necessary for grounding of equipment installed in these corrosive or wet locations [*NEC* 680.7(A) or IRC E4205.1].

In the same manner, **flexible cords** used in these locations are required to contain an EGC that is an insulated copper conductor sized in accordance with *NEC* Table 250.122 or IRC Table E3908.13, not smaller than 12 AWG. If a flexible cord is utilized to facilitate the removal of equipment or disconnection of equipment for maintenance or repair, the flexible cord must terminate in a grounding-type attachment plug having a fixed grounding contact member [*NEC* 680.7(B) or IRC E4205.1].

NEC Article 680 and IRC Chapter 42 contain specific grounding (connection to an equipment grounding conductor) requirements throughout the article (chapter). The following is a list of equipment that is required to be specifically grounded:

1. Through-wall lighting assemblies and underwater luminaires, other than those low-voltage lighting products listed for the purpose not requiring grounding [*NEC* 680.23(F)(2) or IRC E4205.1]
2. All electric equipment located within 1.5 m (5 ft) of the inside wall of an indoor spa or hot tub, with an exception for electrical equipment listed for operation at the low-voltage contact limit or less [*NEC* 680.43(F)], all electrical equipment located within a fountain or within 1.5 m (5 ft) of the inside wall of a fountain, other than listed low-voltage luminaires not requiring grounding [*NEC* 680.54(A)(1)], or a therapeutic tub [*NEC* 680.62(D)].
3. All electric equipment associated with the recirculation system of an indoor spa or hot tub, with an exception for electrical equipment listed for operation at the low-voltage contact limit or less [*NEC* 680.43(F)], a fountain [*NEC* 680.54(A)(2)], or a therapeutic tub [*NEC* 680.62(D)].
4. The grounding terminals of junction boxes, transformer enclosure, or other enclosures in the supply circuit to a wet-niche or no-niche luminaire and the field-wiring chamber of a dry-niche luminaire are required to be connected to the equipment grounding terminal of the panelboard [*NEC* 680.24(F) or IRC E4205.1].
5. A cord-connected pool pump filter of a storable pool must incorporate an approved system of double insulation or its equivalent and must be provided with a means for the termination of an equipment grounding conductor for the connection to the internal and nonaccessible non-current-carrying metal parts of the pump [*NEC* 680.31 or IRC E4207.1].

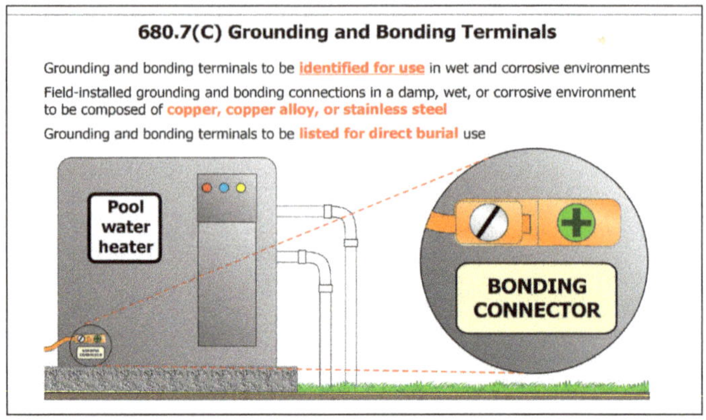

Figure 14.6. Grounding and bonding terminals must be identified for a wet and corrosive environment and listed for direct burial

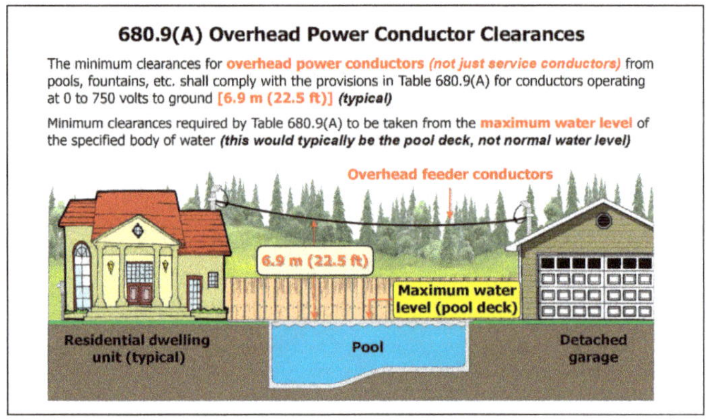

Figure 14.7. Overhead clearances measured from the maximum water level

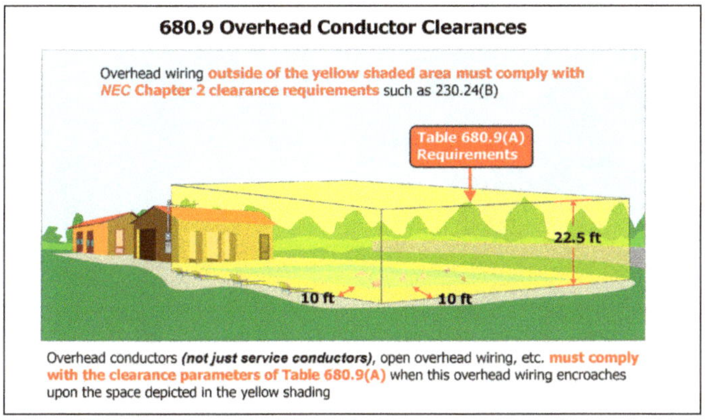

Figure 14.8. Overhead conductor clearances around pools

Table 680.9(A) Overhead Conductor Clearances *(in part)*

Insulated cables, 0-750 Volts to Ground, Supported on and Cabled Together with a Solidly Grounded Bare Messenger or Solidly Grounded Neutral Conductor

Clearance Parameters		
A. Clearance in any direction to the water level, edge of water surface, base of diving platform, or permanently anchored raft	6.9 m	22.5 ft
B. Clearance in any direction to the observation stand, tower, or diving platform	4.4 m	14.5 ft
C. Horizontal limit of clearance measured from inside wall of the pool	This horizontal limit shall extend to the outer edge of the structures listed in A and B of this table but not to less than 3 m (10 ft)	

Table 14.4. Overhead conductor clearances [*NEC* Table 680.9(A) (in part)]

6. A cord-connected storable or portable pump utilized with, but not built-in or permanently attached as an integral part of, a storable or portable immersion pool [*NEC* 680.35(B)]. Panelboards that are not part of the service equipment supply any electrical equipment associated with a fountain [*NEC* 680.54(A)(3)].
7. Panelboards that are not part of the service equipment supply any electrical equipment associated with a fountain [*NEC* 680.54(A)(3)].

Terminals used for equipment grounding and bonding must be identified for use in wet and corrosive environments, such as those found in and around swimming pool locations. Field-installed grounding and bonding connections and terminals used in a damp, wet, or corrosive environment are to be made of copper, copper alloy, or stainless-steel materials and must be listed for direct burial use (*NEC* 680.7(C) and IRC E4205.9).

Later on in this chapter, we will take a close look at specific equipotential bonding requirements for swimming pools, spas, and hot tubs.

CORD- AND PLUG-CONNECTED EQUIPMENT

Fixed or stationary equipment (other than underwater luminaires) for permanently installed pools may be connected to the electrical supply using a flexible cord and plug connection to allow for removal or disconnection during maintenance or repairs. Underwater luminaires for permanently installed pools can be powered by a flexible cord but cannot be connected with a cord and plug. For pools other than storable pools, the length of a flexible cord in this application is limited to 900 mm (3 ft).

Flexible cords used with equipment are required to contain a copper equipment grounding conductor sized based on the rating of the overcurrent device but not smaller than a 12 AWG and must terminate in a grounding-type attachment plug. Note that this rule is specific. Only copper equipment grounding conductors are permitted. These equipment grounding conductors are to be connected to a fixed metal part of the assembly. If connected to a removable part, these removable parts must be mounted on or bonded to a fixed metal part [*NEC* 680.8(A), (B), and (C) and IRC E4202.3].

OVERHEAD CONDUCTOR CLEARANCES FOR POOLS, ETC.

Conductors that run over outdoor bodies of water, such as pools, spas, and hot tubs, must meet the clearance requirements specified in *NEC* Table 680.9(A) or IRC Table E4203.7 and illustrated in *NEC* Figure 680.9(A). Where a minimum clearance from the water level is specified, the measurement is taken from the maximum water level of the specified body of water, which is typically the deck, not the normal water level. These clearances apply to overhead conductors including existing service-drop conductors, branch circuits, feeders, or any other open overhead wiring. Such overhead wiring must meet the minimum distances.

Row (A) of these tables addresses clearances in any direction to the water level, the edge of the water surface, the base of diving platforms, or a permanently anchored raft at pools and similar installations. Row (B) addresses clearances in any direction to towers, diving structures, or observation stands. Row (C) of *NEC* Table 680.9(A) pertains to horizontal clearances and the entire area that extends 3.0 m (10 ft) horizontally measured from inside walls of the pool or similar installations (*NEC* 680.9(A), Table 680.9(A), Figure 680.9(A) or IRC E4203.7, Table E4203.7). The horizontal clearances addressed by *NEC* Table 680.9(A), Row C are addressed in the IRC in the text of E4203.7.

NEC Table 680.9(A) and IRC Table E4103.7 are similar and provide the required minimum clearances from portions of

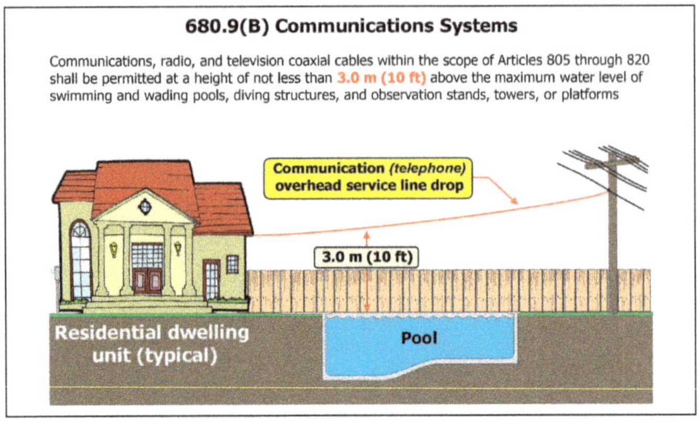

Figure 14.9. Overhead conductor clearances for communication cables

swimming pools and similar installations and adjacent structures for various types of conductors and different voltages.

Communication conductors, community antenna systems, and radio and television coaxial cables complying with *NEC* Chapter 8 and their supporting messengers are permitted to be installed at a height of at least 3.0 m (10 ft) or more above the maximum water level of the swimming and wading pools, diving structures and observation stands, towers or platforms.

A network-powered broadband communication cable is treated as a power cable as far as overhead clearances are concerned. The minimum clearances required for network-powered broadband communication systems conductors are not permitted to be less than 4.4 m (14.5 ft) from any observation stand, tower, or diving platform and not less than 6.9 m (22.5 ft) from the water level, edge of the water surface, base of a diving platform, or permanently anchored raft as indicated in the applicable table heights under the 0-750 volts column [*NEC* 680.9(C), Table 680.9(A), IRC E4203.7 and Table E4203.7].

UNDERGROUND WIRING DEPTHS AND LOCATIONS

The wiring methods underground near pools must be installed in a manner that withstands the conditions unique to the pool environment. The long-standing distance of 1.5 m (5 ft) *(removed for the 2017 NEC and 2018 IRC)* around a pool that underground wiring restrictions were subject to was revived for the 2020 *NEC* (2021 IRC) and applies to all the acceptable and identified wiring methods that are listed for direct burial in these areas where exposure to a corrosive environment is a concern. Identified wiring methods in these underground areas around pools are rigid metal conduit, intermediate metal conduit, rigid polyvinyl chloride conduit, and reinforced thermosetting resin conduit. The following are also identified wiring methods in underground areas around pools when they are listed for direct burial use: jacketed

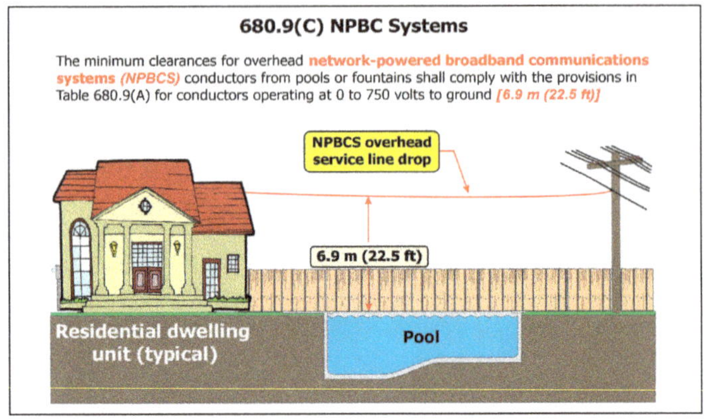

Figure 14.10. Overhead conductor clearances for NPBC cable

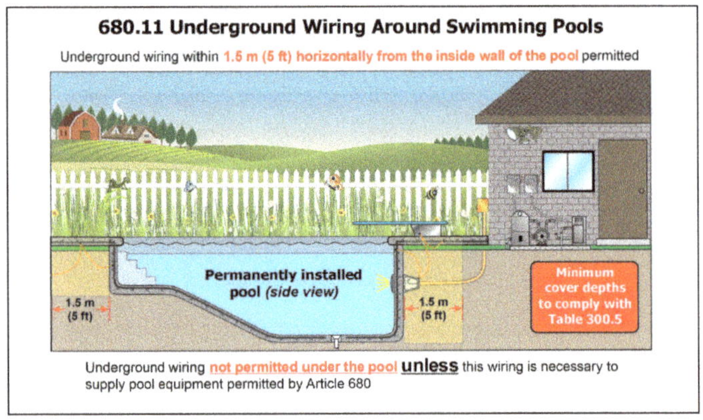

Figure 14.11. All underground wiring within 1.5 m (5 ft) of a pool must comply with *NEC* 680.11 or IRC E4203.8

Type MC cable, liquidtight flexible nonmetallic conduit, and liquidtight flexible metal conduit.

Underground wiring installed within 1.5 m (5 ft) of a pool must be installed in a complete system between outlets, junction boxes, or splice points [not simply sleeved for the portion within 1.5 m (5 ft) of the pool]. Underground wiring is not permitted to be installed under the pool unless this wiring is necessary to supply pool equipment. The minimum burial depth cover requirements for these pool wiring methods will be facilitated by *NEC* Table 300.5 or IRC Table E3803.1 [*NEC* 680.11(A) and (B) and IRC E4203.8].

ELECTRIC POOL WATER HEATERS

Electric pool water heaters incorporating resistive heating elements are required to be subdivided into loads not exceeding 48 amperes. This is a requirement for manufacturers of electric pool water heaters. Overcurrent protection must not be rated at more than 60 amperes.

The branch-circuit conductor ampacity and the rating or setting of the overcurrent protective devices cannot be less than 125 percent of the total load that is included on the water heater nameplate (*NEC* 680.10(A) or IRC E4206.12).

Electrically powered swimming pool heat pumps and chillers using the circulating water system and providing

Photo 14.4. Underground wiring around pools is required to comply with *NEC* Table 300.5(A) or IRC Table E3803.1

Photo 14.6. Gas-fired pool water heaters generally require GFCI protection

Photo 14.5. Electric pool water heaters incorporating resistive heating elements are required to be subdivided into loads not exceeding 48 amperes

Photo 14.7. Chlorine and other pool chemicals can severely deteriorate electrical equipment *(Photo courtesy of David Williams of Michigan)*

heating, cooling, or both must be listed and rated for their intended use. The ampacity of the branch-circuit conductors and the rating or setting of overcurrent protective devices must be sized to comply with the equipment nameplate (*NEC* 680.10(B) or IRC E4206.12).

GFCI protection is not required for electric water heater installations as the risk of electric shock is considered adequately mitigated through the use of proper grounding provisions and the listing installation requirement for the use of current collectors on the input and output side of the electric heater. These current collectors have proven reliable to prevent unacceptable levels of current in the pool in the event of corrosion of the immersed heating elements of an electric pool heater.

GAS-FIRED WATER HEATERS

Branch circuits or feeders that serve gas-fired heaters for permanently installed swimming pools and outdoor spas that will operate at voltages above the low-voltage contact limit are required to be provided with ground-fault circuit-interrupter protection (GFCI) for personnel. As stated earlier, GFCI protection is not required for electric water heater installations in part due to the use of current collectors on the input and output side of an electric heater. However, these current collectors are not present with a gas-fired swimming pool heater. The typical 125-volt branch circuit to a gas-fired water heater is susceptible to a loss of current and a ground-fault condition as much as any other piece of electrical equipment (*NEC* 680.28).

EQUIPMENT ROOMS AND PITS

If a room, vault, or pit is provided for the installation of electric equipment for pools and similar items, that area must be equipped with adequate drainage to prevent water accumulation during normal pool equipment operation or filter maintenance procedures, as described in NEC 680.12. Pool equipment installed in designated equipment rooms, vaults, or pits must be suitable for the type of environment in accordance with NEC 300.6, which addresses protection against corrosion and deterioration. Chemicals such as chlorine can cause severe corrosive and deteriorating effects on electrical connections, equipment, and enclosures when stored and kept in the same vicinity. Adequate ventilation of indoor spaces, such as equipment and storage rooms, can help reduce the likelihood of corrosive vapors accumulating. While indoor pool equipment rooms are uncommon for dwelling units, they can be found in larger dwelling units.

Provisions at *NEC* 680.12(B) call for at least one GFCI-protected 125-volt, 15- or 20-ampere receptacle on a general-purpose circuit to be located within an equipment room for pools, spas, fountains, etc. If more than one receptacle outlet is supplied or any other receptacles are supplied by branch circuits rated 150 volts or less to ground within a pool equipment room, GFCI protection will also be mandated for those receptacle outlets. Requiring at least one GFCI-protected receptacle within an equipment room lessens the potential hazard of employing an extension cord run from a receptacle outlet elsewhere that is not GFCI-protected when the need for electricity is needed while working on or servicing the equipment room where water is typically present on the floor during normal operation or maintenance.

CORROSIVE ENVIRONMENTS

Swimming pool and spa equipment are often subject to deteriorating chemicals such as chlorine, especially in rooms or pits. The equipment and wiring methods in these locations are subject to failure due to exposure to these corrosive chemicals. Chlorine and other pool chemicals can severely deteriorate the electrical connections of conductors, accelerating the rust and deterioration of the metal parts in electrical pool equipment. These chemicals are often stored, mixed, and dispensed in the same room or area as the pool's electrical equipment. A "Corrosive Environment" (as applied to a swimming pool, spa, or hot tub environment) is defined in NEC Article 100 and IRC E4201.2 as "areas or enclosures without adequate ventilation where electrical equipment is located and pool sanitation chemicals are stored, handled, or dispensed."

The *NEC* provides informational notes after this definition stating that sanitation chemicals and pool water are considered to pose a risk of corrosion *(gradual damage or destruction of materials)* due to the presence of oxidizers *(e.g., calcium hypochlorite, sodium hypochlorite, bromine, chlorinated isocyanurates)* and chlorinating agents that release chlorine when dissolved in water. These informational notes go on to indicate that adequate ventilation of indoor equipment rooms can reduce the likelihood of the accumulation of corrosive vapors. Chemicals such as chlorine cause severe corrosive and deteriorating effects on electrical connections, equipment, and enclosures when stored and kept in the same vicinity.

The wiring methods in these corrosive areas are required to be suitable for use in corrosive environments. One method the AHJ can depend on for suitability for use in corrosive environments is to demand that the wiring methods be listed and identified for use in such areas. Rigid metal conduit, intermediate metal conduit, rigid polyvinyl chloride conduit, reinforced thermosetting resin conduit, and liquidtight flexible nonmetallic conduit are wiring methods that are considered to be resistant to the corrosive environment as specified in *NEC* 680.14 and IRC E4202.2. IRC Table E4202.1 contains permitted wiring methods for corrosive environments along with locations and purposes for said wiring methods.

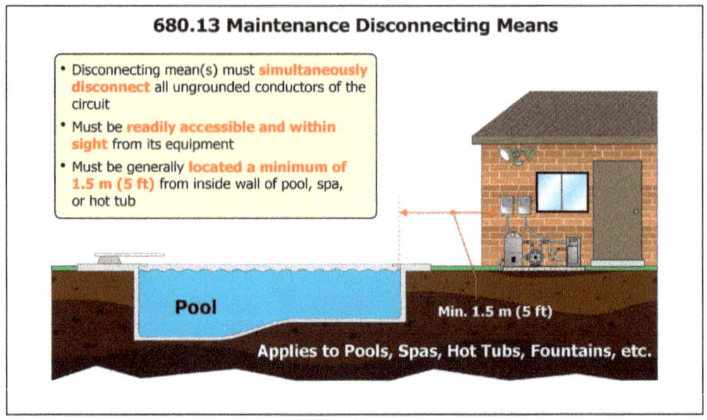

Figure 14.12. Equipment disconnecting means being readily accessible from and within sight of the equipment

Photo 14.8. Pool equipment disconnecting means. Note: sufficient working space is required at disconnect per *NEC* 110.26 or IRC E3405.1

MAINTENANCE DISCONNECTING MEANS

One or more disconnect switches or other means are required to simultaneously disconnect utilization equipment associated with bodies of water, such as pools, spas, hot tubs, fountains, and hydromassage bathtubs. This requirement does not apply to luminaires. The disconnecting means that each piece of equipment must be able to disconnect all ungrounded (hot) conductors. Each disconnecting means is required to be readily accessible, and it must be located within sight of the equipment it serves. This disconnecting means is typically used by individuals servicing the equipment. Service personnel may not have ready access to the branch-circuit protective devices, which are often located elsewhere on the premises (typically inside the dwelling unit). Readily accessible is

Photo 14.9. At least one receptacle outlet is required in the vicinity of a permanently installed swimming pool

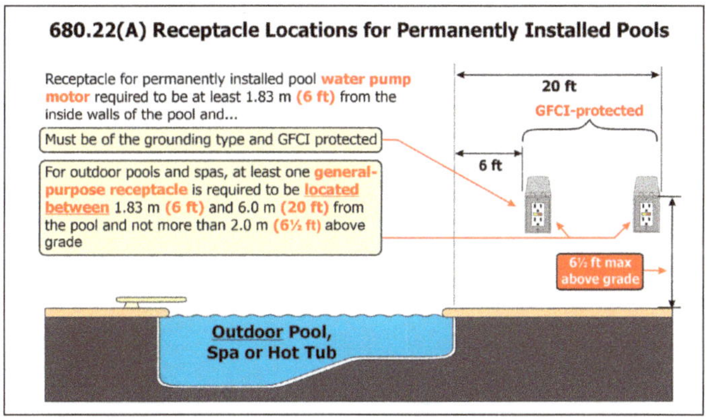

Figure 14.13. Receptacle outlet location requirements for permanently installed pools

defined in NEC Article 100 and IRC Chapter 35 as being "capable of being reached quickly for operation, renewal, or inspections without requiring those to whom ready access is requisite to take actions such as to use tools (other than keys), to climb over or under, to remove obstacles, or to resort to portable ladders, and so forth." These disconnecting mean(s) are required to be located at least 1.5 m (5 ft) horizontally from the inside walls of a pool, spa, or hot tub unless separated from the open water by a permanently installed barrier that provides at least a 1.5 m (5 ft) reach path. This disconnecting means is required by the Code for servicing the equipment and is not necessarily intended for emergency purposes (NEC 680.13 or IRC E4203.3).

Requirements in the *Code* for an emergency shutoff or control switch for the purpose of stopping the motor(s) that provide power to the recirculation system and jet system for spas and hot tubs are provided in *NEC* 680.41(A) and IRC E4208.4. These emergency shutoffs do not apply to one-family dwellings but would be required for two-family or multi-family dwellings. This emergency shutoff must be located not less than 1.5 m (5 ft), adjacent to and within sight of the spa or hot tub, and be readily accessible to users.

RECEPTACLE REQUIREMENTS AND LOCATIONS FOR PERMANENTLY INSTALLED SWIMMING POOLS

In this section, we will take a look at the requirements for receptacle outlets located around and associated with permanently installed pools (Part II of *NEC* Article 680). These receptacle outlet requirements are not applicable to other parts of *NEC* Article 680 (such as storable pools, spas, fountains, etc. unless specifically referenced in those other parts. Receptacles that provide power for circulation water-pump motors for permanently installed pools are required to be at least 1.83 m (6 ft) away from the inside walls of a pool. This circulation water-pump motor receptacle must be of a grounding type and protected by a ground-fault circuit interrupter (GFCI) [*NEC* 680.22(A)(2) or IRC E4203.1.1].

Where a flexible cord-and-attachment plug is used to connect a pump motor to this receptacle outlet installed as indicated above, the cord length is not permitted to exceed 900 mm (3 ft) in length; it must include a copper equipment grounding conductor not smaller than 12 AWG; and be provided with a grounding-type attachment plug (*NEC* 680.8 or IRC E4202.3.1). This outlet and all outlets supplying pool pump motors that are supplied on branch circuits rated 150 volts or less to the ground and 60 amperes or less, single-phase and 3-phase, would be required to be GFCI protected. Notice the word "outlet" is used here, meaning that this GFCI provision would apply to a pool pump motor that is cord-and plug-connected or hard-wired [*NEC* 680.21(A)(2), 680.21(C) or IRC E4203.1.4].

As stated earlier, an exception allows listed low-voltage motors not requiring grounding, with ratings not exceeding the low-voltage contact limit that is supplied by listed transformers or power supplies, to be installed without GFCI protection [NEC 680.21(C) Exception].

For permanently installed pools, at least one 125-volt, 15- or 20-ampere receptacle must be installed in the vicinity of the pool. This receptacle must be GFCI protected on a general-purpose branch circuit and must be located not closer than 1.83 m (6 ft) and not farther than 6.0 m (20 ft) from the inside wall of the pool. This required receptacle must be located not more than 2.0 m (6½ ft) above the same floor, platform, or grade on which the pool is installed. All other receptacles must be located not less than 1.83 m (6 ft) from the inside walls of a pool. This required receptacle applies whether the permanently installed pool is constructed at either a new or existing structure or dwelling unit(s). All 15- and 20-ampere, 125-volt receptacles located within 6.0 m (20 ft) of the inside walls of a pool must be protected by a ground-fault circuit interrupter [*NEC* 680.22(A)(1), (A)(3), and (A)(4) or IRC E4203.1.2, E4203.1.3, and E4203.1.4].

Other outlets, such as for remote control, signaling, fire alarm, and communications, are to be located not less than 3.0 mm (10 ft) from the inside walls of the pool. [*NEC* 680.22(A)(3) or IRC E4203.5].

As stated earlier, at least one GFCI-protected 125-volt, 15- or 20-ampere receptacle on a general-purpose circuit is required to be located within an equipment room associated with pools, spas, fountains, etc. If more than one receptacle outlet is supplied or any other receptacles are supplied by branch circuits rated 150 volts or less to ground within an equipment room, GFCI protection will also be required [*NEC* 680.12(B)].

It is important to clarify that where receptacle locations involve taking measurements, the distances are required to be the shortest path that a supply cord of an appliance that is connected to the receptacle would follow without penetrating any floor, wall, ceiling, doorway with hinged or sliding door, window opening, or other effective permanent barrier [*NEC* 680.22(A)(5) or IRC E4203.1].

LUMINAIRES, LIGHTING OUTLETS AND CEILING FANS

New Outdoor Installations

Generally, luminaires, lighting outlets, and ceiling-suspended (paddle) fans cannot be installed over outdoor pools or over any area extending 1.5 m (5 ft) horizontally from the inside walls of a pool unless they are located 3.7 m (12 ft) or more above the maximum water level [*NEC* 680.22(B)(1) or IRC E4203.4.1].

GFCI Protection in Adjacent Areas

Where luminaires and lighting outlets are installed in the area extending between 1.5 m (5 ft) and 3.0 m (10 ft) horizontally from the inside walls of a pool, they must be protected by a ground-fault circuit interrupter (GFCI). GFCI protection is not required where they are installed, at least 1.5 m (5 ft) away, 1.5 m (5 ft) above the maximum water level, and are rigidly attached to the structure adjacent to or enclosing the pool [*NEC* 680.22(B)(4) or IRC E4203.4.6].

Existing Installations

Existing luminaires and lighting outlets that are located less than 1.5 m (5 ft) measured horizontally from the inside walls of a pool must be installed so they are at least 1.5 m (5 ft) above the surface of the maximum water level. Such luminaires must be rigidly attached to the existing structure, and GFCI protected [*NEC* 680.22(B)(3) or IRC E4203.4.4].

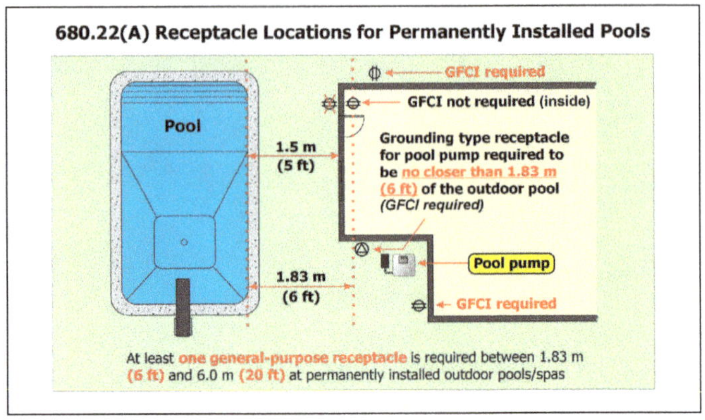

Figure 14.14. Receptacle outlet for permanently installed pool water-pump motor is permitted not less than 1.83 m (6 ft) from water's edge (GFCI protection required)

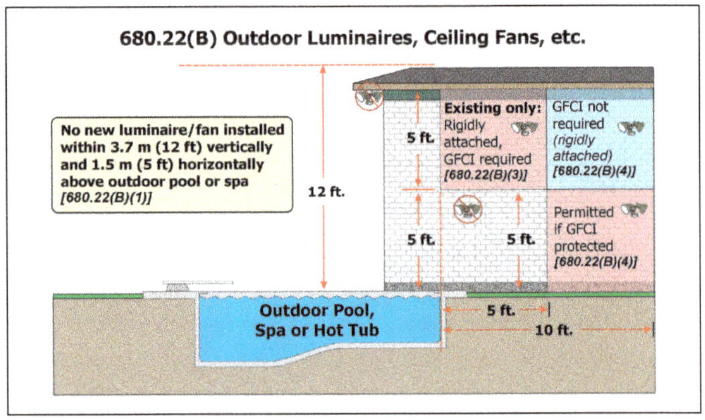

Figure 14.15. Luminaire and fan locations for outdoor pool, spa, or hot tub

Cord-Connected Luminaires

Any cord-connected luminaire installed within 4.9 m (16 ft), measured radially, from any point on the pool water surface, must meet the same requirements stipulated for other types of cord- and plug-connected equipment associated with pools, spas, and hot tubs. This cord-and-plug connection facilitates the removal or disconnection of the luminaire for maintenance or repair. For other than storable-type pools, this flexible cord cannot be longer than 900 mm (3 ft) in length. This cord must also contain a copper equipment grounding conductor, not smaller than 12 AWG, and terminate in a grounding-type attachment plug [NEC 680.22(B)(5), 680.8, or IRC E4202.3].

Indoor Installations

The *Code* permits certain new luminaires or ceiling-suspended paddle fan(s) to be installed closer to an indoor pool than to an outdoor pool, but specific conditions must be met. Luminaires must be of a totally enclosed type. The ceiling-suspended (paddle) fans are required to be identified for use beneath ceiling structures such as porches or patios (possibly

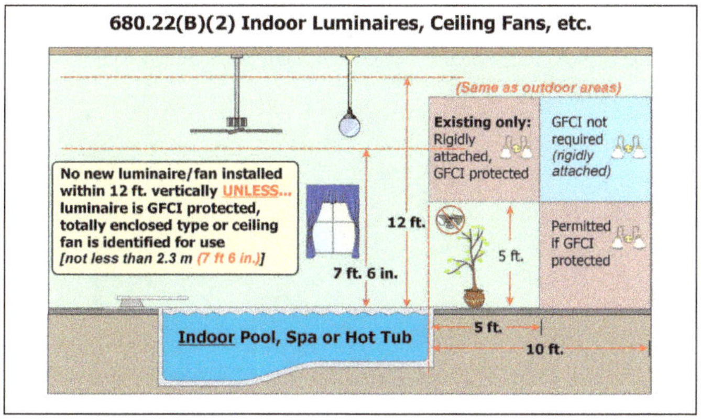

Figure 14.16. Luminaire and fan locations for outdoor pool, spa, or hot tub

Photo 14.10. Certain low-voltage lighting systems are permitted within 1.5 m (5 ft) of pools, spas, etc.

damp locations). The distance from the bottom of the luminaire or ceiling-suspended (paddle) fan to the maximum water level cannot be less than 2.3 m (7 ft 6 in.). The following additional condition must also be met: a ground-fault circuit interrupter is required to be installed in the branch circuit supplying the luminaire(s) or ceiling-suspended (paddle) fan(s) [*NEC* 680.22(B)(2) or IRC E4203.4.2].

Low-Voltage Luminaires

Some specific low-voltage luminaires are permitted to be located less than 1.5 m (5 ft) from the inside walls of a permanently installed pool under certain conditions. These low-voltage luminaires must be of the type that does not require connection to an equipment grounding conductor. They cannot exceed the voltage limitations defined in the definition of "Low Voltage Contact Limit" at *NEC* Article 100 or IRC E4201.2. These luminaires must also be supplied by listed transformers or power supplies that comply with *NEC* 680.23(A)(2) or IRC E4206.1 for transformers or power supplies listed, labeled, and identified for swimming pool and spa use [*NEC* 680.22(B)(6) or IRC E4203.4.3].

NEC Article 411 for "Low-Voltage Lighting" also allows low-voltage lighting systems to be installed in close proximity to swimming pools if "permitted by *NEC* Article 680" [*NEC* 411.6(B)]

Listed low-voltage gas-fired luminaires, decorative fireplaces, fire pits, and similar equipment using low-voltage ignitors that do not require grounding and are supplied by listed transformers or power supplies with outputs that do not exceed the low-voltage contact limit are also permitted to be located less than 1.5 m (5 ft) from the inside walls of the pool. This will be discussed in greater detail later in this chapter (see Photo 14.25).

UNDERWATER LUMINAIRES

Underwater lighting is permitted by the *Code* to be supplied either directly by the branch circuit or through a suitable, listed isolated-winding transformer or power supply meeting applicable provisions as indicated below.

Where installed below the normal water level of the permanently installed pool, underwater luminaires must meet the following requirements:

Supplied through a Transformer

Transformers and power supplies used for the supply of underwater luminaires, together with the transformer or power supply enclosure, are required to be listed, labeled, and identified for swimming pool and spa use. The transformer or power supply must incorporate either a transformer of the isolated-winding type having an ungrounded secondary and grounded metal barrier between the primary and secondary windings or a transformer that incorporates an approved system of double insulation between the primary and secondary windings.

These underwater luminaire(s) must also be designed such that when the luminaire is properly installed per *Code* but without a ground-fault circuit interrupter, there is no shock hazard with any likely combination of fault conditions during normal use. This does not include relamping procedures [*NEC* 680.23(A)(1) and (A)(2) or IRC E4206.1 and E4206.4].

Supplied by a Branch Circuit

When supplying underwater luminaires operating at voltages greater than the low-voltage contact limit, the Code specifies that a ground-fault circuit interrupter must be installed in the branch circuit. This precaution will ensure that there is no shock hazard during lamping, relamping, or servicing. The origins of ground-fault circuit interrupter protection for the *NEC* can be traced back to this original GFCI requirement for underwater lighting at swimming pools. The first GFCI requirement in the entire *NEC* was introduced in the 1968 edition of the *NEC* at 680-4(g)(2), with a definition for a "ground-fault circuit interrupter" immediately following.

The ground-fault circuit interrupter will protect against a shock hazard with any likely fault-condition combination that involves a person in a conductive path from any ungrounded part of the branch circuit or the luminaire to ground. Compliance with this requirement can only be obtained by the use of a listed underwater luminaire and by installation of a listed ground-fault circuit interrupter in the branch circuit or a listed transformer or power supply for luminaires operating at not more than the low voltage contact limit. [*NEC* 680.23(A)(3) and (A)(8) or IRC E4206.4].

The low voltage contact limit for swimming pool environment luminaires is defined as a voltage not exceeding 15 volts (RMS) for sinusoidal AC, 21.2 volts peak for nonsinusoidal AC, 30 volts for continuous DC, and 12.4 volts peak for DC that is interrupted at a rate of 10 to 200 Hz [*NEC* Article 100 or Chapter 35}.

No underwater luminaires are permitted to be installed where they operate on supply circuits over 150 volts between conductors. This would not prohibit the use of luminaires such as electric-discharge luminaires where the supply circuit voltage does not exceed 150 volts between conductors on the supply side, but the luminaire operating voltage is higher than 150 volts on the output side [*NEC* 680.23(A)(4) or IRC E4206.4.1].

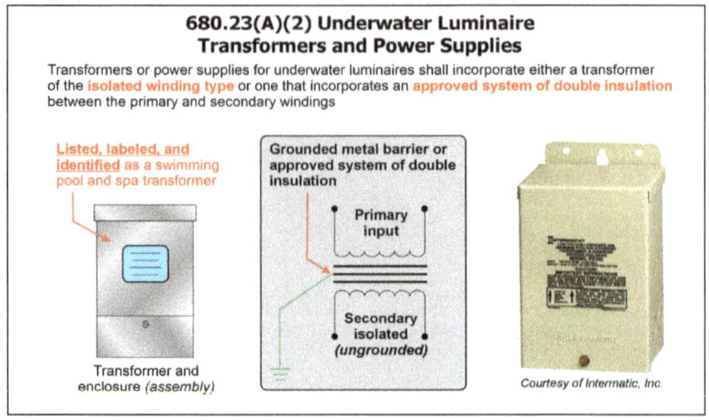

Figure 14.17. Isolated-winding transformers to be listed as swimming pool and spa transformers with ungrounded secondary

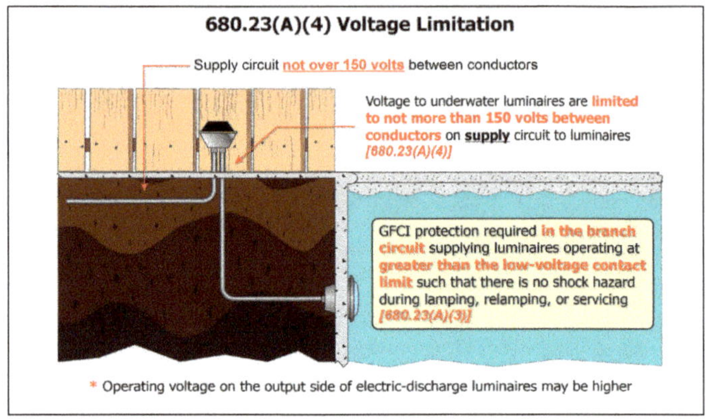

Figure 14.18. Underwater luminaire voltage limitations: The junction box for wet-niche locations is to be listed as a swimming pool junction box.

UNDERWATER LUMINAIRE MOUNTING REQUIREMENTS

Where mounted in pool walls, underwater luminaires must generally be installed so the top of the luminaire lens is at least 450 mm (18 in.) below the normal water level of the pool. Lesser depths are permitted for luminaires specifically listed and identified for use at a depth of not less than 100 mm (4 in.) below the normal water level of the pool [*NEC* 680.23(A)(5) or IRC E4206.4.2]. Where a luminaire is installed facing in an upward position, it must have its lens adequately guarded to prevent contact by any person, or it must be listed for use without a guard. This guarding may be by some identified wire grill or mesh [*NEC* 680.23(A)(6) or IRC E4206.4.2].

Underwater luminaires that depend on submersion for their safe operation must have protection against overheating built into the luminaire. This precaution protects against the hazards of overheating when they are not submerged. Compliance with the preceding requirements is obtained by using listed underwater luminaires, carefully following the installation instructions, and by installation of a GFCI protective device in the branch circuit or a listed swimming pool and spa transformer or power supply that limits the operating voltage to the low-voltage contact limits [*NEC* 680.23(A)(7) and (A)(8) or IRC E4206.4]. For additional details, see the information found in the product safety standards noted at the end of this chapter.

WET-NICHE LUMINAIRES

By far, the most common underwater luminaire employed at dwelling unit swimming pools is the wet-niche luminaire. By its very definition (see *NEC* Article 100 and Chapter 35), a wet-niche luminaire requires a forming shell to be installed for the mounting of all wet-niche underwater luminaires. Where mounted in a permanently installed pool or fountain structure, a wet-niche luminaire must be installed in the forming shell so that the luminaire will be surrounded by water. The forming shell must also be equipped with provisions for conduit entries. Where in contact with the pool water, metal parts of the luminaire and forming shell must also be of brass or other approved corrosion-resistant metal. The forming shell used with nonmetallic conduit systems must have provisions for terminating an 8 AWG copper

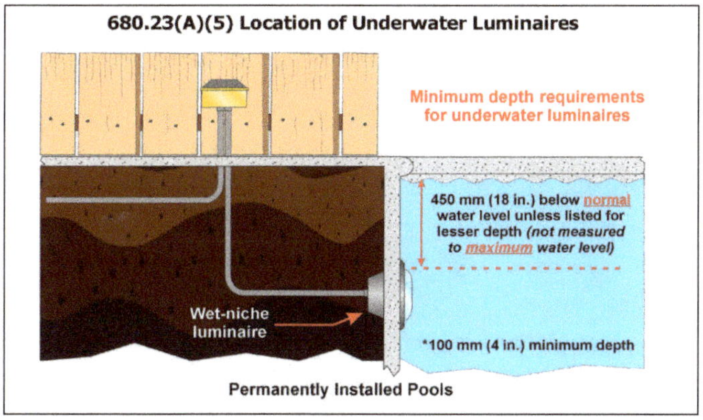

Figure 14.19. Underwater luminaire water depth and listing requirements

Photo 14.11. Underwater luminaires generally required to be installed 450 mm (18 in.) below the normal water level

Photo 14.12. Wet-niche luminaire removed from the forming shell (rough stages)

bonding jumper. This requirement would not apply to listed low-voltage systems that are not required to be grounded [*NEC* 680.23(B)(1) or IRC E4206.5].

The conduit must extend from the forming shell to a listed swimming pool junction box or enclosure (deck box). The junction box is required to be located not less than 100 mm (4 in.) above the ground level or pool deck or not less than 200 mm (8 in.) above the maximum water level, whichever provides the greatest elevation. The measurement is taken to the inside bottom of the box. The junction or deck box must also be located so it is not less than 1.2 m (4 ft) from the inside wall of the pool unless separated from the pool by a solid fence, wall, or other permanent barrier. A flush mount deck box may be used with lighting systems that operate at the low-voltage contact limit or less. Conduit installed from the forming shell to the junction box or other enclosure must be rigid metal, intermediate metal, liquidtight flexible nonmetallic, or rigid polyvinyl chloride conduit (PVC). Rigid metal or intermediate metal conduit must be made of brass or other approved corrosion-resistant metal, such as stainless steel [*NEC* 680.23(B)(2); 680.24(A)(2); or IRC E4206.5, E4206.9].

Where a nonmetallic conduit [liquidtight flexible nonmetallic or rigid polyvinyl chloride conduit (PVC)] is used, an 8 AWG insulated copper bonding jumper is required to be installed along with the luminaire's flexible cord assembly so it can be terminated on a suitable lug in the forming shell, and in the junction box, transformer enclosure, or ground-fault circuit interrupter enclosure. The 8 AWG insulated copper bonding jumper is not required to be used with a listed low-voltage lighting system that does not require grounding. Check the manufacturer's instructions for details on installation requirements [*NEC* 680.23(B)(2)(b) or IRC E4205.5].

The 8 AWG bonding jumper must be encapsulated with a listed potting compound at the point of termination within the forming shell. This compound provides protection from corrosion and other deteriorating effects caused by the pool water. For other than listed low-voltage lighting systems not requiring grounding, wet-niche luminaires supplied by a flexible cord or cable assemblies must have all their exposed non-current-carrying metal parts grounded by an insulated copper equipment grounding conductor that is an integral part of the cord or cable. This equipment grounding conductor must be connected to an equipment-grounding terminal in the supply junction box, transformer enclosure, or other enclosure. Also, this equipment grounding conductor for cords cannot be smaller than the supply conductors and never smaller than 16 AWG [*NEC* 680.23(B)(3) and (B)(4) or IRC E4206.5].

The end of the flexible-cord jacket and the flexible-cord conductor terminations located within a wet-niche luminaire are required to be covered with, or encapsulated in, a suitable potting compound. This will help to prevent the siphoning of water from the forming shell through the cord jacket or its contained conductors back to the junction box. The manufacturer typically encapsulates this termination at the time the cord is attached to the luminaire assembly.

The luminaire must be bonded to and secured to the forming shell by a positive locking device that will ensure a low-resistance contact. A special tool is typically required in order to remove the luminaire from the forming shell.

Bonding is not required for certain types of luminaires listed for the application. Such luminaires have no metal parts other than those, such as wires or lamp filaments, that are specifically intended to carry current [NEC 680.23(B)(5) or IRC E4206.5].

Underwater wet-niche luminaires include markings that will indicate the proper housing(s) with which they are to be used. Also, the luminaire housings are marked to indicate the appropriate luminaires with which the housings are to be used.

Luminaires must be removable from the underwater forming shell for relamping or normal maintenance. The luminaire maintenance location must be accessible by maintenance personnel without their entering or going into the pool water to change the bulb or perform maintenance. This can typically be achieved by installing a specified length of flexible cord inside the forming shell to allow the luminaire to be brought to the surface and placed on the deck or other dry location while the relamping occurs [*NEC* 680.23(B)(6) and IRC E4206.5.1].

Provisions were added to the 2020 *NEC* and the 2021 IRC for spas that can be drained so the luminaire(s) can be placed on the spa bench area for servicing without having to take the luminaire all the way to the deck (see Photo 14.16). For servicing wet-niche luminaires in spa locations with luminaires installed low in the foot well of the spa, the luminaire is only required to reach the bench location, where the spa can be drained to make the bench location dry for maintenance and repair. As stated earlier, wet-niche luminaires installed in permanently installed swimming pools are typically required to be installed in such a manner that they can be completely removable from the water to the deck area for inspection, relamping, or other maintenance. The top of bench areas of a spa that can be drained below the bench area serves the same function as the deck of a pool, with no need to take the spa luminaire all the way to the deck in order to change a light bulb when a dry bench can serve the same purpose.

Listed wet-niche luminaires are typically provided with a factory-installed, permanently attached flexible cord that extends at least 3.7 m (12 ft) outside the luminaire enclosure. This will permit the removal of the luminaire from the forming shell so it can be lifted onto the pool or spa deck, or bench area for servicing without lowering the water level or disconnecting it from the branch-circuit conductors.

Luminaires with longer cords are available for installation where the deck or junction box or enclosure is located such that a 3.7 m (12 ft) length of cord will not permit the luminaire to be removed from the forming shell and placed on the deck for servicing.

To avoid possible cord damage, any cord length in excess of that necessary for servicing should be trimmed from the supply end (junction box) rather than being stored in the forming shell. See also the UL Product Safety Standards Information and the UL Product Spec (formally UL White Book) under the category Luminaires and Forming Shells (WBDT).

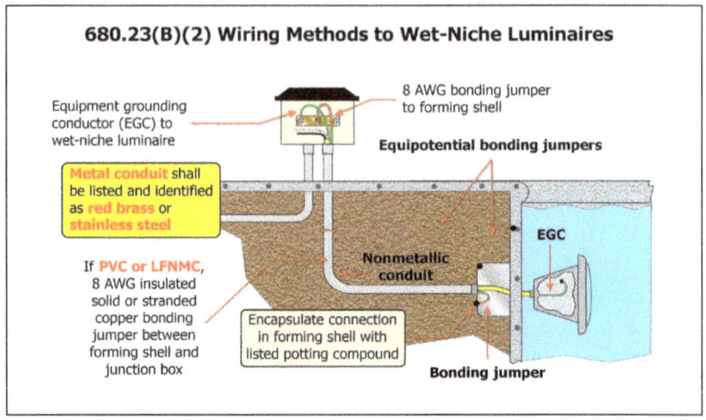

Figure 14.20. Wiring methods and bonding requirements for wet-niche luminaires

Photo 14.13. Red brass rigid metal conduit (shown) permitted as a wiring method to an underwater luminaire considered to be a corrosive environment

Photo 14.14. Equipotential bonding grid bonding jumper to nonmetallic wet-niche forming shell

DRY-NICHE LUMINAIRES

Even though the majority of the underwater luminaires at dwelling-unit permanently installed pools will be wet-niche

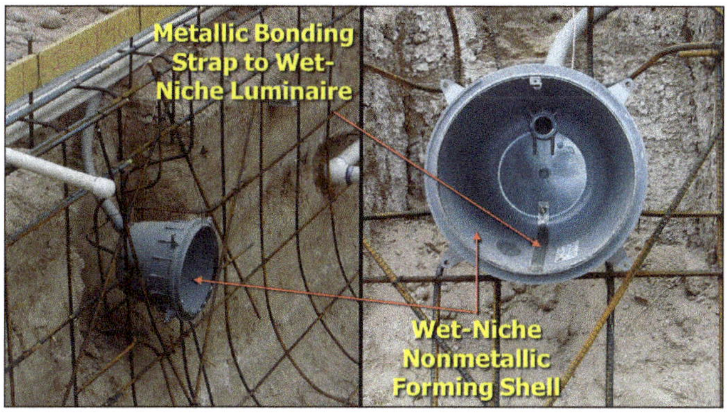

Photo 14.15. Nonmetallic wet-niche forming shell with metallic bonding strip to future metal wet-niche luminaire

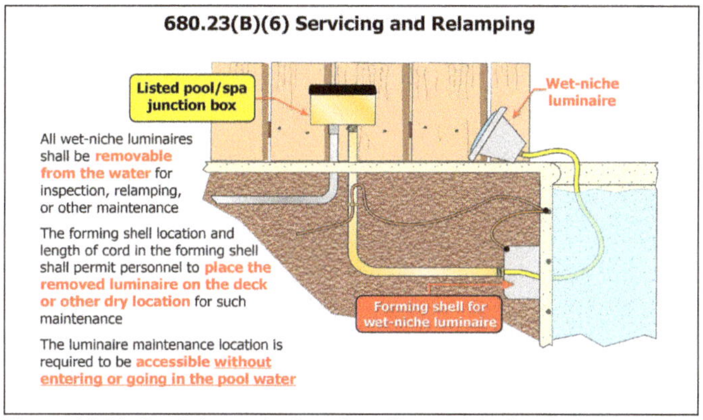

Figure 14.21. Servicing and relamping of wet-niche luminaires

Photo 14.16. Luminaires in spas are only required to be accessible from the bench area for servicing

types, there are other types available. One such type is a dry-niche luminaire. A dry-niche luminaire is intended to be installed on the floor or wall of a pool, spa, or fountain in a niche that is sealed against the entry of pool water. A dry-niche luminaire is required to be provided with a provision for drainage of water. Other than listed low-voltage luminaires not requiring grounding, it must also have a means for accommodating one equipment grounding conductor for each conduit entry.

A specific junction box is not required for dry-niche luminaires. Where a junction box is used, it does not have to be elevated or located above the deck of the pool where the luminaire is specifically identified for the purpose [*NEC* 680.23(C)(1) and (2) or IRC E4206.6].

NO-NICHE LUMINAIRES

Another type of underwater luminaire is a no-niche luminaire. A no-niche luminaire is intended to be installed above or below the water without a niche. These types of luminaires must be listed for the purpose and installed according to the requirements for wet-niche luminaires.

Underwater luminaires of the no-niche type are intended to be installed on the walls of built-in, below-ground spas or swimming pools. These luminaires are required to be installed with the top of the lens not less than 450 mm (18 in.) below the normal water level unless otherwise listed and marked. These luminaires are designed to be mounted to a bracket that is permanently secured in or on the wall where the luminaire will be completely surrounded by water and are provided with a factory-installed, permanently attached flexible cord that extends at least 3.7 m (12 ft) outside the luminaire enclosure.

The flexible cord permits the luminaire to be removed from the mounting bracket and lifted to the pool or spa deck for servicing without lowering the water level or disconnecting the luminaire from the branch-circuit conductors. Luminaires with longer cords are available for installations where the junction box or splice enclosure is so located that a 3.7 m (12 ft) long cord will not permit luminaire removal from the mounting bracket and placement on the deck for servicing.

To avoid possible cord damage, any cord length in excess of that necessary for servicing should be trimmed from the supply end (junction box) rather than stored between the luminaire and wall [*NEC* 680.23(D) or IRC E4206.7]. See also the UL Product Safety Standards Information and the UL Product Spec (formally UL White Book) under the category Luminaires and Forming Shells (WBDT).

THROUGH-WALL LIGHTING ASSEMBLIES

The term *through-wall lighting assembly* is defined in *NEC* Article 100 and Chapter 35 as "A lighting assembly intended for installation above grade, on or through the wall of a pool, consisting of two interconnected groups of components separated by the pool wall." These assemblies are required to be equipped with a threaded entry or hub or a nonmetallic hub for the purpose of accommodating the termination of the supply conduit. A through-wall lighting assembly must include an insulated copper equipment grounding conductor and comply with the construction requirements of *NEC* 680.23(B)(3) or IRC E4205.4. It must also meet all the requirements for underwater luminaires as described at *NEC* 680.23 or IRC E4205. Where connection to a forming shell is specified, the connection must be to the conduit termination point. Even though these rules for through-wall lighting assembly type underwater luminaires are found in Part II (Permanently Installed Pools) of *NEC* Article 680, these assemblies are extremely rare for a permanently installed pool. These assemblies are typically found in association with a storable pool, which we will cover later in this chapter.

BRANCH-CIRCUIT WIRING METHODS FOR UNDERWATER LUMINAIRES

Branch-circuit wiring on the supply side of deck or junction boxes and enclosures and connected to conduits run to underwater wet-niche, dry-niche, or no-niche luminaires installed in corrosive environments (as defined at *NEC* Article 100 or IRC Chapter 35) are required to be listed and identified for use in such areas and comply with *NEC* 680.14 or IRC E4202.2. Rigid metal conduit, intermediate metal conduit, liquidtight flexible nonmetallic conduit, reinforced thermosetting resin conduit, or rigid polyvinyl chloride conduit (PVC) are wiring methods considered suitable for use in corrosive environments.

Where installed in a noncorrosive environment, such as on or within an attic of a dwelling unit being routed to the pool area, any *NEC* Chapter 3 or IRC Chapter 38 wiring method identified for such use can be used. In noncorrosive environments, wiring methods such as electrical metallic tubing are permitted to be used to protect conductors where they are installed on or within a building. A branch circuit or feeder run inside a dwelling unit in a noncorrosive location could include such wiring methods as nonmetallic-sheathed cable, electrical nonmetallic tubing (ENT), Type MC cable, or Type AC cable.

Liquidtight flexible metal conduit or liquidtight flexible nonmetallic conduit can be used to connect to power supplies or transformers supplying pool luminaires. The total length is limited to 3.0 m (10 ft), and any one length is limited to 1.8 m (6 ft). Liquidtight flexible nonmetallic conduit, Type B (LFNC-B), is permitted in lengths longer than 1.8 m (6 ft) [*NEC* 680.23(F)(1), Exception and 680.14 or IRC Table E4202.1].

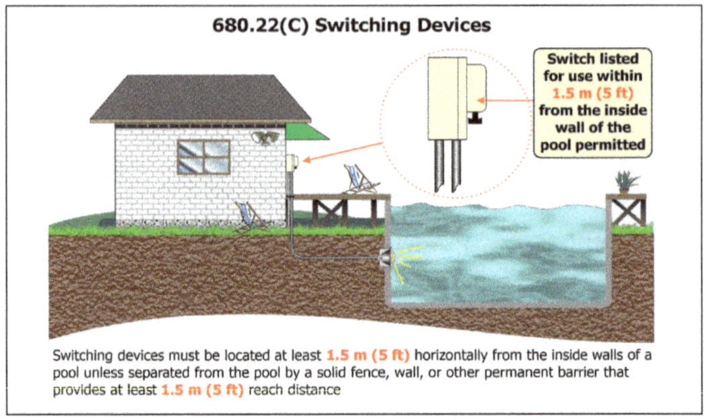

Figure 14.22. Switching devices are generally required to be installed at least 1.5 m (5 ft) from the inside walls of a pool

EQUIPMENT GROUNDING FOR UNDERWATER LUMINAIRES

Through-wall lighting assemblies, wet-niche, dry-niche, or no-niche luminaires must be connected to an insulated copper equipment grounding conductor. This equipment grounding conductor requirement does not apply to listed low-voltage luminaires, which do not require grounding provisions. This equipment grounding conductor is required to be run with the branch-circuit conductors and sized based on the rating of the circuit overcurrent protective device.

This equipment grounding conductor can never be smaller than 12 AWG. An exception applies to the equipment grounding conductor on the secondary side of a transformer. The equipment grounding conductor installed between the wiring chamber of the secondary winding of a transformer and a junction box must be sized in accordance with the transformer's secondary overcurrent device in this circuit. See *NEC* Table 250.122 or IRC Table E3908.13 for the minimum size of the equipment grounding conductor.

Equipment grounding conductors used in conjunction with underwater luminaire circuits are generally to be installed without splice or joint. Two conditions exist to permit such splice or joint. One condition permits the equipment grounding conductor to be spliced on grounding terminals where the same branch circuit supplies more than one luminaire. The second condition permits the equipment grounding conductor to be spliced on grounding terminals located on equipment in the same circuit as the underwater luminaire.

For example, an underwater luminaire is protected by a ground-fault circuit interrupter and controlled by a clock-operated switch. The equipment grounding conductor is permitted to terminate on equipment-grounding terminals on the ground-fault circuit interrupter and the clock-operated switch enclosure, as well as on the deck box that connects to the conduit that runs to the luminaire equipment grounding conductor [*NEC* 680.23(F)(2)(a) and (b) or IRC E4205.4 Exceptions 1 and 2].

SWITCHING DEVICES

Switching devices of any type (including maintenance

Photo 14.17. Listed swimming pool and spa junction boxes

disconnects) must be located at least 1.5 m (5 ft) from the inside walls of a pool. Lesser distances are permitted where switching devices are separated from the pool by a permanent barrier, such as a solid fence or wall that provides at least a 1.5 m (5 ft) reach distance. This rule applies to equipment disconnects, circuit breakers, and fusible switches, as well as to switches installed for other purposes, such as controlling a luminaire. The *Code* recognizes that switches listed for use within the 1.5 m (5 ft) limitation are permitted. An example would be a factory-installed switch that is part of a listed spa assembly [*NEC* 680.22(C) or IRC E4203.2].

JUNCTION BOXES AND ELECTRICAL ENCLOSURES FOR TRANSFORMERS OR GFCIS

Where a conduit is connected to a junction box or enclosure used for a transformer, ground-fault circuit interrupter, or a similar device that extends directly to an underwater forming shell, the box or enclosure is required to meet several requirements in the *Code*. First, it must be listed, labeled, and identified as a swimming pool or spa junction box (deck box). The junction or deck box must be equipped with threaded entries or hubs or a nonmetallic hub. The box must be made of copper, brass, suitable plastic, or other approved corrosion-resistant material. It is also required to be manufactured such that electrical continuity is maintained between every connected metal conduit and the equipment-grounding terminals by means of copper, brass, or other approved corrosion-resistant metal that is integral to the box [*NEC* 680.24(A)(1) or IRC E4206.9.1].

JUNCTION BOX LOCATION(S)

Where the luminaire operates over the low voltage contact limit, the junction boxes are required to be located not less than 100 mm (4 in.) above the ground level or pool deck and not less than 200 mm (8 in.) above the maximum pool water level, whichever provides the greatest elevation. The measurement is to be taken from the inside bottom of the junction (deck) box. The junction or deck box is also required to be located at least 1.2 m (4 ft) from the inside wall of the pool. Less than 1.2 m (4 ft) of separation is permitted where the deck or junction box is separated from the pool by a solid fence, wall, or other permanent barrier.

The *Code* permits a flush deck box to be used for lighting systems operating at the low-voltage contact limit or less. An approved potting compound must be used to fill the box to prevent the entrance of moisture. The flush deck box must be located at not less than 1.2 m (4 ft) from the inside wall of the pool. Where the underwater luminaires operate at the low voltage contact limit or less, the junction box location is permitted to comply with these flush deck box requirements. Potting compounds specifically designed to be used in these junction boxes are available from the junction box manufacturers. These potting compounds do not harden solidly, which permits the potting compound to be removed when the connections require servicing [*NEC* 680.24(A)(2) or IRC E4206.9].

Where junction boxes and enclosures are installed above the grade of the finished deck around the pool, they cannot be located in the walkway or deck unless additional protection from physical damage is provided. For example, protection may be accomplished by locating the junction boxes and enclosures under diving boards or adjacent to fixed structures such as a wooden fence. [*NEC* 680.24(C) or IRC E4206.9.3]

Junction boxes, transformer enclosures, and ground-fault circuit-interrupter enclosures that are connected to a conduit extending directly to a forming shell or mounting bracket of a no-niche luminaire must be provided with a number of equipment grounding conductor terminals that will provide for at least one more grounding terminal than the number of conduit entries [*NEC* 680.24(D) or IRC E4206.9.4].

Where the flexible cord of an underwater luminaire terminates within a junction box, transformer or power supply enclosure, ground-fault circuit interrupter, or other enclosure, it must be provided with a strain relief [*NEC* 680.24(E) or IRC E4206.9.5].

The junction box, transformer enclosure, or other enclosure in the supply circuit that is run to a wet-niche or no-niche luminaire and the field wiring chamber of a dry-niche luminaire must be grounded by means of the equipment-grounding terminal of the panelboard. This terminal must be directly connected to the panelboard enclosure [*NEC* 680.24(F) or IRC E4205.6].

EQUIPOTENTIAL BONDING REQUIREMENTS

Equipotential bonding requirements are an important and

unique protective method employed to increase the safety of the users of bodies of water such as pools, spas, and hot tubs. Bonding connects together the conductive elements of the pool structure, nearby metallic objects, and electrical equipment enclosures. Bonding is required to eliminate voltage gradients (rises) in the pool area. This is simply an attempt to keep all these conductive elements at the same voltage potential. When metallic parts are bonded together, they eliminate differences in voltage potential that may exist between the individual conductive parts and thus reduce the shock hazard. Equipotential bonding is required for pools with or without associated electrical equipment related to the pool.

It should be clearly understood that if used, the 8 AWG or larger solid copper bonding conductor discussed below is not required to be extended or attached to any remote panelboard, service equipment, or any grounding electrode. The sole purpose of this bonding conductor is to eliminate any voltage gradients or differences in voltage potential with metal parts and equipment in and around the pool area. The purpose here is not to "ground" the pool structure. How much more grounded can you get than this huge concrete and metal structure buried in the earth with thousands of pounds of water in it? This bonding (connected together) is referred to as equipotential bonding throughout *NEC* Article 680 and IRC Chapter 42 [*NEC* 680.26(A) and (B) or IRC E4204.1, E4204.2].

This is a good point for exploring the different purposes of bonding and grounding and developing a clear understanding of the two terms. Bonding *(connected together)* serves to equalize voltage potential differences in order to reduce shock hazards between different parts of the pool by connecting potentially conductive parts together. Grounding *(connected to earth)* involves providing a low-impedance, ground-fault return path from the equipment required to be grounded back to the source of the electrical system, typically through the equipment grounding conductor routed with the branch-circuit conductors. This path facilitates the operation of the fuse or circuit breaker to allow it to remove the faulted condition. In short, bonding is connecting together, while grounding is connecting to earth. Grounding and bonding work together to provide safety from shock hazards, which is so important. It should be clearly understood that grounding is not the function of this equipotential bonding described at *NEC* 680.26 or IRC E4204.

Due to the fact that several types of electrical equipment commonly used with pools, hot tubs, or spas that are bonded together must also be grounded, an interconnection between the grounded (neutral) conductor and bonding grid exists. This interconnection, usually at the electrical service equipment, is often remote and is not intended to play a part in equipotential bonding.

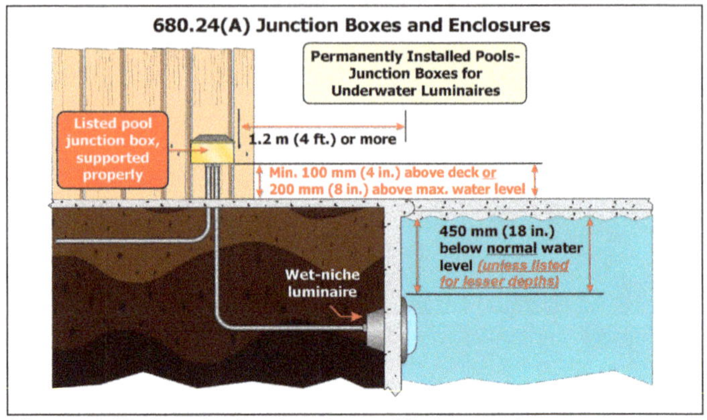

Figure 14.23. Junction box location and listing requirements

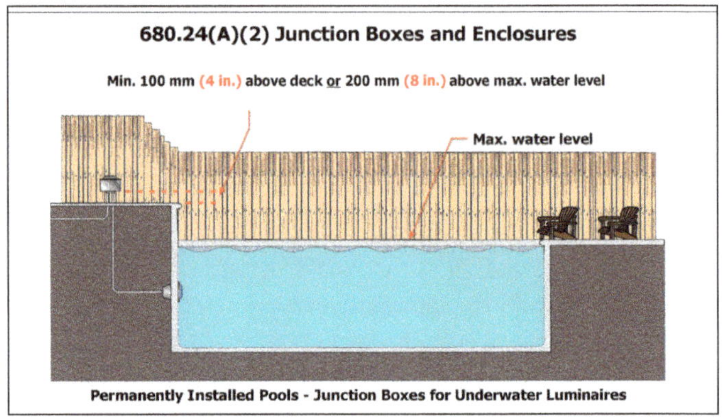

Figure 14.24. Junction box location at split-level pool

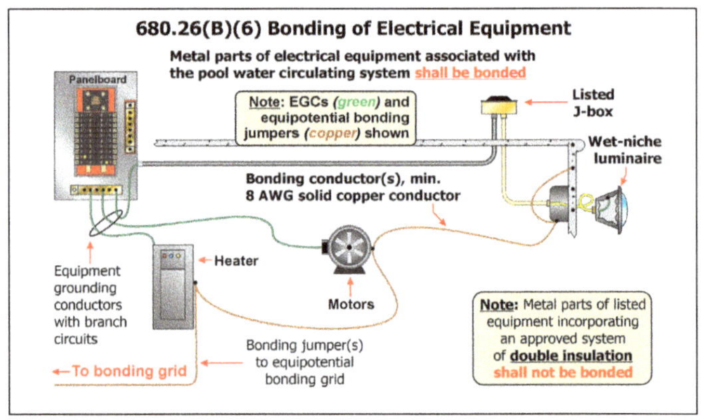

Figure 14.25. Bonding and grounding of pool-associated components and equipment.

The use of the structures and structural reinforcing steel of an in-ground swimming pool as described in *NEC* 680.26(B)(1) and (B)(2) or IRC E4204.2 is prohibited from being intentionally used as a grounding electrode. Even if someone might purposely try and use the structural reinforcing steel of an in-ground swimming pool as a grounding electrode for a remote panelboard for pool equipment, it has never been the intent of the *Code* to use an in-ground pool as a grounding electrode in this manner. Adding the additional requirement at *NEC* 250.52(B) and IRC E3608.7 to prohibit the use of

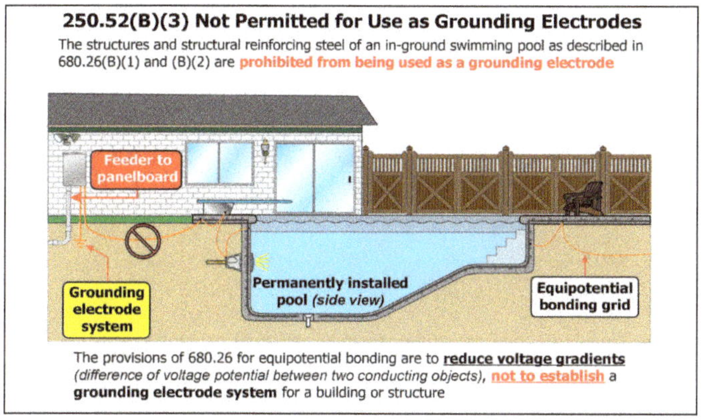

Figure 14.26. Structural components or rebar of an in-ground swimming pool are not permitted to be used as a grounding electrode

the metal components of an in-ground swimming pool as a grounding electrode is an important clarification to point out the difference between grounding and bonding. The equipotential bonding requirements of *NEC* 680.26 and IRC E4204 are to reduce voltage gradients (difference of voltage potential between two conducting objects) and not to create a grounding electrode system for a building or structure.

Swimming pools, spas, and hot tubs present special shock hazards to people. The water is mixed with chemicals, which often increase conductivity. People are in direct contact with the water and come in contact with grounded objects. Electrical equipment is used to circulate and heat the water and for other related functions such as illumination. The *Code* requirements, if carefully followed, will reduce the electric shock hazards to an acceptable level. One of these measures is the bonding together of conductive portions of the pool and metal parts of electrical equipment associated with the pool. Once again, the goal is to provide a means of equalizing the voltage potential of all equipment and parts. If this is done properly, there will be no current flow between these parts at different voltage potentials that would be harmful to personnel (people). The required bonding (connected together) is accomplished by connecting all the conductive parts together by adequately sized and properly connected copper bonding conductor(s) or other recognized equipotential bonding methods.

The *Code* describes the basic bonding requirements and provides a list of items required to be bonded together. These bonding requirements can be found at *NEC* 680.26(B) or IRC E4204.2. Included are the following conductive parts that must be bonded together:

1. All metal parts of the pool structure, including the reinforcing metal of the ***conductive pool shell***, coping stones, and ***perimeter surface*** deck, must be bonded together. The usual steel tie wires are considered to be suitable for bonding the reinforcing steel or rebar together, and welding, brazing, or special clamping is not required. These tie wires must be made tight. Reconstructed pool shells are also required to comply with the equipotential bonding requirements. Reinforcing rebar steel effectively insulated by an encapsulating, nonconductive compound at the time of manufacture is not required to be bonded. If this is the case, the *Code* requires provisions to be made for an alternate means to eliminate voltage gradients. This alternate means to structural reinforcing steel, as well as the components and make-up of the conductive pool shell and the perimeter surfaces, will be discussed in greater detail later in this chapter [*NEC* 680.26(B)(1) and (B)(2) or IRC E4204.2].
2. All ***metallic components*** or parts of the pool structure, including reinforcing metal not addressed above at *NEC* 680.26(B)(1)(a) or IRC IRC E4204.2, must be bonded together. As stated earlier, reinforcing steel that is encapsulated with a nonconductive compound is not be required to be bonded [*NEC* 680.26(B)(3) or IRC E4204.2.3].
3. All metal-forming shells of wet-niche luminaires and mounting brackets of no-niche luminaires for ***underwater lighting*** must be bonded together. Listed low-voltage lighting systems with nonmetallic forming shells are not required to be bonded. Check the manufacturer's instructions for installation details to ensure compliance with listing requirements [*NEC* 680.26(B)(4) or IRC E4204.2.4].
4. All ***metal fittings*** within or attached to the pool structure must be bonded together. Isolated parts that are not over 100 mm (4 in.) in any dimension and do not penetrate into the pool structure more than 25 mm (1 in.) are exempted from the bonding requirement. Metallic pool cover anchors intended for insertion in a concrete or masonry deck surface, 25 mm (1 in.) or less in any dimension and 51 mm (2 in.) or less in length, and metallic pool cover anchors intended for insertion in a wood or composite deck surface, 51 mm (2 in.) or less in any flange dimension and 51 mm (2 in.) or less in length, are not required to be bonded as well [*NEC* 680.26(B)(5) or IRC E4204.2.5].
5. Metal parts of ***electrical equipment*** associated with the pool water circulating system, including pump motors and pool heaters, and metal parts of equip-

ment associated with pool covers, including electric motors, must be bonded together. It should be noted that metal parts of listed equipment incorporating an approved system of double insulation, such as a double-insulated pool pump motor, are prohibited from being bonded to the equipotential bonding grid [*NEC* 680.26(B)(6) or IRC E4204.2.6].

6. Metal-sheathed cables and raceways, metal piping, metal fences, metal awnings, metal doors, and all ***fixed metal parts*** must be bonded together. This would include fixed metal parts that are within 1.5 m (5 ft) horizontally of the inside walls of the pool and within 3.0 m (12 ft) above the maximum water level of the pool or any observation stands, towers or platforms, or from any diving structures, that are not separated from the pool by a permanent barrier [*NEC* 680.26(B)(7) or IRC E4204.2.7].
7. The ***pool water*** itself must also be bonded together with the previously mentioned items. An intentional bond of a minimum conductive surface area of 5800 mm^2 (9 $in.^2$) is required to be in direct contact with the pool water. This bonding can be accomplished with any of the conductive parts that are required to be bonded, as described above. This could include such things as a metal pool ladder or railing where at least 9 square inches of the ladder is in contact with the water. A metal-forming shell of a wet-niche luminaire can also satisfy this pool water bonding requirement. Listed products to accomplish this bonding of the pool water are commercially available as well (see Photo 14.19) [*NEC* 680.26(C) or IRC E4204.3].

Photo 14.18. Bonded parts such as metallic ladders, luminaire housings, and rails with 9 in^2 in contact with the pool water can serve to bond the pool water

Photo 14.19. Listed product for bonding of pool water (Product made by Perma-Cast Co.)

For pool water heaters rated at more than 50 amperes and having specific instructions regarding bonding and grounding, only those parts designated to be bonded are required to be bonded, and only those parts designated to be grounded are required to be grounded [*NEC* 680.26(B)(6)(b) or IRC E4204.2.6].

As stated previously, accessible metal parts of listed equipment incorporating an approved system of double insulation and providing a means for grounding internal nonaccessible, non-current-carrying metal parts are not permitted to be bonded by direct connection to the equipotential bonding grid. The equipment grounding conductor run with the supply conductors to the double-insulated motor is the permitted means that should be used to ground the internal nonaccessible metal parts of a double-insulated motor.

As previously stated, where a double-insulated water pump is installed, a bonding connection to the equipotential bonding grid is not permitted. However, the *Code* does require the installation of an 8 AWG solid copper bonding conductor that is of sufficient length to make a bonding connection to a future replacement motor in the event that the replacement motor is not a double-insulated type motor. The bonding conductor is to be extended from the equipotential bonding grid to an accessible location in the vicinity of the motor that could possibly be replaced in the future. Where there is no connection between the bonding grid and the equipment grounding conductor for the premises, the bonding conductor is required to be connected to the equipment-grounding conductor of the existing pump motor circuit. This connection could be made at the motor termination compartment or in a suitable junction box or other enclosure [*NEC* 680.26(B)(6)(a) or IRC E4204.2.6].

EQUIPOTENTIAL BONDING GRID

All the metallic structural components and other metal fittings, wiring methods, and equipment must be connected together (bonded) to form what is commonly referred to as an *equipotential bonding grid.* When structural steel (rebar) is involved in the makeup of the deck steel and/or the belly

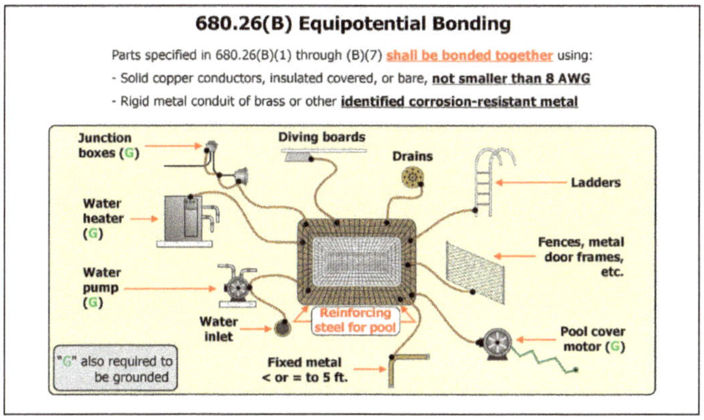

Figure 14.27. The equipotential bonding grid forms an equipotential bonding plane in and around the pool area

steel, this mass of structural steel performs as the equipotential bonding grid. This structural steel or equipotential bonding grid is divided into two (2) distinctive parts by the *Code*: the ***conductive pool shell (belly steel)*** [*NEC* 680.26(B)(1) or IRC E4204.2(1)] and the ***perimeter surfaces (deck steel)*** [*NEC* 680.26(B)(2) or IRC E4204.2(2)]. For the conductive pool shell, cast-in-place or poured concrete, pneumatically applied or sprayed concrete, and concrete block (including painted or plastered coatings) are all considered conductive materials due to water permeability and porosity. This would apply to reconstructed pool decks or shells as well. Vinyl liners and fiberglass composite shells are not included in these bonding requirements and are considered to be nonconductive materials.

For the perimeter surfaces, the equipotential bonding grid is considered to extend under paved walking surfaces for a distance of 1 m (3 ft) horizontally beyond the inside walls of the pool unless separated from the pool by a permanent wall or building 1.5 m (5 ft) or more in height. The deck or perimeter surface does include unpaved surfaces as well as poured concrete and other types of paving and is required to be attached to the pool reinforcing steel or copper conductor grid of the conductive pool shell at a minimum of four (4) points uniformly spaced around the perimeter of the pool. For nonconductive pool shells, bonding at four points is not a requirement. The components and make-up of the conductive pool shell and the perimeter surfaces will be discussed in greater detail later in this chapter.

The bonding connection between the various components of the equipotential bonding grid can be made with a solid copper conductor or with rigid metal conduit of brass or other identified corrosion-resistant metal. If a conductor is used, it may be insulated, covered, or bare, but it cannot be smaller than 8 AWG. This bonding conductor is not required to be extended or attached to remote panelboards, service equipment, or any grounding electrodes. This bonding conductor is, again, simply for bonding metal components together; it is not intended for the purpose of establishing a connection to earth or a ground-fault return path.

The connection(s) must be made in accordance with *NEC* 250.8 or IRC E3406.14 by such things as listed pressure connectors or clamps that are labeled as suitable for the purpose and are made of stainless steel, brass, copper, or copper alloy. Where pressure-type connectors are used, be certain they are listed and sized to accommodate the 8 AWG copper bonding conductor and listed for attachment to the steel reinforcing bar or whatever is being bonded. Consult the connector manufacturer's specifications for the proper model connector. Some clamps or connectors are suitable for use with pipe only or with reinforcing rods only. Some of the clamps or connectors are identified for use with a certain size or diameter pipe only, such as ½ in. – 1 in. diameter trade sizes only. This will also ensure a secure connection on what are often widely different sizes of reinforcing bars or rods and the 8 AWG bonding conductor.

If the connection is covered with concrete or earth, use a wire pressure connector that is listed and identified as being suitable for direct earth burial or concrete encasement, whichever applies.

The process of exothermic welding is an excellent way to bond the required bonding conductor(s) to reinforcing rods or bars. If the exothermic welding process is used, it is critical to use the proper form or mold. A proper form will be sized for both the reinforcing rod and the bonding conductor to be used. Also, be certain all the manufacturers' instructions are followed. Too "hot" a load may ruin the connection. Safe work practices and proper personal protective equipment are essential to prevent serious burns.

EQUIPOTENTIAL BONDING GRID – CONDUCTIVE POOL SHELL

Where the deck steel and/or the belly steel is comprised of structural steel (rebar), this mass of structural steel performs as the equipotential bonding grid. This equipotential bonding grid is comprised of two (2) different and distinct parts. The first of these two that we will discuss is the ***conductive pool shell.*** This conductive pool shell is sometimes referred to as the ***belly steel.*** The conductive pool shell is to be comprised of the following elements:

Unencapsulated structural reinforcing steel of a conductive pool, where the reinforcing steel is bonded together by the usual steel tie wires or the equivalent. If the structural reinforcing steel is encapsulated in a nonconductive compound or if unencapsulated structural reinforcing steel is not available or utilized at a conductive pool shell, a copper conductor grid system is required to be installed as an alternative to unencapsulated structural reinforcing steel.

If installed, this copper conductor grid system would need to comply with and consist of (1) through (4) below:

(1) Be constructed of minimum 8 AWG bare solid copper conductors bonded to each other at all points of crossing (see *NEC* 250.8 or IRC E3406.14 for bonding connection provisions).

"(2) Conform to the contour of the pool shell.

(3) Be arranged in a 300 mm (12 in.) by 300 mm (12 in.) network of conductors in a uniformly spaced perpendicular grid pattern with a tolerance of 100 mm (4 in.).

(4) Be secured within or under the pool no more than 150 mm (6 in.) from the outer contour of the pool shell.

[See *NEC* 680.26(B)(1)(a) and (b) or IRC E4204.2(1)].

EQUIPOTENTIAL BONDING GRID – PERIMETER SURFACES

The second element of the equipotential bonding grid that consists of unencapsulated structural reinforcing steel is the perimeter surfaces or deck steel, as it is sometimes referred to. The perimeter surfaces are comprised of the following elements.

Unencapsulated structural reinforcing steel, which is for the perimeter surface, can consist of such things as structural metal rebar or welded wire mesh. Where structural reinforcing steel is encapsulated in a nonconductive compound or where unencapsulated structural reinforcing steel is not available or utilized, a couple of alternative methods must be employed. These alternative means consist of either a "*copper ring*" system or a "*copper grid*" system being utilized. The *copper ring system* consists of the following requirements being met:

(1) At least one minimum 8 AWG bare solid copper conductor shall be provided.

(2) The conductor(s) shall follow the contour of the perimeter surface.

(3) The copper ring is permitted to be spliced but only when employing listed splicing devices or utilizing the exothermic welding process.

(4) The copper ring must be installed 450 to 600 mm (18

Photo 14.20. The equipotential bonding grid includes both the conductive pool shell (belly steel) and the perimeter surface (deck steel)

Photo 14.21. Conductive pool shell, sometimes referred to as the belly steel.

Photo 14.22. Conductive perimeter surface, sometimes referred to as the deck steel

to 24 in.) from the inside walls of the pool.

(5) The copper ring is required to be secured within or under the perimeter surface 100 mm to 150 mm (4 in. to 6 in.) below the subgrade.

[See *NEC* 680.26(B)(2)(a) and (b) or IRC E4204.2(2)].

Where structural reinforcing steel is not available or is encapsulated in a nonconductive compound, and a *copper grid system* is chosen as the alternative method of bonding the perimeter surface, the following requirements must be met:

(1) Be constructed of 8 AWG solid bare copper and be ar-

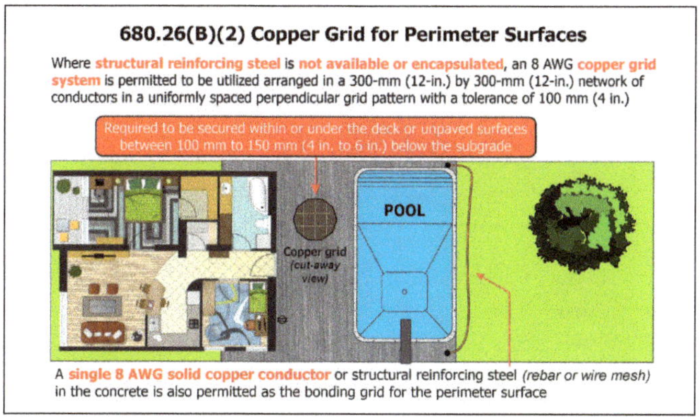

Figure 14.28. Copper grid system permitted for bonding of perimeter surfaces

Photo 14.23. Bonding conductor (8 AWG solid copper) to pool pump motor and water heater

ranged in a 300 mm (12 in.) by 300 mm (12 in.) network of conductors in a uniformly spaced perpendicular grid pattern with a tolerance of 100 mm (4 in.).

(2) Follow the contour of the perimeter surface extending 1 m (3 ft) horizontally beyond the inside walls of the pool.

(3) Use only listed splicing devices or exothermic welding for splices.

(4) Be secured within or under the deck or unpaved surfaces between 100 mm to 150 mm (4 in. to 6 in.) below the subgrade.

UNDERWATER AUDIO EQUIPMENT INSTALLATION(S)

All underwater audio equipment and similar installations must be listed and specifically identified for use as underwater audio equipment. Each underwater speaker must be mounted in an approved metal forming shell, the front of which is enclosed by a captive metal screen or equivalent that is bonded to and secured to the forming shell by a positive locking device that assures a low-resistance contact. Typically, a special tool will be required so that it can be opened for installation or servicing of the speaker. The forming shell must be installed in a recess in the wall or floor of the pool. It is interesting to note that the forming shell for the underwater audio equipment is specifically required to be a metal type. The rigid metal conduit or intermediate metal conduit of brass or other identified corrosion-resistant metal, liquidtight flexible nonmetallic conduit (LFNC-B), rigid polyvinyl chloride (PVC) conduit, or reinforced thermosetting resin conduit is required to extend from the forming shell to a listed swimming pool junction box (deck box) or other enclosure. The junction box must meet the location requirements of the *Code*, which were covered earlier in this chapter in the sections on pool-associated luminaires.

Where rigid polyvinyl chloride (PVC) conduit, reinforced thermosetting resin conduit, or liquidtight flexible nonmetallic conduit is used as the wiring method from the deck or junction box or enclosure to the metal forming shell, an 8 AWG insulated copper bonding conductor must be installed within this conduit with provisions for terminating in the forming shell and in the junction box. This conductor may be solid or stranded. The termination of the 8 AWG copper bonding jumper in the forming shell must be covered with, or encapsulated in, a suitable listed potting compound to protect the connection from the possible deteriorating effect of pool water.

The forming shell and metal screen must be made of brass or other approved corrosion-resistant metal [*NEC* 680.27(A) or IRC E4206.10].

WIRING METHODS FOR PERMANENTLY INSTALLED POOL-ASSOCIATED MOTORS

The feeder or branch circuit wiring methods permitted for use in a permanently installed pool, spa, or hot tub motor will depend on the type of environment in which the motor is installed. Wiring methods installed in a corrosive environment, as defined in NEC Article 100 or IRC Chapter 35, must be listed and identified for use in such areas and must consist of rigid metal conduit, intermediate metal conduit, rigid polyvinyl chloride conduit, and reinforced thermosetting resin conduit. These wiring methods are considered resistant to corrosive environments [NEC 680.14 or IRC E4202.2]. Type MC cable listed for a corrosive environment is also an acceptable wiring method for a corrosive location [NEC 680.21(A)(1) or IRC E4202.2].

All motors associated with permanently installed pools and installed in a corrosive environment are required to be connected to an equipment-grounding conductor. This equipment grounding conductor is to be sized based on the rating of the overcurrent protective device (fuse or circuit breaker typically) ahead of the branch circuit supplying the motor in conjunction with *NEC* Table 250.122 or IRC Table E3908.13. This equipment-grounding conductor must be an insulated copper conductor and is not permitted to be smaller than 12 AWG [*NEC* 680.7(A) or IRC E4205.5].

Where the wiring method is installed in noncorrosive environments, such as on or inside the dwelling unit, branch circuits would need to comply with the general requirements in NEC Chapter 3 or IRC Chapter 38. This could be a nonmetallic-sheathed cable (Type NM cable) in dry locations.

The Code defines a corrosive environment pertaining to a pool location as "areas or enclosures without adequate ventilation, where electrical equipment is located, and pool sanitation chemicals are stored, handled, or dispensed." [See NEC Article 100 or IRC E4201.2]. This could include areas where pool sanitation chemicals are stored, handled, or dispensed, as well as confined areas under decks adjacent to these areas. A corrosive environment or area could also include areas with circulation pumps, automatic chlorinators, filters, open areas under decks adjacent to or abutting the pool structure, and similar locations.

Where flexible connections are needed, such as at or adjacent to a motor, liquidtight flexible metal (LFMC) or liquidtight flexible nonmetallic conduit (LFNC) with listed fittings, or Type MC cable listed and suitable for the use (perhaps a corrosive environment) is a permitted wiring method for pool pump motors [*NEC* 680.21(A)(1) or IRC Table E4202.1].

Double-insulated pump motors are being used with permanently installed pools on a more frequent basis. Double-insulated pump motors have been used for storable pools for many years without any reported problems. A double-insulated pump motor may be cord- and plug-connected. For pools other than storable, the cord length cannot exceed 900 mm (3 ft). The listed double-insulated swimming pool pump motor incorporates an approved system of double insulation. The internal, non-current-carrying metal parts are grounded via the branch-circuit equipment grounding conductor. The equipment grounding conductor must be an insulated copper conductor not smaller than 12 AWG. The cord must terminate in a grounding-type attachment plug. This type of pump motor should not have an external bonding lug, even though some manufacturers will include the exterior bonding lug, which serves no significant function [*NEC* 680.7, 680.8, 680.21(B) or IRC E4202.3(1), E4202.4].

As noted earlier in this chapter, outlets that supply all pool pump motors on branch circuits rated 150 volts or less to

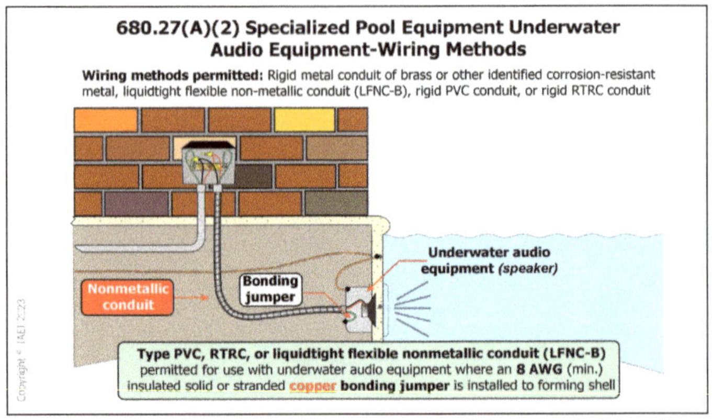

Figure 14.29. Underwater audio equipment and forming shell bonding requirements

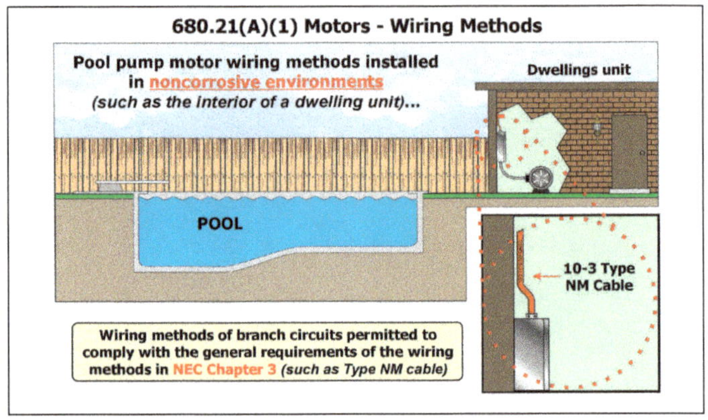

Figure 14.30. Wiring methods for motors in a noncorrosive environment *(interior of a dwelling unit)*

the ground and 60 amperes or less, single- or 3-phase, are required to be provided with Class A GFCI protection. This would include both receptacle outlets (cord-and plug-connected) and direct connection or hard-wired connection. A receptacle outlet supplying a pool pump motor is required to be located at least 1.83 m (6 ft) from the inside walls of a pool. This circulation pump motor receptacle outlet in this location must be a grounding-type receptacle and must be GFCI-protected [*NEC* 680.5(B), 680.21(C), 680.22(A)(2) or IRC E4202.4, E4203.1.1].

SWIMMING POOL PANELBOARDS IN OUT BUILDINGS

Grounding at the disconnecting means supplied by a feeder(s) or branch circuit(s) and located in a remote building(s) or structure(s) is covered in as covered in Chapter 15, where the methods for complying with the *Code* rules are reviewed [*NEC* 250.32(B) or IRC E3607.3]. At a separate building or structure, an equipment grounding conductor (EGC) must be installed with the supply conductors (feeder or branch circuit). This EGC must be connected to the building or structure disconnecting means and to the required established

Photo 14.24. Liquidtight flexible nonmetallic conduit is employed as a wiring method for a pool pump motor where necessary to employ flexible connections

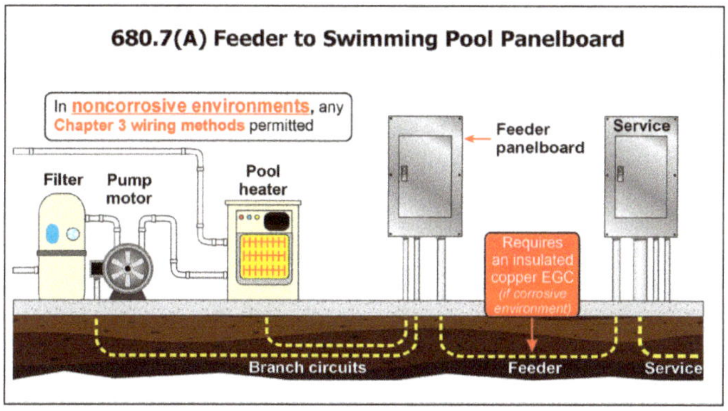

Figure 14.31. Feeder to swimming pool panelboard

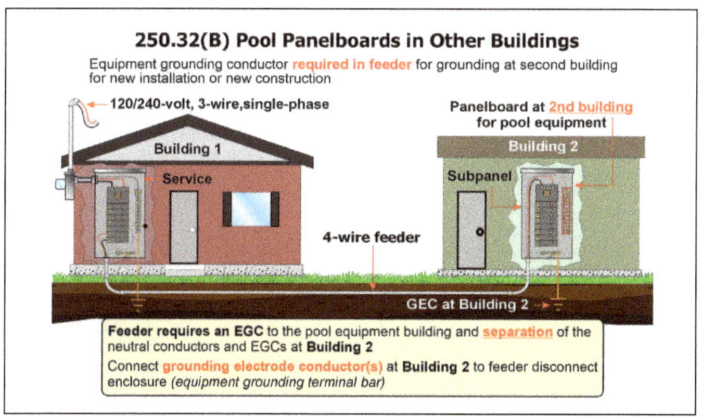

Figure 14.32. The pool equipment panelboard located at the second building requires EGC installed in the second building

grounding electrode(s) at the separate building or structure [see *NEC* 250.32(A) or IRC E3607.3 for the requirement for establishing a grounding electrode system at outbuildings supplied with a feeder or branch circuit]. This equipment grounding conductor is used for grounding or bonding of equipment, structures, or frames of appliances required to be grounded or bonded at a separate building or structure. Any installed grounded conductor to the separate building or structure must be isolated from and is not permitted to be connected to the equipment grounding conductor terminal bar or the grounding electrode(s). The equipment grounding conductor that is run with the feeder or branch circuits must be insulated (if installed in a corrosive environment) and connected to the equipment-grounding terminal of the panelboard at the separate building or structure. This requirement prevents the neutral conductor and equipment grounding conductor from connecting together in parallel at the separate building or structure.

For installations made in compliance with previous editions of this *Code* that permitted such connection *(existing installations only)*, the grounded (neutral) conductor run with the supply conductors to the separate building or structure was permitted to be connected to the building or structure disconnecting means and to the grounding electrode(s) and can be used for grounding or bonding of equipment, structures, or frames required to be grounded or bonded. As stated above, this situation would normally require an equipment grounding conductor for this purpose for anything other than an existing installation with an existing feeder with no equipment grounding conductor.

Even in existing installations, specific conditions must be employed before the grounded (neutral) conductor can be used for grounding at a separate building or structure. One of these conditions is that no equipment grounding conductor has been run with the supply to the existing building or structure. Grounding the neutral conductor is permitted only where there are no continuous metallic paths of any kind between the two buildings. Common metallic water pipes, metallic gas piping, CATV cable shields, and telephone grounding conductors are examples of potential shared metallic paths. A third condition involves ground-fault protection of equipment that has not been installed on the supply side of the feeder(s). Ground-fault protection of equipment is a rare condition for the dwelling unit environment [*NEC* 250.32(B)(1) or IRC E3607.3].

OTHER TYPES OF EQUIPMENT OPERATED AROUND PERMANENTLY INSTALLED POOLS

Electrically Operated Pool Covers for Permanently Installed Pools

An electrically operated pool cover is a piece of motor-driven equipment specifically designed to cover and uncover the water surface of a pool by means of a flexible sheet or rigid frame.

Electric motors, controllers, and wiring generally must be located at least 1.5 m (5 ft) from the inside wall of the pool. They are permitted to be at a lesser distance when separated from the pool by a permanent barrier such as a wall or cover. The device that controls the operation of the cover motor must be located so that the operator has a complete view of the pool. Electric motors must be of the totally enclosed type where they are installed below grade level. Motors that are part of listed systems with ratings not exceeding the low-voltage contact limit and supplied by listed transformers or power supplies [complying with *NEC* 680.23(A)(2) or IRC E4206.1] are permitted to be located less than 1.5 m (5 ft) from the inside walls of the pool.

The branch circuit that supplies the electric motor and controller is required to be protected by a ground-fault circuit interrupter unless the motor is part of listed systems with ratings not exceeding the low-voltage contact limits supplied by listed transformers or power supplies that comply with *NEC* 680.23(A)(2) or IRC E4206.1 [*NEC* 680.27(B)(1) and (2) or IRC E4206.11].

Pool Area Heating and Deck Area Heating for Permanently Installed Pools

All permanently installed pool deck areas, including a covered pool, where electrically operated comfort heating units are to be installed within 6.0 m (20 ft) of the inside wall of the pool must comply with the following requirements:

1. Unit heaters must be of the totally enclosed or guarded type and are to be rigidly mounted to the structure. Unit heaters cannot be mounted over the pool or within the area extending 1.5 m (5 ft) horizontally from the inside walls of a pool.

2. Permanently wired radiant heaters must be properly guarded and securely fastened to the heater mounting device(s). A permanently wired radiant heater is not permitted to be installed over a pool or within the area extending 1.5 m (5 ft) horizontally from the inside walls of the pool. This type of heater must be mounted at least 3.7 m (12 ft) vertically above the pool deck unless it is otherwise specifically approved for a lower height.

3. Radiant heating cables are not permitted to be embedded in or below the pool deck

[See *NEC* 680.27(C) or IRC E4206.13].

Gas-Fired Water Heaters for Permanently Installed Pools

As discussed earlier in this chapter, branch circuits or feeders that serve gas-fired swimming pool and spa water heaters that will operate at voltages above the low-voltage contact limit are required to be provided with ground-fault circuit-interrupter protection for personnel (*NEC* 680.28).

OTHER EQUIPMENT LOCATED IN CLOSE PROXIMITY TO THE POOL AREA

Electrical and other equipment (other than traditional pool pump motors and controllers) with ratings exceeding the low-voltage contact limit must be located at least 1.5 m (5 ft) horizontally from the inside walls of a pool unless separated from the pool by a solid fence, wall, or other permanent barrier. Of particular concern is electric power production equipment such as generators, solar photovoltaic (PV) systems, fuel cell systems, wind systems, and energy storage systems being installed in close proximity to a pool. With this power-producing equipment, not only is there a concern about metal parts being at different voltage potentials, but a real concern is also a shock hazard associated with this type of equipment located too close to a pool [*NEC* 680.22(E) or IRC E4203.5].

LOW-VOLTAGE, GAS-FIRED LUMINAIRES, DECORATIVE FIREPLACES, FIRE PITS, AND SIMILAR EQUIPMENT

Listed low-voltage gas-fired luminaires, decorative fireplaces, fire pits, and similar equipment using low-voltage ignitors are permitted to be located less than 1.5 m (5 ft) from the inside walls of the pool. These types of gas-fired units do not require grounding and are supplied by listed transformers or power supplies with outputs that do not exceed the low-voltage contact limit as defined in *NEC* Article 100 or IRC Chapter 35. These transformers or power supplies must comply with *NEC* 680.23(A)(2) or IRC E4206.1, which demands that the transformers or power supplies used to supply underwater luminaires be listed, labeled, and identified for swimming pool and spa use. These transformers or power supplies must also incorporate an isolated winding type transformer with an ungrounded secondary that has a grounded metal barrier between the primary and secondary windings or one that incorporates an approved system of double insulation between the primary and secondary windings. Junction boxes and electrical enclosures for these transformers or power supplies supplying this type of equipment are required to comply with the provisions of *NEC* 680.24 or IRC E4206.9.1.

Metallic equipment associated with such listed low-voltage gas-fired units must be bonded to the equipotential bonding grid in accordance with *NEC* 680.26(B) or IRC E4204.2. Any metallic gas piping for this type of unit must be bonded

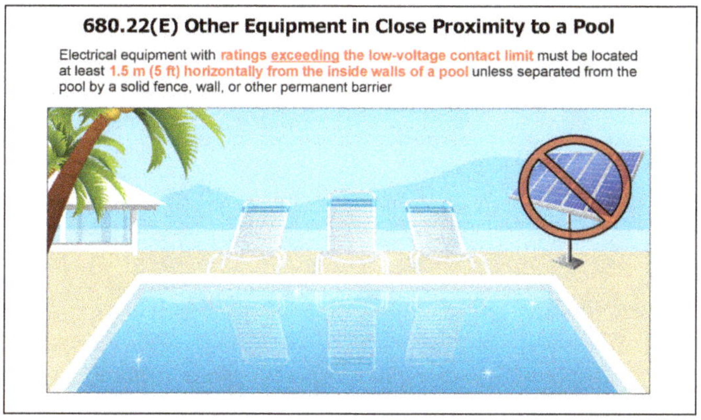

Figure 14.33. Other equipment (such as a PV array) must be located at least 1.5 m (5 ft) horizontally from the inside walls of a pool

Photo 14.25. Low-voltage decorative fire displays within 1.5 m (5 ft) from the inside walls of a pool

in accordance with the requirements in *NEC* 250.104(B) or IRC E3609.7 (bonding of other metal piping) and *NEC* 680.26(B)(7) or IRC E4204.2(7) (bonding of fixed metal parts). See requirements of *NEC* 680.22(B)(7) or IRC E4203.4.7

STORABLE POOLS, STORABLE SPAS, AND STORABLE HOT TUBS

A storable swimming, wading, or immersion pool, or storable/portable spa and hot tub by definition, is a pool installed entirely on or above the ground that is intended to be stored when not in use and is designed for ease of relocation, regardless of water depth. A very significant change occurred to the definitions of a *permanently installed pool* and a *storable pool* for the 2023 *NEC* (2024 IRC) *(see sidebar for the definitions recorded earlier in this chapter)*. The previous definitions relied on the depth of the pool to determine its definition. For example, the previous definition of a *storable pool* indicated that it was a pool constructed on or above the ground and capable of holding water to a maximum depth of 1.0 m (42 in.). Storable pool manufacturers have been producing storable pools with wall heights of 1.2 m (48 in.) for over 40 years without any negative impact on electrical safety. Today, it is very common to see storable pools that exceed the defined previously 1.0 m (42 in.) water depth but qualify in every other way as a storable pool. From an electrical safety standpoint, these storable pools capable of holding water at a depth greater than 1.0 m (42 in.) *(not meeting the previous definition of a storable pool)* are no different than a pool that has a maximum 1.0 m (42 in.) water depth *(meeting the previous definition of a storable pool)*. Providing requirements for electrical safety around a storable pool is not impacted if the maximum water depth is greater than 1.0 m (42 in.).

A pool, spa, or hot tub with a nonmetallic, molded polymeric or inflatable fabric wall, regardless of its dimensions, installed entirely on or above the ground is generally considered to be a storable pool, spa, or hot tub as well. The rules for a permanently installed pool *(Part II of NEC Article 680)* are not applied to a storable pool *(Part III of NEC Article 680)* unless specifically referenced in Part III of *NEC* Article 680. It is critical for both the installer and the AHJ to apply the definitions and to know the difference between a permanently installed pool and a storable pool before applying the *Code* rule to either or both of these types of pools.

PUMP MOTORS FOR STORABLE POOLS, STORABLE SPAS, OR STORABLE HOT TUBS

A cord-connected pool filter pump motor used for a storable pool, storable spa, or storable hot tub must incorporate an approved system of double insulation or its equivalent. This equipment must be provided with a means for the termination of an equipment grounding conductor for the connection to the internal and nonaccessible, non-current-carrying metal parts of the pump motor.

An equipment grounding conductor that is run with the power-supply conductors within the flexible cord must terminate in a grounding-type attachment plug having a fixed grounding contact member [*NEC* 680.7(B) or IRC E4207.1]. Cord-connected pool pumps for storable pools must have ground-fault circuit-interrupter protection that is an integral part of the attachment plug or located in the power supply cord within 300 mm (12 in.) of the attachment plug [*NEC* 680.31 or IRC E4207.1].

Listed pump motors for storable swimming pools, storable spas, and storable hot tubs are permitted to have a minimum 7.5 m (25 ft) supply cord. This cord allowance is permitted by *NEC* 680.8(A) and the product standard for storable pools. The product safety standard for storable pool pumps, UL 1081 (Standard for Swimming Pool Pumps, Filters, and Chlorinators) along with the UL Product Spec (formally UL

White Book) (Category WCSX), also requires that these listed pump motors be marked with the word "CAUTION" and the following or equivalent statement: "This Pump Is for Use with Storable Pools Only — Do Not Use with Permanently-Installed Pools."

Some swimming pool pump motors have 900 mm (3 ft) supply cords. These pump motors are not intended to be used for storable pools but are intended for permanent connection only with a permanently installed pool. They are required to be marked with the word "CAUTION" and the following or equivalent statement: "This Pump Is for Use with Permanently Installed Pools Only — Do Not Use with Storable Pools." An extension cord should never be used for a storable pool, storable spa, or storable hot tub pump motor.

All electric equipment, including power-supply cords, used with storable pools, storable spas, and storable hot tubs is required to be protected by ground-fault circuit interrupters [NEC 680.32, 680.5(B) or IRC E4207.2]. All 125-volt through 250 volts, 60 ampere or fewer receptacles located within 6.0 m (20 ft) of the inside walls of a storable pool, storable spa, or storable hot tub must be GFCI protected. This measurement is made by taking the shortest path an extension or supply cord connected to the receptacle would follow without piercing a wall, floor, ceiling, doorway with hinged or sliding door, window opening, or other effective permanent barrier (*NEC* 680.32 or IRC E4207.2).

Photo 14.26. Typical storable swimming pool

Photo 14.27. Photos show a storable/portable spa or hot tub (*left*) and a storable swimming pool (*right*)

LUMINAIRES FOR STORABLE POOLS, STORABLE SPAS, OR STORABLE HOT TUBS

A luminaire installed in or on the wall of a storable pool, storable spa, or storable hot tub must be part of a cord- and plug-connected lighting assembly where the luminaire operates within the low voltage contact limits. A lighting assembly operating at over the low voltage contact limits (without a transformer or power supply) and with the luminaire lamp(s) operating at not over 150 volts shall be permitted to be cord-and-plug-connected where the assembly is listed as an assembly for the purpose. These are typically a through-wall lighting assembly. There are two types of luminaires for this application. The first type of luminaire operates within the low-voltage contact limit. This assembly is required to have:

1. no exposed metal parts;
2. a luminaire lamp that is suitable for use at the supplied voltage;
3. an impact-resistant polymeric lens, luminaire body, and transformer enclosure;
4. a transformer of power supply with a primary rating not over 150 volts; and
5. be listed as a portable pool luminaire assembly.

The second type of luminaire for a storable pool, storable spa, or storable hot tub is permitted to operate over the low-voltage contact limit but not over 150 volts. This type of assembly is required to have:

1. no exposed metal parts;
2. luminaire body and lens made from an impact-resistant polymeric;
3. a ground-fault circuit interrupter with open neutral conductor protection provided as an integral part of the assembly;
4. a luminaire lamp permanently connected to the ground-fault circuit interrupter with open-neutral protection;
5. luminaire that meets the requirements for underwater luminaires for permanently installed pools (GFCI protection, lamping, and relamping and servicing requirements, voltage limitations, wall mounting locations, etc.); and
6. be listed as a portable pool luminaire assembly.

See *NEC* 680.33, 680.33(A) and (B), and 680.23(A) or

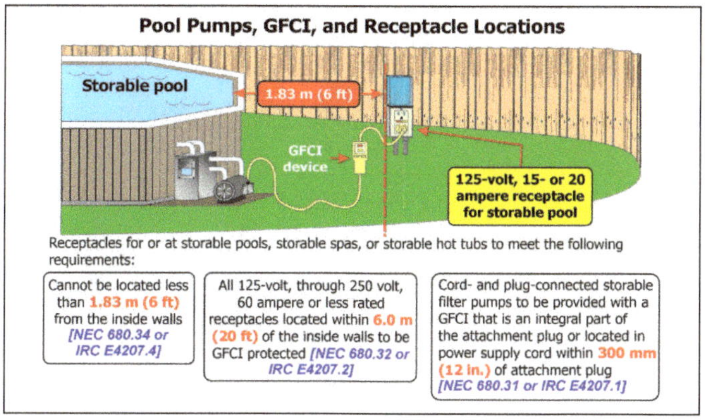

Figure 14.34. Receptacle outlet requirements at storable pools, storable spas, or storable hot tubs

IRC E4207.3 and E4206.1 for *Code* requirements for luminaires for storable pools, storable spas and storable hot tubs.

RECEPTACLE LOCATIONS FOR STORABLE POOLS, STORABLE SPAS, OR STORABLE HOT TUBS

Receptacles are not permitted to be located within 1.83 m (6 ft) of the inside walls of a storable pool, storable spa, or storable hot tub. This measurement is made by taking the shortest path an extension or supply cord connected to the receptacle would follow without piercing a wall, floor, ceiling, doorway with a hinged or sliding door, window opening, or other effective permanent barrier.

It should be noted that unlike with permanently installed pools, a 125-volt, 15- or 20-ampere receptacle outlet that is required to be located 1.83 m (6 ft) to 2.0 m (20 ft) from a permanently installed pool is not required for a storable pool [*NEC* 680.34 or IRC E4207.4].

SPAS, HOT TUBS AND EQUIPMENT ASSEMBLIES

Definitions for spa or hot tub, self-contained spa or hot tub, and for packaged spa or hot tub equipment assembly are provided in the *Code* at *NEC* Article 100 or IRC Chapter 35. Generally, a spa or hot tub is not designed or intended to have its contents drained or discharged after each use. Hot tubs are usually self-contained units that sit on the ground or a platform, and spas may be self-contained units that sit on the ground or a platform and are often referred to as "portable spas" or are built into the ground. Spas and hot tubs installed indoors are required to comply with Part I (general) and Part IV (spas and hot tubs) of Article 680 of the *NEC*. Spas and hot tubs installed outdoors are typically required to comply with Part I and Part II (permanently installed pools) of Article 680 of the *NEC*.

EMERGENCY SWITCH FOR SPAS AND HOT TUBS

An emergency switch is not required by the Code for a spa or a hot tub associated with a single-family dwelling. This would be a requirement for installations at two-family dwellings and other non-dwelling unit installations, such as hotels. Two-family dwellings and non-residential spas and hot tubs are required to be provided with an emergency shut-off switch, which may be a control switch. This switch is required to stop the pump motor(s) that provide power to the recirculation and jet system. The switch must be clearly labeled and readily accessible to spa or tub users, located at least 1.5 m (5 ft) away but adjacent to and within sight of the spa or hot tub. This requirement provides a means to quickly shut off the pump motor(s). Should a user become stuck in the suction opening of the circulation system, other individuals have the capability of stopping the pump(s) (NEC 680.41 or IRC E4208.4). Even though an emergency switch is not required by the Code for a single-family dwelling spa or a hot tub, it is not prohibited. Some spas or hot tub builders and/or homeowners have voluntarily chosen to install an emergency switch at their single-family dwelling spa or hot tub.

SPAS OR HOT TUBS INSTALLED OUTDOORS

Spas or hot tubs installed outdoors must generally comply with the provisions of the *Code* that apply to permanently installed pools. This significant requirement basically means that all the rules for receptacles, luminaires, grounding, and bonding that apply to a permanently installed swimming pool also apply to spas and hot tubs that are installed outdoors. Two alternatives to this general rule allow for a listed packaged units utilizing a factory-installed remote panelboard to be connected with liquidtight flexible metal conduit or liquidtight flexible nonmetallic conduit installed per its listing and the manufactures specifications to the spa or hot tub enclosure. The other alternative permits a listed packaged units to be cord- and plug-connected with a cord not longer than 4.6 m (15 ft) when protected by a ground-fault circuit interrupter [*NEC* 680.42(A)(1) and (A)(2) or IRC E4202.2 and Table E4202.1].

Self-contained spas typically have a control panel or panelboard contained in the control and heating equipment that is installed under the skirt of the spa. The *Code* simplifies the electrical connection of this type of spa, which is typically wired with a conduit that emerges from underground as it permits a liquidtight flexible conduit to be used for this type of electrical connection. Both liquidtight flexible metal conduit and liquidtight flexible nonmetallic conduit are permitted for this use [*NEC* 680.42(A)(1) or IRC Table E4202.1].

The metal-to-metal connection to a common frame or base may accomplish bonding for outdoor spas and hot tubs. Other conductive elements would have to be bonded using the techniques required for outdoor pools as described at *NEC* 680.26(B) or IRC E4204.1 through E4204.3. The metal bands or hoops used to secure the wooden staves of a hot tub are not required to be bonded by an equipotential bonding grid. The equipotential bonding requirements for perimeter surfaces of a permanently installed pool shall not apply to a listed self-contained spa or hot tub installed for aboveground use. The spa or hot tub must also not be identified as suitable only for indoor use. This type of installation must be in accordance with the manufacturer's instructions and must be located on or above grade. To avoid the equipotential bonding grid requirements of a permanently installed pool, the top rim of the spa or hot tub must be at least 710 mm (28 in.) above all perimeter surfaces that are within 760 mm (30 in.), measured horizontally from the spa or hot tub. The height of nonconductive external steps for entry to or exit from the self-contained spa cannot be used to reduce or increase this rim height measurement [*NEC* 680.42(B) and IRC E4204.4 and E4204.2(2) Exception No. 1].

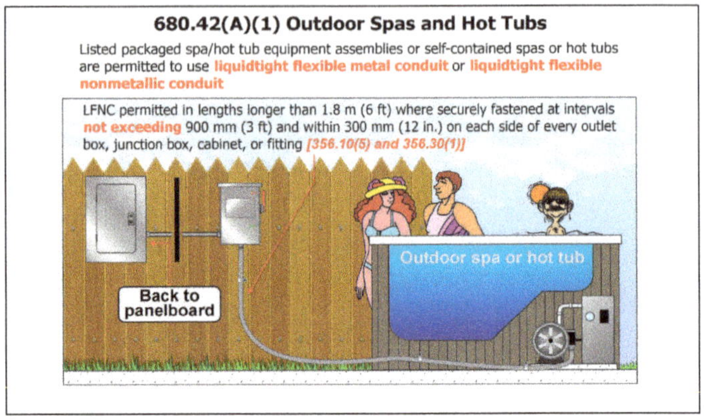

Figure 14.35. LFMC and LFNC are permitted as wiring methods to listed packaged spas or hot tubs

Photo 28. Typical storable hot tub installed outdoors

SPAS OR HOT TUBS INSTALLED INDOORS

Spas or hot tubs installed indoors must conform to the requirements of *NEC* Article 680, Parts I and II, except as specifically modified by *NEC* 680.43. They are permitted to be connected by any of the general wiring methods in Chapter 3 of *NEC* or Chapter 38 of the IRC. Where rated at 20 amperes or less, listed packaged units are permitted to be cord- and plug-connected to facilitate their removal or disconnection for maintenance and repair.

This is significantly different from the limited types of wiring methods permitted for use with spas and hot tubs installed outdoors. In addition, some of the requirements for indoor spas and hot tubs may overlap with those in other sections of the *Code*. For example, the *Code* requires that all 125-volt through 250-volt receptacles supplied by a single-phase branch circuit rated 150 volts or less to ground installed in bathrooms to have GFCI protection [*NEC* 210.8(A)(1) or IRC E3902.1]. For indoor spas and hot tubs, the *Code* requires GFCI protection for all receptacles rated 125 volts through 250 volts, 60 amperes or less, that are located within 3.0 m (10 ft) of the inside walls of the indoor spa or hot tub. If an indoor spa or hot tub were to be installed in a bathroom, compliance with the more stringent GFCI requirements of the bathroom would need to be enforced (*NEC* 680.43(A)(2) or IRC E4202.3(4), E4203.1.5, E4203.1.6).

The equipotential bonding requirements for perimeter surfaces of a permanently installed pool described at *NEC* 680.26(B)(2) do not apply to a listed self-contained spa or hot tub installed indoors above a finished floor [*NEC* 680.43, Ex. No. 2]

RECEPTACLES IN THE VICINITY OF INDOOR SPAS OR HOT TUBS

At least one 125-volt, 15- or 20-ampere rated receptacle on a general-purpose branch circuit is required to be located not less than 1.83 m (6 ft) and not more than 3.0 m (10 ft) from the inside wall of an indoor spa or hot tub.

Receptacles installed to comply with wall-spacing requirements (existing or new) and all existing receptacles in the dwelling unit must be located at least 1.83 m (6 ft) away from the inside walls of an indoor spa or hot tub. As stated earlier, receptacles rated at 125 volts through 250 volts and 60 amperes or less located within 3.0 m (10 ft) of the inside walls of an indoor spa or hot tub must be protected by a ground-fault circuit interrupter.

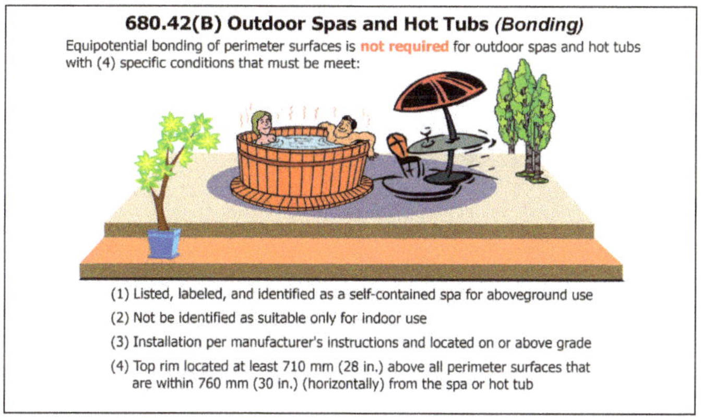

Figure 14.36. Bonding requirements for outdoor spas and hot tubs

Photo 14.29. Indoor spas and hot tubs have their own rules for receptacles, switches, luminaires, and bonding and grounding

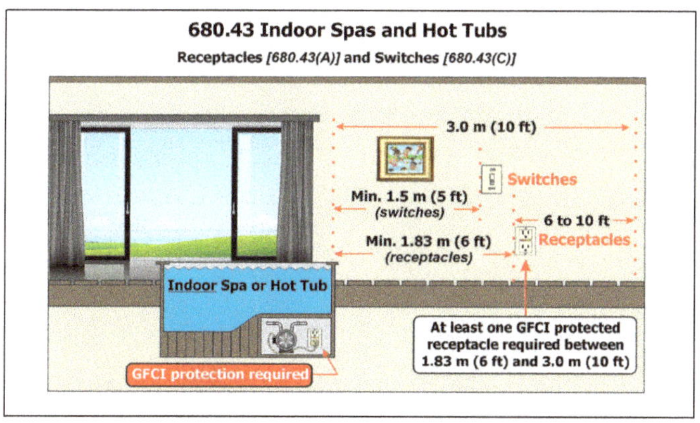

Figure 14.37. Receptacle and switch requirements for indoor spas and hot tubs

The distances given in this rule must be measured as the shortest path the supply cord of an appliance connected to the receptacle would follow without piercing a floor, wall, ceiling, doorway with a hinged or sliding door, window opening, or other effective permanent barriers.

Receptacles that provide power for an indoor spa or hot tub that is cord- and plug-connected must not exceed 150 volts to ground and also be protected by a ground-fault circuit interrupter [*NEC* 680.43(A)(1), (2), (3), and (4) or IRC E4203.1, E4203.1.5, and E4203.1.6].

WALL SWITCHES IN THE VICINITY OF INDOOR SPAS OR HOT TUBS

Wall switches are required to be located at least 1.5 m (5 ft), measured horizontally, from the inside walls of an indoor spa or hot tub [*NEC* 680.43(C) or IRC E4203.2].

LUMINAIRES, LIGHTING OUTLETS, AND CEILING-SUSPENDED FANS NEAR INDOOR SPAS OR HOT TUBS

Luminaires, lighting outlets, and ceiling-suspended (paddle) fans located over an indoor spa or hot tub or within 1.5 m (5 ft) from the inside walls of an indoor spa or hot tub are generally required to be not less than 2.3 m (7 ft 6 in.) above the maximum water level and must be protected by a ground-fault circuit interrupter.

Luminaires, lighting outlets, and ceiling-suspended (paddle) fans located 3.0 m (12 ft) or more above the maximum water level are not required to be protected by a ground-fault circuit interrupter. Requirements for luminaires, lighting outlets, and ceiling-suspended (paddle) fans located over and in close proximity to an indoor spa or hot tub are addressed at *NEC* 680.43(B)(1) or IRC E4203.4.5.

Where protected by a ground-fault circuit interrupter, luminaires are permitted to be installed less than 2.3 m (7 ft 6 in.) over an indoor spa or hot tub if they are either of the following types:

1. recessed luminaires with a glass or plastic lens and a nonmetallic or electrically isolated metal trim, suitable for use in damp locations or
2. surface-mounted luminaires with a glass or plastic globe and a nonmetallic body or a metallic body isolated from contact. Such luminaires are required to be suitable for use in damp locations.

Underwater luminaires for spas and hot tubs are required to be installed to meet the same requirements for those underwater luminaires installed in permanently installed pools [*NEC* 680.43(B)(2); 680.23 or 680.33; or IRC E4206.4].

BONDING OF INDOOR SPAS AND HOT TUBS

The following metal parts of an indoor spa or hot tub are required to be bonded together:

1. all metal fittings within or attached to the spa or hot tub structure (excluding small conductive surfaces not likely to become energized);

2. metal parts of any electric equipment associated with the spa or hot tub water circulating system, including pump motors, unless part of a listed, labeled, and identified self-contained spa or hot tub;

3. metal raceway and metal piping located within 1.5 m (5 ft) of the inside walls of the spa or hot tub and not separated from the spa or hot tub by a permanent barrier; and

4. all metal surfaces located within 1.5 m (5 ft) of the inside walls of a spa or hot tub and not separated from the spa or hot tub area by a permanent barrier. Small conductive surfaces, such as air and water jets and drain fittings, are not likely to become energized, and when not connected to metallic piping, towel bars, mirror frames, or similar non-electrical equipment, they need not be bonded.

5. non-current-carrying metal parts of electrical devices and controls not associated with the spas or hot tubs that are located within 1.5 m (5 ft) from such units [*NEC* 680.43(D) or IRC E4204.5].

Small conductive surfaces not likely to become energized (such as air and water jets and drain fittings), where not connected to metallic piping, towel bars, mirror frames, and similar nonelectrical equipment, are not be required to be bonded. [*NEC* 680.43(D) Exception].

METHODS OF BONDING INDOOR SPAS AND HOT TUBS

All metal parts associated with an indoor spa or hot tub must be bonded by any of the following methods:

1. interconnection of threaded metal piping and fittings;

2. metal-to-metal mounting on a common frame or base or

3. an insulated, covered, or bare 8 AWG solid copper bonding jumper

[See *NEC* 680.43(D) and (E) or IRC E4204.5.1].

As stated earlier, the equipotential bonding requirements for perimeter surfaces of a permanently installed pool described at *NEC* 680.26(B)(2) are not to be applied to a listed self-contained spa or hot tub installed indoors above a finished floor [*NEC* 680.43, Ex. No. 2].

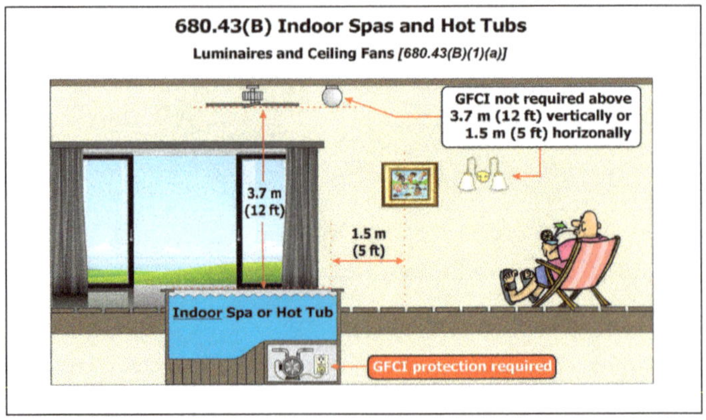

Figure 14.38. Mounting height requirements for luminaire and ceiling fan for indoor spas and hot tubs *(without GFCI protection)*

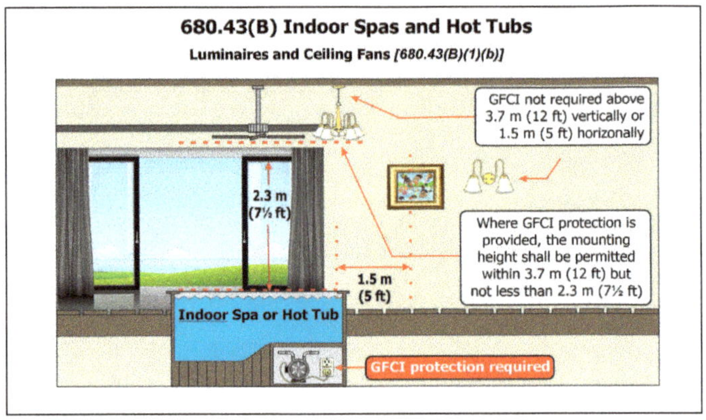

Figure 14.39. Mounting height requirements for luminaire and ceiling fan for indoor spas and hot tubs *(with GFCI protection)*

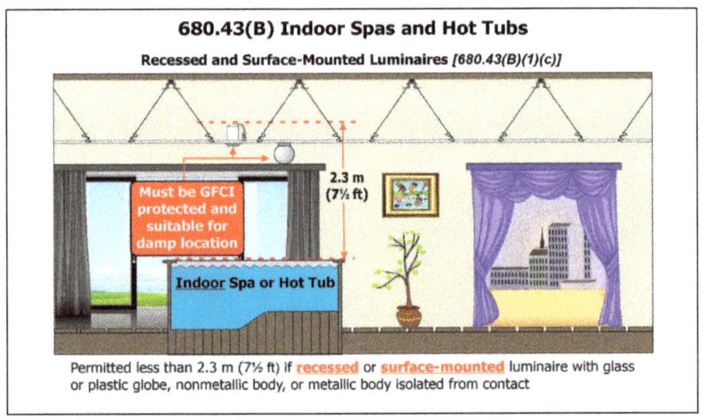

Figure 14.40. Recessed and surface-mounted luminaire requirements for indoor spas and hot tubs

GROUNDING OF INDOOR SPAS AND HOT TUBS

Equipment associated with an indoor spa or hot tub is required to be grounded by connection to the equipment grounding conductor associated with the supply branch circuit or feeder. This includes all electric equipment located within 1.5 m (5 ft) of the inside wall of the spa or hot tub and also electric equipment associated with the circulating system of the spa or hot tub [*NEC* 680.43(F) or IRC E4205.1].

Electrical equipment listed for operation at the low-voltage contact limit or less and supplied by a transformer or power supply that complies with *NEC* 680.23(A)(2) are not required to be connected to the equipment grounding conductor *[see exception to NEC 680.43(F)]*.

METHODS OF GROUNDING INDOOR SPAS AND HOT TUBS

The *Code* requires all electric equipment associated with hot tubs and spas to be grounded by connection to an equipment grounding conductor and to be connected by the general wiring methods. Grounding and bonding terminals are to be identified for use in wet and corrosive environments. Field-installed grounding and bonding connections in a damp, wet, or corrosive environment must be composed of copper, copper alloy, or stainless steel. These terminals will also need to be listed for direct burial use where appropriate as well [*NEC* 680.7, 680.8(C), 680.42(A) or IRC E4202.3, E4205.7, E4205.9].

Listed packaged spa or hot tub equipment assemblies or self-contained spas or hot tubs utilizing a factory-installed or assembled control panel or panelboard are permitted to use flexible connections such as a flexible cord. As stated earlier, for spas and hot tubs, cord-and-plug connections with a cord not longer than 4.6 m (15 ft) are permitted where protected by a ground-fault circuit interrupter. Where a spa or hot tub is connected with a flexible cord, the equipment grounding conductor must be connected to a fixed metal part of the assembly [*NEC* 680.43, Ex. No. 1, 680.8(C), 680.42(A) or IRC E4202.3, E4205.7].

WATER HEATERS FOR INDOOR SPAS OR HOT TUBS

All electric spa or hot tub water heaters are required to be listed. The heating elements incorporating resistive heating elements must be subdivided into loads not exceeding 48 amperes and cannot be protected at more than 60 amperes. The ampacity of the branch-circuit conductors and the rating or setting of overcurrent protective devices cannot be less than 125% of the total load of the marked nameplate rating (*NEC* 680.43, 680.10, or IRC E4208.2).

Branch circuits or feeders serving the gas-fired spa and hot tub water heaters operating separately from the spa or hot tub they serve and operating at voltages above the low-voltage contact limit are required to be ground-fault circuit interrupter (GFCI) protected [*NEC* 680.44(C)].

GROUND-FAULT CIRCUIT-INTERRUPTER PROTECTION OF SPAS AND HOT TUBS

Outlet(s) that supply spas or hot tubs are generally required to be protected by a ground-fault circuit interrupter. This requirement applies to a self-contained spa or hot tub, a packaged spa or hot tub equipment assembly, and a field-installed spa or hot tub with a heater load of 50 amperes or less (*NEC* 680.44, 680.5(B) or IRC E4208.1).

Both indoor and outdoor installations require GFCI protection. GFCI protection may be provided at the receptacle outlet by a GFCI circuit breaker or a self-contained GFCI receptacle or device.

The *Code* excludes those units from the field-supplied GFCI protection at the outlet where a listed self-contained unit or a listed packaged equipment assembly is marked to indicate that GFCI protection is integral to the equipment and protects all electrical parts within the unit or assembly. (Some examples of electrical parts that are required to be protected are pumps, air blowers, heaters, lights, controls, and sanitizer generator wiring.) It is essential to check for proper GFCI protection at the time the unit is connected in the field. Accessory parts may have been added after factory assembly.

Prior to the 2011 edition of the *NEC*, users of the *Code* were told that the electrical equipment for a combination pool/hot tub or spa assembly that is commonly bonded is not required to have separate or additional GFCI protection for the spa or hot tub [see *NEC* 680.44(C) of the 2008 *NEC*]. These units are typically in-ground combination pools and spas that have a common equipotential bonding grid, water circulation system, and common GFCI protection. This statement concerning no additional GFCI protection for combination pool/hot tub or spa assemblies was removed from the 2011 *NEC* as it was considered common knowledge. One would have to assume that this is still common knowledge today.

HYDROMASSAGE BATHTUBS IN DWELLING UNITS

Hydromassage bathtubs are required to comply with Part VII of *NEC* Article 680 only. They are not required to comply with other parts of the article, such as Part I (General Requirements), as do permanently installed pools or spas and hot tubs. The general requirements for hydromassage bathtubs are addressed in Section E4209 in the IRC. A hydromassage bathtub is a permanently installed bathtub that is equipped with a recirculating piping system, pump, and associated equipment to circulate the bath water. The tub is installed with a permanent supply and drain connection to the dwelling's plumbing system. It is designed so it can accept, circulate, and discharge water upon each use [see *NEC* Article 100 or IRC Chapter 35].

Electric equipment, such as luminaires, switches, receptacles, and other electric equipment, located in the same room and not directly associated with a hydromassage bathtub, is required to be installed in accordance with the general rules of the *NEC* Chapters 1 through 4, covering the installation of that type of equipment located in bathrooms (*NEC* 680.72 or IRC E4209.3).

Since a hydromassage bathtub must also comply with the general rules for standard bathtubs, the rules in the *Code* for equipment in the tub area would be applicable. These general rules prohibit any part of cord-connected luminaires, chain-, cable-, or cord-suspended luminaires, lighting track, pendants, or ceiling-suspended (paddle) fans from being located within a zone measured 900 mm (3 ft) horizontally and 2.5 m (8 ft) vertically from the top of the hydromassage bathtub rim. Luminaires located within this zone have to be listed for damp locations or wet locations if subject to shower spray. [*NEC* 410.10(D) or IRC E4003.11].

These general rules would also generally prohibit receptacle outlets from being installed within a zone measured 900 mm (3 ft) horizontally and 2.5 m (8 ft) vertically from the top of the bathtub rim or shower stall threshold (see exceptions). As with luminaires, this identified zone is all-encompassing and includes the space directly over the tub or shower stall. In small bathrooms with physical dimensions less than this required zone, an exception will permit the bathroom receptacle(s) required by *NEC* 210.52(D) or IRC E3901.6 to be installed opposite the bathtub rim or shower stall threshold on the farthest wall within the room [*NEC* 406.9(C) and exception or IRC E4002.11 and exception].

Hydromassage bathtubs and their associated electric components are required to be on an individual branch circuit(s) (branch circuit that supplies only one utilization equipment) and protected by a readily accessible ground-fault circuit interrupter (GFCI) device. This means that the GFCI device, such as a GFCI receptacle, cannot be located under the tub skirt, which would render this GFCI device "not readily accessible." GFCI protection is required for all 125-volt, single-phase receptacles not exceeding 30 amperes within 1.83 m (6 ft) of the hydromassage bathtub. This requirement covers those receptacles not otherwise covered in the *Code* for dwelling unit bathrooms (*NEC* 680.71 or IRC E4209.2).

ACCESS TO HYDROMASSAGE BATHTUB EQUIPMENT

Electrical equipment (including air and water pump motors) associated with the hydromassage bathtub must be accessible

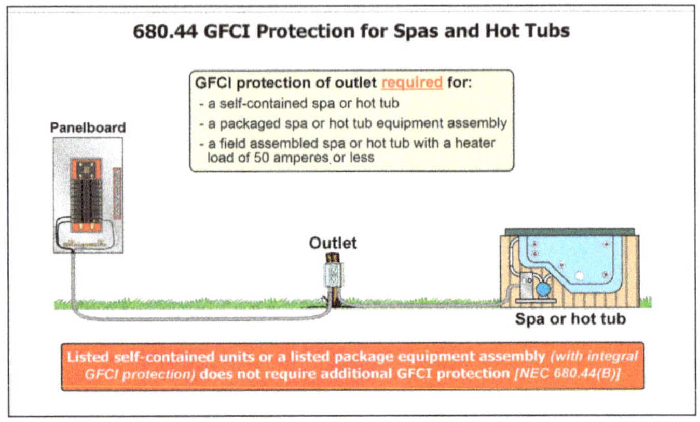

Figure 14.41. Ground-fault circuit-interrupter protection is required for spas and hot tubs

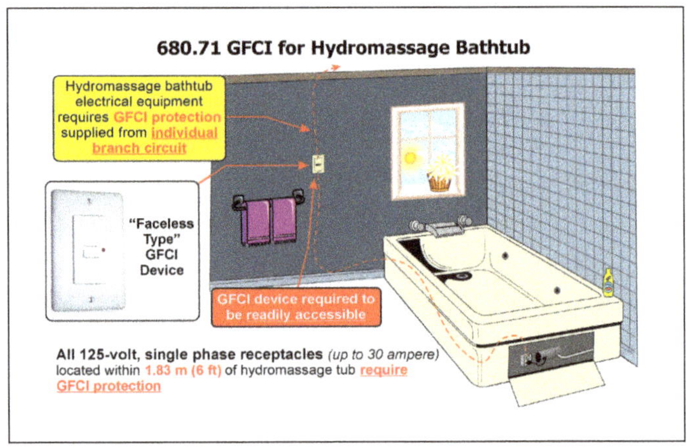

Figure 14.42. GFCI protection required for hydromassage bathtub

without damage to the building structure or finish. If the tub skirt is going to be used as the access, this would prohibit the skirt from being permanently cemented or sealed in place. Caulking would be permitted as caulking can be removed without damage to the building structure or finish. Sufficient access to internal parts such as the pump, pump motor, heater, or control panel is required. Dimensions and location of the access opening may be part of the installation instructions or the local building code. This access opening typically has a minimum of 300 mm (12 in.) by 300 mm (12 in.) in width and height, but again, check your local building code. Where the hydromassage bathtub is cord- and plug-connected, along with the supply receptacle installed under the tub skirt and accessible only through a service access opening, the receptacle shall be installed so that its face is within direct view and not more than 300 mm (1 ft) from the opening. A receptacle protected by a readily accessible GFCI device *(with the GFCI device located in a readily accessible location somewhere other than under the tub skirt)* can be installed under the tub skirt (*NEC* 680.73 or IRC E4209.4).

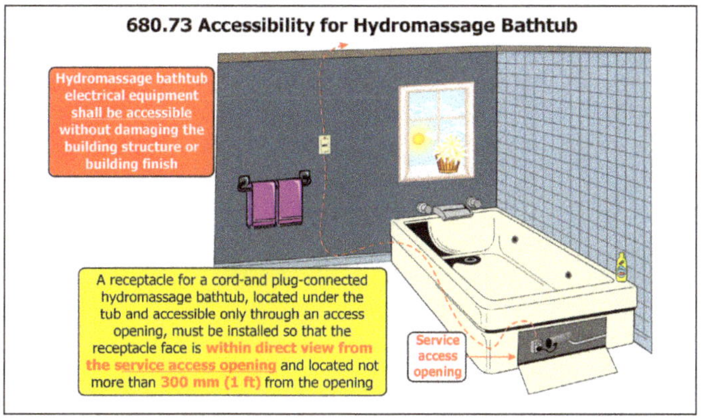

Figure 14.43. Access is required for hydromassage bathtub motors and associated electrical equipment

Photo 14.30. Access is required to hydromassage bathtub electrical equipment

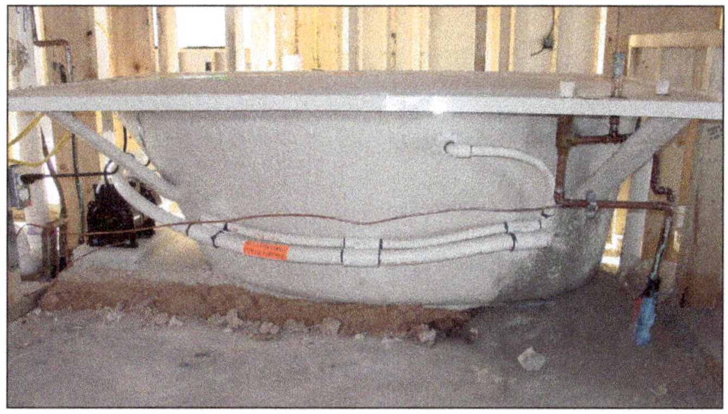

Photo 14.31. Typical hydromassage bathtub at rough-in stage, showing the 8 AWG solid copper bonding jumper

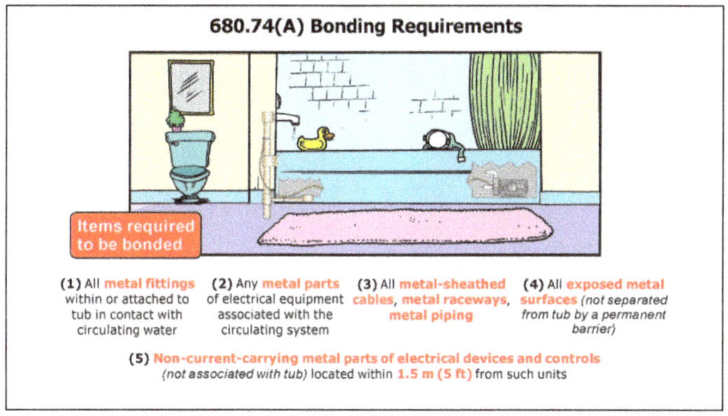

Figure 14.44. Metal water piping and metal parts associated with the hydromassage bathtub are required to be bonded together with an 8 AWG solid copper bonding conductor

BONDING FOR HYDROMASSAGE BATHTUBS

NEC Section 680.74(A) or IRC E4209.5 requires the following parts associated with a hydromassage bathtub to be bonded together:

1. All metal fittings within or attached to the tub structure that are in contact with the circulating water
2. Any metal parts of electrical equipment associated with the tub water circulating system *(pump and blower motors, etc.)*
3. All metal-sheathed cables, metal raceways, and metal piping located within 1.5 m (5 ft) of the inside walls of the tub and not separated from the tub by a permanent barrier
4. All exposed metal surfaces located within 1.5 m (5 ft) of the inside walls of the tub (not separated from the tub by a permanent barrier)
5. Non-current-carrying metal parts of electrical devices and controls that are not associated with the hydromassage tubs and that are located within 1.5 m (5 ft) from such units

An exception relieves small conductive surfaces not likely to become energized from the bonding requirements of *NEC* 680.74(A) or IRC E4209.5. This would include such things as air and water jets, supply valve assemblies, drain fittings not connected to metallic piping, towel bars, mirror frames, and similar nonelectrical equipment not connected to the metal framing of the hydromassage tub. A second exception pardons double-insulated motors and blowers from bonding requirements for hydromassage tubs. A third exception exempts small conductive surfaces of electrical equipment that are not likely to become energized by the bonding requirements for hydromassage bathtubs. This would include such things as mounting straps or yokes of listed light switches or receptacles that are grounded [*NEC* 680.74(A) Ex. No. 1, Ex. No. 2, and Ex. No. 3 or IRC E4209.5 Ex. No. 1, Ex. No. 2, and Ex. No. 3].

All metal parts related to a hydromassage tub that must be bonded are required to be bonded together using a minimum size bonding jumper of 8 AWG, which consists of a solid copper conductor. This bonding jumper may be insulated, covered, or bare. Its purpose is to connect the various non-current-carrying conductive parts of the tub assembly. In all installations, including those with a double-insulated motor, a means must be provided to ground internal, inaccessible, non-current-carrying metal parts. Typically, this is achieved by the equipment grounding conductor of the branch circuit for the double-insulated motor. If an existing double-insulated motor is replaced with a non-double-insulated motor, the required bonding jumper must be long enough to connect to the new, non-double-insulated pump motor and serve as a bonding means. When using a double-insulated circulating pump motor, this bonding jumper should be terminated to the equipment grounding conductor of the branch circuit. The purpose of the 8 AWG or larger solid copper bonding jumper is to achieve equipotential bonding in the area of the hydromassage bathtub and does not need to be extended or connected to a remote panelboard, service equipment, or any grounding electrode (NEC 680.74(B) or IRC E4209.6).

Photo 14.32. Bonding of metal water piping associated with hydromassage bathtubs

Photo 14.33. Metal parts of electrical equipment, metal piping systems, and pump motors associated with hydromassage bathtubs are required to be bonded together

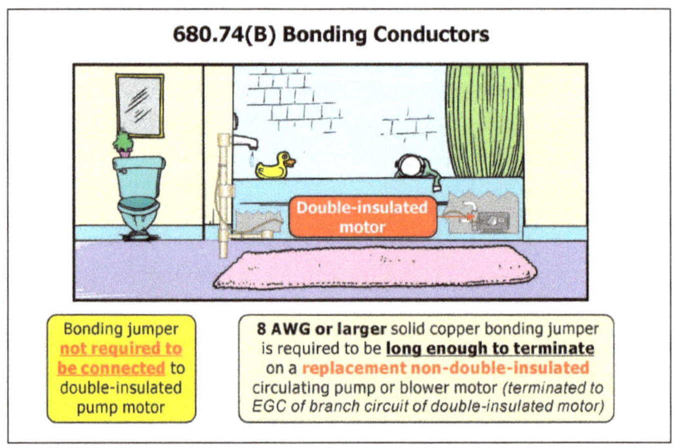

Figure 14.45. Bonding provisions must be made for non-double-insulated hydromassage replacement motors

PRODUCT SAFETY STANDARD REQUIREMENTS

It is advised to refer to the following electrical product safety guide card information and product safety standards for additional guidance on the proper installation, operation, and use of electrical equipment covered in this chapter. The four-letter code in parentheses indicates the product category in the Underwriters Laboratories (UL) Product Spec (formerly UL White Book).

- Electrical Equipment for Use in Ordinary Locations (AALZ)
- Flexible Metal Conduit (DXUZ)
- Flexible Metal Conduit, Liquid-tight (DXHR)
- Flexible Nonmetallic Conduit, Liquid-tight (DXOQ)
- Flexible Cord (ZJCZ)
- Intermediate Ferrous Metal Conduit (DYBY)
- Rigid Ferrous Metal Conduit (DYIX)
- Rigid Nonferrous Metallic Conduit (DYWV)
- Rigid Nonmetallic PVC Conduit (DZYR)
- Electrical Metallic Tubing (FJMX)
- Electrical Metallic Tubing Fittings (FKAV)
- Electrical Nonmetallic Tubing (FKHU)
- Electrical Nonmetallic Tubing Fittings (FKKY)
- Submersible Luminaires (IFEV)
- Ground-Fault Circuit Interrupters (KCXS)
- Grounding and Bonding Equipment (KDER)
- Panelboards (QEUY)
- Swimming Pool and Spa Equipment (WABX)
- Blowers, Swimming Pool and Spa Equipment (WAGN)
- Controls, Swimming Pool and Spa Equipment (WAWU)
- Luminaires and Forming Shells (WBDT)
- Heaters, Swimming Pool and Spa Equipment (WBRR)
- Hot Tub and Spa Equipment Assemblies (WBYQ)
- Junction Boxes, Swimming Pool and Spa Equipment (WCEZ)
- Ozone Generators, Swimming Pool and Spa Equip-

Equipment Grounding Conductor (EGC) Connections and
Wiring Methods from Service to Permanently Installed Pool Equipment

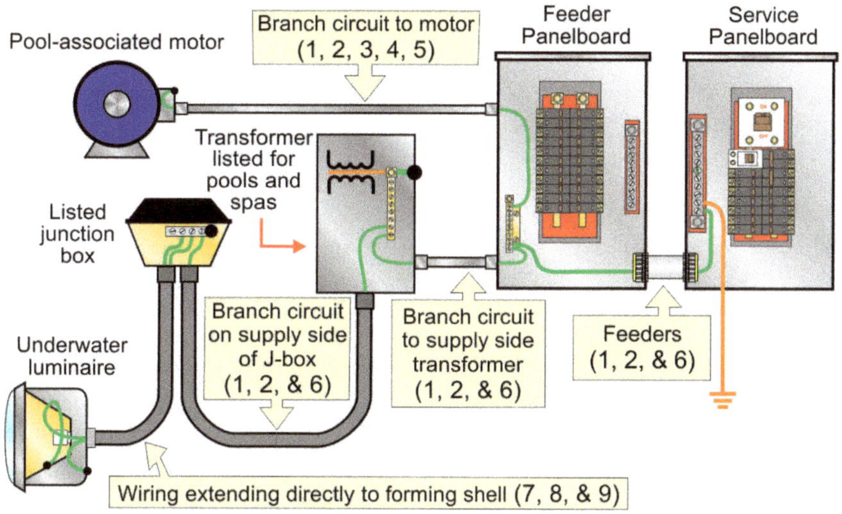

Legend:

1. **Where installed in a corrosive environment**; Rigid metal conduit, intermediate metal conduit, rigid polyvinyl chloride PVC conduit, and reinforced thermosetting resin conduit with an insulated copper EGC sized in accordance with *NEC* Table 250.122 or IRC Table E3908.13 but not smaller than 12 AWG.
2. **Where installed in noncorrosive environments**, branch circuits or feeders to comply with the general requirements in *NEC* Chapter 3 or IRC Chapter 38.
3. For motors installed in a **corrosive environment**; Wiring methods of List Item No. 1 or shall be Type MC cable listed for the environment with an insulated copper EGC sized in accordance with *NEC* Table 250.122 or IRC Table E3908.13 but not smaller than 12 AWG.
4. For a motor application; Liquidtight flexible metal or liquidtight flexible nonmetallic conduit with listed fittings permitted where flexibility required.
5. Flexible cord permitted for connections at motors.
6. Where feeders or branch circuits on the supply side of enclosures and junction boxes connected to conduits run to underwater luminaire are installed in a **corrosive environment**; Wiring methods of List Item No. 1 or liquidtight flexible nonmetallic conduit with an insulated copper EGC sized in accordance with *NEC* Table 250.122 or IRC Table E3908.13 but not smaller than 12 AWG.
7. Wet-niche or no-niche luminaire supplied by flexible cord; Insulated copper EGC that is an integral part of the cord and not smaller than the supply circuit conductors and not smaller than 16 AWG.
8. For underwater luminaires; Rigid metal, intermediate metal, liquidtight flexible nonmetallic or rigid polyvinyl chloride PVC conduit. Metal conduit must be of brass or other corrosion-resistant metal.
9. For underwater luminaires; Where a nonmetallic conduit is used, an 8 AWG insulated solid or stranded copper bonding jumper shall be installed in conduit unless a listed low-voltage lighting system not requiring grounding is used.

Figure 14.46. Equipment grounding conductor connections and permitted wiring methods from service equipment to pool equipment

ment (WCKA)

- Potting Compounds, Swimming Pool and Spa Equipment (WCRY)
- Pumps, Swimming Pool and Spa Equipment (WCSX)
- Self-Contained Spas (WCZW)
- Swimming Pool and Spa Transformers (WDGV)
- Water Treatment Equipment (WDLC)
- Wire, Thermoset-Insulated (ZKST)
- Wire, Thermoplastic-Insulated (ZLGR)

REFERENCES

1. *National Electrical Code* and *NEC* are registered trademarks of the National Fire Protection Association, Inc., Quincy, MA 02169. This reprinted material is not the official position of the National Fire Protection Association, which is represented only by the standard in its entirety.

2. *International Residential Code for One- and Two-Family Dwellings* (2025). Copyright © International Code Council, Inc., June 2023.

3. UL, Northbrook, IL, *UL Guide Information for Electrical Equipment Directory* (UL Product iQ® material) unless indicated otherwise.

REVIEW QUESTIONS

1. A low-voltage contact limit is a voltage not exceeding which of the following values:

A. 15 volts (RMS) for sinusoidal ac and 21.2 volts peak for nonsinusoidal ac
B. 30 volts for continuous dc
C. 12.4 volts peak for DC that is interrupted at a rate of 10 to 200 Hz
D. All of the above

2. The maximum voltage between conductors supplying current to a luminaire installed below a pool surface is _____ volts.

A. 120
B. 150
C. 208
D. 250

3. A junction box connected to a conduit that extends directly to a forming shell of a no-niche luminaire must be provided with a number of grounding terminals that must be at least ____ more than the number of conduit entries.

A. One
B. Two
C. Three
D. Four

4. For permanently installed swimming pools, which of the following statements is NOT true about GFCI protection of luminaires or ceiling-suspended (paddle) fans installed around pools?

A. Luminaires installed more than 3.7 m (12 ft) above the pool are not required to have GFCI protection.
B. Existing luminaires installed between 1.5 m and 3.0 m (5 ft and 10 ft) above the pool and less than 1.5 m (5 ft) from the pool do not require GFCI protection.
C. Ceiling-suspended (paddle) fans installed more than 2.3 m (7 ft 6 in.) above an indoor pool do not require GFCI protection.
D. Luminaires installed in the area extending between 1.5 m (5 ft) and 3.0 m (10 ft) horizontally and not less than 1.5 m (5 ft) above the maximum water level and rigidly attached to the structure adjacent to or enclosing the pool, do not require GFCI protection.

5. A flush deck box is permitted for an underwater pool luminaire operating on a lighting system at the low-voltage lighting system or less if which of the following conditions are met:

A. The flush deck box is located not less than 600 mm (2 ft) from the inside wall of the pool.
B. The flush deck box is located not less than 1.2 m (4 ft) from the inside wall of the pool.
C. An approved potting compound is used to fill the box to prevent the entrance of moisture.
D. Both B. and C.

6. Which one of the following is the minimum size bonding conductor permitted to be used for equipotential bonding at a permanently installed swimming pool installation?

A. 8 AWG stranded copper conductor
B. 8 AWG solid copper conductor
C. 6 AWG solid copper conductor
D. 4 AWG stranded copper conductor

7. Where structural reinforcing steel is not available or is encapsulated in a nonconductive compound and a copper grid system is chosen as the alternative method of bonding the perimeter surface, which of the following requirements must be met:

A. Be secured within or under the deck or unpaved surfaces between 100 mm to 150 mm (4 in. to 6 in.) below the subgrade.
B. Be constructed of 8 AWG solid bare copper and be arranged in a 300 mm (12 in.) by 300 mm (12 in.) network of conductors in a uniformly spaced perpendicular grid pattern with a tolerance of 100 mm (4 in.).
C. Follow the contour of the perimeter surface extending 1 m (3 ft) horizontally beyond the inside walls of the pool.
D. All of the above

8. A storable swimming or wading pool is constructed on or above the ground that is intended to be stored when not in use, is designed for ease of relocation, and is capable of holding water to a maximum depth of what dimension?

A. regardless of water depth (no maximum water depth given)
B. 900 mm (36 in.)
C. 1.0 m (42 in.)
D. 1.2 m (48 in.)

9. Wall switches are required to be located at least what distance, measured horizontally, from the inside walls of an indoor hot tub?

A. 900 mm (3 ft)
B. 1.2 m (4 ft)
C. 1.5 m (5 ft)
D. 3.0 m (10 ft)

10. Receptacles, including those that supply power to a water-pump motor are required to be located at least what distance from the inside walls of a permanently installed swimming pool?

A. 1.52 m (5 ft)
B. 1.83 m (6 ft)
C. 3.0 m (10 ft)
D. 6.0 m (20 ft)

11. The specified parts of the equipotential bonding grid for a permanently installed pool can be bonded together using solid copper conductor(s) not smaller than ____ AWG or with rigid metal conduit of brass or other identified corrosion-resistant metal.

A. 12
B. 10
C. 8
D. 6

12. A dry-niche luminaire must be provided with a means for drainage of water and for other than listed low-voltage luminaires not requiring grounding, a means for accommodating one equipment grounding conductor for each conduit entry.

True _____ False _____

13. Other equipment (other than traditional pool pump motors and controllers), such as a ground-mounted PV array with ratings exceeding the low-voltage contact limit, must be located at least __________ horizontally from the inside walls of a pool unless separated from the pool by a solid fence, wall, or other permanent barrier.

A. 900 mm (3 ft)
B. 1.2 m (4 ft)
C. 1.5 m (5 ft)
D. 3.0 m (10 ft)

14. For other than listed low-voltage luminaires not requiring grounding, a metal raceway cannot be solely depended upon for equipment grounding purposes between a metal forming shell and the junction box of a through-wall lighting assembly, wet-niche, dry-niche, or no-niche luminaire.

True _____ False _____

15. Junction boxes extending to underwater pool luminaires are required to be located not less than what distance horizontally from the inside wall of the pool where there is no barrier separation?

A. 900 mm (3 ft)
B. 1.2 m (4 ft)
C. 1.5 m (5 ft)
D. 1.8 m (6 ft)

16. Which of the following conduits is NOT permitted to be used between a wet-niche luminaire forming shell and the pool junction box?

A. electrical metallic tubing (EMT)
B. rigid polyvinyl chloride (PVC)
C. rigid red brass
D. liquidtight flexible nonmetallic

17. A listed cord-and-plug-connected pool pump incorporating an approved system of double insulation that provides a means for grounding only the internal and nonaccessible, non-current-carrying metal parts of the pump shall be connected to any wiring method ______________ ____________________ that is suitable for the location.

A. recognized by *NEC* or the IRC
B. recognized in *NEC* Chapter 3 or IRC Chapter 38
C. listed for a wet location
D. approved by the AHJ

18. Transformers and power supplies used for the supply of underwater luminaires, together with the transformer or power supply enclosure, are required to be __________________.

A. listed
B. listed, labeled and identified for swimming pool and spa use
C. listed for the purpose
D. approved for use with swimming pools are spas

19. Which of the following self-contained spas, hot tubs, or packaged spas, or hot tub equipment assemblies is NOT required to have any GFCI protection for personnel?

A. A self-contained spa or hot tub is marked as having integral GFCI protection.
B. A packaged spa or hot tub equipment assembly is marked as having integral GFCI protection.
C. A field-installed spa or hot tub is supplied by a branch circuit rated 150 volts or less to ground and with a heater load of 60 amperes or less.
D. A field-installed spa or hot tub is supplied by a branch circuit rated 150 volts or less to ground and with a heater load of 100 amperes.

20. Which of the following is NOT a permitted wiring method for a permanently installed swimming pool pump motor installed in a corrosive environment associated with a single-family dwelling?

A. Rigid polyvinyl chloride (PVC) conduit
B. Liquidtight flexible nonmetallic conduit containing a bare 12 AWG equipment grounding conductor
C. Rigid metal conduit containing an insulated 12 AWG equipment-grounding conductor
D. Rigid nonmetallic conduit containing an insulated 12 AWG equipment-grounding conductor

21. A pool pump motor for a permanently installed pool supplied by an outlet that is directly connected (hard wired) from a 30-amperes, 240-volt, single-phase branch circuit does NOT require Class A ground-fault circuit-interrupter protection for personnel.

True _____ False _____

22. Where a double-insulated water-pump motor is installed for a permanently installed swimming pool, a solid 8 AWG copper conductor that is of sufficient length to make a bonding connection to a replacement motor shall be extended from the equipotential bonding means to an accessible point in the motor vicinity.

True ______ False ______

23. Hydromassage bathtubs and their associated electric components are required to be on a/an __________ branch circuit(s) and protected by a __________ ground-fault circuit interrupter.

A. dedicated, readily accessible
B. individual, readily accessible
C. individual, accessible
D. new, Class A

24. Under certain conditions, some specific low-voltage luminaires are permitted to be located less than 1.5 m (5 ft) from the inside walls of a permanently installed pool but must be of the type that does not require ________________ .

A. maintenance
B. corrosion protection
C. grounding (connection to an equipment grounding conductor)
D. luminaire protection

25. Areas or enclosures without adequate ventilation, where electrical equipment is located and pool sanitation chemicals are stored, handled, or dispensed is the definition of which of the following terms?

A. Corrosive Condition
B. Hazardous (Classified) Location
C. Corrosive Environment
D. Pool Equipment Room or Areas

26. All metallic parts of a swimming pool structure, including reinforcing metal, are required to be bonded. Where reinforcing steel is ___________________________, the reinforcing steel is not required to be bonded.

A. wrapped in electrical tape
B. encapsulated with a nonconductive compound
C. buried in the deck
D. encapsulated with a reflective paint

27. Branch circuits or feeders that serve gas-fired swimming pool and spa water heaters that will operate at voltages above __________ are required to be provided with ground-fault circuit-interrupter protection for personnel.

A. 1 milliamp
B. the high-voltage contact limit
C. the low-voltage contact limit
D. the medium-voltage contact limit

28. Feeders and branch circuits installed in a corrosive environment or a wet location are required to contain an equipment grounding conductor (EGC) that is an __________ copper conductor not smaller than __________.

A. unsulated, 12 AWG
B. approved, 8 AWG
C. insulated, 12 AWG
D. covered, 8 AWG

29. Structural reinforcing steel for an in-ground swimming pool shall be used as part of the grounding electrode system.

True _____ False _____

30. Electrically powered swimming pool heat pumps and chillers using the circulating water system and providing heating, cooling, or both must be listed and rated for their intended use. The ampacity of the branch-circuit conductors and the rating or setting of overcurrent protective devices must be sized to comply with __________________________ .

A. the authority having jurisdiction
B. the equipment nameplate
C. 125 percent of the load
D. industry standards

31. Electrically powered pool lifts must be __________ in accordance with *NEC* 680.26(B)(5) and (B)(7).

A. grounded
B. bonded
C. installed
D. connected

32. At least ________ GFCI-protected 125-volt, 15- or 20-ampere receptacle on a general-purpose circuit is required to be located within a pool equipment room.

A. one
B. two
C. three
D. receptacle not required

33. For servicing wet-niche luminaires in spa locations with luminaires installed low in the foot well of the spa, the luminaire is only required to reach the bench location, where the spa can be drained to make the bench location dry for maintenance and repair.

True _____ False _____

15 Branch Circuits and Feeders Run to Accessory Buildings

Although the terms *outbuilding* and *accessory buildings* are both used in the *Code*, these terms are not defined in the *Code*. These terms can be associated with any building that is detached from a one- or two-family dwelling and that serves a support or ancillary function to the dwelling unit. Examples of an outbuilding include a garage, workshop, cabana, gazebo, water pump house, boathouse, shed, pool pump house, or similar structures. Article 100 of the *NEC* defines a "Building" as "A structure that stands alone, or that is separated from adjoining structures by firewalls." In Chapter 35, the IRC defines an "Accessory Structure" as "a structure that is accessory to and incidental to that of the dwelling(s), and that is located on the same lot."

A "Building" is defined in Chapter 35 as "any one-and two-family dwelling or townhouse, or portion thereof, used or intended to be used for human habitation, for living, sleeping, cooking, or eating purposes, or any combination thereof, or any accessory structure."

Article 225 of the *NEC* covers outside branch circuits and feeders. An outside circuit could supply equipment on or near the dwelling. A branch circuit to a pole-mounted yard light is one example. Or it could supply a separate building or structure located on the premises. A feeder used to supply a detached garage is another example. The rules in *NEC* Article 225 apply to both underground and overhead branch circuits and feeders.

Where the IRC does not provide specific Code rules for outside branch circuits or feeders supplying separate buildings or structures on a residential property, the requirements of the NEC must be applied to the electrical installation by default, as indicated in IRC E3401.2.

LOAD CALCULATIONS AND OVER-CURRENT PROTECTION

Outdoor branch-circuit and feeder loads are required to be calculated in accordance with the ampacity tables and requirements of *NEC* 310.15 based on loads as determined under *NEC* 220.10 and Part III of Article 220 (see Chapter 2 for additional information on performing load calculations). The IRC generally provides specific feeder sizing requirements in Section E3704, and specifics are provided in E3704.2, E3704.3, and Table E3704.2(1). Section E3705 provides specifics as to conductor sizing and maximum allowable overcurrent protection for conductors. These rules are consistent with those provided in Articles 215, 220, and 310 of the *NEC*.

Outdoor branch circuits and feeder conductors must be protected against overcurrent. The conductor ampacities are permitted to be determined as specified in the "Allowable Ampacity Tables" of *NEC* Article 310 or IRC Chapter 37 to determine the minimum ampacity of outside branch circuits and feeders.

MINIMUM SIZE OF CONDUCTORS

Open individual conductors are permitted for use in overhead spans between buildings, structures, or poles. For spans up to 15 m (50 ft), these overhead conductors must be at least 10 AWG copper or 8 AWG aluminum. Where an overhead span is longer than 15 m (50 ft), a minimum size of 8 AWG copper or 6 AWG aluminum is required. Smaller conductors are permitted where a messenger wire supports the overhead span [NEC 225.6(A)]. The span is measured from one support to the next. The permitted support for an overhead circuit is a suitable building or structure, such as a pole. The Code does not permit live vegetation, such as trees, to support overhead conductor spans (NEC 225.26).

The typical wiring method utilized for overhead spans is individual open conductors connected to screw-in knobs, triplex or quadruplex cables. Duplex, triplex, and quadruplex cables are defined in NEC Article 100 as part of a Messenger-Support Wiring system. They consist of factory-assembled and twisted insulated conductors that are spiraled around a bare support cable. The bare support cable, or messenger, typically has a steel alloy core for additional strength. The messenger is often used as the equipment grounding conductor of the circuit. Type UF cable (NEC Article 340, wiring method) is also permitted for use in overhead spans, provided it is specifically marked and identified as sunlight resistant and installed to conform to the requirements for messenger-supported wiring.

CONDUCTOR COVERING AND WIRING METHODS ON BUILDINGS AND OUTBUILDINGS

Open wiring installed on insulators must be insulated for the nominal voltage where it is located within 3.0 m (10 ft) of any building or outbuilding (NEC 225.4). The term "Insulated Conductor" is defined in NEC Article 100 and IRC Chapter 35 as "a conductor encased within material of composition and thickness that is recognized by the Code as electrical insulation."

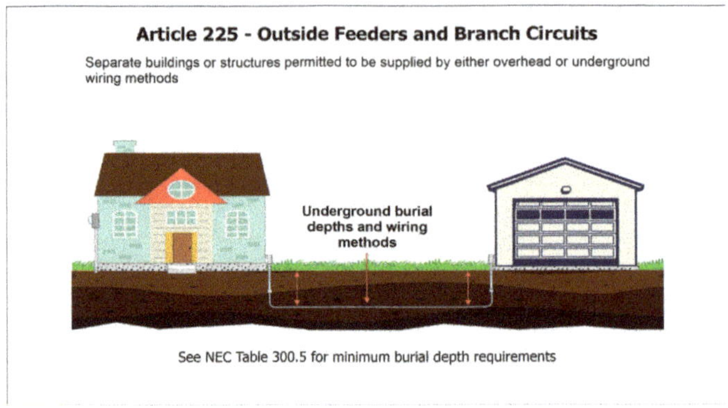

Figure 15.1. Feeders or branch circuits are permitted to supply separate structures (outbuildings) by overhead or underground wiring methods

Conductors within cable assemblies or raceways must be of the thermoset (rubber-insulated) or thermoplastic type. Where installed in wet locations, conductors must be suitable for wet locations. Examples of insulation types identified for a wet location used at dwelling units would be Type THW-2, Type THHW, Type XHHW-2, etc. Where these conductors are exposed to the direct rays of the sun, they must also be listed for sunlight resistance or listed and marked "sunlight resistant." Conductors used for direct-burial applications must be of a type identified for such use, such as Type UF cable [*NEC* 310.10(A), (B), (C), (D), and (E) or IRC Table E3801.4].

Where installed on the surfaces of one- and two-family dwellings or outbuildings, outside branch circuits and feeders typically consist of one or more of the following types (*NEC* 225.10 or IRC E3801.4):

1. Open wiring on insulators
2. Multi-conductor cable assemblies
3. Type MC cable assemblies
4. Type UF cable assemblies
5. Type MI cable assemblies
6. Type SE cable assemblies
7. Type TC-ER cable assemblies
8. Messenger-supported wiring
9. Rigid metal conduit, intermediate metal conduit, rigid polyvinyl chloride (PVC) conduit, or reinforced thermosetting resin conduit (RTRC)
10. Electrical metallic tubing, flexible metal conduit, liquidtight flexible metal conduit, and liquidtight flexible nonmetallic conduit

Photo 15.1. Typical quadruplex cables used for outside overhead services and feeders

11. In wireways and auxiliary gutters

Outside branch circuits and feeders are also permitted to be in cable trays, as a cablebus, and installed as a busway, but these items are not typically found in dwelling unit construction.

Outside branch circuits and feeder wiring installed on buildings, structures, or poles must be protected against physical damage. Mechanical protection for conductors is required when the AHJ determines they are at risk of physical damage. All raceways installed on the exterior surfaces of buildings must be suitable for use in a wet location and designed to drain [NEC 225.20, 225.22]. Furthermore, all conductors within a raceway that are above ground in a wet location must meet the insulation designator for a wet location as specified in NEC 310.4, 300.9, and IRC E3802.8.

OVERHEAD CIRCUITS LEAVING OR ENTERING A BUILDING

Where overhead outside branch circuits and feeders leave or enter a building, they must comply with NEC 230.52, 230.54 [see NEC 225.11], or IRC E3605.9. These rules outline the permitted methods for feeder or branch-circuit conductors located outside the structure. NEC 230.52 permits individual open conductors to enter a building through roof bushings or the wall in an upward slant through individual, noncombustible, nonabsorbent insulating tubes. Under today's construction methods, this method is less likely to be used than the weatherhead wiring method permitted by NEC 230.54 or IRC E3605.9.

A conduit is usually installed from the last junction or pull box or runs from the panelboard to a location outside the building. The conduit on the outside of the building is often referred to as a mast. The rules for outside feeders or branch circuits are similar to those for an overhead service. The mast ends in a gooseneck or weatherhead. A weatherhead or gooseneck is installed to prevent water from entering the conduit. If the mast is used to support the overhead conductors, it must be of adequate strength or be supported by braces or guy wires. The overhead conductors can place a considerable strain on the mast due to the dead weight of the overhead cables. Wind or ice can produce additional strain. All fittings on a raceway mast must be identified for use with that mast. Only the overhead feeder or branch circuit's conductors are permitted to be attached to the mast. Cables for other purposes, such as internet, telephone, or CATV, require independent support (NEC 225.17 or IRC E3604.5).

Type SE cable (service-entrance cable) is permitted to be used as a feeder between buildings. A weatherhead or "gooseneck" is installed to prevent water from entering the cable assembly where Type SE cable connects to the overhead span at the buildings. A gooseneck resembles an "R" and is formed by placing a 180-degree bend on the cable with a small loop at the top of the cable and secured to the rising cable with tape. The conductors are taped where they leave the outer jacket of the cable. The Code still recognizes the use of a gooseneck, which many electrical inspectors often require in areas adjacent to the seashore. Experience has shown that an improperly installed raintight service head may lead to a situation where rain, melting snow, or ice will dribble down and into the enclosure it is connected to.

A weatherhead or gooseneck in a service-entrance cable assembly is required to be located "above the point of attachment" of the outside branch circuit and feeder conductors to the building or outbuilding. Where it is impracticable to locate the weatherhead or gooseneck above the point of attachment, it is permitted to be installed at a distance of not more than 600 mm (2 ft) away from the point of attachment.

To prevent the entrance of moisture into conductors, a drip loop must be formed on all individual outside branch circuits and feeder conductors, either below the level of the raintight service head or below the level of the termination of the service-entrance cable sheath.

Conductors from inside the building connect to the overhead feeder or branch circuit at the gooseneck or weatherhead. Connections at this location are made with connectors that are suitable for the purpose. Mechanical connectors such as "split bolts" and irreversible crimp connectors are typically used to make these connections. It is important to follow the manufacturer's instructions and use the proper tool when crimping the connector. The connector must be suitable for the type of conductor materials used. It is common to see copper building wire connected to an overhead cable with aluminum conductors at this location. A connector rated for copper-to-aluminum connections must be used if this is the case. Any exposed conductive part of the connector must be insulated with weather-resistant tape. The thickness of the tape insulation on the splice should be equal to that of the insulation on the conductor.

Raceways installed on the exterior of buildings or other structures must be arranged to drain. Such raceways must also be listed and approved for use in wet locations. Selecting the appropriate type of raceway fitting for the location is a crucial consideration. The interior of a raceway installed in a wet location is considered a wet location as well (see NEC 300.9), so properly identified conductors will be a consideration. Raceway fittings are designed to provide enhanced protection from elements such as rain and ice (NEC 225.22 or IRC E3605.8).

A cable assembly must be supported and held securely in place by straps or other approved means. *NEC* 230.51 gives the proper spacing of cable supports. In some cases, the requirements in *NEC* 230.51 are more restrictive than would be found in the respective cable's *NEC* article. For example, *NEC* 230.51 requires that service cables be supported at intervals of not more than 750 mm (30 in.). The requirements for securing and supporting the same type of SE cable on the inside of the building are given at 338.10(B)(4), which references *NEC* 334.30, which would permit SE cable to be secured and supported at intervals of 1.4 m (4½ ft) (*NEC* 225.21).

The point of attachment to a building and an outbuilding must be located so as to provide the minimum required clearances for the conductors. Under no circumstances is the point of attachment permitted to be less than 3.0 m (10 ft) above the finished grade (NEC 225.16(A), 230.26, or IRC E3604.3).

The means of attachment for the outside branch circuits and feeders to a building or outbuilding is required to be made by fittings identified for use with service conductors. Open conductors must be attached to fittings identified for use with service conductors or to noncombustible, nonabsorbent insulators securely attached to the building or other structure. The method of support for the insulator or insulator racks must consider the possibility of weak or rotted structural members. It may be adequate to screw the insulator directly into new structural lumber. With an existing building, common practice is to install a threaded rod through the building and into the attic or other space. The rod is then backed with solid lumber and/or a steel backing plate.

CLEARANCE FOR OVERHEAD CONDUCTORS

Outside branch circuits and feeders that are run as spans of open conductors or as open multi-conductor cables are required to be installed at the following heights:

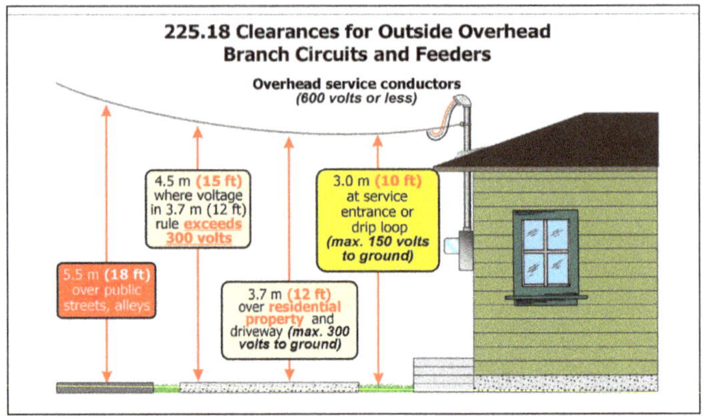

Figure 15.2. Overhead vertical clearances from grade for outdoor overhead branch circuits and feeder conductors

1. **3.0 m (10 ft)** above finished grade, sidewalks, or from any platform or projection from which they might be reached where the supply conductors are limited to 150 volts to ground and accessible to pedestrians only
2. **3.7 m (12 ft)** over residential driveways
3. **4.5 m (15 ft)** for those areas listed in the 3.7-m (12-ft) classification where the voltage exceeds 300 volts to ground (this is rare for residential applications)
4. **5.5 m (18 ft)** over public streets, alleys, roads, and parking areas subject to truck traffic, and other types of land, such as cultivated, grazing, forest, and orchard, traversed by vehicles (see Chapter 4 of this text for illustrations on clearances of overhead conductors over the ground and roofs)
5. **7.5 m (24.5 ft)** over railroad tracks

These clearances are similar to the clearances for overhead service conductors at *NEC* 230.24(B) and IRC E3604.2.2. These overhead clearances for outside branch circuits and feeders are found at *NEC* 225.18.

CLEARANCES BETWEEN BUILDINGS AND STRUCTURES FROM OVERHEAD CONDUCTORS

Similar to the requirements for overhead service conductors, a vertical clearance of not less than 2.6 m (8 ft 6 in.) is required to be maintained from the surface of a roof to the outside branch circuit and feeder conductors. This vertical clearance above the roof level is required to be maintained for a distance of not less than 900 mm (3 ft) in all directions from the edge of the roof. These clearance requirements for outside branch circuits and feeders can be found at *NEC* 225.19. Six exceptions or conditions to this 2.6 m (8 ft 6 in.) rule are commonly applied to residential dwellings as follows:

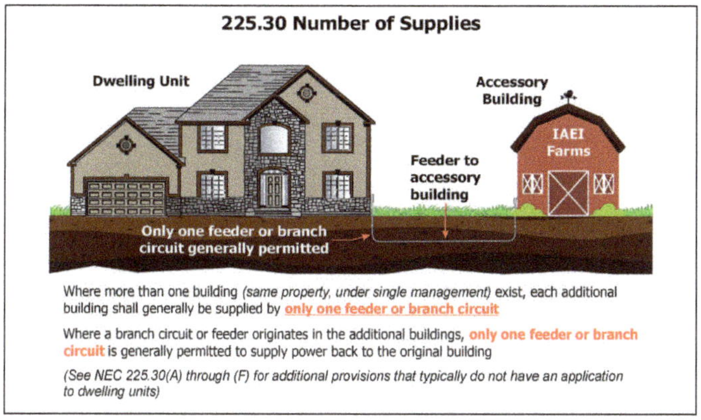

Figure 15.3. One feeder is generally permitted to accessory buildings

1. Where the voltage between the outside branch-circuit and feeder conductors is not over 300 volts and where the roof has a slope of 100 mm (4 in.) in 300 mm (12 in.) or greater, a reduction in clearance to 900 mm (3 ft) is permitted [*NEC* 225.19(A), Ex. No. 2].
2. Where the voltage between the outside branch-circuit and feeder conductors is not over 300 volts, a reduction in the minimum required clearance above an overhanging portion of a roof to not less than 450 mm (18 in.) is permitted if not more than 1.8 m (6 ft) of the conductors pass above the roof overhang. They are terminated at a through-the-roof raceway or approved support. The overhang portion of the roof is limited to 1.2 m (4 ft) horizontally in this case. The requirement for maintaining the vertical clearance of 900 mm (3 ft) from the edge of the roof does not apply to the final conductor span(s) where the conductors are attached to the side of a building [*NEC* 225.19(A) Ex. No. 3 and Ex. No. 4)].
3. Vertical, diagonal, and horizontal clearances for outside branch circuits and feeders cannot be less than 900 mm (3 ft) from a chimney, radio, and television antenna and other nonbuilding structures [*NEC* 225.19(B)].
4. Horizontal clearances from buildings to overhead conductors or cables cannot be less than 900 mm (3 ft) [*NEC* 225.19(C)].
5. Final spans of outside branch circuits or feeders attached to the building they supply or from which they are fed are permitted to be attached to the building. The final span must be kept 900 mm (3 ft) away from the sides and bottom of windows, doors, porches, balconies, ladders, stairs, fire escapes, or similar locations [*NEC* 225.19(D)(1)].
6. A vertical clearance is required from porches, platforms, or other surfaces from which overhead conductors may be reached. The final span above or within 900 mm (3 ft) horizontally from the porch deck, platform, or similar location is required to meet the minimum clearances above ground specified by the *Code*. For example, this would require that a final drop to a building be located a minimum of 3.0 m (10 ft) over a deck, porch, or balcony [*NEC* 225.19(D)(2), 225.18].

Overhead branch-circuit or feeder conductors are also not permitted to be installed beneath openings through which materials may be moved and are not to be installed where the conductors will obstruct entrance to these building openings [*NEC* 225.19(D)(3)]. The IRC only covers overhead conductor spans for service conductors, so defaulting to the *NEC* would be applicable here. As stated previously, these overhead clearances for outside branch circuits and feeders are similar to the requirements for overhead service conductors. A preview of Figures 4.2 through 4.7 in Chapter 4 would be appropriate for comparison.

UNDERGROUND OUTSIDE BRANCH CIRCUITS AND FEEDERS

Underground branch circuits and feeders are commonly used to extend circuits between the main dwelling unit and outbuildings. Circuits run from dwelling unit buildings to supply lighting equipment used for driveway, area, or landscape lighting are also commonly run underground. These circuits must comply with installation requirements for underground wiring. The general requirements are found at *NEC* 300.5 or IRC E3803.1. These rules include but are not limited to required burial depths, types of conductors or cables, cables under buildings, protection from damage, splices and taps, and so forth. Other requirements include backfill, raceway seals, bushings, and single conductors. See the end of Chapter 11 in this publication for additional specific information on the installation of underground conductors (*NEC* Chapter 3 and IRC Chapter 38).

NUMBER OF SUPPLIES TO SEPARATE BUILDINGS AND STRUCTURES

The number of supplies to each building is generally limited to one feeder or branch circuit. This requirement applies when more than one building is located on the same property and is under single management. A multi-wire branch circuit is considered a single circuit for the application of this rule.

If more than one branch circuit is desired or needed at the additional building or structure, a feeder to the additional building or structure would be the solution. A sub-panelboard or feeder panelboard would then be installed at the second building or structure, supplying multiple branch circuits originating at this sub- or feeder panelboard.

Where a branch circuit or feeder originates in the additional building(s) or structure(s), only one feeder or branch circuit is generally permitted to supply power back to the original building or structure. There are special conditions that would permit additional supplies or feeders, such as for a fire pump, but these special conditions typically do not apply to one- and two-family dwellings or their accessory structures (*NEC* 225.30).

DISCONNECTING MEANS REQUIRED AT EACH SEPARATE BUILDING AND STRUCTURE

The *Code* requires each building or other structure served to be provided with a disconnecting means at the building or other structure (*NEC* 225.31). The permitted locations of the disconnecting means can be either inside or outside the building or structure being served. This required disconnecting means must be at a readily accessible location nearest to the point of entrance of the supply conductors. The provisions of *NEC* 230.6 *(conductors considered outside the building)* are permitted to be utilized when applying this section. As a result, a feeder located under a 50 mm (2 in.) concrete slab is considered outside the building. The IRC does not specifically address the requirements for disconnecting means at separate buildings or structures.

NUMBER AND GROUPING OF DISCONNECTS

The *Code* requires that the disconnecting means for and located at the separate building or structure consist of not more than six switches or circuit breakers mounted in a single enclosure or a group of separate enclosures. There can be no more than six disconnecting means grouped at any one location. The *Code* defines disconnecting means as "a device, or group of devices, or other means by which the conductors of a circuit can be disconnected from their source of supply" (*NEC* Article 100 or IRC Chapter 35). Where multiple disconnects are used as the feeder disconnect for a separate building or accessory structure, the two to six disconnecting means must be grouped at one common location [*NEC* 225.33 and 225.34].

If more than one disconnecting means is utilized at the separate building or structure, the permission to allow the two to six disconnecting means to consist of not more than six switches or six circuit breakers mounted in a single enclosure or a group of separate enclosures is different than what is permitted for service equipment by *NEC* 230.71(B) or IRC E3601.7. If service equipment consists of two to six disconnecting means, those service disconnecting means can only be installed in two to six separate enclosures *(not in a single enclosure)*.

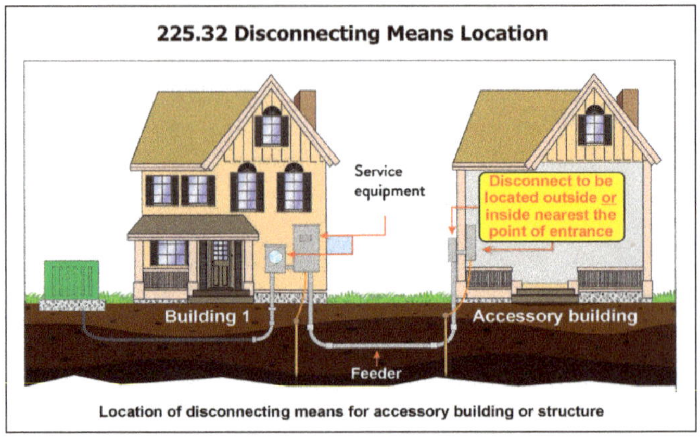

Figure 15.4. Disconnecting means required at separate buildings located outside or inside nearest the point of entrance of the feeder conductors

This rule applies regardless of whether one or several enclosures are installed. For example, up to six enclosures with a single breaker or set of fuses in each enclosure can be installed at one location and grouped together. Alternatively, a single panelboard is permitted to have up to six disconnecting means installed and serve as the disconnecting means. Manufacturers' requirements may restrict this practice. A typical residential-style panelboard will require a main breaker to be installed when the panelboard is used as the building's disconnecting means. This type of panelboard will be marked as "Suitable for Use as Service Equipment" when a suitable main breaker and hold-down kit is installed. The main breaker is a plug-in-type circuit breaker of the appropriate ampere rating. The hold-down kit is an additional fastener that secures the circuit breaker to the internal bus. The Code requires the use of a hold-down kit whenever a plug-in-type circuit breaker is back-fed [NEC 110.3(B) and 408.36(D) or IRC E3706.5].

TYPE OF DISCONNECTING MEANS

The *Code* requires the disconnecting means for a separate building or structure supplied by a feeder to be suitable for use as service equipment when applied in accordance with *NEC* 250.32(B)(1), Exception No. 1. This is the rule that permits an existing grounded (neutral) conductor to be used as the ground-fault return path to existing buildings or

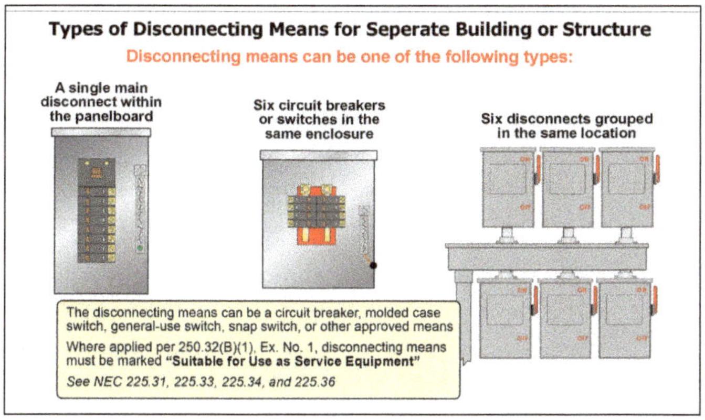

Figure 15.5. Feeder supply disconnecting means types for separate building or structure

structures. This also generally relates to the construction of the disconnect switch or panelboard. It will have a grounded (neutral) conductor terminal bar that can be bonded to the enclosure, which will permit the grounded (neutral) system conductor to be connected to a grounding electrode conductor and the disconnecting means or panelboard enclosure. This same panelboard, when listed and marked "Suitable for Use as Service Equipment," will have the capability of isolating or "floating" the grounded (neutral) bar when used as a feeder panelboard. Electrical equipment suitable for use as service equipment is listed and marked either "Suitable for Use as Service Equipment" or "Suitable for Use Only as Service Equipment."

This equipment provides the grounding and bonding capabilities as mentioned above. It should also be equipped with either a main disconnect or a label that indicates suitability when used as service equipment when a main breaker is installed. Some equipment may be marked "Suitable for Use as Service Equipment When Not More Than Six Disconnecting Means are Installed." The construction of such a panelboard will restrict the number of main disconnects that can be installed (*NEC* 225.36 and IRC E3606.4).

Even though the equipment is generally required to be marked as service equipment, it is not to say that the conductors that originate from a service-supplied electrical system in one building and supply a second building or structure on the premises are service conductors. By definition, they are either feeder or branch-circuit conductors.

For a garage or outbuilding located at one- or two-family dwellings, a circuit breaker, molded case switch, general-use switch, snap switch, or other approved means can be used as the required disconnecting means. Two single-pole, three-way, or four-way switches or breakers capable of individual operation are permitted as the disconnecting means for a multi-wire branch circuit. One pole is required for each ungrounded conductor. The switch handles would have to be equipped with handle ties or a master handle. A more practical solution may be a two-pole switch. Remember, a disconnecting means for the feeder supplying an existing separate building might have to be suitable for use as service equipment [*NEC* 225.33(B) and 225.36].

RATING OF THE DISCONNECTING MEANS

In all cases, the rating of the disconnecting means that a separate building or structure must be capable of supporting at least the same load as the one to be supplied. Parts I and II of NEC Article 220 are used to determine the minimum load for branch circuits. Feeder size is determined by applying Parts III and IV of NEC Article 220. Where the branch circuit or feeder disconnecting means consists of more than one switch or circuit breaker, combining the ratings of all the switches or circuit breakers to determine the rating of the disconnecting means is permitted. In no case shall the minimum rating be less than:

1. 15 amperes for a single branch circuit.
2. 30 amperes for a feeder that supplies one or two two-wire branch circuits.
3. 60 amperes for a feeder that supplies more than two branch circuits.

Chapter 2 contains additional information on *NEC* Article 220 and the requirements for calculating the load on a branch circuit or feeder (*NEC* 225.33, 225.39).

EMERGENCY DISCONNECT(S) FOR DWELLING UNITS

New provisions were added to the 2023 edition of the NEC at 225.41, calling for an emergency disconnect at all one-and two-family dwelling units supplied by an outside branch circuit or feeder. This emergency disconnect rule pertaining to a dwelling unit supplied by a branch circuit or feeder is very similar to the emergency disconnect(s) rules required for a dwelling unit supplied by a service at NEC 230.85 or IRC E3601.8.

This emergency disconnecting means is required to be installed in a readily accessible outdoor location on or within sight of the dwelling unit. The emergency disconnecting means must have a short-circuit current rating equal to or greater than the available fault current. If more than one disconnecting means is provided for this emergency disconnect(s), the disconnecting means would be required to be grouped together [*NEC* 225.41(A)].

Where equipment for isolation of other energy source systems (such as a PV system) is not located adjacent to the required branch circuit or feeder emergency disconnect, a plaque or directory identifying the location of all equipment for isolation of other energy sources would be required to be located adjacent to this required emergency disconnecting means [*NEC* 225.41(B)].

This required branch circuit or feeder disconnecting means is required to be marked "EMERGENCY DISCONNECT." This marking has to comply with NEC 110.21(B) (Field-Applied Hazard Markings). This section requires the marking to be of sufficient durability to withstand the environmental conditions and to warn of hazards using effective words, colors, symbols, or any combination thereof. This marking requirement stipulates that the marking must be permanently affixed to the equipment and not handwritten. This marking or label is to be located on the outside front of the disconnect enclosure with a red background and white lettering. The letters are to be at least 13 mm (½ in.) in height.

SURGE PROTECTION REQUIRED

Where a feeder supplies a dwelling unit, a surge-protective device (SPD) is required to be installed in or adjacent to the distribution equipment that is connected to the load side of the feeder containing branch circuit overcurrent protective device(s) as surge protection is most effective when located closest to the branch circuits. This is a new 2023 *NEC* requirement located at *NEC* 225.42. This is similar to the rules for surge protection for services supplying a dwelling unit at *NEC* 230.67 or IRC E3606.5.

This feeder SPD is required to be of the Type 1 or Type 2 SPD type and have a nominal discharge current rating of not less than 10 kA. Where the distribution equipment supplied by a feeder is replaced at existing dwelling units, all of the requirements of this SPD section must be applied as well.

It should be noted that surges can be generated from multiple sources, including, but not limited to, lightning, electric utilities, and equipment utilization.

EQUIPMENT INSTALLATION IN ACCESSORY BUILDINGS

Most commonly, items such as food freezers, receptacles, power tools, luminaires, and water pumps are used in an accessory building or outbuilding. Additionally, other equipment includes materials, fittings, devices, tools, and similar items used as part of or in connection with an electrical installation. Refer to Chapter 13 for specific information and requirements for luminaires, receptacles, and switches, and Chapter 7 for appliance requirements. If the accessory building or outbuilding is a detached garage with electrical power, the same basic requirements that apply to an attached garage—such as at least one 125-volt, 15- or 20-ampere rated receptacle per vehicle bay, switch lighting outlet, etc.—also apply to this detached garage. See Chapter 11 for electrical requirements for garages and accessory buildings.

Photo 15.2. The accessory building is supplied by a feeder from the main dwelling unit *(the feeder can be seen entering the building at the front left corner of the accessory building)*

GROUNDING REQUIREMENTS IN SEPARATE BUILDINGS OR STRUCTURES (OUTBUILDINGS)

Outbuildings must comply with grounding requirements similar to that of the main structure. For new installations, the feeder or branch circuit must provide a means to ground equipment in the outbuilding, and an equipment grounding conductor must be run with the supply conductors. By exception only, the grounded (neutral) conductor may be used for this purpose in certain limited cases for existing buildings and existing installations only.

The outbuilding must have a grounding electrode system as described at *NEC* 250.52(A), or IRC E3608.1 established at each building or structure. A "made" grounding electrode system (ground rod, etc.) must be established if there is no established grounding electrode system (concrete-encased electrode, etc.) present and formed through the natural construction of the building at the outbuilding where supply conductors are brought to the building (*NEC* 250.32 or IRC E3607.3).

FEEDER AND BRANCH CIRCUIT EQUIPMENT —GROUNDING MEANS

The Code generally requires feeders to include or provide an equipment grounding conductor, which may be installed as an individual bare, covered, or insulated conductor. The Code also permits some metallic wiring methods, such as

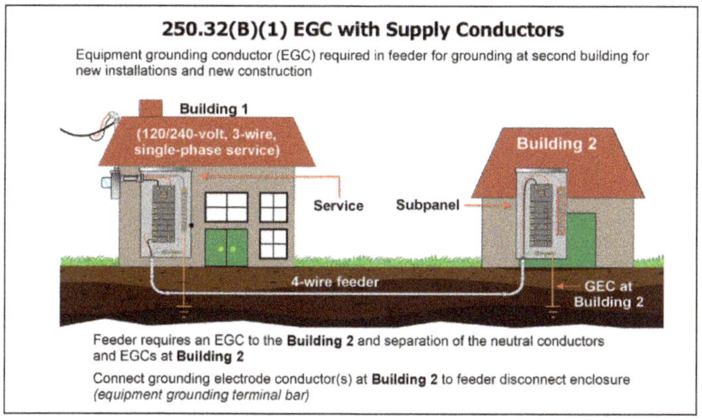

Figure 15.6. Grounding at the second building using the feeder equipment grounding conductor

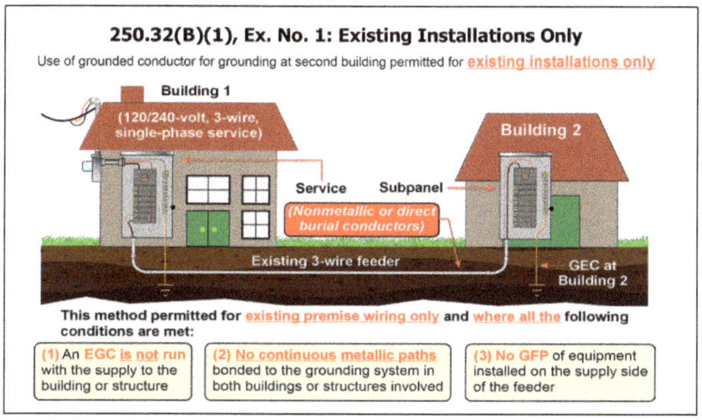

Figure 15.7. Grounding at the existing second building using the grounded (neutral) conductor at separate buildings (existing building and existing installations only)

rigid or intermediate metal conduit, to be used as a suitable equipment grounding means. When the metal conduit is used as an equipment grounding conductor, special consideration must be given to the potential effects that corrosion may have on the installation. The Code requires equipment to be suitable for the environment in which it is installed. The wiring methods that are recognized as equipment grounding conductors are identified in either NEC 250.118 or IRC E3908.9. Note that some of the listed wiring methods have conditions, such as limited length, that would restrict their use as an equipment grounding conductor in a lengthy feeder.

The equipment grounding conductor is required to be connected to a terminal bar or suitable lug that is directly connected to the enclosure for the feeder panelboard or building disconnecting means. The grounded (neutral) conductor is connected to the neutral terminal bar that is insulated from or isolated electrically from the enclosure. This is often referred to as "floating the neutral." The equipment grounding conductor that is run with the feeder or branch circuit to the additional building or structure is to be used for grounding equipment and connected to one or more grounding electrodes established at the additional building or structure. The equipment grounding conductor is sized based on the rating of the overcurrent device on the line side of the feeder or branch circuit (*NEC* 250.122 or IRC E3908.13).

The grounded (neutral) circuit conductor cannot be used to ground equipment in the second building or outbuilding where an equipment grounding conductor is so installed. The *Code* prohibits the grounded (neutral) circuit conductor from being connected to the equipment grounding conductor, grounding electrode, or grounded metal enclosures at the outbuilding or separate structure on the load side of the service disconnecting means [*NEC* 250.24(B) or IRC E3607.2].

GROUNDING USING THE GROUNDED CONDUCTOR (FOR EXISTING BUILDINGS ONLY)

The grounded (neutral) circuit conductor connection to the grounding electrode(s) and metal enclosures is permitted for grounding at existing installations only at an outbuilding when all of the following conditions are met:

1. An equipment grounding conductor is not run with the existing branch circuit or existing feeder conductors to the additional building or structure.
2. There are no continuous metallic paths bonded to the grounding systems at both buildings or structures involved. Continuous metallic paths can be established by metal water pipes and conduits or other pipes as well as by the shield in communications, television, or similar cables.
3. Ground-fault protection of equipment has not been installed on the supply side of the feeder(s) (ground-fault protection of equipment would be a rare occurrence at a dwelling unit).

Where these conditions are not present at the existing additional building or structure, it is permitted to connect the grounded (neutral) conductor to the grounding electrode conductor and the equipment grounding conductors at the disconnecting means for the existing outbuilding or separate structure. Where this is done, the grounding electrode system installation is similar to the grounding of the electrical system at the service. Note that the supply to these additional existing buildings or structures is not a service but is either a feeder or branch circuit. As a result, the rules for bonding on the line side of the service equipment do not apply at the disconnecting means for these existing buildings or structures. It is also important to note that the use of the grounded (neutral) conductor for grounding purposes at the second building or structure is only applicable for existing installations at existing buildings only [*NEC* 250.32(B)(1) Ex. No. 1 or IRC E3607.3.2].

GROUNDING ELECTRODE SYSTEM FOR A SEPARATE BUILDING OR STRUCTURE (OUTBUILDING)

Where separate buildings or structures are supplied by a feeder(s) or branch circuit(s), the grounded system in each building or structure must have one or more grounding electrodes installed (*NEC* 250.32(A) or IRC E3607.3). The point of connection of the grounding electrode conductor is the same location as the equipment grounding conductor that is part of the supply circuit. The grounding electrode conductor is connected to the metal enclosure of the building disconnecting means and the equipment grounding conductor. If the grounded (neutral) supply conductor is used to ground equipment in the outbuilding at an existing installation of an existing building, then the grounding electrode conductor is connected to the AC system grounded (neutral) circuit conductor, and both are bonded to the metal enclosure of the building disconnecting means (*NEC* 250.32(A) or IRC E3607.3).

A grounding electrode is not required to be installed at a separate outbuilding or structure where the building or structure is supplied by only one branch circuit, and the branch circuit includes an equipment grounding conductor for grounding the non-current-carrying parts of all equipment [*NEC* 250.32(A), Exception or IRC E3607.3, Exception]. *NEC* 225.30 clarifies that multi-wire branch circuits are considered single-branch circuits when applying the requirements for a number of supplies.

A grounding electrode conductor connects the equipment grounding terminal bar and equipment grounding conductors connected to it to one or more grounding electrodes that make up the grounding electrode system at a separate building or structure.

SIZING GROUNDING ELECTRODE CONDUCTOR(S)

The size of the grounding electrode conductor(s) must be determined from NEC Table 250.66 or IRC Table E3603.4, based on the size of the largest feeder or branch-circuit conductor(s) supplying the building or structure. It is important to note that the grounding electrode conductor to the grounding electrode(s) must be installed to meet all applicable installation criteria for grounding electrode conductors, including protection from physical damage and protection from magnetic fields when installed in metallic enclosures. The size and location of the grounding electrode conductor

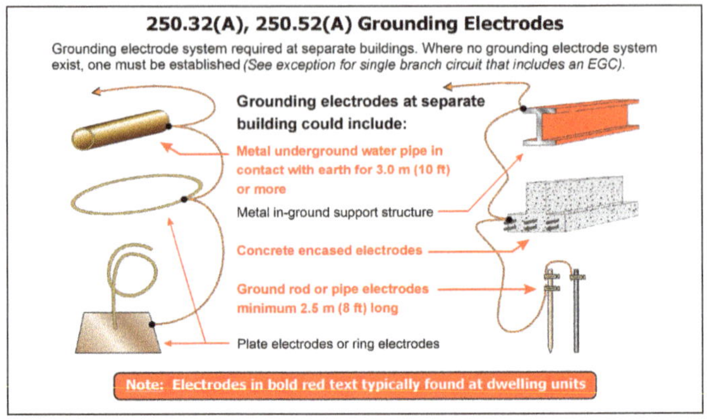

Figure 15.8. Grounding electrode system is generally required at separate buildings or structures

may require physical protection (NEC 250.64(B) or IRC E3610.2).

GROUNDING ELECTRODES AT SEPARATE BUILDINGS OR STRUCTURES (OUTBUILDINGS)

The *Code* requires a separate building or structure to have a grounding electrode or grounding electrode system installed to meet the requirements of *NEC* 250.50 or IRC E3608.1. A grounding electrode system must be established if there is no existing grounding electrode system at the outbuilding and supply conductors are brought to the building. The only exception is a single branch-circuit supply, where the branch-circuit supply circuit includes an equipment grounding conductor.

The *Code* refers to various types of electrodes that, if present at the outbuilding, must be used. Examples of electrodes that may be present in an outbuilding include underground metal water pipes, concrete-encased electrodes, and pipe or rod electrodes. If no electrodes are present, an electrode, such as a driven ground rod, must be installed. For a more complete discussion on grounding electrodes, see Chapter 5.

BONDING GROUNDING ELECTRODES TOGETHER

The *Code* requires that where multiple electrodes are present, they must be bonded together (*NEC* 250.50, 250.58, or IRC E3607.6 and E3608.1). This requirement is important not only for the electrical power distribution systems on a building or structure but also for any grounding electrodes installed, used, or both for limited energy systems such as internet, telephone, or cable TV lines in or on the building or structure.

Grounding electrodes at separate buildings are not required to be bonded to the grounding electrode at the building where the service is located. Situations often arise where more than one building is installed on a common foundation *(see NEC Article 100 or IRC Chapter 35 for the definition of "Building")*. Where this situation exists, the *Code* does not require the separate electrodes for the separate services on each building to be bonded together.

It is common for multiple buildings to be located on the same property. If supplied with electricity, all such buildings are very likely to have their own grounding electrode system. Where a service is located on one building, and a feeder from that building supplies another building on the premises, a close review of the system will show that the two grounding electrode systems at each building or structure are, in fact, bonded together. The grounding electrode systems at Building No. 1 and Building No. 2 are connected together by the equipment grounding conductor contained in the feeder or, in some existing installations, by the grounded conductor.

REFERENCES

1. *National Electrical Code* and *NEC* are registered trademarks of the National Fire Protection Association, Inc., Quincy, MA 02169. This reprinted material is not the official position of the National Fire Protection Association, which is represented only by the standard in its entirety.

2. *International Residential Code for One- and Two-Family Dwellings* (2024). Copyright © International Code Council, Inc., June 2023.

3. UL, Northbrook, IL, *UL Guide Information for Electrical Equipment Directory* (UL Product iQ® material) unless indicated otherwise.

REVIEW QUESTIONS

1. The rating of the disconnecting means for a separate building of structure must be at least equal to the load to be supplied. Part I and II of *NEC* Article 220 is used to determine the minimum load for branch circuits. Feeder size is determined by applying Parts III and IV of *NEC* Article 220.

True _____ False _____

2. For 1000 volts ac, nominal, or less, overhead copper conductors sized at 10 AWG not exceeding _______________ can be used for overhead spans to supply an outbuilding without the use of a messenger wire?

A. 15 m (50 ft)
B. 18 m (60 ft)
C. 23 m (75 ft)
D. 25 m (80 ft)

3. Individual outside overhead branch circuits and feeder conductors at dwelling units cannot be smaller than 10 AWG copper or 8 AWG aluminum unless supported by a messenger wire.

True _____ False _____

4. An emergency disconnect is required to be installed at all one- and two-family dwelling units supplied by an outside branch circuit or feeder. This emergency disconnecting means is required to be installed _________________ _______________________________________.

A. in a readily accessible location
B. outdoors
C. on or within sight of the dwelling unit
D. all of the above

5. Outside wiring installed in the open on insulators is required to be insulated for the nominal voltage where they are installed within ________________ of any building or outbuilding?

A. 3.0 m (10 ft)
B. 4.5 m (15 ft)
C. 5.5 m (18 ft)
D. 6.0 m (20 ft)

6. Outside conductors within cable assemblies or raceways (excluding Type MI cable) must be of the rubber-covered or thermoplastic type.

True _____ False _____

7. Where outside branch circuits or feeders leave or enter a building, they are required to be equipped with a service head. These conductor/cable assemblies are permitted to be formed in a ____________ and taped with a self-sealing weather-resistant thermoplastic.

A. camelback
B. horseshoe
C. gooseneck
D. "P" trap

8. A service head or gooseneck in a suitable cable assembly is generally required to be located above the point of attachment of the outside branch circuit and feeder conductors to the building or outbuilding.

True _____ False _____

9. Where it is impracticable to locate the service head above the point of attachment of a service cable assembly, the service head location is permitted to be installed not more than what distance away from the point of attachment?

A. 600 mm (2 ft)
B. 900 mm (3 ft)
C. 1.2 m (4 ft)
D. 1.8 m (6 ft)

10. Raceways installed on exterior surfaces of buildings must be listed or approved for use in a wet location and installed so that they will be arranged to drain.

True _____ False _____

11. Outside branch circuits and feeders rated at not over 1000 volts, nominal that are run as spans of open conductors are required to be installed not less than ________________ above finished grade, sidewalks, or from any platform or projection from which they might be reached where the supply conductors are limited to 150 volts to ground and accessible to pedestrians only?

A. 2.5 m (8 ft)
B. 2.7 m (9 ft)
C. 3.0 m (10 ft)
D. 3.7 m (12 ft)

12. Where raceway-type masts are installed, hubs used with conduit that serves as the mast are required to be:

A. approved
B. identified
C. listed
D. rigidly attached

13. Open conductors shall be supported on glass or porcelain knobs, racks, brackets, or strain insulators.

True _____ False _____

14. Without employing any exceptions, a vertical clearance of not less than ____________ is required to be maintained from the surface of a flat roof for outside branch-circuit and feeder conductors?

A. 900 mm (3 ft)
B. 1.5 m (5 ft)
C. 2.0 m (6 ft 6 in.)
D. 2.6 m (8 ft 6 in.)

15. Where the voltage between the outside branch circuit and feeder conductors is not more than 300 volts, and where the roof has a slope of not less than 100 mm (4 in.) in 300 mm (12 in.) or greater, a reduction in clearance to _______________ is permitted?

A. 300 mm (1 ft)
B. 450 mm (1½ ft)
C 600 mm (2 ft)
D. 900 mm (3 ft)

16. The vertical, diagonal, and horizontal clearances for outside branch circuits and feeders cannot be less than _______________ from a chimney or radio and television antenna?

A. 300 mm (1 ft)
B. 450 mm (1½ ft)
C. 600 mm (2 ft)
D. 900 mm (3 ft)

17. Individual open outside branch circuit and feeder conductors and cables cannot be installed within _______________ of the final grade level that will permit personal contact where the voltage does not exceed 150 volts to ground and accessible to pedestrians only.

A. 3.0 m (10 ft)
B. 3.7 m (12 ft)
C. 4.3 m (14 ft)
D. 4.9 m (16 ft)

18. Generally, lamps used for outdoor lighting must be installed so that they are located above all energized conductors or other electric utilization equipment.

True _____ False _____

19. Live vegetation, such as trees, are not permitted to be used for the support of overhead outside branch-circuit and feeder conductor spans unless the overhead span is being used for temporary wiring.

True _____ False _____

20. A residential outside branch circuit run underground under a driveway in rigid PVC conduit must have a minimum cover of _______________ with the branch circuit rated at 120 volts, protected by a 15-ampere overcurrent device, and is GFCI protected.

A. 150 mm (6 in.)
B. 300 mm (12 in.)
C. 450 mm (18 in.)
D. 600 mm (24 in.)

21. Generally, without applying any special conditions, a 60-ampere feeder and a separate 15-ampere branch circuit for lighting are both permitted to supply a residential single-family detached garage from the service at the main structure.

True ______ False ______

22. Which of the following is NOT a permitted disconnecting means for an outbuilding located on residential property?

A. A set of three-way or four-way switches
B. A main breaker in a panelboard marked"Suitable for Use as Service Equipment."
C. A main lug panelboard
D. A fusible switch marked "Suitable for Use as Service Equipment."

23. For _______________, the grounded conductor run with the supply feeder to an outbuilding or structure is permitted to be connected to the outbuilding or structure disconnecting means and to the grounding electrode(s) and shall be used for grounding purposes.

A. new and existing premises wiring systems
B. existing premises wiring systems only
C. new building smaller than 500 sq. ft.
D. new premises wiring systems only

24. Where a branch circuit or feeder originates in the additional outbuilding(s) or structure(s), only __________ feeder or branch circuit is permitted to supply power back to the original building or structure without applying any special conditions.

A. one
B. two
C. three
D. none can be brought back to the original building

25. Which of the following is TRUE concerning the disconnecting means for an additional outbuilding(s) or structure(s) supplied by a feeder from the main structure?

A. Must consist of not more than six switches or six circuit breakers
B. The disconnecting means must be mounted in a single enclosure or in a group of separate enclosures
C. The disconnecting means must be grouped in one location
D. All of the above

26. Where a feeder supplies a dwelling unit, a surge-protective device (SPD) is required to be installed in or adjacent to the distribution equipment that is connected to the load side of the feeder containing branch circuit overcurrent protective device(s). This feeder SPD is required to have a nominal discharge current rating of not less than __________.

A. 5 kA
B. 8 kA
C. 10 kA
D. 12 kA

16 Existing Electrical Installations and Wiring

MODIFYING CIRCUITS AND EQUIPMENT IN EXISTING DWELLINGS

In many cases, electrical wiring with adequate capacity that was installed properly at the time the structure was built has served existing one- and two-family dwellings well for many years. Unfortunately, many electrical systems are abused during their life. Often, branch circuits and the service are overloaded as new or additional equipment is placed into service in the dwelling. It is common to find electric ranges, clothes dryers, electric water heaters, and electric heating equipment in older dwellings that are being supplied by existing electrical branch circuits that do not have adequate capacity for additional loads.

All too often, untrained and inexperienced personnel modify existing electrical systems or circuits. Much of this work is done unsafely, mainly due to a lack of understanding of the applicable Code rules. Sometimes, these modifications are made to cut corners and save the cost of hiring a trained and professional electrician to do the job correctly. In many areas, electrical inspection of modifications or additions to the wiring systems of existing dwellings is required but often goes unnoticed. These inspections are essential to help ensure safety. In such cases, an electrical permit and inspection are needed for the work.

EXTENSION OF EXISTING BRANCH CIRCUITS

Older one- and two-family homes were often wired with Type NM (Romex) cable, Type AC (BX) cable assemblies, or, in homes built before the 1960s, knob-and-tube wiring that used circuit conductors with Type R (rubber) insulation. In existing single- and two-family homes, this type of insulation is often old, dried out, and very brittle, which can cause serious problems when disturbed, such as when replacing a luminaire, receptacle, or light switch. This is especially true where circuits have been overloaded or operated at higher temperatures than the insulation was rated for.

Extending an existing branch circuit involves connecting the current conductors to new ones. Splicing new conductors into the existing junction or device box can often cause additional issues. Some of these issues may pose serious fire or electric shock risks. Often, the existing junction or device box is too small to accommodate the extra conductors for the new circuit extension, or the circuit's capacity is insufficient for the new load.

Extensions or modifications to existing branch circuits might require arc-fault circuit-interrupter (AFCI) protection to be added to those circuits. This topic will be covered in more detail later in this chapter.

MINIMUM BOX SIZE

When adding conductors to existing boxes, care must be taken to ensure that the existing box has enough space for both the current conductors and the new ones being added. NEC 314.16 or IRC E3905.12 provides the requirements for determining the minimum size of the box needed to hold all conductors, clamps, devices, etc., that will be contained in that particular box.

For example, an existing 75 mm x 50 mm x 65 mm (3 x 2 x 2½ in.) metal box has a 14-2 (with ground) Type NM cable supplying a duplex receptacle in the box, which has internal cable clamps. By reference to *NEC* 314.16(B) or IRC E3905.12.1, it is found that the conductors, clamps, and devices are counted as six conductors. *NEC* Table 314.16(B)(1) or IRC Table E3905.12.2.1 requires a conductor allowance of 32.8 cm^3 (2.00 cu. in.) for each 14 AWG conductor. The duplex receptacle counts as two conductors and the internal clamps as one. The first four equipment grounding conductors are grouped together and counted as one.

NEC Table 314.16(A) or IRC Table E3905.12.1 allows six 14 AWG conductors in the box. To add one more 14-2 (with ground) AWG cable, a larger box that can accommodate at least eight 14 AWG conductors must be used. The minimum size standard (metal) box for this installation is 75 mm x 50 mm x 90 mm (3 x 2 x 3½ in.). When replacing the existing box, a nonmetallic box with a capacity of 262 cm^3 (16 cubic inches) or more must be chosen.

Chapter 12 provides detailed information on box fill calculations to determine the appropriate box size for electrical installations. Please see Chapter 12, for more details on this topic.

INCREASING BOX SIZE

While adding an extension ring to a box will increase the

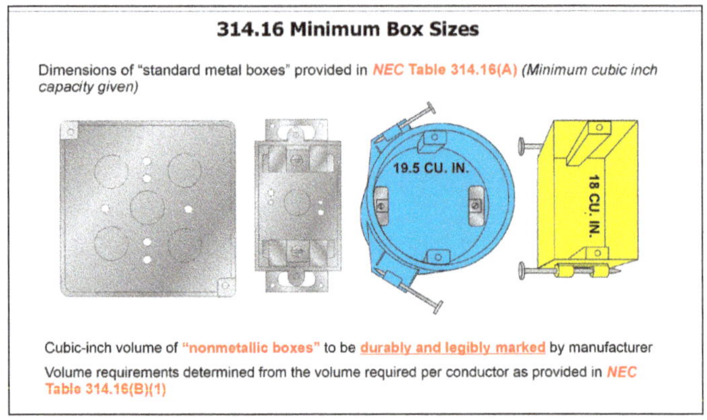

Figure 16.1. Box volume and conductor fill can be determined by *NEC* Table 314.16(A) and *NEC* Table 314.16(B)(1)

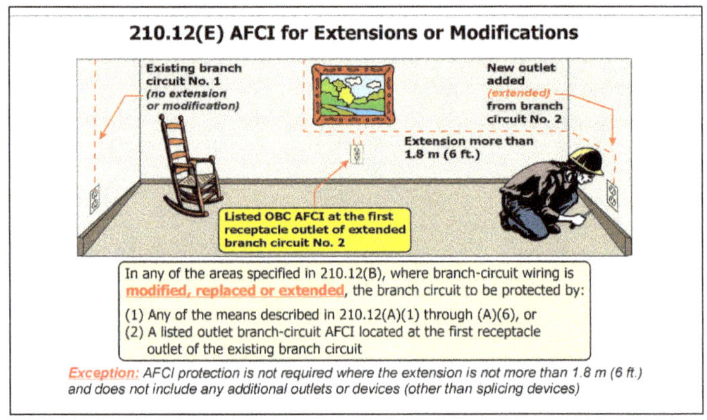

Figure 16.2. Extensions or modifications to existing wiring require AFCI protection in specific areas of a dwelling unit

cubic inch capacity of an existing outlet, device, or junction box, it becomes probable that the existing contained conductors will then be too short to be handled properly. At times, a compression sleeve or splice should be used to extend the conductors before adding an extension ring. Compression sleeves are available with and without insulation. Ensure the proper compression tool is used to crimp the sleeve. A twist-on wire connector can also be used to extend the conductors, but it requires more space in the box.

When adding an extension ring isn't practical, it may be necessary to remove the existing box and replace it with one that has more cubic inches of space. If replacing the box isn't feasible, adding a domed box side can sometimes be an option. When marked with its cubic inch capacity, this capacity can be added to that of the box. This approach can provide the needed cubic inch capacity for extra conductors or devices.

AFCI PROTECTION FOR BRANCH CIRCUIT EXTENSIONS OR MODIFICATIONS

In any room or area of an existing dwelling, as referenced at NEC 210.12(B) or IRC E3902.20, such as family rooms,

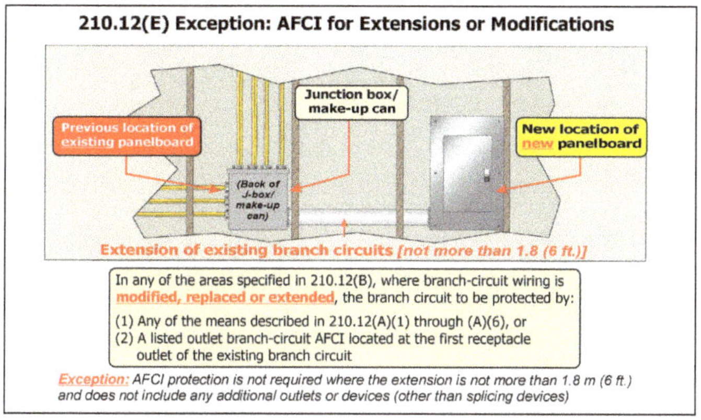

Figure 16.3. AFCI protection is not required for extension of modification to existing branch circuits of less than 1.8 m (6 ft)

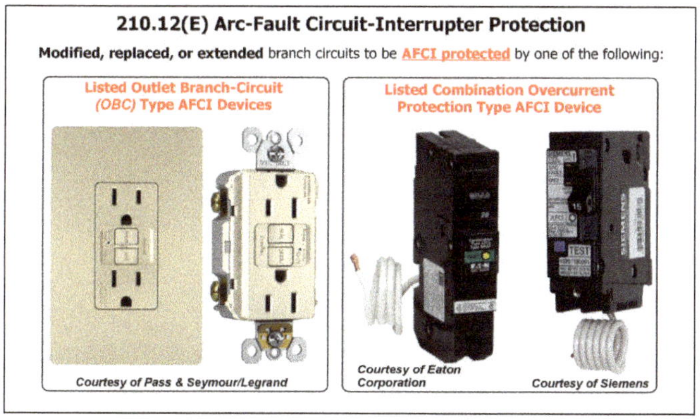

Photo 16.1. AFCI protection for modified, replaced, or extended existing branch circuits can be delivered by a listed AFCI outlet branch-circuit device or a combination AFCI overcurrent device

living rooms, bedrooms, kitchens, or sunrooms, where branch-circuit wiring is modified, replaced, or extended (and outlets or devices are added), the existing branch circuit must be protected by either a listed combination AFCI circuit breaker at the circuit's start or a listed outlet branch-circuit AFCI at the first receptacle outlet of the existing branch circuit [NEC 210.12(E) or IRC E3902.21]. This setup allows listed outlet branch-circuit AFCI devices to be installed at the first receptacle of the existing branch circuit, helping address common issues that can arise during renovations when the circuit is extended or modified.

Particularly in older existing dwelling units, it is often not possible to install a listed combination AFCI overcurrent device on the existing service or panelboard without performing a complete overhaul or replacing the panelboard with a newer one that is compatible with today's listed combination AFCI overcurrent devices.

When using listed outlet branch-circuit AFCI devices, the first receptacle outlet is designated as the location to avoid confusion if the first outlet on the existing branch circuit happens to be a switch, luminaire, or smoke alarm outlet. If there are no receptacle outlets on the branch circuit that are to be extended, a listed outlet branch-circuit AFCI device must be installed at the first receptacle outlet of the extension to the existing branch circuit. AFCI protection for the new extended wiring will include both series and parallel AFCI protection, while the existing branch circuit will only have upstream series-type AFCI protection.

An existing panelboard that can accept the current combination of AFCI devices allows the installer to choose whether to install an outlet branch-circuit AFCI at the first receptacle outlet or to install a listed combination AFCI overcurrent device at the start of the branch circuit.

An exception to AFCI "Branch Circuit Wiring Extensions, Modifications, or Replacements" specifies what qualifies as a "dwelling unit branch circuit extension" and clarifies that existing branch circuit conductors can be extended up to 1.8 m (6 ft.) without AFCI protection, provided no additional outlets or devices are installed. This 1.8 m (6 ft) limit excludes conductors inside an enclosure, panelboard cabinet, or junction box from being counted. Sometimes, a service upgrade is necessary at existing dwelling units.

This action may sometimes involve relocating a panelboard, even a short distance, from the original location, such as moving it from a clothes closet to an acceptable area away from combustible materials. This situation can easily cause the existing branch circuit conductors to be too short to reach their new location at the new panelboard. Often, these branch circuits are spliced and extended using the old panelboard cabinet as a junction box with the busbars removed. This is a good example of an extension or modification that is exempted from AFCI protection with this exception to NEC 210.12(E) or IRC E3902.22, as long as no new outlets or devices are added.

SUPPORT OF OLD-WORK STYLE BOXES

Outlet and device boxes must be supported in compliance with *NEC* 314.23 or IRC E3906.8. For existing installations, *NEC* 314.23(C) or IRC E3906.8.3 permits a flush installation in existing covered or finished surfaces where clamps, anchors, or fittings provide adequate support. These types of boxes are often referred to as "*old-work*" style boxes. Box manufacturers provide a variety of mechanisms to support flush boxes from the existing building surface. Some are in the shape of a "W" bracket on the outside of the back of the box, having a center bolt or screw with its head inside the box. After the conductors are inserted inside the box, the entire assembly is passed through the opening. The "W" bracket snaps into place behind the wall surface and secures the box to the wall or ceiling covering by tightening the screw. See Figure 16.4 and Photo 16.2 for examples of these types of old work boxes and clamps. Other devices, like captive wings or arms, are designed into nonmetallic old-work boxes [some people refer to these as "snap-in," "pop-in," or "cut-in" style boxes] and swing out behind the finished surface. Specifically designed sheet metal straps, sometimes referred to as "battleships" or "Madison clamps," are often used to support metal or nonmetallic boxes that have ears from existing structural surfaces.

The side view of the strap resembles a navy battleship with smokestacks. The metal box with ears is mounted from outside the wall, and the plaster ears rest against the surface of the exterior wall. A "battleship" clamp is inserted through the small gap between the side of the box and the opening on each side of the box. The ends of the clamp are bent inward to secure the box against the wall. A word of caution is appropriate here.

When using these "battleship clips," carefully bend them back tightly against the box to prevent accidental contact with the terminals of the enclosed device. Electricians have learned through experience to insulate energized wiring connections with vinyl or friction electrical tape to avoid accidental contact with metal boxes during installation of this type of support.

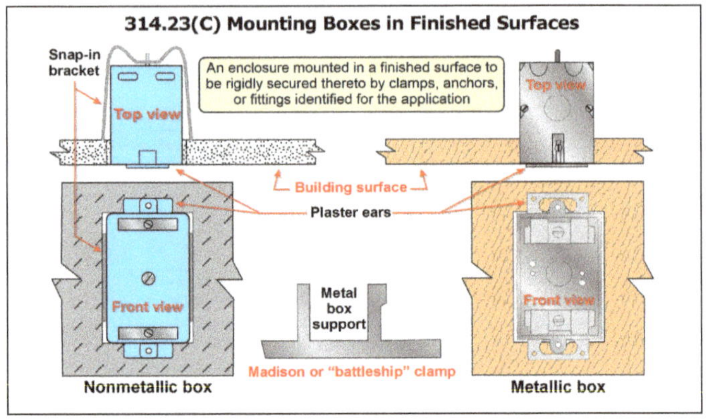

Figure 16.4. Device or outlet box supported by wallboard or wall covering

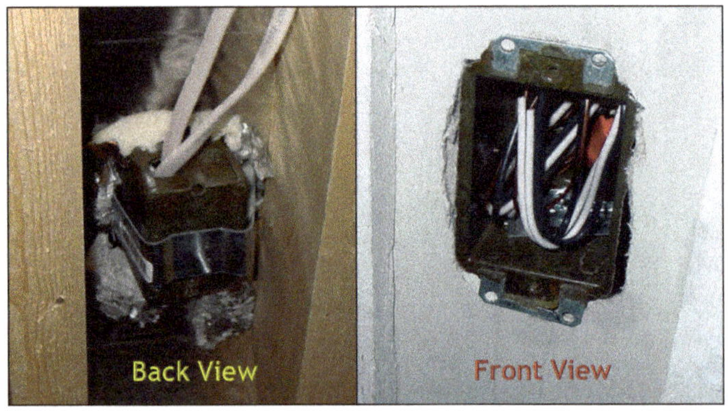

Photo 16.2. Front and back view of a typical "old-work" or "snap-in" style box

EXTENDING TYPE NM CABLES WITH SMALL EQUIPMENT GROUNDING CONDUCTORS

In older, existing homes, some older Type NM cables used for branch circuits were manufactured with conductor insulation of white and black, and a bare equipment grounding conductor that was smaller than the circuit conductors. In many areas around the country, these smaller equipment grounding conductors were "back wrapped," meaning that they were wrapped around the end of the NM cable jacket with four or five turns and then secured under the cable clamp rather than being connected inside the box. This type of termination is not uncommon in existing one- and two-family dwelling electrical systems.

However, this installation violates NEC 250.148 or IRC E3908.14, which require that equipment grounding conductors be connected within or to the box with suitable devices and used solely for this purpose. All equipment grounding conductors must be connected together inside the box to ensure continuity of the grounding system downstream in the branch circuit. A grounding screw or a listed grounding device, such as a grounding clip, must be used to establish the connection between the equipment grounding conductor and the box.

Where a branch-circuit extension originates from an existing device, outlet, or junction box, any smaller-sized (16 AWG) bare equipment grounding conductors should be treated carefully so they will not break. Connecting the existing cable conductors to a new Type NM cable assembly that has a larger sized (14 or 12 AWG) equipment grounding conductor does not create a conflict or *Code* violation. For the most part, new *Code* rules are not retroactive.

Another way to extend the equipment grounding conductor is by using a listed grounding device. The new equipment grounding conductor can be attached to the existing metal box using a clip or ground screw. This is likely the best approach for this situation. The UL Product Safety Standard requires Type NM to have an equipment grounding conductor sized in accordance with NEC 250.122 or IRC Table E3908.13.

KNOB-AND-TUBE WIRING

NEC Article 394 covers knob-and-tube type wiring systems. Knob and tube wiring is a method of electrical wiring that uses copper wires covered in insulation, running through porcelain knobs and tubes. The copper conductors are pro-

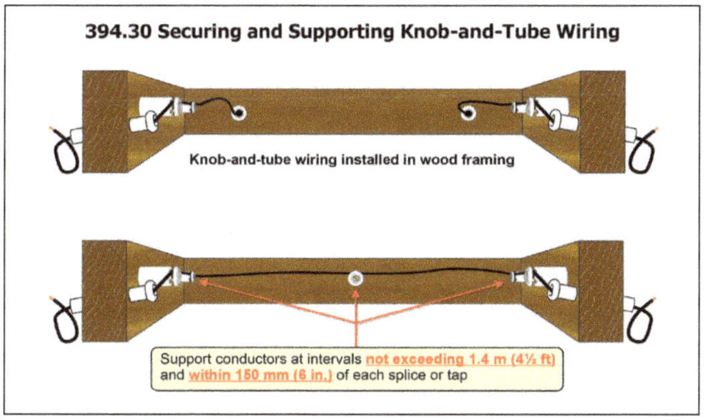

Figure 16.5. Knob-and-tube wiring installed through rafters with running boards installed to provide protection

Photo 16.3. Typical knob-and tube type wiring is sometimes found in older dwelling units

tected by flexible cloth insulating sleeving called loom. Early insulation was made from asphalt-saturated cotton cloth; later, rubber became more common. This wiring method was one of the first types used for wiring all buildings (not just dwelling units), gaining popularity in the 1930s and continuing through the early 1950s. Like the cable assemblies mentioned earlier, Knob and tube wiring may have conductors with insulation that are old and brittle. This information is included here to provide electricians and inspection authorities with additional insight into the specific types of problems they are likely to encounter in older homes with knob and tube wiring.

Many older one- and two-family dwellings have had these systems in place since their construction. These wiring systems may still be safe. However, testing may be needed to rule out past abuse, such as protecting the conductors with a higher-than-allowable fuse or circuit breaker (over-fusing) or rodents chewing on the insulation over time. The use of a Megger or meg-ohm meter is one method used to test for insulation integrity.

The fact that branch circuits lack an equipment-grounding conductor should not pose a safety risk when the receptacles only supply two-wire appliances or lamps. Clearly, the metal boxes used at lighting and receptacle outlets are not grounded. The electrician should handle these boxes carefully. Any work on the box, devices, or circuit should only be done after turning off the circuit at the source and testing it to verify it is off.

As mentioned earlier, concealed knob-and-tube wiring was a method that used knobs (porcelain cleats), tubes, and flexible nonmetallic tubing (loom) to protect and support single insulated conductors (see definition at NEC Article 100). These conductors were hidden within the hollow spaces of walls and ceilings in buildings. Concealed knob-and-tube wiring must comply with NEC Article 394 and the relevant provisions of other articles, particularly NEC Article 300.

USES PERMITTED FOR KNOB-AND-TUBE WIRING

Concealed knob-and-tube wiring can be installed in the hollow spaces of walls, ceilings, unfinished attics, and roof spaces. It is only permitted for extensions of existing installations. Other uses are only allowed with special permission (*the written consent of the authority having jurisdiction*) [NEC 394.10].

Installations in unfinished attics and roof spaces must comply with NEC 394.23. If the unfinished attic or roof space is accessible via a stairway or permanent ladder, conductors should be installed along the side of or through bored holes in floor joists, studs, or rafters. Conductors run through bored holes in the joists, studs, or rafters, and those 2.1 m (7 ft) or less above the floor or floor joists must be protected by sturdy running boards that extend at least 25 mm (1 in.) on each side of the conductors.

Running boards must be securely attached. In unfinished attic and roof spaces inaccessible by stairs or permanent ladders, knob-and-tube conductors must be installed along the sides of or through drilled holes in floor joists, studs, or rafters. Buildings completed before wiring installation and with headroom less than 900 mm (3 ft) at all points are not required to have wiring run along the sides of or through drilled holes in floor joists, studs, or rafters.

USES NOT PERMITTED FOR KNOB-AND-TUBE WIRING

Concealed knob-and-tube wiring cannot be installed in the hollow spaces of walls, ceilings, and attics where such spaces are insulated by loose, rolled, or foamed-in-place insulating material that envelops the conductors (NEC 394.12). This rule requires additional planning to ensure a safe installation when insulating the wall cavities and attics of older homes for energy conservation purposes.

Some inspection authorities use NEC 90.4 to allow insulation to be installed around knob-and-tube wiring. This should only be done after the wiring has been inspected and confirmed to be in good condition. Additionally, standard overcurrent protection such as circuit breakers or Type S fuses is installed to prevent fusing conductors with a higher ampacity than permitted by NEC 240.4(D) for smaller conductors.

GENERAL REQUIREMENTS FOR KNOB-AND-TUBE WIRING

Conductors used for knob-and-tube branch-circuit wiring must meet the specifications listed in NEC Table 310.4(1). The ampacity of knob-and-tube branch-circuit wiring is determined by NEC 310.15. Generally, NEC Table 310, which covers "Allowable Ampacities of Single Insulated Conductors In Free Air," is appropriate. Note the asterisks following 14, 12, and 10 AWG sizes at NEC Table 310.17, which refer to a note at the bottom of the table. Unless otherwise permitted by NEC 240.4, the overcurrent protection for 14 AWG is limited to 15 amperes, 20 amperes for 12 AWG, and 30 amperes for 10 AWG copper [NEC 240.4(D) or IRC E3705.5.3]. Motor circuits are an example where these ampacity tables may be used.

According to NEC 394.30, knob-and-tube conductors must be securely supported on noncombustible, nonabsorbent insulating materials and must not touch any other objects. These supports should be installed as follows:

1. Within 150 mm (6 in.) of each side of each tap or splice, and
2. At intervals not exceeding 1.4 m (4½ ft)

To support conductors in dry locations, NEC 394.30(A) allows pulling conductors through hollow spaces if each conductor is individually protected by flexible nonmetallic tubing (loom). This tubing must be in continuous lengths between supports, between boxes, or between a support and a box.

Where solid knobs are used, conductors must be securely fastened to them with tie wires that have insulation matching that of the conductor. Typically, split ceramic knobs with a base and cap, featuring a captive mounting nail, are utilized [NEC 394.30(B)].

A clearance of at least 75 mm (3 in.) must be maintained between conductors, and at least 25 mm (1 in.) between the conductor and the surface it passes over. When space is too tight to meet these minimum clearances, conductors must be individually enclosed in flexible nonmetallic tubing (loom). The loom must be continuous from the last support or box to the terminal point. Limited space is typically found at meter enclosures, panelboards, outlets, and switch points [NEC 394.19(A) and (B)].

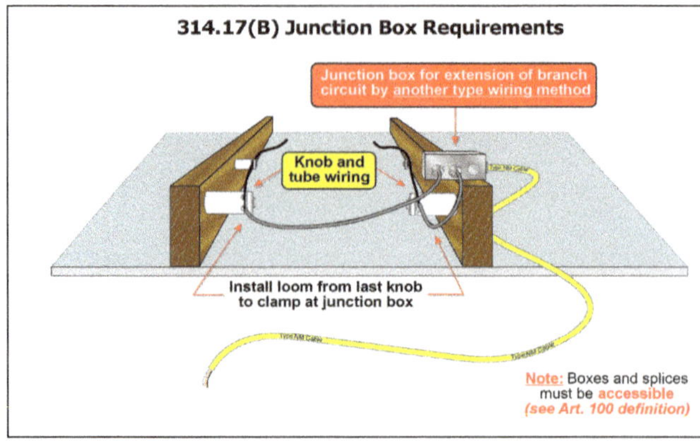

Figure 16.6. Junction box installed with knob-and-tube wiring using loom at transition points

Where installed through holes in structural members such as walls, floors, and wood cross members, conductors must comply with NEC 398.17 for open wiring on insulators. Ceramic tubes are installed in holes bored into framing members at a slight angle to prevent the tubes from falling out and to keep the conductor in contact with the wooden framing member. When knob-and-tube conductors pass through wood cross members in plastered partitions, they must be protected by noncombustible, nonabsorbent insulating tubes extending at least 75 mm (3 in.) beyond the wooden member.

Clearance from piping, exposed conductors, and similar elements is required to comply with NEC 398.19(C) for clearances from other exposed conductors, piping, and related components.

SPLICES IN KNOB-AND-TUBE WIRING

All splices must be soldered unless approved devices such as split bolts or wire connectors are used. In-line or strain splices are not allowed (NEC 394.56). Typically, splices or taps in knob-and-tube wiring are made near a knob because rigid support is necessary within 150 mm (6 in.) [NEC 394.30(A)(1)]. This rigid support helps prevent any strain from affecting the splice or tap connection.

Boxes are not required where conductors for knob-and-tube wiring are spliced or tapped. This does not violate the requirement in NEC 300.15, which generally mandates a box for conductor splice points, as knob-and-tube wiring is not specified in NEC 300.15 as needing a box. All boxes installed for knob-and-tube wiring must comply with NEC Article 314 [NEC 314.17(B) and (C)]. Switches used with knob-and-tube wiring must be installed to meet the requirements outlined in NEC 404.4 for damp and wet locations and 404.10(B) for mounting.

CHANGING WIRING METHODS FROM EXISTING KNOB-AND-TUBE WIRING

A junction box must be installed where two different wiring methods meet to properly convert an existing knob-and-tube wiring system and allow an extension of a wiring method such as Type NM nonmetallic-sheathed cable for continuing or extending an existing branch circuit. During remodels of dwelling units, existing knob-and-tube wiring circuits that are to remain in service are typically "picked up" or supplied through a run of the new wiring method being used, whether it is a cable or raceway. This junction or splice can be made at the end or in the middle of a run of existing knob-and-tube wiring. Experience shows that making the connection at the end of a run provides reasonable safety.

Step 1. Verify that 120 volts exist between the conductors to be extended. Their size will typically be 14 or 12 AWG copper.

Step 2. Determine which one is the ungrounded conductor or "hot leg." It will be necessary to use some form of voltage tester with one lead attached to a grounded object, such as a grounded metal water pipe. Voltage testers that indicate the presence of voltage by recognizing the magnetic field without a direct connection to an equipment ground are also useful.

Step 3. After identifying the ungrounded (hot) conductor of the circuit, remove the branch circuit fuse or turn the circuit breaker off at the panelboard. Check to make sure the circuit is not energized.

Step 4. Install a junction box to be used for the splice in an accessible location near the knob-and-tube wiring circuit to be supplied. Using a nonmetallic junction box is a common alternative that will avoid inductive heating problems. If a metal box is used, steps need to be taken to eliminate heating surrounding metal by induction, such as cutting a slot between the knockouts through which the individual conductors enter into the box [*NEC* 300.20(B). Perhaps a better solution would be to avoid any heating of conductors and metal by installing a cable connector large enough so both the hot and neutral conductors contained inside the loom enter the box through the same connector.

Step 5. Place looms over each conductor from the last knob and routes the conductors into the junction box, leaving an adequate length for making the needed connections. For metal boxes, *NEC* 314.17(B)(1) requires that flexible tubing (loom) must extend from the last support (knob) and be secured to the box. Cable connectors suitable for Type NM cable can be used to connect the conductor and loom to the box. Where a nonmetallic junction box is used, the conductors are required to enter through individual holes with the loom extending not less than 6 mm (¼ in.) inside the box.

Conductors of a type listed in NEC Table 310.4(1) may be used when it is necessary to add new or additional conductors because the existing knob-and-tube wiring is too short. This includes conductors with Type TW, THW, THHN, and XHHW insulation. In such cases, place the loom over the new conductor between the last knob and the junction box.

As mentioned earlier, splices for knob-and-tube wiring are not required to be in a box (NEC 300.15). Using a properly sized split-bolt connector is ideal for splicing the existing knob-and-tube wiring. Add insulation over the splice with a snap-on insulator cover or vinyl electrical tape.

Step 6. Install the Type NM cable or other new wiring method for the box, make the connections, and install the box cover. If the Type NM cable is from a branch-circuit panelboard to supply power to the knob-and-tube wiring, the equipment grounding conductor can be used to ground a metal box.

Step 7. Ground the metal junction box if required by *NEC* 250.86. Note that Exception No. 1 permits metal enclosures, such as junction boxes, not to be connected to an equipment grounding conductor where they are added to existing installations of knob-and-tube wiring that does not have an equipment ground if they are in runs of less than 7.5 m (25 ft), are free from probable contact with ground, grounded metal, metal lath or other conductive material, and if guarded against contact by persons. *NEC* 250.110(1) requires that equipment likely to become energized be connected to an equipment grounding conductor where within 2.5 m (8 ft) vertically or 1.5 m (5 ft) horizontally of grounded metal objects and subject to contact by persons.

REPLACING EXISTING LUMINAIRES

Branch circuits in existing homes are often wired with rubber or other conductor insulation rated at 60°C. When replacing an existing luminaire in a dwelling unit, the new luminaire is usually designed and identified for connection to a wiring method with 90°C insulation. If the branch circuit conductors don't have a temperature rating suitable for the new luminaire, luminaire tap conductors with appropriate temperature ratings should be installed from the luminaire to the existing branch circuit outlet box. This offers an alternative to replacing all the existing conductors with ones that have the proper temperature rating. This method can also be used when the luminaire is supplied with a fixture whip.

Tap conductors with insulation that will withstand the temperature encountered in the luminaire are permitted to be run from the point where the conductors are spliced to the fixture wires to an outlet box placed at least 300 mm (12 in.) away from the luminaire. These tap conductors are required to be in a suitable raceway or within Type AC or Type MC cable that is at least 450 mm (18 in.) long but not more than 1.8 m (6 ft) in length.

Examples of suitable raceways include types such as electrical nonmetallic tubing, flexible metallic tubing, flexible metal conduit, and liquid- tight flexible metal or liquid- tight flexible nonmetallic conduit (NEC 410. 410.117).

The electrical installer, inspector, and designer must always consider the purpose served by this 450 mm (18 in.) to 1. 1.8 m (6 ft) whip: that it is meant to provide conductors with a high- temperature rating to safely withstand the heat generated in the luminaire. It also functions to limit heat transfer to conductors with insulation that has a lower temperature rating.

The wording used in the Code indicates that the outlet box must be placed at least 300 mm (12 in.) away from the luminaire. This does not mean that the box can be left unsecured or floating. The Code requires all outlet boxes to be rigidly and securely fastened in place.

In some cases, the size of the luminaire opening is too narrow to see the box, let alone inspect the wiring inside it. The diameter of the circular opening in a ceiling can be as small as 115 mm (4 ½ in.). Many electrical inspectors would likely consider such a junction box as inaccessible. The method for making luminaire tap connections should not be used unless the outlet box itself is accessible through some other means. An open space above the box, such as an attic, soffit, or crawl space, may be used to access the box. Alternatively, the junction box can be positioned so that its cover is flush with the ceiling, making the connections accessible through the cover.

Outlet boxes are considered accessible when located above a suspended or dropped ceiling where the box can be properly secured. Ceiling framing members are permitted to support the box, which can be fastened to these members with mechanical means such as bolts, screws, rivets, or approved clips. Support wires that do not provide secure support should not be the sole support for outlet boxes. Support wires and their fittings, along with the ceiling grid, are allowed to support the box, and support wires must be secured at both ends.

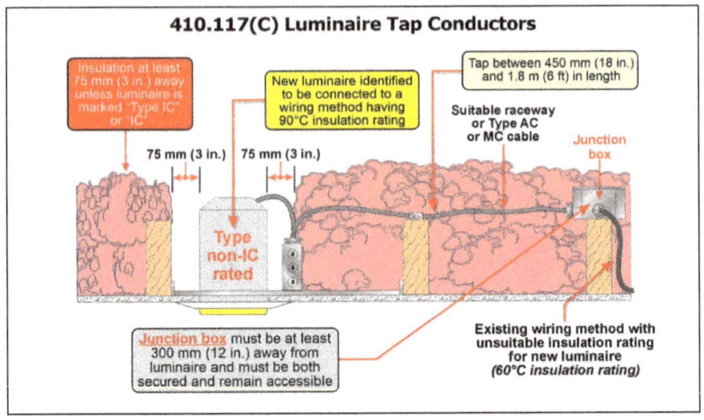

Figure 16.7. Installation of tap conductors to new luminaires

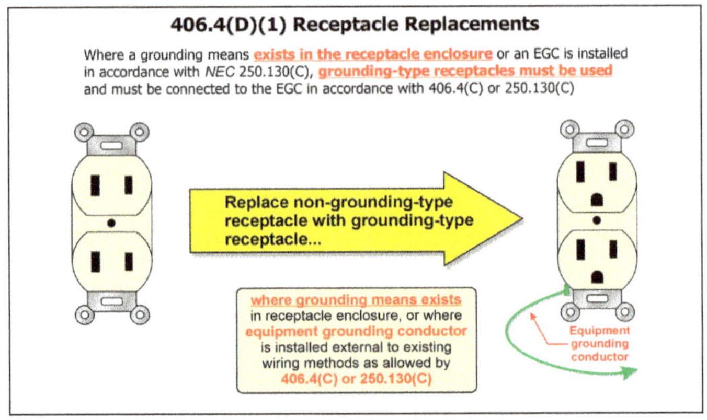

Figure 16.8. Receptacle replacement options where grounding means exist in the outlet box

REPLACING NON-GROUNDING TYPE RECEPTACLES

Under current Code requirements, receptacles installed on 15- and 20-ampere branch circuits must be grounding types [NEC 406.4(A) or IRC E4002.2]. This means the Code also requires that receptacles and cord connectors with grounding contacts have those contacts effectively grounded. Effective grounding is achieved by connecting the receptacle to a grounded enclosure or an equipment grounding conductor. Often, older wiring systems lack grounded enclosures or equipment grounding conductors. Therefore, the Code allows a replacement receptacle to be either non-grounding or to have a grounding terminal that remains ungrounded. When replacing ground-fault or arc-fault circuit-interrupter (GFCI or AFCI) type receptacles, they must be installed in a readily accessible location. These conditions are outlined in NEC 406.4(A) through (D).

Where a Grounding Means Exists in the Box

When a grounding means exists in the box or an equipment grounding conductor is installed, grounding-type receptacles must be used, and their grounding terminal must be connected to the equipment grounding conductor [NEC

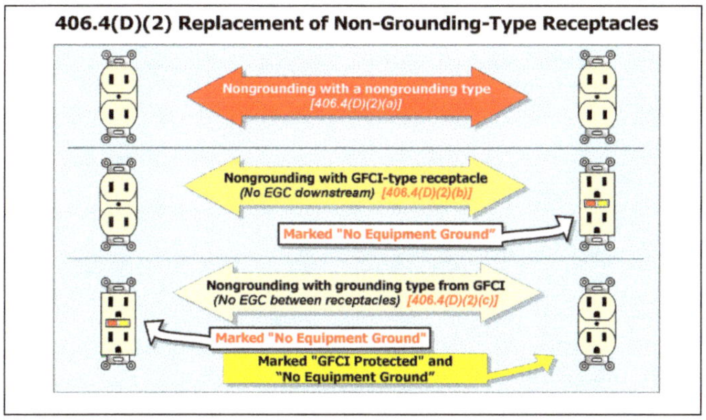

Figure 16.9. Replacement of receptacle options where a grounding means does not exist in the outlet box

406.4(A), 406.4(D)(1) or IRC E4002.2]. Refer to the section on "Grounding of Branch-Circuit Extensions" later in this chapter for more information on this topic.

Where a Grounding Means Does Not Exist in the Box

When a grounding means or equipment grounding conductor is absent in the receptacle enclosure, typically a device box, there are three options for replacing receptacles in NEC 406.4(D)(2): (a) a non-grounding receptacle can replace the existing non-grounding receptacle; (b) a ground-fault circuit-interrupter (GFCI) receptacle can replace the non-grounding receptacle; or (c) a non-grounding receptacle can be replaced with a grounding-type receptacle if supplied through a GFCI.

Some extra requirements apply when replacing an existing non-grounding-type receptacle with a grounding-type.

1. Where a ground-fault circuit interrupter-type receptacle is used to replace a non-grounding-type receptacle, the replacement receptacle must be marked "No Equipment Ground" on the receptacle or its cover plate. This marking is to inform the user that even though a grounding terminal is present in the device, an equipment grounding conductor is not connected to the device. This fact may be important, as some equipment requires a grounding connection in order to function properly. An example is a surge protection device commonly used to protect sensitive electronic equipment. This type of equipment is often found in dwellings and includes computers, many types of televisions, and audio equipment. An equipment grounding conductor is not permitted to be connected from the GFCI receptacle to any outlet supplied by the GFCI.
2. An existing non-grounding-type receptacle can be replaced with a grounding-type one that is protected on the supply side by a GFCI device. GFCI protection can be either a circuit breaker or a receptacle-type device. Grounding-type receptacles installed under this provision must be marked with the words "GFCI Protected" and "No Equipment Ground" on the receptacle or the cover plate. This identification is meant to inform the user that only a GFCI device protects the grounding-type receptacle and that an equipment-grounding conductor is not connected to the receptacle.

It is not permitted to install an equipment grounding conductor from a GFCI-type receptacle to other downstream receptacles where an equipment grounding conductor is not supplied to the enclosure for the GFCI device. This is to prevent the false impression that an equipment grounding terminal of the receptacle is functional.

The design of a GFCI device is such that it will function to protect people from shock hazards associated with the use of electrical equipment supplied by the GFCI device, even though it is not connected to an equipment grounding conductor. This is due to its ability to sense the imbalance of current when a part of the current passing through the device (usually the "hot" conductor) does not return through the device (usually the neutral conductor). It has been accurately described as an "imbalance detector" (see Figure 10.1 in Chapter 10 for a graphic illustration of the principles of a GFCI device).

Where a GFCI device is installed in an outlet box as a replacement for a non-grounding-type receptacle, care must be taken when testing the operation of the device. The "push to test button" on the receptacle will properly test the operation of the device due to the internal connections of the circuitry. Likewise, a "push to test" of a circuit breaker that may be installed in the panelboard ahead of the receptacle will properly test the operation of the GFCI circuit protection features. However, using a GFCI plug-in-type tester will not cause the GFCI-type receptacle or circuit breaker to operate. That is due to the design of the tester, which passes current over the equipment-grounding circuit path. Since an equipment grounding conductor is not present, there is no path for the test current to flow over.

Extra precautions are necessary when testing GFCI-protected outlets with a plug-in tester, especially if metal cover plates are installed on the outlets. This applies to both indoor and outdoor settings. As noted earlier, the design of the plug-in GFCI tester is to send a small current through the equipment-grounding conductor path. Since there is no equipment-grounding conductor, the tester energizes the metal cover plate. A shock risk arises if the person testing provides a path for the current to flow. Always follow the tester manufacturer's instructions when using it on circuits without an equipment-grounding conductor.

GROUNDING OF BRANCH-CIRCUIT EXTENSIONS

Where no equipment grounding conductor exists at older dwelling units, it is possible to install a new equipment grounding conductor to an existing non-grounded outlet. *NEC* 250.130(C) indicates that a conductor installed to provide an equipment grounding conductor for an existing circuit is permitted to be connected to any of the following:

1. Any accessible point on the grounding electrode system. Examples of permitted connection points are the underground metal water pipe within 1.52 m (5 ft) of where it enters the building or an exposed portion of a driven ground rod.
2. Any accessible point on the grounding electrode conductor
3. The equipment grounding terminal bar within the enclosure where the branch circuit for the receptacle originates
4. An equipment grounding conductor that is part of another branch circuit that originates from the enclosure where the branch circuit for the receptacle or branch circuit originates
5. For grounded systems, the grounded (neutral) service conductor is within the service equipment enclosure. This is accomplished by connecting the equipment grounding conductor to the neutral bar inside the service panelboard.
6. For ungrounded systems, the grounding terminal bar within the service equipment enclosure *(this would be a rare situation for a dwelling unit)*

For several Code cycles now, the Code has prohibited grounding receptacle grounding terminals from being connected to interior metal water pipes that are bonded to the service equipment enclosures or neutral conductors at the service. The NEC Code-Making Panel took this measure to reduce the use of interior metal water piping for equipment grounding or bonding, as repairs with nonmetallic materials can interrupt the equipment-grounding path.

Permission to use an equipment grounding conductor that is not run with the circuit conductors is the exception, not

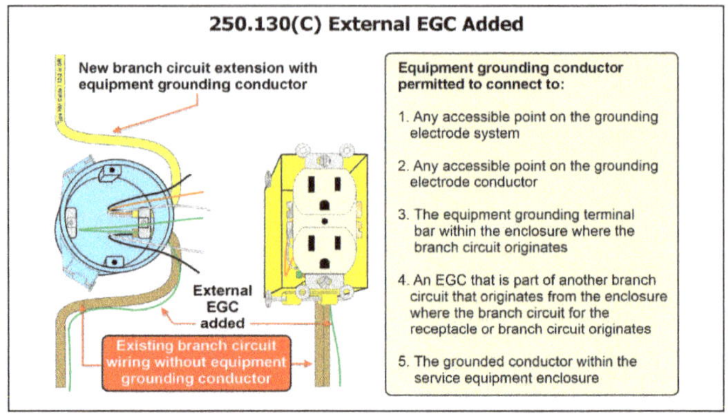

Figure 16.10. Equipment grounding conductor for branch-circuit extensions or grounding receptacles

the rule! Generally, NEC 250.134 requires that the equipment grounding conductor be run with the circuit conductors. NEC 300.3(B) or IRC E3406.7 also requires this. This is an effort to keep all the circuit conductors, including the equipment grounding conductor, together to maintain the circuit impedance as low as possible. Exception No. 1 to NEC 250.134 allows the equipment grounding conductor, installed from the equipment-grounding terminal of the receptacle to one of the locations mentioned above, to be routed separately from the branch-circuit conductors.

The Code requires that metal enclosures, other than service enclosures, be grounded. Usually, this grounding is done by connecting to the equipment grounding conductor that runs with the circuit conductors.

REPLACEMENT WHERE GFCI PROTECTION IS REQUIRED

The code requirements are generally not retroactive. However, NEC 406.4(D)(3) mandates that an existing non-GFCI receptacle must comply with GFCI protection requirements for 15- and 20-ampere, 125-volt receptacles at the time of replacement if that location would now require GFCI protection under current codes. First, review the list of locations where GFCI protection is mandated. Most of these locations for one- and two-family dwellings are listed in Section 210.8(A) of the NEC or Section E3902 of the IRC. If the existing receptacle is in a location now requiring GFCI protection, such as a kitchen countertop or bathroom receptacle, the replacement must also be GFCI-protected.

This requirement applies regardless of whether the original receptacle had GFCI protection. GFCI protection can be provided either by a GFCI receptacle installed at the location where protection is needed or by a GFCI-type circuit breaker protecting the circuit that supplies the receptacle(s). An example of when a GFCI-type circuit breaker might be used is when replacing the receptacle type is impractical, such as

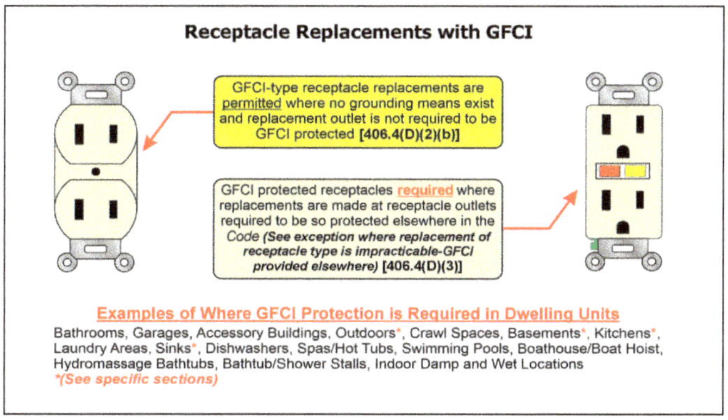

Figure 16.11. Receptacle replacement where GFCI is now required by the *Code* predicates the use of GFCI protection

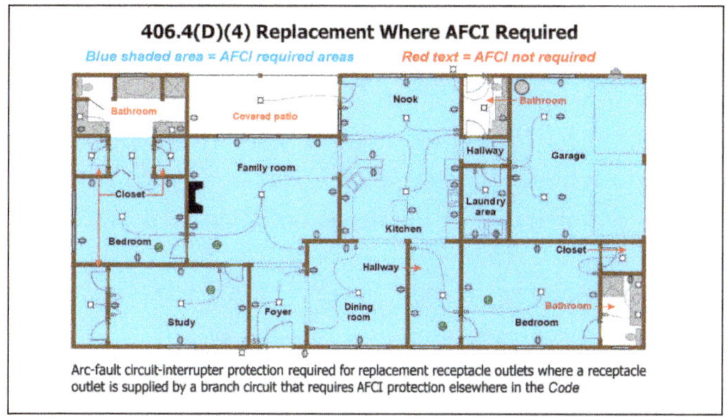

Figure 16.12. Receptacle replacement where AFCI is now required by the *Code* requires AFCI protection on that branch circuit

when the existing outlet box size cannot accommodate a GFCI receptacle. In such cases, the existing receptacle can be replaced with a new one of the same type, with GFCI protection supplied by a circuit breaker, and the new receptacle marked "GFCI Protected" and "No Equipment Ground," in accordance with NEC 406.4(D)(2) [see NEC 406.4(D)(3), Exception]. Refer to Chapter 10 for locations where GFCI protection for receptacles is required.

In this case, the existing receptacle could be replaced with a new receptacle of the existing type, with GFCI protection provided in the circuit breaker form, and the new receptacle is marked "GFCI Protected" and "No Equipment Ground," in accordance with *NEC* 406.4(D)(2) [see *NEC* 406.4(D)(3), Exception]. See Chapter 10 for locations where GFCI protection for receptacles is required.

It is important to understand that this requirement for GFCI protection applies whether or not a grounding means exists in the box. If a grounding means exists in the box, or one is added in accordance with NEC 250.130(C), the grounding terminal of the GFCI device must be grounded. If a grounding means does not exist in the box, installing a GFCI device is allowed under the conditions of NEC 406.4(D)(2), which were discussed earlier in this chapter.

REPLACEMENT WHERE AFCI PROTECTION IS REQUIRED

Similar to the ground-fault circuit interrupter (GFCI) requirement for replacement receptacle outlets, the Code also requires arc-fault circuit interrupter (AFCI) protection for replacement receptacle outlets supplied by branch circuits that need AFCI protection elsewhere in the Code [NEC 406.4(D)(4)]. This provides essential AFCI protection in older, existing homes, as data from the US Consumer Product Safety Commission shows that over 90 percent of electrical fires occur in homes over 10 years old. Before this requirement, AFCI protection in older homes was usually not provided unless circuit breakers were replaced or during a service upgrade, even though previous editions of the Code did not mandate the use of AFCI-type circuit breakers when replacing circuit breakers individually or during a complete upgrade, such as a service upgrade. AFCI protection was only required at an existing home when a new branch circuit was added or existing branch circuits were extended or modified to serve receptacle outlets in areas that need AFCI protection, like a bedroom, living room, den, etc.

The language in this AFCI requirement for existing dwellings allows AFCI protection to be provided either by an AFCI circuit breaker or by an AFCI outlet branch-circuit (OBC) receptacle when replacing an existing receptacle. Both the AFCI outlet branch-circuit devices and the AFCI circuit breaker protect all receptacles and outlets downstream on the branch circuit. The UL standard for AFCIs (UL 1699) covers outlet branch-circuit (OBC) type AFCI receptacles. An OBC-type AFCI receptacle offers series arcing protection upstream and downstream from its installation point, as well as parallel arc protection upstream (backward) of that point.

In previous editions of the NEC, there were two exceptions for the requirement to provide AFCI protection at existing replacement receptacles. These two exceptions were added in the 2017 NEC. The first exception covered situations where an existing two-wire receptacle (without an equipment grounding conductor) is replaced, and it is not possible to install an equipment grounding conductor. In this case, if the panelboard supplying the branch circuit does not offer a listed AFCI combination type overcurrent device, there was no method (at the time of the 2017 NEC revisions) to meet the AFCI protection requirement for replacement receptacles without this exception.

A GFCI receptacle is necessary to comply with NEC 406.4(D)(2)(b) and NEC 406.4(D)(3) when replacing a non-grounding-type receptacle. At the same time, NEC

406.4(D)(4)(1) requires a listed outlet branch circuit (OBC) type AFCI receptacle. Without this exception, the installer faced uncertainty about which Code rule to follow. Since a receptacle that offers both GFCI and AFCI (dual function) protection simultaneously is not required, Exception No. 1 to 406.4(D)(4) offers a solution for potential conflicts in Code rules.

For the 2020 NEC, Ex. No. 1 to NEC 406.4(D)(4) was removed because it is no longer necessary or relevant. Commercially available devices, such as dual-function AFCI/GFCI receptacle outlets, are easily accessible and can meet the main requirement, making the exception unnecessary.

It should also be noted that the exception for 210.12(E) does not apply to replacement receptacles. This AFCI exception excludes AFCI protection where extensions of existing conductors are no more than 1.8 m (6 ft) and does not cover any additional outlets or devices. Again, this exception does not apply to replacement receptacles (NEC 406.4(D)(4), Exception).

REPLACEMENTS - WHERE TAMPER-RESISTANT OR WEATHER-RESISTANT RECEPTACLES ARE REQUIRED

Listed tamper-resistant receptacles are required for replacement outlets where the original installation mandates tamper-resistant features elsewhere in the Code, except when a non-grounding receptacle is replaced with another non-grounding one [NEC 406.4(D)(5)]. This requirement closely resembles the existing ground-fault circuit interrupter (GFCI) requirement for replacement outlets and the arc-fault circuit interrupter (AFCI) protection for replacement receptacles. A similar requirement applies to weather-resistant receptacles at NEC 406.4(D)(6).

Substantiation for introducing tamper- resistant receptacles into the NEC showed that an average of 2, 400 children are injured each year by inserting foreign objects (such as paper clips and keys) into energized electrical receptacles. Since the number of existing dwellings far exceeds new residential construction, this tamper- resistant receptacle requirement has a greater safety impact for small children than if it applied only to new constructions.

This rule requires replacing existing non- tamper- resistant receptacles with tamper- resistant ones during repairs at locations where receptacles are mandated to be tamper- resistant by NEC 406. 12 or IRC E 4002. 14. Now, these Code updates mandate tamper- resistant receptacles at all 15- and 20- ampere, 125- and 250- volt non- locking receptacles in dwelling units, boathouses, mobile and manufactured homes, including attached and detached garages, accessory buildings,

Figure 16.13. Tamper-resistant receptacles are required for *ALL* replacement receptacles at dwelling units

and common areas. It is important to note that before the 2023 NEC (2024 IRC), tamper- resistant receptacles were only required in areas specified by NEC 210. 52 (Dwelling Unit Receptacle Outlets) and 550.13. 13 (mobile homes) or IRC E 3901. 1. Refer to Chapter 13 for additional details on the functions of tamper- resistant receptacles.

Additionally, not only are listed tamper- resistant receptacles required for 15- and 20- ampere, 125- volt non- locking receptacles at dwelling units, but such receptacles are also mandated for 250- volt non- locking receptacles in those areas.

This replacement requirement applies to receptacles where weather- resistant models are specified elsewhere in the code [NEC 406. 4 (D)(6)]. Weather- resistant receptacles are necessary at all 15- and 20- ampere, 125- and 250- volt non-locking receptacles installed in damp or wet locations.

Without weather- resistant receptacles in these settings, deterioration and other damage can adversely affect non-weather- resistant models, often causing the faces to become brittle and break. Receptacles in damp and wet locations are frequently exposed to moisture, UV rays, and temperature fluctuations. Non- weather- resistant receptacles are not designed or rated to withstand these conditions.

This requirement for weather-resistant receptacles mandates their use in existing buildings and structures when replacing receptacles in damp and wet locations. Without this requirement, a standard non-weather-resistant receptacle could be installed, risking the same failures as the one being replaced [NEC 406.4(D)(6)].

REPLACEMENT OF SWITCHES WHERE NO GROUNDING MEANS EXIST IN THE BOX

Snap switches, dimmers, and control switches are generally required to be properly grounded by connecting them to an

Photo 16.4. Weather-resistant receptacle is required in damp or wet locations for replacement receptacles

equipment grounding conductor [NEC 404.9(B) or IRC E4001.11.1]. Additionally, snap switches, dimmers, and control switches must include a means to bond a metal faceplate to the equipment grounding conductor if a metal faceplate is installed during installation or at a later time. These switches and faceplates must be connected to an equipment grounding conductor using one of the following methods:

(1) The switch is mounted with metal screws to a metal box or metal cover that is connected to an equipment grounding conductor or to a nonmetallic box with integral means for connecting to an equipment grounding conductor.
(2) An equipment grounding conductor or equipment bonding jumper is connected to an equipment grounding termination of the snap switch.

The Code includes three exceptions to the grounding requirement for snap switches, dimmers, and control switches. The first exception applies to snap switches for replacement purposes only, where no equipment grounding means exists within the box or where the existing wiring method does not include or provide an equipment grounding conductor. Several conditions apply to this first exception.

A snap switch that is wired under the provisions of Exception No. 1 to *NEC* 404.9(B) or IRC E4001.11.1 Ex. No. 1 and located within 2.5 m (8 ft) vertically, or 1.5 m (5 ft) horizontally, of ground or exposed grounded metal objects must be provided with faceplates of nonconducting noncombustible material with nonmetallic attachment screws (unless the switch mounting yoke is nonmetallic) or shall be protected by a ground-fault circuit interrupter.

Exception No. 2 addresses listed switch kits and listed switch assemblies constructed with nonmetallic yokes and equipped with nonmetallic faceplates. These switches are not designed for, and do not have provisions for, attaching a metallic faceplate to them. Since only a nonmetallic faceplate can be attached to these switches, it is only logical to exempt these switches from grounding requirements.

The nonmetallic faceplates of these listed kits or assemblies cannot be installed on any other type of device, and these devices do not have mounting means to accept other types of faceplates.

Exception No. 3 speaks to snap switches with integral nonmetallic enclosures. These switch enclosures must meet the requirements of NEC 300.15(E) or IRC E3905.1.3 (Integral Enclosure). These switches incorporate integral nonconductive (non-metallic) faceplates and cannot inherently interchange faceplates, metal or otherwise. With the design of these nonmetallic devices, there is no need to provide a means to connect a metal faceplate to an EGC.

See NEC 404.9(B) or IRC E4001.11.1, as well as the exceptions to these rules, for specific information on grounding general-use snap switches, dimmers, and control switches.

REPLACING LUMINAIRES WHERE NO GROUNDING MEANS EXISTS IN THE BOX

The Code requires that luminaires be grounded by connection to an equipment grounding conductor in accordance with Part V of NEC Article 410 [NEC 410.40], NEC 410.42, and the Exception and IRC E4003.3. Additionally, exposed conductive surfaces of luminaires must be connected to an equipment grounding conductor or be separated from all live parts and other conducting surfaces by a listed double insulation system.

The exception also permits small, isolated luminaires' accessory parts, such as lamp tie wires, mounting screws, clips, and decorative bands on glass, spaced at least 38 mm (1½ in.) from any lamp terminal, not to be grounded by connection to an equipment grounding conductor. This exception to NEC 410.42 also prevents the connection of portable luminaires with a polarized attachment plug to an equipment grounding conductor.

Manufacturers of luminaires with exposed conductive surfaces typically provide a means for connecting an equipment grounding conductor to such luminaires. *NEC* 410.44 (with exceptions) requires luminaires and equipment that require connection to an equipment grounding conductor to be mechanically connected to an equipment grounding conductor.

NEC 410.44, Ex. No. 1 allows replacement luminaires to be connected to an equipment grounding conductor in the same way as replacement receptacles, as permitted by NEC 250.130(C), which authorizes adding a separate equipment grounding conductor. The luminaire must then comply with 410.42, which requires a connection to an equipment grounding conductor or being separated from all live parts and other conducting surfaces by a listed double insulation system.

In older homes, the existing wiring may not include a grounding means at the outlet box. The Code offers an alternative method of installing a separate equipment grounding conductor to enable the installation of luminaires that must be grounded (connected to an equipment grounding conductor). Essentially, the same provisions are allowed for both luminaire and receptacle outlets where no grounding means exists at the outlet box. NEC 250.130(C) permits installing an equipment grounding conductor from the lighting outlet to one of the connection points listed below.

1. Any accessible point on the grounding electrode system
2. Any accessible point on the grounding electrode conductor
3. The equipment grounding terminal bar where the branch circuit originates
4. An equipment grounding conductor that is part of another branch circuit that originates from the enclosure where the branch circuit for the receptacle or branch circuit originates
5. For grounded systems, the grounded service conductor is within the service equipment enclosure. The connection is made to the neutral terminal bar.

A second exception to NEC 410.44 for replacement luminaires permits no connection to an equipment grounding conductor when none exists at the outlet, provided the replacement luminaire is GFCI protected or lacks exposed conductive parts. This exception resembles the one for receptacles and snap switches supplied from older circuits without an equipment grounding conductor.

GROUNDING COOKING APPLIANCES AND CLOTHES DRYERS ON EXISTING BRANCH CIRCUITS

The Code requires that the metal frames of ranges and dryers be grounded. A new installation must use the equipment grounding conductor that is required to be run with the branch circuit conductors as the grounding method, as specified by NEC 250.140(A). NEC 250.140(B) allows the frames of existing ranges, wall-mounted ovens, counter-mounted cooking units, clothes dryers, and outlet or junction boxes that are part of the supply circuit to be

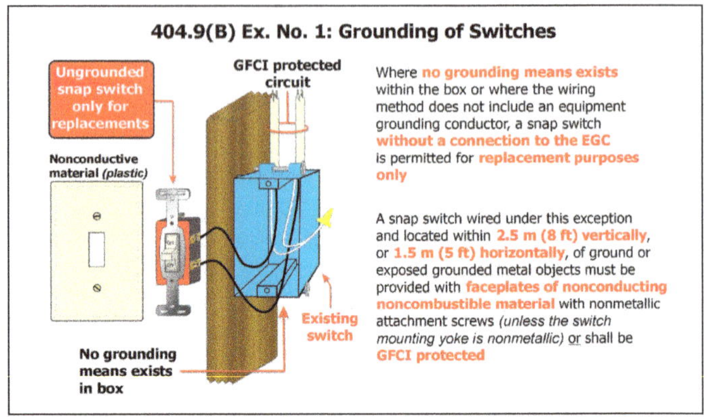

Figure 16.14. One method for the grounding requirements for replacement snap switches where no grounding means exists in the box

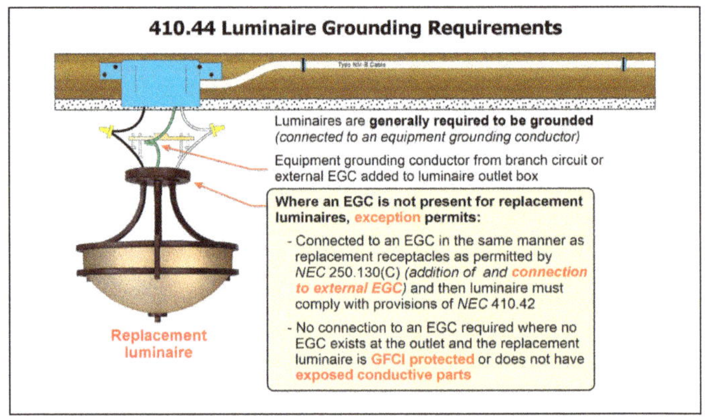

Figure 16.15. Grounding is generally required for replacement luminaires

grounded to the grounded (neutral) conductor of an existing branch circuit. It is very important to remember that using the grounded conductor for grounding is only allowed for existing installations with existing branch circuit conductors. There are four key conditions or qualifiers that must be applied to these existing installations:

1. The supply circuit is 120/240-volt, single-phase, 3-wire; or 208Y/120-volt derived from a 3-phase, 4-wire, wye-connected system.
2. The grounded (neutral) conductor is not smaller than 10 AWG copper or 8 AWG aluminum.
3. The grounded (neutral) conductor is insulated, or the grounded (neutral) conductor is uninsulated and part of a Type SE service-entrance cable, and the branch circuit originates at the service equipment.
4. Grounding contacts of receptacles furnished as part of the equipment are bonded to the equipment. (Note: Receptacle outlets are no longer furnished on listed electric ranges).

The first words of Item 4, "The grounded (neutral) conductor is insulated," permit the use of three-conductor Type NM

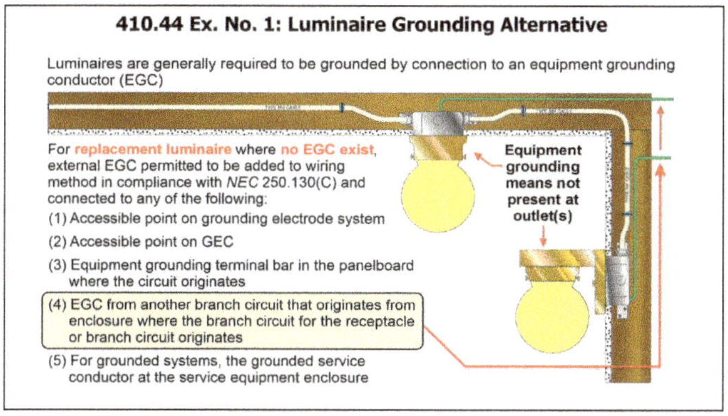

Figure 16.16. Separate equipment grounding conductor permitted to be installed where a grounding means does not exist at the luminaire outlet box

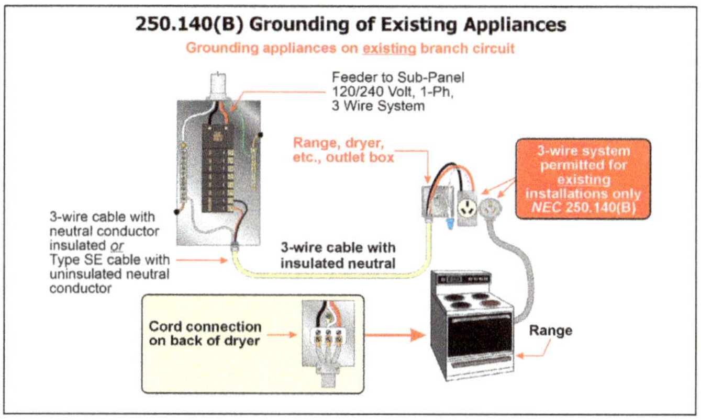

Figure 16.17. Grounding the frame of existing ranges and dryers permits the use of the grounded conductor for grounding under specific conditions

cable (one without a bare equipment-grounding conductor). Where the branch circuit originates at a panelboard on the load side of service, the grounded circuit conductor must be an insulated conductor. This helps to prevent the neutral conductor from being accidentally bonded to the equipment grounding conductor or the panelboard enclosures.

Ranges and dryers in mobile homes and recreational vehicles are not permitted to use the grounded (neutral) conductor-to-ground equipment. So, the exception described above does not apply to circuits supplying ranges and dryers in structures covered by the Scopes of Articles 550, Mobile Homes, and 551, Recreational Vehicles. For the grounding of new ranges and dryers, see Chapter 7.

LOADS FOR ADDITIONS TO EXISTING INSTALLATIONS: STANDARD CALCULATIONS

NEC 220.16(A) covers loads for structural additions to existing dwelling units or a previously unwired portion of an existing dwelling unit. These loads are required to be computed in accordance with *NEC* 220.14 (other loads). These loads must also comply with *NEC* 220.41 and 220.5(C).

Loads for new circuits or extended circuits in previously wired dwelling units must be calculated based on volt-amperes per square foot, including the specific loads for each outlet. This process mirrors the method used for estimating loads in a new home. According to NEC 220.41, the Code specifies that the minimum unit load for dwelling units should not be less than 33 volt-amperes per square meter (3 volt-amperes per square foot). Additionally, NEC 220.5(C) states that the calculated floor area does not include open porches or unfinished spaces that are not suitable for future use as habitable or occupiable areas.

OPTIONAL CALCULATIONS

NEC 220.83 allows an optional calculation for extra loads in dwelling units that are currently supplied by an existing 120/240-volt or 208Y/120-volt, 3-wire service or feeder. The required load calculations are as follows:

- The nameplate load of appliances that are fastened in place, permanently connected, or located on a specific circuit, ranges, wall-mounted ovens, counter-mounted cooking units, clothes dryers, water heaters, and space heaters must be used.
- After making the load calculation, a 100% demand factor from *NEC* 220.83(A) is applied to the first 8 KVA of load. A demand factor of 40% is then applied to the remainder of the load where additional air-conditioning equipment or electric space-heating equipment is ***not*** going to be installed.
- Use a larger connected load of air-conditioning and space-heating units, but not both (*NEC* 220.61).

The calculation is similar for a dwelling unit where air-conditioning or electric space-heating equipment will be added or installed [NEC 220.83(B)]. The first step is to include (at 100%) the larger load of either the air-conditioning or central electric space-heating equipment (including fewer than four separately controlled space-heating loads). The smaller of these two loads is considered a noncoincident load per NEC 220.61, and only the larger load is calculated since both loads will not operate simultaneously. Then, add the "other loads." These include lighting at 3 volt-amperes per square foot; 1500 volt-amperes for each 2-wire small appliance branch circuit; 1500 volt-amperes for each laundry branch circuit; and any range, wall-mounted oven, counter-mounted cooking unit, or other appliances that are permanently connected or fastened in place, rated according to their nameplate.

"Other loads" include 1500 volt-amperes for each 20-ampere appliance branch circuit; 1500-volt-amperes for the laundry circuit, lighting, and general receptacle/portable appliance load at 3 volt-amperes per square foot; household range or wall-mounted oven and counter-mounted cooking unit at their nameplate ratings, and all other appliances fastened in place.

BRANCH CIRCUITS REQUIRED

It is common practice to require compliance with the current edition of the Code for areas or rooms being remodeled or for rooms or spaces being added to an existing dwelling. For example, if a utility room for laundry equipment is added to the structure, a 20-ampere circuit for laundry appliances is generally required by the inspection authority. Additionally, a switch-controlled lighting outlet should be installed. However, when a circuit is added to an existing kitchen for a small appliance, such as a microwave oven, it is not typical to require the entire kitchen to be rewired to meet the current Code. Usually, the added circuit is considered an improvement over the existing wiring system, thereby enhancing safety.

EDISON-BASE TYPE PLUG FUSES

It is common to find Edison-type fuses used for overcurrent protection in much older homes. These Edison-based fuses are round, screw-in types often called plug fuses. They come in 15-, 20-, and 30-amp sizes and are screwed directly into the electrical panel or fuseholder. Plug fuses typically control 120-volt circuits that supply lighting and receptacles in older homes. Inside, they have a metal alloy ribbon that carries the current of the branch circuit. In case of a ground fault or overload, the ribbon melts and opens the circuit. When this happens, the fuse must be replaced. During electrical remodels, electricians sometimes remove these fuses, and a copper or copper alloy penny might fall out of the socket. Placing a penny in the fuseholder bypasses the overcurrent protection and can cause a fire.

NEC 240.51 covers the classification for plug fuses of the Edison-base type. They are classified as not over 125 volts and at 30 amperes or less. The use of an Edison-base type of plug fuse is restricted to replacements in existing installations where there is no evidence of overfusing or tampering.

The most common misuse of this type of fuse occurs when they are oversized, such as a 30-amp (green) fuse used to protect 15- or 20-ampere-rated conductors. An Edison-base fuseholder is only allowed to be installed where it is designed to accept Type S fuses using adapters (NEC 240.52).

Where there is evidence of overfusing or tampering, a Type S fuse must be installed. It must be of the plug type and classified as not exceeding 125 volts and rated for 0 to 15 amperes, 16 to 20 amperes, or 21 to 30 amperes. A Type S fuse with an ampere rating as specified here is not permitted to be interchangeable with a lower ampere rating. A Type S fuse adapter must be designed so that it cannot be used in any fuseholder other than a Type S fuseholder or a fuseholder with a Type S adapter inserted.

Type-S adapters are used to prevent the use of the wrong size fuse and to ensure proper overcurrent protection. A special "can-opener type" tool is required to remove an adapter. Attempting to remove a Type-S adapter with other tools can usually cause damage to the Edison-base fuse socket (NEC 240.53).

Optional Calculation for Additional Loads in Existing Dwelling Units

Where Additional A/C or Electric Space-Heating WILL NOT be Installed *[220.83(A)]*	
Loads (in kVA)	**Percent of Load**
First 8 kVA of load at:	100
Remainder of load at:	40

Where Additional A/C or Electric Space-Heating WILL be Installed *[220.83(B)]*	
Loads	**Percent of Load**
A/C equipment	100
Central electric space heating	100
Less than four separately controlled space-heating units	100
First 8 kVA of load at:	100
Remainder of load at:	40

Table 16.1. See *NEC* 220.83(A) and *NEC* 220.83(B)

Photo 16.5. Older style service equipment showing old Edison-base fuses

Photo 16.6. Type S fuse and adapter

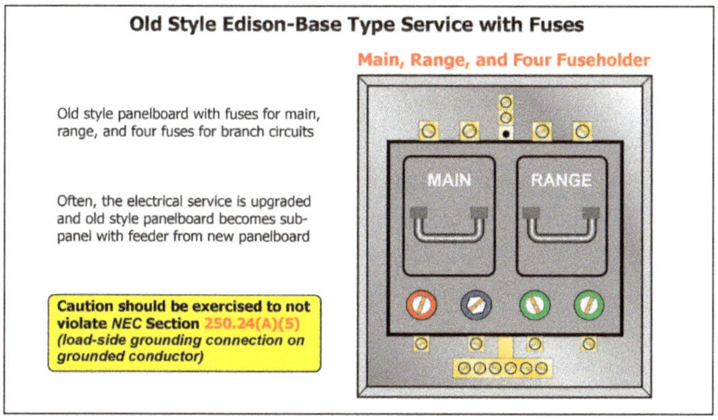

Figure 16.18. "Main, range, and four" overcurrent protection found in many older existing dwelling units

OVERCURRENT PROTECTION OF EXISTING PANELBOARD

The general rule in NEC 408.36 or IRC E3706.3 is that a panelboard must be individually protected on the supply side by no more than two main overcurrent devices. Exception No. 2 to NEC 408.36 allows existing panelboard installations to operate without individual protection (of no more than two main overcurrent devices) when such panelboards are used as service equipment for a single residential occupancy. This exception permits the continued use of the old-style "six-disconnect split-bus panelboard" in existing dwelling units.

SUPPLYING OLD SERVICE EQUIPMENT

Installing a new service on an existing home typically involves replacing all the equipment up to and including the service panelboard. Sometimes, it is possible to re-feed the existing service equipment with a new feeder. The existing service now becomes a feeder panelboard or subpanel. The existing service panel board cannot always be used as a feeder panelboard. Where the existing service equipment uses fuses, it was common for the manufacturer to bond the grounded (neutral) terminal board to the enclosure at the factory. In this case, the existing service equipment cannot be converted to a feeder panel and supplied for the new service equipment. This type of feeder installation would be in violation of NEC 250.24(B), 250.142, and 408.40, or IRC E3908.7. The grounded (neutral) conductor is not generally permitted to be used to ground equipment on the load side of the service disconnecting means. There are a few exceptions to this rule, none of which apply in this case. Thus, the grounded system conductor cannot be connected to an enclosure of a feeder panelboard where both the service equipment and the feeder panelboard are in the same building. In other words, the grounded (neutral) conductors would need to be isolated or "floating" from the equipment grounding conductor terminal bar and the metal cabinet of the existing panelboard.

Some types of panelboard construction allow a bonding jumper or bar to be removed between the grounded (neutral) conductor terminal bar and the equipment grounding conductor terminal bar, according to the manufacturer's design. In this case, the existing panelboard can be converted into a feeder panelboard. It will also be necessary to remove the existing grounding electrode conductor from the grounded (neutral) conductor terminal bar and connect it to the equipment grounding conductor terminal bar. The grounding electrode conductor can be connected to the new service if it is of the correct size. A thorough inspection of the connections at each electrode is required. The clamps or fittings may have corroded over time and might need cleaning or replacement (NEC 250.8, 250.12, or IRC E3908.18).

When the existing grounding electrode conductor is too short to reach the new service location, the Code allows splicing using one of two methods: (a) the exothermic welding process or (b) an irreversible compression-type connector that is listed and designed for this purpose. It is essential to follow the manufacturer's instructions for proper tools and procedures with both methods. See NEC 250.64(C) or IRC E3610.1 for splicing a grounding electrode conductor.

OTHER *CODE* REQUIREMENTS WHEN UPDATING OR REPLACING EXISTING SERVICE EQUIPMENT AT EXISTING DWELLING UNITS

It is common to upgrade or replace existing services in older homes. This can occur due to damaged service equipment during weather events or because more capacity is required than the current service can provide. Often, when remodeling or adding to an existing home, a bigger, modern electrical service is required.

In any of these cases, there are a few locations in the Code that require other upgrades to the electrical service when existing electrical systems are replaced or upgraded.

Emergency Disconnect(s)

When replacing existing service equipment at a dwelling unit, all requirements of NEC 230.85 or IRC E3601.8 must be followed. This section mandates installing an emergency disconnecting means in a readily accessible outdoor location on or within sight of the dwelling. If only meter sockets, service entrance conductors, or related raceways and fittings are replaced, the requirements for the emergency disconnecting means do not need to be applied [NEC 230.85(C) and

Exception or IRC E3601.8].

Refer to NEC 230.85 or IRC E3601.8 for comprehensive requirements on emergency disconnects for dwelling units. This includes details on the location, rating, grouping, disconnect type, identification of other isolated disconnects, and marking criteria. Additionally, see Chapter 4 for further information on outside emergency disconnects for dwelling units.

Surge Protection

Where existing service equipment is replaced at an existing dwelling unit, all of the requirements of *NEC* 230.67 or IRC E3606.5 must be applied [*NEC* 230.67(D) or IRC E3606.5.3]. This section(s) calls for whole-house surge protection, with the surge protection device (SPD) being an integral part of the service equipment or located immediately adjacent thereto. The SPD is not required to be located at the service equipment if surge protection is located at each next-level distribution equipment downstream toward the load. This next level of surge protection is not common in dwelling units.

The SPD at dwelling units must be either a Type 1 or Type 2 SPD. This SPD must have a nominal discharge current rating of at least 10 kA. Refer to NEC 230.67 or IRC E3606.5 for detailed information on whole-house surge protection for dwelling units. Also, refer to Chapter 4 for complete information on surge protection for dwelling units.

REFERENCES

1. *National Electrical Code* and *NEC* are registered trademarks of the National Fire Protection Association, Inc., Quincy, MA 02169. This reprinted material is not the official position of the National Fire Protection Association, which is represented only by the standard in its entirety.

2. *International Residential Code for One- and Two-Family Dwellings* (2021). Copyright © International Code Council, Inc., June 2023.

3. UL, Northbrook, IL, *UL Guide Information for Electrical Equipment Directory* (UL Product iQ® material) unless indicated otherwise.

REVIEW QUESTIONS

1. Older one- and two-family dwellings could be wired with Type NM (Romex) and Type AC (BX) cable assemblies, or, in older homes built prior to the 1960s, knob-and-tube wiring was installed that included circuit conductors with Type R (rubber) insulation.

True _____ False _____

2. In older, existing homes, some older Type NM cables used for common branch circuits were manufactured with conductor insulation of white and black and a bare equipment grounding conductor that was two sizes smaller than the circuit's ungrounded conductors.

True _____ False _____

3. Concealed knob-and-tube wiring is a wiring method that uses knobs, tubes, and flexible nonmetallic tubing (loom) for the protection and support of single-insulated conductors.

True _____ False _____

4. To properly convert a system of existing knob-and-tube wiring to an extension of Type NM nonmetallic-sheathed cable to be used for the continuation of a branch circuit, a ____________ must be installed at the junction of the two different wiring methods.

A. branch circuit extender
B. junction box
C. panelboard
D. auxiliary gutter

5. Where no equipment grounding conductor is present at an existing receptacle, the installation of an equipment-grounding conductor connected 7.5 m (25 ft) inside the building to a properly bonded isolated interior metal water pipe can be run to each grounding terminal of grounding-type receptacles.

True _____ False _____

6. Loads for new circuits or extended circuits in previously wired dwelling units are required to be computed in accordance with *NEC* 220.14.

True _____ False _____

7. Plug fuses of the Edison-base type are classified at not over 125 volts and at __________ and below.

A. 20 amperes
B. 30 amperes
C. 40 amperes
D. 50 amperes

8. The use of an Edison-base type of plug fuse is restricted to replacements in existing installations where there is no evidence of overfusing or tampering.

True _____ False _____

9. An Edison-based fuseholder is permitted to be installed only where it is made to accept ____________ by the use of adapters.

A. Type B fuses
B. Type J fuses
C. Type S fuses
D. a circuit breaker

10. Generally, a panelboard is required to be protected by an overcurrent protective device having a rating not greater than that of the panelboard. For existing panelboards at dwelling units, individual protection shall not be required for a panelboard used as service equipment for an individual residential occupancy.

True _____ False _____

11. Ground-fault circuit-interrupter (GFCI) protected receptacles must be provided where replacements are made at existing receptacle outlets that are required to be so protected elsewhere in the *NEC* or IRC.

True _____ False _____

12. ____________________ shall be provided where replacements are made at receptacle outlets that are required to be so protected or designed elsewhere in the *NEC* or IRC.

A. New receptacles
B. Listed tamper-resistant receptacles
C. Weather-resistant receptacles
D. Both B. and C.

13. In any of the areas of an existing dwelling that is listed in *NEC* 210.12(B) or IRC E3902.17 where branch-circuit wiring is ____________, the branch circuit is required to be AFCI protected and can be AFCI protected by a listed combination AFCI overcurrent device located at the origin of the branch circuit, or a listed outlet branch-circuit AFCI located at the first receptacle outlet of the existing branch circuit.

A. modified
B. replaced
C. extended
D. all of the above

14. Branch circuit extensions for a panelboard relocation from a clothes closet to another location can be made without AFCI protection being installed to the existing extended branch circuits as long as the extensions do not exceed ____________________.

A. 600 mm (24 in.)
B. 900 mm (3 ft)
C. 1.8 m (6 ft)
D. 3.0 m (10 ft)

15. If a receptacle outlet located in any areas specified in *NEC* 210.12(B) or IRC E3902.17 is replaced, a replacement receptacle at this outlet is required to be which one of the following:

A. A listed outlet branch-circuit type arc-fault circuit-interrupter receptacle
B. A receptacle protected by a listed outlet branch-circuit type arc-fault circuit-interrupter type receptacle
C. A receptacle protected by a listed combination type arc-fault circuit-interrupter type circuit breaker
D. Any of the above

16. Where existing service equipment is replaced at an

existing dwelling unit, all of the requirements for an emergency disconnecting means must be applied. This emergency disconnecting means must be installed in a ______________________________ on or within sight of the dwelling unit.

A. readily accessible location
B. outdoor location
C. dry location
D. Both A. and B.

17. Where existing service equipment is replaced at an existing dwelling unit, all of the requirements for surge protection must be applied. This calls for whole-house surge protection with the surge protection device (SPD) being ______________________________.

A. installed in a readily accessible location
B. an integral part of the service equipment
C. located immediately adjacent thereto
D. Either B. or C.

17 Low Voltage and Limited Energy Systems

This chapter covers electrical systems that are not the same as typical power and lighting circuits. Subjects covered include wiring methods, grounding requirements, and conductor requirements for remote-control, signaling, and power-limited circuits; fire alarm systems; communication circuits; and community antenna television systems.

It is essential to become familiar with the regulations, installation methods, and equipment in order to apply the rules to one- and two-family dwellings correctly. Much of the material in *NEC* Articles 724, 725, 760, 800, 805, 820, 830, 840, and IRC Chapter 43 is similar. All involve limited energy systems. Common topics include installation techniques, wiring methods, and grounding and bonding requirements.

A new *NEC* Article 800, titled "General Requirements for Communications Systems," was added to Chapter 8 for the 2020 *NEC*. The previous *NEC* Article 800 (Communications Circuits) was moved to Article 805 to make room for this new Chapter 8 article. Article 800 combines common requirements previously found in *NEC* Articles 800 *(now Article 805)* for communications circuits, 820 for community antenna television and radio distribution systems, 830 for network-powered broadband communications systems, and 840 for premises-powered broadband communications systems into a "general" article that applies to all these low voltage articles unless modified by the forenamed articles. This would include such things as mechanical execution of work, abandoned cables, spread of fire or products of combustion, and temperature limitations of wires and cables. In previous editions of the *Code*, if the user of the *Code* saw a change in one of the *NEC* Chapter 8 articles, it was a good bet that the same change would be occurring in the other *NEC* Chapter 8 articles as well.

The scope statement of Article 800 (NEC 800.1) is similar to that of NEC Article 300. The text emphasizes that NEC

Article 800 is a general article and clarifies that its requirements are modified by subsequent Articles in NEC Chapter 8. Since this general article does not include requirements for NEC Article 810 (Radio and Television Equipment), it is necessary to list the four articles to which the general requirements apply. This new article will serve as a placeholder for redundant requirements in Articles 805, 820, 830, and 840. This article for communication circuits greatly improves clarity and usability while eliminating redundant requirements across NEC Chapter 8.

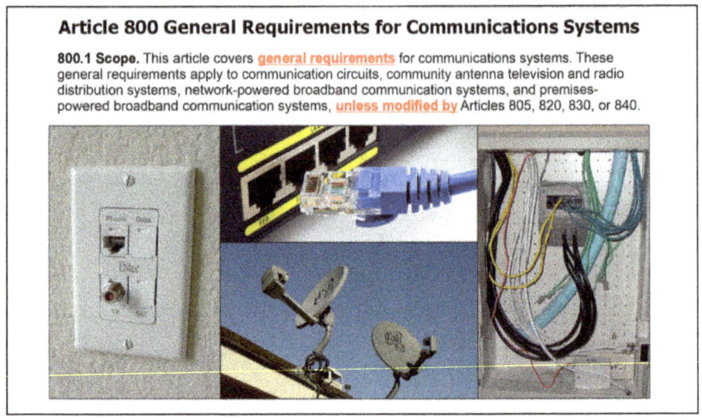

Figure 17.1. *NEC* Article 800 provides general requirements common to *NEC* Articles 805, 820, 830, and 840

Photo 17.1. Typical communication, Internet, and cable TV wiring and systems entering a dwelling unit building

LOW VOLTAGE AND LIMITED ENERGY SYSTEMS

The *NEC* specifies minimum requirements for installing Class 1 circuits in NEC Article 724, while Class 2 and Class 3 circuits are covered in NEC Article 725. Generally, only Class 2 control circuits are used in dwelling units, which is why Chapter 43 of the IRC focuses solely on Class 2 circuit requirements for homes. Consequently, the NEC must be followed as the default standard, since the IRC does not address other types of low-voltage and limited-energy circuits and wiring used in dwelling units.

Fire alarm circuits and wiring must comply with the minimum requirements of NEC Article 760 and any applicable provisions of NFPA 101, Life Safety Code, and NFPA 72, National Fire Alarm and Signaling Code. Chapter 10 provides information related to fire and smoke detection equipment and systems installed in dwelling units (see NFPA 72 and IRC R314). Any installation of optical fiber cables and wiring must meet the minimum requirements in NEC Article 770 (Optical Fiber Cables).

Requirements for communications systems and circuits fall under the scope of *NEC* Article 805. Radio and television equipment requirements are provided in *NEC* Article 810. Community antenna television and radio distribution systems fall under the requirements of *NEC* Article 820. Network-powered broadband communications systems rules are located in *NEC* Article 830, and premises-powered optical fiber-based broadband communications systems are covered by *NEC* Article 840. Where any of these systems are installed in dwelling units, they must meet the minimum requirements of these *NEC* Articles.

CLASS 1 POWER-LIMITED CIRCUITS

The requirements for Class 1 remote-control, signaling, and power-limited circuits are outlined in NEC Article 724. This is a new section introduced in the 2023 NEC. The scope of this new article specifies that it covers Class 1 circuits, including power-limited Class 1 remote-control and signaling circuits that are not part of a device or utilization equipment.

A ***Class 1 Circuit*** is defined as "the portion of the wiring system between the load side of the Class 1 power source and the connected equipment" [*NEC* Article 100]. Class 1 circuits must be supplied from a source with a rated output of no more than 30 volts and 1000 volt-amperes [NEC 724.40]. As mentioned earlier, Class 1 circuits are generally not used in dwelling units.

CLASS 2 AND CLASS 3 POWER-LIMITED CIRCUITS

NEC Article 725 covers Class 2 and Class 3 remote-control, signaling, and power-limited circuits that are not an integral part of a device or appliance (*NEC* 725.1 or IRC E4301.1). These circuits are characterized by usage and electrical power limitations that differentiate them from electric light and power circuits. For this reason, the *Code* permits alternative requirements that are different and less restrictive than the wiring method requirements for power and lighting circuits. Differences include minimum conductor sizes, ampacity correction factors, overcurrent protection, insulation requirements, and wiring methods and materials.

Photo 17.2. Typical Class 2 control circuits and wiring found at dwelling units (cable TV, Cat 5 internet, security wiring, etc.)

CODE DEFINITIONS AND TYPICAL APPLICATIONS

Let's review some of the low-voltage definitions we need to know for this chapter.

A ***remote-control circuit*** is any electrical circuit that manages another circuit using a relay or similar device. Examples include control circuits that operate a motor's magnetic starter, a relay that activates a furnace, start or stop an A/C unit, adjust lighting levels with a dimmer panel, or open and close a garage door.

A ***signaling circuit*** is any electrical circuit that energizes signaling equipment. Examples include doorbells, buzzers, smoke alarms, fire or burglar alarms, and similar devices. A Class 1 Circuit is the portion of the wiring system between the load side of the Class 1 power source and the connected equipment. The voltage and power limitations of the Class 1 source are specified in NEC 724.40. Note that Class 1 circuits were moved from NEC Article 725 to a new NEC Article 724 for the 2023 NEC.

A ***Class 2 Circuit*** is the part of the wiring system between the load side of a Class 2 power source and the connected equipment. Due to its power limitations, a Class 2 circuit emphasizes safety from a fire initiation standpoint and provides acceptable protection from electric shock.

A ***Class 3 Circuit*** is the section between the load side of a Class 3 power source and the connected equipment. Due to its power limitations, a Class 3 circuit emphasizes fire safety. Because it allows higher levels of voltage and current than Class 2, additional safety measures are put in place to prevent electric shock hazards.

A ***Class 4 Circuit*** runs between the load side of a Class 4 transmitter and the Class 4 receiver or utilization equipment, as appropriate. Thanks to active monitoring and control of voltage and current, a Class 4 circuit prioritizes fire safety and provides acceptable protection against electric shock. Class 4 fault-managed power systems are addressed in NEC Article 726. However, NEC 726.12 indicates that Class 4 power systems are not allowed in dwelling units.

Most low-voltage system installations in dwelling units are Class 2 systems. The source of these systems is usually a Class 2 transformer. Such systems include doorbells, thermostats, and garage door openers. A Class 2 power supply could also be found in computer or audio equipment intended for consumer use.

REMOTE-CONTROL, SIGNALING, AND POWER-LIMITED CIRCUITS

Remote-control, signaling, or power-limited circuits are commonly used in dwellings. These types of circuits are classified as Class 1, Class 2, or Class 3 circuits. Class 1 circuits are either line voltage, such as 120 volts, or low voltage, such as 30 volts, but have higher power capability than Class 2 circuits. Class 1 circuits are not usually utilized in the typical dwelling unit.

Class 3 circuits are higher-voltage but low-amperage circuits. Table 11(A) in Chapter 9 of the *NEC* indicates that these circuits are permitted to have a voltage of up to 150 volts and a current capacity of up to 1 ampere. Because these Class 3 circuits present a shock hazard, they are typically not commonly used in dwelling units as well.

The most common remote control and signaling circuits used in one- and two-family dwellings are Class 2 circuits. A Class 2 power source can supply only a limited amount of electrical power. For this reason, a Class 2 circuit is considered to have a limited ability to initiate a fire and provide acceptable protection from electric shock. Table 11(A) in Chapter 9 of the *NEC* gives the voltage and current limitations for Class 2 alternating-current circuits.

Class 2 circuit power supplies are normally inherently protected from overcurrents by using transformers marked "Class 2." These transformers may be located within equipment enclosures or separately mounted on an outlet box, which is commonly done for doorbell circuits (see photo 17.3 and Figure 17.2). Class 2 systems generally are not required to be grounded.

WIRING METHODS FOR CLASS 2 CIRCUITS

The *Code* requires Class 2 and Class 3 circuits to be installed in a neat and workmanlike manner. Cables must be supported by the building structure in a manner that prevents damage to the cable during normal building use [NEC 725.24 and IRC E4304.5].

There are a few additional requirements for the installation of Class 2 circuit cables. They are to be installed in compliance with *NEC* 800.133 (Installation of Communications

Wires and Cables and CATV-Type Coaxial Cables) and 800.154 (Applications of Listed Communications Wires, Cables, and Raceways, and Listed Cable Routing Assemblies) or IRC Chapter 43. The rules in *NEC* Chapter 3 for installing power circuits do not apply to cables installed under the rules of *NEC* Article 725 or IRC Chapter 43 due to the organization and arrangement of the *Code* provided in *NEC* 90.3.

ELECTRIC LIGHT OR POWER CIRCUITS AND CLASS 2 CIRCUITS

Cables and conductors of Class 2 circuits generally cannot be placed in the same raceway, compartment, outlet box, device, or junction box or similar fitting with conductors of electric light, power circuits, or Class 1 circuits unless specific conditions apply [NEC 725.136(A) or IRC E4304.1]. These specific conditions or exceptions are spelled out at NEC 725.136(B) through (I) or IRC E4304.1 Exceptions.

One provision that allows Class 2 cables and conductors to be installed alongside conductors of electric light or power, Class 1, or non-power circuits, is when they are separated by a barrier [NEC 725.136(B) or IRC E4304.1, Exception No. 1].

Another provision applies only to conductors within enclosures. The general rule does not apply where the power conductors in outlet boxes, junction boxes, or similar fittings or compartments are introduced to connect to the same equipment connected to the Class 2 circuits. One of two additional conditions must be met:

1. the electric light or power, Class 1, or other non-power-limited circuit conductors are required to be routed within the enclosure to maintain a minimum of 6-mm (¼ in.) separation from the cable or conductors of the Class 2 circuit or
2. the power circuit conductors operate at 150 volts or less and the Class 2 circuits are installed using a CL3, CL3R, CL3P, or permitted substitute and any Class 3 conductors that extend beyond the jacket are separated by a minimum of 6 mm (¼ in.) or by a nonconductive sleeve or nonconductive barrier from all other conductors [*NEC* 725.136(D) or IRC E4304.1, Exception No. 2].

There are several common examples of the application of these exceptions in typical dwelling units. Residential lighting control panels are becoming more common. Often, these systems require that power and control cables be brought to the same switch location. A conventional device box is not equipped with barriers, so one of the noted exceptions must

Photo 17.3. Class 2 transformer installed at a location remote from panelboard (in attic space)

Photo 17.4. Class 2 control circuit and electric power for electric-fired fireplace in dwelling unit brought to the same device box

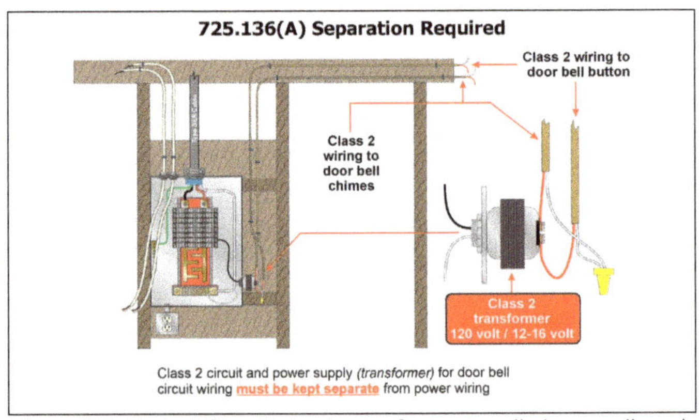

Figure 17.2. Typical Class 2 circuit transformer installed at panelboard for doorbell in dwelling

be used to separate the power wiring and Class 2 control wiring safely. Installation of an electric-fired fireplace often results in a similar situation. The control circuit for ignition and the power circuit for the warm air circulation fan is often brought to the same device box (see photo 17.4).

Raceways cannot typically be used as a means of support for Class 2 circuit conductors (*NEC* 725.24 and 300.11(C) and (D) or IRC E4304.5). However, installations that comply with *NEC* 300.11(C)(2) are permitted to be supported by power supply raceways. This *NEC* section, along with IRC E4304.5, permits the Class 2 circuit cable to be supported

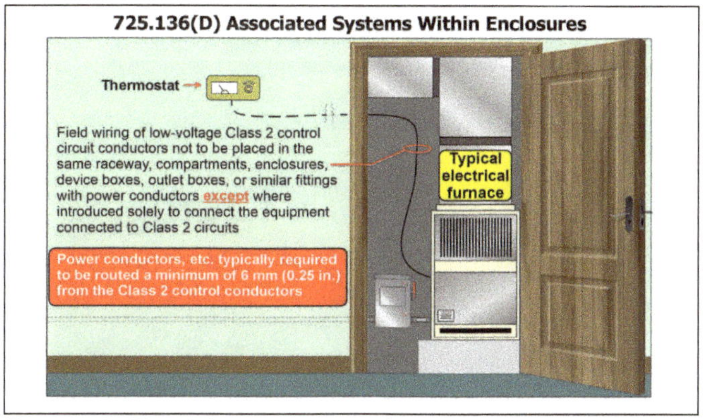

Figure 17.3. Class 2 thermostat wiring for heating system controls

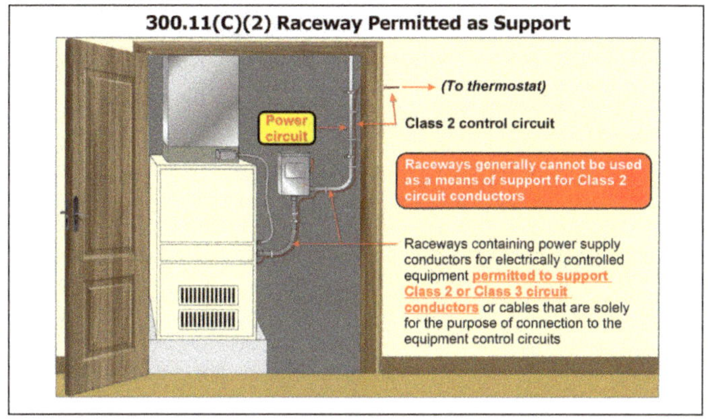

Figure 17.4. Support of Class 2 control wiring by raceway

Photo 17.5. Proper separation of Class 2 wiring, coaxial cable, and power conductors at typical dwelling unit

Photo 17.6. Rough wiring in the garage includes power in the ceiling for the door opener and Class 2 circuit wiring for the door opener remote-control circuit

by a raceway where the raceway supplies the same piece of equipment controlled by the Class 2 circuit. Some examples of this application are furnaces, boilers, AC units, or evaporative coolers. The power wiring often runs through a raceway in order to protect the conductors from physical damage, and the Class 2 control cable is then taped or strapped to the exterior of that raceway (see Figure 17.4).

Caution should be exercised when installing Class 2 conductors in the same raceway, boxes, enclosures, etc., as power wiring, as *NEC* 300.3(C)(1) requires all conductors to have an insulation rating equal to at least the maximum circuit voltage applied to any conductor within the enclosure, cable, or raceway. It is not impossible, but not common, to find Class 2 conductors with the same insulation rating as power wiring.

The *Code* generally requires a 50 mm (2-in.) separation between Class 2 circuits and conductors of electric light, power, or non-power-limited circuits. There are also exceptions that permit a lesser separation. Remember that the chance of interference decreases as the separation distance increases [*NEC* 725.136(I) or IRC E4304.2].

Class 2 circuit conductors are permitted to be run within the same cable with communications circuits. However, the Class 2 circuits are required to be classified as communications circuits and must meet the requirements of Part V of *NEC* Article 805 for communications circuits [*NEC* 725.139(D)(1) or IRC E4304.3].

The rules and exceptions in this section also apply to Class 3 circuits. Class 3 circuits are not commonly used in dwelling units. However, commercial-grade audio equipment typically employs a Class 3 power supply. Such systems are sometimes installed in dwellings; one example is a sophisticated home theater system.

Jacketed cables of Class 2 circuits are permitted to be installed in the same enclosure or raceway with jacketed cables of power-limited fire alarm systems (*NEC* Article 760), optical fiber cables (*NEC* Article 770), communication circuits (*NEC* Article 805), CATV and radio distribution cables (*NEC* Article 820), or network-powered broadband communication cables (*NEC* Article 830) as long as the Class 2 circuits comply with the applicable parts of the *NEC* article of the other jacketed cable [*NEC* 725.139(E) or IRC E4304.4].

CLASS 2 CONDUCTORS AND CABLES WITHIN BUILDINGS

Class 2 conductors and cables used as wiring within a dwelling unit are required to be listed as being resistant to the spread of fire in accordance with *NEC* 800.154 and 800.179 or IRC E4303.2. The typical Class 2 cable used within one- and two-family dwellings is Type CL2X. Other cables that can be substituted as indicated in *NEC* Table 722.135(E) or IRC Table E4303.2 include Types CM, CMX, and CL3. *NEC* Article 722 is a new article introduced in the 2023 *NEC* titled *Cables for Power-Limited Circuits and Fault-Managed Power Circuits*. The scope of this new article indicates that the article covers the general requirements for the installation of single- and multiple-conductor cables used in Class 2 and Class 3 power-limited circuits, power-limited fire alarm (PLFA) circuits, and Class 4 fault-managed power circuits.

POWER OVER ETHERNET (PoE)

Power over Ethernet (PoE) is one of the fastest-growing and most popular low-voltage systems. It is becoming an integral part of our daily lives, driven by ever-evolving technology. PoE can be best described as a technology for wired Ethernet local area networks (LANs) that allows the electrical current needed for device operation to be transmitted through data cables instead of power cords. This reduces the number of wires needed for network installation. Common uses of PoE include voice-over IP phones. The use of PoE means these phones have a single connection to a wall outlet and can be remotely powered down, much like older analog systems. Another common application for PoE is networked surveillance cameras, which allow for quick deployment and easy repositioning. Wireless Wi-Fi and Bluetooth applications, along with radio-frequency identification (RFID) readers, are increasingly common in residential settings.

Prior to 2017, the Code did not address the installation of PoE. With its growing popularity and increased installations, the authority having jurisdiction (AHJ) and installers had no guidelines to ensure safe installation. Currently, PoE is covered by Code requirements in NEC 725.60, 725.144, and 840.160 that govern these systems. Installing the listed 4-pair communications cables for a communications circuit, or replacing other cables with 4-pair communications cables as per NEC Table 722.135(E), must comply with NEC 725.144 and NEC Table 725.144 [see NEC 840.160]. Additionally, NEC 840.160 specifies that listed communication cables, which carry the communications circuit, are also permitted to carry circuits for powering listed communications equipment. The power source for these cables and equipment must be rated and listed according to NEC 840.170(C). This section states that the power source for circuits supplying power over communications cables to remote equipment must be limited according to Table 11(B) in NEC Chapter 9 for voltage sources up to 60 volts DC and must be listed as specified in one of the following:

Power over Ethernet *(PoE)* Cable *(NEC Default)*

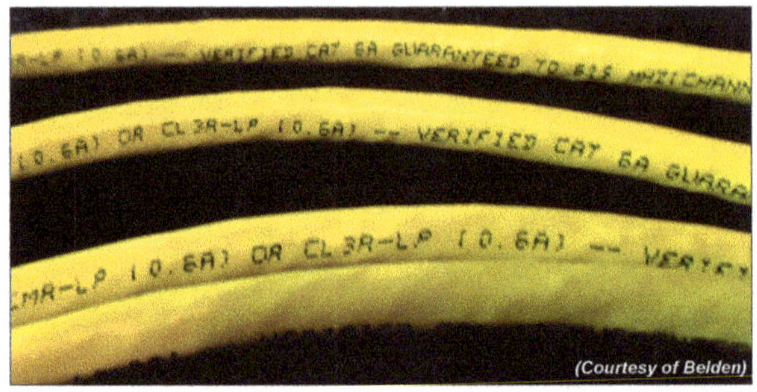

Type CL3R-LP cable listed to carry the marked current (0.6 A) for use with PoE equipment

Figure 17.5. Listed Type CL3R-LP cable *(note that the cable is marked and identified to carry 0.6 amperes of current)*

***NEC* Table 725.144**

Table 725.144: Ampacities of Each Conductor in Amperes in 4-Pair Class 2 or Class 3 Balanced Twisted-Pair Cables Based on Copper Conductors at an Ambient Temperature of 30°C (86° F) with All Conductors in All Cables Carrying Current, 60°C (140°F), 75°C (167°F), and 90°C (194°F) Rated Cables

AWG	Number of 4-Pair Cables in a Bundle																	
	1-7			8-19			20-37			38-61			62-91			92-192		
	Temperature Rating			Temperature Rating			Temperature Rating			Temperature Rating			Temperature Rating			Temperature Rating		
	60°C	75°C	90°C	60°C	75°C	90°C	60°C	75°C	90°C	60°C	75°C	90°C	60°C	75°C	90°C	60°C	75°C	90°C
26	1.00	1.23	1.42	0.71	0.87	1.02	0.55	0.68	0.78	0.46	0.57	0.67	0.45	0.55	0.64	NA	NA	NA
24	1.19	1.46	1.69	0.81	1.01	1.17	0.63	0.78	0.91	0.55	0.67	0.78	0.46	0.56	0.65	0.40	0.48	0.55
23	1.24	1.53	1.78	0.89	1.11	1.28	0.77	0.95	1.10	0.66	0.80	0.93	0.58	0.71	0.82	0.45	0.55	0.63
22	1.50	1.86	2.16	1.04	1.28	1.49	0.77	0.95	1.11	0.66	0.82	0.96	0.62	0.77	0.89	0.53	0.63	0.72

Note 1: For bundle sizes over 192 cables, or for conductor sizes smaller than 26 AWG, ampacities shall be permitted to be determined by qualified personnel under engineering supervision.

Note 2: Where only half of the conductors in each cable are carrying current, the values in the table shall be permitted to be increased by a factor of 1.4.

Informational Note No. 1: Elevated cable temperatures can reduce a cable's data transmission performance. For information on practices for 4-pair balanced twisted pair cabling, see TIA-TSB-184-A and 6.4.7, 6.6.3, and Annex G of ANSI/TIA-568-C.2, which provide guidance on adjustments for operating temperatures between 20°C and 60°C.

Informational Note No. 2: The per-contact current rating of connectors can limit the maximum allowable current below the ampacity shown in Table 725.144.

Table 17.1. Reproduction of *NEC* Table 725.144 for bundling of PoE cables

(1) A power source must be listed as specified in 725.60(A)(1), (A)(2), (A)(3), or (A)(4). Power sources cannot have output connections paralleled or interconnected unless they are listed for such interconnection.
(2) A power source must be listed as communications equipment for limited-power circuits.

For the 2017 NEC, NEC 840.160 stated that when power supplied over a communications cable to communications equipment exceeds 60 watts, communication cables and the power circuit must comply with NEC 725.144, where communications cables are used instead of Class 2 and Class 3 cables. In the 2020 NEC, the 60-watt trigger was removed because it was not relevant to 725.144. Sixty watts was simply the wattage of the highest standardized PoE at the time of the 2017 NEC.

As mentioned earlier, over the past decade or so, power-over-local area networking (LAN) cable technologies like PoE have become a practical powering option for many applications. Device manufacturers are creating more ad-

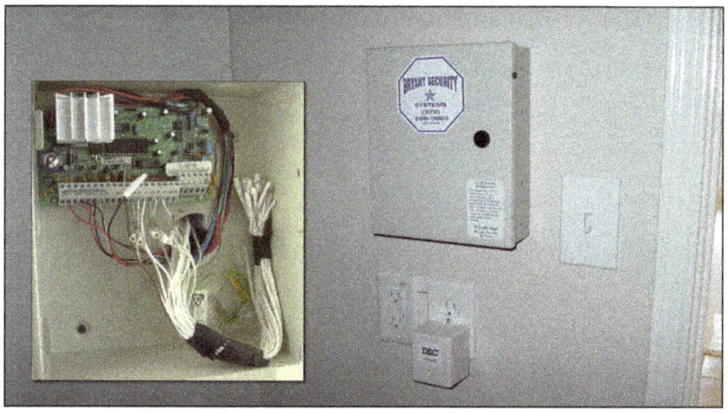

Photo 17.7. Typical security/fire alarm systems that are becoming familiar in dwelling units

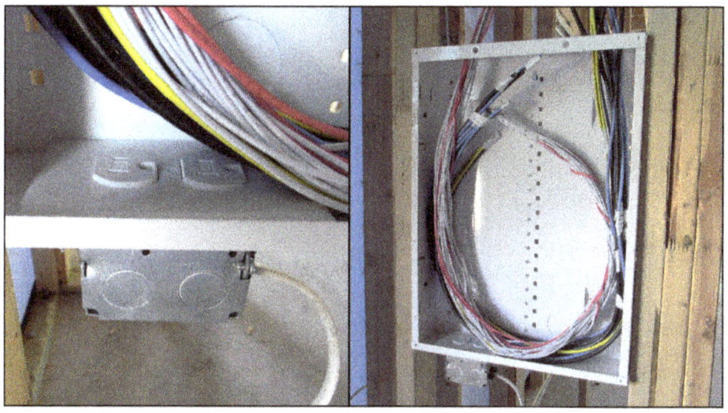

Photo 17.8. Communication circuits have grown in recent years to include such things as telephone, fiber-optic cable TV, and high-speed Internet cable circuits.

vanced equipment that requires higher power levels. As these power levels increase, the heat generated inside the cable also rises, especially when cables are bundled together. The extra heat from increased current could push the cables beyond their rated temperatures. To address this issue, a new type of "limited-power" cable was added to the Code at NEC 725.144 to simplify cable selection and installation considerations. The "-LP" cable designation indicates that the cable has been tested to carry the specified current under typical worst-case installation conditions without exceeding the cable's temperature rating. The requirements at NEC 725.144 and Table 725.144 consider large bundle sizes, high ambient temperatures, and environmental factors such as enclosed spaces or conduits. These cables are usually 4-pair "Category" type cables originally designed for data and communication transmission. However, advancements in technology and device design have led to these cables being increasingly used to supply low voltage (less than 60 volts DC) and limited power alongside data and communication signals.

FIRE ALARM SYSTEMS

The Code covers the installation of wiring and equipment for fire alarm systems, including all circuits controlled and powered by the fire alarm system at NEC Article 760 (NEC 760.1). A fire alarm system circuit can be either a non-power-limited fire alarm system (NPLFA) or a power-limited fire alarm system (PLFA). What distinguishes between the two? Among other things, an NPLFA's output voltage is limited to no more than 600 volts, which is nominal [NEC 760.41(A)]. A PLFA's voltage and power are limited by a listed power source that complies with NEC 760.121(A).

The *Code* does not require that a fire alarm system be installed in a dwelling. If there is a requirement for a fire alarm system at dwelling units in your area, it comes from the locally enforced building code or locally adopted ordinance. The *Code* contains requirements for wiring the fire alarm system but does not include requirements for designing or operating the system.

For further information on the selection, installation, operation, and maintenance requirements of fire warning equipment for use in dwelling units, see the most recent edition of NFPA 72-2022, the National Fire Alarm and Signaling *Code*.

Where a fire alarm system is installed in a damp or wet location, *Code* requirements for exposure to deteriorating agents and protection against corrosion are required to be followed [*NEC* 310.10(B), (C), and (F), 110.11, 300.5(B), 300.6, 300.9, and 760.3(C) or IRC E3404.5, E3803.10, and E3802.8]. Fire alarm cables used in a wet location must be listed for use in wet locations or have a moisture-impervious metal sheath.

It has become common to install combination fire and burglar alarm systems in homes to provide protection against both dangers. These systems can trigger an alarm locally (inside the home), use relay contacts or data communication gateways, or connect over telephone lines to a remote monitoring service. Many companies offer alarm monitoring and troubleshooting services, such as checking for low battery backup. They send alerts they verify as genuine to the local police or fire department for immediate response. Generally, these systems require a special permit from local authorities for continuous fire and burglary monitoring. These requirements can vary and may not be needed in some areas. Check with your local building inspector and fire/police departments to learn the rules in your area.

BRANCH CIRCUIT FOR FIRE ALARM CIRCUITS

A dedicated branch circuit supplying the fire alarm equipment(s) and no other loads is required for the power source for both power-limited fire alarm (PLFA) circuits and non-

Photo 17.9. A dedicated branch circuit is required for fire alarm equipment and cannot be supplied by GFCI or AFCI in the branch circuit

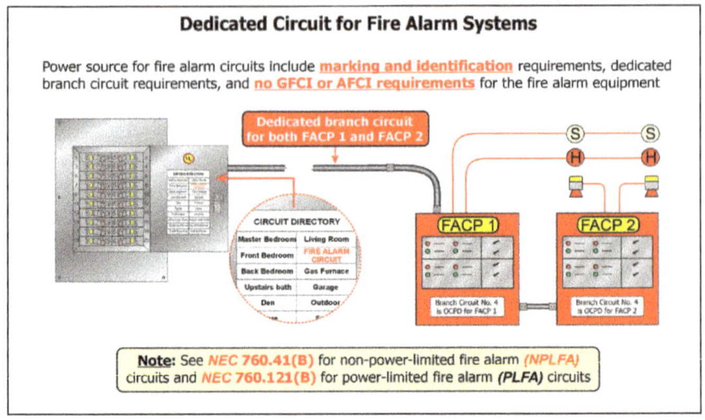

Figure 17.6. A dedicated branch circuit is required for fire alarm equipment(s)

power-limited fire alarm circuits (NPLFA). Since a fire alarm system might have multiple control panels (fire detection and alarm notification, sprinkler water flow, elevator shutdown, door release, fire doors, damper control, and fan shutdown, etc.) with only a small load demand on a circuit, one branch circuit may feed more than one control panel as long as there is adequate capacity on the circuit to do so. The location of the branch-circuit overcurrent protective device feeding the fire alarm control panel(s) is to be marked on said control panel(s). This will facilitate quicker response time when troubleshooting the fire alarm branch circuit so the downtime of the fire alarm system is kept to a minimum. Detailed marking and identification requirements such as red lettering indicating "FIRE ALARM CIRCUIT," accessibility to qualified personnel, marking not to obscure the manufacturer's markings, etc., are also required. This dedicated branch circuit cannot be supplied through ground-fault circuit interrupters or arc-fault circuit interrupters. The fire alarm branch-circuit disconnecting means is permitted to be secured in the "on" position [NEC 760.41(A) and (B) and 760.121(B)].

CLASSIFICATION OF FIRE ALARM CIRCUITS

Fire alarm circuits refer to the portion of the wiring system between the load side of the overcurrent device or the power-limited supply and the connected equipment for all circuits powered and controlled by the fire alarm system. These circuits are categorized as either "non-power-limited fire alarm circuits (NPLFA)" or "power-limited fire alarm circuits (PLFA)" [NEC Article 100]. Power-limited fire alarm circuits are typically installed in residences and often sourced from listed PLFA or Class 3 24-volt power supplies. All fire alarm circuits must adhere to NEC Article 760, Part I. Additionally, power-limited fire alarm circuits are required to comply with NEC Article 760, Part III.

FIRE ALARM CIRCUIT IDENTIFICATION

To help prevent unintentional interference with fire alarm circuits during testing and servicing, they must be "specifically identified" at all terminal and junction points. The method of identification is up to the installer. Often, a cable with a red jacket connected to enclosures painted red is the preferred way to identify fire alarm circuits. In some cases, engraved labels or other identification methods may be appropriate (NEC 760.30).

WIRING METHODS FOR PLFA CIRCUITS

For power-limited fire alarm circuits (PLFA), the conductors and equipment installed on the supply (line) side of overcurrent-protective devices, transformers, or current-limiting devices must also comply with the relevant rules in NEC Article 760, Part III, and NEC Chapter 3. An overcurrent device must protect a transformer or other device supplied by power-supply conductors rated no more than 20 amperes (NEC 760.127).

Power-limited fire alarm circuit conductors and cables installed at dwelling units as described in NEC 722.179 must be installed as specified below:

1. Where installed in raceways or ***exposed on the surface of ceiling and sidewalls or "fished" in concealed spaces***, cable splices or terminations are required to be made in listed fittings, boxes, enclosures, fire alarm devices, or utilization equipment. Where installed exposed, cables are required to be adequately supported and installed in such a way that maximum protection against physical damage is afforded by building construction, such as baseboards, door frames, ledges, etc. Where located within 2.1 m (7

ft) of the floor, cables are required to be securely fastened in an approved manner at intervals of not more than 450 mm (18 in.). See *NEC* 760.130(B)(1)
2. Where ***passing through a floor or wall***, cables must be installed in metal raceways or rigid nonmetallic conduit to a height of 2.1 m (7 ft) above the floor unless adequate protection can be afforded by building construction such as detailed in point 1 above, or unless an equivalent solid guard is provided. See *NEC* 760.130(B)(2).
3. Cables specified in *NEC* Chapter 3 and meeting the requirements of 722.179(A)(15)(a) and (A)(15)(b) are permitted to be installed in ***nonconcealed spaces*** where the exposed length of cable does not exceed 3 m (10 ft).

Additional requirements for installing power-limited fire alarm circuits are located at NEC 760.133, referencing the requirements of Parts I and II of NEC Article 722 and NEC 760.136 through NEC 760.143. Power-limited circuit conductors generally cannot be installed in any cable, compartment, enclosure, outlet box, raceway, or similar fitting that contains conductors for electric light, power, Class 1, or non-power-limited fire alarm circuits. Several exceptions are provided for conditions such as installing power-limited conductors in enclosures like control panels, where they must terminate to operate [NEC 760.136(A) through (H)].

Power-limited fire alarm circuit conductors are generally required by *NEC* 760.136(H) to be separated at least 50 mm (2 in.) from conductors of any electric light, power, Class 1, or non-power-limited fire alarm circuit conductors. Two conditions exist that allow this rule to be relaxed.

Condition 1 excludes PLFA cables from the separation requirements when either (a) all the conductors for electric light, power, Class 1, or non-power-limited fire alarm circuits, or (b) all the power-limited fire alarm circuit conductors are installed in raceways or in metal-sheathed, metal-clad, nonmetallic-sheathed, or Type UF cables.

Condition 2 excludes PLFA cables from separation requirements when all the electric light, power, Class 1, or non-power-limited fire alarm circuit conductors are permanently separated from all the power-limited fire alarm circuit conductors by a continuous and securely fixed nonconductor, such as porcelain tubes or flexible tubing, in addition to the wire's insulation.

Conductors of one or more Class 2 circuits are permitted within the same cable, enclosure, raceway, or cable routing assembly with conductors of power-limited fire alarm circuits, provided the insulation of the Class 2 circuit conductors in the cable, enclosure, or raceway is at least that required by the power-limited fire alarm circuits [*NEC* 760.139(B)].

The *Code* does not permit electrical conduits or raceways to be used as a means of support for power-limited fire alarm circuit conductors (*NEC* 760.143). Unlike with Class 2 circuits, no exception in 760.143 permits the use of electrical conduits or raceways to support PLFA circuits or conductors as with Class 2 circuits in reference to *NEC* 300.11(C)(2) or IRC E4304.5.

Power-limited fire alarm circuits, as well as non-power-limited fire alarm circuits that extend beyond one building and run outdoors, are required to meet the installation requirements of Parts II, III, and IV of *NEC* Article 805 and are required to meet the installation requirements of Part I of *NEC* Article 300. (*NEC* 760.32).

PLFA (POWER-LIMITED FIRE ALARM) SYSTEM CONDUCTORS

Power-limited fire alarm systems can be installed similarly to non-power-limited fire alarm systems [see NEC 760.130 and 760.130(A)]. Additionally, power-limited fire alarm conductors and cables described in NEC 722.179 are allowed. NEC 760.154 and Table 760.154 indicate that Types FPLP, FPLR, and FPL are permitted for use in dwellings. Allowed substitutes include Types CMP, CMR, CMG, and CM, according to NEC Table 760.154(A).

Fire alarm conductors must be copper and can be either stranded or solid. Conductors in a multi-conductor cable are allowed to be no smaller than 26 AWG, and individual conductors are not smaller than 18 AWG. Conductors of 26 AWG are only permitted when spliced with a connector listed as suitable for 26 to 24 AWG or larger conductors that are terminated on equipment, or where the 26 AWG conductors are terminated on equipment listed as suitable for 26 AWG conductors [NEC 760.142, 760.49(A), and 760.130(A)].

COMMUNICATIONS CIRCUITS

Article 805 of the NEC covers communication circuits, including devices such as telephones, optical fiber cables for Internet service delivery, and outside wiring for fire and burglar alarm systems installed in one- and two-family homes. The NEC provides some key rules for installing communication systems. Equipment connected to a communication system must be listed for that purpose and installed following the instructions in NEC 805.18 and 805.170. Separation, grounding, and bonding rules are also essential for the safe installation of these systems.

MECHANICAL EXECUTION OF WORK

The Code also requires that communication circuits and equipment be installed neatly and professionally. Cables installed exposed on the surface of ceilings and walls must be supported by the building's structural components so that they are not damaged during normal building use. Acceptable securing methods include hardware such as straps, staples, cable ties, and hangers. Nonmetallic cable ties and other nonmetallic cable accessories used to secure and support communication cables in spaces used for environmental air (return air plenums) must be listed as having low smoke and heat release properties (NEC 800.24(A) and NEC 110.12).

IN RACEWAYS, BOXES, AND CABLES

Communication wires and cables are allowed to be installed in any raceway listed in NEC Chapter 3, following its requirements. They are also permitted to be installed in the listed plenum, riser, or general-purpose communications raceways, chosen according to the installation requirements outlined in NEC 800.113 for communication wires, cables, and raceways. Additionally, these cables may be installed in cable routing assemblies in compliance with NEC 800.110(C) [NEC 800.110].

The provisions of NEC 800.133(A)(1) allow communication cables to be installed in the same raceway, box, or enclosure with cables of these types:

1. Class 2 and Class 3 remote-control, signaling, and power-limited circuits in accordance with NEC 645.5(E)(2) or Parts I and II of NEC Article 725.
2. Power-limited fire alarm systems following Parts I and III of NEC Article 760.
3. Nonconductive and conductive optical fiber cables complying with Parts I and V of NEC Article 770.
4. Communication circuits conforming to Parts I and IV of NEC Articles 800 and 805.
5. Community antenna television and radio distribution systems in accordance with Parts I and V of NEC Articles 800 and 820.
6. Low-power, network-powered broadband communications circuits as per Parts I and V of NEC Articles 800 and 830.

CIRCUITS RUN IN THE SAME CABLE ASSEMBLY

Class 2 and Class 3 circuits are covered by NEC Article 725. The IRC discusses Class 2 circuits in Chapter 43. A Class 1 circuit cannot be run in the same cable as a communica-

Photo 17.10. Type CMR is permitted as a substitute wiring method for PLFA circuits at dwelling units

tions circuit. Conductors for Class 2 and Class 3 circuits are allowed in the same listed communications cable with communications circuits [NEC 800.133(A)(2)]. According to IRC E4304.3, Class 2 circuits are permitted in the same cable with communications circuit conductors if the cable is a listed communications cable installed per the requirements of Part V of Article 800 of the NEC. These cables must be listed as communications cables. This information is found at 725.139(D)(1) of the NEC.

According to NEC 800.113(L)(1)(9), listed hybrid power and communications cables are allowed in dwelling units. The listing requirements typically permit these cables where the power cable is a listed Type NM or NM-B, conforming to Part III of NEC Article 334, and the communications cable is a listed Type CM cable. In such cases, the jackets on the listed NM or NM-B and listed Type CM cables must be rated for at least 600 volts, and the hybrid cable must be listed as resistant to the spread of fire.

ELECTRIC LIGHT OR POWER CIRCUITS AND COMMUNICATION CIRCUITS

Communications wires, cables, and CATV-type coaxial cables cannot be installed in any raceway, compartment, outlet box, junction box, or similar fitting that contains conductors of electric light or power circuits, or Class 1 circuits, non-power-limited fire alarm circuits, or medium-power network-powered broadband communications circuits [NEC 800.133(A)(3)]. There are a few exceptions to this general rule that might typically apply to dwelling units.

The first exception indicates that the general separation requirement does not apply when the conductors of electric light or power, Class 1, or non-power-limited fire alarm circuits are separated from all communication circuits and CATV-type coaxial cables by a permanent barrier or a listed divider.

The second exception applies only to conductors inside enclosures. The general separation rule does not apply when conductors in outlet boxes, junction boxes, or similar fittings or compartments are used solely for the power supply to communications and/or coaxial system distribution equipment. Electric light or power conductors, Class 1, or non-power-limited fire alarm circuit conductors must be routed within the enclosure to keep a minimum of 6 mm (¼ in.) separation from communications circuit conductors and CATV-type coaxial cables.

The Code usually requires a 50 mm (2 in.) gap between communication circuits and CATV-type coaxial cables, as well as conductors of electric light, power, or other non-power-limited circuits [NEC 800.133(B)]. An exception allows a smaller separation if either all the electric light and power circuits or all the communication cables and CATV-type coaxial cables are run in separate raceways or in separate metal-sheathed, metal-clad, or nonmetallic-sheathed cables [NEC 800.133(B) Exception No 1]. A common example of this exception is running cables horizontally through bored or punched holes in framing members. The wiring methods for power or lighting in dwelling units typically involve nonmetallic-jacketed cables such as Type NM-B cables. Similarly, communication circuit wiring usually uses nonmetallic cables like Type CM. Although this practice can be code-compliant, it is not recommended, as electrical noise from the power circuit may interfere with the communication circuit. Standard cabling guidelines generally suggest maintaining a 300 mm (12 in.) separation to reduce interference.

Raceways cannot be used to support communications and CATV coaxial wires and cables. Communication cables or wires and CATV coaxial cables are not allowed to be strapped, taped, or otherwise attached to the outside of any conduit or raceway for support. It is a violation to use the service mast or any other raceway containing typical branch circuits or feeders to support communication or CATV-type coaxial service drops, wires, or cables. An exception allows overhead (aerial) spans of communication wires and cables to be attached to the outside of a raceway-type non-service mast designed for supporting such wires and cables. [see NEC 800.133(C) and exception].

COMMUNICATIONS WIRES AND CABLES WITHIN BUILDINGS

Communication wires and cables used as wiring inside a building must be listed as fire-resistant according to NEC 800.179(E)(2). The common communication cable used in one- and two-family homes is usually Type CMX. Other cables that can be substituted, as shown in NEC Table 805.154, include Types CMP, CMR, CMG, and CM.

Regarding communications wiring in "other space used for environmental air" and the "spread of fire or products of combustion," installations of cables, communications raceways, cable routing assemblies in hollow spaces, vertical shafts, and ventilation or air-handling ducts must be made so that the possible spread of fire or products of combustion will not be substantially increased. Openings around penetrations of cables, communications raceways, and cable routing assemblies through fire-resistant-rated walls, partitions, floors, or ceilings are required to be firestopped using approved methods to maintain the fire resistance rating [NEC 800.26]. Also, see the comments in this chapter under *NEC* Article 725 and the requirements of *NEC* 300.22(C) for wiring in other spaces used for environmental air spaces (plenums).

REQUIRED COMMUNICATIONS OUTLET AT DWELLING UNITS

For new construction, at least one communication outlet must be installed within the dwelling in a readily accessible location and connected to the service provider demarcation point. Given today's demand for electronic technology such as high-speed Internet access and optical fiber cable television, a typical dwelling unit often has several communication circuits and outlets, but at minimum, one is required for new construction (NEC 805.156).

GROUNDING REQUIREMENTS FOR COMMUNICATIONS CIRCUITS

The metallic part of the cable sheath in the supply communication cable is generally required to be grounded or bonded as close as practically possible to the point where the conductors enter the building or be interrupted by an insulating joint or an equivalent device [NEC 805.93(A)]. For one- and two-family dwellings, the primary protector bonding conductor or grounding electrode conductor must be as short as practically possible and not longer than 6.0 m (20 ft). An exception allows a separate communication ground rod to be installed if it is not possible to maintain the maximum 6.0 m (20 ft) length for the grounding electrode conductor. When a separate communication ground rod is used, it must be connected (bonded) to the building's power grounding electrode system [NEC 800.100(A)(4) and Exception].

INTERSYSTEM BONDING TERMINATION

A means to connect all limited energy communication systems, grounding, and bonding conductors must be supplied in accordance with NEC 250.94(A) or (B) or IRC E3609.3.

This method of bonding these communication systems together is called an "intersystem bonding termination" (IBT). This term is defined in both NEC Article 100 and IRC Chapter 35. An Intersystem Bonding Termination is a device that provides a connection point for intersystem bonding conductors for communication systems to the grounding electrode system. It is a critical safety measure to prevent voltage potential differences between communication systems and the power system.

As renewable energy systems like photovoltaic (PV) and wind generation expand, proper grounding and bonding become crucial for safety. These procedures are increasingly challenging due to changes in building practices, such as the rise of flush construction and the use of nonmetallic conduits for services. Often, in new buildings, the grounding electrode(s), the metallic service raceway, and the grounding electrode conductor are installed behind walls and are not accessible. This makes achieving the necessary intersystem grounding and bonding connections more difficult. Therefore, planning and establishing an intersystem bonding termination early in a construction project is extremely important.

The intersystem bonding termination requirement sets a specific, well-defined point for connecting intersystem bonding conductors and grounding electrode conductors to a list of terminals designated as grounding and bonding equipment. The termination must have enough capacity to support multiple communication systems like telecom, satellite, and CATV on-site, but it must include at least three terminals [see NEC 250.94(A)(2) or IRC E3609.3.1(2)].

Specified locations for these termination points for the intersystem bonding and grounding conductor termination are a key part of this requirement. When installed at service equipment, this termination must be securely mounted and electrically connected to the meter enclosure, service equipment enclosure, an exposed nonflexible metallic service raceway, or located at the grounding electrode conductor. When attached to a service or metering equipment enclosure, it must not interfere with opening the enclosure. If it is not connected to the meter socket enclosure, each intersystem termination must be electrically connected to the equipment grounding terminal bar in the service equipment or connected to the grounding electrode conductor. The conductor size used for this connection must not be smaller than 6 AWG copper.

The intersystem bonding termination must be accessible for connection and inspection. These grounding conductors also establish the necessary link between communication and power systems (intersystem bonding). As a result of this requirement for a dedicated termination means, fewer challenges should be faced by energy system installers and technicians in completing the required intersystem bonding and grounding connections for these critical communication systems.

When an accessory building or an additional structure has electrical power supplied via a feeder or branch circuit, an intersystem bonding termination must be established at the disconnecting means for that second building, similar to the IBT established for the primary building or structure [NEC

Figure 17.7. Bonding of the communications grounding electrode and the service power grounding electrode is required if greater than 6.0 m (20 ft) apart

Photo 17.12. Grounding is required for satellite dishes, radio and TV antennas, and other equipment

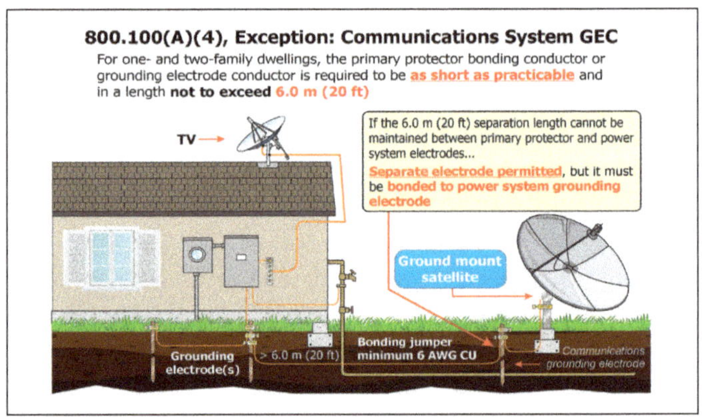

Figure 17.7 Bonding of communications grounding electrode and the service power grounding electrode required if greater than 6.0 m (20 ft) apart

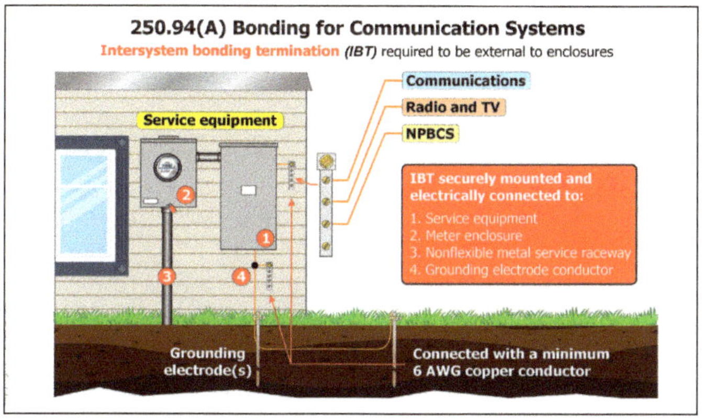

Figure 17.8. A listed intersystem bonding termination device is required to be provided external to enclosures and to provide a minimum of three termination points

Photo 17.13. Typical listed intersystem bonding termination device used for grounding of communication systems

250.94(A)(4)(b) or IRC E3609.3.1].

If a dwelling or structure served has an intersystem bonding termination, the intersystem grounding electrode conductor(s) must be connected to the intersystem bonding termination [NEC 800.100(B)(1)]. If the building or structure served lacks an intersystem bonding termination, as may be the case in an older existing dwelling unit, then the communications bonding conductors or grounding electrode conductors are required to be connected to the nearest accessible location on one of the following:

1. the building grounding electrode system as covered in *NEC* 250.50 or IRC E3608.1;
2. the grounded interior metal water piping system within 1.5 m (5 ft) of where the water pipe enters the building as covered in *NEC* 250.52(A)(1) or IRC E3608.1.1;
3. the power service accessible means external to the enclosures as covered in *NEC* 250.94(A) Exception;
4. a nonflexible metallic power service raceway;
5. the service equipment enclosure;
6. the grounding electrode conductor or the grounding electrode conductor metal enclosure of the power service or
7. the grounding electrode conductor or grounding electrode of a second building or structure disconnecting means that it is grounding to an electrode as covered at *NEC* 250.32 or IRC E3607.3.

See *NEC* 800.100(B)(2) for these provisions.

Where no intersystem bonding termination exists at an existing building or structure, "an accessible means external to enclosures for connecting intersystem bonding and grounding electrode conductors is permitted at the service equipment and at the disconnecting means for any additional buildings or structures by at least one of the following methods:

(1) An exposed nonflexible metallic raceway(s)
(2) An exposed grounding electrode conductor
(3) An approved means for the external connection of a copper or other corrosion-resistant bonding or grounding electrode conductor to the grounded raceway or equipment."

See NEC 800.100(B)(3) and 250.94(A) for exceptions to the provisions just stated or for the above requirements when no intersystem bonding termination is present at an existing building or structure. New construction requires an intersystem bonding termination as described in NEC 250.94(A) or (B) or IRC E3609.3.

While NEC 250.94(A) and IRC E3609.3.1 provide details on using a listed intersystem bonding termination device for connecting communication systems, the Code also acknowledges "other means" of connecting communication bonding conductors at NEC 250.94(B] or IRC E3609.3.2 titled, "Other Means." This alternative connection method allows for connections to an aluminum or copper busbar that is at least 6 mm thick × 50 mm wide (1/4 in. thick × 2 in. wide) and long enough to accommodate at least three terminations for communication systems, in addition to other connections or bonding jumpers. If used, the busbar must be securely fastened and installed in an accessible location. Lugs used with this busbar must be listed for grounding and bonding. If aluminum busbars are used, the installation cannot be in direct contact with masonry or the earth or subject to corrosive conditions. When used outdoors, aluminum or copper-clad aluminum grounding electrode conductors must not be terminated within 450 mm (18 in.) of the ground, in accordance with NEC 250.64(A).

This method is commonly used in commercial and multifamily mixed-use buildings. Although this "other means" of a common busbar is rare for dwelling units, it is acceptable to use this approach in such settings. Many commercial buildings often employ a common grounding terminal bar for connecting multiple electrodes and bonding other systems, such as water piping, building steel, and internal antenna systems, to name a few. These shared busbars also facilitate easy connection to other systems, like communication networks, satellite dish systems, and broadband systems. This "other means" of terminating systems enables the installer to bond various conductors for overall system bonding, including items like corrugated stainless steel tubing (CSST) gas piping, which is still prohibited from being terminated on a listed intersystem bonding device.

An exception to *NEC* 250.94(A) and (B) or IRC E3609.3.1 gives relief from providing a means for connecting intersystem bonding conductors where communications systems are not likely to be used. This would include such things as outhouses, chicken coops, or garden sheds.

Where a separate electrode is installed to ground the communication system, it must be bonded to the power grounding electrode system in that building [*NEC* 800.100(D)]. The bonding jumper required to connect the two electrodes of the different systems together must not be smaller than 6 AWG copper. This bonding requirement is important in limiting potential differences between the communications systems and other associated wiring systems (see Figure 17.7).

RADIO AND TELEVISION EQUIPMENT

NEC Article 810 covers radio and television receiving equipment, as well as amateur radio transmitting and receiving equipment, including amateur and citizen band radios. It also includes various types of antennas, such as wire-strung, multi-element, vertical rod, flat, or parabolic antennas, along with the wiring and cabling that connect them to the equipment. However, the article does not cover equipment and antennas used for coupling carrier current to power line conductors [NEC 810.1].

The Code requires outdoor antennas and lead-in conductors to be properly supported. Antennas are not allowed to be attached to the electric service mast. They also cannot be attached to poles or similar structures that carry open electric light or power wires, or trolley wires of over 250 volts between conductors [NEC 810.12].

Outdoor antennas and lead-in conductors from an antenna to a building must not cross over open conductors of electric light or power circuits and should be installed to prevent accidental contact. When antenna lead-in conductors are near

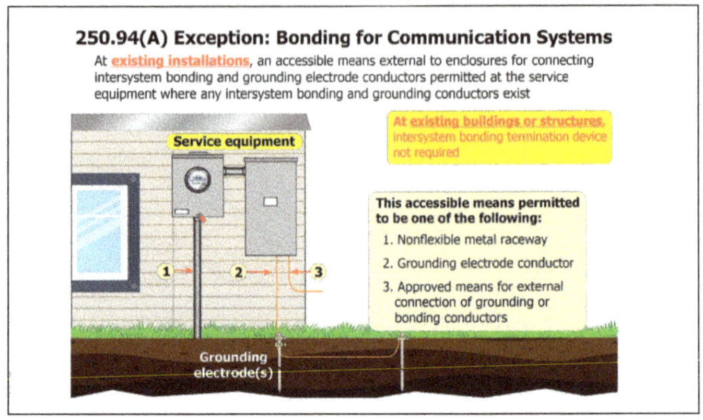

Figure 17.9. Bonding of other systems at existing buildings with no intersystem bonding termination

open electric light or power service conductors with less than 250 volts between them, a clearance of at least 600 mm (2 ft) must be maintained [NEC 810.13].

Each conductor of a lead-in from an outdoor antenna must be equipped with a listed antenna discharge unit unless the cable is enclosed in a continuous metallic shield that is grounded or protected by an antenna discharge unit. The antenna discharge unit should be located outside the building, or if inside, as close as practicable to the point where the conductors enter the building. The antenna discharge unit must be grounded or bonded in accordance with NEC 810.21 [NEC 810.20].

A bonding conductor or grounding electrode conductor, not smaller than 10 AWG copper, 8 AWG aluminum, or 17 AWG copper-clad steel or bronze, is required for grounding as specified by NEC Article 810. The bonding or grounding electrode conductor must be made of copper, aluminum, copper-clad steel, bronze, or similar corrosion-resistant material. Aluminum or copper-clad aluminum conductors cannot be used if they are exposed to corrosive environments or contact masonry or the earth directly. When used outdoors, aluminum or copper-clad aluminum conductors must be installed at least 450 mm (18 in.) away from the ground. The bonding or grounding electrode conductor may be insulated, covered, or bare [NEC 810.21(A), (B), and (H)].

The bonding conductor or grounding electrode conductor must be protected from physical damage or increased in size if protection is lacking. When the bonding or grounding electrode conductor runs in a metal raceway, both ends of the raceway must be bonded to the contained conductor or to the same terminal or electrode to which the bonding or grounding electrode conductor is connected [NEC 810.21(D)].

The bonding conductor or grounding electrode conductor from the antenna and/or antenna discharge unit must be connected to a grounding electrode [*NEC* 810.21(F)]. If the building or structure served has an intersystem bonding

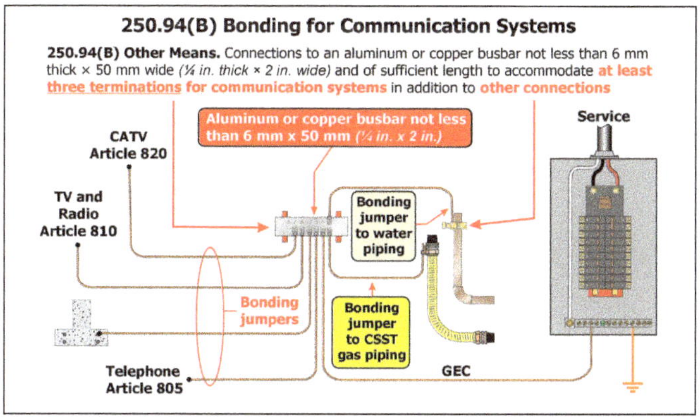

Figure 17.10. Common bus bar allowed is an alternate connection option for communication bonding jumpers and other bonding jumpers

termination, the bonding conductor or grounding electrode conductor is required to be connected to the intersystem bonding termination. If the building or structure served has no intersystem bonding termination (existing building), the bonding conductor or grounding electrode conductor shall be connected to the nearest available electrode. The permitted electrodes are:

1. the building or structure grounding electrode system;
2. the grounded interior metal water piping system within 1.52 m (5 ft) of where the water pipe enters the building;
3. the power service accessible means external to the building, as covered by *NEC* 250.94 or IRC E3609.3;
4. the nonflexible metal power service raceway;
5. the service equipment enclosure; or
6. the grounding electrode conductor or the grounding electrode conductor metal enclosures of the power service.

If the served building or structure has no intersystem bonding termination or grounding means, as mentioned above, an electrode can be installed as outlined in NEC 250.52(A) or IRC E3608.1. Metal underground systems or structures, such as piping systems or underground tanks, can also be used. For example, a metal well casing qualifies. However, underground metal gas piping is not allowed as a grounding electrode as specified at NEC 250.52(B)(1) or IRC E3608.6.

Where a separate grounding electrode is used for the building electric service and the radio and television grounding, both electrodes must be bonded together by a 6 AWG or larger copper conductor [see *NEC* 810.21(J)].

LOCAL COMMUNITY ANTENNA TELEVISION SYSTEMS (CATV)

NEC Article 820 pertains to coaxial cable distribution of radio frequencies commonly used in community antenna television (CATV) systems. Coaxial cable is a cylindrical assembly with a conductor at the center, surrounded by a metallic shield, separated by a dielectric material, and typically covered with an insulating jacket. Generally, coaxial cables installed inside a dwelling must be listed.

Like other low-voltage cables, community antenna television and radio distribution systems must be installed in a neat and workmanlike manner. The building structure must support cables in a manner that prevents damage to the cables from normal building use [NEC 800.24].

Coaxial cable cannot be installed in any raceway, compartment, outlet box, junction box, or other enclosures containing conductors for electric light or power, Class 1 circuits, non-power-limited fire alarm, or medium-power network-powered broadband communications circuits unless:

1. the conductors of the different systems are separated by a permanent barrier or listed divider or
2. the conductors in outlet boxes, junction boxes, or similar fittings or compartments are introduced solely for power supply to the coaxial cable system distribution equipment or for power connection to remote-control equipment. Electric light, power, and Class 1 circuit conductors and non-power-limited fire alarm circuit conductors are required to be routed within the enclosure to maintain a minimum 6 mm (0.25 in.) separation from the coaxial cable.

See *NEC* 800.133(A)(3) and Ex. No. 1 and Ex. No. 2 for the above requirements.

Like other low-voltage conductors, the Code generally requires a 50 mm (2 in.) separation between CATV circuits and conductors of electric light, power, or other non-power-limited circuits. There are also similar exceptions that allow a smaller separation. Keep in mind that the risk of interference decreases as the separation distance increases [NEC 820.133(A)(2)].

Coaxial cables are allowed in the same raceway, boxes, enclosures, or cable routing assemblies with jacketed cables of any of the following:

1. Nonconductive and conductive optical fiber cables
2. Communications circuits
3. Low-power network-powered broadband communications circuits

4. Class 2 and Class 3 remote-control, signaling, and power-limited circuits
5. Power-limited fire alarm systems

See *NEC* 800.133(A)(1) for specific *Code* references for complete *Code* compliance of the above provisions.

Direct-buried coaxial cables must be separated by at least 300 mm (12 in.) from electric service conductors, non-power-limited fire alarm circuit conductors, or Class 1 circuits, unless they are installed in raceways or have metal cable armor. However, coaxial cables are allowed to be directly buried in the same trench as branch-circuit or feeder conductors if they are installed in a raceway or in cable such as Type UF or Type USE [NEC 800.47(B) and Ex. No. 1 and Ex. No. 2].

CATV CABLE LISTING REQUIREMENTS

The CATV cable specifically listed for installation in dwellings is marked CATVX. Other coaxial cables permitted to be substituted for CATVX cables include Types CM, CMP, CATV, CATVP, CMR, CMG, or CATVR. Other permitted substation-type cables are BMR, BM, BLP, BLR, BL, and BLX. [NEC 820.154 and NEC Table 820.154].

Cables installed in a duct, plenum, or environmental air space must be Type CATVP (plenum-rated). Installation in these spaces must comply with NEC 800.3(C) and (D) and NEC 300.22(A) and (C). For "other spaces used for environmental air," such as joist or stud spaces used for return air to the furnace, air conditioner, or air handler, cables are allowed to pass through these spaces perpendicular to (or across) the long dimension without being in a metal raceway. When installed in this manner according to NEC 300.22(C), Exception, the cable is not required to be plenum-rated.

GROUNDING OF CATV SYSTEMS

Coaxial cables entering or attached to buildings must be grounded. When the outer conductive shield of a coaxial cable is grounded, no additional protective devices are necessary. If the coaxial cable enters the building or is terminated outside, the outer shield must be grounded according to NEC 820.100. The bonding or grounding electrode conductor should be as close as possible to the point of entry. It can be insulated, covered, or bare, made of copper or other corrosion-resistant conductive material, stranded or solid, and should not be smaller than 14 AWG or larger than 6 AWG, as specified in [NEC 800.100(A)(3) and (4)].

Lightning protection is the main reason for grounding the CATV cable where it enters the building. The bonding conductor or grounding electrode conductor should be run

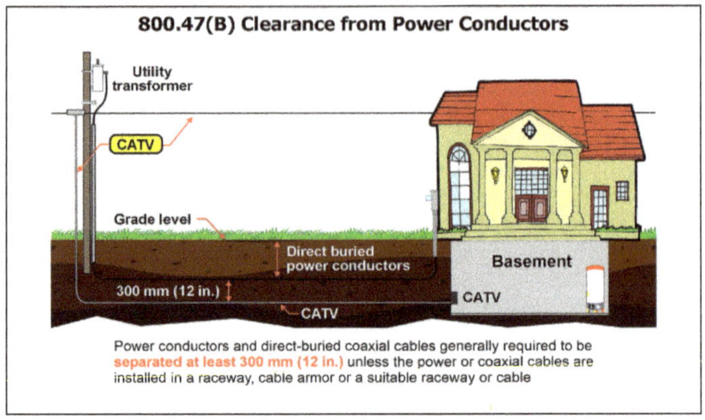

Figure 17.11. Separation of coaxial cables from power conductors

straight as much as possible and protected physically if it is prone to damage. When the bonding or grounding electrode conductor is installed in a metal raceway, both ends of the raceway must be bonded to the enclosed conductor or to the same terminal or electrode that the bonding or grounding conductor is connected to [NEC 800.100(A)(5) and (6)].

The bonding conductor or grounding electrode conductor should be as short as practical. In one- and two-family homes, it must be as short as possible and should not exceed 6.0 m (20 ft). If it is not feasible to keep the conductor within 6.0 m (20 ft), using a separate grounding electrode is allowed. This CATV bonding conductor or grounding electrode conductor must connect to the separate grounding electrode, which must then be bonded to the power grounding system [NEC 800.100(A)(4) and Exception].

GROUNDING ELECTRODE FOR CATV SYSTEMS

If the building or structure served has an intersystem bonding termination, the CATV bonding conductor or grounding electrode conductor must be connected to the intersystem bonding termination NEC 800.100(B)(1)]. If the building or structure served lacks an intersystem bonding termination, the CATV bonding conductor or grounding electrode conductor must be connected to the nearest accessible location, as listed below.

1. the building or structure grounding electrode system as described in NEC 250.50 or IRC E3608.1, including underground metal water pipe, effectively grounded metal frame of a building, concrete-encased electrode, etc.;
2. the grounded interior metal water piping system electrode within 1.5 m (5 ft) of the point of entrance into the building, as outlined in NEC 250.52(A)(1) or IRC E3608.1.1;

820.100 and 800.100 Grounding and Bonding of CATV Cables

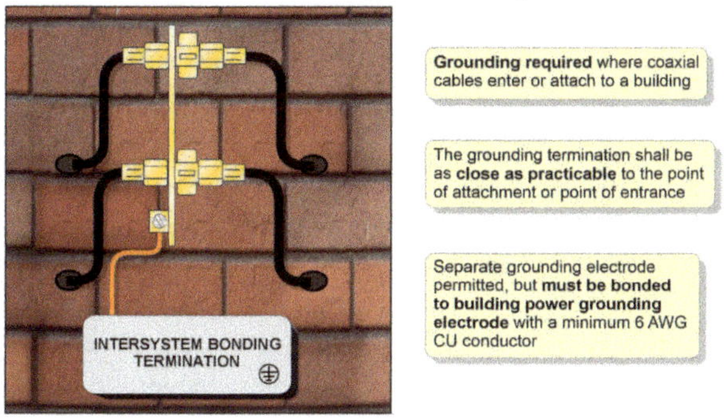

Figure 17.12. Grounding and bonding is required for CATV cables

3. the power service accessible means outside the enclosures, as specified in NEC 250.94(A) Exception;
4. a non-flexible metallic power service raceway;
5. the service equipment enclosure;
6. the grounding electrode conductor or the grounding electrode conductor's metal enclosure of the power service; or
7. the grounding electrode conductor or grounding electrode for a second building or structure disconnecting means, if connected to an electrode according to NEC 250.32 or IRC E3607.3.

See *NEC* 800.100(B)(2) for the above provisions.

Where the building or structure in which the CATV system is installed has no intersystem bonding termination or grounding means, a grounding electrode must be installed. The grounding electrode is permitted to be one of those listed in *NEC* 250.52(A) or IRC E3608.1, including:

1. an underground metal water pipe [250.52(A)(1) or IRC E3608.1.1];
2. the grounded metal frame of the building-[250.52(A)(2)];
3. a concrete-encased electrode [250.52(A)(3) or IRC E3608.1.2]; and
4. a ground ring [250.52(A)(4) or IRC E3608.1.3].

If none of these electrodes are present, an electrode must be installed and used. Additional electrodes allowed for grounding include rods, pipes, or plate electrodes [NEC 800.100(B)(3)]. The Code prohibits using metal underground gas piping, aluminum electrodes, or the structural rebar of a permanently installed inground swimming pool as grounding electrodes, as specified in NEC 250.52(B).

NETWORK-POWERED BROADBAND COMMUNICATIONS SYSTEMS

Network-powered broadband communication systems are a popular and advanced type of communication technology. These systems offer multimedia services, including combinations of voice, data, video, audio, and interactive functions. Typically, they are supplied to premises wiring through a shared cable or wiring that carries multiple frequencies and terminates at a device called a network interface unit (NIU). In NEC Article 100, a network interface unit is defined as "a device that converts a broadband signal into individual voice, audio, video, data, and interactive service signals and provides isolation between the network power and premises signal circuits. These devices often include primary and secondary protectors." The cable supplying the NIU can be coaxial, twisted pair, or a hybrid of coaxial, twisted pair, and optical fiber conductors. The NIU transforms and separates broadband signals into their respective voice, data, and interactive service components.

NEC Article 830 identifies two levels of power sources for these systems, known as low and medium power sources. Low power sources are restricted to a maximum system voltage of 100 volts and a maximum of 100 volt-amperes (VA). Medium power sources are limited to no more than 150 volts and a maximum power of 100 volt-amperes (VA) (see NEC Table 830.15 for specific power limitations).

Direct-current (dc) power sources exceeding 150 volts to ground but not more than 200 volts to ground, with the current to ground limited to 10 mA dc, that meet the current and power limitations for medium-power sources in NEC Table 830.15, are classified as medium-power sources. See UL 60950-21-2007, *Standard for Safety for Information Technology Equipment - Safety - Part 21: Remote Power Feeding*, for listing information on equipment that qualifies as a medium-power source.

Although these power sources are similar to those allowed for Class 3 circuits as covered in NEC Article 725, the installation requirements more closely resemble those of Class 1 circuits or non-power-limited fire alarm circuit requirements. Since the voltage limit for medium power systems is 150 volts, greater spacing and separation provisions are necessary.

The basic broadband communication system powered by a network includes a single supply cable to the NIU that carries both power and signals. The NIU converts the broadband into component signals. Common cables used are coaxial types, with the broadband signal and power on the center conductor. The braided part of the coaxial cable transmits the broadband signal, while a twisted pair carries the power.

Composite optical fiber cables are also used, featuring a pair of conductors for power (see Figure 17.13).

CABLE REQUIREMENTS FOR NPBC SYSTEMS (TYPES, LISTING, AND MARKS)

The Code generally requires the equipment and cables for network-powered broadband communications systems to be listed and suitable for their purpose. Table 17.1 provides a basic overview of the types of broadband communication system cables used in homes and their intended uses.

INSTALLATION REQUIREMENTS FOR NPBC SYSTEMS (SECURE, SUPPORTS, WORKMAN-LIKE MANNER, SEPARATIONS)

Network-powered broadband communication wiring must be installed neatly and professionally. Cables are required to be supported by the building's structural components and secured in a way that prevents damage during normal building use. They must be attached using hardware such as straps, staples, cable ties, hangers, or similar fittings, designed and installed to avoid damaging the cables. When installed in hollow spaces or through framing members, the appropriate clearance from the face of the framing must be maintained, or suitable steel plates must be installed for protection, as specified in NEC 300.4(A), (D), (E), (F), and 300.11 [NEC 800.24]. These cables are not allowed to be supported by or strapped to the outside of any conduit or raceway [NEC 830.133(B)].

Types BMU, BM, or BMR cables should be used for medium-power, network-powered broadband communication systems installed outside and entering buildings. Types BLU and BLX are suitable for low-power systems and circuits (NEC 830.40). The minimum clearances from grade level(s) are specified in NEC 830.44 and align with the requirements of the National Electrical Safety Code (NESC). NEC Table 830.47(A) details the minimum cover requirements for underground installations of network-powered broadband communication wiring systems. These depths are comparable to those specified in NEC Table 300.5 or IRC Table E3803.1. However, network-powered broadband cables must adhere to the minimum depths outlined in NEC Table 830.47(A).

Wiring installed in one- and two-family dwellings is generally required to be either type BM or BLX because wiring in plenums and ducts, or spaces used for transporting environmental air, is restricted. When these conditions are present in dwellings and wiring is installed in such spaces, it must be a type suitable for the location, either BMR or BLP. The separation requirements of NEC 830.133(A) must be followed.

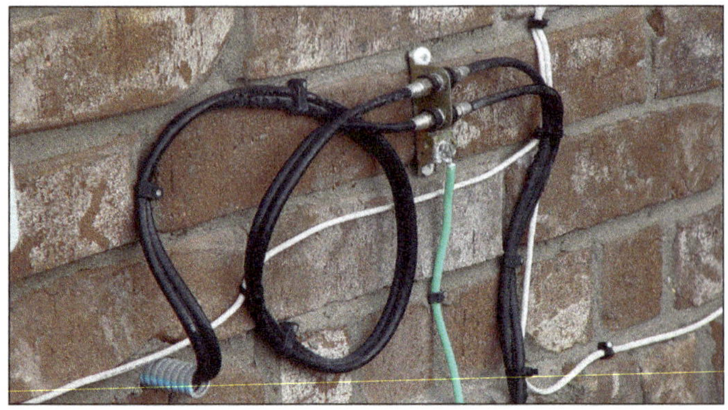

Photo 17.14. Coaxial cable bonding conductor connected to intersystem bonding termination (not shown)

Photo 17.15. Coaxial cable grounding electrode conductor connected to power system grounding electrode conductor where no intersystem bonding termination exists

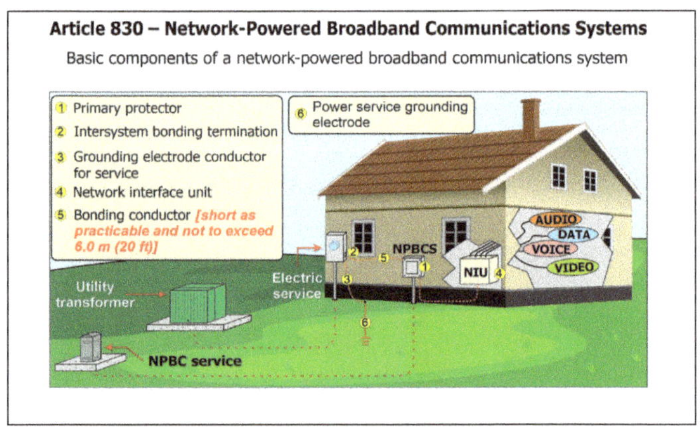

Figure 17.13. Basic components of a network-powered broadband communications system

GROUNDING AND BONDING REQUIREMENTS FOR NPBC SYSTEMS

The grounding and bonding requirements for these systems are similar to those for other communication systems discussed in Chapter 8 of the NEC. The purposes are the same.

The first general requirement is that all metallic enclosures, primary protectors, and metallic members of cables intended

Network-Powered Broadband Communication Cables Typically Used at Dwelling Units

Marking	Type	Intended Use	*NEC* Section
BMR	Medium Power	Use in vertical runs or shafts floor to floor	800.179(B)
BM	Medium Power	General purpose use	800.179(C)
BMU	Medium Power	Outdoor, underground use	830.179(B)
BLP	Low Power	Use in ducts, plenums, other environmental air spaces	800.179(A)
BLX	Low Power	Limited use outside, in dwellings, or in raceways	800.179(D), 830.40(B)
BLU	Low Power	Outdoor, underground use	830.179(C)

Network-powered broadband communication cables are permitted to be wired with any of the substitutions as indicated below *[see NEC Table 830.154]*:

Cable Type	Permitted Cable Substitutions
BM	BMR
BLP	CMP, CL3P
BLR	CMP, CL3P, CMR, CL3R, BLP, BMR
BL	CMP, CMR, CM, CMG, CL3P, CL3R, CL3, BMR, BM, BLP, BLR
BLX	CMP, CMR, CM, CMG, CMX, CL3P, CL3R, CL3, CL3X, BMR, BM, BLP, BRP, BL

Table 17.1. Network-Powered Broadband Communication Cables

to be grounded must be grounded in accordance with the rules in NEC 800.100. Rigid metal conduit (RMC) or intermediate metal conduit (IMC) containing network-powered broadband communications entrance cable must be connected by a bonding conductor or grounding electrode conductor to a grounding electrode according to 800.100(B) [see NEC 800.49].

The bonding conductor or grounding electrode conductor must be insulated, covered, or bare; solid or stranded; copper or another corrosion-resistant conductive material. It must not be smaller than 14 AWG, but it does not need to be larger than 6 AWG. This conductor should be installed in as straight a line as possible, and if physical damage is a concern, it must be properly protected. In dwelling units, the length of this conductor is limited to 6.0 m (20 ft) and should be kept as short as possible. If a length of 6.0 m (20 ft) is not feasible, installing a separate electrode is permitted, along with a minimum 6 AWG bonding jumper between the power system electrode and the additional grounding electrode for the NPBC system [NEC 800.100(A)(1) through (6)].

If the building or structure served has an intersystem bonding termination, the grounding conductor must be connected to it. If there is no intersystem bonding termination (such as at an existing building), the bonding conductor or grounding electrode conductor for these systems should be connected at one of the nearest accessible locations, such as the:

1. building or structure grounding electrode system as covered in *NEC* 250.50 or IRC E3608.1. This includes underground metal water pipes, the grounded metal frame of a building, concrete-encased electrodes, etc.;
2. grounded metal water piping system electrode, within 1.5 m (5 ft) from its point of entrance to the building, as covered in 250.52(A)(1) or IRC E3608.1.1;
3. power service accessible means external to the enclosures as covered in *NEC* 250.94 Exception;
4. nonflexible metallic power service raceway;
5. service equipment enclosure;
6. grounding electrode conductor or the grounding electrode conductor metal enclosure of the power service; or
7. A grounding electrode conductor or grounding electrode for a building or structure disconnecting means in accordance with *NEC* 250.32 or IRC E3607.3.

The connections of the bonding conductor or grounding electrode conductor to grounding electrodes must be suitable for the application [see UL Category (KDSH)]. Grounding electrode connections should be made with equipment appropriate for that specific purpose, and when separate electrodes are installed, they are required to be bonded together in a manner that is common to the power service electrode(s). When bonding or grounding is necessary, the devices used to connect a shield, a sheath, or non-current-carrying metallic members of a cable to a bonding conductor or grounding electrode conductor must be listed or be part of the listed equipment [NEC 800.180].

This chapter offers an overview of most of the requirements for installing network-powered broadband communication systems in dwelling units. For installation criteria not covered here, see NEC Article 830 for the relevant requirements and provisions.

PREMISES-POWERED BROADBAND COMMUNICATIONS SYSTEMS

Premises-powered broadband communication systems are covered by the NEC in Article 840. These systems are broadband communication systems that do not rely on a network for power, and NEC Article 840 addresses the applications involved in these types of non-network-powered communication systems. A basic premises-powered broadband communication system typically consists of an optical fiber, twisted pair, or coaxial cable connected to the premises, supplying a broadband signal to a network terminal. This terminal converts the broadband optical signal into electrical signals such as those used for traditional telephone, video, high-speed Internet, and interactive services. Powering the network terminal and devices is usually done through a premises power supply, which might be built into the terminal or provided separately. To maintain communication during a power outage, a battery backup or uninterruptible power supply (UPS) is typically included in the power system.

NEC Article 840 addresses the specific installation issues related to premises-powered broadband communication systems. These differ from network-powered broadband systems and, therefore, need a separate NEC article.

The IRC does not provide details about network-powered or premises-powered broadband communication system installations in dwellings, so, by default, the requirements outlined in the NEC must be applied.

WIRING IN AIR-HANDLING SPACES

It is best to avoid running any wiring through ducts or other spaces that carry environmental air (plenums) because of the strict wiring methods required in these areas. In dwelling units, return air ducts and plenums are typically made of prefabricated round or rectangular metal ducts used for moving environmental air. The term plenum is defined in Article 100 of the NEC and Chapter 35 of the IRC. In homes, plenums are usually fabricated chambers designed solely to transmit environmental air. For example, a plenum chamber connects the ducts to a forced-air furnace.

Generally, the NEC requires that Class 2 wiring in plenums be listed as CL2P. The "P" suffix usually indicates the cable is suitable for installation in plenums [NEC 722.135(A), 722.179(A)(1) or IRC E4303.2.1]. Other cables, as shown in NEC Table 722.135(E) or IRC Table E4303.2, Cable Uses, and Permitted Substitutions, may be used as substitutes for the CL2P cable. An exception to this requirement allows listed wires and cables to be installed in a wiring method suitable for use in a plenum, in accordance with NEC 300.22. IRC Table E4303.2 covers only Class 2 wiring substitutions and is derived from NEC Table 722.135(E). Chapter 43 of the IRC is limited to Class 2 circuits and wiring. Refer to Chapter 9 for more information on wiring in air-handling spaces.

Photo 17.16. External network terminal at dwelling unit located near service metering equipment

Photo 17.17. Closeup of outdoor external network terminal

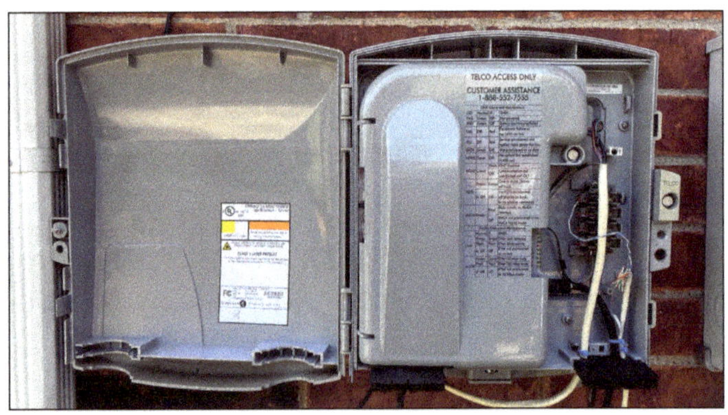

Photo 17.18. Closeup of open outdoor external network terminal

PRODUCT SAFETY STANDARD REQUIREMENTS

It is recommended to consult the following electrical product safety guide card information and safety standards for additional guidance on properly installing, operating, and using electrical equipment covered in this chapter. The four-letter code in parentheses indicates the product category in the Underwriters Laboratories directories.

- Electrical Equipment for Use in Ordinary Locations (AALZ)
- Power Supplies, General Purpose (QQFU)
- Power-Limited Fire-Alarm Cable (HNIR)
- Household Fire-Warning System Units (UTLQ)
- Control Units and Accessories, Household System Type (UTOU)
- Household Fire-Warning System Units (UTLQ)
- Detectors, Automatic Fire (UPLV)
- Detectors, Heat, Automatic Fire (UQGS)
- Microwave and Cable Communication Equipment (POFV)
- Telecommunications Equipment (WYIE)
- Telephone Appliances and Equipment (WYQQ)
- Power Limited Circuit Cable (QPTZ)
- Communication, Coaxial and Broadband Cable Assemblies (DUNH)

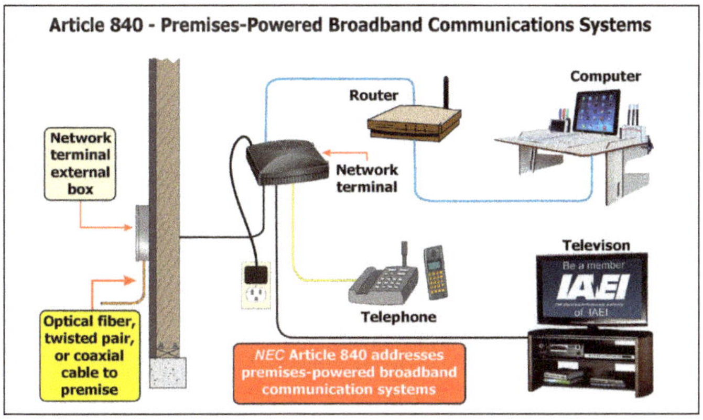

Figure 17.14. Premises-powered broadband communications systems

- Communications Cable (DUZX)
- Community Antenna Television Cable (DVCS)
- Network-Powered Broadband Communications Cables (PWIP)

REFERENCES

1. *National Electrical Code* and *NEC* are registered trademarks of the National Fire Protection Association, Inc., Quincy, MA 02169. This reprinted material is not the official position of the National Fire Protection Association, which is represented only by the standard in its entirety.

2. *International Residential Code for One- and Two-Family Dwellings* (2021). Copyright © International Code Council, Inc., June 2023.

3. UL, Northbrook, IL, *UL Guide Information for Electrical Equipment Directory* (UL Product iQ® material) unless indicated otherwise.

REVIEW QUESTIONS

1. A Class 2 thermostat cable extending through a dwelling unit plenum space must generally be Type _____.

   A. CL2
   B. CL2P
   C. CL2R
   D. CL2X

2. *NEC* Article 725 and Chapter 43 of the IRC cover Class 2 remote-control, signaling, and power-limited circuits that are an integral part of a device or of utilization equipment.

   True _____ False _____

3. Jacketed cables of Class 2 circuits are permitted to be installed in the same enclosure or raceway with jacketed cables of communication circuits as long as the Class 2 circuits comply with the applicable NEC article of the other jacketed cable.

   True _____ False _____

4. The location of the branch-circuit overcurrent protective device supplying a fire alarm control panel(s) is required to be marked on the fire alarm control panel(s).

   True _____ False _____

5. A Class 2 circuit is classified as the portion of the wiring system between the _______ side of a Class 2 power source and the connected equipment.

   A. line
   B. load
   C. line and load
   D. neutral

6. Due to their power limitations, both Class 2 and Class 3 circuits are considered to provide safety from a fire initiation standpoint.

   True _____ False _____

7. Where an accessory building or additional building has electrical power supplied by a feeder or branch circuit, an intersystem bonding termination is required to be established at the ____________________ for that building.

   A. first outlet
   B. entrance
   C. disconnecting means
   D. metallic water piping system

8. Listed hybrid power and communications cables are permitted where the power cable is a listed Type NM or NM-B, conforming to Part III of *NEC* Article 334, and the communications cable is a listed _______ cable. In this case, the jackets on the listed NM or NM-B and listed _________cables are required to be rated for 600 volts minimum, and the hybrid cable is listed as being resistant to the spread of fire.

   A. Type CMR
   B. Type CM
   C. Type CMP
   D. Type CMX

9. In a dwelling unit, cables such as CATV cables are permitted to be installed in joist or stud space used as "other space used for environmental air" ____________________ _______________.

   A. so long as no more than 1.8 m (6 ft) of cable is in the space.
   B. so long as it passes through the joist or stud space perpendicular to the long dimension of such spaces.
   C. so long as it is stapled securely along the studs or joists and no more cable is installed than necessary.
   D. so long as it passes through the joist or stud space in the long dimension of such spaces.

10. An intersystem bonding termination (IBT) for connecting intersystem bonding and grounding conductors required for communication systems is required to be provided external to enclosures at the service equipment and at the disconnecting means for any additional buildings or structures that are supplied by a feeder or branch circuit. The intersystem bonding termination is also required to be:

A. capable of connection of not less than two intersystem bonding conductors.
B. accessible for connection and inspection.
C. capable of connection of not less than three intersystem bonding conductors.
D. both B and C

11. Communications circuits are not permitted to be installed in the same listed communication cable with Class 2 or 3 circuits.

True _____ False _____

12. Power-limited fire alarm (PLFA) system conductors installed in dwellings are typically which of the following types of conductors (without considering cable substitutions):

A. Type CMX, CL3X or CL2X
B. Type CL3P, CL2R or CL3
C. Type MPG, FPLR or FPL
D. Type FPLP, FPLR or FPL

13. Communications circuits are required to be installed in a neat and workmanlike manner.

True _____ False _____

14. If the building or structure served has an intersystem bonding termination, the bonding conductor or grounding electrode conductor for a network-powered broadband communications system shall be connected to the intersystem bonding termination. If the building or structure (with grounding means) served has no intersystem bonding termination (such as at an existing building), the bonding conductor or grounding electrode conductor for this system can be connected at which of the nearest accessible locations:

A. grounded metal water piping system electrode
B. nonflexible metallic power service raceway
C. service equipment enclosure
D. any of the above

15. A raceway cannot be used as a means for supporting Class 2 circuit conductors unless the raceway contains the electric power conductors used to supply the same electrically controlled equipment.

True _____ False _____

16. Communications circuit cables that enter a dwelling or terminate on the outside of a dwelling are generally not required to be grounded.

True ______ False ______

17. *NEC* Article ______ covers the installation of wiring and equipment of fire alarm systems, including all circuits controlled and powered by the fire alarm system.

A. 725
B. 727
C. 760
D. 800

18. A fire alarm circuit is the portion of the wiring system between the load side of the overcurrent device, or the power-limited supply, and all connected equipment of all circuits powered and controlled by the fire alarm system and is classified as "non-power-limited" or "power-limited."

True ______ False ______

19. To help prevent unintentional signals on fire protective signaling circuits while they are being tested and serviced, they must be specifically __________ at all terminal and junction locations.

A. grouped
B. identified
C. marked
D. listed

20. A "Network Interface Unit" (NIU) is defined as "a device that converts a broadband signal into component voice, audio, video, data, and interactive services signals and provides isolation between the network power and the premises signal circuits. These devices often contain primary and secondary protectors."

True ______ False ______

21. Where a non-power-limited fire protective signaling circuit is extended aerially (overhead) beyond one building, it is required to meet the installation requirements of *NEC* Article 800 Parts II, III, and IV and meet the installation requirements of Part I of NEC Article 300.

True ______ False ______

22. For new construction, a minimum of _______ communication outlet(s) is required to be installed within the dwelling at a readily accessible location and cabled to the service provider demarcation point.

A. 1
B. 2
C. 3
D. 4

23. Premises-powered broadband communications systems are systems that typically consist of an optical fiber, twisted pair, or coaxial cable to the premises supplying a broadband signal to a network terminal that converts the broadband signal into component electrical signals, such as traditional telephone, video, high-speed Internet, and interactive services typically powered through a premises power supply.

True ______ False ______

24. The circuit extending from the communications utility's or service provider's serving terminal or tap up to and including the network interface unit (NIU) is the definition of a:

A. Community Antenna Television and Radio Circuit
B. Network-Powered Broadband Communications Circuit
C. Optical Fiber Cable Circuit
D. Premises-Powered Broadband Communications Circuit

25. The bonding conductor or grounding electrode conductor for a network-powered broadband communications cable shall be as short as practicable. In one-family and multifamily dwellings, the bonding conductor or grounding electrode conductor shall be as short as practicable, not to exceed what distance?

A. 2.5 m (8 ft)
B. 3.0 m (10 ft)
C. 6.0 m (20 ft)
D. Unlimited length

26. __________ can best be described as a technology for wired Ethernet local area networks (LANs) that allows the electrical current necessary for the operation of devices to be carried by the data cables rather than by power cords.

A. Smart wiring
B. Power over Internet
C. Power over Ethernet
D. Power over optics

27. The "-LP" cable designation indicates that the cable has been evaluated to carry the marked current under reasonable worst-case installation scenarios without exceeding the __________ rating of the cable.

A. amperage
B. voltage
C. insulation
D. temperature

28. An exception gives relief from providing a means for connecting intersystem bonding conductors where communications systems are not likely to be used.

True ______ False ______

Appendix A Cross Reference from the NEC 2023 to the IRC 2021

***NEC* Reference**	***Description in Both Model Code Documents NEC / IRC***	**IRC Reference**
ARTICLE 90 - Introduction		
90.2(C)	*NEC* Scope - Covered / IRC Scope - Covered	E3401.2
90.2(D)	*NEC* Scope - Not Covered / IRC Scope - Not Covered	E3401.3
ARTICLE 100 - Definitions		
Article 100	Electrical Definitions	Chapter 35
ARTICLE 110 - Requirements for Electrical Installations		
110.2	Approval of Conductors & Equipment / Materials & Components	E3403.1
110.3(A)(1)	Suitability for Installation / Listing and Labeling	E3403.3
110.3(B)	Installation and Use / Listing and Labeling or Identified for a Use	E3403.3
110.4	Voltages	E3404.1
110.5	Conductors / Conductor Material	E3406.2
110.9	Interrupting Rating / Minimum 10KVA Rating for Faults in IRC	E3404.2
110.10	Circuit Impedance & Other Characteristics / Circuit Characteristics	E3404.3
110.12(A)	Unused Openings / General Application	E3404.6
110.12(A)	Unused Openings / Boxes, Conduit Bodies & Fittings	E3906.4
110.12(A), 408.7	Unused Openings / Cabinets and Panelboards	E3907.5
110.12(B)	Integrity of Electrical Equipment & Connections / Integrity of Electrical Equipment	E3404.7
110.12(C)	Cables and Conductors Installed Exposed	*NEC Default*
110.13(A)	Mounting *(Electrical Equipment)*	E3404.8
110.14	Electrical Connections / Aluminum and Copper Connections	E3406.8
110.14	Fine Stranded Conductors	E3406.9
110.14(A)	Terminals	E3406.10
110.14(B)	Splices	E3406.11
110.14(C)	Temperature Limitations for Conductors	E3705.4
110.14(C)(1)	Equipment Provisions / Temperature Limitations	E3603.5
110.14(C)(1)(a)(1)	Conductors Rated 60ºC	E3705.4.1
110.14(C)(1)(b)(1)	Conductors Rated 75ºC	E3705.4.2
110.14(C)(2)	Separate Connector Provisions / Separately Installed Pressure Connectors	E3705.4.3
110.14(D)	Terminal Connection Torque Requirements / Installation	E3406.12
110.17	Servicing and Maintenance of Equipment	*NEC Default*
110.20	Reconditioned Equipment	*NEC Default*
110.21(A)	Manufacturer's Marking / Equipment Identification	E3404.11
110.21(A)(2)	Reconditioned Equipment *(Marking Requirements)*	E3404.11(2)
110.21(B)	Field-Applied Hazard Marking	E3404.12
110.22	Identification of Disconnecting Means	E3404.13
110.25	Lockable Disconnecting Means *(for Disconnects, etc.)*	E4101.8
110.26	Spaces About Electrical Equipment / Working Space and Clearances – General	E3405.1
110.26	Spaces About Electrical Equipment / Working Space – Panelboards	E3606.2
110.26(A)	Working Space	E3405.1
110.26(A)(1), (A)(2), and (A)(3)	Depth, Width, & Height of Working Space	E3405.2
Table 110.26(A)(1)	Working Space Distances	E3405.2
110.26(A)(6)	Grade, Floor. Or Working Platform *(Clear, Level and Flat)*	*NEC Default*
110.26(B), 240.24(D), (E) & (F)	Clear Spaces / Location of Working Spaces & Equipment	E3405.5
110.26(C)	Entrance to and Egress from Working Space	E3405.6
110.26(C)(1)	Minimum Required / Access and Entrance to Working Space	E3405.6
110.26(D)	Illumination	E3405.7

NEC **Reference**	***Description in Both Model Code Documents NEC / IRC***	**IRC Reference**
110.26(E)(1)(a)	Working Space *(cont.)* Indoor Dedicated Electrical Space / Dedicated Panelboard Space	E3405.3
110.26(E)(2)(b) and (c)	Outdoor Dedicated Electrical Space / Dedicated Panelboard Space	E3405.4
110.27(A)	Live Parts Guarded Against Accidental Contact / Energized Parts Guarded Against Accidental Contact	E3404.9
110.27(B)	Prevent Physical Damage *(Electrical Equipment)*	E3404.10
110.28, Table 110.28	Enclosure Types and Selection *(Electrical Equipment)*	E3404.4, Table E3404.4
110.29	In Sight From *[Visible and not More than 15 m (50 ft)]*	Chapter 35
ARTICLE 200 – Use and Identification of Grounded Conductors		
200.2(B)	Continuity of Grounded Conductor	E3406.13
200.4	Neutral Conductor	*NEC Default*
200.6(A), 200.6(B), 310.6(A)	Means of Identifying Grounded Conductors / Grounded Conductors	E3407.1
200.9	Identification of Terminals *(Polarized Connections for Grounded Conductor)*	E3407.4.2
200.10	Identification of Terminals *(Grounded Conductor)*	E3407.4
200.10(A)	Device Terminals *(Grounded Conductor)*	E3407.4.1
200.10(B)	Receptacles, Plugs, and Connectors *(Grounded Conductor)*	E3407.4.2
200.10(C)	Screw Shells / Polarization of Luminaires *(Grounded Conductor)*	E4003.13.1
200.11	Polarity of Connections *(Grounded Conductor)*	*NEC Default*
ARTICLE 210 – Branch Circuits		
210.4	Multiwire Branch Circuits	E3701.5
210.4(B)	Disconnecting Means	E3701.5.1
210.4(D), 200.4(B)	Grouping	E3701.5.2
210.5(A), (B), and (C)	Identification for Branch Circuits	E3407
210.6(A)	Branch-Circuit Voltage Limitations for Dwelling Units	E3702.1
210.7	Multiple Branch Circuits Supplying Devices or Equipment	*NEC Default*
210.8(A)	Ground-Fault Circuit-Interrupter Protection for Dwelling Units	E3902
210.8	Measurement for GFCI Protection	E3902..19
210.8	Location of GFCI *(Readily Accessible Location)*	E3902.15
210.8(A)(1)	Bathroom Receptacles	E3902.1
210.8(A)(2)	Garage and Accessory Building Receptacles	E3902.2
210.8(A)(3)	Outdoor Receptacles	E3902.3
210.8(A)(4)	Crawl Space Receptacles	E3902.4
210.8(A)(5)	Basement Receptacles	E3902.5
210.8(A)(6)	Kitchen Receptacles	E3902.6
210.8(A)(7)	Areas with sinks and permanent provisions for food preparation, beverage preparation, **or** cooking	*N 3902.7*
210.8(A)(8)	Sinks *(top inside edge of bowl) [Receptacles within 1.8 m (6 ft)]*	E3902.8
210.8(A)(9)	Boathouse Receptacles	E3902..15
210.8(A)(10)	Bathtub and Shower Stalls *[Receptacles within 1.8 m (6 ft)]*	E3902..9
210.8(A)(11)	Laundry Areas	E3902.,10
210.8(A)(12)	Indoor Damp and Wet Locations	E3902..11
210.8(A), Ex. No. 1, 426.28	Snow-Melting and Deicing Equipment Protection *(GFCI Exception)*	E4101.7 , 3902.18
210.8(A), Ex. No. 2	Permanently Installed Fire/Burglar Alarm *(GFCI Exception)*	E3902.5, .18
210.8(A), Ex. No. 3	Listed weight-supporting ceiling receptacles (WSCR) with compatible weight-supporting attachment fittings (WSAF) installed for the support ceiling luminaires or ceiling-suspended fans *(GFCI Exception)*	*3902.18*
210.8(A), Ex. No. 4	Factory-installed receptacles *(not readily accessible)* mounted internally to bathroom exhaust fan assemblies *(GFCI Exception)*	*3902.18*
210.8(C)	Crawl Space Lighting Outlets	E3902.4

NEC Reference	Description in Both Model Code Documents *NEC / IRC*	IRC Reference
210.8(D), 422.5(A)	Ground-Fault Circuit-Interrupter Protection for Dwelling Units *(cont.)* Specific Appliances *(Dishwashers, Microwave Ovens, Electric Ranges, Etc.)*	*3902.12*
210.8(E), 210.63	Equipment Requiring Servicing	*3902.13*
210.8(F)	GFCI Protection for Outdoor Outlets *(A/C Disconnects)*	*3902.14*
210.11(A)	Number of Branch Circuits *(Minimum Number Required)*	E3703.6
210.11(B)	Load Evenly Proportioned Among Branch Circuits / Branch-Circuit Load Proportioning	E3703.7
210.11(C)(1), 210.52(B)(1) and (B)(2)	Small-Appliance Branch Circuits / Kitchen and Dining Area Receptacles	E3703.2
210.11(C)(2)	Laundry Branch Circuits	E3703.3
210.11(C)(3)	Bathroom Branch Circuits	E3703.4
210.11(C)(4)	Garage Branch Circuits	E3703.5
210.12	Arc-Fault Circuit-Interrupter Protection	E3902..21
210.12	Location of AFCI Device *(Readily Accessible Location)*	E3902..20
210.12(A)	Means / Methods for Providing AFCI Protection	E3902.21
210.12(B)	Rooms/Areas Requiring AFCI Protection	E3902..21
210.12(B), Ex. No. 1	Individual Branch Circuit Supplying a Fire Alarm System *(Installed in a Metal Raceway System) (AFCI Exception)*	E3902., Ex.
210.12(B), Ex. No. 2	Individual Branch Circuit Supplying Outlet for Arc Welder *(Expires Jan. 1, 2025) (AFCI Exception)*	*NEC Default*
210.12(E)	AFCI Protection for Branch Circuit Extensions, Modifications, or Replacements	E3902..22
210.18	Branch-Circuit Ampere Rating	E3702.2
210.19(A)(1) and (2)	Branch Circuit Minimum Ampacity and Size / Branch-Circuit and Feeder Ampacity	E3701.2
210.19(B)	Branch Circuits with More than One Receptacle	E3701.4
210.19(C)	Household Ranges and Cooking Appliances / Minimum Branch Circuit for Ranges	E3702.10.1
210.20(A)	Continuous and Noncontinuous Loads / Continuous load	E3701.2.1
210.21(B)(1)	Single Receptacle on an Individual Branch Circuit / Single Receptacle	E4002.1.1
210.21(B)(3)	Receptacle Ratings / Two or More Receptacles	E4002.1.2
210.22	Permissible Loads, Individual Branch Circuits	*NEC Default*
210.23(A)	10-Ampere Branch Circuits	*3702.3*
210.23(A)(1)	Loads Permitted for 10-Ampere Branch Circuits	*3702.3*
210.23(A)(2)	Loads Not Permitted for 10-Ampere Branch Circuits	*3702.3*
210.23(B)	15- and 20-Ampere Branch Circuits / Branch Circuits Serving Multiple Loads or Outlets	E3702.4, E3702.6
210.23(C)	30-Ampere Branch Circuits	E3702.5
210.23(D)	40- and 50-Ampere Branch Circuits	*NEC Default*
210.23(E)	Branch Circuits Larger Than 50 Amperes	*NEC Default*
210.24	Branch-Circuit Requirement - Summary	E3702.15
210.25(A)	Dwelling Unit Branch Circuits	E3702.15
210.25(B)	Common Area Branch Circuits	E3702.15
210.50(C)	Appliance Receptacle Outlets *[Installed Within 1.8 m (6 ft) of Appliance]*	E3901.5
210.52	Required Outlets - Dwelling Unit Receptacle Outlets / Receptacle Outlets - General	E3901.1
210.52(A)	General Provisions / General Purpose Receptacle Distribution	E3901.2
210.52(A)(1)	Spacing (Receptacle Outlets)	E3901.2.1
210.52(A)(2)	Wall Space (Receptacle Outlets)	E3901.2.2
210.52(A)(3)	Floor Receptacles	E3901.2.3
210.52(A)(4)	Countertop and Similar Work Surface Receptacle Outlets *(Not Part of Wall Space Requirements)*	E3901.2.4
210.52(B), 210.11(C)(1)	Small Appliances	E3703.2, E3901.3
210.52(B)(1)	Receptacle Outlets Served / Small Appliance Receptacles	E3901.3
210.52(B)(2)	No Other Outlets / Other Outlets Prohibited	E3901.3.1

NEC Reference	*Description in Both Model Code Documents NEC / IRC*	IRC Reference
210.52(B)(3)	*(Required Outlets)* Small Appliances *(cont.)* Kitchen Receptacle Requirements / Limitations	E3901.3.2
210.52(C)	Countertop and Work Surface Receptacles *(Kitchen, Pantries, Dining Rooms, Breakfast Rooms, Etc.)*	E3901.4
210.52(C)(1)	Wall Spaces *(Kitchen Countertop Receptacles)*	E3901.4.1
Figure 210.52(C)(1)	Determination of Area Behind a Range, Counter-Mounted Cooking Unit, or Sink	Figure E3901.4.1
210.52(C)(2)	Island and Peninsular Countertops and Work Surfaces *(Optional)*	E3901.4.2
210.52(C)(2)	*(Provisions to be Provided at Island or Peninsula for Future Addition of Receptacle Outlets)*	*NEC Default*
210.52(C)(3)	Receptacle Outlet Location *(On or Above but not Below Countertop or Work Surface)*	E3901.4.3
210.52(D)	Bathrooms *(Receptacle Outlets)*	E3901.6
210.52(E)	Outdoor Outlets *(Receptacle Outlets)*	E3901.7
210.52(E)(3)	Balconies, Decks, and Porches *(Receptacle Outlets)*	E3901.7
210.52(F)	Laundry Areas *(Receptacle Outlets)*	E3901.8
210.52(G)	Basements, Garages, and Accessory Buildings *(Receptacle Outlets)*	E3901.9
210.52(G)(1)	Garages *(Receptacle Outlets)*	E3901.9
210.52(G)(2)	Accessory Buildings *(Receptacle Outlets)*	E3901.9
210.52(G)(3)	Basements *(Receptacle Outlets)*	E3901.9
210.52(H)	Hallways *(Receptacle Outlets)*	E3901.10
210.52(I)	Foyers *(Receptacle Outlets)*	E3901.11
210.63	Equipment Requiring Servicing *[Receptacle Outlet Required Within 7.5 m (25 ft) of Equipment]*	E3901.12
210.63(A)	Heating, Air Conditioning, and Refrigeration Equipment	E3901.12
210.63(B)(1)	Indoor Service Equipment	*NEC Default*
210.63(B)(2)	Indoor Equipment Requiring Dedicated Equipment Spaces	*NEC Default*
210.70(A)	Lighting Outlets Required - Dwelling Units / Lighting Outlets - General	E3903.1
210.70(A)(1)	Habitable Rooms *(Lighting Outlets)*	E3903.2
210.70(A)(2)	Additional Locations *(Lighting Outlets) (Hallways, Stairways, etc.)*	E3903.3
210.70(C)	Storage or Equipment Spaces *(Lighting Outlets)*	E3903.4
ARTICLE 215 - Feeders		
215.2(A)	Feeders – Minimum Rating and Size - General	E3701.2
215.2(A)(1)	Continuous and Noncontinuous Loads	E3701.2.1
215.2(A)(2)	Ampacity Adjustment or Correction Factors	E3701.2.2
215.3	Feeders – Overcurrent Protection / Overcurrent Protection Required	E3705.5
215.6	Feeders – Equipment Grounding Conductor	E3704.6
215.9	Feeder GFCI Protection *(in lieu of Branch Circuit GFCI Protection)*	*NEC Default*
215.12	Identification of Feeders	E3407
215.15	Barriers Required for Equipment Supplied by Feeders	*NEC Default*
215.18(A) through (E)	Surge Protection Required for Dwelling Units Suppled by a Feeder	*NEC Default*
220.40, 310.12(B) and (C)	Feeders – Minimum Rating and Size – *(Not Required to be Larger than Service Conductors)*	E3704.1
ARTICLE 220 – Branch-Circuit, Feeder, and Service Calculations		
220.5	Calculations	E3404.1
220.5(A), 110.4	Voltages	E3404.1
220.5(C)	Floor Area *(Dwelling Unit Calculated Floor Area)*	E3704.4
220.10	Branch Circuit Load Calculations	E3702, E3704.4
220.11(A)	Branch Circuits Serving Motor-Operated and Combination Loads	E3702.8
220.11(B)	Branch-Circuit Inductive and LED Lighting Loads	E3702.9
220.14(B)	Electric Dryers and Electric Cooking Appliances in Dwelling Units / Branch-Circuit Load for Ranges and Cooking Appliances	E3702.10
220.16(A)	Loads for Additions to Existing Installations *(Dwelling Units)*	*NEC Default*
220.40	Feeder and Service Load Calculations - General	E3704.2

NEC Reference	*Description in Both Model Code Documents NEC / IRC*	IRC Reference
	Feeder and Service Load Calculations *(cont.)*	
220.41	Minimum Unit Load for Dwelling Units / Lighting and General Use Receptacle Load *(3 Volt-Amperes/ft^2)*	E3704.4
220.45	General Lighting / Feeder Loads	E3704.2
Table 220.45	Lighting Load Demand Factors / Feeder Load Calculations	Table E3704.2(1)
220.52(A)	Small Appliance Circuit Load	Table E3704.2(1)
220.52(B)	Laundry Circuit Load	Table E3704.2(1)
220.53	Appliance Load	Table E3704.2(1)
220.54	Electric Clothes Dryer	Table E3704.2(1)
220.55	Electric Ranges and Other Cooking Appliances	Table E3704.2(1)
Table 220.55	Demand Factors and Loads for Household Electric Ranges, Wall-Mounted Ovens, Etc.	Table E3704.2(2)
220.57	Electric Vehicle Supply Equipment (EVSE) Load	*Table E3704.2(1)*
220.60	Noncoincident Loads	Table E3704.2(1)
220.61(A)	Feeder or Service Neutral Load / Feeder Neutral Load	E3704.3
220.61(B)	Permitted Reductions *(70% Demand Factor)*	E3704.3
220.82(A)	Optional Feeder and Service Load Calculations / Minimum Service Load Calculations	E3602.2, Table E3602.2
220.83	Feeder and Service Load Calculations for Existing Dwelling Units	*NEC Default*
220.85	Feeder and Service Load Calculations for Two Dwelling Units	*NEC Default*
ARTICLE 225 – Outside Branch Circuits and Feeders		
225.4	Conductor Insulation *[Insulated Within 3.0 m (10 ft) of Building]*	*NEC Default*
225.6(A)	Conductor Size and Support – Overhead Spans	*NEC Default*
225.10	Wiring on Buildings *(or Other Structures)*	*NEC Default*
225.18	Clearance for Overhead Conductors and Cables	*NEC Default*
225.19	Clearances from Buildings *(Overhead Spans)*	*NEC Default*
225.19(A)	Above Roofs	*NEC Default*
225.19(C)	Horizontal Clearances *[Not Less than 900 mm (3 ft)]*	*NEC Default*
225.19(D)(1)	Clearance from Windows, Doors, Porches, Balconies, Etc. *[Not Less than 900 mm (3 ft)]*	*NEC Default*
225.26	Vegetation *(Trees)* as Support *(Not Permitted)*	*NEC Default*
Article 225 – Part II	Buildings or Other Structures Supplied by a Feeder(s) or Branch Circuit(s)	*NEC Default*
225.30	Number of Supplies	*NEC Default*
225.31	Disconnecting Means	*NEC Default*
225.31(B)	Location *(Disconnecting Means)*	*NEC Default*
225.33	Maximum Number of Disconnects	*NEC Default*
225.34	Grouping of Disconnects	*NEC Default*
225.36	Type of Disconnecting Means	*NEC Default*
225.39	Rating of Disconnect	*NEC Default*
225.39(C)	One-Family Dwelling *(Min. 100 Amperes)*	*NEC Default*
225.41(A), (B), and (C)	Emergency Disconnects *(Location, Rating, Markings, Etc.)*	*NEC Default*
225.42(A) through (E)	Surge Protection *(Location, Type, Rating, Etc.)*	*NEC Default*
ARTICLE 230 – Services		
230.1	Scope	E3601.1
230.2	Number of Services	E3601.2
230.3	One Building or Other Structure Not to Be Supplied Through Another	E3601.3
230.6	Conductors Considered Outside the Building	*NEC Default*
230.7	Other Conductors in Raceway or Cable *(Prohibited)*	E3601.4
230.8	Raceway Seal	E3601.5
230.9, 230.9(A)	Clearances on Buildings *(Overhead Service Conductors)*	E3604.1
230.9(B)	Vertical Clearances	E3604.2
230.9(C)	Building Openings	*NEC Default*
230.10	Vegetation as Support *(Not Permitted)*	*NEC Default*

***NEC* Reference**	***Description in Both Model Code Documents NEC / IRC***	**IRC Reference**
Article 230 – Part II	Overhead Service Conductors	E3604
230.23	Size and Ampacity *(Overhead Service Conductors)*	*NEC Default*
230.24	Clearances / Vertical Clearances *(Overhead Service Conductors)*	E3604.2
230.24(A)	Above Roofs	E3604.2.1
230.24(A), Ex. No. 1	*(Subject to Pedestrian or Vehicular Traffic)*	E3604.2.1 Ex. 1
230.24(A), Ex. No. 2	*[Slope of 100 mm in 300 mm (4 in. in 12 in.) or Greater]*	E3604.2.1 Ex. 2
230.24(A), Ex. No. 3	*(Overhanging Portion of the Roof)*	E3604.2.1 Ex. 3
230.24(A), Ex. No. 4	*(Service Drop Attached to the Side of a Building)*	E3604.2.1 Ex. 4
230.24(A), Ex. No. 5	*(Roof Area is Guarded or Isolated)*	E3604.2.1 Ex. 5
230.24(B)	Vertical Clearance for Overhead Service Conductors / Vertical Clearance from Grade	E3604.2.2
230.24(B)(1)	3.0 m (10 ft) - *(Service Entrance to Buildings and Lowest Point of the Drip Loop)*	E3604.2.2(1)
230.24(B)(2)	3.7 m (12 ft) – *(Over Residential Property and Driveways – Not Exceeding 300 Volts to Ground)*	E3604.2.2(2)
230.24(B)(3)	4.5 m (15 ft) - *(Over Residential Property and Driveways – Exceeding 300 Volts to Ground)*	*NEC Default*
230.24(B)(4)	5.5 m (18 ft) – *(Over Public Streets, Alleys, Roads, Parking Areas Subject to Truck Traffic)*	E3604.2.2(3)
230.24(C)	Clearance from Building Openings *[NEC 230.9(C)]*	*NEC Default*
230.24(D)	Clearance from Swimming Pools, Etc. *[NEC 680.9]*	*NEC Default*
230.24(E)	Clearance from Communication Wires and Cables *[NEC 800.44(A)(4)]*	*NEC Default*
230.26	Point of Attachment *(Overhead Service Conductors)*	E3604.3
230.27	Means of Attachment *(Overhead Service Conductors)*	E3604.4
230.28	Service Masts as Supports *(Overhead Service Conductors)*	E3604.5
230.28(A)	Strength *(...of adequate strength...)*	E3604.5.1
230.28(B)	Attachment *(no coupling located above last point of securement)*	E3604.5.2
230.29	Supports Over Buildings	E3604.6
Article 230 – Part III	Underground Service Conductors	E3605
230.30(A)	Insulation *(to be Insulated)*	E3605.1
230.31	Size and Ampacity	*NEC Default*
230.32	Protection Against Damage	E3605.4
230.33	Spliced Conductors	E3605.3
Article 230 – Part IV	Service-Entrance Conductors	E3605
230.40	Number of Service-Entrance Conductor Sets	*NEC Default*
230.40, Ex. No. 3	Service-Entrance Conductors for Accessary Building	*NEC Default*
230.41	Insulation of Service-Entrance Conductors	E3605.1
230.42	Minimum Size and Ampacity	E3602.1, E3602.2
230.43	Wiring Methods for Service-Entrance Conductors	E3605.2
230.46	Spliced and Tapped Conductors	E3605.3
230.50	Protection Against Physical Damage	E3605
230.50(A)	Underground Service Entrance Conductors	E3605.4
230.50(B)	All Other Service-Entrance Conductors	E3605.5
230.51	Mounting Supports	E3605.7
230.53	Raceways to Drain	E3605.8
230.54	Overhead Service Locations *(Service-Entrance Conductors)*	E3605.9
230.54(A)	Service Head / Rain-Tight Service Head	E3605.9.1
230.54(B)	Service-Entrance Cable Equipped with Service Head or Gooseneck Service Cable, Service Head or Gooseneck	E3605.9.2
230.54(E)	Service Heads and Goosenecks Above Service Drop or Overhead Service Attachment / Service Head Location	E3605.9.3
230.54(D)	Secured	E3605.9.7
230.54(E)	Separately Bushed Openings	E3605.9.4

NEC Reference	*Description in Both Model Code Documents NEC / IRC*	IRC Reference
230.54(F)	Overhead Service Locations *(Service-Entrance Conductors) (cont.)* Drip Loops	E3605.9.5
230.54(G)	Arranged that Water Will Not Enter Service Raceway or Equipment / Conductor Arrangement	E3605.9.6
Article 230 – Part V	Service Equipment - General	E3606
230.62	Service Equipment – Enclosed or Guarded / Service Equipment Enclosures	E3606.1
230.62(C)	Barriers *(in Service Equipment)*	*NEC Default*
230.66, 230.70(B)	Marking of Service Equipment and Disconnects	E3601.6.1, E3606.4
230.67(A) through (E)	Surge Protection *(Service Equipment) (Location, Type, Rating, Etc.)*	E3606.5
Article 230 – Part VI	Service Equipment - Disconnecting Means	E3601
230.70	General / Service Disconnect Required	E3601.6
230.70(A)(1)	Readily Accessible Location / Service Disconnect Location	E3601.6.2
230.70(A)(2)	Bathrooms *(Not to be Installed In)*	E3601.6.2
230.70(B), 230.66	Marking of Service Equipment and Disconnects	E3601.6.1, E3606.4
230.71(A)	Max. Number of Service Disconnects - General	*NEC Default*
230.71(B)	Two to Six Service Disconnecting Means / Maximum Number of Disconnects	E3601.7
230.71(B)(1)	Separate enclosures *(Must be Located in…)*	E3601.7.1
230.72	Grouping of Disconnects	E3601.7
230.79	Rating of Service Disconnecting Means	E3602.1, E3602.3
230.79(A)	One-Circuit Installation *(Not less than 15 Amperes)*	E3603.2, Ex. 1
230.79(B)	Two-Circuit Installation *(Not less than 30 Amperes)*	E3603.2, Ex. 2
230.79(C)	One-Family Dwellings / Ampacity of Ungrounded Conductors *(Not less than 100 Amperes, 3-Wire)*	E3602.1
230.79(D)	All Others / Ampacity of Ungrounded Conductors *(Not less than 60 Amperes)*	E3602.1
230.82	Equipment Permitted to be Connected to the Supply Side of Service Disconnecting Means	*NEC Default*
230.85(A) through (E)	Emergency Disconnects at Dwelling Units *(Location, Rating, Markings, Etc.)*	E3601.8
Article 230 – Part VII	Service Equipment - Overcurrent Protection	E3603
230.90	Where Required / Overload Protection	E3603.3
230.90(A)	Ungrounded Conductors	E3603.3.1
230.90(B)	Not in Grounded Conductor	E3603.3.2
230.91	Location *(Overcurrent Protection)*	E3603.3.3
ARTICLE 240 - Overcurrent Protection		
240.4	Protection of Conductors / Overcurrent Protection Required	E3705.5
240.4(B)	Devices Rated 800 Amperes or Less / Overcurrent Devices of the Next Higher Rating	E3705.5.2
240.4(D)	Small Conductors	E3705.5.3, Table E3705.5.3
240.4(G)	Overcurrent Protection for Specific Conductor Applications / Air Conditioning and Heat Pump Equipment	E3705.5.4
240.4(H), 310.12	Dwelling Unit Service and Feeder Conductors	*NEC Default*
240.5	Protection of Flexible Cords, Flexible Cables, and Fixture Wires / Cords	E3705.5.1
240.5(B)(1)	Supply Cord of Listed Appliance or Luminaire *(Overcurrent Protection)* / Loading and Protection (Flexible Cords)	E3909.2
240.6	Standard Ampere Ratings / Fuses and Fixed-Trip Circuit Breakers	E3705.6
Table 240.6(A)	Standard Ampere Ratings for Fuses and Inverse Time Circuit Breakers	Table E3705.6
240.7	Listing Requirements	*NEC Default*
240.15(A)	Overcurrent Device Required / Overcurrent Protection Required	E3705.5
240.16	Interrupting Rating	*NEC Default*
Article 240 – Part II	Location *(Overcurrent Protection Device)*	E3705
240.21	Location in Circuit / Overcurrent Protection Required	E3705.5

NEC Reference	*Description in Both Model Code Documents NEC / IRC*	IRC Reference
240.21(B)	Location *(Overcurrent Protection Device) (cont.)* Feeder Taps Rules *(Location of Overcurrent Protection)*	*NEC Default*
240.24	Location of Overcurrent Devices in or on Premises	E3705.7
240.24(A)	Accessibility *[Readily Accessible and not More than 2.0 m (6 ft 7 in.) Above the Floor or Working Platform]*	E3705.7(1) and (6)
240.24(B)	Occupancy *(Ready Access to Each Occupant)*	E3705.8
240.24(C)	Not Exposed to Physical Damage	E3705.7(2)
240.24(D)	Not in Vicinity of Easily Ignitible Material *(Clothes Closets)*	E3705.7(3)
240.24(E)	Not Located in Bathrooms	E3705.7(4)
240.24(F)	Not Located Over Steps *(Stairways)*	E3705.7(5)
240.81	Indicating *(Circuit Breakers) ("Up" Position of Handle to be "On" Position)*	E4001.3
ARTICLE 250 – Grounding and Bonding		
Article 100	Effective Ground-Fault Current Path *(Definition)*	Chapter 35
250.4(A)(5)	Effective Ground-Fault Current Path *(Description)*	E3908.4
250.4(A)(5)	Earth as a Ground-Fault Current Path *(Not Permitted)*	E3908.5
250.8	Connection of Grounding and Bonding Equipment *(to EGCs, etc.)*	E3406.14
250.8(A)	Permitted Methods	E3406.14.1
250.8(B)	Methods Not Permitted *(Fittings Dependent Solely on Solder)*	E3406.14.2
250.10	Protection of Ground Clamps and Fittings	E3611.6
250.12	Clean Surfaces *(For Equipment to be Grounded or Bonded)*	E3908.18, E3611.7
250.20(B)(1)	AC Systems of 50 Volts to 1000 Volts *(To be Grounded)*	E3607.1
250.24(A)	System Grounding Connections / System Service Ground *(GEC Connected to the Grounded Service Conductor at each Service)*	E3607.1
250.24(A)(1)	General *(GEC Connections)* / Location of Grounding Electrode Conductor Connection *(GEC Connection to be Made at Service)*	E3607.2
250.24(B)	Load-Side Grounding Connections / Load-Side Grounded Conductor Neutral *(No Grounded Conductor Connections on Load Side of Service)*	E3607.2, E3908.7
250.24(C)	Main Bonding Jumper *(Connection at Service)*	E3607.5
250.24(D)	Grounded Conductor Brought to Service Equipment	*NEC Default*
250.24(E)	Grounding Electrode Conductor *(Connection at Service)*	E3607.4
250.28(A) through (D)	Main Bonding Jumper *(Sizing and Installation Requirements)*	E3609.5
250.32	Buildings or Structures Supplied by Feeder(s) or Branch Circuit(s)	E3607.3
250.32(A)	Grounding Electrode *(Required to be Installed at 2nd Building)*	E3607.3
250.32(A), Exception	Exception *(Single Branch Circuit)*	E3607.3 Exception
250.32(B)(1)	Supplied by a Feeder or Branch Circuit / Equipment Grounding Conductor *(EGC Required)*	E3607.3.1
250.32(B)(1) Ex. No. 1	Exception: Grounded Conductor, Existing Premises *[Use of Grounded Conductor for Grounding (Existing Installations Only)]*	E3607.3.2
250.50	Grounding Electrode System	E3608.1
250.52	Grounding Electrodes	E3608
250.52(A)	Electrodes Permitted for Grounding	E3608.1
250.52(A)(1)	Metal Underground Water Pipe	E3608.1.1
250.52(A)(2)	Metal In-Ground Support Structure(s)	*NEC Default*
250.52(A)(3)	Concrete-Encased Electrode	E3608.1.2
250.52(A)(4)	Ground Rings	E3608.1.3
250.52(A)(5)	Rod and Pipe Electrodes	E3608.1.4
250.52(A)(6)	Other Listed Electrodes / Other Electrodes	E3608.1.6
250.52(A)(7)	Plate Electrodes	E3608.1.5
250.52(A)(8)	Other Local Metal Underground Systems or Structures	*NEC Default*
250.52(B)	Not Permitted for Use as Grounding Electrodes	E3608
250.52(B)(1)	Metal Underground Gas Piping System	E3608.6
250.52(B)(2)	Aluminum Electrodes	E3608.5
250.52(B)(3)	Structural Reinforcing Steel - Swimming Pools	E3608.7
250.53	Grounding Electrode System Installation *(Requirements)*	E3608

NEC **Reference**	***Description in Both Model Code Documents NEC / IRC***	**IRC Reference**
250.53(A)	Grounding Electrode System Installation *(Requirements) (cont.)* Rod, Pipe and Plate Electrodes	E3608.3
250.53(A)(2)	Supplemental Electrode Required *(For Rods, Pipes and Plates)*	E3608.4
250.53(A)(2) Exception	Exception *(Single Rod, Pipe or Plate of 25 Ohms or Less)*	E3608.4 Exception
250.53(A)(4)	Rod and Pipe Electrodes / Installation *[Driven at least 2.44 m (8 ft)]*	E3608.1.4.1
250.53(B)	Electrode Spacing	E3608.4
250.53(C)	Bonding Jumper *(Connecting Grounding Electrodes)*	E3608.2
250.53(D)	Metal Underground Water Pipe / Installation	E3608.1.1.2
250.53(D)(2)	Supplemental Electrode Required *(For Water Pipe)*	E3608.1.1.2
250.58	Common Grounding Electrode	E3607.6
250.64	Grounding Electrode Conductor Installation	E3610
250.64(A)	Aluminum or Copper Clad Aluminum Conductors / Securing and Protection Against Physical Damage	E3610.2
250.64(B)	Securing and Protection Against Physical Damage	E3610.2
250.64(C)	Continuous *(One Continuous Length)*	E3610.1
250.64(E)	Raceways and Enclosures for Grounding Electrode Conductors	E3610.3
250.64(F)	Installation to Electrode(s)	E3610.1
250.66	Size of AC Grounding Electrode Conductor / Grounding Electrode Conductor Size	E3603.4
Table 250.66	Grounding Electrode Conductor for AC Systems / Service Conductor and Grounding Electrode Conductor Sizing	Table E3603.4
250.66(A)	Connections to Rod, Pipe, or Plate Electrodes *(6 AWG Copper)*	Table E3603.4, Footnote f
250.66(B)	Connections to Concrete-Encased Electrodes *(4 AWG Copper)*	Table E3603.4, Footnote f
250.68	Grounding Electrode Conductor and Bonding Jumper Connection to Grounding Electrodes	E3611
250.68(A)	Accessibility	E3611.2
250.68(B)	Effective Grounding Path	E3611.3
250.68(C)	Grounding Electrode Conductor Connections	E3611.4
250.68(C)(1)	Metallic Water Pipe and Structural Metal / Interior Metal Water Piping *[Connection Within 1.52 m (5 ft) from entrance to building]*	E3611.4
250.68(C)(3)	Rebar Concrete-Encased Electrode Extension	E3611.5
250.70	Methods of Grounding and Bonding Conductor Connection to Electrodes	E3611.1
250.86	Other Conductor Enclosures or Raceways / Metal Enclosures	E3908.1
Article 250 – Part V	Bonding	E3609
250.90	General	E3609.1
250.92(A)	Bonding of Equipment for Services	E3609.2
250.92(B)	Method of Bonding at the Service	E3609.4
250.94(A)	Bonding for Communication Systems *(Intersystem Bonding Termination)*	E3609.3.1
250.94(B)	Bonding for Communication Systems *(Other Means)*	E3609.3.2
250.96(A)	Bonding Other Enclosures	E3908.19
250.102	Grounded Conductors, Bonding Conductors and Jumpers	E3609.5
250.102(C), Table 250.102(C)(1)	Size – Supply Side Bonding Jumper / Sizing Supply-Side Bonding Jumper and Main Bonding Jumper,	E3609.5
Table 250.102(C)(1)	Size – Grounded Conductor, Main Bonding Jumper, System Bonding Jumper, and Supply-Side Bonding Jumper	*NEC Default,* Table E3603.4
250.102(D)	Size - Equipment Bonding Jumper on Load Side of Overcurrent Device	E3908.20
250.104	Bonding of Piping Systems and Exposed Structural Metal	E3609
250.104(A)	Metal Water Piping Bonding	E3609.6
250.104(B)	Bonding Other Metal Piping *(Including Gas Piping)*	E3609.7
250.104(C)	Bonding of Structural Metal	*NEC Default*
Article 250 – Part VI	Equipment Grounding and Equipment Grounding Conductors (EGC)	E3908
250.109	Metal Enclosures *(as EGCs)*	E3908.1

NEC Reference	Description in Both Model Code Documents *NEC / IRC*	IRC Reference
250.110	Equipment Grounding and EGCs *(cont.)* Equipment Fastened in Place (Fixed) or Connected by Permanent Wiring Methods	E3908.2
250.112	Specific Equipment Fastened in Place (Fixed) or Connected by Permanent Wiring Methods	E3908.3
250.114(3)	Equipment Connected by Cord-and-Plug *(Residential Occupancies)*	*NEC Default*
250.118	Types of Equipment Grounding Conductors	E3908.9
250.118(5)	Listed Flexible Metal Conduit *(as EGC)*	E3908.9.1
250.118(6)	Listed Liquidtight Flexible Metal Conduit *(as EGC)*	E3908.9.2
250.118(B)	(EGC) Uses Not Permitted / Prohibited Use	E3610.4
250.118(B)(1)	Prohibited as GEC	E3610.4
250.118(B)(2)	Metal Frame of Building Prohibited as EGC	*NEC Default*
250.119	Identification of Wire-Type Equipment Grounding Conductors	E3407.2
250.120	Equipment Grounding Conductor Installation	E3908.12
250.122	Size of Equipment Grounding Conductors	E3908.13
Table 250.122	Minimum Size EGCs for Grounding Raceways and Equipment / Equipment Grounding Conductor Sizing	Table E3908.13
Article 250 – Part VII	Methods of Equipment Grounding Conductor Connections	E3908.11
250.130	Equipment Grounding Conductor Connections	E3908.11
250.134	Equipment Fastened in Place or Connected by Permanent Wiring Methods (Fixed) – Grounding	E3908.10
250.138	Cord-and-Plug-Connected Equipment	*NEC Default*
250.140	Frames of Ranges and Clothes Dryers	*NEC Default*
250.140(B)	Use of Grounded Circuit Conductor for Grounding Existing Frames Of Appliances *(Existing Installations Only)*	*E3908.8.1*
250.142	Use of Grounded Circuit Conductor for Grounding Equipment	E3908
250.142(A)	Supply-Side Equipment / Line-Side Grounded (Neutral) Conductor	E3908.6
250.142(B)	Load-Side Equipment *(Grounded Conductor not to be Used)*	E3908.8
250.142(B), Ex. No. 1	Exception for Existing Installations of Existing Frames of Ranges, Dryers, Etc.	*NEC Default*
250.146	Connecting Receptacle Grounding Terminal to an EGC	E3908.15
250.148	Continuity of EGCs and Attachment in Boxes	E3908.14
250.148(A)	Connectors and Splices *(All Spliced EGCs in Box to be Connected Together)*	E3908.14
250.148(B)	EGC Continuity *(Not dependent on Device for Continuity of EGC)*	E3908.14
250.148(C)	Metal Boxes	E3908.16
250.148(D)	Nonmetallic Boxes	E3908.17
ARTICLE 300 – General Requirements for Wiring Methods and Materials		
Table 300.1(C)	Metric Designators and Trade Sizes	*NEC Default*
300.3(B), 300.5(I)	Conductors of the Same Circuit *(Same Raceway, Etc.)*	E3803.8
300.3(B)(1), 300.5(I), Ex. No. 1	Paralleled Installations	E3803.8 Exception
300.4	Protection Against Physical Damage	Chapter 38
300.4(A)	Cables and Raceways Through Wood Members	E3802.1, Table E3802.1
300.4(A)(1)	Bored Holes	Table E3802.1
300.4(A)(2)	Notches in Wood	Table E3802.1
300.4(B)	Nonmetallic-Sheathed Cables and Electrical Nonmetallic Tubing Through Metal Framing Members	Table E3802.1
300.4(D)	Cables and Raceways Parallel to Framing Members and Furring Strips	Table E3802.1
300.4(F)	Cable & Raceways Installed in Shallow Grooves	Table E3802.1
300.4(G)	Fittings *(Insulated and Identified Fittings)*	E3906.1.1
300.5	Underground Installations	E3803
300.5(A)	Minimum Cover Requirements	E3803.1
Table 300.5	Minimum Cover Requirements, 0 to 1000 Volts ac, 1500 Volts dc, Nominal, Burial in Millimeters (Inches)	Table E3803.1

NEC Reference	*Description in Both Model Code Documents NEC / IRC*	IRC Reference
300.5(B)	Underground Installations *(cont.)* Wet locations *(interior of enclosures or raceways installed underground to be considered a wet location)*	E3803.10
300.5(C)	Underground Cable Under Buildings *(generally required to be in a raceway)*	E3803.11
300.5(D)	Protection from Damage	E3803.3
300.5(D)(1)	Emerging from Grade	E3803.3
300.5(D)(2)	Conductors Entering Buildings *(protected to the point of entrance)*	E3803.3
300.5(D)(3)	Service Conductors / Warning Ribbon	E3803.2
300.5(E)	Splices and Taps	E3803.4
300.5(F)	Backfill	E3803.5
300.5(G)	Raceway Seals	E3803.6
300.5(H)	Bushing	E3803.7
300.5(I)	Conductors of the Same Circuit / Single Conductors	E3803.8
300.5(J)	Earth Movement	E3803.9
300.6(C)(1)	Nonmetallic Equipment Exposed to Sunlight *(Sunlight Resistant)*	*E3802.9*
300.7(A)	Raceways Exposed to Different Temperatures *(Sealing Requirements)*	E3802.7
300.9	Raceways in Wet Locations Above Grade *(interior of raceways to be considered a wet location)*	E3802.8
300.10	Electrical Continuity of Metal Raceways and Enclosures	E3904.1
300.11	Securing and Supporting	E3904.3
300.11(A)	Secured in Place / Securing and Supporting	E3904.3
300.11(C)	Raceways Used as Means of Support	E3904.4
300.11(D)	Cables Not Used as a Means of Support / Prohibited Means of Support	E3904.3.1
300.12	Mechanical Continuity—Raceways and Cables	E3904.2
300.13(A)	Mechanical and Electrical Continuity - Conductors (General) / Continuity *(Conductors)*	E3406.11.1
300.14	Length of Free Conductors at Outlets, Junction Boxes, and Switch Points / Length of Conductor for Splice or Termination *[150 mm (6 in.)]*	E3406.11.3
300.15	Boxes, Conduit Body or Fitting - Where Required	E3905.1
300.15(B)	Equipment	E3905.1.1
300.15(C)	Protection	E3905.1.2
300.15(E)	Integral enclosure	E3905.1.3
300.15(F)	Fitting	E3905.1.4
300.15(G)	Direct Buried Conductors and Cables / Buried Conductors	E3905.1.5
300.15(J)	Luminaires	E3905.1.6
300.17	Number and Size of Conductors in Raceway / Conduit and Tubing Fill	E3904.6
300.18	Raceway installations *(Complete Runs)*	E3904.5
300.21	Spread of Fire or Products of Combustion / Penetrations of Fire-Resistance-Rated Assemblies	E3402.2, E3402.3
300.22(C)	Other Spaces Used for Environmental Air (Plenums) / Air-Handling Stud Cavity and Joist Spaces	E3904.7
300.22(C), Exception	Air-Handling Stud Cavity *(permitted in spaces perpendicular to the long dimension)*	E3904.7
300.26	Remote-Control and Signaling Circuits Classification	*NEC Default*
ARTICLE 310 Conductors for General Wiring		
310.3	Conductors	E3406
310.3(A)	Minimum Size of Conductors	E3406.3
310.3(B)	Conductor Material	E3406.2
310.3(C)	Stranded Conductors	E3406.4
310.3(D)	Insulated / Individual Conductor Insulation	E3406.5
310.4	Conductor Construction and Applications	E3406.5
Table 310.4(1)	Conductor Applications and Insulations Rated 600 Volts (*conductors permitted to be rated up to 1000 volts if listed and marked*)	*NEC Default*
310.6	Conductor Identification	E3407

NEC **Reference**	***Description in Both Model Code Documents NEC / IRC***	**IRC Reference**
310.6(A), 200.6	Conductor Identification *(cont.)* Grounded Conductors	E3407.1
310.6(B), 250.119	Equipment Grounding Conductors	E3407.2.2.2
310.6(C)	Ungrounded conductors	E3407.3
310.10	Uses Permitted *(Conductors)* / Electrical Conductors and Connections	E3406
310.10(A)	Dry Locations	E3406.5
310.10(B)	Dry and Damp Locations	E3406.5
310.10(C)	Wet Locations	E3406.5
310.10(D)	Locations Exposed to Direct Sunlight	E3605.6
310.10(E)	Direct-Burial Conductors *(Identified for Such Use)*	*NEC Default*
310.10(G)	Conductors in Parallel	E3406.6
310.12	Single-Phase Dwelling Services and Feeders *(Sizing)*	E3603.1
310.12(A)	Services / Ungrounded Service Conductors	E3603.1.1
310.12(B)	Feeders / Ungrounded Feeder Conductors	E3603.1.2
310.12(C)	Feeder Ampacities / Feeder Size Relative to Service Size	E3603.1.3
310.12(D)	Grounded Conductors	E3603.1.4
Table 310.12(A)	Single-Phase Dwelling Services and Feeders	Table E3603.1.1
310.14	Ampacities for Conductors Rated 0 Volts - 2000 Volts / Conductor Sizing and Overcurrent Protection	E3705* **BC & Feeders*
310.14(A)	General	E3705.1*
310.14(A)(2)	Selection of Ampacity *(Conductors)*	E3701.3*
310.14(A)(3)	Temperature Limitation of Conductors	E3603.5
310.15	Ampacity Tables	E3705
310.15(B)(1)	General *(Ambient Temperature Correction Factors)*	E3705.2
Table 310.15(B)(1)(1)	Ambient Temperature Correction Factors Based on 30°C (86°F)	Table E3705.2
310.15(B)(2)	Rooftops *(Raceways or Cables Installed on Rooftops)*	*NEC Default*
310.15(C)(1)	Adjustment Factors *(More Than Three Current-Carrying Conductors)*	E3705.3
Table 310.15(C)(1)	Adjustment Factors for More Than Three Current-Carrying Conductors / Conductor Proximity Adjustment Factors	Table E3705.3
310.15(E)	Neutral Conductor *(Current-Carrying Conductor)*	*NEC Default*
310.15(F)	Grounding or Bonding Conductors *(Not to be Counted)*	*NEC Default*
310.16	Ampacities of Insulated Conductors in Raceway, Cable, or Earth (Directly Buried)	E3705.1
Table 310.16	Ampacities of Insulated Conductors with Not More Than Three Current-Carrying Conductors in Raceway, Cable, or Earth (Directly Buried) *[Based on Ambient Temperature of 30°C (86°F)]*	Table E3705.1
ARTICLE 312 – Cabinets, Cutout Boxes, and Meter Socket Enclosures		
312.2	Damp and Wet Locations *[6 mm (¼ in.) airspace between the enclosure and the wall]*	E3907.2
312.3	Position in Wall	E3907.3
312.4	Repairing Noncombustible Surfaces	E3907.4
312.5(A)	Openings to be Closed *(Conductor Entry)*	E3907.7
312.5(C)	Cables *(Secured to Cabinet or Enclosure)*	E3907.8
312.6(A)	Deflection of Conductors – Width of Wiring Gutter	E3907.9
Table 312.6(A)	Minimum Wire-Bending Space at Terminals (Wiring Gutter)	Table E3907.9.1(2)
312.6(B)	Deflection of Conductors - Wire-Bending Space at Terminals	E3907.9
Table 312.6(B)(2)	Minimum Wire-Bending Space at Terminals	Table E3907.9.1(1)
312.8(A)	Switch and Overcurrent Device Enclosures with Splices, Taps, and Feed-Through Conductors	E3907.1.1
312.8(B)	Power Monitoring Equipment or Energy Management Equipment	E3907.1.2
312.10	Screws or Other Fasteners *(Entering Wiring Spaces in Boxes)*	E3906.8.2.1
ARTICLE 314 – Outlet, Device, Pull, and Junction Boxes; Conduit Bodies; Fittings; and Handhole Enclosures		
314.2	Round Boxes	*NEC Default*
314.3	Nonmetallic Boxes	E3905.3

NEC Reference	*Description in Both Model Code Documents NEC / IRC*	IRC Reference
314.4	Metal Boxes	E3905.2
Article 314 – Part II	Installation *(Outlet Boxes, Etc.)*	E3905
314.15	Damp or Wet Locations	E3905.11
314.16	Number of Conductors in Outlet, Device, Junction Boxes, Conduit Bodies	E3905.12
314.16(A)	Box Volume Calculations	E3905.12.1
314.16(A)	Barriers *(Installed in Boxes)*	E3905.12.1
314.16(A)(1)	Standard Boxes *(Metal Boxes)*	E3905.12.1.1
314.16(A)(2)	Other Boxes *(Nonmetallic Boxes)*	E3905.12.1.2
Table 314.16(A)	Metal Boxes / Maximum Number of Conductors in Metal Boxes	Table E3905.12.1
314.16(B)	Box Fill Calculations	E3905.12.2
314.16(B)	Barriers *(Each Space in Box with Barriers Calculated Separately)*	E3905.12.2
Table 314.16(B)(1)	Volume Allowance Required per Conductor	Table E3905.12.2.1
314.16(B)(1)	Conductor Fill	E3905.12.2.1
314.16(B)(2)	Clamp Fill	E3905.12.2.2
314.16(B)(3)	Support Fittings Fill	E3905.12.2.3
314.16(B)(4)	Device or Equipment Fill	E3905.12.2.4
314.16(B)(5)	Equipment Grounding Conductor Fill	E3905.12.2.5
314.16(B)(6)	Terminal Block Fill	*NEC Default*
314.16(C)	Conduit Bodies	E3905.12.3
314.16(C)(1)	General	E3905.12.3
314.16(C)(2)	With Splices, Taps or Devices	E3905.12.3.1
314.16(C)(3)	Short Radius Conduit Bodies *(Not to Contain Splices, Taps, Etc.)*	E3905.12.3.2
314.17	Conductors Entering Boxes, Conduit Bodies or Fittings	E3906.1
314.17(A)	Opening to be Closed / Openings	E3906.2
314.17(B)	Boxes and Conduit Bodies / Metal Boxes and Conduit Bodies	NEC Default
314.17(B)(1)	Conductors Entering Through Individual Holes or Through Flexible Tubing *(Securing to Box)*	*NEC Default*
314.17(B)(2)	Conductors Entering Through Cable Clamps / Nonmetallic-Sheathed Cable Entering Boxes *(Securing to Box)*	E3905.3.1
314.17(B)(2), Exception	Exception *(Single-Gang Nonmetallic Box and Type NM Cable)*	E3905.3.1, Ex.
314.19	Boxes Enclosing Flush Devices or Flush Equipment	E3905.5
314.20	Flush-Mounted Installations *[6 mm (¼ in.) set back noncombustible material]*	E3906.5
314.21	Repairing Noncombustible Surfaces *[Min. 3 mm (⅛ in.) Gap]*	E3906.6
314.22	Surface Extensions	E3906.7
314.23	Supports *(For Boxes)*	E3906.8
314.23(A)	Surface Mounting	E3906.8.1
314.23(B)	Structural Mounting	E3906.8.2
314.23(B)(1)	Nails and Screws	E3906.8.2.1
314.23(B)(2)	Braces	E3906.8.2.2
314.23(C)	Mounting in Finished Surfaces	E3906.8.3
314.23(E)	Raceway Supported Enclosures Without Devices, Luminaires or Lampholders	E3906.8.4
314.23(F)	Raceway Supported Enclosures with Devices, Luminaires or Lampholders	E3906.8.5
314.23(G)	Enclosures in Concrete or Masonry	E3906.8.6
314.24	Dimensions of Boxes / Minimum Depth of Boxes for Outlets, Devices, and Utilization Equipment	E3905.4
314.25	Covers and Canopies	E3906.9
314.25(A)	Nonmetallic or Metal Covers and Plates / Covers and Plates	E3906.10
314.25(B)	Exposed Combustible Wall or Ceiling Finish	E3906.11
314.27	Outlet Boxes	E3905
314.27(A)	Boxes at Luminaire or Lampholder Outlets	E3905.6
314.27(A)(1)	Vertical Surface Outlets *(Max. Weight)*	E3905.6.1

NEC Reference	*Description in Both Model Code Documents NEC / IRC*	IRC Reference
314.27(A)(2)	Outlet Boxes at Luminaire or Lampholder Outlets *(cont.)* Ceiling Outlets *(Max. Weight)*	E3905.6.2
314.27(B)	Floor Boxes	E3905.7
314.27(C)	Boxes at Ceiling-Suspended (Paddle) Fan Outlets / Boxes at Fan Outlets *(Outlet Boxes Mounted in Ceilings of Habitable Rooms)*	E3905.8
314.27(D)	Utilization Equipment	E3905.9
314.27(E)	Weight-Supporting Ceiling Receptacles (WSCR) and Weight-Supporting Attachment Fittings (WSAF)	E3905.6.3
314.28	Pull and Junction Boxes and Conduit Bodies *(Minimum Size)*	*NEC Default*
314.28(A)(1)	Straight Pulls *(Eight times the trade size of the largest raceway)*	*NEC Default*
314.28(A)(2)	Angle or U Pulls, or Splices *(Six times the trade size of the largest raceway)*	*NEC Default*
314.29	Boxes, Conduit Bodies, and Handhole Enclosures to be Accessible / Conduit Bodies and Junction, Pull and Outlet Boxes to be Accessible	E3905.10
Wiring Method Articles ***(Article 320 – Armored Cable: Type AC, Article 330 – Metal-Clad Cable: Type MC, Article 334 - Nonmetallic Sheathed Cable: Type NM, NMC, and NMS, Article 348 - Flexible Metal Conduit: Type FMC, Etc.)***		
.6 of Particular Article	Listing Requirements	*NEC Default*
.10 of Particular Article	Uses Permitted	Table E3801.4
.12 of Particular Article	Uses Not Permitted	Table E3801.4
.15 of Particular Article	Exposed Work / Exposed Cable	E3802.3
.17 of Particular Article	Cable Installed Through or Parallel to Framing Members	E3802.2.2, Table E3802.1
.23 of Particular Article	Cables in Accessible Attics	E3802.2
.24 of Particular Article	Bending Radius / Bends	E3802.5
.30 of Particular Article	Securing and Supporting	E3904.3
.40 of Particular Article	Boxes and Fittings	*NEC Default*
.80 of Particular Article	Ampacity *(IRC - Type NM Cable Only)*	E3705.4.4
320.30(A), 330.30(A), 334.30, 338.10(B), 340.10(4)	Cable Securing Means	E3802.6
Article 334 - Nonmetallic Sheathed Cable: Type NM, NMC, and NMS		
334.6	Listing Requirements	*NEC Default*
334.10	Uses Permitted	Table E3801.4
334.12	Uses Not Permitted	Table E3801.4
334.15	Exposed Work / Exposed Cable	E3802.3
334.15(A)	To Follow Surface / Surface Installation	E3802.3.1
334.15(B)	Protection from Physical Damage	E3802.3.2
334.15(C)	In Unfinished Basements and Crawl Spaces	E3802.4
334.17	Through or Parallel to Framing Members	E3802.2.2
334.19	Cable Entries *[Sheath of cable to extend not less than 6 mm (¼ in.) beyond any cable clamp or cable entry]*	*NEC Default*
334.23	In Accessible Attics	E3802.2
334.24	Bending Radius / Bends *(not less than five times the diameter of the cable)*	E3802.5
334.30	Securing and Supporting	E3904.3
334.40	Boxes and Fittings	*NEC Default*
334.80	Ampacity / Conductors of Type NM Cable *(Installed in Insulation, etc.)*	E3705.4.4
334.108	Equipment Grounding Conductor *(Required to Include an EGC)*	E3908.9.3
ARTICLE 400 - Flexible Cords and Cables ****(IRC E3909 only references flexible cords for appliances)***		
400.4, Table 400.4	Types *(Flexible Cords and Cables)*	*NEC Default*
Table 400.4	Flexible Cords and Cables	*NEC Default*
400.5	Ampacities for Flexible Cords and Cables	E3909.2
Table 400.5(A)(1)	Ampacity for Flexible Cords and Flexible Cables / Maximum Ampere Load for Flexible Cords	Table E3909.2
400.10	Uses Permitted	E3909.1
400.10(A)	Uses / Where Permitted	E3909.1

NEC Reference	*Description in Both Model Code Documents NEC / IRC*	IRC Reference
400.10(B)	Flexible Cords and Cables *(Uses Permitted) (cont.)* Attachment Plugs	E3909.4
400.12	Uses Not Permitted / Where Permitted	E3909.1
400.13	Splices	E3909.3
400.17	Protection from Damage	*NEC Default*
ARTICLE 404 – Switches		
404.1	Scope *(Wireless switches and control equipment and not covered)*	*NEC Default*
404.2	Switch Connections	E4001.9
404.2(A)	Three-Way and Four-Way Switches	E4001.9
404.2(B)	Grounded Conductors	E4001.8
404.2(C)	Switches Controlling Lighting Loads *(Grounded Conductor at Switch)*	E4001.15
404.4	Damp or Wet Locations	E4001.7
404.4(C)	Switches in Tub or Shower Spaces *(Generally not Permitted)*	E4001.7
404.5	Time Switches and Similar Devices	E4001.4
404.7	Indicating *[Up Position Shall be the (On) Position]*	E4001.3
404.8	Accessibility and Grouping	E4001.6
404.8(A)	Location / Access	E4001.6
404.8(B)	Voltage Between Adjacent Devices *(Shall not Exceed 300 Volts Between Adjacent Devices)*	*NEC Default*
404.8(C)	Multipole Snap Switches	E4001.13
404.9	Provisions for General-Use Snap Switches, Dimmers, and Control Switches	E4001.11
404.9(A)	Faceplates / Snap Switch Faceplates	E4001.11
404.9(B)	Grounding / Faceplate Grounding	E4001.11.1
404.9(C)	Construction	*NEC Default*
404.10	Mounting of Snap Switches, Dimmers, and Control Switches	E4001.10
404.10(B)	Box Mounted	E4001.10
404.11	Circuit Breakers as Switches	*NEC Default*
404.12	Grounding of Enclosures	E4001.5
404.14	Rating and Use of Snap Switches / Rating and Application of Snap Switches	E4001.1
404.14(A)	Alternating-Current General-Use Snap Switch	E4001.1
404.14(C)	CO/ALR Snap Switches	E4001.2.1
404.14(D)	Snap Switch Terminations *(Push-in terminals to be used with not greater than 15 amperes branch circuits connected to 14 AWG solid copper conductors)*	*E4001.2*
404.14(F)	Dimmer and Electronic Control Switches	E4001.12
404.14(G)	Cord-and-Plug-Connected Loads	E4001.14
ARTICLE 406 – Receptacles, Cord Connectors, and Attachment Plugs (Caps)		
406.3	Receptacle Rating and Type	E4002
406.3(A)	Receptacles *(Listed and Marked with Voltage and Ampere Rating)*	*NEC Default*
406.3(B)	Rating / Rating and Type	E4002.1
406.3(C)	CO/ALR Receptacles	E4002.3.1
406.3(D)	Receptacle Terminations	*4002.3*
406.3(F)	Controlled Receptacle Marking	*NEC Default*
406.3(G)	Receptacle with USB Charger	*NEC Default*
406.4	General Installation Requirements	E4002
406.4(A)	Grounding Type	E4002.2
406.4(B)	Connection to Equipment Grounding Conductor *(EGC)*	*NEC Default*
406.4(C)	Methods of Connection to Equipment Grounding Conductor *(EGC)*	*NEC Default*
406.4(D)	Replacements	*NEC Default*
406.4(D)(1)	Grounding-Type Receptacles	*NEC Default*
406.4(D)(2)	Non-Grounding-Type Receptacles	*NEC Default*
406.4(D)(3)	GFCI Protection *(GFCI for Replacement Receptacles)*	*NEC Default*
406.4(D)(4)	AFCI Protection *(AFCI for Replacement Receptacles)*	*NEC Default*

NEC Reference	*Description in Both Model Code Documents NEC / IRC*	IRC Reference
406.4(D)(5)	General Installation Requirements *(Replacement Receptacles) (cont.)* Tamper-Resistant Receptacles *(TR for Replacement Receptacles)*	*NEC Default*
406.4(D)(6)	Weather-Resistant Receptacles *(WR for Replacement Receptacles)*	*NEC Default*
406.4(D)(7)	Controlled Receptacles *(Controlled Receptacles for Replacement Receptacles)*	*NEC Default*
406.4(E)	Cord- and Plug-Connected Equipment	*NEC Default*
406.4(G)	Protection of Floor Receptacles	*NEC Default*
406.5	Receptacle Mounting	E4002
406.5(A)	Boxes That Are Set Back / Receptacles Mounted in Boxes	E4002.6
406.5(B)	Boxes That Are Flush / Receptacles Mounted in Boxes	E4002.6
406.5(C)	Receptacles Mounted on Covers	E4002.7
406.5(D)	Position of Receptacle Faces	E4002.5
406.5(E)	Receptacles in Countertops	E3901.6, E3902.7, E4002.15
406.5(F)	Receptacles in Work Surfaces	E4002.15
406.5(G)	Receptacle Orientation	E4002
406.5(G)(1)	Countertops and Work Surfaces / Receptacle Position *(Not to be Installed in a Face-Up Position)*	E4002.16
406.5(G)(2)	Under Sinks *(Not to be Installed in a Face-Up Position)*	*NEC Default*
406.5(H)	Receptacles in Seating Areas and Other Similar Surfaces	*NEC Default*
406.5(I)	Exposed Terminals	E4002.13
406.5(J)	Voltage Between Adjacent Devices *(Shall not Exceed 300 Volts Between Adjacent Devices)*	*NEC Default*
406.6	Receptacle Faceplates (Cover Plates)	E4002.5
406.6(B)	Grounding / Faceplates *(Metal Faceplates)*	E4002.4
406.6(D)	Receptacle Faceplate *(Cover Plates)* with Integral Night Light and/or USB Charger	*NEC Default*
406.9	Receptacles in Damp or Wet Locations	E4002
406.9(A)	Damp locations *(Listed Weather-Resistant Type Receptacles)*	E4002.8
406.9(B)	Wet Locations	E4002.9, E4002.10
406.9(B)(1)	15- and 20-Ampere Receptacles in a Wet Location *(Listed Weather-Resistant Type Receptacles)*	E4002.9
406.9(B)(2)	Other Receptacles / Other Receptacles in Wet Locations	E4002.10
406.9(C)	Bathtub and Shower Space *(Receptacles not to be Installed Within Tub or Shower Zone)*	E4002.11
406.9(C) Ex. No. 1	Receptacle for Hydro-Massage Tub Permitted *(NEC 680.73)*	*4002.11 Ex1*
406.9(C) Ex. No. 2	Receptacle in Bathroom with less than the Required Zone Permitted	E4002.11, Ex.2
406.9(C) Ex. No. 3	Weight-Supporting Ceiling Receptacles (WSCR) in Damp Locations Permitted *[NEC 410.10(D)]*	*4002.11 Ex 4*
406.9(C) Ex. No. 4	Single Receptacle for Personal Hygiene Device Permitted	*NEC Default*
406.9(D)	Flush Mounting with Faceplate	E4002.12
406.12	Tamper-Resistant Receptacles *(All Receptacles in Dwelling Units)*	E4002.14
406.12 Exception	Receptacles Located More than 1.7 m (5 ft) Above the Floor, Receptacle Located in Appliance Space, Etc.	E4002.14, Ex.
ARTICLE 408 – Switchboards, Switchgear, and Panelboards		
408.3	Support and Arrangement of Busbars and Conductors	*NEC Default*
408.4	Descriptions Required	E3706.2
408.4(A)	Circuit Directory or Circuit Descriptions / Panelboard Circuit Identification	E3706.2
408.6	Short-Circuit Current Rating *(Not less than Available Fault Current)*	*NEC Default*
408.7	Unused Openings *(For Circuit Breakers and Switches)*	E3907.5
Article 408 – Part III	Panelboards	E3706
408.30	General / Panelboard Rating	E3706.1

NEC Reference	*Description in Both Model Code Documents NEC / IRC*	IRC Reference
	Panelboards *(cont.)*	
408.36	Overcurrent Protection / Panelboard Overcurrent Protection	E3706.3
408.36(D)	Back-Fed Devices	E3706.5
408.37	Panelboards in Damp or Wet Locations	E3907.2
408.38	Enclosure *(Panelboards to be Mounted in Cabinets, etc. and shall be Dead-Front)*	*NEC Default*
408.40	Grounding of Panelboards	E4001.5
408.41	Grounded Conductor Terminations *(Generally One Grounded Conductor per Terminal)*	E3706.4
408.43	Panelboard Orientation *(Not to be Installed in the Face-Up or Face-Down Position)*	*NEC Default*
408.54	Maximum Number of Overcurrent Devices	*NEC Default*
408.55, Table 312.6(A), Table 312.6(B)(2)	Wire-Bending Space Within an Enclosure Containing a Panelboard	E3907.9, Table E3907.9.1(1), Table E3907.9.1(2)
408.55(A)	Top and Bottom Wire-Bending Space	E3907.9.1
408.55(B)	Side Wire-Bending Space	E3907.9.2
408.55(C)	Back Wire-Bending Space	E3907.9.3
ARTICLE 410 – Luminaires, Lampholders, and Lamps		
410.5	Live Parts / Energized Parts	E4003.1
410.6	Listing Required *(Luminaires to be Listed)*	E3403.3
410.8	Inspection / Access	E4004.3
410.10	Luminaires in Specific Locations	E4003
410.10(A)	Wet and Damp Locations	E4003.9
410.10(B)	Corrosive Locations	*NEC Default*
410.10(D)	Bathtub and Shower Areas *[900 mm (3 ft) Horizontal by 2.5 m (8 ft) Vertical Zone]*	E4003.11
410.11	Luminaires Near Combustible Material	E4003.2
410.12	Luminaires Over Combustible Material	*NEC* Default
410.16	Luminaires in Clothes Closets / Luminaires in Clothes Closet Storage Space	E4003.12
410.16(A)	Clothes Closet Storage Space / Luminaires in Clothes Closet Storage Space / Luminaires in Clothes Closet Storage Space	E4003.12
Article 100	Definition: Clothes Closet Storage Space / Clothes Closet	E4003.12
Figure 410.16(A)	Clothes Closet Storage Space / Closet Storage Space	Figure E4003.12
410.16(B)	Luminaire Types Permitted *(in Clothes Closets)*	E4003.12
410.16(C)	Luminaire Types Not Permitted *(in Clothes Closets)*	E4003.12
410.16(D)	Location *(in Clothes Closets)*	E4003.12
410.18	Space for Cove Lighting	*NEC* Default
410.22	Outlet Boxes to Be Covered / Outlet Box Covers	E4004.1
410.23	Covering of Combustible Material at Outlet Boxes	E4004.2
410.24	Connection of Electric-Discharge and LED Luminaires	*NEC Default*
410.24(A)	Independent of the Outlet Box	*NEC Default*
410.24(B)	Access to Boxes *(Suitable Openings in the Back of the Luminaire)*	*NEC Default*
410.30	Supports	E4004.4
410.30(A)	General	E4004.4
410.36	Means of Support	E4004.5
410.36(A)	Luminaires Supported by Outlet Boxes	E4004.5
410.36(B)	Suspended Ceilings	*NEC Default*
410.36(G)	Trees	*NEC Default*
Article 410 – Part V	Grounding of Luminaires	E3908, E4003
410.40	Equipment Grounding Conductor *(Connection to)*	E3908.11
410.42	Luminaire(s) with Exposed Conductive Surfaces	E4003.3
410.44	Connection to the Equipment Grounding Conductor	E3908.11

A)	Properly Secured	*NEC Default*
B)	Protection Through Metal	*NEC Default*
C)	Luminaire Stems	*NEC Default*
D)	Splices and Taps	*NEC Default*
E)	Stranding *(Use of Stranded Conductors)*	*NEC Default*
F)	Tension *(Luminaire Weight not to put Tension on Conductors)*	*NEC Default*
A) through (C)	Cord-Connected Lampholders and Luminaires	*NEC Default*
A) through (C)	Luminaires as Raceways	E4003.13.2
	Screw-Shell Type *(For Lampholders Only)*	E4003.4
410 – Part X	Special Provisions for Flush and Recessed Luminaires	E4003
5(B)	Recessed Incandescent Luminaires	E4003.5
5	Clearance and Installation	E4004
5(A)	Clearance from Combustible Material	E4004.8
5(A)(1)	Non-Type IC *(Not Identified for Contact with Insulation)*	E4004.8
5(A)(2)	Type-IC *(Identified for Contact with Insulation)*	E4004.8
5(B)	Clearance from Thermal Installation / Recessed Luminaire Installation *(Thermal Insulation)*	E4004.9
2	Lamp Wattage Marking *(Incandescent Lamp Luminaires)*	E3404.11
410 – Part XII	Special Provisions for Electric-Discharge Lighting Systems of 1000 Volts or Less	E4003, E4004
)(E)(1)	Integral Thermal Protection *(Fluorescent Luminaires)*	E4003.6
)(F)	High-Intensity Discharge Luminaires	E4003.7
)(F)(5)	Metal Halide Lamp Containment	E4003.8
5	Luminaire Mounting	E4004
5(A)	Exposed Components	E4004.6
5(B)	Combustible Low-Density Cellulose Fiberboard	E4004.7
410 – Part XIV	Lighting Track	E4005
)	Installation / Fittings	E4005.1, E4005.2
)(A)	Lighting Track	E4005.1
)(B)	Connected Load	E4005.3
)(C)	Locations Not Permitted / Prohibited Locations	E4005.4
)(D)	Support / Fittings	E4005.2
4	Fastening	E4005.5
5(B)	Equipment Grounding Conductor *(Connection to EGC Required)*	E4005.6
)	Listing of Decorative Lighting *(Holiday Lighting to be Listed)*	*NEC Default*
LE 411 – Low Voltage Lighting		
	Scope *(Includes Class 2 Power Source)*	*NEC Default*
A)	Listed Systems	*NEC Default*
	Voltage Limitations *(not exceed 30 volts ac or 60 volts dc) (Wet Locations: not exceed 15 volts ac or 30 volts dc)*	*NEC Default*
	Low-Voltage Lighting Systems	*NEC Default*
	Specific Location Requirements	*NEC Default*
A)	Walls, Floors, and Ceilings	*NEC Default*
B)	Pools, Spas, Fountains, and Similar Locations	*NEC Default*
	Secondary Circuits	*NEC Default*
A)	Grounding *(Shall not be Grounded)*	*NEC Default*
A) Exception	Secondary Circuits of a Class 2 Power Source *(Listed for Grounding)* Permited to be Grounded	*NEC Default*
B)	Isolation	*NEC Default*

NEC Reference	*Description in Both Model Code Documents NEC / IRC*	IRC Reference
411.7(C)	Low-Voltage Lighting System Secondary Circuits *(cont.)* Bare Conductors	*NEC Default*
411.8	Branch Circuit *(Max. 20-Ampere Branch Circuit)*	*NEC Default*
ARTICLE 422 - Appliances		
422.5	GFCI Protection for Appliances	*NEC Default*
422.5(A)	General *(Class A GFCI Protection Required)*	*NEC Default*
422.5(A)(6), 210.8(D)(6)	Sump Pumps	*NEC Default*
422.5(A)(7), 210.8(D)(7)	Dishwashers / Kitchen Dishwasher Branch Circuit *(See NEC 210.8(D) for Other Specific Appliances Requiring GFCI Protection)*	E3902.12
422.5(B)	Type and Location *(GFCI Protection) (Readily Accessible, Etc.)*	*NEC Default*
422.6	Listing Required	E3403.3
422.10	Branch Circuits	E3702.2
422.10(A)	Individual Branch Circuits *(Supplying Appliances)*	*NEC Default*
422.10(B)	Branch Circuits Supplying Two or More Loads	*NEC Default*
422.11	Overcurrent Protection	E4101.4
422.11(A)	Branch-Circuit Overcurrent Protection	E3702.2
422.11(B)	Household-Type Appliances with Surface Heating Elements	*NEC Default*
422.11(E)	Single Non-Motor-Operated Appliance	E4101.4.1
422.11(G)	Motor-Operated Appliances	*NEC Default*
422.12	Central Heating Equipment / Branch Circuits for Central Heating	E3703.1
422.13	Storage-Type Water Heaters / Branch Circuits Serving Heating Loads	E3702.10
422.16	Flexible Cords *(Appliances)*	E4101.3, Table E4101.3
422.16(A)	General / Flexible Cords	
422.16(B)	Specific Appliances	
422.16(B)(1)	Electrically Operated Kitchen In-Sink Waste Disposers / Disposal	
422.16(B)(2)	Built-in Dishwashers and Trash Compactors	
422.16(B)(2)(5)	Receptacle for a Built-In Dishwasher to be Located in the Space Adjacent to the Space Occupied by the Dishwasher	
422.16(B)(3)	Wall-Mounted Ovens and Counter-Mounted Cooking Units	
422.16(B)(4)	Range Hoods and Microwave Oven/Range Hood Combinations *(Individual Branch Circuit Required)*	
422.18	Ceiling-Suspended (*Paddle*) Fans	E4101.6
422.18(A)	Support	E4101.6
422.18(B)	Location	E4003.11
422.19	Space for Conductors *(Canopies)*	*NEC Default*
422.20	Outlet Boxes to be Covered	*NEC Default*
422.21	Covering of Combustible Material at Outlet Boxes	*NEC Default*
422.22	Utilizing Separable Attachment Fittings	*NEC Default*
Article 422 – Part III	Disconnecting Means *(Appliances)*	E4101.5
422.30	General	E4101.5
422.31	Disconnection of Permanently Connected Appliances	Table E4101.5
422.31(A)	Rated at Not over 300 Volt-Amperes or ⅛ Horsepower	Table E4101.5
422.31(B)	Appliances Rated over 300 Volt-Amperes	Table E4101.5
422.31(C)	Motor-Operated Appliances Rated over ⅛ Horsepower	Table E4101.5
422.33	Disconnection of Cord-and-Plug-Connected or Attachment Fitting-Connected Appliances	Table E4101.5
422.33(A)	Separable Connector or an Attachment Plug (or Attachment Fitting) and Receptacle	*NEC Default*
422.33(B)	Connection at the Rear Base of a Range	*NEC Default*
422.34	Unit Switch(es) as Disconnecting Means	Table E4101.5
422.35	Switch and Circuit Breaker to be Indicating	E4101.5
422.60(A)	Nameplate Marking	*NEC Default*
422.62	Appliances Consisting of Motors and Other Loads	*NEC Default*

NEC Reference	*Description in Both Model Code Documents NEC / IRC*	IRC Reference
ARTICLE 424 – Fixed Electric Space-Heating Equipment		
424.4	Branch Circuits	E3703.1
424.4(A)	Branch-Circuit Requirements	E3703.1
424.4(B)	Branch Circuit Conductor Sizing *(Continuous Load)*	E3702.10
424.6	Listed Equipment	E3403.3
424.10	Installation-General *(Factory-Installed Receptacle Outlets)*	E3901.1
424.12	Locations	*NEC Default*
424.12(A)	Exposed to Physical Damage	*NEC Default*
424.12(B)	Damp or Wet Locations	*NEC Default*
424.13	Spacing from Combustible Materials	E4101.2
424.19	Disconnecting Means	E4101.5
424.19(A)	Heating Equipment with Supplementary Overcurrent Protection	Table E4101.5
424.19(B)	Heating Equipment Without Supplementary Overcurrent Protection	Table E4101.5
424.19(C)	Unit Switch(es) as Disconnecting Means	Table E4101.5
424.20	Thermostatically Controlled Switching Devices	Table E4101.5
424.22	Overcurrent Protection	*NEC Default*
424.28(A)	Marking Required *(Nameplate)*	*NEC Default*
Article 424 – Part V	Electric Space-Heating Cables	*NEC Default*
424.38	Area Restrictions	*NEC Default*
424.38(A)	Extending Beyond the Room or Area	*NEC Default*
424.38(B)	Uses Not Permitted	*NEC Default*
424.44	Installation of Cables in Concrete or Poured Masonry Floors	*NEC Default*
424.44(E)	GFCI Protection / Electrically Heated Floors	E3902.14
ARTICLE 426 - Fixed Outdoor Electric Deicing and Snow-Melting Equipment		
426.28	Ground-Fault Protection of Equipment / Snow-Melting and Deicing Equipment Protection *(GFCI Not Required, GFP Required)*	E4101.7
ARTICLE 430 – Motors, Motor Circuits, and Controllers		
430.6	Conductor Ampacity and Motor Rating Determination	*NEC Default*
430.6(A)(1)	Table Values	*NEC Default*
430.6(A)(2)	Nameplate Values	*NEC Default*
430.14	Location of Motors *(Adequate Ventilation)*	*NEC Default*
Article 430 – Part II	Motor Circuit Conductors	*NEC Default*
430.21	General	*NEC Default*
430.22	Single Motor / Branch Circuits Serving a Single Motor	E3702.7
430.22(G)	Conductors for Small Motors	*NEC Default*
430.24	Several Motors or a Motor(s) and Other Load(s) / Branch Circuits Serving Motor Operated and Combination Loads	E3702.6
Article 430 – Part III	Motor and Branch-Circuit Overload Protection	*NEC Default*
430.31	General	*NEC Default*
430.32	Continuous-Duty Motors	*NEC Default*
430.42	Motors on General-Purpose Branch Circuits	*NEC Default*
Article 430 – Part IV	Motor Branch-Circuit Short-Circuit and Ground-Fault Protection	*NEC Default*
430.52	Rating or Setting for Individual Motor Circuit	*NEC Default*
Table 430.52(C)(1)	Maximum Rating or Setting of Motor Branch-Circuit Short-Circuit and Ground-Fault Protective Devices	*NEC Default*
430.58	Rating of Circuit Breaker	*NEC Default*
Article 430 – Part IX	Disconnecting Means	*NEC Default*
430.102	Location	*NEC Default*
430.109	Type *(Disconnecting Means)*	*NEC Default*
Article 430 – Part XIII	Grounding *(of Motors)*	*NEC Default*
430.245	Method of Grounding *(Connection to an EGC)*	*NEC Default*
Table 430.247 – Table 430.250	Full-Load Currents in Amperes *(Motor Tables)*	*NEC Default*

NEC Reference	*Description in Both Model Code Documents NEC / IRC*	IRC Reference
ARTICLE 440 – Air-Conditioning and Refrigerating Equipment		
440.4	Marking on Hermetic Refrigerant Motor-Compressors and Equipment	*NEC Default*
440.4(A)	Hermetic Refrigerant Motor-Compressor Nameplate *(Nameplate Requirements)*	*NEC Default*
440.6	Ampacity and Rating / Branch Circuits for Air-Conditioning and Heat Pump Equipment	E3702.12
440.9	Grounding and Bonding *(Rooftop Equipment Requires Wire-Type EGC)*	*NEC Default*
Article 440 – Part II	Disconnecting Means	*NEC Default*
440.12	Rating and Interrupting Capacity	*NEC Default*
440.13	Cord-Connected Equipment	*NEC Default*
440.14	Location *(Disconnecting Means)*	Table E4101.5
Article 440 – Part III	Branch-Circuit Short-Circuit and Ground-Fault Protection	*NEC Default*
440.22	Application and Selection	*NEC Default*
440.22(A)	Rating or Setting for Individual Motor-Compressor	*NEC Default*
Article 440 – Part IV	Branch-Circuit Conductors	E3702
440.32	Single Motor-Compressor	*NEC Default*
440.35	Multimotor and Combination-Load Equipment / Branch Circuits for Air-Conditioning and Heat Pump Equipment	E3702.12
Article 440 – Part VII	Provisions for Room Air Conditioners	*NEC Default*
440.60	General	*NEC Default*
440.61	Grounding *(Connection to an EGC)*	*NEC Default*
440.62	Branch-Circuit Requirements / Branch Circuits Serving Room Air Conditioners	E3702.13
440.62(A)	Room Air Conditioner as a Single Motor Unit / Branch Circuits Serving Room Air Conditioners	E3702.13
440.62(B)	Where No Other Loads Are Supplied	E3702.13.1
440.62(C)	Where Lighting Units or Other Appliances Are Also Supplied	E3702.13.2
440.63	Disconnecting Means	*NEC Default*
440.64	Supply Cords *(Max. Length)*	*NEC Default*
440.65	Protection Devices *(LCDI, AFCI, HDCI)*	*NEC Default*
ARTICLE 555 – Marinas, Boatyards, and Noncommercial Docking Facilities, Etc.		
555.10	Signage *(Shock Hazard Signage Required)*	*NEC Default*
555.14	Equipotential Planes and Bonding of Equipotential Planes	*NEC Default*
555.15	Replacement of Equipment *(Comply with Current Code)*	*NEC Default*
555.33(A) through (C)	Receptacles	*NEC Default*
555.34(A) and (B)	Wiring Methods and Installation	*NEC Default*
555.35(A) through (D)	Ground-Fault Protection of Equipment (GFPE) and Ground-Fault Circuit interrupters (GFCI)	*NEC Default*
555.35(A)	Feeders *(GFPE Rated not More Than 100 mA)*	*NEC Default*
555.35(B)	Branch Circuits	*NEC Default*
555.35(B)(1)	Receptacles Providing Shore Power *(GFPE Rated not More Than 30 mA)*	*NEC Default*
555.35(B)(2)	Outlets for Other Than Shore Power *(GFCI Protection Required)*	*NEC Default*
555.35(C)	Boat Hoists *(GFCI Protection Required)*	E3902.16
555.36(A) through (C)	Disconnecting Means for Shore Power Connection(s)	*NEC Default*
555.36(C)	Emergency Electrical Disconnect *(For Shore Power to Boats)*	*NEC Default*
555.37(A) through (E)	Equipment Grounding Conductor *(Equipment to Be Connected to EGC)*	*NEC Default*
555.38(A) and B)	Luminaires	*NEC Default*
ARTICLE 625 – Electric Vehicle Power Transfer Systems		
625.17(A) through (D)	Cords and Cables	*NEC Default*
625.17(A)	Power-Supply Cord	*NEC Default*
625.17(C)	Overall Cord and Cable Length	*NEC Default*
625.22	Personnel Protection System	*NEC Default*
625.40	Electric Vehicle Branch Circuit *(Individual Branch Circuit)*	E3702.14

NEC Reference	*Description in Both Model Code Documents NEC / IRC*	IRC Reference
625.41	Electric Vehicle Power Transfer Systems *(cont.)* Overcurrent Protection	*NEC Default*
625.42	Rating *(Considered a Continuous Load)*	E3702.14
625.43	Disconnecting Means	*NEC Default*
625.48	Interactive Equipment *(Bidirectional Power Feed, Etc.)*	*NEC Default*
625.54	GFCI Protection for Personnel *(All Receptacles for EV Charging)*	*NEC Default*
Article 625 – Part IV	Wireless Power Transfer Equipment	*NEC Default*
625.101	Grounding	*NEC Default*
625.102(A) through (E)	Installation	*NEC Default*
ARTICLE 680 – Swimming Pools, Fountains, and Similar Installations		
Article 680 – Part I	General	E4201
680.1	Scope	E4201.1
Article 100	Definitions *(Related to Pools, Spas, etc.)*	Chapt 35
680.4	Inspection After Installation *(AHJ can inspect after Initial Installation)*	*NEC Default*
680.5	Ground-Fault Circuit-Interrupter *(GFCI)* and Special Purpose Ground-Fault Circuit-Interrupter *(SPGFCI)* Protection **(**Note to Reader**: SPGFCI will not be addressed further as the conditions of SPGFCI do not exist at dwelling units)*	E4206.2
680.5(A)	General *(GFCI or SPGFCI in addition to NEC 210.8)*	E4206.2
680.5(B)	150 Volts or Less to Ground *(Class A GFCI Protection)*	E4206.2
680.5(C)	Above 150 Volts to Ground *(SPGFCI Protection)*	*NEC Default*
680.6	Listing Requirements *(Equipment to be Listed)*	E3403.3
680.7	Grounding and Bonding	E4205, E4205.1
680.7(A)	Feeders and Branch Circuits	E4205.1
680.7(B)	Cord and Plug Connections *(Must Contain an EGC)*	, E4202.3 E4205.2
680.7(C)	Terminals *(Identified for a Wet Location)*	E4205.3
680.8	Cord and Plug Connected Equipment / Flexible Cords	E4202.3
680.8(A)	Length	E4202.3(1)
680.8(B)	Equipment Grounding	E4202.3(1)
680.8(C)	Construction / Cord-Connected Equipment	E4205.7
680.9	Overhead Conductor Clearances	E4203.7
680.9(A)	Power Conductors *[6.9 m (22.5 ft)]*	E4203.7
680.9(B)	Communications Systems *[3.0 m (10 ft)]*	E4203.7
680.9(C)	Network-Powered Broadband Communications Systems *[6.9 m (22.5 ft)]*	E4203.7
Table 680.9(A)	Overhead Conductor Clearances	Table E4203.7
Figure 680.9(A)	Clearances from Pool Structures	*NEC Default*
680.10	Electric Pool Water Heaters Incorporating Resistive Heating Elements and Electrically Powered Swimming Pool Heat *Pump* and Chillers	*NEC Default*
680.10(A)	Electric Pool Water Heaters	E4206.12
680.10(B)	Electrically Powered Swimming Pool Heat Pumps and Chillers	*NEC Default*
680.11, Table 300.5	Underground Wiring *[Within 1.5 m (5 ft) Horizontally from the Inside Wall of the Pool]*	E4203.8
680.11(A)	Underground Wiring *(Wiring Methods)*	E4203.8
680.11(B)	Wiring Under Pools	E4203.8
680.12	Equipment Rooms, Vaults, and Pits	*NEC Default*
680.12(B)	Receptacles *(At least one GFCI-protected receptacle to be located within an equipment room)*	E4203.1.7
680.13	Maintenance Disconnecting Means / Disconnecting Means	E4203.3
680.14	Corrosive Environments / Wiring Methods in Corrosive Environments	E4202.1
680.14(A)	Wiring Methods *(Installed in a Corrosive Environment)*	E4202.2, Table E4202.1

680.14(B)	Other Equipment *(Installed in a Corrosive Environment)*	*NEC Default*
***NEC* Reference**	***Description in Both Model Code Documents NEC / IRC***	**IRC Reference**
Article 680 – Part II	Permanently Installed Swimming Pools	*NEC Default*
680.20	General *(Comply with Part I and Part II of NEC Article 680)*	*NEC Default*
680.21	Motors	E4203
680.21(A)	Wiring Methods	E4202.1, Table E4202.1
680.21(B)	Double Insulated Pool Pumps	E4202.4
680.21(C)	Gound-Fault Protection *(Motors) (Comply with NEC 680.5)*	E4203.1.4
680.21(D)	Pool Pump Motor Replacement *(GFCI Protection Required)*	*NEC Default*
680.22(A)	Receptacles	E4203.1
680.22(A)(1)	Required Receptacle, Location *[At Least One Receptacle Between 1.83 m (6 ft) and 6.0 m (20 ft) from Pool]*	E4203.1.3
680.22(A)(2)	Circulation and Sanitation System *(Receptacles)* / Location	E4203.1.1
680.22(A)(3)	Other Receptacles/ Location	E4203.1.1
680.22(A)(4)	GFCI Protection *[Within 6.0 m (20 ft) of Pool]*	E4203.1.3
680.22(A)(5)	Measurements / Receptacle Outlets	E4203.1
680.22(B)	Luminaires, Lighting Outlets, and Ceiling Suspended (Paddle) Fans	E4203.4
680.22(B)(1)	New Outdoor Installation Clearances / Outdoor Location	E4203.4.1
680.22(B)(2)	Indoor Clearances / Indoor Locations	E4203.42
680.22(B)(3)	Existing Installations / Existing Lighting Outlets and Luminaires	E4203.4.4
680.22(B)(4)	GFCI Protection in Adjacent Areas	E4203.4.6
680.22(B)(5)	Cord-and-Plug-Connected Luminaires	*NEC Default*
680.22(B)(6)	Low-Voltage Luminaires	E4203.4.3
680.22(B)(7)	Low-Voltage Gas-Fired Luminaires, Decorative Fireplaces, Fire Pits, and Similar Equipment	E4203.4.7
680.22(B)(8)	Measurements / Luminaires and Ceiling Fans	*NEC Default*
680.22(C)	Switching Devices	E4203.2
680.22(D)	Other Outlets *(Remote Control, Fire Alarm, Communications, etc.)*	E4203.5
680.22(E)	Other Equipment *(PV Arrays, etc.)*	E4203.6
680.23	Underwater Luminaires	E4206.4
680.23(A)(1)	Luminaire Design Normal Operation / Underwater Luminaires	E4206.4
680.23(A)(2)	Transformers and Power Supplies	E4206.1
680.23(A)(3)	GFCI Protection, Lamping, Relamping, and Servicing / Underwater Luminaires	E4206.4
680.23(A)(4)	Voltage Limitation / Maximum Voltage	E4206.4.1
680.23(A)(5)	Location, Wall-Mounted Luminaires / Luminaire Location	E4206.4.2
680.23(B)	Wet-Niche Luminaires	E4206.5
680.23(B)(1)	Forming Shells	E4206.5
680.23(B)(2)	Wiring Extending Directly to the Forming Shell / Nonmetallic Conduit	Table E4202.1, E4205.3
680.23(B)(3)	Equipment Grounding Provisions for Cords / Flexible Cords	E4205.4
680.23(B)(4)	Luminaire Grounding Terminations	E4206.5
680.23(B)(5)	Luminaire Bonding	E4206.5
680.23(B)(6)	Servicing *(Removable from Water)*	E4206.5.1
680.23(C)	Dry-Niche Luminaires	E4206.6
680.23(D)	No-Niche Luminaires	E4206.7
680.23(E)	Through-Wall Lighting Assembly	E4206.8
680.23(F)	Branch-Circuit Wiring	Table E4202.1
680.23(F)(1)	Wiring Methods	E4202.1, E4202.2
680.23(F)(2)	Equipment Grounding / Luminaires and Related Equipment	E4205.2
680.23(F)(3)	Conductors / Wiring on Load Side of Ground-Fault Circuit-Interrupters and Transformers	E4206.3
680.24	Junction Boxes and Electrical Enclosures for Transformers or Ground-Fault Circuit Interrupters	E4206.9

680.24(A)	Junction Boxes *(Listed as a Swimming Pool Junction Box)*	E4206.9.1
***NEC* Reference**	***Description in Both Model Code Documents NEC / IRC***	**IRC Reference**
680.24(B)	Junction Boxes and Electrical Enclosures for Transformers or Ground-Fault Circuit Interrupters *(Permanently Installed Swimming Pools) (cont.)* Other Enclosures	E4206.9.2
680.24(C)	Protection / Protection of Junction Boxes and Enclosures	E4206.9.3
680.24(D)	Grounding Terminals *(One More than Number of Conduit Entries)*	E4206.9.4
680.24(E)	Strain Relief	E4206.9.5
680.24(F)	Grounding / Luminaires and Related Equipment	E4205.4
680.26	Equipotential Bonding *(for Permanently Installed Swimming Pools)*	E4204
680.26(A)	Performance *(Reduce Voltage Gradients)*	E4204.1
680.26(B)	Bonded Parts *(Connections and Parts Required to be Bonded)*	E4204.2
680.26(B)(1)	Conductive Pool Shell *(Belly Steel)*	E4204.2(1)
680.26(B)(2)	Perimeter Surface *(Deck Steel)*	E4204.2(2)
680.26(B)(3)	Metallic Components *(Reinforcing Metal, Etc.)*	E4204.2(3)
680.26(B)(4)	Underwater Lighting	E4204.2(4)
680.26(B)(5)	Metal Fittings *(Ladders, Railings, Etc.)*	E4204.2(5)
680.26(B)(6)	Electrical Equipment *(Pool Pumps, Etc.)*	E4204.2(6)
680.26(B)(7)	Fixed Metal Parts *(Metal Fences, Etc.)*	E4204.2(7)
680.26(C)	Pool Water	E4204.3
680.27	Specialized Pool Equipment	E4206
680.27(A)	Underwater Audio Equipment	E4206.10
680.27(A)(1)	Speakers	E4206.10.1
680.27(A)(2)	Wiring Methods	E4206.10.2
680.27(A)(3)	Forming Shell and Metal Screen	E4206.10.3
680.27(B)	Electrically Operated Pool Covers	E4206.11
680.27(B)(1)	Motors and Controllers	E4206.11
680.27(B)(2)	Protection *(GFCI Protected Branch Circuit)*	E4206.11
680.27(C)	Deck Area Heating / Pool Area Heating	E4206.14
680.27(C)(1)	Unit Heaters	E4206.14.1
680.27(C)(2)	Permanently Wired Radiant Heaters	E4206.14.2
680.27(C)(3)	Radiant Heating Cables Not Permitted / Prohibited	E4206.14.3
680.28	Gas-Fired Water Heater *(GFCI Protection)*	*NEC Default*
Article 680 – Part III	Storable Pools, Storable Spas, Storable Hot Tubs, and Storable Immersion Pools / Storable Pools, Storable Spas, and Storable Hot Tubs	E4207
680.30	General *(Comply with Part I and Part III of NEC Article 680)*	
680.31	Pumps	E4207.1
680.32	Ground-Fault Circuit-Interrupters (GFCI) Protection *(Required)*	E4207.2
680.33	Luminaires	E4207.3
680.33(A)	Within the Low Voltage Contact Limit	E4207.3.1
680.33(B)	Over the Low Voltage Contact Limit but Not over 150 Volts	E4207.3.2
680.34	Receptacle Locations *[Not Located Within 1.83 m (6 ft) of Pool]*	E4207.4
680.35(A) through (G)	Storable and Portable Immersion Pools	
Article 680 – Part IV	Permanently Installed and Self-Contained Spas and Hot Tubs and Permanently Installed Immersion Pools / Spas and Hot Tubs	E4208
680.40	General *(Comply with Part I and Part IV of NEC Article 680)*	*NEC Default*
680.41	Location of Other Equipment	*NEC Default*
680.41(B)	Equipment Exceeding the Low-Voltage Contact Limit	*NEC Default*
680.42	Outdoor Installations *(Generally Comply with Part I and Part IV of NEC Article 680)*	E4203
680.42(A)	Flexible Connections	Table E4202.1
680.42(B)	Bonding / Bonding of Outdoor Hot Tubs and Spas	E4204.4
680.42(C)	Underwater Luminaires	*NEC Default*
680.43	Indoor Installations	E4203

680.43(A)	Receptacles / Indoor Locations	E4203.1.5
NEC **Reference**	***Description in Both Model Code Documents NEC / IRC***	**IRC Reference**
680.43(A)(1)	Receptacles *(Indoor Spas and Hot Tubs, Etc.) (cont.)* Location *[Not Located Within 1.83 m (6 ft) of Spa, Hot Tub, Etc.]*	E4203.1.5
680.43(A)(2)	GFCI Protection for Receptacles / Indoor GFCI Protection	E4203.1.6
680.43(A)(3)	Protection, Spa or Hot Tub Supply Receptacle	E4208.1
680.43(A)(4)	Measurements	E4203.1
680.43(B)	Installation of Luminaires, Lighting Outlets and Ceiling Suspended Paddle Fans / Luminaires, Equipment, and Ceiling Fans	E4203.4.2
680.43(B)(1)	Elevation	E4203.4.2
680.43(B)(2)	Underwater Applications	*NEC Default*
680.43(C)	Switches *[Located at least 1.5 m (5 ft) from Spa, Etc.]*	E4203.2
680.43(D)	Bonding / Bonding of Indoor Spas and Hot Tubs	E4204.5
680.43(E)	Methods of Bonding / Methods	E4204.5.1
680.43(F)	Grounding	E4205.1
680.43(G)	Underwater Audio Equipment	E4208.3
680.44	GFCI Protection / Ground-Fault Circuit-Interrupters	E4208.1
680.44(A)	General *(Self-Contained Spas and Hot Tubs) (GFCI Protection)*	E4208.1
680.44(B)	Listed Units *(Self-Contained GFCI Protection)*	E4208.1
680.44(C)	Gas-Fired Water Heaters	*NEC Default*
Article 680 – Part V	Fountains	*NEC Default*
680.50	General *(Comply with Part I and Part V of NEC Article 680)*	*NEC Default*
680.50(A)	Additional Requirements *(Water Common to a Pool to Comply with Part II of NEC Article 680)*	*NEC Default*
680.50(B)	Location of Equipment Exceeding the Low-Voltage Contact Limit	*NEC Default*
680.51(A) through (G)	Luminaires, Submersible Pumps, and Other Submersible Equipment	*NEC Default*
680.52(A) and (B)	Junction Boxes and Other Enclosures	*NEC Default*
680.54(A) and (B)	Grounding and Bonding	*NEC Default*
680.54(C)	Equipotential Bonding of Splash Pads	*NEC Default*
680.55(A) and (B)	Methods of Grounding	*NEC Default*
680.56(A) through (D)	Cord-and-Plug-Connected Equipment	*NEC Default*
680.58	GFCI Protection for Adjacent Receptacle Outlets	*NEC Default*
680.59	GFCI Protection for Permanently Installed Nonsubmersible Pumps	*NEC Default*
Article 680 – Part VII	Hydromassage Bathtubs	E4209
680.70	General *(Comply with Part VII ONLY of NEC Article 680)*	E4201.1
680.71	Protection / Ground-Fault Circuit-Interrupters	E4209.2
680.72	Other Electric Equipment	E4209.3
680.73	Accessibility *(Accessible Without Damaging the Building Finish)*	E4209.4
680.74	Bonding / Bonded Parts	E4209.5
680.74(A)	General *(Parts to be Bonded)*	E4209.5
680.74(B)	Bonding Conductor / Method of Bonding	E4209.6
Article 680 – Part VIII	Electrically Powered Pool Lifts	*NEC Default*
680.80	General *(Comply with Part VIII of NEC Article 680 [Except Where Other Parts are Specifically Referenced)]*	*NEC Default*
680.81	Equipment Approval	*NEC Default*
680.82	Protection *(GFCI Protection)*	*NEC Default*
680.83	Equipotential Bonding	*NEC Default*
680.84	Switching Devices and Receptacles	*NEC Default*
680.85	Nameplate Marking	*NEC Default*
ARTICLE 690 – Solar Photovoltaic (PV) Systems		
690.4(A) through (G)	General Requirements	*NEC Default*
690.4(B), 110.3(B)	Equipment *(Listing Requirements for PV Equipment)*	E3403.3
690.4(C)	Qualified Persons	*NEC Default*
Article 690 – Part II	Circuit Requirements	*NEC Default*

690.7(A) through (D)	Maximum Voltage	*NEC Default*
***NEC* Reference**	***Description in Both Model Code Documents NEC / IRC***	**IRC Reference**
690.8(A) through (D)	Circuit Requirements *(PV Circuits) (cont.)* Circuit Sizing and Current	*NEC Default*
690.9(A) through (D)	Overcurrent Protection	*NEC Default*
~~690.10~~	Stand-Alone System *(See NEC Article 710)*	*NEC Default*
690.11	AFCI Protection (dc)	*NEC Default*
690.12(A) through (D)	Rapid Shutdown of PV Systems on Buildings	*NEC Default*
Article 690 – Part III	Disconnecting Means	*NEC Default*
690.13(A) through (E)	PV System Disconnecting Means	*NEC Default*
690.15(A) through (D)	Disconnecting Means for Isolating Photovoltaic Equipment	*NEC Default*
Article 690 – Part IV	Wiring Methods and Materials for PV Systems	*NEC Default*
690.31(A) through (G)	Wiring Systems *(Wiring Methods for PV Systems)*	*NEC Default*
690.31(A)	Wiring Systems *(Conductor Ampacity, Etc.)*	*NEC Default*
690.31(B)	Identification and Grouping	*NEC Default*
690.31(C)	Cables *(Single-Conductor Cables, Small-Conductor Cables, Etc.)*	*NEC Default*
690.31(D)	Direct-Current (dc) Circuits on or in Buildings	*NEC Default*
690.31(D)(2)	Marking and Labeling Required	*NEC Default*
690.32	Component Interconnections	*NEC Default*
690.33(A) through (D)	Mating Connectors	*NEC Default*
690.33(C)	Type *(To be Listed and Identified for Intermatability)*	*NEC Default*
Article 690 – Part V	Grounding and Bonding of PV Systems	*NEC Default*
690.41	PV System DC Circuit Grounding and Protection *(for PV Systems)*	*NEC Default*
690.41(A)	PV System DC Circuit Grounding Configurations	*NEC Default*
690.41(B)	DC Ground-Fault Detector-interrupter (GFDI) Protection	*NEC Default*
690.41(B)(1)	Ground-Fault Detection	*NEC Default*
690.43(A) through (D)	Equipment Grounding and Bonding *(for PV Systems)*	*NEC Default*
690.45	Size of EGCs *(for PV Array and Support Structures)*	*NEC Default*
690.47(A) and (B)	Grounding Electrode System *(for PV Systems)*	*NEC Default*
Article 690 – Part VI	Source Connections *(for PV Systems)*	*NEC Default*
690.56	Identification of Power Sources *(See NEC 705.10)*	*NEC Default*
690.59	Connection to Other Sources *(See Parts I and II of NEC 705)*	*NEC Default*
Article 694 - Wind Electric Systems		
694.7(A) through (G)	Construction and Maintenance *(Installed Only by Qualified Persons)*	*NEC Default*
694.7(B)	Equipment *(Must be Listed or Field-Evaluated)*	E3403.3
694.7(D)	Overvoltage Protection *[Surge Protective Devices (SPD)]*	*NEC Default*
694.7(E)	Receptacles *(GFCI Protection Required)*	*NEC Default*
Article 694 – Part II	Circuit Requirements	*NEC Default*
694.10(A) through (C)	Maximum Voltage	*NEC Default*
694.12(A) and (B)	Circuit Sizing and Current	*NEC Default*
694.15(A) through (C)	Overcurrent Protection	*NEC Default*
Article 694 – Part III	Disconnecting Means	*NEC Default*
694.20	All Conductors	*NEC Default*
694.22	Additional Provisions	*NEC Default*
694.22(A)	Disconnecting Means	*NEC Default*
694.22(C)	Requirements for Disconnecting Means	*NEC Default*
694.23(A) and (B)	Turbine Shutdown	*NEC Default*
Article 694 – Part IV	Wiring Methods	*NEC Default*
694.30(A) through (C)	Permitted Wiring Methods	*NEC Default*
Article 694 – Part V	Grounding and Bonding	*NEC Default*
694.40(A) and (B)	Equipment Grounding and Bonding	*NEC Default*
Article 694 – Part VI	Marking	*NEC Default*
694.54	Identification of Power Sources *(Marking)*	*NEC Default*

Article 694 – Part VII	Connection to Other Sources	*NEC Default*
694.60	Identified Interactive Equipment	*NEC Default*

***NEC* Reference**	***Description in Both Model Code Documents NEC / IRC***	**IRC Reference**
694.66	Connection to Other Sources *(Wind Electric Systems) (cont.)* Operating Voltage Range	*NEC Default*
ARTICLE 702 - Optional Standby Systems		
Article 702 – Part I	General	*NEC Default*
702.5(A) through (D)	Interconnection or Transfer Equipment	*NEC Default*
702.5(B)	Meter-Mounted Transfer Switches	*NEC Default*
702.7(A) through (C)	Signs	*NEC Default*
Article 702 – Part II	Wiring	*NEC Default*
702.10	Wiring Optional Standby Systems	*NEC Default*
702.11	Portable Generator Grounding	*NEC Default*
702.12	Outdoor Generator Sets	*NEC Default*
ARTICLE 722 – Cables for Power-Limited Circuits and Fault-Managed Power Circuits		
Article 722 – Part I	General	*NEC Default*
722.3(O)	Specific Requirements *[Class 2 and Class 3 cables (Part II of NEC Article 725)]*	*NEC Default*
722.24	Mechanical Execution of Work *(Neat and Workmanlike Manner)*	*NEC Default*
722.24(B)	Support of Cables	E4304.5
722.25	Abandoned Cables	*NEC Default*
722.135(A) through (I)	Installation of Cables	E4304.5
Table 722.135(B)	Installation of Listed Cables in Buildings	*NEC Default*
Article 722 – Part II	Listing Requirements	*NEC Default*
722.179	Listing and Marking of Cables	E4303.2
722.179(A)	Listing of Cables	E4303.2
722.179(A)(3)	General-Purpose Cable	*NEC Default*
722.179(A)(9)	Limited Power (LP) Cable	*NEC Default*
722.179(A)(10)	Undercarpet Cable	*NEC Default*
722.179(A)(14)	Class 2 and Class 3 Cable Voltage and Temperature Ratings	*NEC Default*
722.179(B)	Marking	E4303.2.5
Table 722.179(B)	Cable Type Marking	Table E4303.2.5
ARTICLE 725 – Class 2, and Class 3 Power-Limited Circuits		
Article 725 – Part I	General	*NEC Default*
725.1	Scope	E4301.1
725.24	Mechanical Execution of Work *(Neat and Workmanlike Manner)*	E4304.5
725.30	Class 2 and Class 3 Circuit Identification	*NEC Default*
725.31	Safety-Control Equipment	*NEC Default*
Article 725 – Part II	Class 2 and Class 3 Circuits	Chapter 43
725.60	Power Sources for Class 2 and Class 3 Circuits	E4302.1
725.60(A)	Power Source	E4302.1
725.121(B)	Interconnection of Power Sources	E4302.2
725.60(C)	Marking	*NEC Default*
725.127	Wiring Methods on Supply Side of Class 2 or Class 3 Power Source	E4303.1
725.130(A) and (B)	Wiring Methods and Materials on Load Side of the Class 2 or Class 3 Power Source	E4303.2
725.136(A) through (I)	Separation from Electric Light, Power, Class 1, Non–Power-Limited Fire Alarm Circuit Conductors, and Medium-Power Network-Powered Broadband Communications Cables / Separation from Other Conductors	E4304.1
725.139(A) through (F)	Installation of Conductors of Different Circuits in the Same Cable, Enclosure, Cable Tray, Raceway, or Cable Routing Assembly / Class 2 Circuits with Communications Circuits	E4304.3
725.144, Table 725.144	Bundling of Cables Transmitting Power and Data	*NEC Default*
Article 725 – Part III	Listing Requirements	*NEC Default*
725.160	Listing and Marking of Equipment for Power and Data Transmission	*NEC Default*

ARTICLE 760 – Fire Alarm Systems		
760.3(A)	Spread of Fire or Products of Combustion *(see NEC 300.21)*	*NEC Default*
***NEC* Reference**	***Description in Both Model Code Documents NEC / IRC***	**IRC Reference**
760.24	Fire Alarm Systems *(cont.)* Mechanical Execution of Work *(Neat and Workmanlike Manner)*	*NEC Default*
760.25	Abandoned Cables	*NEC Default*
760.30	Fire Alarm Circuit Identification	*NEC Default*
Article 760 – Part II	Non–Power-Limited Fire Alarm (NPLFA) Circuits	*NEC Default*
760.41(A)	Power Source	*NEC Default*
760.41(B)	Branch Circuit *(Not Supplied Through GFCI or AFCI Circuits)*	*NEC Default*
Article 760 – Part III	Power-Limited Fire Alarm (PLFA) Circuits	*NEC Default*
760.121(A)	Power Source	*NEC Default*
760.121(B)	Branch Circuit *(Not Supplied Through GFCI or AFCI Circuits)*	*NEC Default*
760.130	Wiring Methods and Materials on Load Side of PLFA Power Source	*NEC Default*
760.130(A)	NPLFA Wiring Methods and Materials	*NEC Default*
760.130(B)	PLFA Wiring Methods and Materials	*NEC Default*
760.136(A) through (H)	Separation from Electric Light, Power, Class 1, NPLFA, and Medium-Power Network-Powered Broadband Communications Circuit Conductors	*NEC Default*
760.139(A) through (E)	Installation of Conductors of Different PLFA Circuits, Class 2, Class 3, and Communications Circuits in the Same Cable, Enclosure, Cable Tray, Raceway, or Cable Routing Assembly.	*NEC Default*
760.154	Applications of Listed PLFA Cables	*NEC Default*
Table 760.154	Applications of Listed PLFA Cables in Buildings	*NEC Default*
ARTICLE 800 – General Requirements for Communications Systems		
Article 800 – Part I	General	*NEC Default*
800.24	Mechanical Execution of Work *(Neat and Workmanlike Manner)*	*NEC Default*
800.25	Abandoned Cables	*NEC Default*
800.26	Spread of Fire or Products of Combustion	*NEC Default*
Article 800 – Part II	Wires and Cables Outside and Entering Buildings	*NEC Default*
800.44(A) through (E)	Overhead (Aerial) Wires and Cables *(Entering Buildings)*	*NEC Default*
800.47(A) and (B)	Underground Systems Entering Buildings	*NEC Default*
800.49	Metallic Entrance Conduit Grounding	*NEC Default*
800.53	Separation from Lightning Conductors	*NEC Default*
Article 800 – Part III	Grounding Methods	*NEC Default*
800.100	Cable and Primary Protector Bonding and Grounding	*NEC Default*
800.100(A)	Bonding Conductor or Grounding Electrode Conductor	*NEC Default*
800.100(B)	Electrode *(Connections, Installations, etc.)*	*NEC Default*
800.100(B)(1)	In Buildings or Structures with an Intersystem Bonding Termination	*NEC Default*
800.100(B)(2)	In Buildings or Structures with Grounding Means	*NEC Default*
800.100(B)(3)	In Buildings or Structures Without an Intersystem Bonding Termination or Grounding Means	*NEC Default*
Article 800 – Part IV	Installation Methods Within Buildings	*NEC Default*
800.113	Installation of Cables Used for Communications Circuits, Communications Wires, Cable Routing Assemblies, and Communications Raceways	*NEC Default*
800.113(L)	One- and Two-Family Dwellings	*NEC Default*
Table 800.113	Cables Used for Communications Circuits	*NEC Default*
800.133(A) through (C)	Installation of Communications Wires and Cables and CATV-Type Coaxial Cables	*NEC Default*
800.154	Applications of Listed Communications Wires, Cables, Raceways and Listed Cable Routing Assemblies	*NEC Default*
Table 800.154(a)	Applications of Listed Communications Wires, Cables, and Network-Powered Broadband Communications System Cables in Buildings	*NEC Default*
Article 800 – Part V	Listing Requirements	*NEC Default*

800.179(A) through (I)	Wires and Cables *(Plenum Cables, General-Purpose Cables, Etc.)*	*NEC Default*
800.180	Grounding Devices *(NEC Chapter 8 Circuits)*	*NEC Default*
***NEC* Reference**	***Description in Both Model Code Documents NEC / IRC***	**IRC Reference**
ARTICLE 805 – Communications Circuits		
805.18	Installation of Equipment	*NEC Default*
Article 805 – Part II	Wires and Cables Outside and Entering Buildings	*NEC Default*
805.47	Underground Communications Wires and Cables Entering Buildings - Underground Block Distribution	*NEC Default*
805.50	Circuits Requiring Primary Protectors	*NEC Default*
805.50(C)	Entering Buildings	*NEC Default*
Article 805 – Part III	Protection *(Communications Circuits)*	*NEC Default*
805.90(A)	Application *(Protective Devices) (Listed Primary Protector)*	*NEC Default*
805.93	Grounding, Bonding, or Interruption of Non-Current-Carrying Metallic Sheath Members of Communications Cables	*NEC Default*
Article 805 – Part IV	Installation Methods Within Buildings	*NEC Default*
805.154	Substitutions for Listed Communications Cables	*NEC Default*
Table 805.154	Cable Substitutions	*NEC Default*
805.156	Dwelling Unit Communications Outlet *(Minimum One per Dwelling)*	*NEC Default*
Article 805 – Part V	Listing Requirements	*NEC Default*
805.170	Protectors *(Shall be Listed for Electrical Connection to a Communications Network)*	E3403.3
805.173	Drop Wire and Cables	*NEC Default*
ARTICLE 820 - Community Antenna Television and Radio Distribution Systems		
820.15	Power Limitations *(not over 60 Volts)*	*NEC Default*
Article 820 – Part III	Protection	*NEC Default*
820.93(A) through (D)	Grounding of the Outer Conductive Shield of Coaxial Cables	*NEC Default*
Article 820 – Part IV	Grounding Methods	*NEC Default*
820.100	Cable Bonding and Grounding	*NEC Default*
820.100(B)	Shield Protection Devices	*NEC Default*
820.103	Equipment Grounding	*NEC Default*
Article 820 – Part V	Installation Methods Within Buildings	*NEC Default*
820.154	Substitutions of Listed CATV Cables	*NEC Default*
Table 820.154	Coaxial Cable Uses and Permitted Substitutions	*NEC Default*
ARTICLE 830 – Network-Powered Broadband Communication Systems		
830.15	Power Limitations *(Low- or Medium-Power Sources)*	*NEC Default*
Table 830.15	Limitations for Network-Powered Broadband Communications Systems	*NEC Default*
Article 830 – Part II	Cables Outside and Entering Buildings	*NEC Default*
830.44	Overhead (Aerial) Cables	*NEC Default*
830.44(A)	On Poles and In-Span or Above Roofs	*NEC Default*
830.44(B)	Clearance from Ground	*NEC Default*
830.44(C)	Over Pools	*NEC Default*
830.44(D)	Final Spans	*NEC Default*
830.44(E)	Between Buildings	*NEC Default*
830.44(F)	On Buildings	*NEC Default*
830.47	Underground Network-Powered Broadband Communications Cables Entering Buildings	*NEC Default*
830.47(A)	Protection from Physical Damage	
Table 830.47(A)	NPBCS Minimum Cover Requirements	*NEC Default*
830.47(B)	Pools *(Cables Under Pools) (Comply with NEC 680.11)*	*NEC Default*
Article 830 – Part III	Protection *(Network-Powered Broadband Communication Systems)*	*NEC Default*
830.90	Primary Electrical Protection	*NEC Default*
Article 830 – Part IV	Grounding Methods	*NEC Default*
830.93	Grounding or Interruption of Metallic Members of Network-Powered Broadband Communications Cables	*NEC Default*
830.93(A)	Entering Building	*NEC Default*

830.93(B)	Terminating Outside of the Building	*NEC Default*
***NEC* Reference**	***Description in Both Model Code Documents NEC / IRC***	**IRC Reference**
Article 830 – Part V	Network-Powered Broadband Communication Systems *(cont.)* Installation Methods Within Buildings	*NEC Default*
830.133	Installation of Network-Powered Broadband Communications Cables and Equipment	*NEC Default*
830.133(B)	Support of Network-Powered Broadband Communications Cables	*NEC Default*
830.154	Substitutions of NPBC Cables	*NEC Default*
Table 830.154	Cable Substitutions	*NEC Default*
830.179	Network-Powered Broadband Communications Equipment and Cables *(Listing Requirements)*	E3403.3
ARTICLE 840 – Premises-Powered Broadband Communication Systems		
Article 840 – Part II	Cables Outside and Entering Buildings	*NEC Default*
840.47	Underground Wires and Cables Entering Buildings	*NEC Default*
Article 840 – Part III	Protection	*NEC Default*
840.93	Grounding or Interruption	*NEC Default*
840.94	Premises Circuits Leaving the Building	*NEC Default*
Article 840 – Part IV	Grounding Methods	*NEC Default*
840.101	Premises Circuits Not Leaving the Building	*NEC Default*
840.101(A)	Coaxial Cable Shield Grounding	*NEC Default*
840.101(B)	Communications Circuit Grounding	*NEC Default*
840.101(C)	Network Terminal Grounding	*NEC Default*
840.102	Premises Circuits Leaving the Building	*NEC Default*
Article 840 – Part VI	Premises Powering of Communications Equipment over Communications Cables	*NEC Default*
840.160	Powering Circuits	*NEC Default*
Article 840 – Part VII	Listing Requirements	*NEC Default*
840.170(A) through (D)	Equipment and Cables	*NEC Default*

Appendix B

Appendix B

RESIDENTIAL ELECTRICAL INSPECTIONS CHECKLISTS (2023 NEC)

Key Questions for Residential Pre-Inspections

1. What wiring methods will be used, and are the wiring methods suitable for the conditions?
2. Will multiwire branch circuits be used?
3. Are the small-appliance branch circuits located and installed properly *(kitchens, dining room receptacles, etc.)*?
4. What sizes of service equipment, service-entrance conductors, and feeder conductors are required?
5. Is the service overhead or underground?
6. What type(s) of grounding electrode(s) are present?
7. What size(s) of grounding electrode conductor(s) are required?
8. Where is mechanical equipment located, and what types of equipment will be installed?

RESIDENTIAL ROUGH-IN INSPECTION CHECKLIST

Residential Rough-In Inspection: General Requirements (All Areas)			
✓	Item	*NEC* Reference	Inspection Activity
	1.	Chapter 3, Chapter 7, and Chapter 8	Check wiring methods *(usually cable assemblies)* for support and suitability for the conditions.
	2.	300.4(A)(1) and Exceptions	Check cable installation through or parallel to framing members for 32 mm (1¼ in.) clearance or protective steel plates. Protective steel plates must be min. 1.6 mm (1/16 in.) thick *(unless listed for lesser thickness)* and of the appropriate length and width.
	3.	300.22(C) and Exception	Verify wiring methods located within other spaces used for environmental air is only crossing perpendicular to the long dimension of such spaces.
	4.	314.15, 314.16, 314.27, 422.18	Check boxes for suitability of use *(Sized properly for conductor fill, double conductor allowance for devices that require more than one opening, etc.)*.
	5.	300.15, 314.27, 314.29	Verify that boxes are installed in accessible locations for all junction and outlet boxes.
	6.	314.17(B), 314.17(B)(2), and 314.17(B)(2) Exception	Check that cables are secured to boxes. *[Where single gang nonmetallic boxes are used, and cable is fastened within 200 mm (8 in.) of the box, securing to the box is not required]*.
	7.	314.17(B)(1), 314.17(B)(2)	Flexible tubing or cable sheath shall extend not less than 6 mm (¼ in.) inside the metallic or nonmetallic boxes and beyond any cable clamp.
	8.	314.16(A) and (B)	Check boxes for proper conductor fill and box fill calculations.
	9.	314.20	Check positioning of boxes that are intended to be flush with combustible and noncombustible finished surfaces.
	10.	250.8(A) and (B), 250.86, 250.146(A) and (C), 250.148(C)	Check for splicing devices on all equipment grounding conductors within boxes and bonding connections to metal boxes.
	11.	250.118, 250.122	Check equipment grounding conductors for suitability and size.
	12.	314.27(B), and (C)	Check boxes used in floors, or for support of ceiling fans, for listing.
	13.	314.27(C)	Verify where outlet boxes mounted in the ceilings of habitable rooms in locations acceptable for the installation of a ceiling-suspended (paddle) fan, that the outlet box or outlet box system is listed for sole support of a ceiling-suspended (paddle) fan or that the ceiling-suspended (paddle) fan is capable of being supported from the structural framing.
	14.	410.116(A), 410.116(B)	Check recessed luminaires for proper clearances from combustibles and thermal insulation *(typically non-IC rated luminaires)*.
	15.	* Applicable Energy Code	Verify recessed luminaires comply with the applicable energy codes.

✓	Residential Rough-In Inspection: General Requirements (All Areas) *(cont.)*		
✓	Item	*NEC* Reference	Inspection Activity
	16.	* Applicable Building Code *(IRC R314)*	* Verify smoke alarm locations and requirements. **(The applicable Building Code typically covers smoke alarm requirements).*
	17.	334.80, 334.112, Table 310.15(C)(1)	Check Type NM cable where installed in bored holed with fire- or draft-stopping for possible correction factors for de-rating of conductors. Same applies for Type NM cable installed in thermal insulation without maintaining spacing between conductors.
	18.	404.2(C), 404.22	Check wall switch locations for the presence of a grounded conductor *(where conditions require grounded conductor to be installed).*
	19.	300.5(C)	Verify that any underground cables and conductors installed under a building are installed in a raceway.
	20.	760.41(B), 760.121(B)	Check for an individual branch circuit being provided for NPLFA and PLFA fire alarm systems. This branch circuit cannot be provided through GFCI or AFCI devices.
	21.	210.12(B)	AFCI protection required on all 120 volt, single-phase, 10-, 15-, and 20-ampere branch circuits that supply outlets or devices in dwelling unit kitchens, family rooms, dining rooms, living rooms, parlors, libraries, dens, bedrooms, sunrooms, recreation rooms, closets, hallways, laundry areas, or similar rooms or areas and protected by a listed combination-type AFCI device. *(This would typically be checked and verified on final inspection but could have impact on rough-in inspection.)*
	22.	**Note to User*	*GFCI and AFCI protection is typically enforced on the final inspection. Check Final Inspection Checklist for these requirements.*
	23.	210.23(A), 310.3(A)	Verify the presents of any 10-ampere branch circuits. The minimum size conductors for dwelling unit branch circuits is 14 AWG copper or 12 AWG aluminum or copper-clad aluminum.

Residential Rough-In Inspection: Kitchen			
✓	Item	*NEC* Reference	Inspection Activity
	1.	210.52(A) and (C), 210.52(C)(2)	Check spacing of receptacle outlets for walls, countertops, and work surfaces including areas behind corner-mounted ranges and sinks.
	2.	210.11(C)(1), 210.52(B)	Verify that a minimum of two 20-A small-appliance branch circuits is used for receptacle outlets in kitchen, pantry, dining room, and breakfast room.
	3.	210.70(A), 210.52(B)(2), 210.70(A)(1) Ex. No. 1	Verify that a lighting outlet controlled by a listed wall-mounted control device *(located near an entrance to the room)* is provided and supplied on a general lighting circuit *(not small-appliance branch circuit). (Receptacle outlet cannot be used as lighting outlet in kitchen)*
	4.	210.3, 210.22, 210.23, 210.52(B)(1) Ex. No. 2, 422.10	Verify that properly sized circuits have been provided for specific kitchen appliances, such as refrigerators, dishwashers, disposals, ranges, cooktops, trash compactors, and the like.
	5.	210.52(B)(1), 210.52(B)(2) and Exceptions	Verify that the two or more small-appliance branch circuits have no other outlets other than receptacle outlets in the kitchen, pantry, dining room, and breakfast room, as well as an electric clock receptacle and electric loads associated with gas-fired appliances.
	6.	210.52(B)(3), 210.11(C)(1)	Verify that the kitchen countertop receptacle outlets are supplied by no fewer than two small-appliance branch circuits.
	7.	210.52(C), 210.52(A)(1)	Verify that receptacle outlets serving kitchen countertops and work surfaces are not also being used to serve as receptacle outlets for wall spacing requirements.
	8.	210.52(B)(3)	In dwellings with more than one kitchen, no small-appliance branch circuit can serve more than one kitchen.
	9.	422.16(B)(4)	If a range hood or an over-the-range microwave oven with integral range hood is to be cord-and-plug connected, verify that receptacle is supplied by an individual branch circuit.

Residential Rough-In Inspection: Kitchen *(cont.)*			
✓	Item	*NEC* Reference	Inspection Activity
	10.	210.52(C)(2), 300.5(B), 310.10(C)	If a kitchen island or peninsular will be present at the time of final inspection, verify that a proper wiring method is present as well. If this wiring method is installed underground (under the slab), verify that the wiring method is identified for a wet location and any enclosed conductors are identified for a wet location as well.

Residential Rough-In Inspection: Dining Room			
✓	Item	*NEC* Reference	Inspection Activity
	1.	210.52(A)(1) and (A)(2)	Check receptacle outlets for proper spacing.
	2.	210.52(B)(1)	Verify that all required receptacle outlets are supplied by small-appliance branch circuits.
	3.	210.70(A)(1), 210.52(B)(2)	Check for a lighting outlet controlled by a listed wall-mounted control device on a general lighting circuit.

Residential Rough-In Inspection: Bathrooms			
✓	Item	*NEC* Reference	Inspection Activity
	1.	210.52(D)	Verify that at least one receptacle outlet(s) is installed on a wall or partition adjacent to and within 900 mm (36 in.) of each basin. *(Receptacle not required to be mounted on wall or partition if it is intended to be installed on the side or face of basin cabinet and not more than 300 mm (12 in.) below countertop).*
	2.	406.9(C) and Exceptions	Verify that no receptacles are installed within a zone measured 900 mm (3 ft) horizontally and 2.5 m (8 ft) vertically from the top of the bathtub rim or shower stall threshold *(see exceptions for smaller bathrooms, personal hygiene devices, etc.)*. Space below the zone is also included.
	3.	406.9(C), Ex. No. 4	If an electronic toilet or personal hygiene device *(such as an electronic bidet seat)* will be present at final inspection, verify that the single receptacle outlet will be readily accessible and not located in the space between the toilet and the bathtub or shower area.
	4.	210.11(C)(3) and Ex., 210.23(B)(2)	Verify that receptacle outlets are supplied by dedicated 20-A branch circuit(s). *[Where a 20-ampere circuit supplies a single bathroom, other outlets (lighting, etc.) within the same bathroom shall be permitted to be supplied by this circuit].*
	5.	680.71	Where a hydromassage tub will be installed, verify that pump motor is supplied by a GFCI protected, individual branch circuit *(GFCI device must be installed in a readily accessible location)*.
	6.	680.73	Where a hydromassage tub will be installed, make sure electrical equipment is accessible without damaging the building finish *(access opening)*.
	7.	680.73	Where a hydromassage tub will be installed, and where the hydromassage tub will be cord- and plug-connected with the supply receptacle accessible only through a service access opening, make sure the receptacle face will be within direct view and not more than 300 mm (1 ft) from the opening.
	8.	680.74 and Exceptions	Where a hydromassage tub will be installed, verify bonding of electrical equipment and grounded metal parts. Verify that copper bonding jumper is long enough to terminate on a non-double-insulated replacement motor.
	9.	210.70, 210.70(A)(1) Ex. No. 1	Check for a lighting outlet controlled by a listed wall-mounted control device *(located near an entrance to the room)* on a general lighting circuit *(unless exception in Item 3. above is employed). (Receptacle outlet cannot be used as lighting outlet in bathrooms.)*
	10.	240.24(E)	Verify that overcurrent devices *(panelboards)*, other than supplementary overcurrent protection, are not located in bathrooms.

Residential Rough-In Inspection: Other Habitable Rooms (Bedrooms, Family Rooms, Dens, Etc.)			
✓	Item	*NEC* Reference	Inspection Activity
	1.	210.52, 210.52(A), 210.52(2), 210.70(A)(1) Ex. No. 1	Check receptacle outlets for proper spacing. *(Switched receptacles do not count as required receptacle for wall spacing) (Shall have ½ of a duplex receptacle not switched to be acceptable for wall spacing).*
	2.	210.70(A)(1)	Check for lighting outlets controlled by a listed wall-mounted control device *(including switched receptacles).*
	3.	210.12, 210.12(A)	AFCI protection required on all 120 volt, single-phase, 10-, 15-, and 20-ampere branch circuits that supply outlets or devices in dwelling unit kitchens, family rooms, dining rooms, living rooms, parlors, libraries, dens, bedrooms, sunrooms, recreation rooms, closets, hallways, laundry areas, or similar locations or areas and protected by a listed combination-type AFCI device. *(This would typically be checked and verified on final inspection but could have impact on rough-in inspection.) (AFCI device required to be readily accessible.)*
	4.	210.23(A), 310.3(A)	Verify the presents of any 10-ampere branch circuits. The minimum size conductors for dwelling unit branch circuits is 14 AWG copper or 12 AWG aluminum or copper-clad aluminum.

Residential Rough-In Inspection: Hallways and Foyers			
✓	Item	*NEC* Reference	Inspection Activity
	1.	210.70(A)(2)	Check for at least one lighting outlet controlled by a listed wall-mounted control device *(or automatic-, remote-, or centrally controlled).*
	2.	210.52(H)	Verify that hallways with a continuous length of 3.0 m (10 ft) or more have at least one receptacle outlet to serve that hallway.
	3.	300.3(C)(1), 725.136(A)	Verify doorbell chimes location and wiring method, as well as the voltage of the conductors *(chimes typically located in hallways). (Separation is generally required between 120-volt wiring and Class 2 wiring.)*
	4.	210.52(I)	Verify that foyers *[with an area that is greater than 5.6 m^2 (60 ft^2)]* that are not part of a hallway have a receptacle(s) located in each wall space 900 mm (3 ft) or more in width.

Residential Rough-In Inspection: Stairways			
✓	Item	*NEC* Reference	Inspection Activity
	1.	210.70(A)(2)(1)	Check for at least one lighting outlet controlled by a listed wall-mounted control device *(or automatic-, remote-, or centrally controlled).*
	2.	210.70(A)(2)(3)	Verify that listed wall-mounted control devices are provided at each floor level, landing level, and entry ways where there are six or more risers between levels.
	3.	240.24(F)	Verify that overcurrent devices *(located in panelboards)* are not being located and installed over steps of a stairway *(stairway landings would be acceptable with proper OCPD heights and working space clearances).*

Residential Rough-In Inspection: Closets			
✓	Item	*NEC* Reference	Inspection Activity
	1.	410.16	Where luminaires are installed, check clearances between luminaires and designated closet storage spaces.
	2.	240.24(D), 110.26(B)	Verify that overcurrent devices *(panelboards)* will not be located in the vicinity of easily ignitible material, such as in clothes closets.

Residential Rough-In Inspection: Laundry Area			
✓	Item	*NEC* Reference	Inspection Activity
	1.	210.52(F)	Verify that at least one receptacle outlet is installed for the laundry area equipment.

Residential Rough-In Inspection: Laundry Area *(cont.)*			
✓	Item	*NEC* Reference	Inspection Activity
	2.	210.11(C)(2)	Verify that a dedicated 20-A circuit supplies the laundry outlet(s) and no other outlets other than laundry outlet(s).
	3.	210.50(C)	Check for an appliance receptacle outlet located within 1.8 m (6 ft) of the intended appliance location.
	4.	210.8(A)(11)	Verify that all 125-volt through 250-volt receptacles installed in the laundry area will be provided with GFCI protection. *(GFCI protection device required to be readily accessible.) (Typically enforced on Final Inspection)*
	5.	422.10, 250.134, 250.138, 250.140, 220.54	For other appliances such as 240-volt clothes dryers *(if used)*, check for proper sized branch-circuit conductors, including equipment grounding conductors.
	6.	210.11(C)(2), 210.70(A)(1)	Verify that lighting outlets for the laundry area are supplied from general lighting circuits.
	7.	210.70(A)(1), 210.70(A)(1), Ex. No. 1	Check for a lighting outlet controlled by a listed wall-mounted control device *(located near entrance to the room)* on a general lighting circuit. *(Receptacle outlet cannot be used as lighting outlet in laundry areas.)*

Residential Rough-In Inspection: Attached Garages and Detached Garages or Accessory Buildings with Electric Power			
✓	Item	*NEC* Reference	Inspection Activity
	1.	210.52(G)(1)	For attached garages and in each detached garage with electric power, verify that at least one receptacle outlet is provided in each vehicle bay and not more than 1.7 m (5½ ft) above the floor.
	2.	210.11(C)(4), 210.52(G)(1)	Need at least one 120-volt, 20-ampere branch circuit installed to supply garage receptacle outlets. This circuit shall have no other outlets with the exception of outdoor receptacle outlets. Additional branch circuits rated 15 amperes or greater permitted to serve receptacle outlets other than those required by 210.52(G)(1).
	3.	210.70(A)(2)(1) & (2)	For attached garages and in each detached garage with electric power, check for at least one wall-switch-controlled *(or automatic-, remote-, or centrally controlled)* lighting outlet.
	4.	625.40, 625.54	Verify that each outlet installed for the purpose of electric vehicle supply equipment (EVSE) greater than 16 amperes or 120 volts is supplied by an individual branch circuit and that this outlet(s) is GFCI protected. Each of these circuit(s) shall have no other outlets.
	5.	210.52(G)(2), 210.8(A)(2)	Verify that at least one receptacle outlet is installed in each accessory building with electric power and that GFCI protection is provided.

Residential Rough-In Inspection: Basements and Attics			
✓	Item	*NEC* Reference	Inspection Activity
	1.	210.52(G)(3)	Verify that at least one receptacle outlet is provided in each portion of an unfinished basement areas in addition to any receptacles installed for specific equipment.
	2.	210.63, 210.63(A), 210.8(E)	Verify that a GFCI protected receptacle outlet is installed within 7.5 m (25 ft) *(provided for servicing mechanical equipment)*, on the same level as the HACR equipment.
	3.	422.12, 422.12, Ex. 1 and Ex. 2	Verify that individual branch circuits are supplied for any central heating equipment provided. *(Auxiliary equipment and permanently connected air conditioning equipment permitted to be connected to same branch circuits.)*
	4.	210.70(C)	Verify that a wall-switch-controlled lighting outlet or a lighting outlet containing a switch is provided at the normal entrance to equipment requiring servicing *(lighting outlet needs to be at or near the equipment)*.

Residential Rough-In Inspection: Basements and Attics *(cont.)*			
✓	Item	*NEC* Reference	Inspection Activity
	5.	320.23, 330.23, 334.23, 334.15, 334.15(C), 320.15	Check accessible attics, attic entrances, scuttle holes, basements, and crawl spaces, for clearances from or protection of cable assemblies.
	6.	334.15(B)	Verify that Type NM cable is protected from physical damage where necessary *(passing through a floor, etc.)*.
	7.	210.8(A)(4), 210.8(C)	Verify that receptacles and lighting outlets in crawl spaces are provided with GFCI protection.

Residential Rough-In Inspection: Outdoors			
✓	Item	*NEC* Reference	Inspection Activity
	1.	210.52(E)(1)	Check for at least two receptacle outlets, readily accessible while standing at grade level and located not more than 2.0 m (6½ ft) above grade level, one each at the front and back of the dwelling.
	2.	210.52(E)(3)	All balconies, decks, and porches *[that are within 102 mm (4 in.) horizontally of the dwelling unit]* required to have at least one receptacle accessible from the balcony, deck, or porch. This outlet(s) shall not be located more than 2.0 m (6½ ft) above the balcony, deck, or porch walking surface.
	3.	210.63, 210.63(A), 210.8(E)	Verify that a GFCI protected receptacle outlet is installed within 7.5 m (25 ft) and provided for servicing mechanical equipment and on the same level as the mechanical equipment.
	4.	334.12(A)(9), 334.12(B), 334.15(B), 300.5(B), 300.9	Verify cable assemblies *(such as Type NM cable)* are properly sleeved through cement, concrete or brick. *(Type NM and NMS not allowed in wet or damp locations. The interior of conduit installed in a wet location is considered a wet location [outdoors or beneath a slab (underground)].*
	5.	210.70(A)(2)(2)	Check for exterior lighting outlets at outdoor entrances or exits with grade-level access controlled by a listed wall-mounted control device *(or remote, central, or automatic controlled). (vehicle door in a garage not considered an entrance or exit)*
	6.	300.5(B), 300.5(E), 300.9, 310.10(C), 110.14(B), 310.4, 310.3(C)	Verify that conductors and any splices or terminations installed outdoors are suitable for a wet location.

RESIDENTIAL SERVICE, FEEDER, AND GROUNDING INSPECTIONS CHECKLISTS

Note to Reader: *The following inspection items are typically preformed as part of either the rough-in or final inspection. They have been broken out into a separate checklist for clarity and emphases.*

Residential Service, Feeder, and Grounding Inspections: Services			
✓	Item	*NEC* Reference	Inspection Activity
	1.	Article 220, 220.82(A), 230.42, 230.42(B), 230.79	Review the calculation of service load and determine the minimum size of service conductors.
	2.	310.12(A) through (D), Table 310.12, Table 310.16, Annex D, Example D7	Verify that service conductors and/or service-entrance conductors and the main power feeder are sized properly per the ampacity values and adjustment factors of 310.12 *(83% reduction permitted)*.
	3.	230.66(A) and (B), 230.70(B), 230.70(C)	Verify that service equipment is identified as suitable for use as service equipment and listed as such.
	4.	230.67	Verify that the service supplying the dwelling unit is provided with a Type 1 or Type 2 surge-protective device (SPD) that is an integral part of the service equipment or is located immediately adjacent thereto *(rated not less than 10 kA)*.
	5.	230.62(C)	Verify that barriers are placed in all service equipment such that no uninsulated, ungrounded service busbar or service terminal is exposed to inadvertent contact by personal while servicing load terminations.

Residential Service, Feeder, and Grounding Inspections: Services *(cont.)*			
✓	**Item**	***NEC* Reference**	**Inspection Activity**
	6.	230.85(A) through (E)	Verify that an outdoor, readily accessible emergency disconnect has been installed with all service conductors terminating in a disconnecting means having a short-circuit current rating equal to or greater than the available fault current. This outdoor emergency disconnect(s) must be marked or labeled properly *[see 230.85(E) for marking details]*.
	7.	230.70(A)(1), 230.70(A)(2), 230.91	Verify that service disconnecting means are located at a readily accessible location either outside or inside nearest the point of entrance of the service conductors. (*not in bathrooms*)
	8.	230.71, 230.71(B), 230.72	If more than one service disconnect is employed, verify that service disconnects are grouped together, with no more than six in any one location, located in separate enclosures.
	9.	110.26(A) through (F), 230.70(A)(1), 230.91, 240.24(A) through (F)	Check for proper accessibility, working clearances, and dedicated electrical spaces around service equipment.
	10.	240.24(A) through (F)	For overcurrent devices used with services, verify that they are located in acceptable locations *(not exposed to physical damage, not located in clothes closets, bathrooms, or over steps, etc.)*.
	11.	230.30(B), 230.43, 230.50, 230.51	Check underground service conductor and service-entrance wiring methods for suitability, support, and protection from physical damage.
	12.	230.54	Check for proper drip loop and listed weatherhead on overhead services.
	13.	230.26, 230.28, 230.54(C)	Verify that the point of attachment is of adequate strength and will provide required support.
	14.	230.24	Verify that overhead service conductors are not readily accessible and have proper clearances above roofs, final grade, and driveways.
	15.	230.10	Verify that vegetation *(such as trees)* is not being used for support of overhead service conductors or service equipment.
	16.	230.24(D), 680.9(A), Table 680.9(A), Figure 680.9(A)	Verify that overhead service conductors have proper clearances above and around swimming pools, fountains, and similar aquatic installations.
	17.	230.28	Check service masts for adequate strength and support.
	18.	230.29	Where a substantial structure supporting service conductors over a roof is metal, it shall be bonded by a bonding jumper to the grounded overhead service conductor.
	19.	230.9(A) through (C)	Check for proper clearance of service conductors from building openings, doors, porches, balconies, or similar locations.
	20.	230.32, 230.50(A), 300.5, Table 300.5, 338.12(B)	Check underground service conductors for proper depth, fill, protection, marking, and allowances for ground movement *(USE cable not permitted for interior wiring)*.

Residential Service, Feeder, and Grounding Inspections: System Grounding			
✓	**Item**	***NEC* Reference**	**Inspection Activity**
	1.	250.50, 250.52(A)	Determine which grounding electrodes are present and verify that they are bonded together to form a grounding electrode system. This may require an inspection of the foundation footings *before the footing is poured* to verify if the components of a concrete-encased electrode are present.
	2.	250.52(B)	Verify that no prohibited electrodes are being used in the grounding electrode system *(metal underground gas piping, aluminum electrodes, or metal structure of an in-ground swimming pool)*.
	3.	250.52(A)(5) and (A)(7), 250.53(A), 250.53(B)	Check any rod, pipe, or plate electrodes for proper size, type, and installation.
	4.	250.53(A)(2), 250.53(D)(2)	Verify that all single metal underground water piping, rods, pipes, and plate electrodes are supplemented with additional electrodes.

Residential Service, Feeder, and Grounding Inspections: System Grounding *(cont.)*			
✓	**Item**	***NEC* Reference**	**Inspection Activity**
	5.	250.68(C)(1)	Verify that grounding electrode conductors and/or bonding jumpers are connected to interior metal water piping *(that qualifies as a grounding electrode)* within the first 1.52 m (5 ft) of the point of entrance in the building.
	6.	250.118(B)	Verify that no equipment grounding conductor is being used in a dual role as a grounding electrode conductor *(see exception)*.
	7.	250.64(C)	Verify that grounding electrode conductors are generally installed in one continuous length without a splice.
	8.	250.64(E)	Where grounding electrode conductors are installed in ferrous metal raceways *(for physical protection, etc.)*, verify that the ferrous metal raceways are bonded to and electrically continuous with the enclosed GECs.
	9.	250.64(F), 250.66, 250.66(A) through (C), Table 250.66	Check grounding electrode conductor(s) and bonding jumpers for proper sizing and installation.
	10.	250.8, 250.10, 250.68, 250.70	Check grounding electrode conductor connections to grounding electrodes for proper type, for protection, and for accessibility.
	11.	250.24(C), 250.28, Table 250.102(C)(1)	Verify that the main bonding jumper is installed and of the proper size and type.
	12.	250.104(A), 250.68(B)	Verify metal water piping systems are bonded, bonding jumpers are properly sized, and continuity around removable devices is assured.
	13.	250.104(B)	Verify that all metal piping systems such as metal gas piping are bonded together, and that bonding jumpers are properly sized. *(The equipment grounding conductor for the circuit that may energize the piping shall be permitted to serve as the bonding means).*
	14.	250.80, 250.92,	Verify that service raceways and enclosures are properly grounded and bonded.
	15.	250.94(A) and (B)	Verify existence of intersystem bonding termination device or other means for connecting intersystem bonding and grounding conductors required for communication systems.
	16.	310.12(D), 250.24(C), Table 250.102(C)(1)	Verify the proper sizing and installation of the grounded *(neutral)* conductor.

Residential Service, Feeder, and Grounding Inspections: Feeders and Panelboards			
✓	**Item**	**NEC Reference**	**Inspection Activity**
	1.	210.19, 215.2, Article 220, 310.14, 334.80, 338.12(B)	Review the calculation of feeder and branch circuit loads and verify that the proper conductors or cables are being used. Verify that cables and conductors are properly sized and rated, for ampacity, ambient temperatures and length of conductors.
	2.	110.14(D)	Verify that an approved means was used to achieve the indicated torque values where tightening torque values for terminal connections are indicated on equipment or in installation instructions provided by the manufacturer of electrical equipment *(panelboards, etc,)*.
	3.	408.2(A)	Check of any reconditioned equipment *(panelboards cannot be reconditioned)*.
	4.	408.43	Ensure that no panelboard(s) are installed in the face-up or face-down position.
	5.	408.37, 312.2	Make sure that any panelboards installed outdoors in damp or wet locations are placed or equipped so as to prevent moisture or water from entering and accumulating within the cabinet and mounted so there is at least 6-mm (¼-in.) airspace between the enclosure and the wall or other supporting surface. Enclosures installed in wet locations shall be weatherproof.
	6.	310.12(B)	When sizing the main power feeder, verify that this main power feeder supplies <u>all</u> loads that are part of or associated with the dwelling unit.

Residential Service, Feeder, and Grounding Inspections: Feeders and Panelboards *(cont.)*			
✓	Item	NEC Reference	Inspection Activity
	7.	Article 220, 240.6(A), 408.30, 408.36, 408.58	Verify proper marking, ratings, and overcurrent protection of panelboards.
	8.	110.26, 110.26(E), 240.24	Check for proper location, accessibility, working clearances, and dedicated electrical spaces around panelboards and overcurrent devices.
	9.	210.11(A)	Verify that at least the minimum number of overcurrent devices and branch circuits have been provided.
	10.	210.23(A), 310.3(A)	Verify the presents of any 10-ampere branch circuits. The minimum size conductors for dwelling unit branch circuits is 14 AWG copper or 12 AWG aluminum or copper-clad aluminum.
	11.	408.4(A)	Check for proper circuit directory and identification of all overcurrent devices *(spare positions that contain unused overcurrent devices shall be marked accordingly)*.
	12.	408.6	Make sure all panelboards have a short-circuit current rating not less than the available fault current.
	13.	110.12(A), 408.7, 312.5(A)	Check for unused cable openings and open circuit breaker spaces in panelboard fronts or cabinets *(must be properly closed)*.
	14.	408.36, 408.54	Check panelboards for excessive number of circuits and circuit provisions *(per manufacturer's specifications and listing)*.
	15.	240.24(F)	Verify that overcurrent devices *(located in panelboards)* are not being located and installed over steps of a stairway *(stairway landings would be acceptable with proper OCPD heights and working space clearances)*.
	16.	110.26(B), 240.24(D)	Verify that overcurrent devices *(panelboards)* are not located in the vicinity of easily ignitible material, such as in clothes closets.
	17.	240.24(E)	Verify that overcurrent devices *(panelboards)*, other than supplementary overcurrent protection, are not located in bathrooms.
	18.	408.55	Verify that the wire-bending space within an enclosure containing a panelboard is in conformance.
	19.	250.24(B), 250.142(B), 310.3(D)	Verify that the grounded conductor of a feeder circuit is insulated and isolated from equipment grounding conductors and grounded enclosures.
	20.	215.6, 250.32(B), 250.118, 250.122, 408.40	Verify that panelboards are grounded by an appropriate and properly sized equipment grounding conductor(s).
	21.	408.41	Verify that each grounded conductor terminates within the panelboard on an individual terminal that is not also used for another conductor.
	22.	250.24(D)(1), Table 250.102(C)(1)	Verify the proper sizing of the grounded *(neutral)* conductor for all feeders.

RESIDENTIAL FINAL INSPECTIONS CHECKLISTS

Residential Final Inspection: General Requirements (All Areas)			
✓	Item	NEC Reference	Inspection Activity
	1.		Check for correction of any deficiencies noted on previous inspections.
	2.	110.14(D)	Verify that an approved means was used to achieve the indicated torque values where tightening torque values for terminal connections are indicated on equipment or in installation instructions provided by the manufacturer of electrical equipment *(receptacles, panelboards, etc,)*.
	3.	210.2, 240.2(A) and (B), 406.2, 410.2, 408.2(A), 800.2	Check of any reconditioned equipment *(GFCI devices, AFCI devices, circuit breakers, receptacles, luminaires, panelboards, communication equipment, etc. cannot be reconditioned)*.
	4.	110.14, 110.14(A) through (C), 404.14(C), 406.3(C)	Verify that conductor terminations, splicing methods, and temperature limitations are compatible with conductor materials.

✓	Item	NEC Reference	Inspection Activity
Residential Final Inspection: General Requirements (All Areas) ***(cont.)***			
	5.	230.85(A) through (E)	Verify that an outdoor, readily accessible emergency disconnect has been installed with all service conductors terminating in a disconnecting means having a short-circuit current rating equal to or greater than the available fault current. This outdoor emergency disconnect(s) must be marked or labeled properly *[see 230.85(E) for marking details]*.
	6.	210.23(A), 310.3(A)	Verify the presents of any 10-ampere branch circuits. Verify loads permitted *(lighting outlets, etc.)* and loads not permitted *(receptacle outlets, etc.)* on a 10-ampere branch circuit. The minimum size conductors for dwelling unit branch circuits is 14 AWG copper or 12 AWG aluminum or copper-clad aluminum.
	7.	314.20	Check positioning of boxes intended to be flush with combustible and noncombustible finished surfaces.
	8.	406.5, 406.6	Check for proper positioning of receptacles and faceplates on vertical surfaces *(walls)*.
	9.	314.21	Check for gaps around outlet boxes in walls *[maximum 3 mm (⅛ in.) gap]*.
	10.	*Various NEC requirements*	Verify that all receptacle outlets and luminaires are functioning properly.
	11.	250.146, 250.148, 404.9(B), 406.6(B)	Verify that receptacles are bonded to metal boxes and receptacles, switches, and metal faceplates are grounded.
	12.	406.12 and Exception	Verify that all 15- and 20-ampere, 125- and 250-volt nonlocking-type receptacles in the dwelling unit are tamper-resistant type receptacles *(including their attached and detached garages and accessory buildings) (See exception)*.
	13.	210.8(A)(7)	Verify GFCI protection for all 125-volt through 250-volt receptacles installed in areas with a sink and permanent provisions for food preparation, beverage preparation, or cooking *(this differs from a kitchen)*.
	14.	210.8, 210.8(A)(8)	Verify GFCI protection for all 125-volt through 250-volt receptacles installed within 1.8 m (6 ft) of all sinks *(including the kitchen sink)*. This measurement is to be taken from the top inside edge of the bowl of the sink and measured as the shortest path the power supply cord of an appliance connected to the receptacle could take without piercing a floor, wall, ceiling, or fixed barrier.
	15.	210.8(A)(12)	Check for GFCI protection for all 125-volt through 250-volt receptacles installed in indoor damp and wet locations.
	16.	210.8, 210.12	Verify that all devices that provide GFCI and AFCI protection are installed in a readily accessible location *(just the providing device required to be readily accessible)*.
	17.	210.12(A)(1) through (A)(6), 210.12(B)	Check for listed combination-type AFCI protection on all 120-volt, single-phase, 10-, 15-, and 20-ampere branch circuits that supply outlets and devices in kitchens, family rooms, dining rooms, living rooms, parlors, libraries, dens, bedrooms, sunrooms, recreation rooms, closets, hallways, laundry areas, or similar locations or areas. AFCI devices to be installed in readily accessible locations.
	18.	200.10, 200.11, 406.10, 410.51	Check polarity of devices and luminaries.
	19.	410.6	Verify that all luminaires, lampholders, and retrofit kits are listed.
	20.	314.23, 314.27(A)	Check for outlet boxes at luminaire vertical surface and ceiling mounted locations and that the box is designed for the support of luminaires and lampholders. Check for proper fastening and support of luminaire boxes.
	21.	314.27(C)	Verify where outlet boxes mounted in the ceilings of habitable rooms in locations acceptable for the installation of a ceiling-suspended (paddle) fan, that the outlet box or outlet box system is listed for sole support of a ceiling-suspended (paddle) fan or that the ceiling-suspended (paddle) fan is capable of being supported from the structural framing.

✓	Item	NEC Reference	Inspection Activity
Residential Final Inspection: General Requirements (All Areas) ***(cont.)***			
	22.	314.28(C)	Verify that all junction boxes and conduit bodies are provided with covers compatible with the box.
	23.	314.27(B), (C), and (D)	Check outlet boxes at floor outlets, ceiling-suspended (paddle) fans, and other utilization equipment for listing requirements and suitable installation.
	24.	210.21, 210.24	Verify that device ratings are compatible with circuit and equipment ratings.
	25.	410.116(A) and (B)	Verify that thermal insulation is not installed within 75 mm (3 in.) of recessed luminaire enclosures, wiring compartment, ballast, transformer, LED driver, or power supply except where such luminaire is identified for contact with thermal insulation (Type IC).
	26.	250.8, 250.86, 250.148, 250.148(A) through (C)	Check that all equipment grounding conductors that are spliced or terminated within boxes are connected together and for bonding connections to metal boxes.
	27.	300.15	Check for proper use of connectors and fittings and for protection of cables.
	28.	300.4(F) & (G), 300.5(H), 314.17, 314.102	Check for bushings or equivalent protection for cables entering boxes and other enclosures.
	29.	110.12(A), 314.17(A), 312.5(A), 408.7	Verify that unused openings in boxes and other enclosures are sufficiently closed in a manner substantially equivalent to the wall of the equipment.
	30.	250.110, 250.112, 250.114	Verify that appliances, motors, and other equipment are grounded.
	31.	300.5(A), Table 300.5(A)	Check buried raceways and cables for proper burial depth.
	32.	300.5, 340.10, 334.12(B)(4)	Verify conductors or cables used underground are suitable for the purpose.
	33.	110.3(B)	Check installation of listed equipment for compliance with manufacturer's instructions.
	34.	300.21	Verify that fire rating of assemblies has been restored at electrical penetrations.
	35.	Article 422, Part III	Check for disconnecting means on both permanently connected and cord- and plug-connected appliances.
	36.	Articles 422, 424, 430, and 440	Verify that circuits for mechanical equipment have correct conductor size and overcurrent protection.
	37.	422.12 and Exceptions	Verify that central heating equipment is supplied by an individual branch circuit *(see exceptions)*.
	38.	424.44(E)	Check for GFCI protection for cables installed in any electrically heated floors of bathrooms, kitchens, and in hydromassage bathtub locations.
	39.	760.41(B), 760.121(B)	Verify that NPLFA and PLFA fire alarm systems are not supplied through GFCI or AFCI devices.
	40.	406.3(E), Figure 406.3(E)	Verify that all 125-volt, 15- and 20-ampere receptacles that are controlled by an automatic control device are marked with the controlled symbol and the word "CONTROLLED."

✓	Item	NEC Reference	Inspection Activity
Residential Final Inspection: Kitchen			
	1.	210.52(A) and (C)(1)	Check spacing of receptacles for walls, countertops, and work surfaces, and area behind corner-mounted ranges and sinks.
	2.	210.11(C)(1), 210.52(B)(3)	Verify that a minimum of two 20 ampere small-appliance branch circuits are used for kitchen countertop receptacle outlets.
	3.	210.52(B)(1), 210.52(B)(2) and Exceptions	Verify that small-appliance branch circuits are used for receptacle outlets only, located in kitchen, dining room, breakfast rooms, pantry, and similar areas *[no other outlets permitted on small-appliance branch circuits (with exceptions)]*.

Residential Final Inspection: Kitchen *(cont.)*			
✓	**Item**	**NEC Reference**	**Inspection Activity**
	4.	210.52(B)(1), 210.52(B)(1), Ex. No. 2	Verify that refrigeration equipment is supplied by a small-appliance branch circuit *(or exception applied)*.
	5.	210.52(B)(1), Ex. No. 2	Verify that *(by exception)* specific appliance equipment *(such as the dishwasher or refrigerator)* is supplied by an individual 15- or 20-ampere rated branch circuit.
	6.	210.52(C), 210.52(C)(1)	For the base countertops, verify that receptacle outlets are installed so that no point along the wall line is more than 600 mm (24 in.) measured horizontally from a receptacle outlet in that space. Verify that any countertop or work surface that measures 300 mm (12 in.) or wider is served by at least one small-appliance receptacle outlet.
	7.	210.52(C)(2), 300.5(B), 310.10(C)	If receptacle outlet(s) are installed to serve a kitchen island or peninsular *(this is optional)*, verify proper installation and location. If no receptacle outlet is present, verify that a proper wiring method is present for future addition of receptacle outlet(s). If this wiring method is installed underground *(under the slab)*, verify that the wiring method is identified for a wet location and any enclosed conductors are identified for a wet location as well.
	8.	210.52(C)(3)(1)	Verify that all receptacle outlets installed to serve kitchen countertops or work surfaces are installed on or above, but not more than 500 mm (20 in.) above, a countertop or work surface *(kitchen receptacle outlets prohibited from being installed below the countertop or work surface it is intended to serve)*.
	9.	210.52(C)(3)(2) and (3)	Verify that any receptacle outlet assembly installed to serve a kitchen countertop or work surface is installed properly and that said assembly is listed for use in a countertop or work surface.
	10.	406.5(G)(1)	Check that no receptacle outlets are installed in a face-up position in or on countertop or work surfaces *(unless listed for countertop or work surface applications). (This is not limited to kitchen countertop or work surfaces.)*
	11.	210.52(C)(3)	Ensure that receptacle outlets rendered not readily accessible by appliances fastened in place or appliance garages will not be considered as the required outlet(s) to serve kitchen countertops or work surfaces.
	12.	210.52(C), 210.52(A)(1)	Verify that receptacle outlets serving kitchen countertops and work surfaces are not also being used to serve as receptacle outlets for wall spacing requirements.
	13.	210.8(A)(6), Article 100 Definition: Kitchen	Verify that all 125-volt through 250-volt receptacle outlets installed in the area defined as a kitchen are provided with GFCI protection.
	14.	422.5(A)(7), 210.8, 210.8(D)	Verify that GFCI protection is provided for outlets that supply dishwashers installed in dwelling unit locations. *(GFCI device to be readily accessible)*
	15.	422.16(B)(2)(5)	The receptacle outlet for a cord-and-plug connected built-in dishwasher must be located in the space adjacent to the space occupied by the dishwasher.
	16.	210.12(B)(1), 210.12	Verify that all 120-volt, single-phase, 15- and 20-ampere branch circuits supplying outlets or devices installed in kitchens are AFCI protected. Device providing AFCI protection to be installed in a readily accessible location.
	17.	210.70(A), 210.52(B)(2), 210.70(A)(1) Ex. No. 1	Verify that a lighting outlet controlled by a listed wall-mounted control device is provided and wired on a general lighting circuit *(receptacle outlet cannot be used as lighting outlet in kitchen)*.
	18.	210.23, 422.10, 210.52(B)(1), Ex. No. 2	Verify that properly sized circuits are provided for specific kitchen appliances *(dishwashers, disposals, ranges, cooktops, trash compactors, etc.)*. These appliances can be supplied by an individual 15- or 20-ampere rated branch circuit.
	19.	422.16, Table 400.4	Check for proper type, length, and use of flexible cords for appliance connections.
	20.	406.5(G)(2)	Check cabinet below kitchen sink to verify that no receptacles are installed in a face-up position in the area below a sink *(any sink in the dwelling)*.
	21.	422.16(B)(4)	If a range hood or an over-the-range microwave oven with integral range hood is to be cord-and-plug connected, verify that receptacle is supplied by an individual branch circuit.

Residential Final Inspection: Dining Room			
✓	Item	NEC Reference	Inspection Activity
	1.	210.12, 210.12(B)(3)	Verify that all 120-volt, single-phase, 15- and 20-ampere branch circuits supplying outlets or devices installed in dining rooms are AFCI protected. AFCI devices to be installed in a readily accessible location.
	2.	210.52(A)(1) and (2)	Check receptacle outlets for proper wall spacing.
	3.	210.52(B)(1)	Verify that all required receptacle outlets are supplied by small-appliance branch circuits.
	4.	210.70(A)(1), 210.52(B)(2)	Check for a lighting outlet controlled by a listed wall-mounted control device on a general lighting circuit.

Residential Final Inspection: Bathrooms			
✓	Item	NEC Reference	Inspection Activity
	1.	210.52(D)	Verify that receptacle outlet(s) are installed on a wall or partition adjacent to and within 900 mm (36 in.) of each basin. *[Receptacle not required to be mounted on wall or partition if it is intended to be installed on the side or face of basin cabinet and not more than 300 mm (12 in.) below the top of the basin].* Receptacle outlet assemblies permitted.
	2.	406.9(C) and Exceptions	Verify that no receptacles are installed within a zone measured 900 mm (3 ft) horizontally and 2.5 m (8 ft) vertically from the top of the bathtub rim or shower stall threshold *(see exceptions for smaller bathrooms, personal hygiene devices, etc.)*. Space below the zone is also included.
	3.	406.9(C), Ex. No. 4	If an electronic toilet or personal hygiene device *(such as an electronic bidet seat)* will be present at final inspection, verify that the single receptacle outlet will be readily accessible and not located in the space between the toilet and the bathtub or shower area.
	4.	210.8(A)(1)	Verify that all 125-volt through 250-volt receptacle outlets installed in bathrooms are GFCI protected.
	5.	210.8(A)(10), Article 100 Definitions: Bathroom	Verify that GFCI protection is provided where 125-volt through 250-volt receptacles are installed within 1.8 m (6 ft) of the outside edge of a bathtub or shower stall when installed in areas not defined as a bathroom.
	6.	210.11(C)(3) and Ex., 210.23(B)(2)	Verify that receptacle outlet(s) are supplied by dedicated 120-volt, 20-ampere branch circuit(s). *[Where a 20-ampere circuit supplies a single bathroom, other outlets (lighting, etc.) within the same bathroom shall be permitted to be supplied by this circuit].*
	7.	680.71	Where a hydromassage tub is installed, verify GFCI protection for same and supplied by an individual branch circuit.
	8.	680.74	Where a hydromassage tub will be installed, verify bonding of electrical equipment and grounded metal parts. Verify that copper bonding jumper is long enough to terminate on a non-double-insulated replacement motor.
	9.	680.73	Where a hydromassage tub will be installed, make sure electrical equipment is accessible without damaging the building finish *(access opening)*.
	10.	680.73	Where a hydromassage tub will be installed, and where the hydromassage tub is cord- and plug-connected with the supply receptacle accessible only through a service access opening, make sure the receptacle is within direct view and not more than 300 mm (1 ft) from the opening.
	11.	410.10(D)	Check luminaires and ceiling fan locations over bathtub rim or shower stalls *[Prohibited within a zone 900 mm (3 ft) horizontally and 2.5 m (8 ft) vertically above same]*. Wall-mounted luminaires located in this zone shall be listed for damp locations or listed for wet locations where subject to shower spray.
	12.	210.70(A)(1), 210.70(A)(1) Ex. No. 1	Check for a lighting outlet controlled by a listed wall-mounted control device on a general lighting circuit *[unless 210.11(C)(3), Ex. is applied]. (Receptacle outlet cannot be used as lighting outlet in bathrooms).*

Residential Final Inspection: Other Habitable Rooms (Bedrooms, Family Rooms, Dens)			
✓	Item	NEC Reference	Inspection Activity
	1.	210.52(A)(1) and (2)	Check receptacle outlets for proper spacing *[no point measured horizontally along the floor line of any wall space more than 1.8 m (6 ft) from a receptacle outlet]*. Any wall 600 mm (2 ft) or wider requires receptacle outlet(s).
	2.	210.70(A)(1) and 210.70(A)(1), Ex. No. 1	Check for lighting outlets controlled by a listed wall-mounted control device *(including switched receptacles)*.
	3.	210.12(B)	Check for arc-fault circuit-interrupter *(AFCI)* protection for all branch circuits serving outlets and devices in kitchens, family rooms, dining rooms, living rooms, parlors, libraries, dens, bedroom, sunrooms, recreation rooms, closets, hallways, laundry areas, or similar locations or areas. AFCI devices to be installed in a readily accessible location.

Residential Final Inspection: Hallways and Foyers			
✓	Item	NEC Reference	Inspection Activity
	1.	210.70(A)(2)	Check for at least one lighting outlet controlled by a listed wall-mounted control device *(or automatic-, remote-, or centrally controlled)*.
	2.	210.52(H)	Verify that hallways that are continuous for 3.0 m (10 ft) or more in length have at least one receptacle outlet.
	3.	300.3(C)(1), 725.136(A)	Verify doorbell chimes location and wiring method, as well as voltage rating of the conductors *(chimes typically located in hallways)*. *(Class 2 conductors not to be installed in same enclosure as power conductors.)*
	4.	210.52(I)	Verify that foyers *[with an area that is greater than 5.6 m² (60 ft²)]* that are not part of a hallway have a receptacle(s) located in each wall space 900 mm (3 ft) or more in width.
	5.	210.12(B)	Check for arc-fault circuit-interrupter *(AFCI)* protection for branch circuits supplying outlets or devices in hallways or similar locations or areas. AFCI protection device to be installed in a readily accessible location.

Residential Final Inspection: Stairways			
✓	Item	NEC Reference	Inspection Activity
	1.	210.70(A)(2)	Check for at least one lighting outlet controlled by a listed wall-mounted control device *(or automatic-, remote-, or centrally controlled)*.
	2.	210.70(A)(2)(3)	Verify that listed wall-mounted control device are provided at each floor level and landing level that includes a stairway entry way where there are six or more risers.
	3.	240.24(F)	Verify that panelboards (*overcurrent devices*) are not located in a stairway over steps *(landings would be acceptable with proper OCPD heights, and working clearances)*.
	4.	210.12(B)	Check for arc-fault circuit-interrupter (*AFCI*) protection for branch circuits supplying outlets or devices in hallways and similar locations or areas *(stairways)*. This protection device shall be in a readily accessible location.

Residential Final Inspection: Laundry Area			
✓	Item	NEC Reference	Inspection Activity
	1.	210.52(F)	Verify that at least one receptacle outlet is installed for the laundry area designated for the installation of laundry equipment.
	2.	210.8(A)(11)	Verify GFCI protection for all 125-volt through 250-volt receptacle outlets installed within the laundry area.
	3.	210.11(C)(2)	Verify that a dedicated 20-ampere circuit supplies the laundry receptacle outlet(s) and no other outlets.
	4.	210.11(C)(2)	Verify that lighting outlets for the laundry area are supplied from general lighting circuits.

Residential Final Inspection: Laundry Area *(cont.)*			
✓	Item	NEC Reference	Inspection Activity
	5.	210.70(A)(1), 210.70(A)(1), Ex. No. 1	Check for a lighting outlet controlled by a listed wall-mounted control device *(located near entrance to the room)* on a general lighting circuit. *(Receptacle outlet cannot be used as lighting outlet in laundry areas.)*
	6.	210.12(B)	Verify AFCI protection for all for branch circuits supplying 125-volt, 15- and 20-ampere outlets and devices installed within the laundry area. AFCI protection device to be installed in a readily accessible location.
	7.	210.50(C)	Check for an appliance receptacle outlet within 1.8 m (6 ft) of the intended laundry appliance location.
	8.	210.21, 210.24, 250.140	Check for proper receptacle rating based on branch-circuit rating, including receptacles for electric clothes dryers *(if provided)*.
	9.	250.140, 250.134, 250.138	Verify 4-prong outlet *(equipment grounding conductor provided)* for 240-volt clothes dryer in new installations.

Residential Final Inspection: Closets			
✓	Item	NEC Reference	Inspection Activity
	1.	410.11, 410.16(A) through (D)	If closet luminaire is installed, verify that luminaire is acceptable type for installation in a closet *(lighting outlet not required in clothes closet)*.
	2.	410.16(A), (B), and (C)	If closet luminaire is installed, check clearances between luminaires and storage spaces *(Open, partially open, or pendant luminaires are not permitted)*.
	3.	240.24(D)	Verify that no overcurrent devices *(panelboards)* have been installed in the vicinity of easily ignitible materials *(such as in a clothes closet)*.
	4.	210.12(B)	Verify AFCI protection for branch circuits supplying all 125-volt, 15- and 20-ampere outlets and devices installed within closets. AFCI protection device to be installed in a readily accessible location.

Residential Final Inspection: Basements and Attics			
✓	Item	NEC Reference	Inspection Activity
	1.	210.52(G)(3)	Verify that at least one receptacle outlet is provided in each separate unfinished basement area in addition to any receptacles installed for specific equipment.
	2.	210.63(A), 210.8(E)	Verify that a GFCI protected receptacle outlet is installed within 7.5 m (25 ft) and provided for servicing mechanical equipment, on same level as equipment.
	3.	210.8, 210.8(A)(5), 210.8(A), Ex. No. 2	Verify that GFCI protection is provided for all 125-volt through 250-volt receptacle outlets in basements and unfinished portions of basements *(other than receptacles for permanently installed fire/burglar alarms)*. GFCI devices to be installed in a readily accessible location.
	4.	422.5(A)(6)	Check for GFCI protection for any installed sump pumps.
	5.	422.12, 422.12, Ex. No. 1 422.12, Ex. No. 2	Verify that individual branch circuits are supplied for central heating equipment, if any. *(Auxiliary equipment and permanently connected air conditioning equipment permitted to be connected to same branch circuits.)*
	6.	210.70(C)	Verify that a lighting outlet controlled by a wall switch or listed wall-mounted control device *(located at the entrance to equipment requiring servicing)* or a lighting outlet containing a switch is provided. Lighting outlet should be located at or near equipment.
	7.	320.23, 330.23, 334.23, 334.15, 330.15	Check accessible attics, attic entrances, and scuttle holes for clearances from or protection of cable assemblies.
	8.	210.8(C)	Verify GFCI protection provided for crawl space lighting outlets *(not exceeding 120 volts)*

Residential Final Inspection: Attached Garages and Detached Garages or Accessory Buildings with Electric Power			
✓	**Item**	**NEC Reference**	**Inspection Activity**
	1.	210.52(G)(1)	For garages, verify that at least one receptacle outlet is provided in each vehicle bay and located not more than 1.7 m (5½ ft) above the floor.
	2.	210.11(C)(4), 210.52(G)(1)	Need at least one 120-volt, 20-ampere branch circuit installed to supply garage receptacle outlets. This circuit shall have no other outlets with the exception of outdoor receptacle outlets. Additional branch circuits rated 15 amperes or greater permitted to serve receptacle outlets other than those required by 210.52(G)(1).
	3.	210.8(A)(2)	Verify that GFCI protection is provided for all 125-volt through 250-volt receptacles installed in garages and accessory buildings.
	4.	210.70(A)(2)(1) and (2)	For garages, verify that a lighting outlet controlled by a listed wall-mounted control device is provided interior, and a lighting outlet controlled by a listed wall-mounted control device is also provided to illuminate the exterior side of outdoor entrances or exits with grade level access.
	5.	625.40, 625.54	Verify that each outlet installed for the purpose of electric vehicle supply equipment (EVSE) greater than 16 amperes or 120 volts is supplied by an individual branch circuit and that this outlet(s) is GFCI protected. Each of these circuit(s) shall have no other outlets.
	6.	625.54	Verify that all receptacles installed for the connection of electric vehicle charging are provided with GFCI protection for personnel.
	7.	210.52(G)(2)	At least one receptacle outlet shall be installed in each accessory building with electric power.

Residential Final Inspection: Outdoors			
✓	**Item**	**NEC Reference**	**Inspection Activity**
	1.	210.52(E)(1)	Check for at least two GFCI protected receptacle outlets, accessible at grade level, one each located at the front and back of the dwelling.
	2.	210.52(E)(3)	Verify that all balconies, decks, and porches *[located within 102 mm (4 in.) horizontally of the dwelling unit]* have at least one receptacle outlet that is accessible from the balcony, deck, or porch. This receptacle outlet shall not be located more than 2.0 m (6½ ft) above the balcony, deck, or porch walking surface.
	3.	210.63(A), 210.8(E)	Verify that a GFCI protected, 125-volt, single-phase, 15- or 20-ampere receptacle outlet is installed within 7.5 m (25 ft) provided for servicing mechanical equipment, on same level as the equipment.
	4.	230.85(A) through (E)	Verify that an outdoor, readily accessible emergency disconnect has been installed with all service conductors terminating in a disconnecting means having a short-circuit current rating equal to or greater than the available fault current. This outdoor emergency disconnect(s) must be marked or labeled properly *[see 230.85(E) for marking details]*.
	5.	210.8(A)(3), 406.4(D)(3), 426.28	Verify that all 125-volt through 250-volt outdoor receptacle outlets are GFCI protected *(except for a receptacle that is not readily accessible and supplied by circuits for deicing or snow-melting equipment)*.
	6.	210.8(F) and 210.8(F) Ex. No. 1 and Ex. No. 2	Check that all outdoor outlets *(including outlets for AC units)* supplied by single-phase branch circuits rated 150 volts to ground or less, 50 amperes or less are provided with GFCI protection *(see exceptions)*.
	7.	406.9(A)	Verify that receptacles installed outdoors in a location protected from the weather or in other damp locations have an enclosure for the receptacle that is weatherproof when the receptacle is covered *(attachment plug cap not inserted and receptacle covers closed)*.
	8.	406.9(B)(1)	Verify that all 15- and 20-ampere, 125- and 250-volt receptacles installed in a wet location *(such as outdoors)* are provided with a cover identified as “extra duty” that is weatherproof whether or not the attachment plug cap is inserted or not *(in-use covers or outlet box hood covers)*.

✓	Item	NEC Reference	Inspection Activity
	Residential Final Inspection: Outdoors *(cont.)*		
	9.	406.9(A), 406.9(B)(1), 406.4(D)(6)	Verify that all 15- and 20- ampere, 125- and 250- volt non-locking receptacles installed in damp or wet locations are of the listed weather-resistant type.
	10.	406.12 and Exception, 406.4(D)(5)	Verify that all 15- and 20- ampere, 125- and 250- volt non-locking receptacles installed in outdoor locations are of the listed tamper-resistant type *(including attached and detached garages and accessory buildings)*.
	11.	210.70(A)(2)(2)	Check for exterior lighting outlet(s) at outdoor entrances or exits with grade-level access controlled by a listed wall-mounted control device *(or remote-, central-, or automatic-controlled). (The garage vehicle door is not considered an outdoor entrance or exit, luminaires optional for this location.)*
	12.	300.15, 314.27(A)	Check for boxes at exterior luminaire locations.
	13.	410.10(A), 110.3(B)	Verify that luminaires installed outdoors are suitable for damp or wet locations.
	14.	410.10(A)	Verify that luminaires installed in wet locations are installed such that water cannot enter or accumulate in wiring compartment.
	15.	314.15	If approved drainage openings for boxes or conduit bodies listed for use in damp or wet locations are installed in the field, verify that they are not smaller than 3 mm (⅛ in.) and not larger than 6 mm (¼ in.).
	16.	Article 411, 411.3, 411.4	If low-voltage lighting systems are installed, verify that they and listed and installed as required per *NEC* Article 411.

✓	Item	NEC Reference	Inspection Activity
	Residential Final Inspection: Service Equipment, Feeders, and Panelboards		
	1.	See "Residential Service, Feeder, and Grounding Inspections" checklist	Verify that all checklist items from "Residential Service, Feeder, and Grounding Inspections" checklist are in compliance.
	2.	Article 250	Review bonding and grounding requirements if not completed during previous inspections.
	3.	230.66(A) and (B), 230.70(B), 230.70(C)	Verify service equipment is marked and identified as being suitable for use as service equipment.
	4.	230.67(A) through (E)	Verify that the service supplying the dwelling unit is provided with whole-house surge-protective device (SPD) that is an integral part of the service equipment or is located immediately adjacent thereto.
	5.	230.85(A) through (E)	Verify that an outdoor, readily accessible emergency disconnect has been installed with all service conductors terminating in a disconnecting means having a short-circuit current rating equal to or greater than the available fault current. This outdoor emergency disconnect(s) must be marked or labeled properly *[see 230.85(E) for marking details]*.
	6.	310.6(A), (B), and (C), 215.12(A), (B) and (C), 200.6, 250.119	Ensure that all service and feeder conductors are properly identified by insulation color or markings.
	7.	200.2(B)	Verify that the continuity of the grounded conductor does not depend on a connection to a metallic enclosure, raceway, or cable armor.
	8.	210.4(B) and (D), 200.4(B)	Verify grouping of all multiwire branch circuit conductors for readily identification *(unless grouping is obvious)*. Verify simultaneous disconnection for all multiwire branch circuits.
	9.	240.4, 110.14(C), 310.14, 210.20, 210.22, 215.3, 230.42, 240.6(A)	Check overcurrent devices for compatibility with conductors *(terminals, ratings, and ampacities)*.
	10.	408.43	Ensure that no panelboard(s) are installed in the face-up or face-down position.

Residential Final Inspection: Service Equipment, Feeders, and Panelboards *(cont.)*			
✓	**Item**	**NEC Reference**	**Inspection Activity**
	11.	230.62(C)	Verify that barriers are placed in all service equipment such that no uninsulated, ungrounded service busbar or service terminal is exposed to inadvertent contact by personal while servicing load terminations with the service disconnect in the open position.
	12.	110.22, 230.70(B), 408.4(A)	Check for proper circuit directory and identification of all overcurrent devices and disconnects *(spare positions that contain unused overcurrent devices shall be marked accordingly).*
	13.	110.12(A), 408.7, 312.5(A)	Ensure there are no open spaces in panelboard fronts or cabinets.
	14.	300.3(C)(1), 725.136(A)	Verify doorbell chimes location and wiring method, as well as voltage rating of the conductors *(chimes typically located in hallways, but transformer could be located at panelboard). (Class 2 conductors not to be installed in same enclosure as power conductors.)*
	15.	408.36(D)	Verify that any backfed overcurrent devices are secured in place by an additional fastener that requires other than a pull to release the device from the mounting means on the panelboard bus.
	16.	422.31(A), (B), and (C), 110.25	If a circuit breaker serves as the disconnecting means for an appliance where the circuit breaker is not within sight from the appliance and is capable of being locked in the open position, verify that the provision for locking or adding a lock to the disconnecting means is installed on or at the circuit breaker and remains in place with or without the lock installed.
	17.	408.37, 312.2	For panelboards installed outdoors, verify that panelboard cabinet is weatherproof and listed for damp/wet locations.
	18.	312.2	For panelboards installed outdoors, check for minimum 6 mm (¼ in.) air gap between enclosure and wall. Also, check for fittings listed for a wet location where raceways and cables entering above the level of uninsulated live parts.
	19.	250.80, 250.92,	Verify that metal service raceways and metal enclosures are properly grounded and bonded.
	20.	250.94(A) and (B)	Verify the presence of an intersystem bonding termination device or other means, for connecting intersystem bonding and grounding conductors required for communication systems.

Answers

CHAPTER 1

	Answer	*NEC* Section	IRC Section
1.	False (1826)	Ohm's Law	Ohm's Law
2.	True	Ohm's Law	Ohm's Law
3.	False (Ampere)	Ohm's Law	Ohm's Law
4.	True	Ohm's Law	Ohm's Law
5.	True	Ohm's Law	Ohm's Law
6.	A.	Ohm's Law	Ohm's Law
7.	True	Ohm's Law	Ohm's Law
8.	True	90.2(A)	(*NEC* Default)
9.	B.	110.3(C) and Informational Note	(*NEC* Default)
10.	True	Article 100, Definitions — Premises Wiring Premises Wiring (System)	Chapter 35, Definitions — Premises Wiring (System)
11.	B.	110.2	E3403.1
12.	D.	Article 100, Definitions — Location(s)	Chapter 35, Definitions — Location(s)
13.	C.	90.4(B), Article 100, Definitions	Chapter 35, Definitions
14.	C.	110.14(A)	E3406.10
15.	B.	240.4(D)(6)	Table E3705.5.3
16.	D.	Chapter 9, Table 8	(*NEC* Default)
17.	C.	Table 310.4(1)	(*NEC* Default)
18.	B.	Table 310.16	Table E3705.1
19.	C.	Table 310.4(1)	(*NEC* Default)
20.	B.	Table 310.16	Table E3705.1
21.	B.	Table 310.4(1), 310.10(A) and (C)	(*NEC* Default)
22.	False	310.15(C)(1)(b)	E3705.3, Ex. No. 1
23.	False	310.3(A), 240.4(D)	E3406.3
24.	A.	310.15(B)(1), 310.16(3), Table 310.16	E3705.2, Table E3705.2
25.	D.	310.16(3), Table 310.16	Table E3705.1 and Table E3705.2
26.	B.	Table 310.16 and Table 310.15(B)(1)(1)	Table E3705.1 and Table E3705.2
27.	B.	110.14(C)(1)(a)(1)	E3705.4.1
28.	A.	110.14(C)(1)(b)(1)	E3705.4.2
29.	C.	310.15(C)(1), Table 310.15(C)(1)	E3705.3, Table E3705.3
30.	C.	110.14(B)	E3406.11
31.	C.	110.12(A)	E3404.6
32.	B.	110.11	E3404.5
33.	A.	110.12(B)	E3404.7
34.	C.	90.4(C), *NEC* 100, *Definitions - Special Permission*	R104.10
35.	C.	110.2, Article 100, Definitions Approved	Chapter 35, Definitions
36.	B.	Article 100, Definitions — Locations, Textbook	Chapter 35, Definitions
37.	D.	312.2	E3907.2
38.	A.	UL Product Spec (ZMVV), Text in book	*(NEC Default)*

39.	C.	Table 310.4(1), Text in book, Table 1.3 in book	*(NEC Default)*
40.	B.	Table 310.15(C)(1), 334.80, Text in book	Table E3705.3, E3705.4.4

CHAPTER 2

	Answer	*NEC* Section	IRC Section
1.	C.	*NEC* 100, Definitions — Habitable Room	Chapter 35, Definitions — Habitable Room
2.	B.	220.41 E3704.4	
3.	B.	220.41(1)	*(NEC Default)*
4.	A.	210.11(C)(1), 210.52(B)(1)	E3703.2, E3901.3
5.	B.	210.11(C)(2), 210.52(F)	E3703.3
6.	D.	220.52(A) and (B)	Table E3704.2(1)
7.	True	Article 100, Definitions — Branch Circuit	Chapter 35, Definitions — Branch Circuit
8.	True	210.52(B)(1) and (B)(2)	E3901.3, E3901.3.1
9.	True	Article 100, *Definitions— Branch Circuit, Individual*	Chapter 35, *Definitions — Branch Circuit, Individual*
10.	True	Article 100, Definitions — Feeder	Chapter 35, Definitions — Feeder
11.	True	220.5(A)	(*NEC* Default)
12.	C.	220.54	Table E3704.2(1)
13.	C.	210.11, 220.41 (See examples in Informative Annex D)	Table E3704.2(1)
14.	False	220.52(A) and (B), 220.41	Table E3704.2(1)
15.	B.	220.5(C)	E3704.4
16.	True	220.85	(*NEC* Default)
17.	D.	310.12(A) and (B)	E3603.1.1, E3603.1.2
18.	C.	310.12(A) and (B)	E3603.1.1, E3603.1.2

CHAPTER 3

	Answer	*NEC* Section	IRC Section
1.	B.	300.4(G)	E3906.1.1
2.	A.	312.2	E3907.2
3.	B.	312.6(B)(1), Table 312.6(A)	E3907.9, Table E3907.9.1(2)
4.	False	312.2, 408.37	E3907.2
5.	A.	312.3	E3907.3
6.	C.	312.4	E3907.4
7.	True.	408.41	E3706.4
8.	C.	408.36, Ex. No. 1	*(NEC Default)*
9.	True	408.40	*(NEC Default)* and E4001.5
10.	D.	110.12(A)	E3404.6
11.	False	312.6(B)(2), Ex. No. 2, Table 312.6(A)	*(NEC Default)*
12.	False	312.6(C), 300.4(G)	E3906.1.1
13.	True	312.6(C), 300.4(G)	E3906.1.1
14.	D.	312.8(A)(2)	E3907.1.1(2)
15.	True	404.4(A), 312.2	E3907.2

16.	D.	312.6(C); 300.4(G)	E3907.6, E3906.1.1
17.	True	404.6(A)	(*NEC* Default)
18.	C.	404.8(A), 240.24(A)	E4001.6, E3705.7(1) and (6)
19.	True	250.24(B)	E3607.2
20.	D.	408.4(A)	E3706.2
21.	True	408.30	E3706.1
22.	A.	408.4(A)	E3706.2
23.	B.	408.36	E3706.3
24.	C.	110.12(A)	E3404.6
25.	True	408.36(D)	E3706.5
26.	True	408.40	*(NEC Default)* and E4001.5
27.	B.	312.2	E3907.2
28.	A.	312.8(A)(4)	E3907.1.1(3)
29.	B.	230.62(C)	*(NEC Default)*
30.	A.	Chapter 3 text in book	Chapter 3 text in book
31.	C.	404.7	E4001.3
32.	False	404.7, 240.81	E4001.3

CHAPTER 4

	Answer	*NEC* Section	IRC Section
1.	B.	Article 100, Definitions – Service Conductors, Overhead	Chapter 35, Definitions – Service Conductors, Overhead
2.	True	230.28(A)	E3604.5.1
3.	D.	230.9(A)	E3604.1
4.	D.	230.24(A)	E3604.2.1
5.	C.	230.43	E3605.2, Table E3801.4
6.	D.	230.26	E3604.3
7.	B.	230.24(B)(2)	E3604.2.2(2)
8.	False	230.43	E3605.2, Table E3801.4
9.	False	230.79(C) (100 amperes)	E3602.1
10.	C.	230.90(A)	E3603.3
11.	C.	230.54(C), Exception	E3605.9.3, Exception
12.	D.	230.50(B)(1)	E3605.5
13.	True	408.36, Ex. No. 1	(*NEC Default*)
14.	C.	230.24(A), Ex. No. 5	E3604.2.1, Ex. No. 5
15.	C.	230.6	(*NEC Default*
16.	False	230.2	E3601.2
17.	B.	110.26(A)(2)	E3405.2
18.	B.	230.24(A), Ex. No. 2	E3604.2.1, Ex. No. 2
19.	True	230.40, Ex. No. 4; 210.25	(*NEC* Default)
20.	D.	230.85	E3601.8
21.	D.	310.12 and Table 310.12	E3603.1 and Table E3603.1.1
22.	A.	230.29	E3604.6
23.	D.	230.67	E3606.5
24.	A.	Article 100, Definitions – Supplementary Overcurrent Protective Device	(*NEC Default*
25.	B.	230.85(E)	E3601.8

26.	True	230.67(D)	E3606.5.3
27.	B.	230.71(B)(1), 230.72(A)	E3601.7
28.	C.	230.46	E3605.3
29.	B.	230.24(A), Ex. No. 2	E3604.2.1, Ex. No. 2

CHAPTER 5

	Answer	*NEC* Section	IRC Section
1.	D.	Article 100, Definitions – Bonding Jumper, Supply-Side	Chapter 35, Definitions – Bonding Jumper, Supply-Side
2.	A.	250.64(A)(3)	E3610.2(3)
3.	C.	250.66, Table 250.66	E3603.4, Table E3603.4
4.	D.	250.66, Table 250.66	E3603.4, Table E3603.4
5.	False	Article 100, Definitions — Grounded (Grounding)	Chapter 35, Definitions — Grounded (Grounding)
6.	True	250.24(B), 250.142(B)	E3607.2, E3908.7, E3908.8
7.	C.	250.52(A)(3)	E3608.1.2
8.	D.	250.28(D)(1), Table 250.102(C)(1)	E3607.5, E3609.5, Table E3603.4
9.	B.	250.102(E)(2)	E3908.21.2
10.	B.	250.4(A)(5)	E3908.5
11.	False	250.52(A)(1)	E3608.1.1
12.	A.	250.112 and 250.112(L)	(*NEC Default*)
13.	B.	250.94(A)(2), 250.94(B)	E3609.3.1(2), E3609.3.2
14.	D.	250.94(A)(4)	E3609.3.1(4)
15.	A.	250.28(B)	(*NEC Default*)
16.	True	250.53(A)(2)	E3608.4
17.	B.	250.52(B)(1)	E3608.6
18.	True	250.32(B)(1)	E3607.3.1
19.	D.	250.8(A), 250.8(B)	E3406.14.1, E3406.14.2
20.	False	250.68(C)(3)(c)	E3611.5
21.	D.	250.53(A)(2), Exception	E3608.4, Exception
22.	A.	250.64(B)(2)	E3610.2
23.	D.	250.28(A)	(*NEC Default*)
24.	C.	250.53(A)(4)	E3608.1.4.1
25.	True	250.64(A)(2)	E3610.2(2)
26.	D.	250.52(A)(5)(1)	E3608.1.4(1)
27.	C.	250.104(B)	E3609.7
28.	C.	250.52(A)(3)	E3608.1.2
29.	False	250.64(C)	E3610.1, Exception
30.	C.	250.92(B)	E3609.4
31.	True	250.64(D)(1)	(*NEC Default*
32.	False	250.32(A), Exception	E3607.3, Exception
33.	D.	250.68(C)(3)	E3611.5
34.	C.	250.52(A)(2)	(*NEC Default*)
35.	B.	Article 100, Definitions – Likely to Become Energized	Chapter 35, Definitions – Likely to Become Energized

CHAPTER 6

	Answer	*NEC* Section	IRC Section
1.	True	Article 100, Definitions — Feeder	Chapter 35, Definitions – Feeder
2.	False	215.3 (Part I of *NEC* Article 240 & IRC Chapter 37)	E3705.5
3.	B.	250.122(B), Chapter 9, Table 8	E3908.13
4.	True	310.12(C)	E3704.1
5.	True	215.6	E3704.6
6.	C.	215.2(A)(1)	E3701.2.1
7	True	Article 338 (also see text in the book)	(See text in the book)
8.	C.	338.2, 338.10(B)(4)(b)	E3801.2, Table E3801.4
9.	False	338.10(B)(1)	(*NEC Default*)
10.	True	(See text in this book)	(See text in this book)
11.	D.	(See text in this book)	(See text in this book)
12.	B.	240.21	E3705.5
13.	True	240.21(B)(1)	(*NEC Default*)
14.	True	250.24(B)	E3607.2, E3908.7
15.	B.	Chapter 9, Table 1	Table E3907.8
16.	True	352.12(B)	(*NEC Default*)
17.	A.	352.44	(*NEC Default*)
18.	D.	348.42, 350.42	(*NEC Default*)
19.	True	338.12(B)(1), 310.4, Table 310.4(1)	(*NEC Default*)
20.	D.	300.4(A) and (D), 338.10(B)(4)(a), 334.17	E3802.2.2, Table E3802.1
21.	C.	314.28(A)(1)	(*NEC Default*)
22.	C.	Annex C, Table C.9	E3904.6, Table E3904.6(9)
23.	A.	356.30(1)	(*NEC Default*)
24.	False	250.24(B)	E3607.2, E3908.7

CHAPTER 7

	Answer	*NEC* Section	IRC Section
1.	D.	422.31(A)	E4101.5, Table E4101.5
2.	C.	422.30	E4101.5, Table E4101.5
3.	D.	422.16(B)(2)(2)	E4101.3, Table E4101.3
4.	D.	422.13, 422.10(A)(3)	E3702.10, E3701.2.1
5.	A.	422.31(A) and (C)	E4101.5, Table E4101.5
6.	True	406.5(G)(2)	(*NEC Default*)
7.	False	422.16(B)(1)	E4101.3, Table E4101.3
8.	True	422.16(B)(3)	(NEC Default)
9.	C.	422.16(B)(2)(1)	E4101.3, Table E4101.3
10.	C.	422.31(B)	E4101.5, Table E4101.5
11.	B	422.31(C), 430.110(A)	E4101.5, Table E4101.5
12.	D.	314.27(C)	E3905.8].
13.	True	422.17	E4101.2

14.	C.	422.12	E4101.5, Table E4101.5
15.	True	250.114	(*NEC Default*)
16.	True	422.16(A)(1)	E3909.1
17.	True	422.33(B)	(*NEC Default*
18.	B.	422.13, 422.10(A)(3)	E3702.10, E3701.2.1
19.	True	422.11(A)	E4101.4.1
20.	True	422.16(B)(1), (2), and (4), Exceptions	(*NEC Default*)

CHAPTER 8

	Answer	*NEC* Section	IRC Section
1.	False	424.20(A)	E4101.5, Table E4101.5
2.	D.	424.94	(*NEC Default*
3.	C.	424.4(A)	E3702.10(2)
4.	True	424.46	(*NEC Default*)
5.	B.	424.41(A)	(*NEC Default*)
6.	A.	424.41(F)	(*NEC Default*)
7.	B.	424.39	(*NEC Default*)
8.	D.	424.11	(*NEC Default*)
9.	True	Article 100 – Heating Panel Set	(*NEC Default*)
10.	True	210.52 Informational Note	(*NEC Default*)
11.	A.	424.19(A)(2)(2)	E4101.5, Table E4101.5
12.	D.	424.36	(*NEC Default*)
13.	D.	424.48	(*NEC Default*)
14.	D.	424.38(B), 424.44	(*NEC Default*)
15.	D.	424.44(E)	E3902.14
16.	D.	424.58	(*NEC Default*)
17.	C.	424.101(A)	(*NEC Default*)
18.	True	210.52	(*NEC Default*)
19.	B.	424.13	

CHAPTER 9

	Answer	*NEC* Section	IRC Section
1.	A.	440.14, Article 100, Definitions — In Sight From	E4101.5, Table E4101.5
2.	D.	440.63	(*NEC Default*)
3.	B.	440.32	(*NEC Default*)
4.	False	440.14	Table E4101.5, Note a.
5.	True	440.63	E4101.5, Table E4101.5
6.	True	440.4(A)	(*NEC Default*)
7.	True	440.14	Table E4101.5, Note a.
8.	B.	440.62(C)	E3702.12.2
9.	C.	210.63 and 210.63(A)	E3901.12
10.	C.	440.22(A) and 440.22(A), Ex. No. 2	(*NEC Default*)

11.	D	440.64	*(NEC Default)*
12.	D	440.65	*(NEC Default)*
13.	B.	440.12(B)(2)	*(NEC Default)*

CHAPTER 10

	Answer	*NEC* Section	IRC Section
1.	D.	210.8(A)	E3902
2.	D.	210.8(A)	E3902
3.	False	210.52(B)(1), Ex. No. 2	E3901.3, Ex. No. 2
4.	B.	Article 100, Definitions - GFCI, Informational Note	*(NEC Default)*
5.	B.	210.8	E3902.15
6.	D.	210.8(E), 210.63(A), 210.8(A)(11), 210.8(F), Ex. No. 1	*(NEC Default)*)
7.	D.	210.8(A)(1) and Ex. No. 4	E3902.1
8.	True	210.8(A)(2)	E3902.2
9.	B.	210.12(B), Ex. No. 1	E3902.17, Ex.
10.	C.	210.8(A) and Ex. No. 1	E3902.3, Ex, E4101.7
11.	False	210.8(A)(5) and Ex. No. 3	*(NEC Default)*
12.	False	210.8(A)(2)	E3902.2
13.	C.	210.8(A)(8)	E3902.7
14.	B.	210.8(A)(8)	E3902.7
15.	D.	210.12(A)	E3902.17
16.	D.	Local Building Codes	R314.3
17.	True	210.8(A)(12)	E3902.10
18.	D.	406.12(1)	E4002.14
19.	B.	680.71	E4209.2
20.	D.	406.12, Exception to (1) through (10)	E4002.14, Exception
21.	C.	210.8(A)(10)	E3902.8
22.	False	210.8(11)	E3902.9
23.	D.	210.12(E), Exception	E3902.18, Exception
24.	True	422.5(A)(7)	E3902.11
25.	D.	210.8	E3902.15
26.	C.	210.8(C)	E3902.4
27.	True.	210.8(A)(9)	E3902.12
28.	B.	Article 100, Definitions	Chapter 35, Definitions
29.	D.	210.8(A)(5)	E3902.5
30.	B.	210.8(A)(6)	E3902.6
31.	D.	210.8(D)	*(NEC Default)*

CHAPTER 11

	Answer	*NEC* Section	IRC Section
1.	D.	210.52(C)	E3901.4.1
2.	A.	210.52(A)(2)(1)	E3901.2.2
3.	A.	210.50(C)	E3901.5

4.	False	210.70(A)(2)(1)	E3903.3
5.	B.	210.21(B)(2), Table 210.21(B)(2)	E3702.3
6.	A.	210.21(B)(1)	E4002.1.1
7.	C.	210.23(A)(1) and (A)(2)	
8.	True	210.52(A)(1)	E3901.2.1
9.	B.	314.17(B)(2), Exception	E3905.3.1, Exception
10.	C.	210.23(C)	E3702.4
11.	True	404.9(B), Ex. No. 2	E4001.11.1, Ex. No. 2
12.	B.	210.70(A)(2)(2)	E3903.3
13.	B.	210.19(C)	E3702.9.1
14.	C.	210.52(C)(3)(1)	E3901.4.3(1)
15.	False	210.52(C)(2)	E3901.4.2(1)
16.	False	210.52(C)(3)	E3901.4.3(3)
17.	C.	210.4(B)	E3701.5.1
18.	D.	210.6(A)(1)	E3702.1
19.	D.	300.5, Table 300.5	E3803.1, Table E3803.1
20.	A.	300.14	E3406.11.3
21.	D.	300.4(D)	E3802.2.2, Table E3802.1
22.	C.	406.9(A) and (B)	E4002.8 and E4002.9
23.	B.	300.5, Table 300.5 — Column 4	E3803.1, Table E3803.1
24.	True	250.8(A)	E3406.14.1
25.	B.	334.80	E3705.4.4
26.	D.	300.5(D)(1)	E3803.3
27.	B.	210.52(C)(1)	E3901.4.1
28.	True	340.12(9)	E3802.3.3, Table E3801.4
29.	C.	210.52(I)	E3901.11
30.	True	310.3(A)	E3406.3
31.	True	210.52(H)	E3901.10
32.	False	310.3(A), 402.6	(NEC Default)
33.	B.	300.14	E3406.11.3
34.	B.	210.11(C)(4) and Ex. No. 1	E3703.5 and Exception
35.	False	210.8(C), 210.12(B)	E3902.4, E3902.17
36.	True	210.52(C)(3)	E3901.4.3
37.	A	210.52(G)(1)	E3901.9
38.	True	210.52(C)(2)	E3901.4.2(1)
39.	A.	210.52(G)(2)	E3901.9
40.	B.	210.52(E)(3)	E3901.7 (*NEC Default)*

CHAPTER 12

	Answer	*NEC* Section	IRC Section
1.	C.	314.16(A), Table 314.16(A)	E3905.12.1, Table E3905.12.1
2.	B.	314.16(B)(2)	E3905.12.2.2
3.	True	314.16(C)(3)	(NEC Default)
4.	D.	314.16(B)(1), Table 314.16(B)(1)	E3905.12.2.1, Table E3905.12.2.1
5.	C.	314.16(B)(1), Table 314.16(B)(1)	E3905.12.2.1, Table E3905.12.2.1

6.	C.	110.12(A)	E3404.6
7.	False	314.23(B)(2)	E3906.8.2.2
8.	D.	314.27(C), 422.18	E3905.8, E4101.6
9.	B.	314.16(B), 314.16(A)(1), Table 314.16(A), Table 314.16(B)(1)	E3905.12.2, E3905.12.1.1, Table E3905.12.1, Table E3905.12.2.1
10.	False	314.19	E3905.5
11.	C.	314.16(B)(1), (2), (3), (5), Table 314.16(B)(1)	E3905.12.2, Table E3905.12.2.1
12.	D.	314.27(C)	E3905.8
13.	C.	314.16(B)(1), (5), Table 314.16(B)(1)	E3905.12.2, Table E3905.12.2.1
14.	D.	314.16(A), Table 314.16(A), 314.16(B)(1), 314.16(B)(5), Table 314.16(B)(1) [16.875 cu. in.]	E3905.12.1, Table E3905.12.1, E3905.12.2, Table E3905.12.2.1
15.	C.	314.16(B)(1), (2), (4), (5), Table 314.16(B)(1)	E3905.12.2, Table E3905.12.2.1
16.	C.	314.27(A)(2)	E3905.6.2
17.	True	314.15	E3905.11
18.	B.	314.16(B)(4)	E3905.12.2.4
19.	True	314.20	E3906.5
20.	B.	314.16(B)(2) and (B)(3)	E3905.12.2.2, E3905.12.2.3
21.	True	314.22, Exception	E3906.7, Exception
22.	True	314.24(A)	E3905.4.1
23.	D.	314.24(B)(5)	E3905.4.2(5)
24.	True	314.17(B)(4)	E3905.3.2
25.	A.	314.20	E3906.5
26.	A.	314.23(B)(1)	E3906.8.2.1
27.	C.	314.27(A)(1), Exception	E3905.6.1, Exception
28.	B.	314.16(B)(4)	E3905.12.2.4
29.	A.	300.21	E3402.2, R302.4
30.	B.	300.22(C), Exception	E3904.7
31.	B.	314.25	E3906.9
32.	A.	314.15	E3905.11
33.	D.	314.16(A) and (B)	E3905.12.1 and E3905.12.2
34.	D	Book Text [314.27(E)]	E3905.6.3
35.	A.	314.16(B)(5)	E3905.12.2.5

CHAPTER 13

	Answer	*NEC* Section	IRC Section
1.	D.	410.68	*(NEC Default)*
2.	False	410.56(C)	*(NEC Default)*
3.	B.	410.10(A), Article 100, Definitions — Locations, Damp, Informational Note	E4003.9, Chapter 35, Definitions — Locations, Damp *(NEC Default)*
4.	B.	410.10(D)(1)	E4003.11
5.	True	406.9(C)	E4002.11
6.	A.	410.130(E)(1)	E4003.6
7.	B.	406.9(D)	E4002.12
8.	B.	406.9(C)	E4002.11
9.	True	406.9(C)	E4002.11

10.	True	406.12(1)	E4002.14
11.	D.	406.9(C), Ex. No. 4	E4002.11
12.	A.	410.116(B)	E4004.9
13.	True	404.22 and Exception	E4001.15 and *(NEC Default)*
14.	True	200.10(C), 410.90	E4003.13.1
15.	C.	314.27(A)(2)	E3905.6.2
16.	True	406.10(C)	*(NEC Default)*
17.	C.	406.9(C), Ex. No. 2	E4002.11, Ex. No. 2
18.	B.	406.9(A) and (B)(1)	E4002.8, E4002.9
19.	B.	410.116(A)(1)	E4004.8
20.	True	410.16(A) and (B)	E4003.12
21.	C.	404.2(C) and Exception	E4001.15
22.	True	406.4(D)(6)	(NEC Default)
23.	D.	410.16(B)	E4003.12
24.	D.	410.42, Exception, 410.44, Ex. No. 2	E4003.3 and *(NEC Default)*
25.	A.	406.6(D)	*(NEC Default)*
26.	C.	410.150(A) and (C)(2)	E4005.1, E4005.2, E4005.4
27.	False	410.150(C)(7)	E4005.4
28.	D.	404.10(B)	E4001.10
29.	B.	Article 100, Definitions – Retrofit Kit	(NEC Default)
30.	True	410.116(A)(1) and (A)(2)	E4004.8
31.	True	Book Text	Book Text
32.	B.	406.9(B)(1)	E4002.9 and *(NEC Default)*
33.	True	404.4(C)	E4001.7

CHAPTER 14

	Answer	***NEC* Section**	**IRC Section**
1.	D.	Article 100, Definitions - Low Voltage Contact Limit	E4201.2 - Definitions
2.	B.	680.23(A)(4)	E4206.4.1
3.	A.	680.24(D)	E4206.9.4
4.	B.	680.22(B)	E4203.4
5.	D.	680.24(A)(2)(c)	E4206.9.1
6.	B.	680.26(B)	E4204.2
7.	D.	680.26(B)(2)(c)	E4204.2(2), Ex. No. 2.2.3
8.	C.	Article 100, Definitions - Storable Swimming Pool	E4201.2
9.	C.	680.43(C)	E4203.2
10.	B.	680.22(A)(1), (A)(2), and (A)(3)	E4203.1.1, E4203.1.2, and E4203.1.3
11.	C.	680.26(B)	E4204.2
12.	True	680.23(C)(1)	E4206.6
13.	C.	680.22(E)	E4203.5
14.	True	680.23(F)(2)	E4205.2
15.	B.	680.24(B)(2)(b)	E4206.9.1(5)
16.	A.	680.23(B)(2) and (a) and (b)	E4202.1, Table E4202.1
17.	B.	680.21(B)	E4202.4
18.	B.	680.23(A)(1) and (A)(2)	E4206.1, E4206.4
19.	D.	680.5(B) and 680.44(A) and (B)	E4208.1
20.	B.	680.7(A), 680.14	E4202.1, Table E4202.1, E4205.5

21.	False	680.21(C), 680.5(B)	E4203.1.4
22.	True	680.26(B)(6)(a)	E4204.2.6(6.1)
23.	B	680.71	E4209.2
24.	C	680.22(B)(6)	E4203.4.3
25.	C.	Article 100, Definitions - Corrosive Environment	E4201.2 - Definitions
26.	B.	680.26(B)(3)	E4204.2(3)
27.	C.	680.28	*(NEC Default)*
28.	C.	680.7(A)	E4202.2
29.	False	250.52(B)(3)	E3608.7, E4204.2
30.	B.	680.10(B)	E4202.1, E4205.2
31.	B	680.83	*(NEC Default)*
32.	A.	680.12(B)	E4203.1.7
33.	True	680.23(B)(6)	E4206.5.1

CHAPTER 15

	Answer	*NEC* Section	IRC Section
1.	True	225.39	*(NEC Default)*
2.	A.	225.6(A)	*(NEC Default)*
3.	True	225.6(A)	*(NEC Default)*
4.	D.	225.41 and 225.41(A)	*(NEC Default)*
5.	A.	225.4	*(NEC Default)*
6.	True	225.4	*(NEC Default)*
7.	C.	225.11, 230.54(B), Exception	E3605.9.2, (Service only)
8.	True	225.11, 230.54(C)	E3605.9.3, (Service only)
9.	A.	225.11, 230.54(C), Ex.	E3605.9.3, Ex., (Service only)
10.	True	225.22	E3605.8, (Service only)
11.	C.	225.18(1)	E3604.2.2, (Service only)
12.	B.	225.17(A)	E3604.5.1, (Service only)
13.	True	225.12	*(NEC Default)*
14.	D.	225.19(A)	E3604.2.1, (Service only)*(NEC Default)*
15.	D.	225.19(A), Ex. No. 2	E3604.2.1, Ex. 2, (Service only)
16.	D.	225.19(B)	*(NEC Default)*
17.	A.	225.18(A)	*(NEC Default)*
18.	False	225.25	*(NEC Default)*
19.	False	225.26, 590.4(J) and Exception	*(NEC Default)*
20.	B.	300.5(A), Table 300.5, Column 4	E3803.1, Table E3803.1, Column 4
21.	False	225.30	*(NEC Default)*
22.	C.	225.36, 225.31, and 225.33(B)	*(NEC Default)*
23.	B.	250.32(B)(1), Ex. No. 1	E3607.3.2
24.	A.	225.30	*(NEC Default)*
25.	D.	225.33(A)	*(NEC Default)*
26.	C.	225.42(E)	

CHAPTER 16

	Answer	*NEC* Section	IRC Section
1.	True	See Book- "Extension of Existing Branch Circuits"	See Book- "Extension of Existing Branch Circuits"
2.	True	See Book- "Extending Type NM Cables"	See Book- "Extending Type NM Cables"
3.	True	Article 100, Definitions	*(NEC Default)*
4.	B.	394.10, 300.16(A)	*(NEC Default)*
5.	False	250.130(C), 250.68(C)(1)	AJ107.3, Ex. No. 3
6.	True	220.16(A)	*(NEC Default)*
7.	B.	240.51(A)	*(NEC Default)*
8.	True	240.51(B)	AJ107.3, Ex. No. 2
9.	C.	240.52	*(NEC Default)*
10.	True	408.36, Ex. No. 2	*(NEC Default)*
11.	True	406.4(D)(3)	AJ109.5.3.3
12.	D.	406.4(D)(5) and (D)(6)	*(NEC Default)*
13.	D.	210.12(E)	E3902.18
14.	C.	210.12(E) Ex.	E3902.18, Ex.
15.	D.	406.4(D)(4)	*(NEC Default)*
16.	D.	230.85(C)	E3601.8
17.	D.	230.67(D)	E3606.5.3

CHAPTER 17

	Answer	*NEC* Section	IRC Section
1.	B.	722.135(E), 722.179(B)	E4303.2.1
2.	False	725.1	E4301.1
3.	True	725.139(E)	E4304.4
4.	True	760.41(B), 760.121(B)	*(NEC Default)*
5.	B.	Article 100, Definitions – Class 2 Circuit	E4301.2
6.	True	Article 100, Definitions – Class 2 and 3 Circuit	E4301.2
7.	C.	250.94(A)	E3609.3.1
8.	B.	800.113(L)(1)(9), See Book- "Circuits Run in Same Cable Assembly"	*(NEC Default)*
9.	B.	300.22(C), Exception	E3904.7
10.	D.	250.94(A)	E3609.3.1
11.	False	800.133(A)(1)(1)	E4304.4
12.	D.	760.154 and Table 760.154	*(NEC Default)*
13.	True	800.24	*(NEC Default)*
14.	D.	800.100(B)(2)	*(NEC Default)*
15.	True	725.24, 300.11(C)(2)	E4304.5
16.	False	805.93(A) and (B)	*(NEC Default)*

17.	C.	760.1	*(NEC Default)*
18.	True	Article 100, Definitions	*(NEC Default)*
19.	B.	760.30	*(NEC Default)*
20.	True	Article 100, Definitions	*(NEC Default)*
21.	False	760.32	*(NEC Default)*
22.	A.	805.156	*(NEC Default)*
23.	True	840.1, Informational Note	*(NEC Default)*
24.	B.	Article 100, Definitions	*(NEC Default)*
25.	C.	800.100(A)(4)	*(NEC Default)*
26.	C.	See book text under Power over Ethernet (PoE)	*(NEC Default)*
27.	D.	725.144(B), book text	*(NEC Default)*
28.	True	Exception to 250.94(A) and (B)	Exception to E3609.3.1 and E3609.3.2

Notes

ONE- AND TWO-FAMILY DWELLING

Residential Electrical Systems

TWELFTH EDITION

President/CEO: Rudy Garza

Creative Director / Editor / Cover Design:
Laura Hildreth

Technical Edit and Review:
Mike Ornoski
L. Keith Lofland
Mike Hofkin
Joseph Wages, Jr.

Technical Drawings
Joseph Wages, Jr.
Laura L. Hildreth

Organization Contributions and Support:
National Fire Protection Association (NFPA)
Underwriters Laboratories Inc. (UL)
International Code Council (ICC)

Photos / Illustrations:
Arlington Industries, Scranton, PA
Carlon, Memphis, TN
Cooper Bussmann, St. Louis, MO
Eaton Cutler-Hammer, Beachwood, OH
Greaves Corporation, Guilford, CT
Hunter, Christel, Las Vegas, NV
Hunter, Randy, Las Vegas, NV
IAEI Archives, Richardson, TX
Ideal Industries, Inc, Sycamore, IL
Lofland, L. Keith, Sachse, TX
McGovern, Bill, Dallas, Texas
Pass & Seymour/Legrand, West Hartford, CT
Raco Boxes, South Bend, IN
Steel City Boxes, Memphis, TN
Strongwell, Bristol, VA
Thomas and Betts, Memphis, TN
Thomas, Tim
UL, Northbrook, IL
Wages, Joseph Jr., Celina, TX
Watson, John, Amarillo, TX

www.ingramcontent.com/pod-product-compliance
Lightning Source LLC
LaVergne TN
LVHW070846210926
848482LV00016B/875